Nonparametric Models for Longitudinal Data

With Implementation in R

MONOGRAPHS ON STATISTICS AND APPLIED PROBABILITY

Editors: F. Bunea, P. Fryzlewicz, R. Henderson, N. Keiding, T. Louis, R. Smith, and W. Wong

For more information about this series please visit:
https://www.crcpress.com/Chapman--HallCRC-Monographs-on-Statistics--
Applied-Probability/book-series/CHMONSTAAPP

Nonparametric Models for Longitudinal Data

With Implementation in R

Colin O. Wu
Xin Tian

CRC Press
Taylor & Francis Group
Boca Raton London New York

CRC Press is an imprint of the
Taylor & Francis Group, an **informa** business

A CHAPMAN & HALL BOOK

To Our Parents

To Li-Ping and Enpei

To Lang, Allison, and Frank

Contents

IV   Shared-Parameter and Mixed-Effects Models   301

List of Figures

List of Tables

Preface

Longitudinal studies, which commonly refer to studies with variables repeatedly observed over time, play an important role in biomedical studies, such as long-term epidemiological studies and clinical trials, as well as other scientific areas. A well-known example of longitudinal study in biomedicine is the Framingham Heart Study (www.framinghamheartstudy.org), which began in 1948 and has led to the identification of major cardiovascular disease risk factors, including blood pressure, triglyceride and cholesterol levels. Given the success of early longitudinal studies, most important biomedical studies today contain at least some repeatedly measured variables over time. A partial list of such high-impact biomedical studies includes the Multicenter AIDS Cohort Study (MACS), the National Growth and Health Study (NGHS), the Enhancing Recovery in Coronary Heart Disease Patients (ENRICHD) study, the Coronary Artery Risk Development in Young Adults (CARDIA) study, and the Multi-Ethnic Study of Atherosclerosis (MESA) study. These are all large scale long-term longitudinal studies with thousands of subjects and multiple observations over years of follow-up. Among them, the ENRICHD is a randomized clinical trial, while the rest are all epidemiological studies with at least ten years of follow-up. Tremendous progress in statistical methodology has been made in the past three decades for the development of computational and statistical methods to analyze data from longitudinal studies. Building on the early developments in longitudinal analysis, a major area of statistical methodology is based on parametric or semiparametric statistical models that properly take the intra-subject correlations into account, such as the popular linear and nonlinear mixed-effects models. These statistical methods still dominate the applications and methodological research in longitudinal analysis today.

However, the rapid advancement of computing capability and data storage tools in recent years makes it possible for researchers to efficiently collect, store and transfer a large amount of data in a relatively short period of time. As a result, researchers are able to extensively explore and analyze large longitudinal studies using flexible nonparametric statistical analysis and data mining tools. Although well-designed clinical trials are still believed to be the gold standard for evaluating efficacy in biomedical studies, an extensive exploratory analysis is often useful for guiding the appropriate study questions and hypotheses, which may be tested through a clinical trial.

There has been remarkable development of nonparametric methods for the analysis of longitudinal data for the past twenty-five years. In contrast to parametric or semiparametric methods, nonparametric methods are more flexible and often used in situations where there are no established parametric or semiparametric models for the available data. In exploratory longitudinal analysis, parametric or semiparametric forms of the data distributions are usually completely unknown. When there are a large amount of data available, subjectively chosen parametric or semiparametric models may give inadequate fits to the data and lead to potentially biased conclusions. Thus, an important aspect of nonparametric methods is to provide some flexible tools to describe the temporal trends of the patterns and correlation structures of the data. Since a completely unstructured nonparametric approach may be too general to lead to useful conclusions in practice, most research activities on nonparametric longitudinal analysis are focused on flexible nonparametric models subject to certain scientifically meaningful and practical structural restrictions. When appropriately used, structured nonparametric methods have the advantage of balancing model flexibility with practicability in real applications.

The aim in this book is to provide a summary of recent results of unstructured and structured nonparametric methods for the analysis of longitudinal data. Given our own experience and research interests, our coverage is focused on the statistical methods and theories which are particularly useful for biomedical studies, although in principle these methods may have applications in other scientific areas. We intend to strike a proper balance between methodology, applications and theory. To do this, we include four longitudinal studies, among them, two large epidemiology studies, one large clinical trial and one small-sized study, as motivating examples to illustrate the real applications and scientific interpretations of the statistical methods. Statistical implementation based on R software packages is also presented for the corresponding statistical results, including tables and figures, in each chapter.

For a more application-oriented reader, the methodology and computational part is sufficient for the applications of the statistical methods described in this book. Since longitudinal analysis is still an active area of statistical research, we allocate a sizable portion of the book to cover the theoretical aspects of the methods. Although our coverage of the theoretical results does not include all the methods described in this book, nor does every method have its theoretical property systematically investigated, we intend to cover the important theoretical derivations as much as possible, so that a reader interested in the theoretical aspects of nonparametric longitudinal analysis may gain sufficient background knowledge to the recent developments in the literature. To benefit the readers who are interested in doing some methodological research in this area, such as graduate students and new researchers in statistics and biostatistics, we include in each chapter some discussions of the potential questions and directions for future research. Since almost all the methods presented in this book are grown out of real scientific questions in biomedical studies, we also include in the discussions of each chapter the relevant papers in the biomedical literature to help motivate the statistical procedures and their interpretations. Throughout the book, we hope to send a clear message that most methodological and theoretical developments of nonparametric longitudinal analysis, certainly the ones described in this book, are motivated by some real studies and intended to answer certain important scientific questions, which may not be properly answered by the mathematically simpler parametric or semiparametric approaches.

It would be impossible to complete this book without the help and support from our families, friends and colleagues. We are deeply in debt to many colleagues who at various stages have provided many insightful comments and suggestions to initial drafts of this book. In particular, we are grateful to Mr. John Kimmel, Executive Editor of Statistics, who initiated this book project to the first author during a snow storm in Washington, D.C., in 2011, and encouraged us constantly during the preparation of this book. It is certainly a long journey from its initiation in 2011 to this date – which has certainly brought many ups and downs. Along the way, we have to constantly keep up with the new publications in this area and update the materials accordingly. We greatly appreciate the many excellent comments and suggestions provided by Ms. Robin Lloyd-Starkes (Project Editor), Ms. Sherry Thomas (Editorial Assistant), several anonymous reviewers and the proofreader, which led to significant improvement on the presentation of this book.

We are grateful to many of our colleagues who have collaborated with either one or both of us at various stages of research and publications. These include statistical collaborators, such as John A. Rice, Grace L. Yang, Donald R. Hoover, Chin-Tsang Chiang, Jianhua Z. Huang, Lan Zhou, Gang Zheng, Heejung Bang, Wenhua Jiang, Tianqing Liu, Zhaohai Li, Yuanzhang Li, Mohammed Chowdhury, Xiaoying Yang, Wei Zhang, Qizhai Li, Lixing Zhu, Mi-Xia Wu, Hyunkeun Ryan Cho and Seonjin Kim, and biomedical collaborators, such as Joao A.C. Lima, David A. Bluemke, Kiang Liu, Bharath Ambale-Venkatesh, A. John Barrett, Neal S. Young, Richard W. Childs, Cynthia E. Dunbar, Jan Joseph Melenhorst, Phillip Scheinberg, Danielle Townsley, Minoo Battiwalla, Sawa Ito, Adrian Wiestner, Eva Obarzanek, Michael S. Lauer, Narasimhan S. Danthi, Jared Reis, among many others. We are also grateful to our statistical colleagues at the National Institutes of Health, including Nancy L. Geller, Eric Leifer, Paul S. Albert, Dean Follmann, Jin Qin, Aiyi Liu, and others who shared with us their suggestions and insights into statistical methodology and applications. We apologize for not being able to mention all of the wonderful friends and colleagues who have helped us throughout the preparation of this book. Their comments and suggestions have broadened our perspectives on the nature of longitudinal data and statistical methodology in general.

Both of us have greatly benefited from the many teachers, advisors and mentors throughout our lives and professional careers. Their wisdom and guidance have played an enormously important role in shaping our statistical research careers. Among them, Colin O. Wu particularly thanks his mathematics teacher at Yao-Hua High School (Tianjin, China), Mr. Zon-Hua Liu; his undergraduate and graduate school teachers and advisors at UCLA and UC Berkeley, Professors Shua-Yuan Cheng, Erich L. Lehmann, Lucien Le Cam, Chin Long Chiang, David H. Blackwell, Rudolph Beran, Kjell Doksum, Peter Bickel, and P. Warwick Millar (dissertation advisor); and his senior faculty mentor at Johns Hopkins, Professor Robert J. Serfling. Xin Tian also thanks her dissertation advisors at Rutgers, The State University of New Jersey, Professors Cun-Hui Zhang and Yehuda Vardi. Again, we apologize for not being able to list all our great teachers, advisors and mentors, and we take the opportunity to thank them all. We thank our parents for their love and sacrifices in our upbringings and educational opportunities.

Finally, we express our deepest gratitude and appreciation to our families. Their love, understanding, and patience provided emotional support and encouragement for us to work hard throughout many long days and sleepless nights. We understand this book would not be possible, and our professional careers would not be successful, without their love and support. For Colin O. Wu, his sincere appreciation goes to his wife, Li-Ping Yang, and daughter, Enpei Y. Wu; for Xin Tian, her sincere appreciation goes to her husband, Lang Lin, and children, Allison and Frank. We would like to dedicate this book to them.

Colin O. Wu and Xin Tian

Spring, 2018
Bethesda, Maryland

About the Authors

Colin O. Wu is Mathematical Statistician at the Office of Biostatistics Research, Division of Cardiovascular Sciences, National Heart, Lung and Blood Institute, National Institutes of Health (USA). He is also Adjunct Professor at the Department of Biostatistics, Bioinformatics, and Biomathematics, Georgetown University School of Medicine, and Professorial Lecturer at the Department of Statistics, The George Washington University. He received his Ph.D. in statistics from the University of California, Berkeley, in 1990. His former academic positions include Visiting Assistant Professor at the University of Michigan, Ann Arbor, Associate Professor at the Johns Hopkins University, and Guest Lecturer at University of Maryland, College Park. He has published over 150 research articles in statistics, biostatistics and medical journals. He has been serving as Guest Editor for Statistics in Medicine, Associate Editor for Biometrics, reviewer for the National Science Foundation, member of Data Monitoring Committees and Cardiology Review Panels for the United States Department of Veterans Affairs, and statistical expert of the Sphygmomanometer Committee for the Association for the Advancement of Medical Instrumentation. He is Elected Member of the International Statistical Institute, and Fellow of the American Statistical Association.

Xin Tian is Mathematical Statistician at the Office of Biostatistics Research, Division of Cardiovascular Sciences, National Heart, Lung and Blood Institute, National Institutes of Health. She obtained her Ph.D. in Statistics from Rutgers, The State University of New Jersey, in 2003. Her research interests include design and analysis of clinical trials, statistical genetics, and longitudinal data analysis with structured nonparametric models. She has published over 80 research papers in statistical and biomedical journals and book chapters.

Part I

Introduction and Review

Chapter 1

Introduction

In biomedical studies, interests are often focused on evaluating the effects of treatments, medication dosage, risk factors or other biological and environmental covariates on certain outcome variables, such as disease progression and health status, over time. Because the changes of outcomes and covariates and their temporal patterns within each subject usually provide important information of scientific relevance, longitudinal samples that contain repeated measurements within each subject over time are often more informative than the classical cross-sectional samples, which contain the measurements of each subject at one time point only. Since longitudinal samples combine the characteristics of cross-sectional sampling and time series observations, their usefulness goes far beyond biomedical studies and is often found in economics, psychology, sociology and many other scientific areas.

1.1 Scientific Objectives of Longitudinal Studies

In general, there are two main sampling approaches to obtain longitudinal observations in biomedical studies: (a) a randomized *clinical trial* with pre-specified treatment regimens and repeatedly measured observations (Friedman et al., 2015), and (b) an *epidemiological study*, which is often referred to as an *observational cohort study* (Rosenbaum, 2002). The major difference between a randomized clinical trial, or simply a clinical trial, and an observational cohort study is their designs. In a clinical trial, the selection of the experimental treatment regimens, length of the trial period, visiting times and methods of the measurement process are determined by the study investigators, and the treatment regimens are randomly assigned to the study subjects, although, in some occasions, nonrandomized *concomitant treatments*, or *concomitant interventions* may also be given to some subjects due to ethical and logistical reasons. An observational cohort study, on the other hand, is more complicated, because the risk factors, treatments and the measurement process depend on the participants of the study and are not controlled by the investigators.

In a longitudinal clinical trial, the main scientific objective is to evaluate the efficacy of the pre-specified experimental treatment versus a placebo or standard treatment on the primary outcomes, such as certain health status indicators, over time during the trial period. In many situations, a follow-up period is added at the end of the treatment period, so that time-to-event variables, such as time to hospitalization or death, may be included as a primary outcome in addition to the repeatedly measured health outcomes. In a particular analysis, the trial period may be defined based on the objectives of the analysis. For example, if the objective is to evaluate the treatment effects on the *time-trend* of a health indicator within the treatment period, it is appropriate to consider the treatment period as the trial period. On the other hand, if it is also of interest to consider certain time-to-event variables beyond the treatment period, it is then appropriate to include both the treatment period and the follow-up period into the trial period. Effects of the study treatments may be evaluated through the conditional means, conditional distributions or conditional quantiles of the outcome variables. Although regression models based on conditional means of the outcome variables are by far the most popular methods in the analysis of longitudinal clinical trials, regression methods based on conditional distributions or conditional quantiles are often valuable statistical tools in situations when the outcome variables have highly skewed or distributions that are not easily approximated by normal distributions. In addition to the evaluation of randomized study treatments, important secondary objectives include evaluating the effects of concomitant interventions or other covariates on the time-varying trends of the outcome variables. Regression analyses involving covariates other than the randomized study treatments are often useful for evaluating treatment-covariate interactions or identifying subgroups of patient populations to whom the experimental treatments are efficacious. Because of the randomization, a properly designed clinical trial is viewed as a gold standard to make causal inferences about the efficacy of the study treatments.

In a longitudinal observational cohort study, there are no randomized experimental treatments to be tested, and the main objective is to evaluate the potential associations of various covariates, such as demographic and environmental factors, with the outcome variables of interest and their trends over time. Observational cohort studies are often used for the purpose of data exploration, so that more specific scientific hypotheses may be generated and tested in a future properly designed clinical trial. For this purpose, observational cohort studies often involve large sample sizes as well as large numbers of scientifically relevant variables. Statistical inferences obtained from an observational cohort study are useful for understanding the associations between the covariates and the outcome variables, but may not be sufficient to infer the causal effects. Because the variables are repeatedly measured over time, long-term longitudinal observational cohort studies are useful for understanding the natural progression of certain diseases both on a population-wide level and for certain sub-populations represented by the study subjects. In practice, an observational cohort study usually involves a large number of subjects with sufficient numbers of repeated measurements over time, so that novel findings with adequate statistical accuracy can be obtained from the study. Similar to longitudinal clinical trials, the choices of statistical approaches for the analysis of data from observational cohort studies depend on the scientific objectives, and may involve regression models for the conditional means, conditional distributions and conditional quantiles.

1.2 Data Structures and Examples

1.2.1 Structures of Longitudinal Data

For a typical framework of longitudinal data, the variables are repeatedly measured over time. We denote by t a real-valued variable of time, $\mathscr{T}$ the range of time points such that $t \in \mathscr{T}$, $Y(t)$ a real-valued outcome variable and $\mathbf{X}(t) = \left(X^{(0)}(t), \ldots, X^{(K)}(t) \right)^T$, $K \geq 1$, a R^{K+1}-valued covariate vector at time t. Depending on the choice of origin, the time variable t is not necessarily non-negative. As part of the general methodology, interests of statistical analysis with regression models are often focused on modeling and determining the effects of $\{t, \mathbf{X}(t)\}$ on the population mean, subject-specific deviations from the population mean, conditional distribution or conditional quantiles of $Y(t)$. For n randomly selected subjects with each subject repeatedly measured over time, the longitudinal sample of $\{Y(t), t, \mathbf{X}(t)\}$ is denoted by

$$\left\{ \left(Y_{ij}, t_{ij}, \mathbf{X}_{ij} \right) : i = 1, \ldots, n; \, j = 1, \ldots, n_i \right\},$$

where t_{ij} is the jth measurement time of the ith subject, Y_{ij} and $\mathbf{X}_{ij} = \left(X_{ij}^{(0)}, \ldots, X_{ij}^{(K)} \right)^T$ are the observed outcome and covariate vector, respectively, of the ith subject at time t_{ij}, and $n_i \geq 1$ is the ith subject's number of repeated measurements. Due to various reasons, such as schedule changes or some missed visits, the numbers of repeated measurements n_i are usually not the same in practice, even though the study design ideally calls for the same number of repeated measurements for all the study subjects. The total number of measurements for the study is $N = \sum_{i=1}^{n} n_i$. In contrast to the independent identically distributed (i.i.d.) samples in classical cross-sectional studies, which is equivalent to the situation with $n_i = 1$ for all $i = 1, \ldots, n$, the measurements within each subject are possibly correlated, although the inter-subject measurements are assumed to be independent.

A longitudinal sample is said to have a balanced design if all the subjects have their measurements made at a common set of time points, i.e., $n_i = m$ for some $m \geq 1$ and all $i = 1, \ldots, n$ and $t_{1j} = \cdots = t_{nj}$ for all $j = 1, \ldots, m$. An unbalanced design arises if the design time points $\{t_{ij}; 1 \leq j \leq n_i\}$ are different for different subjects. In practice, unbalanced designs may be caused by the presence of missing observations in an otherwise balanced design or by the random variations of the time design points. In long-term clinical trials or epidemiological studies, study subjects are often assigned to a set of pre-specified "design visiting times," but their actual visiting times could be different because of missing visits or changing visiting times due to various reasons. Under an ideal situation, it is possible to observe balanced longitudinal data from a well-controlled longitudinal clinical trial, because the randomized study treatments and clinical visiting times are determined by the study investigators. However, for various reasons that are out of the investigator's control, most longitudinal clinical trials and nearly all the observational cohort studies have unbalanced longitudinal designs.

1.2.2   Examples of Longitudinal Studies

In order to provide a practical sense of the scope for longitudinal studies, we use four real-life examples, two epidemiological studies and two longitudinal clinical trials, throughout this book to illustrate some typical designing features, scientific objectives, and statistical models for longitudinal data analysis. These examples have different sample sizes, data structures and objectives, and require different statistical approaches.

Example 1. Baltimore Multicenter AIDS Cohort Study

This dataset is from the Baltimore site of the Multicenter AIDS Cohort Study (BMACS), which included 400 homosexual men who were infected by the human immunodeficiency virus (HIV) between 1984 and 1991. Because CD4 cells (T-helper lymphocytes) are vital for immune function, an important component of the study is to evaluate the effects of risk factors, such as cigarette smoking, drug use, and health status evaluated by CD4 cell levels before the infection, on the post-infection depletion of CD4 percent of lymphocytes. Although all the individuals were scheduled to have their measurements made at semi-annual visits, the study has an unbalanced design because the subjects' actual visiting times did not exactly follow the schedule and the HIV infections happened randomly during the study. The covariates of interest in these data include both time-dependent and time-invariant variables. Details of the statistical design and scientific importance of the BMACS data can be found in Kaslow et al. (1987) and Wu, Chiang and Hoover (1998). The BMACS data used in this book included 283 subjects with a total of 1817 observations. The number of repeated measurements ranged from 1 to 14 with a median of 6 and a mean of 6.4. Figure 1.1 presents the longitudinal trajectories for the BMACS data.   □

Example 2. National Growth and Health Study

The National Heart, Lung, and Blood Institute Growth and Health Study (NGHS, also known as the National Growth and Health Study in ClinicalTrials.gov) is a multicenter population-based cohort study aimed at evaluating the racial differences and longitudinal changes in childhood cardiovascular risk factors between 1166 Caucasian and 1213 African American girls during childhood and adolescence. Details of the study have been previously described in NGHS Research Group (NGHSRG, 1992). Up to 10 annual measurements

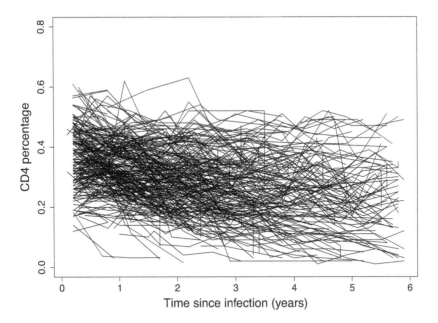

Figure 1.1 *The longitudinal CD4 trajectories for all subjects in BMACS dataset.*

were obtained from the girls followed longitudinally between 9 to 10 years of age (visit 1) at study entry through 18 to 19 years (visit 10). The demographic information, physical measures and cardiovascular risk factors such as blood pressure and lipids levels were obtained during the visits. The body mass index (BMI) defined as weight in *kg* divided by height in m^2 was derived from annual measurements of height and weight. The number of follow-up visits for the Caucasian and African American girls in the study ranged from 1 to 10, with a median of 9 and a mean of 8.2. Figure 1.2(A) shows the BMI and the systolic blood pressure (SBP) for a randomly chosen sample of 150 study participants. Figure 1.2(B)-(C) displays BMI and SBP measurements for three girls from NGHS, from which we can see the individual variations in their longitudinal trajectories. The NGHS data are available for request via the NIH BioLINCC site (https://biolincc.nhlbi.nih.gov/). ☐

Example 3. Enhancing Recovery in Coronary Heart Disease Patients Study

The Enhancing Recovery in Coronary Heart Disease Patients (ENRICHD) study is a randomized clinical trial to evaluate the efficacy of a cognitive behavior therapy (CBT) versus usual cardiological care on survival and depression severity in 2481 patients who had depression and/or low perceived social support after acute myocardial infarction. The primary objective of the study is to determine whether mortality and recurrent myocardial infarction

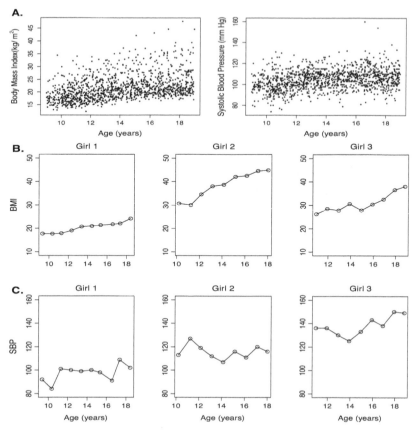

Figure 1.2 *(A) The body mass index and systolic blood pressure of 150 subjects in the NGHS. (B)-(C) The longitudinal BMI and SBP measurements for three girls from the NGHS.*

are reduced by treatment of depression and low perceived social support with cognitive behavior therapy, supplemented with the use of selective serotonin reuptake inhibitor (SSRI) or other antidepressants as needed, in patients enrolled within 28 days after myocardial infarction (MI). The intervention of the trial consists of cognitive behavior therapy initiated at a median of 17 days after the index MI for a median of 11 individual sessions throughout 6 months. Depression severity was measured by the Beck Depression Inventory (BDI) with higher BDI scores indicating worsened depression severity. Group therapy was conducted when feasible, with antidepressants, such as SSRIs, as a pharmacotherapy for patients scoring higher than 24 on the Hamilton Rating Scale for Depression (HRSD) or having a less than 50% reduction in BDI scores after 5 weeks. Antidepressants were also prescribed at the request of the patients or their primary-care physicians, therefore, could be treated

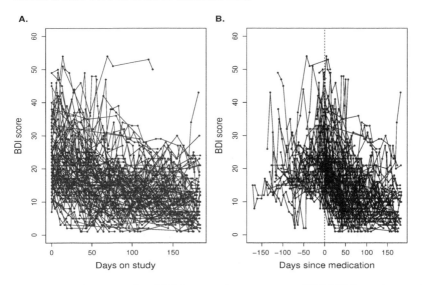

Figure 1.3 *The longitudinal BDI scores for patients in ENRICHD study with horizontal axes as days on study since randomization (A) or days since medication (B).*

as a concomitant treatment in addition to the randomized CBT psychosocial treatment or usual cardiological care specified in the trial design. The main outcome measures consist of a composite primary endpoint of death or recurrent MI, and secondary outcomes including change in BDI for depression or the ENRICHD Social Support Instrument scores for low perceived social support at 6 months. Details of the study design, objectives, and major findings of the trial have been described in ENRICHD (2001, 2003), Taylor et al. (2005) and Bang and Robins (2005), among others.

In addition to the primary objective of the trial, an important question of secondary objective is whether the use of antidepressants has added benefits on the trends of depression (measured by BDI scores) for patients who received pharmacotherapy during the six-month psychosocial treatment period. Because pharmacotherapy was only designed as a concomitant intervention in this trial, the starting time of pharmacotherapy was decided by the patients or their physicians. Unfortunately, since patients in the usual care arm did not have accurate pharmacotherapy starting time and repeated BDI scores recorded within the first six-month period, patients in this arm are not included in the dataset for our analysis. In our data, 92 patients (total 1445 observations) in the psychosocial treatment arm received pharmacotherapy as a concomitant intervention during this period and had clear records of their pharmacotherapy starting time. Among them, 45 started pharmacotherapy at baseline and 47 started pharmacotherapy between 10 and 172 days. In addition, we also included 11 patients in the CBT arm who had record of starting antidepressants before baseline and 454 patients who did not use antidepressants before and during the treatment period. Therefore, this data example is based on 557 depressed patients (total 7117 observations) in the CBT arm. The number of visits for these patients ranges from 5 to 36 and has a median of 12. Figure 1.3 shows the BDI scores of 92 patients with antidepressant starting time recorded in the ENRICHD data. The ENRICHD data are available for request via the NIH BioLINCC site (https://biolincc.nhlbi.nih.gov/).   □

Example 4. The HSCT Data

For patients with hematologic malignancies and life-threatening bone marrow diseases, allogeneic hematopoietic stem cell transplantation (HSCT) has long been recognized as a curative treatment. HSCT is associated with profound changes in levels of various leukocytes and cytokines around the time before and immediately after the transplantation. The HSCT data consists of 20 patients who were transplanted between 2006 and 2009 in a phase II clinical trial at the National Institutes of Health. Patients received a 7-day conditioning preparative regimen including radiation and chemotherapy agents (on days -7 and -1). On day 0, the patients received a CD34+ stem cell-selected HSCT from a Human Leukocyte Antigen (HLA) identical sibling donor. Plasma samples were collected twice weekly from day -8 until 100 days post-transplantation. The database and study design have been described in Melenhorst et al. (2012). Figure 1.4 shows the longitudinal changes in the white blood cell counts of granulocytes, lymphocytes and monocytes and levels of three cytokines, granulocyte colony-stimulating factor (G-CSF), IL-15 and monocyte chemotactic protein-1 (MCP-1) for patients during the pre- and early post-transplantation period. The local polynomial smoothing estimate is superimposed on each scatter plot to show the overall time-trend.   □

1.2.3   Objectives of Longitudinal Analysis

Generally speaking, a proper longitudinal analysis should achieve at least three objectives:

(1) The model under consideration must give an adequate description of the scientific relevance of the data and be sufficiently simple and flexible to be practically implemented. In biomedical studies, an appropriate regression model should give a clear and meaningful biological interpretation and also has a simple mathematical structure.

(2) The methodology must contain proper model diagnostic tools to evaluate the validity of a statistical model for a given dataset. Two important diagnostic methods are confidence regions and tests of statistical hypotheses.

(3) The methodology must have appropriate theoretical and practical properties, and can adequately handle the possible intra-subject correlations of the data. In practice, the intra-subject correlations are often completely unknown and difficult to be adequately estimated, so that it is generally

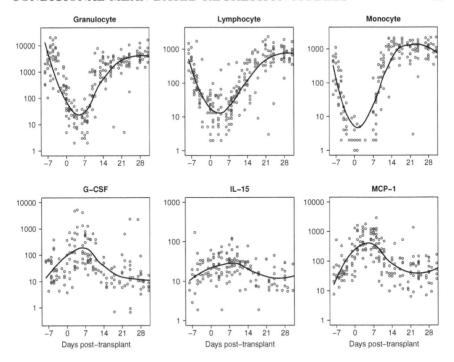

Figure 1.4 *Dynamics of leukocytes and cytokines around the time of stem cell transplantation. The local polynomial smoothing estimators are superimposed on the scatter plots.*

preferred to use estimation and inference procedures that do not depend on modeling the specific correlation structures.   □

We briefly summarize here major classes of regression models for longitudinal analysis, which form the main topics of this book. Detailed estimation and inference procedures based on these models are discussed in the following chapters.

1.3   Conditional-Mean Based Regression Models

This class of models is aimed at characterizing the covariate effects through the conditional mean structures of the response variables given the covariates of interest. These models are appropriate when the outcome variables, i.e., the error terms of the regression model, have nearly symmetric distributions conditioning on the covariates. When the conditional distributions of the outcome variables are skewed, some transformations, such as the logarithmic or Box-Cox transformations, may be applied to the original outcome variables, so that the transformed outcome variables have nearly symmetric conditional distributions given the covariates.

1.3.1 Parametric Models

Naturally, the most commonly used approach in longitudinal analysis is through parametric regression models, such as the generalized linear and nonlinear mixed-effects models. By adopting a parsimonious parametric structure, this class of models can summarize the relationships between the outcome variables and covariates through some simple parameters, so that it has the advantage of having mathematically trackable estimation and inference procedures. The simplest case of these models is the marginal linear model

$$Y_{ij} = \sum_{l=0}^{K} \beta_l X_{ij}^{(l)} + \varepsilon_i(t_{ij}), \tag{1.1}$$

where $\beta_0, \ldots, \beta_K$ are constant linear coefficients describing the effects of the corresponding covariates, $\varepsilon_i(t)$ are realizations of a mean zero stochastic process $\varepsilon(t)$ at t, and $\mathbf{X}_{ij}$ and $\varepsilon_i(t_{ij})$ are independent. Similar to all regression models where a constant intercept term is desired, the choice of $X^{(0)} = 1$ produces a baseline coefficient β_0, which represents the mean value of $Y(t)$ when all the covariates $X^{(l)}(t)$ are set to zero. A popular special case of the error process is to take $\varepsilon(t)$ to be a mean zero Gaussian stationary process. Although (1.1) appears to be overly simplified for many practical situations, its generalizations lead to many useful models which form the bulk of longitudinal analysis.

Estimation and inference methods based on parametric models, including the weighted least squares, the quasi-likelihoods and the generalized estimating equations, have been extensively investigated in the literature. Details of these methods can be found, for example, in Laird and Ware (1982), Pantula and Pollock (1985), Ware (1985), Liang and Zeger (1986), Diggle (1988), Zeger, Liang and Albert (1988), Jones and Ackerson (1990), Jones and Boadi-Boteng (1991), Davidian and Giltinan (1995), Vonesh and Chinchilli (1997), Verbeke and Molenberghs (2000) and Diggle et al. (2002). The main advantage of parametric models is that they generally have simple and intuitive interpretations. User-friendly computer programs are already available in popular statistical software packages, such as SAS, STATA and R. However, these models suffer the potential shortfall of model misspecification, which may lead to erroneous conclusions. At least in exploratory studies, it is often necessary to relax some of the parametric restrictions.

1.3.2 Semiparametric Models

A useful semiparametric regression model, investigated by Zeger and Diggle (1994) and Moyeed and Diggle (1994), is the partially linear model

$$Y_{ij} = \beta_0(t_{ij}) + \sum_{l=1}^{K} \beta_l X_{ij}^{(l)} + \varepsilon_i(t_{ij}), \tag{1.2}$$

where $\beta_0(t)$ is an unknown smooth function of t, β_l are unknown constants and $\varepsilon_i(t)$ and $\mathbf{X}_{ij}$ are defined in (1.1). The objective in (1.2) is to estimate the scalar covariate effects $\{\beta_1, \ldots, \beta_K\}$ while leaving the unknown smooth function $\beta_0(t)$ as a nuisance baseline curve. This model is more general than the marginal linear model (1.1), because $\beta_0(t)$ is allowed to change with t, rather than setting to be a constant over time. But the covariate effects of (1.2) are determined by the linear coefficients β_l, $l = 1, \ldots, K$, which form the parametric components. By including the linear terms of $X_{ij}^{(l)}$, (1.2) is more general than the unstructured nonparametric regression model given below, which involves only (t_{ij}, Y_{ij}) and was studied in the literature by Hart and Wehrly (1986), Altman (1990), Hart (1991), Rice and Silverman (1991), among others.

However, because (1.2) describes the effects of $X_{ij}^{(l)}$ on Y_{ij} through constant linear coefficients, this model is still based on mathematical convenience rather than scientific relevance. For example, there is no reason to expect that the influences of the effects of cigarette smoking and pre-infection CD4 level on the post-infection CD4 cell percent in the BMACS data of Section 1.2.2 are linear and constant throughout the study period. Thus, further generalization of (1.2) is needed in many situations. We review the methods for parametric and semiparametric longitudinal regression models in Chapter 2.

1.3.3 Unstructured Nonparametric Models

To relax the parametric assumptions on the effects of the covariates $\mathbf{X}_{ij}$, a further generalization is to allow the effects of $\mathbf{X}_{ij}$ to be described by nonparametric functions. Although it is possible in principle to model $\{Y_{ij}, t_{ij}, \mathbf{X}_{ij}\}$ through a completely unstructured nonparametric function, such an approach is often impractical due to the well-known problem of "*curse of dimensionality*" in the sense that the resulting covariate effects are difficult to interpret and the estimation and inference procedures are numerically unstable if the dimensionality K of $\mathbf{X}_{ij}$ is high (e.g., Fan and Gijbels, 1996). Thus, without assuming any modeling structures, a completely unstructured nonparametric model for $\{Y_{ij}, t_{ij}, \mathbf{X}_{ij}\}$ is often impractical even for the low-dimensional case of $K = 2$ or 3.

For the unstructured nonparametric models, we only present in this book the special case of estimating the conditional mean $\mu(t) = E[Y(t)|t]$ based on the model

$$Y_{ij} = \mu(t_{ij}) + \varepsilon_i(t_{ij}), \tag{1.3}$$

with longitudinal sample $\{(Y_{ij}, t_{ij}) : i = 1, \ldots, n; j = 1, \ldots, n_i\}$, where, as in (1.1), $\varepsilon_i(t)$ are realizations of a mean zero stochastic process $\varepsilon(t)$ at t. We present three types of smoothing methods for the estimation of $\mu(t)$:

(a) kernel and local polynomial estimators;
(b) basis approximation methods through B-splines;

(c) penalized smoothing spline methods.

The estimation methods in (a) are based on the so-called local smoothing methods in the sense that the estimators of $\mu(t)$ are constructed using the weighted averages of the local observations around the time point t. The estimation methods in (b) and (c), by contrast, are based on global smoothing methods since the data points observed away from the time point t, i.e., global observations, still contribute to the estimators of $\mu(t)$.

We present the local and global smoothing estimation methods for (1.3) in Chapters 3 to 5. The main difference between the smoothing methods of Chapters 3 to 5 and their counterparts with the classical cross-sectional independent and identically distributed (i.i.d.) data is that the statistical properties of the methods of Chapters 3 to 5 are affected by the potential intra-subject correlations of the longitudinal sample $\{(Y_{ij}, t_{ij}) : i = 1, \ldots, n; j = 1, \ldots, n_i\}$. For the special case that there is no intra-subject correlation, the smoothing methods presented in Chapters 3 to 5 are equivalent to their counterparts with cross-sectional i.i.d. data. But, as shown by the examples of Section 1.2.2, this uncorrelated assumption is usually unrealistic in real longitudinal applications.

1.3.4   Structured Nonparametric Models

The problems discussed in the above parametric, semiparametric and unstructured nonparametric modeling approaches motivate the consideration of nonparametric regression models that have certain scientifically interpretable and meaningful structures, which we refer to herein as the "structured nonparametric regression models" or "structural nonparametric regression models." The main idea is to impose some mathematically tractable structures to the model, so that, in order to maintain model flexibility and interpretability, the main parameters of interest are functions or curves, which we refer to as "functional parameters" or "curve parameters."

1. The Time-Varying Coefficient Model

An important class of structured nonparametric regression models is the time-varying coefficient model, which, under the linear structure for covariate effects, has the form

$$Y_{ij} = \mathbf{X}_{ij}^T \boldsymbol{\beta}(t_{ij}) + \varepsilon_i(t_{ij}), \tag{1.4}$$

where $\boldsymbol{\beta}(t) = \left(\beta_0(t), \ldots, \beta_K(t)\right)^T$ is a $(K+1)$-vector of smooth functions (or smooth curves) of t, and $\varepsilon_i(t)$ and $\mathbf{X}_{ij}$ are defined as in (1.1). Because (1.4) assumes a linear model between $Y(t)$ and $\mathbf{X}(t)$ at each fixed time point t, the linear coefficient curve $\beta_l(t)$, $l = 0, \ldots, K$, can be interpreted the same way as in (1.2). In most applications, we assume that Y_{ij} has a baseline population mean at time point t_{ij}, so that $X_{ij}^{(0)}$ is set to $X_{ij}^{(0)} = 1$ and $\beta_0(t)$ represents the intercept at time t. Because all the linear coefficients of (1.4) are functions of t, different linear models may be obtained at different time points.

The model (1.4) is a special case of the general varying-coefficient models discussed in Hastie and Tibshirani (1993). Methods of estimation and inferences based on this class of models have been subjected to intense investigation in the literature. A number of different smoothing methods for the estimation of $\beta(t)$ have been proposed. These include the ordinary least squares kernel and local polynomial methods, the roughness penalized splines, the two-step smoothing methods and the basis approximation smoothing methods. Targeted to specific types of longitudinal designs, each of these methods has its own advantages and disadvantages in practice. We present these smoothing estimation methods for the time-varying coefficient model (1.4) in Chapters 6 to 9.

2. The Shared-Parameter Change-Point Models

In addition to the time-varying linear structured modeling formulation as (1.4), a number of nonparametric extensions and alternatives of (1.4) may be considered in practice depending on the scientific objectives, biological interpretations and data structures of the study. These alternative modeling approaches are needed because, in many studies, the regression models as shown in (1.1) to (1.4) are misspecified, which may lead to biased and erroneous conclusions.

A particularly interesting scenario in longitudinal studies is the presence of a concomitant intervention in the sense that a particular intervention is introduced to a subject because of the unsatisfactory health outcomes observed in the past. In this case, a concomitant intervention cannot be treated as a usual covariate in the regression models (1.1) to (1.4), because a necessary condition for these models to be valid in practice is that the values of the covariates must not depend on the past values of the outcomes. Under the assumption that there is only one concomitant intervention and a subject can switch from "without concomitant intervention" to "with concomitant intervention" only once, a viable way to model the effects of the concomitant intervention and other covariates on the outcome variable is to incorporate the change-point time into the model.

Let S_i be the concomitant intervention change-point time of the ith subject. It is then reasonable to assume that the outcome $Y_i(t)$ of this subject follows different trajectories before and after the change-point time S_i. To do this, we denote by $\mu_0(t, \mathbf{X}_i; \mathbf{a}_i)$ the outcome trajectory at time t before the concomitant intervention, which is determined by the subject-specific parameter vector $\mathbf{a}_i$, and by $\mu_1(t, \mathbf{X}_i; \mathbf{b}_i)$ the change of the outcome trajectory after the concomitant intervention, which depends on the subject-specific parameter vector $\mathbf{b}_i$. The relationship between Y_{ij} and $\{t_{ij}, \mathbf{X}_i, S_i\}$ is described by

$$\begin{cases} Y_{ij} = \mu_0(t, \mathbf{X}_i; \mathbf{a}_i) + \delta_{ij}\mu_1(t, \mathbf{X}_i; \mathbf{b}_i) + \varepsilon_{ij}, \\ (\mathbf{a}_i^T, \mathbf{b}_i^T, S_i)^T \sim \text{Joint Distribution}, \end{cases} \tag{1.5}$$

where $\delta_{ij} = 1_{[t_{ij} \geq S_i]}$ is the indicator of whether the subject is taking concomitant intervention at time t_{ij} and ε_{ij} are the mean zero errors with some covariance structure at time points t_{ij_1} and t_{ij_2} when $j_1 \neq j_2$. In (1.5), the outcome trajectory before the concomitant intervention time S_i, $\mu_0(t, \mathbf{X}_i; \mathbf{a}_i)$, affects the change-point time S_i through the joint distribution of $\mathbf{a}_i$ and S_i. At time points t_{ij} after taking the concomitant intervention, i.e., $\delta_{ij} = 1$, the outcome trajectory becomes $\mu_0(t_{ij}, \mathbf{X}_i; \mathbf{a}_i) + \mu_1(t_{ij}, \mathbf{X}_i; \mathbf{b}_i)$, in which the change-point time S_i affects the value of Y_{ij} through the joint distribution of $\mathbf{b}_i$ and S_i.

Although (1.5) shares the parameters $\mathbf{a}_i$ and $\mathbf{b}_i$ in the trajectory of Y_{ij} and the joint distribution with S_i, hence the name "shared-parameter change-point" model, the terms $\mu_0(t, \mathbf{X}_i; \mathbf{a}_i)$ and $\mu_1(t_{ij}, \mathbf{X}_i; \mathbf{b}_i)$ can be flexible nonparametric curves of t. By specifying the functional forms of $\mu_0(t, \mathbf{X}_i; \mathbf{a}_i)$ and $\mu_1(t_{ij}, \mathbf{X}_i; \mathbf{b}_i)$, (1.5) fits into the framework of structured nonparametric models. We discuss the details of model formulation and interpretations and the estimation methods of (1.5) in Chapter 10.

3. The Nonparametric Mixed-Effects Models

Another useful class of structured nonparametric regression models for a longitudinal sample $\{(Y_{ij}, t_{ij}, \mathbf{X}_{ij}) : i = 1, \ldots, n; j = 1, \ldots, n_i\}$ is a more flexible version of the classical mixed-effects models to be reviewed in Section 2.1. To see why this flexible extension is potentially useful in real applications, we consider the simple case of modeling Y_{ij} as a function of t_{ij} without the covariates $\mathbf{X}_{ij}$. If we denote by $Y_i(t)$ the ith subject's outcome value at time t, we can describe the relationship between $Y_i(t)$ and t by

$$Y_i(t) = \beta_0(t) + \beta_{0i}(t) + \varepsilon_i(t), \tag{1.6}$$

where $\beta_0(t) = E[Y_i(t)]$ is the population-mean curve of t and, for the ith subject, $\beta_{0i}(t)$ is the subject-specific curve and $\varepsilon_i(t)$ is the measurement error at time t. Since the parametric forms of $\beta_0(t)$ and $\beta_{0i}(t)$ are often unknown in practice, a reasonable flexible model is to assume that $\beta_0(t)$ and $\beta_{0i}(t)$ are unknown smooth curves of t, so that the population-mean time-trend of $Y(t)$ can be evaluated by the curve estimates of $\beta_0(t)$, and the subject-specific outcome trajectory of $Y_i(t)$ can be predicted by the estimates of $[\beta_0(t) + \beta_{0i}(t)]$. Here, we use the convention which refers an estimate of the population-mean curve to a curve estimator and an estimate of the subject-specific curve to a curve predictor.

The advantages of using (1.6) include two main aspects:

(a) By decomposing the trajectory of $Y_i(t)$ as the sum of a population-mean curve, a subject-specific deviation curve from the population and a measurement error, (1.6) establishes a clearly interpretable mechanism that can be used to construct a reasonable covariance structure for the repeated measurements.

(b) The predictor of the subject-specific curve over t can be used to track the outcomes of interest at different time points and evaluate the distributions of the outcomes over time.

The above advantages have important applications in real applications when the scientific questions can be better answered by evaluating the correlation structures and tracking the individual subject's outcome trajectories. We describe the details of (1.6), its extensions with time-varying covariates, and the corresponding estimation and inference procedures in Chapter 11. We further describe the estimation methods for distribution functions and longitudinal tracking based on (1.6) and its extensions with time-varying covariates in Chapter 15.

4. Other Modeling Approaches

The structured nonparametric models described above only represent a number of frequently used flexible structural approaches in longitudinal analysis. In real applications, these models are clearly not enough to cover all the potentially important and scientifically interpretable structures. In some circumstances, the models described in this book can be directly extended to meet the practical needs of real studies. In general, however, there are various modeling structures which are beyond the scope of this book and have been developed for various scientific reasons and data structures. In order to maintain focus, we limit the scope of this book to the most frequently used modeling structures.

1.4   Conditional-Distribution Based Models

Beyond the above conditional-mean based models, regression models for conditional distributions are often used in longitudinal analysis. This is particularly true when the distributions of the response variables or their transformed variables are unknown, non-normal or asymmetric. In many longitudinal studies, the scientific objectives cannot be achieved using the conditional-mean based models, and answers to the relevant study questions require appropriate statistical inferences for the conditional distributions. Flexible regression models for conditional distribution functions are then more appropriate tools than the conditional-mean based regression models.

1.4.1   *Conditional Distribution Functions and Functionals*

We present a number of useful formulations of conditional distributions and their functionals. These distribution functions and functionals are useful to show the general patterns of the population. In many instances, they are more informative than only evaluating the conditional means. For example, conditional distributions and their functionals can be used to define a subject's health status and track the disease risks over time.

1. Conditional Distribution Functions

When the outcome variables are discrete, a well-known parametric approach is to consider the generalized linear mixed-effects models (e.g., Molenberghs and Verbeke, 2005), which may also be applied to the discretized versions of continuous outcome variables defined by some prespecified threshold values. When there are no appropriate threshold values or there are no existing transformations for the outcome variables, a natural method for the analysis of $\{Y(t), t, \mathbf{X}(t)\}$ is to directly model the conditional distribution functions

$$P_A(\mathbf{x}, t) = P\big[Y(t) \in A(\mathbf{x}, t)\big|\mathbf{X}(t) = \mathbf{x}, t\big], \tag{1.7}$$

where $A(\mathbf{x}, t)$ is a subset on the real line chosen by the scientific objectives of the analysis. In particular, if $A(\mathbf{x}, t) = (-\infty, y]$ for some real valued y, (1.7) is the conditional cumulative distribution function (CDF)

$$F_t(y|\mathbf{x}) = P\big[Y(t) \le y\big|\mathbf{X}(t) = \mathbf{x}, t\big]. \tag{1.8}$$

The statistical objective of the analysis is to construct estimates and inferences for $P_A(\mathbf{x}, t)$ or $F_t(y|\mathbf{x})$, when these quantities are considered as functions of $\mathbf{x}$ and t. The relationship between $\mathbf{x}$ and $P_A(\mathbf{x}, t)$ or $F_t(y|\mathbf{x})$ shows the covariate effects on the outcome distributions at a given time point t. On the other hand, the covariate effects are possibly time-varying, and the change of $P_A(\mathbf{x}, t)$ or $F_t(y|\mathbf{x})$ as a function of t illustrates the time-trends of $P_A(\mathbf{x}, t)$ or $F_t(y|\mathbf{x})$, respectively.

The applications of $P_A(\mathbf{x}, t)$ or $F_t(y|\mathbf{x})$ in biomedical studies are often originated from evaluating the health status or disease risk levels of an individual or a group of subjects from a chosen population. The set $A(\mathbf{x}, t)$, which defines health status or disease risk levels, is chosen based on the study objectives. In some situations, $A(\mathbf{x}, t)$ is obtained from other studies and treated as known for the current study. In general, $A(\mathbf{x}, t)$ is possibly unknown and may need to be estimated from the same longitudinal sample.

2. Conditional Quantiles and Other Functionals

Various functionals of the conditional distribution functions may also be of interest in longitudinal studies. Choices of these functionals depend on the scientific objectives of the study. A useful functional of the conditional CDF $F_t(y|\mathbf{x})$ is the conditional quantile function given by

$$y_\alpha(t, \mathbf{x}) = F_t^{-1}(\alpha|\mathbf{x}), \tag{1.9}$$

where $F_t^{-1}(\alpha|\mathbf{x})$ is the unique inverse of $F_t(y|\mathbf{x}) = \alpha$ for any $0 < \alpha < 1$ and any given t and $\mathbf{X}(t) = \mathbf{x}$, and $y_\alpha(t, \mathbf{x})$ is the $(100 \times \alpha)$th conditional quantile given $\{t, \mathbf{x}\}$. Other useful functionals include the conditional inter-quantile range

$$\delta_{y_{\alpha_1}, y_{\alpha_2}}(t_1, \mathbf{x}_1; t_2, \mathbf{x}_2) = y_{\alpha_1}(t_1, \mathbf{x}_1) - y_{\alpha_2}(t_2, \mathbf{x}_2) \tag{1.10}$$

for any choices of $\{\alpha_1, t_1, \mathbf{x}_1\}$ and $\{\alpha_2, t_2, \mathbf{x}_2\}$. Conditional quantiles have been used to develop the national guidelines for the diagnosis, evaluation and treatment of high blood pressure in children and adolescents (NHBPEP, 2004).

3. Distribution-Based Tracking

Given the time-varying information provided by the longitudinal observations, an important objective is to evaluate the tracking ability among subjects with certain health status. The scientific values of evaluating the tracking abilities of cardiovascular risk factors in pediatric studies have been discussed by, for example, Kavey et al. (2003), Thompson et al. (2007) and Obarzanek et al. (2010). The concept of tracking ability can be quantified by evaluating whether a subject's health outcome at an earlier time point affects the distribution of the health outcome at a later time point. We discuss a few statistical tracking indices based on the concept of maintaining the relative ranks over time within the population.

Rank-Tracking Probability:
Suppose that there is a pre-defined set of health outcomes $A(t)$ at any given time point t, so that a subject's health outcome at a time point t can be determined by whether $Y(t) \in A(t)$. A simple and direct way to measure the tracking ability of $Y(t)$ at two time points $s_1 < s_2$ is to use the "rank-tracking probability" (RTP) defined by

$$RTP_{s_1,s_2}(A, B) = P\big[Y(s_2) \in A(s_2) \big| Y(s_1) \in A(s_1), \mathbf{X}(s_1) \in B(s_1)\big], \qquad (1.11)$$

where $B(t) \subset R^{K+1}$ is a pre-specified subset for the covariates at time point t. Since $RTP_{s_1,s_2}(A, B)$ is a conditional probability, its values are within $[0, 1]$, and a large value of $RTP_{s_1,s_2}(A, B)$ would suggest that, given $Y(s_1) \in A(s_1)$ and $\mathbf{X}(s_1) \in B(s_1)$, the probability of $Y(s_2) \in A(s_2)$ is large.   □

Interpretations of Rank-Tracking Probability:
The strength of tracking $Y(t)$ is actually measured by the value of $RTP_{s_1,s_2}(A, B)$ relative to the conditional probability of $Y(s_2) \in A(s_2)$ without knowing $Y(s_1) \in A(s_1)$, that is, $P[Y(s_2) \in A(s_2)|\mathbf{X}(s_1) \in B(s_1)]$. If

$$RTP_{s_1,s_2}(A, B) = P\big[Y(s_2) \in A(s_2)\big|\mathbf{X}(s_1) \in B(s_1)\big], \qquad (1.12)$$

then knowing $Y(s_1) \in A(s_1)$ does not increase the conditional probability of $Y(s_2) \in A(s_2)$ given $\mathbf{X}(s_1) \in B(s_1)$. The equation in (1.11) then suggests that $Y(s_1) \in A(s_1)$ has no tracking ability for $Y(s_2) \in A(s_2)$ conditioning on $\mathbf{X}(s_1) \in B(s_1)$. On the other hand, $Y(s_1) \in A(s_1)$ can be defined to have positive or negative tracking value for $Y(s_2) \in A(s_2)$ conditioning on $\mathbf{X}(s_1) \in B(s_1)$, if

$$RTP_{s_1,s_2}(A, B) > P[Y(s_2) \in A(s_2)|\mathbf{X}(s_1) \in B(s_1)] \qquad (1.13)$$

or

$$RTP_{s_1, s_2}(A, B) < P[Y(s_2) \in A(s_2) | \mathbf{X}(s_1) \in B(s_1)] \qquad (1.14)$$

respectively. It clearly follows that $Y(s_1)$ has no tracking ability for $Y(s_2)$ if $Y(s_1)$ and $Y(s_2)$ are conditionally independent given $\mathbf{X}(s_1) \in B(s_1)$. $\qquad\qquad \Box$

Tracking is an extremely useful feature in long-term studies of population changing patterns and can only be evaluated by longitudinal studies with sufficient numbers of repeated measurements. This feature illustrates the major advantage of long-term longitudinal studies over the simpler studies with cross-sectional i.i.d. data or longitudinal studies with small numbers of repeated measurements.

1.4.2 Parametric Distribution Models

If $P_A(\mathbf{x}, t)$ or $F_t(y|\mathbf{x})$ can be determined by a finite dimensional parameter θ within a parameter space Θ, we obtain the parametric models $P_{A, \theta}(\mathbf{x}, t)$ and $F_{t, \theta}(y|\mathbf{x})$ for $P_A(\mathbf{x}, t)$ or $F_t(y|\mathbf{x})$, respectively, with $\theta \in \Theta$. Here, Θ is often taken as an open subset in the Euclidean space. The linear mixed-effects model to be reviewed in Chapter 2.1 is a special case of the parametric models with normal distribution assumptions. Under the parametric modeling assumptions, the conditional distribution functions $P_{A, \theta}(\mathbf{x}, t)$ and $F_{t, \theta}(y|\mathbf{x})$ can be in principle estimated by first estimating the parameter θ using a maximum likelihood procedure and then substituting θ in $P_{A, \theta}(\mathbf{x}, t)$ and $F_{t, \theta}(y|\mathbf{x})$ with its maximum likelihood estimator. Completely specified parametric models $P_{A, \theta}(\mathbf{x}, t)$ and $F_{t, \theta}(y|\mathbf{x})$ lack the much needed flexibility in practice, hence, may lead to biased conclusions when the models are misspecified. These models are not the focus of this book.

1.4.3 Semiparametric Distribution Models

Since the fully parametrized distributions for $P_A(\mathbf{x}, t)$ and $F_t(y|\mathbf{x})$ may not be always available in practice, a relatively more flexible approach is to consider modeling $P_A(\mathbf{x}, t)$ and $F_t(y|\mathbf{x})$ by a semiparametric family through a combination of nonparametric components and finite dimensional parameters. This approach is similar to the semiparametric modeling of Section 1.3.2 for the conditional means. In particular, when $\{Y_i(t) : i = 1, \ldots, n\}$ have normal distributions for any given t, similar semiparametric models for the conditional means, such as the partially linear model (1.2), can also be used to evaluate the conditional distributions of $Y_i(t)$.

Semiparametric models specifically developed for evaluating the conditional distribution functions have also been extensively studied in the literature. A well-known example of semiparametric models in survival analysis is the transformation models, which have been studied, for example, by Cheng, Wei and Ying (1995, 1997). Although the longitudinal data considered in this book (Section 1.2.1) have different structures from the usual time-to-event outcome variables considered in survival analysis, the linear transformation model can be adapted to the current longitudinal data with some modifications. A key feature under the current context is to model the conditional distribution of $Y(t)$ given the covariates $\mathbf{X}(t)$ at any given time point t.

Assume that $F_t(y|\mathbf{x})$ of (1.8) is continuous in y for any given t and $\mathbf{x}$. Let

$$S_t(y|\mathbf{x}) = 1 - F_t(y|\mathbf{x}), \tag{1.15}$$

which is often referred to as the "survival function" in survival analysis when the outcome is a time-to-event variable. Under the context of longitudinal data, a semiparametric linear transformation model for $F_t(y|\mathbf{x})$ is

$$g\{S_t[y|\mathbf{X}(t)]\} = h(y, t) + \mathbf{X}^T(t)\beta, \tag{1.16}$$

where $g(\cdot)$ is a known decreasing link function, $\beta = (\beta_0, \beta_1, \ldots, \beta_K)^T$ is the parameter vector describing the covariate effects, and $h(y, t)$ is an unknown baseline function strictly increasing in y. By leaving $h(y, t)$ to be a nonparametric function of (y, t), (1.16) incorporates both the nonparametric component $h(y, t)$ and the parameter vector β, hence, it leads to a semiparametric model for $F_t(y|\mathbf{x})$. When t is fixed, (1.16) is just the semiparametric linear transformation model studied by Cheng, Wei and Ying (1995, 1997). When t changes across the time range, (1.16) has the added feature of time-trends described by the baseline function $h(y, t)$, while keeping β as a time-invariant multivariate parameter.

Despite the popularity of the transformation models in survival analysis, its extension (1.16) falls short of the model flexibility targeted by this book, because the covariate effect characterized by β is assumed to stay constant for all t. In real biomedical applications, the covariate effects are likely to change with t. This dynamic nature of the covariate effects leads to the consideration of nonparametric regression models for conditional distributions.

1.4.4 Unstructured Nonparametric Distribution Models

Completely unstructured nonparametric models for $P_A(\mathbf{x}, t)$ and $F_t(y|\mathbf{x})$ may also be considered when there are no suitable parametric or semiparameter models available for these functions. Such situations may arise when the potential bias caused by the possible model misspecification is a major concern. Unstructured nonparametric estimation of the conditional CDF with cross-sectional i.i.d. data and certain time series samples has been studied by Hall, Wolff and Yao (1999) based on two kernel smoothing methods, the local logistic estimation method and the adjusted Nadaraya-Watson method.

However, a completely unstructured nonparametric formulation of $P_A(\mathbf{x}, t)$ or $F_t(y|\mathbf{x})$ is determined by both t and the $K+1$ components of $\mathbf{x}$, which can be difficult to estimate if K is large because of the well-known problem of "curse of dimensionality" (e.g., Fan and Gijbels, 1996). Even if the completely unstructured nonparametric estimators of $P_A(\mathbf{x}, t)$ or $F_t(y|\mathbf{x})$ and their inferences are available, the results for $P_A(\mathbf{x}, t)$ or $F_t(y|\mathbf{x})$ are usually difficult to interpret when $K \geq 2$. This drawback severely limits the use of unstructured nonparametric models for $P_A(\mathbf{x}, t)$ or $F_t(y|\mathbf{x})$.

Despite the shortcomings of unstructured nonparametric models, we present in Chapter 12 some useful smoothing methods for estimating the conditional distribution functions and their functionals. These methods are further modified in later chapters to estimate the distribution functions and covariate effects under a number of structured nonparametric models.

1.4.5   Structured Nonparametric Distribution Models

By imposing a functional structure on $F_t(y|\mathbf{x})$, structured nonparametric models can be applied as a useful dimension reduction strategy to alleviate the potential instability associated with the high dimensional nonparametric estimation of $F_t(y|\mathbf{x})$. We describe here a few such structured modeling approaches.

1. The Time-Varying Transformation Models

This structured approach is a direct generalization of (1.16) by generalizing the linear coefficients β to be nonparametric coefficient functions of time

$$\beta(t) = \big(\beta_0(t), \ldots, \beta_K(t)\big)^T, \tag{1.17}$$

so that, by substituting β of (1.16) with $\beta(t)$, the time-varying transformation model for the conditional CDF $F_t\big[y|\mathbf{X}(t)\big] = 1 - S_t\big[y|\mathbf{X}(t)\big]$ is given by

$$g\big\{S_t\big[y|\mathbf{X}(t)\big]\big\} = h(y, t) + \mathbf{X}^T(t)\beta(t). \tag{1.18}$$

Similar to the more restrictive semiparametric model (1.16), the link function $g(\cdot)$ in (1.18) is known. The linear coefficients $\beta_l(t)$, $l = 0, \ldots, K$, describe the time-varying covariate effects on the conditional CDF $F_t\big[y|\mathbf{X}(t)\big]$. Different choices of $g(\cdot)$ also lead to different covariate effects on $F_t\big[y|\mathbf{X}(t)\big]$.

Motivated by the NGHS example of Section 1.2.2, the time-varying transformation model (1.18) was introduced by Wu, Tian and Yu (2010) to evaluate the covariate effects on the time-varying distributions of various cardiovascular risk factors for children and adolescents. The conditional-distribution based regression models are appropriate for pediatric studies because health status and disease risk levels for children and adolescents are often determined by the distributions of risk variables conditioning on age, gender and other covariates. The model (1.18) keeps a reasonable balance between model flexibility and complexity. Details of the estimation methods and application of (1.18) are presented in Chapters 13 and 14.

2. The Mixed-Effects Varying-Coefficient Models

Another class of structured nonparametric models for $F_t(y|\mathbf{x})$, which can be treated as a natural generalization of the parametric family $F_t(y|\mathbf{x})$, is to allow the parameters to be functions of time, i.e., substituting the θ in $F_{t,\theta}(y|\mathbf{x})$ with $\theta(t)$. The resulting distribution function $F_{t,\theta(t)}(y|\mathbf{x})$ then belongs to a parametric family at each fixed time point t. When t changes, the values of the parameters may also change. This modeling approach leads to a class of time-varying parametric models, which includes the time-varying coefficient model as a special case. In particular, if the error term of (1.3) has a mean zero normal distribution with time-varying variance $\sigma^2(t)$, then the time-varying coefficient model (1.3) is equivalent to the time-varying Gaussian model with the conditional mean $\mathbf{X}^T(t)\beta(t)$ given $\mathbf{X}(t)$.

To include a subject-specific effect curve into the model formulation, we may assume that the outcome distribution of the ith subject deviates from the population and its subject-specific deviation from the population is charac-terized by a parameter curve $\theta_i(t)$ at time t. This structured modeling scheme leads to the class of "mixed-effects varying-coefficient models." A simple special case of the mixed-effects varying-coefficient model with a univariate covariate $X_i(t)$ is

$$Y_i(t) = \beta_0(t) + \left[\beta_1(t) + \beta_{1i}(t)\right] X_i(t) + \varepsilon_i(t), \qquad (1.19)$$

where $\beta_0(t)$ and $\beta_1(t)$ are the population-mean intercept and coefficient curves, respectively, $\beta_{1i}(t)$ is the subject-specific deviance curve from the population, and $\varepsilon_i(t)$ is the mean zero error process. Although (1.19) is a conditional-mean based regression model, it can be used to estimate the conditional distributions and their functionals of $Y_i(t)$ at different time points when the distributions of the error process $\varepsilon_i(t)$ are specified.

Various generalizations of (1.19) can be established by including different population-mean and subject-specific coefficient curves. We present in Chap-ter 15 the estimation methods, and applications and generalizations for eval-uating the conditional distributions and their functionals.

1.5 Review of Smoothing Methods

Since this book mainly focuses on the estimation, inferences and applications of nonparametric models (both unstructured and structured nonparametric models), smoothing methods, which have been widely used in nonparametric curve estimation with cross-sectional i.i.d. data, are used throughout the book in conjunction with the modeling structures. In order to gain some useful insights into the commonly used smoothing methods, we briefly review these methods under the simple cross-sectional i.i.d. data

$$\mathscr{Y} = \left\{ (Y_i, t_i)^T : i = 1, \ldots, n \right\} \qquad (1.20)$$

without the presence of $\mathbf{X}(t)$. Suppose that, under the i.i.d. sample $\mathscr{Y}$ for $(Y(t), t)^T$, the statistical interest is to estimate the smooth conditional-mean function $\mu(t) = E\left[Y(t)|t\right]$ under the regression model

$$Y(t) = \mu(t) + \varepsilon(t), \qquad (1.21)$$

where $\varepsilon(t)$ is a mean zero stochastic process with variance σ^2. We summarize below a number of local and global smoothing methods for the estimation of $\mu(t)$, which is assumed to be a smooth function of t. A local smoothing method for $\mu(t)$ refers to the estimation methods using primarily the observations within some neighborhoods of the time t. A global smoothing method for $\mu(t)$ then relies on all the observations. Both local and global smoothing methods have their advantages and disadvantages in real applications.

1.5.1   Local Smoothing Methods

Throughout this book, we consider two kernel-based local smoothing methods: the kernel estimators and the local polynomial estimators. This is because the majority of the local smoothing methods for structured nonparametric models with longitudinal data are developed using kernel-based methods. Other local smoothing methods, such as the locally weighted scatter plot smoothing (LOESS) method, are useful in nonparametric regression, but their theory and applications in longitudinal studies have not been substantially investigated.

1. Kernel Smoothing Methods

The Nadaraya-Watson kernel estimator of $\mu(t)$ can be obtained by minimizing the "local least squares criterion"

$$\ell_K(t) = \sum_{i=1}^{n} [Y_i - \mu(t)]^2 \left(\frac{1}{nh}\right) K\left(\frac{t - t_i}{h}\right),\qquad(1.22)$$

with respect to $\mu(t)$, where $K(\cdot)$ is a kernel function, which is usually taken to be a non-negative probability density function, and $h > 0$ is a bandwidth. Setting the partial derivative of $\ell_K(t)$ with respect to $\mu(t)$ to zero, the Nadaraya-Watson kernel estimator $\widehat{\mu}_K(t)$ of $\mu(t)$ is given by

$$\widehat{\mu}_K(t) = \frac{\sum_{i=1}^{n} \{Y_i K[(t - t_i)/h]\}}{\sum_{i=1}^{n} \{K[(t - t_i)/h]\}}.\qquad(1.23)$$

It follows from (1.22) and (1.23) that $\widehat{\mu}_K(t)$ is a so-called local smoothing estimator of $\mu(t)$, because it is obtained by using the subjects whose t_i are within a neighborhood of t determined by the bandwidth h. This can be seen by considering the special case that $K(\cdot)$ is the density of a uniform distribution. Suppose that $K_U(s)$ is the uniform density on $[-a/2, a/2]$, such that

$$K_U(s) = a^{-1} 1_{[|s| \leq a/2]}\qquad(1.24)$$

for some $a > 0$, where $1_{[A]}$ is the indicator function such that $1_{[A]} = 1$ if A holds, and 0 otherwise. Then, it follows from (1.24) that, for any $h > 0$, $K_U(s)$ is a uniform kernel function, which satisfies

$$\begin{cases} K_U[(t-t_i)/h] = 1/a, & \text{if } t - (ha/2) \leq t_i \leq t + (ha/2); \\ K_U[(t-t_i)/h] = 0, & \text{if } t_i < t - (ha/2) \text{ or } t_i > t + (ha/2). \end{cases}$$

Since any value of $a > 0$ can be used for (1.24), simple choices include $a = 1$ or 2.

The equations (1.22) and (1.24) imply that the local least squares score function $\ell_{K_U}(t)$ under $K_U(\cdot)$ is

$$\ell_{K_U}(t) = \left(\frac{1}{nh}\right) \sum_{i=1}^{n} [Y_i - \mu(t)]^2 \, 1_{[t-(h/2) \leq t_i \leq t+(h/2)]}. \tag{1.25}$$

Consequently, (1.23) gives

$$\widehat{\mu}_{K_U}(t) = \frac{\sum_{i=1}^{n} Y_i \, 1_{[t-(h/2) \leq t_i \leq t+(h/2)]}}{\sum_{i=1}^{n} 1_{[t-(h/2) \leq t_i \leq t+(h/2)]}}. \tag{1.26}$$

Thus, for each given t, $\widehat{\mu}_{K_U}(t)$ is the local average of the Y_i's obtained from the subjects within the neighborhood of t_i's centered at t with radius $h/2$, that is, $|t_i - t| \leq h/2$. The neighborhood shrinks to t when h tends to zero.

When $K(s)$ is not a uniform density but has a bounded support in the sense that $K(s) = 0$ if $|s| > b$ for some constant $b > 0$, then $\widehat{\mu}_K(t)$ given in (1.23) is a weighted local average of the Y_i's within the neighborhood $|t_i - t| \leq h/2$ of t_i's. Some well-known kernel functions with bounded supports include the Epanechnikov kernel

$$K_E(s) = \frac{3}{4}\left(1 - s^2\right) 1_{[|s| \leq 1]}, \tag{1.27}$$

the triangular kernel

$$K_T(s) = \left(1 - |s|\right) 1_{[|s| \leq 1]}, \tag{1.28}$$

the quartic kernel

$$K_Q(s) = \frac{15}{16}\left(1 - s^2\right)^2 1_{[|s| \leq 1]}, \tag{1.29}$$

and the tricube kernel

$$K_C(s) = \frac{70}{81}\left(1 - |s|^3\right)^3 1_{[|s| \leq 1]}. \tag{1.30}$$

Kernel functions with unbounded supports may also be used in practice. A well-known example of the kernel functions with unbounded supports is the Gaussian kernel

$$K_G(s) = \frac{1}{\sqrt{2\pi}} \exp\left(-\frac{s^2}{2}\right). \tag{1.31}$$

The Epanechnikov kernel has been shown to have the optimality property that it minimizes the asymptotic mean squared errors of the kernel estimators under certain mild asymptotic assumptions (Härdle, 1990). Although different kernel functions lead to different local weights for kernel estimators, the asymptotic derivations, simulation studies and practical applications have shown that the theoretical and practical properties of kernel estimators are mostly influenced by the bandwidth choices but not so much by the shapes of the kernel functions (Härdle, 1990; Fan and Gijbels, 1996). This fact is again

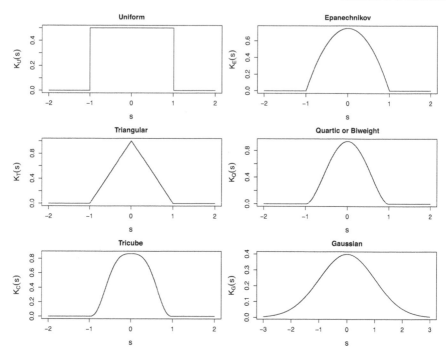

Figure 1.5 *Graphical depiction of the shapes of kernel functions: Uniform (a = 2 in (1.24)), Epanechnikov, Triangular, Quartic, Tricube and Gaussian.*

observed for the smoothing estimators discussed in this book. Thus, in real applications, all the kernel functions shown in this section can be used interchangeably. The main advantages of the Nadaraya-Watson kernel estimators are their computational simplicity and straightforward interpretations. The kernel estimators generally have appropriate theoretical properties, such as consistency and small mean squared errors, when the sample size is large and t is within the interior of its support.

2. Local Polynomial Estimators

A main drawback of the above kernel estimators, as demonstrated in Fan and Gijbels (1996), is that these estimators have excessive biases when t is close to the boundary of its support. Intuitively, the boundary bias is caused by the fact that, when t is a boundary point, the observations are only obtained with t_i at one side of t, which causes the local averages computed by the kernel estimators to be either lower or higher than the true value $\mu(t)$. The local polynomial estimators are a useful local smoothing approach to correct the potential boundary bias associated with the kernel estimators. Suppose that, when t_i is within a small neighborhood of t, $\mu(t_i)$ can be approximated by the

Taylor's expansion

$$\mu(t_i) \approx \sum_{l=0}^{p} \frac{\mu^{(l)}(t)}{l!} (t_i - t)^l = \sum_{l=0}^{p} b_l (t_i - t)^l \tag{1.32}$$

for an integer p. Using the same approach as (1.22) with the approximation of (1.32), the pth order local polynomial estimators can be obtained by minimizing

$$\ell_{p,K}(t) = \sum_{i=1}^{n} \left[Y_i - \sum_{l=0}^{p} b_l (t_i - t)^l \right]^2 \left(\frac{1}{nh} \right) K \left(\frac{t - t_i}{h} \right) \tag{1.33}$$

with respect to b_l, $l = 1, \ldots, p$, where $K(\cdot)$ is the same kernel function as in (1.22), which may be chosen as the ones given in (1.24), (1.27) through (1.30). Comparing (1.32) with (1.33), the minimizer $\widehat{b}_l$ leads to the pth order local polynomial estimator

$$\widehat{\mu}_{L,p}^{(l)}(t) = l! \widehat{b}_l \tag{1.34}$$

of the lth derivative $\mu^{(l)}(t)$ at time point t. The estimator of the entire curve $\mu^{(l)}(\cdot)$ can be obtained by minimizing (1.33) over all the time points within its support. In practice, it is sufficient to compute $\widehat{\mu}_{L,p}^{(l)}(\cdot)$ over a finite number of distinct time points. The pth order local polynomial estimator of $\mu(t)$ is

$$\widehat{\mu}_{L,p}(t) = \widehat{\mu}_{L,p}^{(0)}(t), \tag{1.35}$$

and, by selecting $p = 1$, the local linear estimator of $\mu(t)$ is $\widehat{\mu}_{L,1}(t)$. In all the applications discussed in this book, the main focus is on the estimation of the mean curve $\mu(t)$, rather than its derivatives $\mu^{(l)}(t)$ with $l \geq 1$.

Similar to the kernel estimators, the choice of kernel functions does not have significant influences on the theoretical and practical properties of $\widehat{\mu}_{L,p}(t)$. As discussed in Fan and Gijbels (1996), the main factors affecting the statistical properties of the local polynomial estimators are the bandwidth choices and the degree p of the polynomials. When p increases, the number of parameters used in the approximation (1.32) increases, which, on one hand, may reduce the estimation bias, but, on the other hand, will increase the computational complexity. In practice, the results of minimizing (1.33) with a large p may be numerically unstable, and the practical benefit of using a large p is not significant. For this reason, the local linear estimators with $p = 1$ are often the more popular choices than higher order local polynomial estimators in nonparametric smoothing. $\square$

1.5.2 Global Smoothing Methods

Among a large number of global smoothing methods in the literature, we focus in this book on the methods of basis splines (B-splines) and the penalized smoothing splines. These are the two most extensively studied global smoothing methods in nonparametric regression with longitudinal data. Between these two smoothing methods, the B-splines are more often used than the penalized smoothing splines for various structured nonparametric regression models. By approximating the smooth curves through some basis expansions, the B-splines can be viewed as "extended linear models," which can be readily adapted to many different regression structures. On the other hand, the penalized smoothing spline estimators depend on maximizing some "roughness penalized likelihood functions," which can be computationally difficult for many regression structures.

1. Basis Approximations and B-splines

These smoothing methods approximate the unknown function by expansions of some basis functions and are referred to in the nonparametric statistics literature as the "extended linear models" (Stone et al., 1997; Huang, 1998). Suppose that $\{B_0(t), B_1(t), \ldots\}$ is a set of basis functions, such as polynomial bases, Fourier bases or B-splines, and $\mu(t)$ can be approximated by the expansion

$$\mu(t) \approx \sum_{l=0}^{L} \gamma_l B_l(t), \tag{1.36}$$

where γ_l are the real-valued coefficients. The choices of basis functions depend on the nature and smoothness assumptions of $\mu(t)$. A simple choice is the global polynomial basis $B_l(t) = t^l$. Because the global polynomials can be numerically unstable when the order L is large, we commonly use B-splines, also referred to as polynomial splines, in real applications.

The B-splines are piece-wise polynomials with knots at the boundary and within the interior of the support of t. To give a brief description of the B-splines, we consider q real-valued knots t_j with $t_0 < t_1 < \cdots < t_{q-1}$, where t_0 and t_{q-1} are the boundary knots and the rest are the interior knots. A B-spline curve of degree r is a curve from $[t_r, t_{q-r-1}]$ to the real line composed of a linear combination of B-spline basis $B_{l,r}(t)$ of degree r, such that for $t \in [t_r, t_{q-r-1}]$,

$$S(t) = \sum_{l=0}^{q-r-2} c_l B_{l,r}(t),$$

where c_l are the control points or de Boor points. The B-spline basis is defined in such a way that, for $l = 0, \ldots, q-2$, $B_{l,0}(t) = 1$ if $t_l \leq t < t_{l+1}$ and $B_{l,0}(t) = 0$ otherwise, for $l = 0, \ldots, q-r-2$,

$$B_{l,r}(t) = \frac{t - t_l}{t_{l+r} - t_l} B_{l,r-1}(t) + \frac{t_{l+r+1} - t}{t_{l+r+1} - t_{l+1}} B_{l+1,r-1}(t), \tag{1.37}$$

and $l + r + 1 \leq q - 1$. The choices of $r = 1$, 2 and 3 correspond to the linear, quadratic and cubic B-splines, respectively. For example, if we use a natural cubic spline with $q - 2$ interior knots $\{t_1, \ldots, t_{q-2}\}$, the natural cubic spline is represented by $q-2$ basis functions $\{B_{0,1}(t), \ldots, B_{0,q-2}(t)\}$ defined by

$$\begin{cases} B_{0,1}(t) = 1, \\ B_{0,2}(t) = t, \\ B_{0,(l+2)}(t) = d_l(t) - d_{q-3}(t), \\ d_l(t) = \left[(t-t_l)_+^3 - (t-t_{q-2})_+^3 \right] \Big/ (t_{q-2} - t_l). \end{cases}$$

Here, for $l = 1, \ldots, q-4$, $(t-t_l)_+ = t - t_l$ if $t > t_l$, and $(t-t_l)_+ = 0$ if $t \leq t_l$. Each of the above basis functions has zero second and third derivative outside the boundary knots.

Substituting the approximation (1.36) into (1.21), the observed data $\{(Y_i, t_i)^T : i = 1, \ldots, n\}$ can be approximated by

$$Y_i \approx \sum_{l=0}^{L_n} \gamma_l B_l(t_i) + \varepsilon_i, \tag{1.38}$$

where ε_i, $i = 1, \ldots, n$, are independent error terms with mean zero and variance σ^2 and L_n may increase as n increases. By minimizing the square error

$$\ell_B(t; \gamma) = \frac{1}{n} \sum_{i=1}^{n} \left[\sum_{l=0}^{L_n} \gamma_l B_l(t_i) \right]^2, \tag{1.39}$$

the least squares estimator $\widehat{\gamma} = (\widehat{\gamma}_1, \ldots, \widehat{\gamma}_{L_n})^T$ of $\gamma = (\gamma_1, \ldots, \gamma_{L_n})^T$, if exists, is the unique minimizer of (1.39). The smoothing estimator of $\mu(t)$ based on the basis functions $\{B_0(t), B_1(t), \ldots, B_L(t)\}$ is

$$\widehat{\mu}_B(t) = \sum_{l=0}^{L} \widehat{\gamma}_l B_l(t). \tag{1.40}$$

For most biomedical applications, the linear, quadratic and cubic B-splines are commonly used in practice. Depending on the nature of the specific applications, other basis choices have also been used in the literature.

2. Penalized Smoothing Splines

Another class of global smoothing estimators of $\mu(t)$ is obtained by minimizing a penalized squared error criterion. Let $Q(\mu)$ be a penalty term defined by the smoothness requirement of $\mu(t)$. A smoothing estimator $\widehat{\mu}_\lambda(t)$ of $\mu(t)$ can be obtained by minimizing the penalized least squares criterion

$$L_Q(\mu; \lambda) = \frac{1}{n} \sum_{i=1}^{n} \left[Y_i - \mu(t_i) \right]^2 + \lambda Q(\mu), \tag{1.41}$$

where λ is a smoothing parameter. The degree of smoothing of the estimator

$\mu_\lambda(t)$ depends on the value of λ. A large λ in (1.41) would lead to a smoother $\mu_\lambda(t)$, while a small λ leads to a less smooth $\mu_\lambda(t)$.

In addition to λ, the smoothing estimator is also determined by the smoothness penalty term $Q(\mu)$. Let $\mu''(t)$ be the second derivative of $\mu(t)$ with respect to t. If $Q(\mu) = \int_a^b |\mu''(s)| ds$ is chosen to be the L_1-norm on the interval $[a, b]$, a L_1-penalized smoothing estimator $\widehat{\mu}_Q(t)$ of $\mu(t)$ can be obtained by minimizing

$$L_Q(\mu; \lambda) = \frac{1}{n} \sum_{i=1}^{n} \left[Y_i - \mu(t_i) \right]^2 + \lambda \int_a^b |\mu''(s)| ds. \qquad (1.42)$$

Alternatively, using the L_2 penalized criterion $Q(\mu) = \int_a^b \left[\mu''(s) \right]^2 ds$ for $[a, b]$, a L_2-penalized smoothing estimator $\widehat{\mu}_Q(t)$ of $\mu(t)$ can be obtained by minimizing

$$L_{Q_2}(\mu; \lambda) = \frac{1}{n} \sum_{i=1}^{n} \left[Y_i - \mu(t_i) \right]^2 + \lambda \int_a^b \left[\mu''(s) \right]^2 ds. \qquad (1.43)$$

Intuitively, (1.42) has a large penalizing term when the absolute values of the second derivatives of $\mu(t)$ are large, and (1.43) has a large penalizing term when the squares of the second derivatives of $\mu(t)$ are large. The minimizers of (1.42) and (1.43) are natural cubic splines, as shown, for example, in Green and Silverman (1994) and Eubank (1999).

Other penalized least squares criteria may be constructed by replacing $\mu''(t)$ of (1.42) and (1.43) with $\mu'(t)$ or other roughness penalizing terms. Although the choice of roughness penalizing terms may be ideally selected by the scientific nature of the problem, such a choice is not always available in practice. In real applications, the roughness penalizing terms are often chosen subjectively or by comparing the resulting smoothing estimators. The choice of the smoothing parameter λ is more important than the penalty function $Q(\mu)$ in determining the smoothness and appropriateness of the estimators.

1.6   Introduction to R

The main statistical computing tools we use in this book are the R language and several R packages (R Core Team, 2017). R is a popular language with various statistical applications, including data import, manipulation, graphics and model fitting. We provide examples of R code and outputs to implement the statistical methods discussed in each of the following chapters so that readers with minimal background knowledge can easily try out the models and sample code to fit their own data. We note that these statistical methods are also available or can be implemented in SAS, MATLAB, Python, or other statistical software and languages.

Both R and R packages are free and open-source software, available on the Comprehensive R Archive Network, or CRAN, via the website https://www.r-project.org. It is straightforward to download from CRAN and install a current version of the pre-compiled binary distribution of the R base system and contributed packages. For a comprehensive introduction to R and its applications, we refer to the R manuals that come as part of the R installation, and several comprehensive textbooks, e.g., Venables and Ripley (2002), Dalgaard (2008) and James et al. (2013).

A major advantage of R is that a large number of add-on packages developed by the R Core Team and users are also freely available from CRAN. As of January 2018, CRAN package repository has over 12,000 packages. An R package can be a collection of datasets, functions and documentation, and many R packages provide statistical tools that are widely used among R users. We can benefit and make use of a certain package by installing it from CRAN and then loading it into R with `library()` or `require()`:

```
> install.packages("pkgname")
> library(pkgname)
```

We have implemented the statistical analyses discussed in this book with R version 3.4.3 and compiled the datasets and relevant functions into a R package **npmlda**, which can be obtained from CRAN and GitHub. To install our package from the GitHub repository, we can use `install_github` function in the **devtools** package (Wickham, 2015):

```
> library(devtools)
> install_github("npmldabook/npmlda")
```

The R code for the examples used in this book is available online from the supporting website (https://github.com/npmldabook/rcodes). The authors would appreciate to be informed of any issues, suggestions and improvements on this book.

1.7 Organization of the Book

This book contains two major topics of nonparametric regression models: the conditional-mean based regression models and the conditional-distribution based regression models. To give a clear picture of the model structures, we organize this book into five main parts:

I: *Introduction and Review* (Chapters 1 and 2);

II: *Unstructured Nonparametric Models* (Chapters 3, 4 and 5);

III: *Time-Varying Coefficient Models* (Chapters 6, 7, 8 and 9);

IV: *Shared-Parameter and Mixed-Effects Models* (Chapters 10 and 11);

V: *Nonparametric Models for Distributions* (Chapters 12, 13, 14 and 15).

As an initial building block of the models, Chapter 2 briefly summarizes the main results of the methodology used for longitudinal analysis with parametric and semiparametric regression models, and Chapters 3, 4 and 5 in Part II present the local and global smoothing methods for unstructured nonparametric models.

The main results of the conditional mean based structured nonparametric models are presented in Parts III and IV, where Part III summarizes the smoothing estimation methods for the time-varying coefficient models under a number of different data structures and Part IV presents some extensions of the models in Part III.

Part V summarizes some recent developments in the modeling, estimation and applications of the conditional distribution based structured nonparametric regression models. Two important concepts of these chapters are (a) modeling the changing patterns of conditional distribution functions and their functionals over time, and (b) quantifying and estimating the tracking indices of outcome variables over time based on the conditional distributions. These two concepts demonstrate the advantages of the conditional distribution based models over the conditional mean based modeling schemes for certain objectives of longitudinal analysis.

We attempt to maintain a reasonable balance among methods, theory, applications and implementations. In addition to the methodology and theoretical derivations presented in each chapter, the R packages are used throughout the book to illustrate the applications of the statistical methods. The graphs and tables of each application are accompanied by the corresponding R code.

Chapter 2

Parametric and Semiparametric Methods

We briefly review in this chapter a number of popular parametric and semiparametric models in longitudinal analysis. Because of their simple mathematical structures and interpretations, these models are often the first set of analytical tools to be used in a real study. The estimation and inference methods developed for these models form the foundation of longitudinal analysis. Since the nonparametric models to be presented in this book are flexible extensions of these parametric and semiparametric models, their local and global smoothing estimation methods with different modeling structures are motivated and generalized from the estimation methods in this chapter. We note that, because the topics of parametric and semiparametric longitudinal analysis have been extensively studied in the literature, the models and estimation methods reviewed here only represent a fraction of the available approaches. A more complete account of the most commonly used parametric and semiparametric methods in longitudinal analysis can be found in Fitzmaurice et al. (2009).

2.1 Linear Marginal and Mixed-Effects Models

As a popular approach for modeling the covariate effects on the longitudinal outcome variables, the mixed-effects models generally serve two purposes:

(1) Describe the covariate effects on the mean response profiles.

(2) Describe the subject-specific response profiles.

A regression model serving the first purpose is generally classified as a marginal model or a population-mean model. A regression model serving the second purpose is a random-effects model or a subject-specific model (e.g., Zeger, Liang and Albert, 1988). A mixed-effects model then incorporates both the marginal and random effects. In particular, a linear mixed-effects model is obtained when the marginal and random effects are additive and follow a linear relationship.

2.1.1 Marginal Linear Models

It is convenient to describe the models through a matrix representation. Let the ith subject's responses, time design points, and covariate matrix be

$$\begin{cases} \mathbf{Y}_i &= \left(Y_{i1},\dots,Y_{in_i}\right)^T, \\ \mathbf{t}_i &= \left(t_{i1},\dots,t_{in_i}\right)^T, \\ \mathbf{X}_i &= \begin{pmatrix} 1 & X_{i1}^{(1)} & \cdots & X_{i1}^{(K)} \\ \vdots & \vdots & \vdots & \vdots \\ 1 & X_{in_i}^{(1)} & \cdots & X_{in_i}^{(K)} \end{pmatrix}. \end{cases} \tag{2.1}$$

Note that, in order to allow for an intercept term in the models described below, we set $X_{ij}^{(0)} = 1$ in $\mathbf{X}_i$. The marginal linear model (1.1) for (2.1) is

$$\mathbf{Y}_i = \mathbf{X}_i \beta + \varepsilon_i(\mathbf{t}_i), \tag{2.2}$$

where $\beta = (\beta_0,\dots,\beta_K)^T$ is the vector of linear coefficients with β_0 being the unknown intercept and β_k, $1 \le k \le K$, describing the effect of the kth covariate $X_{ij}^{(k)}$, and $\varepsilon_i(\mathbf{t}_i) = \left(\varepsilon_i(t_{i1}),\dots,\varepsilon_i(t_{in_i})\right)^T$ with $\varepsilon_i(t_{ij})$ being the realization of a mean zero random error process $\varepsilon_i(t)$ at time point $t = t_{ij}$. The within-subject correlation structures of $\varepsilon_i(t_1)$ and $\varepsilon_i(t_2)$ at any two time points $t_1 \ne t_2$ are in general unknown, but may assume to have certain parametric or nonparametric forms.

The model (2.2) is referred to as a marginal model because the conditional mean of $\mathbf{Y}_i$ at $\mathbf{X}_i$ is $\mathbf{X}_i \beta$, so that the β_0 represents the population-mean intercept and the linear coefficients $\beta_1, \dots, \beta_K$ represent the covariate effects. Under the special case that the error term $\varepsilon_i(\mathbf{t}_i)$ of (2.2) is a mean zero Gaussian process with covariate matrix $\mathbf{V}_i(\mathbf{t}_i)$, the responses $\mathbf{Y}_i$ are then independent Gaussian random vectors such that

$$\mathbf{Y}_i \sim N\left(\mathbf{X}_i \beta, \mathbf{V}_i(t_i)\right), \tag{2.3}$$

where $N(\mathbf{a}, \mathbf{b})$ denotes a multivariate normal distribution with mean vector $\mathbf{a}$ and covariance matrix $\mathbf{b}$. A drawback of (2.2) is that the subject-specific relationship between $\mathbf{Y}_i$ and $\mathbf{X}_i$ for the subjects $i = 1, \dots, n$ is not specified by the model.

The covariance structures of (2.2) or its Gaussian model (2.3) are usually influenced by three factors: *random effects*, *serial correlations*, and *measurement errors*. The random effects characterize the stochastic variations between subjects within the population. In particular, we may view that, when the covariates affect the response linearly, some of the linear coefficients may vary from subject to subject. The serial correlations are the results of time-varying associations between different measurements of the same subject. Such correlations are typically positive in biomedical studies, and become weaker as the time interval between the measurements increases. The measurement errors, which are normally assumed to be independent both between and within the subjects, are induced by the measurement process or random variations within the subjects.

2.1.2 The Linear Mixed-Effects Models

This modeling strategy establishes a practical intra-subject correlation structure for the repeated measurements. The subsequent models are capable of predicting the subject-specific outcome trajectories.

Suppose that, for the ith subject, $1 \leq i \leq n$, there is a $[r \times 1]$ vector of explanatory variables $\mathbf{U}_{ij}$ measured at time t_{ij}, which may or may not overlap with the original covariate vector $\mathbf{X}_{ij}$. Using the additive decomposition of random-effects, serial correlations and measurement errors, $\varepsilon_i(t_{ij})$ can be expressed as

$$\varepsilon_i(t_{ij}) = \mathbf{U}_{ij}^T b_i + W_i(t_{ij}) + Z_{ij}, \qquad (2.4)$$

where b_i is the $[r \times 1]$ random vector with multivariate normal distribution $N(\mathbf{0}, \mathbf{D})$, $\mathbf{D}$ is a $[r \times r]$ covariate matrix with (p, q)th element $d_{pq} = d_{qp}$, $W_i(t_{ij})$ for $i = 1, \ldots, n$ are independent copies of a mean zero Gaussian process whose covariance at time points t_{ij_1} and t_{ij_2} is $\rho_W(t_{ij_1}, t_{ij_2})$, and Z_{ij} for $i = 1, \ldots, n$ and $j = 1, \ldots, n_i$ are i.i.d. random variables with $N(0, \tau^2)$ distribution. In general, Z_{ij} does not have to be a normal random variable.

The specification of (2.2) and (2.4) gives the linear mixed-effects model, which was first studied by Laird and Ware (1982),

$$\begin{cases} \mathbf{Y}_i &= \mathbf{X}_i \beta + \mathbf{U}_i b_i + \delta_i, \\ \delta_i(t_{ij}) &= W_i(t_{ij}) + Z_{ij}, \\ \delta_i &= \left(\delta_i(t_{i1}), \ldots, \delta_i(t_{in_i}) \right)^T, \\ \mathbf{U}_i &= \text{the } [n_i \times r] \text{ matrix whose } j\text{th row is } \mathbf{U}_{ij}^T, \end{cases} \qquad (2.5)$$

where the population-mean parameter β represents the influence of $\mathbf{X}_i$ on the population means of the response profile, the subject-specific parameter b_i describes the variation of the ith subject from the population conditioning on the given explanatory variable $\mathbf{U}_i$, and δ_i represents the error term. Conditioning on $\mathbf{X}_i$ and $\mathbf{U}_i$, (2.5) implies that $\mathbf{Y}_i$ for $i = 1, \ldots, n$ are independent Gaussian vectors such that

$$\mathbf{Y}_i \sim N\left(\mathbf{X}_i \beta, \mathbf{U}_i \mathbf{D} \mathbf{U}_i^T + \mathbf{P}_i + \tau^2 \mathbf{I}_i \right), \qquad (2.6)$$

where $\mathbf{P}_i$ is the $[n_i \times n_i]$ covariance matrix whose (j_1, j_2)th element is $\rho_W(t_{ij_1}, t_{ij_2})$ and $\mathbf{I}_i$ is the $[n_i \times n_i]$ identity matrix.

Useful special cases of (2.5) can be derived from the variance-covariance structure of (2.4). A number of the commonly seen special cases include:

(a) The classical linear models with cross-sectional i.i.d. data is a special case of (2.4) where $\varepsilon_i(t_{ij})$ are only affected by the measurement errors Z_{ij}.

(b) When neither the random effects nor the measurement errors are present, the error term is of pure serial correlation $\varepsilon_i(t_{ij}) = W_i(t_{ij})$. Moreover, if $W_i(t_{ij})$ are from a mean zero stationary Gaussian process, the covariance of $\varepsilon_i(t_{ij_1})$ and $\varepsilon_i(t_{ij_2})$ can be specified by

$$Cov[\varepsilon_i(t_{ij_1}), \varepsilon_i(t_{ij_2})] = \sigma^2 \rho\left(|t_{ij_1} - t_{ij_2}|\right), \tag{2.7}$$

where σ is a positive constant and $\rho(\cdot)$ is a continuous function. Useful choices of $\rho(\cdot)$ include the exponential correlation $\rho(s) = \exp(-as)$ for some constant $a > 0$ and the Gaussian correlation $\rho(s) = \exp(-as^2)$, among others.

(c) When $\varepsilon_i(t_{ij})$ are affected by a mean zero stationary Gaussian process and a mean zero Gaussian measurement error, the variance of Y_{ij} is $\sigma^2 \rho(0) + \tau^2$, while the covariance of Y_{ij_1} and Y_{ij_2}, for $j_1 \neq j_2$, is $\sigma^2 \rho(|t_{ij_1} - t_{ij_2}|)$, for some $\sigma > 0$, $\tau > 0$ and continuous correlation function $\rho(\cdot)$.

(d) When there is no serial correlation, the intra-subject correlations are only induced by the random effects, so that $\mathbf{P}_i$ is not present in (2.6). □

2.1.3   Conditional Maximum Likelihood Estimation

In the literature, estimation and inference procedures for linear models with longitudinal data are primarily developed based on the mixed-effects model (2.5) with the Gaussian distribution assumption for (2.4). The Gaussian assumption simplifies the computation of likelihood-based estimation and inference procedures. Further details of the estimation and inference methods summarized here can be found in Verbeke and Molenberghs (2000).

Suppose that the variance-covariance matrix $\mathbf{V}_i(\mathbf{t}_i)$ of (2.4) is determined by a R^q-valued parameter vector α. Let $\mathbf{V}_i(\mathbf{t}_i; \alpha)$ be the variance-covariance matrix parametrized by α. The log-likelihood function for (2.3) under the Gaussian distribution assumption is

$$\begin{cases} L(\beta, \alpha) & = & c + \sum_{i=1}^n \left[-(1/2) \log|\mathbf{V}_i(\mathbf{t}_i; \alpha)| \right. \\ & & \left. -(1/2) (\mathbf{Y}_i - \mathbf{X}_i \beta)^T \mathbf{V}_i^{-1}(\mathbf{t}_i; \alpha) (\mathbf{Y}_i - \mathbf{X}_i \beta) \right], \\ c & = & \sum_{i=1}^n \left[-(n_i/2) \log(2\pi) \right]. \end{cases} \tag{2.8}$$

For a given α, (2.8) can be maximized by

$$\widehat{\beta}(\alpha) = \left\{ \sum_{i=1}^n \left[\mathbf{X}_i^T \mathbf{V}_i^{-1}(\mathbf{t}_i; \alpha) \mathbf{X}_i \right] \right\}^{-1} \left\{ \sum_{i=1}^n \left[\mathbf{X}_i^T \mathbf{V}_i^{-1}(\mathbf{t}_i; \alpha) \mathbf{Y}_i \right] \right\}, \tag{2.9}$$

which is referred to as the conditional maximum likelihood estimator (CMLE). It can be verified by direct calculation that, under (2.5), $\widehat{\beta}(\alpha)$ is an unbiased estimator of β. Direct calculation also shows that the covariance matrix of $\widehat{\beta}(\alpha)$ is

$$
\begin{aligned}
Cov\left[\widehat{\beta}(\alpha)\right] \\
= \left\{\sum_{i=1}^{n}\left[\mathbf{X}_i^T\,\mathbf{V}_i^{-1}(\mathbf{t}_i;\alpha)\mathbf{X}_i\right]\right\}^{-1}\left\{\sum_{i=1}^{n}\left[\mathbf{X}_i^T\,\mathbf{V}_i^{-1}(\mathbf{t}_i;\alpha)\,Cov(\mathbf{Y}_i)\,\mathbf{V}_i^{-1}(\mathbf{t}_i;\alpha)\,\mathbf{X}_i\right]\right\} \\
\times \left\{\sum_{i=1}^{n}\left[\mathbf{X}_i^T\,\mathbf{V}_i^{-1}(\mathbf{t}_i;\alpha)\mathbf{X}_i\right]\right\}^{-1} \\
= \left\{\sum_{i=1}^{n}\left[\mathbf{X}_i^T\,\mathbf{V}_i^{-1}(\mathbf{t}_i;\alpha)\mathbf{X}_i\right]\right\}^{-1}. \qquad (2.10)
\end{aligned}
$$

Note that the second equality sign of (2.10) does not hold when the structure of the variance-covariance matrix is not correctly specified. Further derivation using (2.5), (2.9) and (2.10) shows that $\widehat{\beta}(\alpha)$ has a multivariate Normal distribution,

$$
\widehat{\beta}(\alpha) \sim N\left\{\beta,\left[\sum_{i=1}^{n}\left(\mathbf{X}_i^T\,\mathbf{V}_i^{-1}(\mathbf{t}_i;\alpha)\,\mathbf{X}_i\right)\right]^{-1}\right\}. \qquad (2.11)
$$

When α is known, this result can be used to develop inference procedures, such as confidence regions and test statistics, for β.

2.1.4 Maximum Likelihood Estimation

When α is unknown, as in most practical situations, a consistent estimate of α has to be used. An intuitive approach is to estimate β and α by maximizing (2.8) with respect to β and α simultaneously. Maximum likelihood estimators (MLE) of this type can be computed by substituting (2.9) into (2.8) and then maximizing (2.8) with respect to α. We denote the resulting MLE by $\widehat{\beta}_{ML}$ and $\widehat{\alpha}_{ML}$. The asymptotic distributions of $\{\widehat{\beta}_{ML}, \widehat{\alpha}_{ML}\}$ can be developed using the standard approaches in large sample theory.

Although $\{\widehat{\beta}_{ML}, \widehat{\alpha}_{ML}\}$ has some justifiable statistical properties, as for most likelihood-based methods, it may not be desirable in practice. To see why an alternative estimation method might be warranted in some situations, we consider the simple linear regression with i.i.d. errors and $n_1 = \cdots = n_n = m$,

$$
\mathbf{Y}_i \sim N\left(\mathbf{X}_i\beta, \sigma^2\mathbf{I}_m\right), \qquad (2.12)
$$

where $\mathbf{I}_m$ is the $(m \times m)$ identity matrix. The parameters involved in the model are β and σ. Let $\widehat{\beta}_{ML}$ and $\widehat{\sigma}_{ML}$ be the MLEs of β and σ, respectively, and RSS be the residual sum of squares defined by

$$
RSS = \sum_{i=1}^{n}\left(\mathbf{Y}_i - \mathbf{X}_i\widehat{\beta}_{ML}\right)^T\left(\mathbf{Y}_i - \mathbf{X}_i\widehat{\beta}_{ML}\right).
$$

The MLE of σ^2 is

$$\widehat{\sigma}^2_{ML} = RSS/(nm). \tag{2.13}$$

However, it is well-known that, for any finite n and m, (2.13) is a biased estimator of σ^2. On the other hand, a slightly modified estimator

$$\widehat{\sigma}^2_{REML} = RSS/[nm - (k+1)] \tag{2.14}$$

is unbiased for σ^2. Here, $\widehat{\sigma}^2_{REML}$ is the restricted maximum likelihood estimator (REMLE) for the model (2.12).

2.1.5 Restricted Maximum Likelihood Estimation

This class of estimators was introduced by Patterson and Thompson (1971) for the purpose of estimating variance components in the linear models. The main idea is to consider a linear transformation of the original response variable, so that the distribution of the transformed variable does not depend on β. Let $\mathbf{Y} = \left(\mathbf{Y}_1^T, \ldots, \mathbf{Y}_n^T\right)^T$, $\mathbf{X} = \left(\mathbf{X}_1^T, \ldots, \mathbf{X}_n^T\right)^T$ and $\mathbf{V}$ be the block-diagonal matrix with $\mathbf{V}_i(t_i)$ on the ith main diagonal and zeros elsewhere. Then, with $\mathbf{V}$ parameterized by α, the model (2.3) is equivalent to

$$\mathbf{Y} \sim N\left(\mathbf{X}\beta, \mathbf{V}(\alpha)\right). \tag{2.15}$$

The REMLE of α, the parameter for the variance-covariance matrix in (2.15), is obtained by maximizing the likelihood function of $\mathbf{Y}^* = \mathbf{A}^T\mathbf{Y}$, where $\mathbf{A}$ is a $[N \times (N - k - 1)]$ full rank matrix with $\mathbf{A}^T\mathbf{X} = \mathbf{0}$ and $N = \sum_{i=1}^n n_i$. A specific construction of $\mathbf{A}$ can be found in Diggle et al. (2002, Section 4.5). It follows from (2.15) that $\mathbf{Y}^*$ has a mean zero multivariate Gaussian distribution with covariance matrix $\mathbf{A}^T\mathbf{V}(\alpha)\mathbf{A}$. Harville (1974) showed that the likelihood function of $\mathbf{Y}^*$ is proportional to

$$
\begin{aligned}
L^*(\alpha) \;=\; & \left|\sum_{i=1}^n \mathbf{X}_i^T\mathbf{X}_i\right|^{1/2} \left|\sum_{i=1}^n \mathbf{X}_i^T\mathbf{V}_i^{-1}(t_i;\alpha)\mathbf{X}_i\right|^{-1/2} \prod_{i=1}^n \left|\mathbf{V}_i(t_i;\alpha)\right|^{-1/2} \tag{2.16} \\
& \times \exp\left\{-\frac{1}{2}\sum_{i=1}^n \left[\mathbf{Y}_i - \mathbf{X}_i\widehat{\beta}(\alpha)\right]^T \mathbf{V}_i^{-1}(t_i;\alpha)\left[\mathbf{Y}_i - \mathbf{X}_i\widehat{\beta}(\alpha)\right]\right\}.
\end{aligned}
$$

The REMLE $\widehat{\alpha}_{REML}$ of α maximizes (2.16). The REMLE $\widehat{\beta}_{REML}$ of β is obtained by substituting α of (2.9) with $\widehat{\alpha}_{REML}$. Because (2.16) does not depend on the choice of $\mathbf{A}$, the resulting estimators $\widehat{\beta}_{REML}$ and $\widehat{\alpha}_{REML}$ are free of the specific linear transformations.

The log-likelihood of $\mathbf{Y}^*$, $\log[L^*(\alpha)]$, differs from the log-likelihood $L(\widehat{\beta}, \alpha)$ only through a constant, which does not depend on α, and

$$-\frac{1}{2}\log\left|\sum_{i=1}^n \mathbf{X}_i^T\mathbf{V}_i^{-1}(t_i;\alpha)\mathbf{X}_i\right|,$$

which does not depend on β. Because both REMLE and MLE are based on the likelihood principle, they all have appropriate theoretical properties such as consistency, asymptotic normality and asymptotic efficiency. In practice, neither one is uniformly superior to the other for all the situations. Their numerical values are also computed from different algorithms. For the MLEs, the fixed effects and the variance components are estimated simultaneously. For the REMLEs, only the variance components are estimated.

2.1.6 Likelihood-Based Inferences

The results established in the previous sections are useful to construct inference procedures for β. For the purpose of illustration, only a few special cases are presented here. A more complete account of inferential and diagnostic tools can be found in Zeger, Liang and Albert (1988), Vonesh and Chinchilli (1997), Verbeke and Molenberghs (2000), Diggle et al. (2002), among others.

Suppose that there is a consistent estimator $\widehat{\alpha}$ of α, which may be either the MLE $\widehat{\alpha}_{ML}$ or the REMLE $\widehat{\alpha}_{REML}$. Substituting α of (2.11) with $\widehat{\alpha}$, the distribution of $\widehat{\beta}(\widehat{\alpha})$ can be approximated, when n is large, by

$$\begin{cases} \widehat{\beta}(\widehat{\alpha}) \sim N\left(\beta, \widehat{\mathbf{V}}\right), \\ \widehat{\mathbf{V}} = \left\{ \sum_{i=1}^{n} \left[\mathbf{X}_i^T \mathbf{V}_i^{-1}(\mathbf{t}_i; \widehat{\alpha}) \mathbf{X}_i \right] \right\}^{-1}. \end{cases} \tag{2.17}$$

Suppose that $\mathbf{C}$ is a known $[r \times (k+1)]$ matrix with full rank. It follows immediately from (2.17) that, when n is sufficiently large, the distribution of $\mathbf{C}\widehat{\beta}(\widehat{\alpha})$ can be approximated by

$$\mathbf{C}\widehat{\beta}(\widehat{\alpha}) \sim N\left(\mathbf{C}\beta, \mathbf{C}\widehat{\mathbf{V}}\mathbf{C}^T\right). \tag{2.18}$$

Consequently, an approximate $[100 \times (1-a)]\%$, $0 < a < 1$, confidence interval for $\mathbf{C}\beta$ can be given by

$$\mathbf{C}\widehat{\beta}(\widehat{\alpha}) \pm Z_{1-a/2}\left(\mathbf{C}\widehat{\mathbf{V}}\mathbf{C}^T\right)^{1/2}. \tag{2.19}$$

Taking $\mathbf{C}$ to be the $(k+1)$ row vector with 1 at its lth place and zero elsewhere, the approximate $[100 \times (1-a)]\%$ confidence interval for β_l obtained from (2.19) is given by

$$\widehat{\beta}_l(\widehat{\alpha}) \pm Z_{1-a/2}\widehat{V}_l^{1/2}, \tag{2.20}$$

where $\widehat{V}_l$ is the lth diagonal element of $\widehat{\mathbf{V}}$.

The normal approximation in (2.17) can also be used to construct test statistics for linear statistical hypotheses. Suppose that we would like to test the null hypothesis of $\mathbf{C}\beta = \theta_0$ for a known vector θ_0 against the general alternative that $\mathbf{C}\beta \neq \theta_0$. A natural test statistic would be

$$\widehat{T} = \left[\mathbf{C}\widehat{\beta}(\widehat{\alpha}) - \theta_0\right]^T \left(\mathbf{C}\widehat{\mathbf{V}}\mathbf{C}^T\right)^{-1} \left[\mathbf{C}\widehat{\beta}(\widehat{\alpha}) - \theta_0\right], \tag{2.21}$$

which has approximately a χ^2-distribution with r degrees of freedom, denoted by χ_r^2, under the null hypothesis. A level $(100 \times a)\%$ test based on (2.21) then rejects the null hypothesis when $\widehat{T} > \chi_r^2(a)$ with $\chi_r^2(a)$ being the $[100 \times (1 - a)]$th percentile of χ_r^2. For the special case of testing $\beta_l = 0$ versus $\beta_l \neq 0$, a simple procedure equivalent to (2.21) is to reject the null hypothesis when

$$\left| \widehat{\beta}_l(\widehat{\alpha}) \right| > Z_{1-a/2} \widehat{V}_l^{1/2}, \tag{2.22}$$

where $Z_{1-a/2}$ and $\widehat{V}_l$ are defined in (2.20).

2.2   Nonlinear Marginal and Mixed-Effects Models

We outline in this section a few key features of the nonlinear marginal and mixed-effects models that have already been described in Fitzmaurice et al. (2009, Chapter 5). Because the methods of this book are mainly non-parametric generalizations of the linear regression methods, our aim is to illustrate the differences between the linear and nonlinear approaches. Details of the model formulations, estimation and inference procedures and their applications are referred to Fitzmaurice et al. (2009) and the references therein.

2.2.1   Model Formulation and Interpretation

Nonlinear models generally refer to parametric regression models which cannot be formulated into the framework of (2.2) or (2.5). Thus, by nature, this class of models include a large number of possible functional relationships between $Y(t)$ and $\{t, \mathbf{X}(t)\}$. But, because these are still parametric models, a key feature is that the functional relationships between $Y(t)$ and $\{t, \mathbf{X}(t)\}$ are determined by a set of parameters in a Euclidean space, although the linear relationship between $Y(t)$ and $\mathbf{X}(t)$ are not satisfied. Applications of nonlinear models in biomedical studies can be found, for example, in pharmacokinetics and infectious diseases.

When the objective of the analysis is on the overall population effects of the covariates without considering the effects at the individual level, a nonlinear marginal regression model for $\{Y(t), t, \mathbf{X}(t)\}$ can be written as

$$Y(t) = m\big[t, \mathbf{X}(t); \beta\big] + \varepsilon_i(t), \tag{2.23}$$

where β is an unknown vector of parameters, $m\big[t, \mathbf{X}(t); \beta\big]$ is a nonlinear function of t and $\mathbf{X}(t)$ determined by the parameter vector β, and, as in (2.2), $\varepsilon_i(t)$ is a mean zero random error process. An example of the nonlinear function is the *logistic model*

$$m(t; \beta) = \frac{\beta_1}{1 + \exp[-\beta_3(t - \beta_2)]}, \tag{2.24}$$

where $\beta = (\beta_1, \beta_2, \beta_3)^T$, $\beta_1 > 0$ and $\beta_3 > 0$. Using the matrix notation

$\{\mathbf{Y}_i, \mathbf{t}_i, \mathbf{X}_i\}$ in (2.1), the model (2.24) can be written as

$$
\begin{cases}
\mathbf{Y}_i & = & m(\mathbf{t}_i, \mathbf{X}_i; \beta) + \varepsilon_i(\mathbf{t}_i), \\
m(\mathbf{t}_i, \mathbf{X}_i; \beta) & = & \left(m(t_{i1}, \mathbf{X}_i; \beta), \ldots, m(t_{in_i}, \mathbf{X}_i; \beta) \right)^T, \\
\varepsilon_i(\mathbf{t}_i) & = & \left(\varepsilon_i(t_{i1}), \ldots, \varepsilon_i(t_{in_i}) \right)^T,
\end{cases}
\tag{2.25}
$$

where $\varepsilon_i(t_{ij})$ is the realization of a mean zero random error process $\varepsilon_i(t)$ at time point $t = t_{ij}$. This model is referred to as a marginal model because the parameter vector β describes the relationship between $Y(t)$ and $\mathbf{X}(t)$ for the entire population of interest.

Nonlinear mixed-effects models are formulated by decomposing the effects of $\{t, \mathbf{X}(t)\}$ on $Y(t)$ through two stages, the *individual-level* (or *subject-specific*) modeling and the *population-level* modeling. The first stage, i.e., individual-level modeling, is aimed at describing the trajectory of $Y_i(t)$ through a nonlinear function of $\{t, \mathbf{X}_i(t)\}$ specific to each subject i. The second stage, i.e., population-level modeling, characterizes the differences among individuals across the population. Suppose that, for each subject i, the covariate $\mathbf{X}_i$ is formed by two components: the *"within-subject"* covariates $\mathbf{U}_i$, and the *"between-subject"* covariates $\mathbf{A}_i$. Intuitively, the components of $\mathbf{U}_i$ describe the time-response relationship at the level of the ith individual, and the components of $\mathbf{A}_i$, which do not change over the observation period, characterize the differences between individuals.

The basic nonlinear mixed-effects model can be expressed as

$$
\begin{cases}
\text{Individual-Level Model:} & \mathbf{Y}_i = m(\mathbf{t}_i, \mathbf{U}_i; \theta_i) + \varepsilon_i(\mathbf{t}_i), \\
& Cov(\mathbf{Y}_i | \mathbf{X}_i, b_i) = V_i(\mathbf{X}_i, \beta, b_i, \alpha), \\
\text{Population-Level Model:} & \theta_i = d(\mathbf{A}_i; \beta, b_i),
\end{cases}
\tag{2.26}
$$

where $m(\mathbf{t}_i, \mathbf{U}_i; \theta_i)$ is defined as in (2.25), θ_i is a r-dimensional vector of parameters for some $r \geq 1$ specific to the individual i, $m(\mathbf{t}_i, \mathbf{U}_i; \theta_i)$ and $\varepsilon_i(\mathbf{t}_i)$ are a nonlinear function of time and the vector of errors as in (2.25), respectively, α and β are vectors of fixed-effects parameters, b_i is a vector of random-effects parameters, $V_i(\mathbf{X}_i, \beta, b_i, \alpha)$ is the conditional covariance matrix, and $d(\mathbf{A}_i; \beta, b_i)$ is a r-dimensional function of the "between-subject" covariates $\mathbf{A}_i$. At the individual-level model of (2.26), $\mathbf{Y}_i$ depends on $\mathbf{U}_i$, hence $\mathbf{X}_i$, through the nonlinear function $m(\mathbf{t}_i, \mathbf{U}_i; \theta_i)$, which has a known parametric structure determined by the individual-level parameter θ_i. At the population-level, the subject-specific characteristics described by θ_i depend on $\mathbf{A}_i$, hence $\mathbf{X}_i$, through the known function $d(\mathbf{A}_i; \beta, b_i)$, which is determined by the vector of population-level parameters β and a random variation vector b_i. A common assumption is that b_i has mean zero conditioning on $\mathbf{A}_i$ and its variance-covariance matrix does not depend on $\mathbf{A}_i$, that is,

$$
E(b_i | \mathbf{A}_i) = E(b_i) = \mathbf{0} \quad \text{and} \quad Cov(b_i | \mathbf{A}_i) = Cov(b_i) = \Sigma
\tag{2.27}
$$

for an unstructured covariance matrix Σ. A common choice is to take b_i to be a mean zero multivariate normal random variable, i.e., $b_i \sim N(\mathbf{0}, \Sigma)$. The fixed-effects parameters to be estimated from (2.26) are $\{\beta, \alpha, \Sigma\}$. The random-effects parameter to be estimated from (2.26) and (2.27) is θ_i. These estimated fixed-effects and random-effects parameters are used to predict the subject-specific trajectories of $Y(t)$.

The nonlinear marginal and mixed-effects models (2.25) and (2.26) share three common features with their linear counterparts in (2.2) and (2.5). First, both modeling schemes assume that there are population-mean parameters, i.e., β in these models, which characterizes the fixed covariate effects of the population. Second, both modeling schemes characterize the individual-level (or subject-specific) covariate effects through a vector of random-effects parameters, i.e., b_i in these models. Third, in most practical situations, the distributions of the random-effects parameter vectors $\{b_1, \ldots, b_n\}$ are assumed to be multivariate normal with mean zero. The variance-covariance matrices may be either structured or unstructured. These common features allow the likelihood-based estimation and inference procedures to be used for both the linear and nonlinear models. However, because the individual-level and population-level functions $m(\mathbf{t}_i, \mathbf{U}_i; \theta_i)$ and $d(\mathbf{A}_i; \beta, b_i)$ in (2.26) do not have the simple linear structure, computation of the likelihood-based estimation and inferences for the nonlinear models requires more complex algorithms compared to the linear models.

2.2.2   Likelihood-Based Estimation and Inferences

We only outline the methods for the nonlinear mixed-effects model (2.26), since the marginal model (2.25) can be treated as a special case of (2.26). Under the assumption that the distribution function of $\mathbf{Y}_i$ conditioning on $\mathbf{X}_i$ is known, for example, $\varepsilon_i(\mathbf{t}_i)$ and b_i of (2.26) have multivariate normal distributions, we can write $\gamma = (\beta^T, \alpha^T)^T$ and the log-likelihood function for $\{\gamma, \Sigma\}$ as

$$
\begin{aligned}
\ell(\gamma, \Sigma) &= \log\left[\prod_{i=1}^{n} f_i(\mathbf{Y}_i | \mathbf{X}_i; \gamma, \Sigma)\right] \\
&= \log\left[\prod_{i=1}^{n} \int f_i(\mathbf{Y}_i | \mathbf{X}_i, b_i; \gamma) f(b_i; \Sigma) \, db_i\right],
\end{aligned}
\tag{2.28}
$$

where $f_i(\mathbf{Y}_i | \mathbf{X}_i; \gamma, \Sigma)$ is the ith subject's density function of $\mathbf{Y}_i$ given $\mathbf{X}_i$, $f_i(\mathbf{Y}_i | \mathbf{X}_i, b_i; \gamma)$ is the ith subject's density function of $\mathbf{Y}_i$ conditioning on $\mathbf{X}_i$ and the subject-specific parameters b_i, and $f(b_i; \Sigma)$ is the marginal density function of b_i.

A major obstacle of obtaining the likelihood-based estimation and inferences for the parameters of interest $\{\beta, \alpha\}$ is computation. Maximizing (2.28) with respect to γ and Σ is generally intractable, because the right side integral involves complex nonlinear functions and numerical evaluation of the integral can be computationally intensive. Thus, a number of analytical approximation methods have been proposed in the literature to approximate the log-likelihood function in (2.28), so that the optimization algorithms for computing the approximate maximum likelihood estimators of γ and Σ can be simplified. Because these approximation methods are not used in our nonparametric estimation procedures, we refer to Davidian and Giltinan (1995) for details.

2.2.3 Estimation of Subject-Specific Parameters

An appropriate estimator $\widehat{\theta}_i$ of the subject-specific parameter θ_i can be used to predict the subject's trajectory of $Y_i(t)$ by substituting θ_i in $m(t, \cdot; \theta_i)$ with $\widehat{\theta}_i$. When sufficient data are available for each individual, a simple method to estimate θ_i is to fit the available data from the ith individual to its individual-level model in (2.26). The advantage of estimating θ_i using only the ith subject's individual data is that the estimation does not depend on the model structures. However, this "fitting individual model" approach is often not practical because the numbers of repeated measurements n_i may not be sufficiently large for all $i = 1, \ldots, n$. In situations where not all the subjects have large numbers of repeated measurements, a more practical approach is to pool the information from all n subjects, so that the parameters for the individual-level and population-level models in (2.26) can be estimated simultaneously.

The estimation of $\{\beta, \alpha, \theta_i\}$ using "pooled information" from all n subjects depends on the covariance structures $V_i(\mathbf{X}_i, \beta, b_i, \alpha)$. For simple structures of $V_i(\cdot)$, such as $V_i(\cdot)$ is diagonal or certain nondiagonal with intra-individual variance and covariance parameters, $\gamma = (\beta^T, \alpha^T)^T$ and θ_i can be estimated using the weighted regression method described in Davidian and Giltinan (1995, Chapter 2 and Section 5.2). Other approaches for the estimation and inference of θ_i include "approximate linear mixed-effects model." the *Expectation-Maximization* (EM) algorithm, and the methods based on analytic and numerical approximations to the likelihood. Details on these methods and their implementations have been described in Davidian and Giltinan (1995, Chapters 6 and 7) and Davidian and Giltinan (2003), among others.

2.3 Semiparametric Partially Linear Models

Estimation and inference methods for semiparametric models with longitudinal data have been mostly focused on the partially linear models. Existing results in the literature can be found, for example, in Zeger and Diggle (1994), Moyeed and Diggle (1994), Lin and Ying (2001), Lin and Carroll (2001, 2006), Wang, Carroll and Lin (2005), among others. As discussed in Section 1.3.2, this class of models has been developed to generalize the marginal and mixed-effects linear models. The main objective for these models is on the estimation and inferences of the real-valued parameters while allowing some nonparametric curves as a nuisance component. We summarize here the main approaches described in Zeger and Diggle (1994), Lin and Carroll (2001, 2006) and Wang, Carroll and Lin (2005).

2.3.1   Marginal Partially Linear Models

As discussed in Section 1.3.2, the semiparametric marginal partially linear model (1.2), which has been first investigated by Zeger and Diggle (1994) and Moyeed and Diggle (1994), for the stochastic processes $\{Y(t), t, \mathbf{X}(t)\}$ can be written as

$$
\left\{
\begin{aligned}
Y(t) &= \beta_0(t) + \sum_{l=1}^{K} \beta_l X^{(l)}(t) + \varepsilon(t) \\
&= \beta_0(t) + \mathbf{X}^T(t)\beta + \varepsilon(t), \\
\beta &= (\beta_1, \ldots, \beta_K)^T,
\end{aligned}
\right.
\tag{2.29}
$$

where $\beta_0(t)$ is an unknown smooth function of t, $\varepsilon(t)$ is a mean zero stochastic process with variance $\sigma^2(t)$ and correlation function

$$
\rho(t_1, t_2) = \frac{Cov[\varepsilon(t_1), \varepsilon(t_2)]}{\sigma(t_1)\sigma(t_2)} \quad \text{for any } t_1 \neq t_2,
\tag{2.30}
$$

and $X^{(l)}(t)$, $l = 1, \ldots, K$, and $\varepsilon(t)$ are independent. The correlation structures $\rho(t_1, t_2)$ of $\varepsilon(t)$ distinguish (2.29) with repeatedly measured longitudinal data from its counterpart with cross-sectional i.i.d. data. With a longitudinal sample $\{(Y_{ij}, t_{ij}, \mathbf{X}_{ij}) : i = 1, \ldots, n; j = 1, \ldots, n_i\}$, the errors $\varepsilon_i(t_{ij})$ in (2.29) are independent copies of $\varepsilon(t)$ across the n subjects but with intra-subject correlations specified by (2.30).

A useful special case of $\varepsilon_i(t_{ij})$ is the decomposition

$$
\varepsilon_i(t_{ij}) = W_i(t_{ij}) + Z_{ij},
\tag{2.31}
$$

where $W_i(t)$ are independent copies of a mean zero stationary process $W(t)$ with covariance function at any time points t_1 and t_2

$$
Cov[W(t_1), W(t_2)] = \sigma_W^2 \rho_W(t_1, t_2)
$$

for some σ_W and correlation function $\rho_W(\cdot, \cdot) > 0$, and Z_{ij} are independent and identically distributed measurement errors with mean zero and variance σ_Z^2. The covariance structure of the measurements Y_{ij} for $i = 1, \ldots, n$ and $j = 1, \ldots, n_i$ are

$$
Cov(Y_{i_1 j_1}, Y_{i_2 j_2}) =
\left\{
\begin{aligned}
&\sigma_Z^2 + \sigma_W^2, && \text{if } i_1 = i_2 \text{ and } j_1 = j_2, \\
&\sigma_W^2 \rho_W(t_{i_1 j_1}, t_{i_2 j_2}), && \text{if } i_1 = i_2 \text{ and } j_1 \neq j_2, \\
&0, && \text{otherwise.}
\end{aligned}
\right.
\tag{2.32}
$$

Although the partially linear model (2.29) can be classified as a special case of the time-varying coefficient model (1.4), which is a class of the structured nonparametric models to be discussed in Chapters 6 to 9, the estimation methods of these two types of models are quite different. This is a fact owing to the differences between these two classes of modeling assumptions. In (2.29), the covariate effects are described through the linear coefficients which do not change with time. On the other hand, the covariate effects of (1.4) are unknown curves of t, hence (1.4) is entirely nonparametric, although a linear structure is used at each time t.

When $Y(t)$ is not necessarily a continuous random variable, a generalized marginal partially linear model is to model the conditional distribution of $Y(t)$ through

$$
\begin{cases}
g\{\mu[t, \mathbf{X}(t)]\} &= \beta_0(t) + \sum_{l=1}^{K} \beta_l X^{(l)}(t) \\
&= \beta_0(t) + \mathbf{X}^T(t)\beta, \\
\mu[t, \mathbf{X}(t)] &= E[Y(t)|\mathbf{X}(t)], \\
\beta &= (\beta_1, \ldots, \beta_K)^T,
\end{cases}
\tag{2.33}
$$

where $g(\cdot)$ is a known link function, and $\beta_0(t), \beta_1, \ldots, \beta_K$ are defined in (2.2). With the observations $\{(Y_{ij}, t_{ij}, \mathbf{X}_{ij}) : i = 1, \ldots, n; j = 1, \ldots, n_i\}$, (2.33) can be written as

$$
\begin{cases}
g[\mu(t_{ij}, \mathbf{X}_{ij})] &= \beta_0(t_{ij}) + \sum_{l=1}^{K} \beta_l X_{ij}^{(l)} \\
&= \beta_0(t_{ij}) + \mathbf{X}_{ij}^T \beta, \\
\mu(t_{ij}, \mathbf{X}_{ij}) &= E(Y_{ij}|\mathbf{X}_{ij}), \\
\beta &= (\beta_0, \ldots, \beta_K)^T.
\end{cases}
\tag{2.34}
$$

If $g(\cdot)$ is the identity function, then (2.33) reduces to (2.29), and (2.34) is the expression given in (1.2). Both the models (2.29) and (2.33) describe the conditional expectations of $Y(t)$ through the sum of an unspecified baseline curve $\beta_0(t)$ of t and the linear effects of $\{X^{(1)}(t), \ldots, X^{(K)}(t)\}$ characterized by the coefficients $\{\beta_1, \ldots, \beta_K\}$. When the study objective is to evaluate the effects of the covariates, $\beta_0(t)$ is treated as a nuisance nonparametric component, and the statistical inference is focused on the linear coefficients $\{\beta_1, \ldots, \beta_K\}$.

2.3.2 Mixed-Effects Partially Linear Models

When the subject-specific deviation from the population is also of interest for the analysis, random-effects at the individual level can be built into the model. In this case, we consider that, for the observations $\{(Y_{ij}, t_{ij}, \mathbf{X}_{ij}) : i = 1, \ldots, n; j = 1, \ldots, n_i\}$, there is a subset $\mathbf{U}_{ij} = (U_{ij}^{(1)}, \ldots, U_{ij}^{(K_0)})^T$ of the original covariates $\mathbf{X}_{ij}$, so that $\mathbf{U}_{ij}$ has individual-level effects on Y_{ij} which are specific to the ith subject. Incorporating the subject-specific deviation to (1.2), a semiparametric mixed-effects partially linear model is given by

$$\begin{cases} Y_{ij} &=& \beta_0(t_{ij}) + \sum_{l=1}^{K} \beta_l X_{ij}^{(l)} + \sum_{r=1}^{K_0} b_{ir} U_{ij}^{(r)} + e_{ij}, \\ &=& \beta_0(t_{ij}) + \mathbf{X}_{ij}^T \beta + \mathbf{U}_{ij}^T b_i + e_{ij}, \\ \beta &=& (\beta_1, \ldots, \beta_K)^T, \\ b_i &=& (b_{i1}, \ldots, b_{iK_0})^T, \ b_i \sim N(\mathbf{0}, \mathbf{D}(\gamma_1)), \\ e_i &=& (e_{i1}, \ldots, e_{in_i})^T, \ e_i \sim N(\mathbf{0}, \mathbf{R}(\gamma_2)), \end{cases} \tag{2.35}$$

where $\beta_0(t_{ij})$ and β are the population-level baseline smooth curve and linear coefficients defined in (2.29), b_i is the vector of the mean zero normally distributed subject-specific deviations from the population-level linear coefficients, e_i is the vector of mean zero normally distributed error terms, and $\mathbf{D}(\gamma_1)$ and $\mathbf{R}(\gamma_2)$ are the variance-covariance matrices of b_i and e_i specified by the parameters γ_1 and γ_2, respectively.

The generalized mixed-effects partially linear model can be similarly established by adding the subject-specific parameters into (2.34). Following (2.34) and (2.35), this model can be written as

$$\begin{cases} g\left[\mu(t_{ij}, \mathbf{X}_{ij})\right] &=& \beta_0(t_{ij}) + \sum_{l=1}^{K} \beta_l X_{ij}^{(l)} + \sum_{r=1}^{K_0} b_{ir} U_{ij}^{(r)} \\ &=& \beta_0(t_{ij}) + \mathbf{X}_{ij}^T \beta + \mathbf{U}_{ij}^T b_i, \\ \mu(t_{ij}, \mathbf{X}_{ij}) &=& E(Y_{ij} | \mathbf{X}_{ij}), \\ \beta &=& (\beta_0, \ldots, \beta_K)^T, \\ b_i &=& (b_{i1}, \ldots, b_{iK_0})^T, \ b_i \sim N(\mathbf{0}, \mathbf{D}(\gamma_1)), \end{cases} \tag{2.36}$$

where $\beta_0(t)$ and β are the same as (2.34) and γ_1 is defined in (2.35). By incorporating the subject-specific coefficients b_i, the mixed-effects models (2.35) and (2.36) can be used to predict the subject-specific outcome trajectory for a given individual.

The rest of this section focuses on some of the well-established procedures for the estimation of $\{\beta_0(t), \beta_1, \cdots, \beta_K\}$ as well as the random-effect coefficients b_i. An excellent summary of these estimation methods can be found in Fitzmaurice et al. (2009, Chapter 9).

2.3.3   Iterative Estimation Procedure

We briefly review here an iterative procedure described by Zeger and Diggle (1994) for the estimation of $\beta_0(t), \beta_1, \ldots, \beta_K$ with the marginal partially linear model (2.29). This procedure repeatedly uses a nonparametric smoothing method as described in Section 1.5 and a longitudinal parametric estimation method as described in Section 2.1. This iterative algorithm is a special case of the backfitting algorithm described in Hastie and Tibshirani (1993).

To start, we consider the partially linear model (2.29) whose error term

$\varepsilon(t)$ is obtained from a mean zero Gaussian process with a known correlation function $Cov[\varepsilon(t_1), \varepsilon(t_2); \alpha]$ at any two time points t_1 and t_2, which is defined in (2.30) and determined by a Euclidean space parameter α. Then, by the definition of (2.1), the ith subject's covariance matrix of the repeatedly measured outcome $\mathbf{Y}_i = (Y_{i1}, \ldots, Y_{in_i})^T$ at the time points $\mathbf{t}_i = (t_{i1}, \ldots, t_{in_i})^T$ is $\mathbf{V}_i(\mathbf{t}_i; \alpha)$, where the (j_1, j_2)th element of $\mathbf{V}_i(\mathbf{t}_i; \alpha)$ is $Cov[\varepsilon(t_{ij_1}), \varepsilon(t_{ij_2}); \alpha]$.

Iterative Estimation Procedure:

(a) *Set $\beta_0(t)$ to be an unknown constant, i.e., $\beta_0(t) = \beta_0$, so that the partially linear model (2.29) reduces to the marginal linear model (2.3) with linear coefficients $\beta_0, \beta_1, \ldots, \beta_K$ and covariance matrix $\mathbf{V}_i(\mathbf{t}_i; \alpha)$. The initial estimators of $\beta_0, \beta_1, \ldots, \beta_K$ can be computed by maximizing the log-likelihood function (2.8).*

(b) *Based on the current estimator $\widehat{\beta}_1, \ldots, \widehat{\beta}_K$, calculate the ith subject's residual at time point t_{ij} by*

$$r_{ij}^{(1)} = Y_{ij} - \sum_{l=1}^{k} \widehat{\beta}_l X_{ij}^{(l)}. \tag{2.37}$$

(c) *Treat the residuals $\{r_{ij} : i = 1, \ldots, n; j = 1, \ldots, n_i\}$ as the pseudo-observations and compute the kernel estimator $\widehat{\beta}_0^K(t)$ of $\beta_0(t)$ using (1.23) with Y_{ij} replaced by its residual r_{ij} in (2.37), i.e.,*

$$\widehat{\beta}_0^K(t) = \frac{\sum_{i=1}^{n} \sum_{j=1}^{n_i} r_{ij} K[(t - t_{ij})/h]}{\sum_{i=1}^{n} \sum_{j=1}^{n_i} K[(t - t_{ij})/h]}, \tag{2.38}$$

where $K(\cdot)$ and $h > 0$ are the kernel function and bandwidth.

(d) *Based on the current kernel estimator $\widehat{\beta}_0^K(t)$ computed from (2.38), calculate the residuals*

$$r_{ij}^{(2)} = Y_{ij} - \widehat{\beta}_0^K(t_{ij}). \tag{2.39}$$

(e) *Update the estimators of $\beta_1, \ldots, \beta_K$ by applying the maximum likelihood procedure of (a) to the linear model*

$$r_{ij}^{(2)} = \sum_{l=1}^{K} \beta_l X_{ij}^{(l)} + \varepsilon_{ij}, \tag{2.40}$$

where ε_{ij} has the mean zero Gaussian distribution with intra-subject covariance $\mathbf{V}_i(\mathbf{t}_i; \alpha)$.

(f) *Repeat the steps (b) to (e) until the estimators converge.* □

The likelihood and kernel estimators in the steps (a) and (c) of the above iterative estimation procedure are used for the purpose of illustration. Clearly, other parametric and nonparametric estimation methods can also be used in these steps. A number of the commonly used local and global smoothing methods for estimating the nonparametric curves are described in Chapters 3 to 5. A crucial step in obtaining an adequate kernel estimator for $\beta_0(t)$ is to select an appropriate bandwidth h, while the choice of kernel functions is less important. For smoothing methods other than the kernel estimators, such as splines, this amounts to selecting an appropriate smoothing parameter. We present some of the commonly used smoothing parameter choices in Chapters 3 to 5.

The above iterative estimation procedure has the advantage of being conceptually simple and computationally feasible, since each estimation step is based on well-known estimation methods in the literature. As shown in the next section, it can also be directly generalized to the case with the generalized marginal partially linear model (2.33). However, this iterative estimation procedure ignores the intra-subject correlations in the estimation of $\beta_0(t)$ in step (c). As a result, its estimators of $\beta_1, \ldots, \beta_K$ are not "*semiparametric efficient*" in the sense their asymptotic mean squared errors do not reach the lower bound established in Lin and Carroll (2001). Thus, we briefly mention in the next section some alternative approaches which attempt to take the intra-subject correlations of the data into account. These alternative approaches have their own advantages and disadvantages compared with the above iterative estimation procedure.

2.3.4   Profile Kernel Estimators

Since the methods described here are based on the approach of solving some kernel generalized estimating equations, the estimators are referred to in the literature as the *profile kernel generalized estimating equations* (profile kernel GEE) estimators. We briefly review these methods here, which are summarized in Fitzmaurice et al. (2009, Section 9.6). Details of these methods are described in Lin and Carroll (2001), Wang, Carroll and Lin (2005) and Lin and Carroll (2006).

1. Profile Kernel GEE Estimation Method

This method is a generalization of the above iterative estimation procedure to the generalized marginal partially linear model (2.33). The iteration algorithm contains the following steps:

(a) *Replace β of (2.33) by some preliminary estimators, and then estimate the nonparametric component $\beta_0(t)$ of (2.33) by solving the local polynomial kernel GEE as shown in equation (9.22) of Fitzmaurice et al. (2009, Chapter 9).*

(b) *Substitute $\beta_0(t)$ by its kernel GEE estimator, the parameters β of (2.33)* *are estimated by the profile estimating equation as shown in equation (9.23) of Fitzmaurice et al. (2009, Chapter 9).*

(c) *Both the kernel GEEs and the profile estimating equations are used iteratively until the estimators converge, which lead to the final profile kernel estimators of $\beta_0(t)$ and β.* □

The asymptotic properties of the profile kernel estimators depend on the *working correlation matrix* used in the kernel GEE estimating equation step (a) and the profile estimating equation step (b). A striking result shown by Lin and Carroll (2001) is that, if one accounts for the intra-subject correlation, the profile kernel estimator of $\beta_0(t)$ and β is not "semiparametric efficient" in the sense that its asymptotic mean squared error does not reach an established lower bound. A main reason for this lack of efficiency is because the kernel GEE estimator of $\beta_0(t)$ only uses the local observations obtained within the small neighborhood of t specified by the bandwidth. Consequently, the use of an appropriate working correlation matrix choice does not ultimately lead to semiparametric efficient estimators at the profile estimating equation step.

2. Profile SUR Kernel Estimation Method

In order to improve the asymptotic properties of the profile kernel estimators, this method, as described in Wang, Carroll and Lin (2005), intends to construct "semiparametric efficient" estimators for the Euclidean space parameters in (2.29) and (2.33). The crucial part of this estimation method is to replace the kernel GEE estimator of $\beta_0(t)$ by the *"seemingly unrelated kernel estimator"* referred to as the SUR kernel estimator in Wang (2003). The SUR kernel estimator is obtained by solving a kernel estimating equation, i.e., equation (9.3) of Fitzmaurice et al. (2009, Chapter 9), through an iterative algorithm. The SUR kernel estimator, which does not only rely on local observations around the time point t, places weights on all the observations. Under the identity link function for (2.33), the SUR kernel estimator of $\beta_0(t)$ has a closed-form expression as shown in Lin et al. (2004) and equation (9.7) of Fitzmaurice et al. (2009, Chapter 9). The profile SUR kernel estimators of $\beta_0(t)$ and β are computed by the following steps:

(a) *Substitute $\beta_0(t)$ with its SUR kernel estimator, and estimate β through the profile estimating equation step.*

(b) *Substitute β with its profile estimators, and compute the SUR kernel estimator of $\beta_0(t)$.*

(c) *The above two steps are repeated iteratively until convergence.* □

Compared with the profile kernel GEE estimators, the asymptotic results of Wang, Carroll and Lin (2005) demonstrate that, when $\mathbf{Y}_i$ is normally distributed, the profile SUR kernel estimator of β is asymptotically consistent for any working correlation matrix choice of $\mathbf{V}_i(\mathbf{t}_i; \alpha)$. When the working correlation matrix is the true correlation matrix of the data, the profile SUR kernel estimator of β is semiparametric efficient and reaches the semiparametric efficiency lower bound described in Lin and Carroll (2001). Although the true correlation matrix is unknown in practice, hence, a practical profile SUR kernel estimator of β may not be truly semiparametric efficient for a given longitudinal data, the numerical results of Wang, Carroll and Lin (2005) demonstrate that this estimator, when available, may still be a desirable choice because of its appropriate finite sample statistical properties.

Despite the potentially attractive theoretical advantage of the profile SUR kernel estimator, its implementation in practice is somewhat challenging, because the computation of a profile SUR kernel estimator requires an iterative procedure, except for the Gaussian case, and the properties of the iterative procedure in general settings are still not well understood. Further research is still needed to develop practical algorithms to ensure the implementation of the profile SUR kernel estimators.

3. Likelihood-Based Profile SUR Kernel Estimation Method

This method, which is described in Lin and Carroll (2006), is an extension of the profile SUR kernel estimation method to the semiparametric mixed-effects partially linear models (2.35) and (2.36). Based on the distribution assumptions given in (2.35) and (2.36), we can write the log-likelihood function $\ell[\mathbf{Y}_i; \beta, \gamma, \beta_0(t_{i1}), \ldots, \beta_0(t_{in_i})]$, which is an integral involving the conditional likelihood function $\ell(\mathbf{Y}_i|b_i)$, where $\gamma = (\gamma_1^T, \gamma_2^T)^T$ is the parameter vector determining the covariance matrices in (2.35) and (2.36). The estimation procedure is similar to the profile SUR kernel estimation described above. The specific iteration steps include:

(a) *Given β and γ, estimate $\beta_0(t)$ using the SUR kernel estimator $\widehat{\beta}_0(t; \beta, \gamma)$.*

(b) *Substitute $\beta_0(t_{ij})$ in $\ell[\mathbf{Y}_i; \beta, \gamma, \beta_0(t_{i1}), \ldots, \beta_0(t_{in_i})]$ with the SUR kernel estimators, and estimate β by maximizing the profile log-likelihood*

$$\sum_{i=1}^{n} \ell\left[\mathbf{Y}_i; \beta, \gamma, \widehat{\beta}_0(t_{i1}; \beta, \gamma), \ldots, \widehat{\beta}_0(t_{in_i}; \beta, \gamma)\right]$$

with respect to β and γ.

(c) *The final log-likelihood profile SUR kernel estimators $\widehat{\beta}_{LK}$ and $\widehat{\gamma}_{LK}$ are obtained by repeating the above two steps iteratively until convergence.* □

Similar to the profile SUR kernel estimators for marginal partially linear models, the results of Lin and Carroll (2006) show that the likelihood-based profile SUR kernel estimators $\widehat{\beta}_{LK}$ and $\widehat{\gamma}_{LK}$ have the desired asymptotic properties of consistency and semiparametric efficiency.

2.3.5 Semiparametric Estimation by Splines

The nonparametric baseline curve and the Euclidean space parameters of the marginal and mixed-effects partially linear models (2.29), and (2.33) through (2.36) can also be estimated using some modified spline estimation methods as described in Section 1.5. In the following, we briefly summarize the estimation for the marginal partially linear model (2.29) using B-splines and the roughness penalized smoothing splines. We omit the estimators for the partially linear models (2.33) through (2.36), because they can be similarly computed using these spline approaches. In later chapters, we further generalize these spline methods to the structured nonparametric models.

1. Estimation by B-Splines

The main idea of the B-spline estimation method is to approximate the nonparametric baseline curve $\beta_0(t)$ by some linear expansions of the spline basis functions, so that $\beta_0(t)$ and as well as β can be estimated by the "extended linear models" approach of Stone et al. (1997) and Huang (1998). Approximating $\beta_0(t)$ by the B-spline expansion with a pre-specified integer $L \geq 0$, which may increase as the number of subject n increases,

$$\beta_0(t) \approx \sum_{l=0}^{L} \gamma_l B_l(t), \tag{2.41}$$

where $\{B_0(t), B_1(t), \dots\}$ is a set of B-spline basis functions defined in (1.37), the B-spline approximated model for (2.29) is

$$\begin{cases} Y_{ij} & \approx \sum_{l=0}^{L} \gamma_l B_l(t_{ij}) + \sum_{l=1}^{K} \beta_l X_{ij}^{(l)} + \varepsilon_{ij} \\ & = \mathbf{B}^T(t_{ij})\gamma + \mathbf{X}_{ij}^T \beta + \varepsilon_{ij}, \\ \gamma & = (\gamma_0, \dots, \gamma_L)^T, \\ \mathbf{B}(t_{ij}) & = (B_0(t_{ij}), \dots, B_L(t_{ij}))^T, \\ \beta & = (\beta_1, \dots, \beta_K)^T, \end{cases} \tag{2.42}$$

where ε_{ij} is the error process $\varepsilon(t)$ defined in (2.29) at time point t_{ij}.

Assuming that $\varepsilon_i = (\varepsilon_{i1}, \dots, \varepsilon_{in_i})^T$ has the mean zero Gaussian distribution with covariance matrix satisfying a parametric model denoted by $\mathbf{V}_i(\mathbf{t}_i; \alpha)$, the B-spline estimators of β, γ and α are obtained by minimizing

$$\begin{aligned} L(\beta, \gamma, \alpha) & = \sum_{i=1}^{n} \left\{ [\mathbf{Y}_i - \mathbf{B}^T(\mathbf{t}_i)\gamma - \mathbf{X}_i^T \beta]^T \right. \\ & \left. \times \mathbf{V}_i^{-1}(\mathbf{t}_i; \alpha)[\mathbf{Y}_i - \mathbf{B}^T(\mathbf{t}_i)\gamma - \mathbf{X}_i^T \beta] \right\} \end{aligned} \tag{2.43}$$

with respect to β, γ and α. If (2.43) can be uniquely minimized, the B-spline estimators $\widehat{\beta}$, $\widehat{\gamma}$ and $\widehat{\alpha}$ satisfy

$$L(\widehat{\beta}, \widehat{\gamma}, \widehat{\alpha}) = \min_{\beta, \gamma, \alpha} L(\beta, \gamma, \alpha). \tag{2.44}$$

Based on $\widehat{\gamma} = (\widehat{\gamma}_0, \dots, \widehat{\gamma}_L)^T$ and the B-spline basis functions $\{B_0(t), \dots, B_L(t)\}$, the B-spline estimator of the baseline coefficient curve $\beta_0(t)$ is given by

$$\widehat{\beta}_0(t) = \sum_{l=0}^{L} \widehat{\gamma}_l B_l(t). \tag{2.45}$$

Because (2.42) is the special case of the time-varying coefficient models, we discuss in Chapter 9 the statistical properties of the B-spline estimators for these more general models. Asymptotic properties, such as the asymptotic distributions and semiparametric efficiency, of the B-spline estimators in (2.44) and (2.45) are still not well-understood and require further development.

2. Estimation by Penalized Smoothing Splines

When smoothing splines are used, the estimators are obtained by maximizing some "*roughness penalized likelihood functions.*" For the case of (2.29) with Gaussian errors, i.e., $\varepsilon_i = (\varepsilon_{i1}, \dots, \varepsilon_{in_i})^T$ has the mean zero Gaussian distribution with covariance matrix $\mathbf{V}_i(\mathbf{t}_i; \alpha)$, and $\beta_0(t)$ is twice differentiable with respect to t, the penalized log-likelihood function based on the second derivatives of $\beta_0(t)$ is given by

$$\ell_\lambda[\beta_0(\cdot), \beta, \alpha] = \sum_{i=1}^{n} \Big\{ [\mathbf{Y}_i - \beta_0(\mathbf{t}_i) - \mathbf{X}_i^T \beta]^T \mathbf{V}_i(\mathbf{t}_i; \alpha) \tag{2.46}$$
$$\times [\mathbf{Y}_i - \beta_0(\mathbf{t}_i) - \mathbf{X}_i^T \beta] \Big\} - \lambda \int [\beta_0''(t)]^2 dt,$$

where $\mathbf{Y}_i$, $\mathbf{t}_i$ and $\mathbf{X}_i$ are defined in (2.1), $\beta_0(\mathbf{t}_i) = (\beta_0(t_{i1}), \dots, \beta_0(t_{in_i}))^T$, $\beta_0''(t)$ is the second derivative of $\beta_0(t)$ with respect to t, and $\lambda > 0$ is a positive smoothing parameter. The minimizers of (2.46) are the *penalized smoothing splines* estimators of $\beta_0(\cdot)$, β and α, such that

$$\ell_\lambda[\widehat{\beta}_0(\cdot), \widehat{\beta}, \widehat{\alpha}] = \min_{\beta_0(\cdot), \beta, \alpha} \ell_\lambda[\beta_0(\cdot), \beta, \alpha]. \tag{2.47}$$

Details of the statistical properties of the penalized smoothing splines estimators in (2.47) and their generalizations to the mixed-effects partially linear models can be found in Zhang et al. (1998) and Fitzmaurice et al. (2009, Section 9.6). Similar to the B-splines estimators of (2.44) and (2.45), asymptotic properties, such as the asymptotic distributions and semiparametric efficiency, of the penalized smoothing splines estimators in (2.47) have not been systematically derived and require further development.

2.4 R Implementation

We present a few R functions for fitting the linear mixed-effects models using the BMACS and the ENRICHD examples of Section 1.2. These results can be used to compare with the findings obtained from the more flexible structured nonparametric models to be discussed in Chapters 6 to 11. Although the nonlinear mixed-effects models and the semiparametric partially linear models also have important applications in practice, we omit their implementations here because our focus is on the comparisons of the linear models with the structured nonparametric models.

Several R packages are available to fit the linear mixed-effects models. Among them, the nlme (Pinheiro and Bates, 2000; Pinheiro et al., 2018) and lme4 (Bates et al., 2015) are two widely used and well-documented packages. Both nlme and lme4 can model the intra-subject correlations among repeated measurements using random effects. The nlme package also allows the user to specify complex serial correlation structures. The lme4 package uses more flexible and efficient optimizers to allow for the fitting of singular models, which sometimes happen in the analysis of small to medium sized datasets. We illustrate how to fit the models using functions from these R packages in two examples.

2.4.1 The BMACS CD4 Data

The BMACS CD4 data has been described in Section 1.2. For each observation, the subject's study visit time, cigarette smoking status, age at study enrollment, pre-infection CD4 percentage, and CD4 percentage at the time of visit were recorded. The following R code is used to inspect the data:

```
> data(BMACS)
> str(BMACS)
'data.frame':          1817 obs. of  6 variables:
 $ ID    : int  1022 1022 1022 1022 1022 1022 1022 1049 1049 ...
 $ Time  : num  0.2 0.8 1.2 1.6 2.5 3 4.1 0.3 0.6 1 ...
 $ Smoke : int  0 0 0 0 0 0 0 0 0 0 ...
 $ age   : num  26.2 26.2 26.2 26.2 26.2 ...
 $ preCD4: num  38 38 38 38 38 38 38 44.5 44.5 44.5 ...
 $ CD4   : num  17 30 23 15 21 12 5 37 44 37 ...
> head(BMACS)
    ID Time Smoke   age preCD4 CD4
1 1022  0.2     0 26.25     38  17
2 1022  0.8     0 26.25     38  30
3 1022  1.2     0 26.25     38  23
4 1022  1.6     0 26.25     38  15
```

```
5 1022  2.5      0 26.25     38  21
6 1022  3.0      0 26.25     38  12
```

The BMACS data is already in the *long* format (repeated measurements per subject are listed in separate records or rows). For some datasets, if the data available are stored in the *wide* format (repeated measurements per subject are listed in multiple columns of the same records), they can be easily restructured from *wide* to *long* format using some simple functions such as stack() in the utils package and reshape() in the stats package in the base R distribution or using a flexible data restructuring package tidyr (Wickham and Grolemund, 2017).

For the BMACS data, we are interested in modeling the attenuation in CD4 percentage over time following HIV infection because it is known that the loss of CD4 cells due to HIV leads to AIDS and HIV-related mortality (Kaslow et al.,1987). We first fit a simple linear mixed model that includes a linear trend in time and a random intercept term for each subject,

$$Y_{ij} = \beta_0 + b_{0i} + \beta_1 t_{ij} + \varepsilon_{ij}, \tag{2.48}$$

where Y_{ij} are the CD4 percentages and ε_{ij} are the i.i.d. measurement errors at time t_{ij} with $\varepsilon_{ij} \sim N(0, \sigma^2)$, β_0 and β_1 are the fixed intercept and slope terms, respectively, and $b_{0i} \sim N(0, \sigma_0^2)$ is a normal random intercept term that describes the individual shifts from the common intercept β_0.

The following R commands are used to fit the random intercept model (2.48) and produce the REML estimates for the model parameters. The model formula of the lme function includes both fixed and random effects, where the random argument specifies the random-effect terms followed by |grouping variable to indicate the correlated observations within the same group or subject:

```
> library(nlme)
> CD4fit1 <- lme(CD4~ Time, random=~1|ID, data=BMACS)
> summary(CD4fit1)

Linear mixed-effects model fit by REML
 Data: BMACS
       AIC       BIC      logLik
  12561.91 12583.92 -6276.954

Random effects:
 Formula: ~1 | ID
         (Intercept) Residual
StdDev:    8.824904 6.345293

Fixed effects: CD4 ~ Time
```

```
              Value Std.Error   DF    t-value p-value
(Intercept) 35.37234 0.5966519 1533  59.28471       0
Time         -2.67484 0.1076277 1533 -24.85274       0
 Correlation:
     (Intr)
Time -0.357
```

```
Standardized Within-Group Residuals:
       Min          Q1        Med         Q3        Max
-3.73917541 -0.57204793 -0.04429796  0.56658314  4.54644250
```

```
Number of Observations: 1817
Number of Groups: 283
```

The "random intercept only" model (2.48) implies a compound symmetry covariance structure, that is, a constant variance over time and equal positive correlation between any two measurements from the same subject. In the model (2.48), $Cov(Y_{ij_1}, Y_{ij_2}) = \sigma_0^2$ for $j_1 \neq j_2$, and $Var(Y_{ij}) = \sigma_0^2 + \sigma^2$. Thus, the correlation is $\rho = \sigma_0^2 / (\sigma_0^2 + \sigma^2)$. For CD4 percentage, $\rho = (8.824)^2 / [(8.824)^2 + (6.345)^2] = 0.659$.

We can also introduce a random slope term b_{1i} into the model (2.48), so that

$$Y_{ij} = \beta_0 + b_{0i} + (\beta_1 + b_{1i}) t_{ij} + \varepsilon_{ij}, \tag{2.49}$$

where $(b_{0i}, b_{1i})^T \sim N(0, \Sigma)$ with

$$\Sigma = \begin{pmatrix} \sigma_0^2 & \sigma_{01} \\ \sigma_{01} & \sigma_1^2 \end{pmatrix}$$

represents the individual random deviations from the population-mean intercept and slope. We fit (2.49) using the R commands (the random intercept is included by default):

```
> CD4fit2 <- lme(CD4~ Time, random=~Time |ID, data=BMACS)
> summary(CD4fit2)
```

```
Linear mixed-effects model fit by REML
 Data: BMACS
       AIC       BIC     logLik
  12166.3 12199.32 -6077.148
```

```
Random effects:
 Formula: ~Time | ID
 Structure: General positive-definite,
           Log-Cholesky parametrization
```

```
          StdDev    Corr
(Intercept) 8.845541 (Intr)
Time        3.052241 -0.324
Residual    5.003937
```

```
Fixed effects: CD4 ~ Time
              Value Std.Error   DF   t-value p-value
(Intercept) 35.74864 0.5885714 1533  60.73799       0
Time        -3.08116 0.2353750 1533 -13.09042       0
 Correlation:
     (Intr)
Time -0.435
```

```
Standardized Within-Group Residuals:
        Min           Q1          Med          Q3          Max
-4.20189953 -0.55165113 -0.02258232  0.52023712  4.25186946
```

```
Number of Observations: 1817
Number of Groups: 283
```

Unlike (2.48), which assumes that the subjects have different intercepts at time 0 but the same slope for CD4 percentages, the model (2.49) assumes that the subjects have different intercepts and different slopes. The model (2.49) also implies that the covariance and correlation structures between any two measurements of the same subject depend on their time points t_{ij}, such that,

$$\begin{cases} Cov(Y_{ij_1}, Y_{ij_2}) &= \sigma_0^2 + \sigma_{01}\left(t_{ij_1} + t_{ij_2}\right) + \sigma_1^2 t_{ij_1} t_{ij_2} \quad \text{for } j_1 \neq j_2, \\ Var(Y_{ij}) &= \sigma_0^2 + 2\sigma_{01} t_{ij} + \sigma_1^2 t_{ij}^2 + \sigma^2. \end{cases}$$

Figure 2.1 shows the longitudinal CD4 measurements of two randomly selected subjects, the population-mean CD4 percentage trajectories and the subject-specific CD4 percentage trajectories of these two subjects, which are computed based on the above models (2.48) and (2.49). Comparing the results from these two models, the population-mean (fixed-effects) intercept and slope estimates are similar and both effects are statistically significant.

To examine if the additional random slope term improves the model fitting, we can use anova function that produces a likelihood ratio test to compare the models. The following output shows that the likelihood ratio test is highly significant, which suggests that the model (2.49) is preferable:

```
> anova(CD4fit1, CD4fit2)

        Model df      AIC       BIC    logLik    Test  L.Ratio p-value
CD4fit1   1    4 12561.91 12583.92 -6276.954
CD4fit2   2    6 12166.30 12199.32 -6077.148 1 vs 2 399.6122  <.0001
```

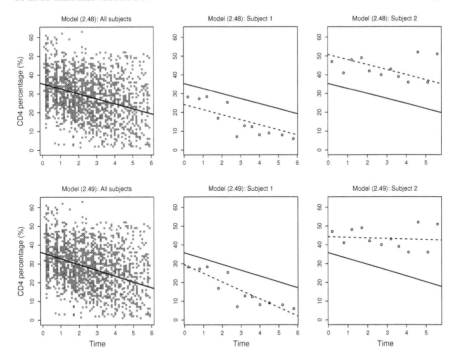

Figure 2.1 *The longitudinal CD4 measurements for all subjects and two randomly selected subjects with the population-averaged and subject-specific regression estimates plotted in solid lines and dashed lines, respectively, based on model (2.48) (random intercept only) and model (2.49) (random intercept and random slope).*

We can also obtain the same fitting results of the model (2.49) using the lmer function from the lme4 package with the following slightly different model formulation:

```
> library(lme4)
> CD4fit2b <- lmer(CD4~ Time + (Time|ID), data= BMACS)
> summary(CD4fit2b )

Linear mixed model fit by REML ['lmerMod']
Formula: CD4 ~ Time + (Time | ID)
   Data: BMACS

REML criterion at convergence: 12154.3

Scaled residuals:
    Min      1Q   Median      3Q      Max
-4.2019 -0.5517 -0.0226  0.5202  4.2519
```

```
Random effects:
 Groups    Name            Variance Std.Dev. Corr
 ID        (Intercept) 78.243    8.846
           Time           9.316    3.052    -0.32
 Residual                25.039    5.004
Number of obs: 1817, groups:  ID, 283

Fixed effects:
             Estimate Std. Error t value
(Intercept)  35.7486     0.5886    60.74
Time         -3.0812     0.2354   -13.09

Correlation of Fixed Effects:
     (Intr)
Time -0.435
```

In addition to the linear trend with time, it may be interested in examining the effects of other covariates on CD4 percentages. The covariates can be added to the linear mixed-effects model. The preCD4 variable is centered first by subtracting the sample mean, which is shown to be a significant factor. Smoking history and age are not significant. These results are obtained using the following R code:

```
> BMACS$preCD4c <- BMACS$preCD4 - mean(BMACS$preCD4)
> CD4fit3<- lme(CD4~ Time + preCD4c + Smoke + age,
                random=~Time|ID, data=BMACS)
> summary(CD4fit3)

Linear mixed-effects model fit by REML
...

Fixed effects: CD4 ~ Time + preCD4 + Smoke + age
                Value Std.Error   DF    t-value p-value
(Intercept)  34.94593 2.3259116 1532  15.024615  0.0000
Time         -3.103432 0.2365191 1532 -13.121275  0.0000
preCD4        0.453743 0.0611912  280   7.415169  0.0000
Smoke         0.687434 1.0333168  280   0.665270  0.5064
age           0.014169 0.0638670 1532   0.221853  0.8245
  ...
```

2.4.2   The ENRICHD BDI Data

The ENRICHD dataset has been described in Section 1.2. For each observation, the subject's ID number, study visit time (in days), BDI score, antidepressant medication use, and the starting time of medication were recorded, and can be seen using the following R commands:

```
> data(BDIdata)
> str(BDIdata)
'data.frame':          7117 obs. of  5 variables:
 $ ID       : int  1 1 1 1 1 1 1 1 1 1 ...
 $ time     : int  0 29 42 47 56 77 83 90 118 125 ...
 $ BDI      : int  25 34 28 29 18 5 19 12 14 18 ...
 $ med      : num  0 0 1 1 1 1 1 1 1 1 ...
 $ med.time : int  30 30 30 30 30 30 30 30 30 30 ...

> BDIdata[BDIdata$ID==1,]

   ID time BDI med med.time
1   1    0  25   0       30
2   1   29  34   0       30
3   1   42  28   1       30
4   1   47  29   1       30
5   1   56  18   1       30
6   1   77   5   1       30
7   1   83  19   1       30
8   1   90  12   1       30
9   1  118  14   1       30
10  1  125  18   1       30
```

The ENRICHD BDI dataset consists of a subset of patients from the psychosocial treatment arm of the randomized ENRICHD trial. The ENRICHD study was designed to evaluate the efficacy of the psychosocial treatment in patients with depression and/or low perceived social support after acute myocardial infarction. Depression severity was measured by Beck Depression Inventory (BDI) score, where higher BDI scores indicate worsened depression. In our analysis, we use the subgroup of 92 patients (1465 observations) with clear records of the pharmacotherapy starting time. We first fit the following linear mixed-effects model involving only time with random intercept and slope

$$Y_{ij} = \beta_{0i} + \beta_{1i} t_{ij} + \varepsilon_{ij}, \tag{2.50}$$

where $\beta_{0i} = \beta_0 + b_{0i}$ and $\beta_{1i} = \beta_1 + b_{1i}$, β_0 and β_1 are the population-mean intercept and slope, b_{0i} and b_{1i} are the random intercept and slope, and Y_{ij} is BDI score for the ith patient at study visit time t_{ij}. The model can be fit by the following R commands, which show a negative slope and suggest a significant effect of the psychosocial treatment that lowered the patients' BDI scores during the clinical trial:

```
# recode time in months
> BDIdata$Tijm <- BDIdata$time*12/365.25
> BDIsub <- subset(BDIdata, med.time >=0 & med.time < 200)
> BDI.Model3 <- lme(BDI ~ Tijm, data=BDIsub, random=~Tijm|ID)
> summary(BDI.Model3)
```

```
Linear mixed-effects model fit by REML
 Data: BDIsub
       AIC      BIC    logLik
  9587.344 9619.073 -4787.672

Random effects:
 Formula: ~Tijm | ID
 Structure: General positive-definite,
            Log-Cholesky parametrization
            StdDev    Corr
(Intercept) 8.206250 (Intr)
Tijm        1.819110 -0.597
Residual    5.466927

Fixed effects: BDI ~ Tijm
                Value Std.Error   DF   t-value p-value
(Intercept) 21.630395 0.9030760 1372 23.951910       0
Tijm        -2.084839 0.2130814 1372 -9.784236       0
 Correlation:
     (Intr)
Tijm -0.621

Standardized Within-Group Residuals:
       Min          Q1          Med          Q3
-4.52027744 -0.49237793 -0.02382951  0.46209759
       Max
 5.02331404

Number of Observations: 1465
Number of Groups: 92
```

However, the model (2.50) does not consider the use of concomitant antidepressant medication. By the trial protocol, in addition to the randomized psychosocial treatments, patients with high baseline depression scores and/or nondecreasing BDI trends were eligible for pharmacotherapy with antidepressants. If the scientific question is whether the antidepressants have added benefits for lowering the BDI scores of the patients undergone this concomitant intervention during the trial, we may incorporate the medication use in the following model. For the ith patient, let s_i, $r_{ij} = t_{ij} - s_i$ and $\delta_{ij} = 1_{[t_{ij} \geq s_i]}$

be the ith patient's starting time of pharmacotherapy, time from initiation of pharmacotherapy, and pharmacotherapy indicator, respectively, at the jth visit. An intuitive model is

$$Y_{ij} = \beta_{0i} + \beta_{1i} t_{ij} + \gamma_{0i} \delta_{ij} + \gamma_{1i} \delta_{ij} r_{ij} + \varepsilon_{ij}, \tag{2.51}$$

where β_{0i}, β_{1i}, γ_{0i} and γ_{1i} are all random coefficients with

$$E\left(\beta_{0i}, \beta_{1i}, \gamma_{0i}, \gamma_{1i}\right)^T = \left(\beta_0, \beta_1, \gamma_0, \gamma_1\right)^T.$$

When $\delta_{ij} = 1$ and $r_{ij} = m$, $\left(\gamma_0 + \gamma_1 m\right)$ describes the mean pharmacotherapy effect at m months since the start of pharmacotherapy. The results of fitting (2.51) can be seen from the following R commands:

```
> BDIsub$Sim <- BDIsub$med.time*12/365.25
> BDIsub$Rijm <- with(BDIsub, med*(Tijm -Sim))
> BDI.model4 <- lme(BDI ~ Tijm+ med + Rijm , data=BDIsub,
                    random=~Tijm+ med + Rijm|ID)
> summary(BDI.model4)

Linear mixed-effects model fit by REML
 Data: BDIsub
       AIC       BIC    logLik
  9482.313 9561.616 -4726.156


Random effects:
 Formula: ~Tijm + med + Rijm | ID
 Structure: General positive-definite,
            Log-Cholesky parametrization
            StdDev    Corr
(Intercept) 8.440818 (Intr) Tijm    med
Tijm        2.641688 -0.469
med         5.811350 -0.361  0.217
Rijm        3.259511  0.118 -0.868 -0.231
Residual    5.118352

Fixed effects: BDI ~ Tijm + med + Rijm
                Value Std.Error    DF   t-value p-value
(Intercept) 23.360014 1.1154194 1370 20.942808  0.0000
Tijm        -0.610008 0.4783029 1370 -1.275360  0.2024
med         -3.582001 1.0000283 1370 -3.581900  0.0004
Rijm        -1.546813 0.5161834 1370 -2.996634  0.0028
 Correlation:
      (Intr) Tijm    med
Tijm -0.356
med  -0.573  0.035
```

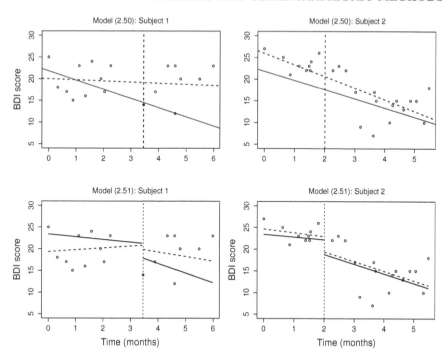

Figure 2.2 *The longitudinal BDI measurements for two randomly selected subjects with the population-averaged and subject-specific regression estimates plotted in solid lines and dashed lines, respectively, based on the models (2.50) and (2.51). For each subject, the starting time of the pharmacotherapy is indicated by a vertical dashed line.*

```
Rijm   0.201 -0.908 -0.132

Standardized Within-Group Residuals:
        Min              Q1             Med              Q3
-4.72295735 -0.49429621 -0.01038197  0.47258323
        Max
 5.24099392

Number of Observations: 1465
Number of Groups: 92
```

In contrast to the model (2.50), the results from (2.51) suggest that the time effect β_1 is no longer significant ($P=0.20$) and the psychosocial treatment is not effective for these patients. However, both γ_0 and γ_1 are statistically significant, indicating a significant decrease in BDI score after the start of the pharmacotherapy. Figure 2.2 shows the fitted population-mean and subject-specific trajectories from the models (2.50) and (2.51).

It is worth noting that (2.51) ignores the correlation between the starting time of pharmacotherapy s_i and the pre-pharmacotherapy depression trends. We revisit this example in Chapter 10, which shows that, through a varying-coefficient modeling approach, (2.51) may lead to potential bias because it ignores the potential effects of the starting time of pharmacotherapy s_i.

2.5 Remarks and Literature Notes

This chapter briefly summarizes a number of parametric and semiparametric regression methods for the analysis of longitudinal data. The parametric models include the linear, generalized linear and nonlinear marginal and mixed-effects models. The semiparametric models are primarily focused on the partially linear marginal and mixed-effects models. The estimation and inference methods for these models include the maximum likelihood and restricted maximum likelihood procedures, and the iteration procedures which combine parameter estimation and nonparametric smoothing methods. Because the linear marginal and mixed-effects models are often the first attempts for longitudinal analysis in a real application, we outline a few details of their estimation and inference methods, including the implementation of the R packages nlme and lme4. These estimation and inference methods provide some useful insights into the more flexible nonparametric estimation methods to be introduced in detail throughout this book.

The analytical approach with repeated measurements data using parametric and semiparametric regression models has a long history in the statistical literature. As noted in Fitzmaurice et al. (2009, Section 1.3), the linear marginal and mixed-effects models are probably the most widely used methods for analyzing longitudinal data, and this approach was popularized by researchers at the U.S. National Institutes of Health (NIH). Given the extensive publications of this subject in the literature, it is difficult to list all or even most of the important publications. Early work of the linear marginal and mixed-effects models, their estimation and inference procedures, and their applications include Patterson and Thompson (1971), Harville (1974), Laird and Ware (1982), Liang and Zeger (1986), Diggle (1988), Zeger, Liang and Albert (1988), among many others. Most of these results are summarized in a number of well-written books, for example, Vonesh and Chinchilli (1997), Verbeke and Molenberghs (2000), Diggle et al. (2002), Molenberghs and Verbeke (2005) and Jiang (2007). The R packages nlme and lme4 for the estimation and inferences with the linear marginal and mixed-effects models are also well developed and well documented in Pinheiro and Bates (2000), Wickham (2014), Bates et al. (2015) and Pinheiro et al. (2017).

Beyond the linear and generalized linear models, theory, methods and applications for the nonlinear marginal and mixed-effects models are nicely summarized in Davidian and Giltinan (1995, 2003). The flexible semiparametric partially linear models summarized in this chapter are described in detail in Zeger and Diggle (1994), Moyeed and Diggle (1994), Lin and Ying (2001), Lin and Carroll (2001, 2006) and Wang, Carroll and Lin (2005). A comprehensive account of the recent developments of longitudinal analysis can be found in Fitzmaurice et al. (2009). But, most of the structured nonparametric models and their estimation and inference methods presented in Part III to Part V, i.e., Chapters 6 to 15, of this book are more recent results that are not included in Fitzmaurice et al. (2009).

Part II

Unstructured Nonparametric Models

Chapter 3

Kernel and Local Polynomial Methods

We present in this chapter the class of kernel-based local smoothing methods for nonparametric curve estimation with unstructured nonparametric regression models. These smoothing methods form a series of building blocks for constructing smoothing estimators when the nonparametric models are subject to structural constraints. In addition, as discussed in Section 2.3, these smoothing estimators can also be used in conjunction with parametric estimation procedures in semiparametric partially linear regression models.

3.1 Least Squares Kernel Estimators

Our statistical objective is to estimate the conditional mean $\mu(t) = E[Y(t)|t]$ of $Y(t)$ from the model

$$Y(t) = \mu(t) + \varepsilon(t), \tag{3.1}$$

based on the sample $\{(Y_{ij}, t_{ij}) : i = 1, \ldots, n; j = 1, \ldots, n_i\}$, where $\mu(t)$ is a smooth function of time t and $\varepsilon(t)$ is a mean zero error term with variance and covariance curves,

$$\begin{cases} Var[Y(t)|t] & = & \sigma^2(t) \ \text{at} \ t \in \mathcal{T}, \\ Cov[Y(t_1), Y(t_2)|t_1, t_2] & = & E\Big\{ [Y(t_1) - \mu(t_1)][Y(t_2) - \mu(t_2)] \Big\} \\ & = & \rho(t_1, t_2) \ \text{at any} \ t_1 \neq t_2 \in \mathcal{T}, \\ \rho(t,t) & \neq & \sigma^2(t) \ \text{for any} \ t \in \mathcal{T}. \end{cases} \tag{3.2}$$

Although there are no parametric conditions assumed for $\mu(t)$, the smoothness assumptions of $\mu(t)$ determine the theoretical and practical properties of its estimators and statistical inferences. We discuss the specific smoothness assumptions of $\mu(t)$ under the asymptotic derivations of Section 3.6.

A natural approach for the nonparametric estimation of $\mu(t)$ is to borrow the smoothing techniques from the cross-sectional i.i.d. data setting, while evaluating the statistical performance of the resulting estimators by taking the potential intra-subject correlations into consideration. A simple method is to use the kernel smoothing method similar to the Nadaraya-Watson type local least squares criterion (1.22) and (1.23), which amounts to estimate

$\mu(t)$ through a weighted average using the measurements obtained within a neighborhood of t defined by a bandwidth.

Let $K(u)$ be a kernel function, which is usually a probability density function as in (1.24) and (1.25) through (1.31), defined on the real line, and let $h > 0$ be a positive bandwidth. A kernel estimator of $\mu(t)$ similar to (1.23) can be obtained by minimizing the local score function

$$\ell_K(t; h, N^{-1}) = \frac{1}{N} \sum_{i=1}^{n} \sum_{j=1}^{n_i} \left[Y_{ij} - \mu(t)\right]^2 K\left(\frac{t - t_{ij}}{h}\right), \tag{3.3}$$

where $N = \sum_{i=1}^{n} n_i$ is the total number of observations for all subjects. The score function (3.3) uses the measurement uniform weight $1/N = 1/\left(\sum_{i=1}^{n} n_i\right)$ on all the observations, and the minimizer of (3.3) leads to the kernel estimator

$$\widehat{\mu}_K(t; h, N^{-1}) = \frac{\sum_{i=1}^{n} \sum_{j=1}^{n_i} Y_{ij} K\left[(t - t_{ij})/h\right]}{\sum_{i=1}^{n} \sum_{j=1}^{n_i} K\left[(t - t_{ij})/h\right]}. \tag{3.4}$$

Because of the measurement uniform weight on each measurement, (3.4) makes no distinction between the subjects that have unequal numbers of repeated measurements. Consequently, subjects with more repeated measurements are used more often than those with fewer repeated measurements.

More generally, given that the measurements of each subject are potentially correlated and the subjects may have different numbers of repeated measurements, a modification of (3.2) may use a subject-specific weight w_i for the ith subject. The local score function then becomes

$$\ell_K(t; h, w) = \sum_{i=1}^{n} \sum_{j=1}^{n_i} w_i \left[Y_{ij} - \mu(t)\right]^2 K\left(\frac{t - t_{ij}}{h}\right), \tag{3.5}$$

where the weights, $w = (w_1, \ldots, w_n)^T$, satisfy $w_i \geq 0$ for all $i = 1, \ldots, n$ with strict inequality for all or some subjects $1 \leq i \leq n$. The minimizer of the $\ell_K(t; h, w)$ of (3.4) leads to the kernel estimator

$$\widehat{\mu}_K(t; h, w) = \frac{\sum_{i=1}^{n} \sum_{j=1}^{n_i} Y_{ij} w_i K\left[(t - t_{ij})/h\right]}{\sum_{i=1}^{n} \sum_{j=1}^{n_i} w_i K\left[(t - t_{ij})/h\right]}. \tag{3.6}$$

An intuitive weight choice other than $1/N$ of (3.3) is to assign the subject uniform weight to each subject, rather than the measurement uniform weight. The resulting kernel estimator then uses $w_i^* = 1/(n n_i)$, and minimizes the local score function

$$\ell_K(t; h, w^*) = \frac{1}{n} \sum_{i=1}^{n} \sum_{j=1}^{n_i} \left\{ \frac{1}{n_i} \left[Y_{ij} - \mu(t)\right]^2 K\left(\frac{t - t_{ij}}{h}\right) \right\}. \tag{3.7}$$

The kernel estimator based on $w^* = \left(1/(n n_1), \ldots, 1/(n n_n)\right)^T$ is

$$\widehat{\mu}_K(t; h, w^*) = \frac{\sum_{i=1}^{n} \left\{ n_i^{-1} \sum_{j=1}^{n_i} Y_{ij} K\left[(t - t_{ij})/h\right] \right\}}{\sum_{i=1}^{n} \left\{ n_i^{-1} \sum_{j=1}^{n_i} K\left[(t - t_{ij})/h\right] \right\}}. \tag{3.8}$$

In (3.8), the subjects with fewer repeated measurements are assigned more weight than those with more repeated measurements.

The effects of weight choices depend on the longitudinal designs and the numbers of repeated measurements. The two most commonly used weight choices, which will be used throughout this book, are

(a) the *subject uniform weight* $w_i^* = 1/(n n_i)$;

(b) the *measurement uniform weight* $w_i^{**} = 1/N$.

We present the theoretical and practical properties of the weight choices in Section 3.6. Other kernel approaches for the estimation of $\mu(t)$ in (3.1) have also been studied by Hart and Wehrly (1986), Müller (1988), Altman (1990) and Hart (1991). Because these methods are based on fixed time design points, we omit the discussion of their details. In practice, however, all these kernel methods are based on the fundamental concept of local least squares with weights determined by a bandwidth and a kernel function, and they generally lead to similar numerical results. This is in contrast to the local and global smoothing methods with structured nonparametric models, i.e., Chapters 6 to 15, where different local and global smoothing methods may produce different smoothing estimators.

3.2 Least Squares Local Polynomial Estimators

The local polynomial estimators of $\mu(t)$ based on the model (3.1) and longitudinal sample $\{(Y_{ij}, t_{ij}) : i = 1, \ldots, n; j = 1, \ldots, n_i\}$ can be derived by extending the approach of (1.32) through (1.35). Similar to the local polynomial methods with cross-sectional i.i.d. data, the local polynomial estimators are motivated by the need to reduce the potential boundary bias associated with the kernel estimators (Fan and Gijbels, 1996), when t_{ij} is close to the boundary of $\mathcal{T}$. Similar to the approximation of (1.32), when t_{ij} is within a small neighborhood of t and $\mu(s)$ is p-times continuously differentiable at t for some $p \geq 1$, the Taylor's expansion of $\mu(t_{ij})$ at t gives

$$\mu(t_{ij}) \approx \sum_{l=0}^{p} \frac{\mu^{(l)}(t)}{l!} \left(t_{ij} - t\right)^l = \sum_{l=0}^{p} b_l \left(t_{ij} - t\right)^l. \tag{3.9}$$

The pth order local polynomial estimator based on (3.9), with the w_i weight for the ith subject, the kernel function $K(\cdot)$ and the bandwidth h, is obtained by minimizing the local score function

$$\ell_{p,L}(t; h, w) = \sum_{i=1}^{n} \sum_{j=1}^{n_i} w_i \left[Y_{ij} - \sum_{l=0}^{p} b_l \left(t_{ij} - t\right)^l\right]^2 K\left(\frac{t - t_{ij}}{h}\right) \tag{3.10}$$

with respect to b_l, $l = 1, \ldots, p$, where $w = \left(w_1, \ldots, w_n\right)^T$.

Choices of the kernel function $K(\cdot)$ include the probability density functions given in (1.24) and (1.27) through (1.31). The minimizer $\widehat{b}_l(t; h, w)$ of $\ell_{p,L}(t; h, w)$ in (3.10) gives the pth order local polynomial estimator

$$\widehat{\mu}_{L,p}^{(l)}(t; h, w) = l! \, \widehat{b}_l(t; h, w) \tag{3.11}$$

of the lth derivative $\mu^{(l)}(t)$ at time point t. The entire curve $\mu^{(l)}(\cdot)$ is obtained by varying t over the range of $\mathscr{T}$. The pth order local polynomial estimator of $\mu(t)$ is

$$\widehat{\mu}_{L,p}(t; h, w) \equiv \widehat{\mu}_{L,p}^{(0)}(t; h, w). \tag{3.12}$$

In particular, the local linear estimator of $\mu(t)$ is $\widehat{\mu}_{L,1}(t; h, w)$, which is the most commonly used local polynomial estimators in real applications. Similar to the kernel estimators, the main factors affecting the statistical properties of the local polynomial estimators are the bandwidth choices, the degree p of the polynomials and the weight choices w_i, while the shape of the kernel function $K(\cdot)$ is less important.

3.3 Cross-Validation Bandwidths

A crucial step in obtaining an adequate kernel or local polynomial estimator of $\mu(t)$ is to select an appropriate bandwidth h. For the smoothing estimation methods other than kernel or local polynomial estimators, such as the B-splines and penalized smoothing splines to be discussed in the next two chapters, this amounts to selecting an appropriate smoothing parameter. Since the repeated measurements within a subject are potentially correlated and the correlation structure is usually completely unknown in practice, a simple procedure for the selection of a data-driven smoothing parameter is to use the "*leave-one-subject-out*" cross-validation (LSCV), which does not depend on the intra-subject correlation structure of the data and can potentially preserve the unknown correlation structure.

3.3.1 The Leave-One-Subject-Out Cross-Validation

Let $\widehat{\mu}_L(t; h, w)$ be a local smoothing estimator of $\mu(t)$, which can be either the kernel estimator $\widehat{\mu}_K(t; h, w)$ of (3.6) or the pth order local polynomial estimator $\widehat{\mu}_{L,p}(t; h, w)$ of (3.12), i.e.,

$$\widehat{\mu}_L(t; h, w) = \widehat{\mu}_K(t; h, w) \quad \text{or} \quad \widehat{\mu}_{L,p}(t; h, w). \tag{3.13}$$

The LSCV bandwidth selection for $\widehat{\mu}_L(t; h, w)$ is carried out with the following three main steps. These steps can be adapted to other estimation methods to compute the corresponding LSCV smoothing parameters in other settings discussed in this book.

Leave-One-Subject-Out Cross-Validation Procedure:

(a) *Compute the "leave-one-subject-out" estimator $\widehat{\mu}_L^{(-i)}(t; h, w)$ based on the remaining data after deleting the entire set of repeated measurements for the ith subject.*

(b) *Predict the ith subject's outcome at time t_{ij} by $\widehat{\mu}_L^{(-i)}(t_{ij}; h, w)$.*

(c) *Define the LSCV score of $\widehat{\mu}_L(t; h, w)$ by*

$$LSCV(h, w) = \sum_{i=1}^{n} \sum_{j=1}^{n_i} w_i \left[Y_{ij} - \widehat{\mu}_L^{(-i)}(t_{ij}; h, w) \right]^2. \tag{3.14}$$

If (3.14) can be uniquely minimized by h_{LSCV} over all positive values of $h > 0$, i.e.,

$$LSCV(h_{LSCV}, w) = \min_{h>0} CV(h, w). \tag{3.15}$$

then h_{LSCV} is defined to be the leave-one-subject-out cross-validated bandwidth of $\widehat{\mu}_L(t; h, w)$ of (3.13). □

The use of h_{LSCV} can be heuristically justified because, by minimizing the cross-validation score $LSCV(h, w)$ of (3.14), it approximately minimizes an average prediction error of $\widehat{\mu}_L(t; h, w)$. In real applications, it is often easy to find out a suitable range of the bandwidths by examining the plots of the fitted curves and then approximate the value of h_{LSCV} through a series of bandwidth choices. This searching method, although somewhat *ad hoc*, may actually speed up the computation and give a satisfactory bandwidth.

3.3.2 A Computation Procedure for Kernel Estimators

Direct minimization of the cross-validation score (3.14) can be computationally intensive, as the algorithm repeats itself each time a new subject is deleted. For the kernel estimator $\widehat{\mu}_K(t; h, w)$, it is possible to use a computationally simpler approach without relying on deleting the subjects one at a time. In this approach, we first define, for $i = 1, \ldots, n$,

$$K_{ij} = K\left(\frac{t - t_{ij}}{h}\right), \quad K_{ij}^* = \frac{w_i K[(t - t_{ij})/h]}{\sum_{i=1}^{n} \sum_{j=1}^{n_i} w_i K[(t - t_{ij})/h]} \quad \text{and} \quad K_i^* = \sum_{j=1}^{n_i} K_{ij}^*,$$

and, then compute $\left[Y_{ij} - \widehat{\mu}_K^{(-i)}(t_{ij}; h, w) \right]$ using the following decomposition,

$$
\begin{aligned}
&Y_{ij} - \widehat{\mu}_K^{(-i)}(t_{ij}; h, w) \\
&= Y_{ij} - \left[\widehat{\mu}_K(t_{ij}; h, w) - \sum_{j=1}^{n_i} (Y_{ij} K_{ij}^*) \right] \left(1 + \frac{K_i^*}{1 - K_i^*} \right) \\
&= \left[Y_{ij} - \widehat{\mu}_K(t_{ij}; h, w) \right] + \sum_{j=1}^{n_i} (Y_{ij} K_{ij}^*)
\end{aligned}
$$

$$-\left[\widehat{\mu}_K\left(t_{ij}; h, w\right) - \sum_{j=1}^{n_i} \left(Y_{ij} K_{ij}^*\right)\right]\left(\frac{K_i^*}{1 - K_i^*}\right) \tag{3.16}$$

$$= \left[Y_{ij} - \widehat{\mu}_K\left(t_{ij}; h, w\right)\right] + \left(\frac{K_i^*}{1 - K_i^*}\right)\left[\frac{\sum_{j=1}^{n_i}\left(Y_{ij} K_{ij}^*\right)}{K_i^*} - \widehat{\mu}_K\left(t_{ij}; h, w\right)\right].$$

The above expression, which was suggested by Rice and Silverman (1991), is specifically targeted to the kernel estimator $\widehat{\mu}_K(t; h, w)$, although different weight choices may be used. When other smoothing methods, such as the local polynomial estimators, are used, the explicit expression at the right side of (3.16) does not generally hold. Thus, for general estimators other than the kernel estimators $\widehat{\mu}_K(t; h, w)$, direct minimization of the cross-validation score (3.14) has to be computed with subjects deleted one at a time.

3.3.3 Heuristic Justification of Cross-Validation

The use of $LSCV(h, w)$ in (3.14) as a risk criterion for $\widehat{\mu}_L^{(-i)}(t_{ij}; h, w)$ can be justified by evaluating the following decomposition of $LSCV(h, w)$,

$$\begin{aligned} LSCV(h, w) &= \sum_{i=1}^{n}\sum_{j=1}^{n_i} w_i \left[Y_{ij} - \mu\left(t_{ij}\right)\right]^2 \\ &+ \sum_{i=1}^{n}\sum_{j=1}^{n_i} w_i \left[\mu\left(t_{ij}\right) - \widehat{\mu}_K^{(-i)}\left(t_{ij}; h, w\right)\right]^2 \\ &+ 2\sum_{i=1}^{n}\sum_{j=1}^{n_i} w_i \left[Y_{ij} - \mu\left(t_{ij}\right)\right]\left[\mu\left(t_{ij}\right) - \widehat{\mu}_L^{(-i)}\left(t_{ij}; h, w\right)\right]. \end{aligned} \tag{3.17}$$

The first term of the right side of (3.17) does not depend on the bandwidths. Since the observations of the ith subject have been deleted for the computation of $\widehat{\mu}_L^{(-i)}(t; h, w)$ and the subjects are assumed to be independent, the expectation of the third term is zero.

Let $ASE\left[\widehat{\mu}_L(\cdot; h, w)\right]$ be the average squared error of $\widehat{\mu}_L(t_{ij}; h, w)$ defined by

$$ASE\left[\widehat{\mu}_L(\cdot; h, w)\right] = \sum_{i=1}^{n}\sum_{j=1}^{n_i} w_i \left[\mu\left(t_{ij}\right) - \widehat{\mu}_L\left(t_{ij}; h, w\right)\right]^2. \tag{3.18}$$

Direct calculation using the definition (3.18) shows that the expectation of the second term of the right side of (3.17) is actually the expectation of $ASE\left[\widehat{\mu}_L^{(-i)}(\cdot; h, w)\right]$, which approximates the expectation of $ASE\left[\widehat{\mu}_L(\cdot; h, w)\right]$ when n is large. Thus, the LSCV bandwidth h_{LSCV} approximately minimizes the average squared error $ASE\left[\widehat{\mu}_L(\cdot; h, w)\right]$. Consistency of a similar LSCV procedure in a different nonparametric regression setting has been shown by Hart and Wehrly (1993). But, under the current setting, the asymptotic properties of h_{LSCV} are still not well understood. Statistical properties of h_{LSCV} are investigated through simulation studies.

3.4 Bootstrap Pointwise Confidence Intervals

Statistical inferences, such as confidence intervals, are usually developed based on either asymptotic distributions of the estimators or bootstrap methods (e.g., Efron and Tibshirani, 1993). Under the context of cross-sectional i.i.d. data, asymptotic distributions are derived by letting the number of subjects n go to infinity. The resulting inferences are reliable at least when the sample size n is large. However, the longitudinal data structure is more complicated because of two reasons. First, since the numbers of repeated measurements n_i, $i = 1, \ldots, n$, are possibly different, the corresponding asymptotic distributions of the estimators may also be different depending on how fast n_i, $i = 1, \ldots, n$, converge to infinity relative to n. It can be seen from the asymptotic properties of Section 3.6 that, in order to get a meaningful asymptotic result, n must converge to infinity, but n_i may be either bounded or converging to infinity along with n. Second, because of the possible intra-correlation structure of the data, which is assumed to be completely unknown, the asymptotic distributions of $\widehat{\mu}_L(t; h, w)$ may be difficult to estimate. Thus, longitudinal inferences that are purely based on the asymptotic distributions of $\widehat{\mu}_L(t; h, w)$ may be difficult to implement in practice. On the other hand, a bootstrap inference can always be constructed based on the available data.

3.4.1 Resampling-Subject Bootstrap Samples

Since the subjects are assumed to be independent, a natural bootstrap sampling scheme is to resample the entire repeated measurements of each subject with replacement from the original dataset. This approach is referred to in the literature (e.g., Hoover et al., 1998) as the *"resampling-subject bootstrap."* This bootstrap scheme can be generally applied to construct confidence intervals for estimators of other nonparametric models in this book. The bootstrap samples are generated using the following steps.

Resampling-Subject Bootstrap:

(a) *Randomly select n bootstrap subjects with replacement from the original dataset, and denote by $\left\{ (Y_{ij}^*, t_{ij}^*) : i = 1, \ldots, n; j = 1, \ldots, n_i \right\}$ the longitudinal bootstrap sample. The entire repeated measurements of some subjects in the original sample may appear multiple times in the new bootstrap sample.*

(b) *Compute the curve estimators, such as the kernel or local polynomial estimator $\widehat{\mu}_L^{boot}(t; h, w)$ of $\mu(t)$ in (3.8) or (3.12), based on the bootstrap sample $\left\{ (Y_{ij}^*, t_{ij}^*) : i = 1, \ldots, n; j = 1, \ldots, n_i \right\}$.*

(c) *Repeat the above two steps B times. Denote the bth, $b = 1, \ldots, B$, bootstrap estimator by $\widehat{\mu}_L^{boot,b}(t; h, w)$, so that B bootstrap estimators*

$$\mathscr{B}_\mu^B(t; h, w) = \left\{ \widehat{\mu}_L^{boot,1}(t; h, w), \ldots, \widehat{\mu}_L^{boot,B}(t; h, w) \right\} \qquad (3.19)$$

of $\mu(t)$ are obtained. □

3.4.2   Two Bootstrap Confidence Intervals

The bootstrap samples generated in (3.19) can be used to construct two types of approximate pointwise confidence intervals for $\mu(t)$, namely the intervals based on percentiles of bootstrap samples and the intervals based on normal approximation.

Approximate Bootstrap Pointwise Confidence Intervals:

(a) **Percentile Bootstrap Intervals.** *Compute the percentiles of the B bootstrap estimators $\mathscr{B}_\mu^B(t; h, w)$ of (3.19). The approximate $[100 \times (1 - \alpha)]th$ percentile bootstrap pointwise confidence interval for $\mu(t)$ is given by*

$$\left(L_{(\alpha/2)}(t), U_{(\alpha/2)}(t)\right), \tag{3.20}$$

where $L_{(\alpha/2)}(t)$ and $U_{(\alpha/2)}(t)$ are the $(\alpha/2)th$ and $(1 - \alpha/2)th$, i.e., lower and upper $(\alpha/2)th$, percentiles of $\mathscr{B}_\mu^B(t; h, w)$, respectively.

(b) **Normal Approximated Bootstrap Intervals.** *Compute the estimated standard error $\widehat{se}_B(t; \widehat{\mu}_L)$ of $\widehat{\mu}_L(t; h, w)$ from the B bootstrap estimators $\mathscr{B}_\mu^B(t; h, w)$. The normal approximated bootstrap pointwise confidence interval for $\mu(t)$ is*

$$\widehat{\mu}_L(t; h, w) \pm z_{(1-\alpha/2)} \widehat{se}_B(t; \widehat{\mu}_L), \tag{3.21}$$

where $z_{(1-\alpha/2)}$ is the $[100 \times (1 - \alpha/2)]th$ percentile of the standard normal distribution and $\widehat{\mu}_L(t; h, w)$ is the smoothing curve estimator computed from the original sample.   □

The percentile bootstrap interval given in (3.20) is a naive procedure, which has the main advantage of not relying on the asymptotic distributions of the curve estimator $\widehat{\mu}_L(t; h, w)$. However, a visual drawback of the percentile interval (3.20) is that $\widehat{\mu}_L(t; h, w)$ is not necessary at the center of the interval (3.20). On the other hand, the normal approximated interval (3.21) is symmetric about $\widehat{\mu}_L(t; h, w)$, which is visually appealing. But, because (3.21) uses normal approximation of the critical values, its accuracy depends on the appropriateness of the normal approximation.

Technically, both (3.20) and (3.21) may lead to reasonable approximations of the actual $[100 \times (1 - \alpha)]\%$ pointwise confidence intervals of $\mu(t)$ if the biases of $\widehat{\mu}_L(t; h, w)$ are small. Ideally, the biases of the estimators need to be adjusted if they are not negligible. Theoretical properties of these bootstrap procedures have not been systematically investigated. But, the practical properties, such as the empirical coverage probabilities and computational feasibility, of these bootstrap procedures have been investigated through a series of simulation studies in the literature, such as Hoover et al. (1998) and Wu and Chiang (2000), among others.

3.4.3 Simultaneous Confidence Bands

We note that the above bootstrap pointwise confidence intervals are only for the inferences of $\mu(t)$ at the given time point t. In most practical situations, such pointwise inferences are sufficient. But, in some studies, the study objectives require the knowledge that there is a high probability that the true regression curve $\mu(t)$ stays within a given band simultaneously for a range of the time values. In such situations, we need to construct a *simultaneous confidence band* of $\mu(t)$ for t within a closed interval $[a, b]$ with some known positive constants $b > a > 0$.

1. Construction of Simultaneous Confidence Bands

We introduce here a straightforward procedure that extends the above pointwise confidence intervals to simultaneous confidence bands for $\mu(t)$ over a given interval $[a, b]$. This procedure is described in Hall and Titterington (1988) for nonparametric regression models with cross-sectional i.i.d. data. Because this is a general approach, it is not limited to the estimators $\hat{\mu}_L(t; h, w)$ and is repeatedly used for other smoothing curve estimators in this book. This simultaneous confidence band procedure requires the following two steps.

Simultaneous Confidence Band Procedure:

(a) Simultaneous Confidence Intervals at Grid Points. *In this step, we partition $[a, b]$ into $M + 1$ equally spaced grid points*

$$a = \xi_1 < \cdots < \xi_{M+1} = b$$

for some integer $M \geq 1$, and construct a set of approximate $[100 \times (1 - \alpha)]\%$ simultaneous confidence intervals $\left(l_\alpha(\xi_r), u_\alpha(\xi_r)\right)$ for $\mu(\xi_r)$, such that

$$\lim_{n \to \infty} P\left\{l_\alpha(\xi_r) \leq \mu(\xi_r) \leq u_\alpha(\xi_r), \text{ for all } r = 1, \ldots, M+1\right\} \geq 1 - \alpha. \quad (3.22)$$

If we apply the Bonferroni adjustment to (3.20) or (3.21), $(l_\alpha(\xi_r), u_\alpha(\xi_r))$ are given by

$$\left(L_{\alpha/[2(M+1)]}\left[\hat{\mu}_L^{boot}(\xi_r; h, w)\right], U_{\alpha/[2(M+1)]}\left[\hat{\mu}_L^{boot}(\xi_r; h, w)\right]\right) \quad (3.23)$$

or

$$\hat{\mu}_L(\xi_r; h, w) \pm z_{1-\alpha/[2(M+1)]} \hat{s}_L^{boot}(\xi_r; h, w), \quad (3.24)$$

respectively. For any $\xi_r \leq t \leq \xi_{r+1}$, we define $\hat{\mu}_L^{(I)}(t; h, w)$ to be the linear interpolation of $\hat{\mu}_L(\xi_r; h, w)$ and $\hat{\mu}_L(\xi_{r+1}; h, w)$, such that

$$\hat{\mu}_L^{(I)}(t; h, w) = M\left(\frac{\xi_{r+1} - t}{b - a}\right)\hat{\mu}_L(\xi_r; h, w) + M\left(\frac{t - \xi_r}{b - a}\right)\hat{\mu}_L(\xi_{r+1}; h, w). \quad (3.25)$$

Similarly, we define $\mu^{(I)}(t)$ to be the linear interpolation of $\mu(\xi_r)$ and

$\mu(\xi_{r+1})$ for any $\xi_r \leq t \leq \xi_{r+1}$. Then $(l_\alpha^{(I)}(t), u_\alpha^{(I)}(t))$ is an approximate $[100 \times (1-\alpha)]\%$ confidence band for $\mu^{(I)}(t)$, such that

$$\lim_{n \to \infty} P\left\{l_\alpha^{(I)}(t) \leq \mu^{(I)}(t) \leq u_\alpha^{(I)}(t), \text{ for all } t \in [a,b]\right\} \geq 1-\alpha, \qquad (3.26)$$

where

$$l_\alpha^{(I)}(t) = M\left(\frac{\xi_{r+1}-t}{b-a}\right) l_\alpha(\xi_r) + M\left(\frac{t-\xi_r}{b-a}\right) l_\alpha(\xi_{r+1}) \qquad (3.27)$$

and

$$u_\alpha^{(I)}(t) = M\left(\frac{\xi_{r+1}-t}{b-a}\right) u_\alpha(\xi_r) + M\left(\frac{t-\xi_r}{b-a}\right) u_\alpha(\xi_{r+1}) \qquad (3.28)$$

are the linear interpolations of $(l_\alpha(\xi_r), l_\alpha(\xi_{r+1}))$ and $(u_\alpha(\xi_r), u_\alpha(\xi_{r+1}))$, respectively.

(b) **Confidence Bands Linking Grid Points.** *To construct the simultaneous confidence bands for $\mu(t)$ for all $t \in [a,b]$, some smoothness conditions have to be assumed. For two commonly used smoothness conditions, we assume that either*

$$\sup_{t \in [a,b]} |\mu'(t)| \leq c_1, \quad \text{for a known constant } c_1 > 0, \qquad (3.29)$$

or

$$\sup_{t \in [a,b]} |\mu''(t)| \leq c_2, \quad \text{for a known constant } c_2 > 0. \qquad (3.30)$$

Then, it can be verified by direct calculation that, for $\xi_r \leq t \leq \xi_{r+1}$,

$$|\mu(t) - \mu^{(I)}(t)| \leq \begin{cases} 2c_1 M\left[(\xi_{r+1}-t)(t-\xi_r)/(b-a)\right], & \text{if (3.29) holds;} \\ (c_2/2)(\xi_{r+1}-t)(t-\xi_r), & \text{if (3.30) holds.} \end{cases}$$

To adjust the simultaneous confidence bands for $\mu^{(I)}(t)$, the approximate $[100 \times (1-\alpha)]\%$ simultaneous confidence bands for $\mu(t)$ are given by

$$\left(l_\alpha^{(I)}(t) - 2c_1 M\left[\frac{(\xi_{r+1}-t)(t-\xi_r)}{b-a}\right], u_\alpha^{(I)}(t) + 2c_1 M\left[\frac{(\xi_{r+1}-t)(t-\xi_r)}{b-a}\right]\right) \qquad (3.31)$$

or

$$\left(l_\alpha^{(I)}(t) - (c_2/2)(\xi_{r+1}-t)(t-\xi_r), u_\alpha^{(I)}(t) + (c_2/2)(\xi_{r+1}-t)(t-\xi_r)\right), \qquad (3.32)$$

when (3.29) or (3.30) holds, respectively. $\qquad\qquad\square$

2. Constructing Simultaneous Confidence Bands in Practice

There are two practical issues for the applications of the confidence bands of (3.31) and (3.32). The first is the Bonferroni adjustment in constructing the simultaneous confidence intervals at the grid points $\{\xi_1, \ldots, \xi_{M+1}\}$

given in (3.23) and (3.24). Since the Bonferroni adjustment usually gives conservative intervals with coverage probability higher than the nominal level of $[100 \times (1 - \alpha)]\%$, the Bonferroni adjusted simultaneous confidence intervals (3.23) and (3.24) are conservative at the equally spaced grid points $\{\xi_1, \ldots, \xi_{M+1}\}$. As a result, the simultaneous confidence band given in (3.31) and (3.32) for $t \in [a, b]$ are also conservative with coverage probabilities higher than the nominal level of $[100 \times (1 - \alpha)]\%$, and their actual coverage probabilities will increase as the number of grid time points $M + 1$ increases. In real applications, we prefer to have a large number for $M + 1$, because the gaps between two adjacent grid points $\xi_r < \xi_{r+1}$, $1 \le r \le M$, are bridged with the inequalities (3.29) and (3.30). Thus, some refinement is often preferred to construct simultaneous confidence intervals at $t \in \{\xi_1, \ldots, \xi_{M+1}\}$, which are less conservative than (3.23) and (3.24). Examples of refined intervals may include adjustments using the inclusion-exclusion identities with more accurate coverage probabilities or other multiple comparison techniques, such as Bretz, Hothorn and Westfall (2011). These refinements, which often involve more intensive computation than the Bonferroni adjustment, can be used in place of the intervals (3.23) and (3.24) when the computation complexity is not an issue.

The second issue is the number and location of the grid points. For simplicity, we assume that the grid points $a = \xi_1 < \cdots < \xi_{M+1} = b$ are equally spaced. But, in real applications, these grid points are not necessarily equally spaced, and their location may be chosen based on the study design and the scientific questions being investigated. The equally spaced grid points can be treated as a default choice when there is no clear indication from the study design where these grid points should be located. The number of grid points $M + 1$ can be selected subjectively by examining the widths of the confidence bands. Theoretical results on the "optimal" choices of location and number of grid points are still not available. Although some heuristic suggestions for choosing M for the simple case of kernel regression with cross-sectional i.i.d. data have been provided by Hall and Titterington (1988), optimal choices of M under the current situation with longitudinal data are not available.

3.5 R Implementation

3.5.1 The HSCT Data

The HSCT data has been described in Section 1.2. For each observational time, the measurements include the patient's study visit time relative to the date of hematopoietic stem cell transplant, three white blood cells or leukocyte counts (granulocytes, lymphocytes and monocytes, in 10^3 cells/μL or K/μL) and multiple cytokines (pg/mL). The R code for examining the data is

```
> library(npmlda)
> str(HSCT)
```

```
'data.frame':   271 obs. of  8 variables:
 $ ID   : int  1 1 1 1 1 1 1 1 1 1 ...
 $ Days : int  -5 -3 0 1 4 7 10 14 17 21 ...
 $ Granu: num  6.962 4.407 0.566 0.253 0.007 ...
 $ LYM  : num  1.705 0.498 0.061 0.097 0.243 ...
 $ MON  : num  0.151 0.025 0.01 NA 0.002 ...
 $ G-CSF: num  30.2 14.3 101 69.5 625.6 ...
 $ IL-15: num  27.6 33.9 60.9 40 74.2 ...
 $ MCP-1: num  35.2 59.4 317.8 230.8 1235.6 ...

> HSCT[HSCT$ID==1,]

   ID Days  Granu   LYM   MON  G-CSF IL-15   MCP-1
1   1   -5  6.962 1.705 0.151  30.19 27.60   35.25
2   1   -3  4.407 0.498 0.025  14.33 33.88   59.36
3   1    0  0.566 0.061 0.010 101.04 60.93  317.81
4   1    1  0.253 0.097    NA  69.54 39.96  230.77
5   1    4  0.007 0.243 0.002 625.56 74.17 1235.62
6   1    7  0.004 0.358 0.004 112.60 82.81 1247.30
7   1   10  0.118 0.401 0.076 643.62 73.33  597.63
8   1   14  3.809 0.478 1.636  48.58 77.25  106.91
9   1   17  2.260 0.377 0.323  30.87 36.94   53.54
10  1   21  7.142 1.283 0.354  20.32 36.33   72.35
11  1   24 10.508 0.710 0.852  11.18 44.43   96.04
12  1   28  8.190 0.525 1.155  10.38 29.19  130.88
13  1   31  7.992 1.621 1.088   8.73 26.96  166.97
14  1   35  3.587 1.087 0.349   9.56 29.51  290.87
```

The summary statistics below show that the three leukocytes have very skewed distributions with some extremely large values, in which we apply log transformations to these three nonzero variables to reduce the skewness of their distributions:

```
> summary(HSCT$Granu)

   Min. 1st Qu.  Median    Mean 3rd Qu.     Max.    NA's
 0.0020  0.0515  1.0880  2.2290  3.3680 20.1800      36

> summary(HSCT$LYM)

   Min. 1st Qu.  Median    Mean 3rd Qu.     Max.    NA's
 0.0020  0.0250  0.0945  0.3315  0.5002  2.3070      43

> summary(HSCT$MON)

   Min. 1st Qu.  Median    Mean 3rd Qu.     Max.    NA's
 0.0010  0.0260  0.3770  0.5909  1.0880  2.1780      78
```

```
> HSCT$Granu.log <- log10(HSCT$Granu)
> HSCT$LYM.log   <- log10(HSCT$LYM)
> HSCT$MON.log   <- log10(HSCT$MON)
```

For a visual representation of the change of leukocytes over time, we generate the scatterplots of the leukocytes against days after transplant in log-scale, shown in Figure 3.1. To better visualize the overall time-trends of the leukocytes during the week before stem cell transplant (this is known as the "conditioning" period) and within 4 weeks after the transplant (known as the "recovery" period), we compute the local means of the log-transformed leukocytes using the kernel smoothing methods in Section 3.1. To demonstration the effects of kernel choices, we use three different kernel functions, namely the Epanechnikov kernel (1.27), the Gaussian kernel (1.31) and the Biweight kernel (1.29). The smoothing estimates of Figure 3.1 show the results based on the Epanechnikov kernel for granulocytes, the Gaussian kernel for lymphocytes, and the Biweight kernel for monocytes. Individual subjects have a median of 15 repeated measurements with a range of 6 to 25.

The solid and dashed lines represent the kernel estimates using the measurement uniform weight (3.4) and the subject uniform weight (3.8). The following `kernel.fit()` function can be used to generate the fitted y values using the Epanechnikov kernel and the subject uniform weight with a given bandwidth `bw`:

```
> Fit.Granu <- with(HSCT[!is.na(HSCT$Granu.log),],
             kernel.fit(sort(unique(Days)),Days, Granu.log,
                   bw=4, Kernel="Ep", Wt=1/ni))
```

Similarly, we can specify different values for the arguments if other choices of kernel function, weights and bandwidth are used.

As shown in Figure 3.1, the kernel estimates with two kinds of weights are very similar, except that there are small differences near the boundary when the number of observations per subject and the related weight may be more influential. Both sets of fitted curves suggest a typical pattern for the changes of leukocytes around the pre- and early post-transplantation period. The leukocyte counts first drop to their lowest levels following the conditioning regimen. Then, the donor cell engraftment and hematopoietic reconstitution occur. The leukocyte counts gradually recover during the first month after the transplantation.

3.5.2 The BMACS CD4 Data

The BMACS CD4 data has been described in Section 1.2 and Section 2.4. We use this dataset to illustrate the local polynomial estimate, bandwidth selection with cross-validation and bootstrap inference. Figure 3.2 depicts the CD4

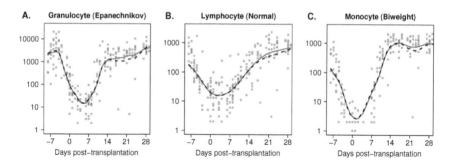

Figure 3.1 *The dynamics of the main leukocyte counts around the time of transplant: (A) Granulocytes, (B) Lymphocytes, and (C) Monocytes. The solid and dashed lines represent the kernel estimates using the measurement uniform weights and the subject uniform weights, with the Epanechnikov, Gaussian and Biweight kernel functions for the three leukocytes, respectively.*

cell percentages at the study visits since seroconversion for the HIV infected men. In R, the local polynomial estimators are implemented through the local (weighted) regression function `loess()` or `lowess()` (Fan and Gijbels, 1996, Section 2.4). It is easy to use `loess()` by specifying the weights, degree of polynomials (commonly 1 or 2) and a smoothing span, which is similar to the bandwidth and is expressed as a proportion of local data points around each value to control the degree of smoothing. Only the tricube kernel function (1.30) is implemented in `loess`. For example, we can use the following commands to produce a local linear fit with span of 0.5 and measurement uniform weights:

```
> fit.linear.5 <- loess(CD4 ~ Time, span=0.5, degree=1,
    data=BMACS)
> Time.int<- seq(0.1,5.9,  by=0.1)
> plot(CD4 ~ Time, data = BMACS, xlab="Yeas since infection",
    ylab="CD4 percentage", ylim=c(0,65),cex=0.7,col="gray50",
    main="Local linear: span=0.5")
> lines(Time.int,
    predict(fit.linear.5,data.frame(Time=Time.int)))
```

Local smoothing estimates of the CD4 percentage over time are shown in Figure 3.2 using the local linear and local quadratic estimators with spans of 0.1 an 0.5. We can see that, for the small value of bandwidth or span=0.1, both local smoothed curves in Figures 3.2(A) and (C) are wiggly because only a small proportion of the points have contributed to the fit. In contrast, the local estimators using the larger span=0.5 yield much smoother curves shown

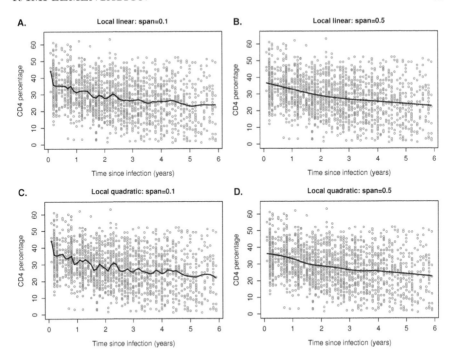

Figure 3.2 *Scatter plots and local smoothing estimates of the CD4 percentage over time since seroconversion for the HIV infected men in BMACS.*

in Figures 3.2(B) and (D). In this case, with relatively larger bandwidth and lower curvature, the local linear and local quadratic fits are very close to each other, and both have captured the declining trend of CD4 percentage over time after HIV infection.

As described in Section 3.3, the bandwidth h may be chosen by the leave-one-subject-out cross-validation (LSCV). Figure 3.3(A) shows the LSCV scores defined in (3.14) against a range of h values from 0.3 to 4.5 years for a local linear fit with the subject uniform weight $1/(nn_i)$, which indicates $h_0 = 0.9$ is the approximate minimizer of the LSCV score. Figure 3.3(B) shows the local linear estimated curve computed using this selected bandwidth with the subject uniform weight $1/(nn_i)$.

The following R functions are used to compute the estimated curve in Figure 3.3(B):

```
# Obtain the number of observations per subject
> Ct <- data.frame(table(BMACS$ID))
> names(Ct)<- c("ID", "ni")
> BMACS<- merge(BMACS, Ct, by= "ID")
```

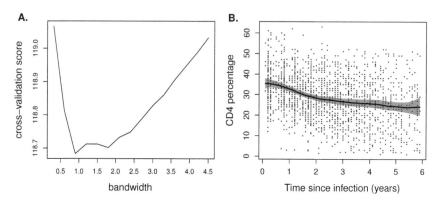

Figure 3.3 *(A) Use of cross-validation to select bandwidth for the local linear fit.*
(B) The local linear fit using cross-validated bandwidth. The gray band represents
the bootstrap 95% pointwise confidence interval computed using percentiles.

```
# LocalLm() is a local linear function with Epanechnikov kernel
> LocalFit.Y  <- with(BMACS, LocalLm(Time.int, Time, CD4,
                      bw=0.9, Wt=1/ni))
```

Next, we use the resampling-subject bootstrap procedure in Sections 3.4.1
and 3.4.2 to compute the pointwise confidence interval of the local linear fit
in Figure 3.3(B). The resulting 95% pointwise confidence interval based on
$B = 1000$ resampling-subject bootstrap samples is computed and shown in
Figure 3.3(B). These results are computed using the following R code:

```
# Generate a resampling-subject bootstrap sample
> IDlist <- unique(BMACS$ID)
> nID <- length(IDlist)
> Bootsample <- function(){
 resample.ID <- sample(IDlist ,nID ,replace=T)
 do.call("rbind", lapply(1:nID,
 function(i) subset(BMACS, ID==resample.ID[i])))}

# Obtain fitted value at a time grid
> LocalLm.Fit<- function(Data, Time.int){
   with(Data, LocalLm(Time.int, Time,CD4,bw=0.9, Wt=1/ni))}

# Compute the 95% CI based on B=1000 bootstrap replicates
> Boot.Fit <-replicate(1000, LocalLm.Fit(Bootsample(),
                        Time.int))
> UpperCI <- apply(Boot.Fit, 1, quantile,.975)
> LowerCI <- apply(Boot.Fit, 1, quantile,.025)
```

```
# plot with local linear fit and 95% CI
> plot(CD4 ~ Time, data = BMACS,
    xlab = "Time since infection(years)",
    ylab = "CD4 percentage", cex=0.3, col="gray70", main="")
> polygon(c(Time.int[1], Time.int, rev(Time.int)),
    c(LowerCI[1], UpperCI, rev(LowerCI)), col="gray60", border=NA)
> lines(Time.int, LocalFit.Y, lwd=2.5, col=1)
```

Although the confidence limits indicate a clear decreasing trend of CD4 percentage over time, their widths become wider near the end of five years since HIV infection. This result reflects more uncertainty and variability of the local linear estimate at the boundary compared to the central area of "time since infection."

3.6 Asymptotic Properties of Kernel Estimators

To provide some insight into the statistical properties of the smoothing estimators in this chapter, we derive in this section the asymptotic properties of the least squares kernel estimators. We focus on the kernel estimators of Section 3.1 because their derivations rely on some basic approaches which can be generalized to the local smoothing estimators in later chapters. The derivations here are focused on the asymptotic expressions of the biases, variances and mean squared errors of $\widehat{\mu}_K(t; h, w)$. Large sample inferences based on $\widehat{\mu}_K(t; h, w)$ can be derived using asymptotic normality with the explicit expressions of its asymptotic means and variances. Because $\widehat{\mu}_K(t; h, w)$ is a linear statistic of Y_{ij}, its asymptotic distributions can be established by checking the triangular array central limit theorem using the expressions of the asymptotic biases and variances.

For mathematical convenience, we assume that the time design points t_{ij} are randomly selected from a distribution function $F(\cdot)$ with density $f(\cdot)$, but n_i, $i = 1, \ldots, n$, are assumed to be nonrandom. Although this design assumption corresponds to a special version of the random designs, it is generally an acceptable assumption for many real settings in longitudinal studies. However, by modifying the notation and several key steps in the derivations, the main theoretical results of this section can be extended to fixed designs or the case that n_i are also random. Since the main purpose of asymptotic derivations is to give some insight into the reliability of the estimation procedures, the asymptotic results established here are useful to guide the practical applications of the kernel estimation procedures in this chapter.

We first derive the asymptotic representations of the mean squared errors and the mean integrated squared errors of $\widehat{\mu}_K(t; h, w)$ for a general weight choice $w = (w_0, \ldots, w_n)^T$. The asymptotic risks for the special cases of $w_i^{**} = N^{-1}$ in (3.4) and $w_i^* = (n n_i)^{-1}$ in (3.8) are further elaborated at the end of this section. These asymptotic results demonstrate that different weight choices may lead to kernel estimators with different asymptotic properties.

3.6.1 Mean Squared Errors

The closeness of $\widehat{\mu}_K(\cdot; h, w)$ to the true curve $\mu(\cdot)$ can be measured in different ways. Suppose that the interest is the adequacy of $\widehat{\mu}_K(t; h, w)$ as an estimator of $\mu(t)$ at a given time point t. A natural measure of the risk of $\widehat{\mu}_K(t; h, w)$ at point t is the mean squared error (MSE) defined by

$$MSE\left[\widehat{\mu}_K(t; h, w)\right] = E\left\{\left[\widehat{\mu}_K(t; h, w) - \mu(t)\right]^2\right\}. \tag{3.33}$$

However, a minor technical difficulty for kernel estimators is that their moments, hence $MSE\left[\widehat{\mu}_K(t; h, w)\right]$ as defined in (3.33), may not exist (e.g., Rosenblatt, 1969), so that modifications of the above mean squared error definitions have to be used.

By a simple reformulation, the kernel estimator $\widehat{\mu}_K(\cdot; h, w)$ of (3.6) can be written as

$$\widehat{\mu}_K(t; h, w) = \frac{\widehat{m}_K(t; h, w)}{\widehat{f}_K(t; h, w)}, \tag{3.34}$$

where

$$\widehat{m}_K(t; h, w) = \sum_{i=1}^{n}\left\{w_i \sum_{j=1}^{n_i}\left[\frac{1}{h}Y_{ij}K\left(\frac{t - t_{ij}}{h}\right)\right]\right\} \tag{3.35}$$

and

$$\widehat{f}_K(t; h, w) = \sum_{i=1}^{n}\left\{w_i \sum_{j=1}^{n_i}\left[\frac{1}{h}K\left(\frac{t - t_{ij}}{h}\right)\right]\right\}. \tag{3.36}$$

It can be shown by applying straightforward algebra to (3.34), (3.35) and (3.36) that

$$\left[1 - d_K(t; h, w)\right]\left[\widehat{\mu}_K(t; h, w) - \mu(t)\right]$$
$$= \left[\widehat{m}_K(t; h, w) - \mu(t)\widehat{f}_K(t; h, w)\right]/f(t), \tag{3.37}$$

where $d_K(t; h, w) = 1 - \left[\widehat{f}_K(t; h, w)/f(t)\right]$. For any interior point t of the support of $f(\cdot)$, it can be shown by the same method used in kernel density estimation with cross-sectional i.i.d. data (e.g., Silverman, 1986), that $d_K(t; h, w) \to 0$ in probability as $n \to \infty$ and $h \to 0$. Then, by (3.34) and (3.37), we have the following approximation,

$$\left[1 + o_p(1)\right]\left[\widehat{\mu}_K(t; h, w) - \mu(t)\right] = f^{-1}(t)\widehat{R}_K(t; h, w), \tag{3.38}$$

where $\widehat{R}_K(t; h, w) = \widehat{m}_K(t; h, w) - \mu(t)\widehat{f}_K(t; h, w)$.

The advantage of using the approximation (3.38) is that, although the $MSE\left[\widehat{\mu}_K(t; h, w)\right]$ as defined in (3.33) may not exist, we can always evaluate the appropriateness of $\widehat{\mu}_K(t; h, w)$ by evaluating the mean squared errors of the approximated term given at the right side of (3.37). Thus, through (3.38), we can define the local risk of $\widehat{\mu}_K(t; h, w)$ at time point t by the following modified MSE

$$MSE\left[\widehat{\mu}_K(t; h, w)\right] = E\left\{\left[f^{-1}(t)\widehat{R}_K(t; h, w)\right]^2\right\}, \qquad (3.39)$$

and the global risk of $\widehat{\mu}_K(\cdot; h, w)$ over a time range by the modified mean integrated squared error (MISE)

$$MISE\left[\widehat{\mu}_K(\cdot; h, w)\right] = \int MSE\left[\widehat{\mu}_K(s; h, w)\right]\pi(s)\,ds, \qquad (3.40)$$

where $\pi(s)$ is a known non-negative weight function whose support is a compact subset in the interior of the support of $f(\cdot)$. It is known in nonparametric regression that, as a "weighted local average," the kernel estimators of $\mu(t)$ have large biases when t is near the boundary of its support. An important reason of using $\pi(t)$ is to reduce the effect of boundary bias.

3.6.2 Assumptions for Asymptotic Derivations

Asymptotic properties for nonparametric estimators with cross-sectional i.i.d. data are generally developed under a set of smoothing assumptions for the unknown curves and the assumption that the sample sizes tend to infinity. Under the context of longitudinal data, the same smoothing assumptions for the unknown curves can still be used, but the asymptotic results may depend on whether the numbers of repeated measurements n_i, $i = 1, \ldots, n$, also tend to infinity in addition to the usual assumption that the number of subjects n tends to infinity. When the numbers of subjects are much larger than the numbers of repeated measurements, one may expect that the asymptotic risks derived under the assumption that n tends to infinity and $\{n_1, \ldots, n_n\}$ are bounded can reasonably approximate the actual risks of the estimators. In such situations, i.e., n tending to infinity while $\{n_1, \ldots, n_n\}$ being bounded, the asymptotic results are simple and can be similarly derived as the classical case with cross-sectional i.i.d. data. In many longitudinal studies, however, the numbers of repeated measurements may not be ignorable relative to the number of subjects, so that a more useful asymptotic assumption should relax the boundedness condition on $\{n_1, \ldots, n_n\}$ and allow n_i to also tend to infinity for some $i = 1, \ldots, n$ as n tends to infinity.

As a result, the asymptotic representations for the mean squared errors, $MSE\left[\widehat{\mu}_K(t; h, w)\right]$ and $MISE\left[\widehat{\mu}_K(\cdot; h, w)\right]$, are derived under the following technical assumptions.

Asymptotic Assumptions:

(a) *For all t, $f(t)$ is continuously differentiable and there is a non-negative constant p, so that $\mu(t)$ is $(p+2)$ times continuously differentiable with respect to t.*

(b) *The variance and covariance of the error process $\varepsilon(t)$ satisfy*

$$\sigma^2(t) = E\left[\varepsilon^2(t)\right] < \infty \quad \text{and} \quad \rho_\varepsilon(t) = \lim_{t' \to t} E\left[\varepsilon(t)\,\varepsilon(t')\right] < \infty. \quad (3.41)$$

Furthermore, $\sigma^2(t)$ and $\rho_\varepsilon(t)$ are continuous for all t.

(c) *The kernel function $K(\cdot)$ has a compact support, and it is a $(p+2)$th order kernel in the sense that it satisfies*

$$\left\{ \begin{array}{rcl} \int u^j K(u)\,du & = & 0 \ \ \text{for all } 1 \le j < p+2, \\[2mm] M_{(p+2)}(K) & = & \int u^{p+2} K(u)\,du \ < \ \infty, \\[2mm] R(K) & = & \int K^2(u)\,du < \infty \quad \text{and} \quad \int K(u)\,du \ = \ 1. \end{array} \right. \quad (3.42)$$

(d) *The weight vector $w = (w_1, \ldots, w_n)^T$ with $w_i \ge 0$ for all $1 \le i \le n$, satisfies*

$$\sum_{i=1}^n \left(w_i\,n_i\right) = 1 \quad \text{and} \quad \sum_{i=1}^n \left(w_i^2\,n_i^2\right) \to 0 \ \ \text{as } n \to \infty. \quad (3.43)$$

(e) *The bandwidth $h > 0$ satisfies*

$$h \to 0, \ nh \to \infty, \quad \text{and} \quad \sum_{i=1}^n \left(w_i^2\,n_i\right)/h \to 0 \quad (3.44)$$

as $n \to \infty$. $\square$

These assumptions are comparable to the ones used for kernel estimation with cross-sectional i.i.d. data. For Assumption (b), we note that in general $\sigma^2(t) \ne \rho_\varepsilon(t)$. The strict inequality between $\sigma^2(t)$ and $\rho_\varepsilon(t)$ holds, for example, when $\varepsilon_{ij} = s(t_{ij}) + W_i$ where $s(t)$ is a mean zero Gaussian stationary process and W_i is an independent white noise (e.g., Zeger and Diggle, 1994). Some of the above assumptions, such as the compactness of the support of $K(\cdot)$ and the smoothness conditions of $f(t)$, $\mu(t)$, $\sigma^2(t)$ and $\rho_\varepsilon(t)$, are made for the simplicity of the derivations. We note that, under Assumption (c) with $p \ge 1$, the kernels satisfying (3.42) belong to the so-called "higher order kernels," which may allow for some parts of the kernel function to take negative values. Although the use of higher order kernels may not seem "natural" as a local smoothing method, they have been shown in the nonparametric regression literature to have bias correction properties (e.g., Jones, 1995). In practice, some non-compactly supported kernels, such as the standard Gaussian kernel (1.31), can provide equally good estimators as well. Asymptotic results analogous to the ones of this section may also be derived when Assumptions (a) to (e) are modified or even weakened.

3.6.3   Asymptotic Risk Representations

Using the right side approximation in (3.38) and the modified MSE (3.39), we define the modified bias and variance (or simply bias and variance) of

$\widehat{\mu}_K(t; h, w)$ by

$$B\left[\widehat{\mu}_K(t; h, w)\right] = f^{-1}(t) E\left[\widehat{R}_K(t; h, w)\right] \tag{3.45}$$

and

$$V\left[\widehat{\mu}_K(t; h, w)\right] = f^{-2}(t) Var\left[\widehat{R}_K(t; h, w)\right], \tag{3.46}$$

respectively. Then, by (3.39), $MSE\left[\widehat{\mu}_K(t; h, w)\right]$ has the decomposition

$$MSE\left[\widehat{\mu}_K(t; h, w)\right] = B^2\left[\widehat{\mu}_K(t; h, w)\right] + V\left[\widehat{\mu}_K(t; h, w)\right]. \tag{3.47}$$

1. Asymptotic Biases, Variances and Mean Squared Errors

The following theorem gives the asymptotic expressions of $B\left[\widehat{\mu}_K(t; h, w)\right]$, $V\left[\widehat{\mu}_K(t; h, w)\right]$, $\text{MSE}\left[\widehat{\mu}_K(t; h, w)\right]$ and $\text{MISE}\left[\widehat{\mu}_K(\cdot; h, w)\right]$.

Theorem 3.1 *Suppose that t is in the interior of the support of $f(\cdot)$ and Assumptions (a) to (e) are satisfied. The following asymptotic expressions hold.*

(a) *When n is sufficiently large,*

$$B\left[\widehat{\mu}_K(t; h, w)\right] = h^{p+2} B_*(t; K, p, \mu, f)\left[1 + o(1)\right], \tag{3.48}$$

where

$$B_*(t; K, p, \mu, f) = M_{(p+2)}(K)\left[\frac{\mu^{(p+2)}(t)}{(p+2)!} + \frac{\mu^{(p+1)}(t) f'(t)}{(p+1)! f(t)}\right],$$

and

$$V\left[\widehat{\mu}(t; h, w)\right] = \left\{h^{-1} \sum_{i=1}^{n}\left(w_i^2 n_i\right) f^{-1}(t) \sigma^2(t) R(K) \tag{3.49}\right.$$

$$\left. + \left[\sum_{i=1}^{n_i}\left(w_i^2 n_i^2\right) - \sum_{i=1}^{n}\left(w_i^2 n_i\right)\right] \rho_\varepsilon(t)\right\}\left[1 + o(1)\right].$$

(b) *The asymptotic representations of $MSE\left[\widehat{\mu}_K(t; h, w)\right]$ and*

$$MISE\left[\widehat{\mu}_K(\cdot; h, w)\right] = \int MSE\left[\widehat{\mu}_K(s; h, w)\right] \pi(s) ds,$$

when $n \to \infty$, are giving by substituting the expressions of $B\left[\widehat{\mu}_K(t; h, w)\right]$ and $V\left[\widehat{\mu}_K(t; h, w)\right]$ in (3.48) and (3.49), respectively, into (3.47).

(c) *If, in addition to Assumption (e), the bandwidth h also satisfies,*

$$\frac{h \sum_{i=1}^{n}\left(w_i^2 n_i^2\right)}{\sum_{i=1}^{n}\left(w_i^2 n_i\right)} \to 0 \quad \text{as } n \to \infty, \tag{3.50}$$

then the asymptotic variance (3.49) reduces to

$$V\left[\hat{\mu}(t; h, w)\right] = \left\{h^{-1}\left[\sum_{i=1}^{n}\left(w_i^2 n_i\right)\right]f^{-1}(t)\,\sigma^2(t)R(K)\right\}\left[1 + o(1)\right], \quad (3.51)$$

Consequently, the asymptotic representations of $MSE\left[\hat{\mu}_K(t; h, w)\right]$ and $MISE\left[\hat{\mu}_K(\cdot; h, w)\right]$ are not affected by the intra-subject correlations. ■

The results of (3.48), (3.49) and (3.51) imply that, in general, the convergence rates of $MSE\left[\hat{\mu}_K(t; h, w)\right]$ and $MISE\left[\hat{\mu}_K(\cdot; h, w)\right]$ depend on whether and how n_i, $i = 0, \ldots, n$, converge to infinity relative to n, and the choice of weight $w = (w_1, \ldots, w_n)^T$ also affects the convergence rates of $MSE\left[\hat{\mu}_K(t; h, w)\right]$ and $MISE\left[\hat{\mu}_K(\cdot; h, w)\right]$.

Proof of Theorem 3.1:

By the definition of $B\left[\hat{\mu}_K(t; h, w)\right]$ in (3.45) and the definition of $\widehat{R}_K(t; h, w)$ in (3.38), it follows from (3.35), (3.36), (3.38), Assumptions (a) and (c), and the change of variables that, when n is sufficiently large,

$$
\begin{aligned}
B\left[\hat{\mu}_K(t; h, w)\right] &= \frac{1}{hf(t)}\sum_{i=1}^{n}\sum_{j=1}^{n_i} w_i E\left\{E\left[\left(Y_{ij} - \mu(t)\right)K\left(\frac{t - t_{ij}}{h}\right)\Big|t_{ij}\right]\right\} \\
&= \frac{1}{hf(t)}\sum_{i=1}^{n}\sum_{j=1}^{n_i}\int w_i\left[E\left(Y_{ij}|t_{ij} = s\right) - \mu(t)\right]K\left(\frac{t - s}{h}\right)f(s)\,ds \\
&= f^{-1}(t)\int\left[\mu(t - hu) - \mu(t)\right]f(t - h_r u)K(u)\,du.
\end{aligned}
$$

Then the expression in (3.48) follows from Assumptions (a), (c) and (d), and the Taylor expansions of $\mu(t - hu)$ and $f(t - hu)$ at $\mu(t)$ and $f(t)$, respectively.

For the asymptotic expression of $V\left[\hat{\mu}_K(t; h, w)\right]$, let $Z_{ij}(t) = Y_{ij} - \mu(t)$ and

$$\left[f^{-1}(t)\widehat{R}_K(t; h, w)\right]^2 = A^{(1)}(t) + A^{(2)}(t) + A^{(3)}(t),$$

where

$$
\begin{aligned}
A^{(1)}(t) &= \left[f(t)h\right]^{-2}\sum_{i=1}^{n}\sum_{j=1}^{n_i}\left[w_i^2 Z_{ij}^2(t)K^2\left(\frac{t - t_{ij}}{h}\right)\right], \\
A^{(2)}(t) &= \left[f(t)h\right]^{-2}\sum_{i=1}^{n}\sum_{j_1 \neq j_2}\left[w_i^2 Z_{ij_1}(t)Z_{ij_2}(t)K\left(\frac{t - t_{ij_1}}{h}\right)K\left(\frac{t - t_{ij_2}}{h}\right)\right], \\
A^{(3)}(t) &= \left[f(t)h\right]^{-2}\sum_{i_1 \neq i_2}\sum_{j_1, j_2}\left[w_{i_1} w_{i_2} Z_{i_1 j_1}(t)Z_{i_2 j_2}(t)K\left(\frac{t - t_{i_1 j_1}}{h}\right)K\left(\frac{t - t_{i_2 j_2}}{h}\right)\right].
\end{aligned}
$$

It remains to evaluate the expectations of $A^{(k)}(t)$ for $k = 1, 2, 3$. Note that,

$$Z_{ij}^2(t) = \left[\mu(t_{ij}) - \mu(t)\right]^2 + 2\left[\mu(t_{ij}) - \mu(t)\right]\varepsilon_{ij} + \varepsilon_{ij}^2,$$

and, by $E\{[\mu(t_{ij}) - \mu(t)]\,\varepsilon_{ij}|t_{ij} = s\} = 0$, $E(\varepsilon_{ij}^2|t_{ij} = s) = \sigma^2(s)$ and

$$E\{[\mu(t_{ij}) - \mu(t)]^2|t_{ij} = s\} = \mu^2(s) - 2\mu(t)\mu(s) + \mu^2(t),$$

it follows that

$$
\begin{aligned}
& E[A^{(1)}(t)] \\
&= [hf(t)]^{-2} \sum_{i=1}^{n} \sum_{j=1}^{n_i} \left\{ w_i^2 \int E[Z_{ij}^2(t)|t_{ij} = s]\, K^2\!\left(\frac{t-s}{h}\right) f(s)\, ds \right\} \\
&= [hf(t)]^{-2} \sum_{i=1}^{n} \sum_{j=1}^{n_i} \left\{ w_i^2 \int \left[\mu^2(s) - 2\mu(t)\mu(s) + \mu^2(t) + \sigma^2(s)\right] \right. \\
&\qquad\qquad \left. \times K^2\!\left(\frac{t-s}{h}\right) f(s)\, ds \right\} \\
&= f^{-2}(t) \sum_{i=1}^{n} \sum_{j=1}^{n_i} \left\{ h^{-1} w_i^2\, \sigma^2(t) R(K) f(t)\, [1 + o(1)] \right\} \\
&= h^{-1} \left[\sum_{i=1}^{n} (w_i^2 n_i) \right] f^{-1}(t)\, \sigma^2(t) R(K)\, [1 + o(1)]. \qquad (3.52)
\end{aligned}
$$

Using similar derivations as those in (3.52), we can show that, by (3.42) and (3.44),

$$
\begin{aligned}
& E[A^{(2)}(t)] \\
&= [hf(t)]^{-2} \sum_{i=1}^{n} \sum_{j_1 \neq j_2 = 1}^{n_i} \left\{ w_i^2 \iint E[Z_{ij_1}(t) Z_{ij_2}(t)|t_{ij_1} = s_1, t_{ij_2} = s_2] \right. \\
&\qquad\qquad \left. \times K\!\left(\frac{t-s_1}{h}\right) K\!\left(\frac{t-s_2}{h}\right) f(s_1) f(s_2)\, ds_1\, ds_2 \right\} \\
&= \left[\sum_{i=1}^{n} (w_i^2 n_i^2) - \sum_{i=1}^{n} (w_i^2 n_i) \right] \rho_\varepsilon(t)\, [1 + o(1)] \qquad (3.53)
\end{aligned}
$$

and

$$E[A^{(3)}(t)] = B^2[\hat{\mu}(t; h, w)]. \qquad (3.54)$$

Then, when n is sufficiently large, the asymptotic variance expression (3.49) follows from (3.46), (3.52) to (3.54), and

$$V[\hat{\mu}(t; h, w)] = \sum_{l=1}^{3} E[A^{(l)}(t)] - B^2[\hat{\mu}(t; h, w)] = E[A^{(1)}(t)] + E[A^{(2)}(t)].$$

The above results give the conclusions in Theorem 3.1(a). Theorem 3.1(b) directly follows from (3.39), (3.40) and (3.47) to (3.49).

To prove Theorem 3.1(c), we first note that $\sum_{i=1}^{n}(w_i^2 n_i^2) > \sum_{i=1}^{n}(w_i^2 n_i)$.

By Assumptions (c) and (d), the asymptotic expressions of $E\left[A^{(1)}(t)\right]$ and $E\left[A^{(2)}(t)\right]$ in (3.52) and (3.53) suggest that

$$E\left[A^{(2)}(t)\right] = O\left[\sum_{i=1}^{n}\left(w_i^2 n_i^2\right)\right] \quad \text{and} \quad E\left[A^{(1)}(t)\right] = O\left[h^{-1}\sum_{i=1}^{n}\left(w_i^2 n_i\right)\right].$$

By Assumption (e) and (3.50), we have $E\left[A^{(2)}(t)\right] = o\left\{E\left[A^{(1)}(t)\right]\right\}$ as $n \to \infty$, so that (3.51) holds. This completes the proof of the theorem. ∎

2. Theoretical Optimal Bandwidths

The theoretically optimal bandwidths can be derived by minimizing the asymptotic expressions of the MSE or MISE of $\widehat{\mu}_K(t;h,w)$ with respect to the choices of h. This result is shown in the next theorem.

Theorem 3.2 *Suppose that t is in the interior of the support of $f(\cdot)$, Assumptions (a) to (e) and (3.50) are satisfied, and $B_*(t; K, p, \mu, f)$ is defined in (3.48). The following conclusions hold:*

(a) *The optimal pointwise bandwidth $h_{opt}(t; w)$, which minimizes the asymptotic expression of $MSE\left[\widehat{\mu}_K(t;h,w)\right]$ for all $h > 0$, is given by*

$$h_{opt}(t; w) = \left\{ \frac{\left[\sum_{i=1}^{n}(w_i^2 n_i)\right] R(K)\sigma^2(t)}{2\,(p+2)\,f(t)\,B_*^2(t; K, p, \mu, f)} \right\}^{1/(2p+5)}. \tag{3.55}$$

(b) *The optimal global bandwidth $h_{opt}(w)$ for the weight w, which minimizes the asymptotic expression of $MISE\left[\widehat{\mu}_K(\cdot; h, w)\right]$ for all $h > 0$, is given by*

$$h_{opt}(w) = \left\{ \frac{\left[\sum_{i=1}^{n}(w_i^2 n_i)\right]\int f^{-1}(s)R(K)\sigma^2(s)\pi(s)\,ds}{2\,(p+2)\int B_*^2(s; K, p, \mu, f)\,\pi(s)\,ds} \right\}^{1/(2p+5)}. \tag{3.56}$$

(c) *The optimal MSE and MISE for $\widehat{\mu}_K(\cdot; h, w)$ are given by*

$$MSE\left[\widehat{\mu}_K\left(t; h_{opt}, w\right)\right]$$
$$= \left[\sum_{i=1}^{n}(w_i^2 n_i)\right]^{(2p+4)/(2p+5)} \tag{3.57}$$
$$\times \left[B_*(t; K, p, \mu, f)\right]^{2/(2p+5)}\left[f^{-1}(t)R(K)\sigma^2(t)\right]^{(2p+4)/(2p+5)}$$
$$\times \left[(2p+4)^{-(2p+4)/(2p+5)} + (2p+4)^{1/(2p+5)}\right]\left[1 + o(1)\right]$$

and

$$MISE\left[\widehat{\mu}_K\left(\cdot; h_{opt}, w\right)\right]$$
$$= \left[\sum_{i=1}^{n}(w_i^2 n_i)\right]^{(2p+4)/(2p+5)}$$

$$\times \left\{ \int B_*^2(s; K, p, \mu, f) \, \pi(s) \, ds \right\}^{1/(2p+5)} \tag{3.58}$$

$$\times \left[\int f^{-1}(s) R(K) \sigma^2(s) \, \pi(s) \, ds \right]^{(2p+4)/(2p+5)}$$

$$\times \left[(2p+4)^{-(2p+4)/(2p+5)} + (2p+4)^{1/(2p+5)} \right] [1 + o(1)],$$

which are the MSE and MISE corresponding to theoretically optimal band-widths $h_{opt}(t; w)$ and $h_{opt}(w)$, respectively. ∎

Proof of Theorem 3.2:

By (3.48), (3.49) and (3.50), the dominating term of the pointwise mean squared error $MSE\left[\widehat{\mu}_K(t; h, w)\right]$ is

$$h^{2(p+2)} B_*^2(t; K, p, \mu, f) + h^{-1} \sum_{i=1}^{n} \left(w_i^2 n_i\right) f^{-1}(t) \sigma^2(t) R(K),$$

so that the optimal pointwise bandwidth $h_{opt}(t; w)$ which minimizes this dominating term satisfies the equation

$$\frac{d}{dh} \left\{ h^{2(p+2)} B_*^2(t; K, p, \mu, f) + h^{-1} \left[\sum_{i=1}^{n} \left(w_i^2 n_i\right) \right] f^{-1}(t) R(K) \sigma^2(t) \right\}$$

$$= \quad 2(p+2) h^{2p+3} B_*^2(t; K, p, \mu, f)$$

$$- h^{-2} \left[\sum_{i=1}^{n} \left(w_i^2 n_i\right) \right] f^{-1}(t) R(K) \sigma^2(t)$$

$$= \quad 0. \tag{3.59}$$

The solution of (3.59) shows that

$$h_{opt}^{2p+5}(t; w) = \frac{\left[\sum_{i=1}^{n} \left(w_i^2 n_i\right) \right] R(K) \sigma^2(t)}{2(p+2) f(t) B_*^2(t; K, p, \mu, f)}, \tag{3.60}$$

which gives the desired result in (3.55).

Substituting the expression of (3.55) into (3.48) and (3.49), the bias and variance of $\widehat{\mu}_K(t; h, w)$ based on $h_{opt}(t; w)$ are

$$B\left[\widehat{\mu}_K(t; h, w)\right] = \left[\sum_{i=1}^{n} \left(w_i^2 n_i\right) \right]^{(p+2)/(2p+5)} \tag{3.61}$$

$$\times (2p+4)^{-(p+2)/(2p+5)} \left[f^{-1}(t) R(K) \sigma^2(t) \right]^{(p+2)/(2p+5)}$$

$$\times \left[B_*(t; K, p, \mu, f) \right]^{1/(2p+5)} [1 + o(1)]$$

and

$$V\left[\widehat{\mu}(t; h, w)\right] = \left[\sum_{i=1}^{n} \left(w_i^2 n_i\right) \right]^{(2p+4)/(2p+5)}$$

$$\times (2p+4)^{1/2p+5} \left[f^{-1}(t) R(K) \sigma^2(t) \right]^{(2p+4)/(2p+5)}$$

$$\times \left[B_*(t; K, p, \mu, f) \right]^{2/(2p+5)} \left[1 + o(1) \right]. \tag{3.62}$$

The mean squared error in (3.57) then follows from (3.61) and (3.62).

For the derivation of (3.58), we first note that, by Theorem 3.1, and the expressions of (3.48) and (3.49), the dominating term of the mean integrated squared error $\text{MISE} \left[\hat{\mu}_K(\cdot; h, w) \right]$ is

$$h^{2(p+2)} \int B_*^2(s; K, p, \mu, f) \pi(s) ds$$

$$+ h^{-1} \left[\sum_{i=1}^n \left(w_i^2 n_i \right) \right] \int f^{-1}(s) \sigma^2(s) R(K) \pi(s) ds.$$

Setting the derivative of the above term with respect to h to zero, the optimal global bandwidth satisfies the equation

$$2(p+2) h^{2p+3} \int B_*^2(s; K, p, \mu, f) \pi(s) ds$$

$$= h^{-2} \left[\sum_{i=1}^n \left(w_i^2 n_i \right) \right] \int f^{-1}(s) \sigma^2(s) R(K) \pi(s) ds. \tag{3.63}$$

The solution of (3.63) gives the expression (3.56).

Substituting the expression of $h_{opt}(w)$ (3.56) into (3.48), (3.49) and the mean integrated squared error $\text{MISE} \left[\hat{\mu}_K(\cdot; h_{opt}, w) \right]$, we have

$$\int \left\{ B \left[\hat{\mu}_K(s; h, w) \right] \right\}^2 \pi(s) ds$$

$$= h_{opt}^{2p+4}(w) \int B_*^2(s; K, p, \mu, f) \pi(s) ds \left[1 + o(1) \right]$$

$$= \left[\sum_{i=1}^n \left(w_i^2 n_i \right) \right]^{(2p+4)/(2p+5)} \tag{3.64}$$

$$\times (2p+4)^{-(2p+4)/(2p+5)} \left[\int f^{-1}(s) R(K) \sigma^2(s) \pi(s) ds \right]^{(2p+4)/(2p+5)}$$

$$\times \left[\int B_*^2(s; K, p, \mu, f) \pi(s) ds \right]^{1/(2p+5)} \left[1 + o(1) \right]$$

and

$$\int V \left[\hat{\mu}(s; h, w) \right] \pi(s) ds$$

$$= \left[\sum_{i=1}^n \left(w_i^2 n_i \right) \right]^{(2p+4)/(2p+5)}$$

$$\times \left[\int f^{-1}(s) R(K) \sigma^2(s) \pi(s) ds \right]^{(2p+4)/(2p+5)}$$

$$\times (2p+4)^{1/2p+5} \left[\int B_*^2(s; K, p, \mu, f) \pi(s) ds \right]^{1/(2p+5)}$$

$$\times \left[1 + o(1) \right]. \tag{3.65}$$

The mean integrated squared error $MISE\left[\widehat{\mu}_K(\cdot; h_{opt}, w)\right]$ in (3.58) is obtained by summing up the right sides of (3.64) and (3.65). This completes the proof for the assertions (a), (b) and (c) of the theorem. ∎

3.6.4   Useful Special Cases

The asymptotic properties demonstrated in Theorems 3.1 and 3.2 are suitable for longitudinal data with general repeated measurements $\{n_i : i = 1, \ldots, n\}$ and weights $w = (w_1, \ldots, w_n)^T$ as long as Assumptions (a) to (e) are satisfied. For useful special cases with further conditions imposed on $\{n_i : i = 1, \ldots, n\}$ and $w = (w_1, \ldots, w_n)^T$, interesting special cases can be deduced from Theorems 3.1 and 3.2. We discuss here the two most commonly used special cases: (a) the subject uniform weight $w^* = (1/(nn_1), \ldots, 1/(nn_n))^T$, and (b) the measurement uniform weight $w^{**} = (1/N, \ldots, 1/N)^T$.

1. Kernel Estimators with Subject Uniform Weight

When $w_i^* = 1/(nn_i)$ is used, we have that $\sum_{i=1}^n (w_i^2 n_i) = \sum_{i=1}^n \left[1/(n^2 n_i)\right]$ and $\sum_{i=1}^n (w_i^2 n_i^2) = 1/n$, and (3.44) and (3.50) imply the following condition for h

$$\frac{1}{n^2 h} \sum_{i=1}^n (1/n_i) \to 0 \quad \text{and} \quad \frac{nh}{\sum_{i=1}^n (1/n_i)} \to 0 \quad \text{as } n \to \infty. \tag{3.66}$$

The following two corollaries, which are direct consequences of Theorems 3.1 and 3.2, summarize the asymptotic expressions for the kernel estimators $\widehat{\mu}_K(t; h, w^*)$ of (3.8).

Corollary 3.1 *Suppose that t is in the interior of the support of $f(\cdot)$, $w^* = (w_1^*, \ldots, w_n^*)^T$ with $w_i^* = 1/(nn_i)$ is used, and Assumptions (a)-(c) and (d) are satisfied. When $n \to \infty$, $B\left[\widehat{\mu}_K(t; h, w^*)\right]$, $V\left[\widehat{\mu}(t; h, w^*)\right]$, $MSE\left[\widehat{\mu}(t; h, w^*)\right]$ and $MISE\left[\widehat{\mu}(t; h, w^*)\right]$ are given as the corresponding terms in Theorem 3.1, such as (3.48) and (3.49), by substituting $\sum_{i=1}^n (w_i^2 n_i^2) = 1/n$ and $\sum_{i=1}^n (w_i^2 n_i)$ with $\sum_{i=1}^n \left[1/(n^2 n_i)\right]$.* ∎

The next corollary shows the optimal convergence rate for the MSE and MISE of $\widehat{\mu}_K(t; h, w^*)$ to converge to zero under the optimal bandwidth choices.

Corollary 3.2 *Under the assumptions of Corollary 3.1 and (3.66), the optimal pointwise bandwidth $h_{opt}(t; w^*)$ and the optimal global bandwidth $h_{opt}(w^*)$ for the weight $w_i^* = 1/(nn_i)$, which minimize $MSE\left[\widehat{\mu}_K(t; h, w^*)\right]$ and $MISE\left[\widehat{\mu}_K(\cdot; h, w^*)\right]$, are given by (3.55) and (3.56), respectively, by substituting $\sum_{i=1}^n (w_i^2 n_i)$ with $\sum_{i=1}^n \left[1/(n^2 n_i)\right]$. The optimal mean squared errors $MSE\left[\widehat{\mu}_K(t; h_{opt}, w^*)\right]$ and $MISE\left[\widehat{\mu}_K(\cdot; h_{opt}, w^*)\right]$ corresponding to the optimal* bandwidths $h_{opt}(t; w^*)$ and $h_{opt}(w^*)$, are given by (3.57) and (3.58), respectively, by substituting $\sum_{i=1}^{n}\left(w_i^2 n_i\right)$ with $\sum_{i=1}^{n}\left[1/\left(n^2 n_i\right)\right]$. ∎

The above two corollaries may be simplified under further asymptotic assumptions. A common longitudinal setting is to assume that the number of repeated measurements are bounded while the number of subjects may tend to infinity. In such situations, $n_i \leq c$ for some constant $c > 0$ as $n \to \infty$, (3.66) reduces to $\lim_{n\to\infty} h = 0$ and $\lim_{n\to\infty} nh = \infty$, and the $\sum_{i=1}^{n}\left[1/\left(n^2 n_i\right)\right]$ used in Corollary 3.2 is then replaced by n^{-1}, so that the optimal convergence rate for $MSE\left[\widehat{\mu}_K(t; h_{opt}, w^*)\right]$ and $MISE\left[\widehat{\mu}_K\left(\cdot; h_{opt}, w^*\right)\right]$ is $n^{(2p+4)/(2p+5)}$.

2. Kernel Estimators with Measurement Uniform Weight

For the weight choice $w_i^{**} = 1/N$, we have $\sum_{i=1}^{n}\left[\left(w_i^{**}\right)^2 n_i\right] = 1/N$ and $\sum_{i=1}^{n}\left[\left(w_i^{**}\right)^2 n_i^2\right] = \left(\sum_{i=1}^{n} n_i^2\right)/N^2$, and (3.44) and (3.50) imply the following condition for h

$$Nh \to 0 \quad \text{and} \quad \frac{h \sum_{i=1}^{n} n_i^2}{N} \to 0 \quad \text{as } n \to \infty. \qquad (3.67)$$

The following two corollaries follow from Theorem 3.1 and Theorem 3.2.

Corollary 3.3 *Suppose that t is in the interior of the support of $f(\cdot)$, $w^{**} = \left(w_1^{**}, \ldots, w_n^{**}\right)^T$, $w_i^{**} = 1/N$, is used, and Assumptions (a)-(c) and (d) are satisfied. When $n \to \infty$, $B\left[\widehat{\mu}_K(t; h, w^{**})\right]$, $V\left[\widehat{\mu}(t; h, w^{**})\right]$, $MSE\left[\widehat{\mu}(t; h, w^{**})\right]$ and $MISE\left[\widehat{\mu}(\cdot; h, w^{**})\right]$ are given by the corresponding terms in Theorem 3.1, such as (3.48) and (3.49), by substituting $\sum_{i=1}^{n}\left(w_i^2 n_i\right)$ with $1/N$.* ∎

As a consequence of the above corollary, the optimal convergence rate for the MSE and MISE of $\widehat{\mu}_K(t; h, w^{**})$ to converge to zero is $N^{-(2p+4)/(2p+5)}$.

Corollary 3.4 *Under the assumptions of Corollary 3.3 and (3.67), the pointwise optimal bandwidth $h_{opt}(t; w^{**})$ and the global optimal bandwidth $h_{opt}(w^{**})$ for $w_i^{**} = 1/N$, which minimize $MSE\left[\widehat{\mu}_K(t; h, w^{**})\right]$ and $MISE\left[\widehat{\mu}_K(\cdot; h, w^{**})\right]$, are given by (3.55) and (3.56), respectively, by substituting $\sum_{i=1}^{n}\left(w_i^2 n_i\right)$ with $1/N$. The optimal mean squared errors $MSE\left[\widehat{\mu}_K(t; h_{opt}, w^{**})\right]$ and $MISE\left[\widehat{\mu}_K(\cdot; h_{opt}, w^{**})\right]$ corresponding to $h_{opt}(t; w^{**})$ and $h_{opt}(w^{**})$, are given by (3.57) and (3.58), respectively, by substituting $\sum_{i=1}^{n}\left(w_i^2 n_i\right)$ with $1/N$.* ∎

When the numbers of repeated measurements are bounded, $n_i \leq c$ for some constant $c > 0$, we have that N/n is bounded, and, by Corollary 3.2, the optimal convergence rate for $MSE\left[\widehat{\mu}_K(t; h_{opt}, w^{**})\right]$ and $MISE\left[\widehat{\mu}_K\left(\cdot; h_{opt}, w^{**}\right)\right]$ is $n^{(2p+4)/(2p+5)}$.

3.7   Remarks and Literature Notes

The estimators described in this chapter are some of the most basic local smoothing methods. These methods are direct extensions of the local smoothing methods from cross-sectional i.i.d. data to longitudinal data. The main advantage of these methods is that they are conceptually simple. As seen from the theoretical derivations of Section 3.6, the asymptotic biases, variances and mean squared errors, including both the MSE and MISE, of these estimators may depend on the number of subjects n as well as the number of repeated measurements n_i. A direct consequence of the local nature of these estimators is that their asymptotic properties are not affected by the correlation structures of the data in most practical situations.

The estimation methods of Sections 3.1 to 3.5 are adapted from Hoover et al. (1998) and Wu and Chiang (2000). The asymptotic results of Section 3.6 are special cases of the results of Wu and Chiang (2000). A noteworthy kernel method for longitudinal data is the SUR kernel method of Wang (2003) described in Section 2.3, which has the advantage of taking the correlation structures of the data into consideration. However, because the SUR kernel method is not easily generalized under the structured nonparametric models to be discussed in the later chapters of this book, we omit its presentation in this chapter and refer its details to Wang (2003). Other omitted topics include the asymptotic properties for the local polynomial estimators, several inference procedures (e.g., Knafl, Sacks and Ylvisaker, 1985; Härdle and Marron, 1991; Eubank and Speckman, 1993) and fast algorithms for computation (e.g., Fan and Marron, 1994). Since the asymptotic results of Section 3.6 are intended to provide some initial insights into local smoothing with repeated measurements data, we choose to present only the simpler case of least squares kernel estimators in this chapter.

Chapter 4

Basis Approximation Smoothing Methods

The estimation of $\mu(t) = E[Y(t)|t]$ in the simple nonparametric regression model (3.1) can also be carried out by an approximation approach using basis expansions for $\mu(t)$ based on the sample $\{(Y_{ij}, t_{ij}) : i = 1, \ldots, n; j = 1, \ldots, n_i\}$, where the time points t_{ij} can be either regularly or irregularly spaced. In contrast to the kernel-based local smoothing methods in Chapter 3, the basis approximation methods belong to the class of "global smoothing methods" because the entire curve $\mu(t)$ within the time range is approximated by a linear combination of a set of chosen basis functions. The coefficients of the linear expansions, which determine the shape of the approximation of $\mu(t)$, are then estimated from the data by finding the "best" fit between the basis approximation of $\mu(t)$ and the data. As an important part of the "global smoothing methods," the basis approximation approach described in this chapter is an extension of the "extended linear models" (Stone et al., 1997; Huang, 1998, 2001 and 2003) to data with intra-subject correlations over time. The methods and theory in this chapter provide useful insights into the mechanism and effects of correlation structures in practical situations. Extensions of the basis approximation methods to more complicated structured nonparametric models with longitudinal data have been extensively studied in the literature, for example, Huang, Wu and Zhou (2002, 2004), Yao, Müller and Wang (2005a, 2005b), among others. We discuss these extensions later in Chapter 9.

4.1 Estimation Method

4.1.1 Basis Approximations and Least Squares

Suppose that there is a set of basis functions $\{B_k(t) : k = 1, \ldots, K\}$ and constants $\{\gamma_k : k = 1, \ldots, K\}$, such that, for any $t \in \mathcal{T}$, $\mu(t)$ can be approximated by the expansion

$$\mu(t) \approx \sum_{k=1}^{K} \gamma_k B_k(t). \tag{4.1}$$

In order to ensure that $\mu(t)$ can be a constant, we assume that, unless specifically mentioned otherwise, $B_1(t) = 1$. Substituting $\mu(t)$ of (3.1) with the right side of (4.1), we approximate (3.1) by

$$Y_{ij} \approx \sum_{k=1}^{K} \gamma_k B_k(t_{ij}) + \varepsilon_{ij}, \tag{4.2}$$

where $\varepsilon_{ij} = \varepsilon_i(t_{ij})$ is the error term of the ith subject at jth visit time t_{ij} for the $\varepsilon(t)$ defined in (3.1). Using the approximation (4.2), the coefficients $\gamma = (\gamma_1, \ldots, \gamma_K)^T$ can be estimated by minimizing the squared error

$$\ell_w(\gamma) = \sum_{i=1}^{n} \sum_{j=1}^{n_i} w_i \left[Y_{ij} - \sum_{k=1}^{K} \gamma_k B_k(t_{ij}) \right]^2, \tag{4.3}$$

where $w = (w_1, \ldots, w_n)^T$ and w_i is a known nonnegative weight for the ith subject, which satisfies $\sum_{i=1}^{n} (n_i w_i) = 1$. Similar to the kernel estimation cases of Chapter 3, useful choices of w_i include the "subject uniform weight" $w^* = (w_1^*, \ldots, w_n^*)^T$ with $w_i^* = 1/(n n_i)$ and the "measurement uniform weight" $w^{**} = (w_1^{**}, \ldots, w_n^{**})^T$ with $w_i^{**} = 1/N$.

If $\ell_w(\gamma)$ can be uniquely minimized, we denote by $\widehat{\gamma} = (\widehat{\gamma}_1, \ldots, \widehat{\gamma}_K)^T$ the least squares estimator of $\gamma = (\gamma_1, \ldots, \gamma_K)^T$ based on (4.3) with weight w. Substituting γ of (4.1) with $\widehat{\gamma}$, the least squares *basis approximation* estimator of $\mu(t)$ based on the basis functions $\{B_k(t) : k = 1, \ldots, K\}$ is

$$\widehat{\mu}_B(t) = \sum_{k=1}^{K} \widehat{\gamma}_k B_k(t). \tag{4.4}$$

Explicit expressions for $\widehat{\gamma}$ and $\widehat{\mu}_B(t)$ can be derived from the following matrix derivations. Let $B(t) = (B_1(t), \ldots, B_K(t))^T$ be the $(K \times 1)$ vector of basis functions. Then, for the ith subject, we denote by $Y_i = (Y_{i1}, \ldots, Y_{in_i})^T$ the $(n_i \times 1)$ vector of n_i outcome observations, $t_i = (t_{i1}, \ldots, t_{in_i})^T$ the $(n_i \times 1)$ vector of time points, $B(t_i) = (B(t_{i1}), \ldots, B(t_{in_i}))^T$ the $(K \times n_i)$ matrix of basis functions, and $W_i = \mathrm{diag}(w_i, \ldots, w_i)$ the $(n_i \times n_i)$ diagonal weight matrix with diagonal elements w_i and 0 elsewhere. The matrix representation of (4.3) can be written as

$$\ell_w(\gamma) = \sum_{i=1}^{n} [Y_i - B(t_i)\gamma]^T W_i [Y_i - B(t_i)\gamma]. \tag{4.5}$$

Suppose that the inverse of $\sum_{i=1}^{n} [B(t_i)^T W_i B(t_i)]$ exists and is unique. Then there is a unique γ which minimizes the right side of (4.5). The least squares basis approximation estimators $\widehat{\gamma}$, which minimizes $\ell_w(\gamma)$ of (4.5), and $\widehat{\mu}_B(t)$, which is obtained from (4.4), are given by

$$\begin{cases} \widehat{\gamma} &= \left\{ \sum_{i=1}^{n} [B(t_i)^T W_i B(t_i)] \right\}^{-1} \left\{ \sum_{i=1}^{n} [B(t_i)^T W_i Y_i] \right\}, \\ \widehat{\mu}_B(t) &= B(t)^T \widehat{\gamma}. \end{cases} \tag{4.6}$$

The linear function space spanned by the basis functions $\{B_k(t) : k = 1, \ldots, K\}$ uniquely determines the basis approximation estimator $\widehat{\mu}_B(t)$. Different sets of basis functions can be used to span the same space and thus give the same estimator $\widehat{\mu}_B(t)$. When different basis functions are used, the corresponding coefficients γ_k, $k = 1, \ldots, K$, and their estimators $\widehat{\gamma}_k$ are also different. For example, both the B-spline basis (also known as polynomial spline basis) and the truncated power basis can be used to span a space of spline functions for $\mu(t)$. But, for the same function $\mu(t)$, the coefficients γ for the two different bases are different; hence, their estimates are also different.

The choice of w_i in (4.3) may influence the theoretical and practical properties of $\widehat{\gamma}$, hence, $\widehat{\mu}_B(t)$. For the "subject uniform weight" $w_i^* = 1/(n n_i)$, each subject is inversely weighted by its number of repeated measurements n_i, so that the subjects with fewer repeated measurements receive more weight than the subjects with more repeated measurements at the corresponding visiting times. For the "measurement uniform weight" $w_i^{**} = 1/N$, all the subjects at all the visiting times receive the same weight, so that the subjects with more repeated measurements are weighted the same as the subjects with fewer repeated measurements. It is conceivable that an ideal choice of w_i may also depend on the intra-subject correlation structures of the data. However, because the actual correlation structures are usually completely unknown in practice, $w_i^{**} \equiv 1/N$ appears to be a practical choice if n_i for all $i = 1, \ldots, n$ are similar, while $w_i^* = 1/(n n_i)$ may be appropriate when n_i are significantly different between different subjects. Some theoretical implications of the choices of w_i are discussed in Section 4.4.

Although any common basis system can be used for function approximation, some basis systems may be more appropriate than others depending on the nature of the data and the scientific questions being investigated. For example, the Fourier basis may be desirable when the underlying functions exhibit periodicity, and global polynomials are familiar choices which can provide good approximations to smooth functions. However, these bases may not be sensitive enough to exhibit certain local features without using a large number K of basis functions. In this respect, B-splines (i.e., polynomial splines) are often desirable. Ideally, a basis should be chosen to achieve an excellent approximation using a relatively small value of K. Some general guidance for choosing basis functions can be found in Chapter 3 of Ramsay and Silverman (2005). All the numerical examples in the R implementation of Section 4.3 are computed using B-spline bases, because they can exhibit local features and provide stable numerical solutions (de Boor, 1978, Ch. II).

4.1.2   Selecting Smoothing Parameters

Once a basis system is chosen, the number of basis functions K in (4.1) is the smoothing parameter for a basis approximation estimator. Similar to the least squares kernel and local polynomial estimators discussed in Chapter 3, the choice of the smoothing parameter K determines the appropriate smoothness of the estimators. When a large number of basis functions is used in (4.1) to approximate $\mu(t)$, the bias of the estimator $\widehat{\mu}_B(t)$ in (4.4) is expected to be small, but the variance of the estimator is expected to be large. On the other hand, a small K leads to smaller variance but larger bias for $\widehat{\mu}_B(t)$.

1. Leave-One-Subject-Out Cross-Validation

For the local smoothing methods, such as the kernel and local polynomial estimators, data-driven smoothing parameters, i.e., bandwidths, can be selected through the "*leave-one-subject-out*" cross-validation (LSCV) procedures described in Section 3.3. The objective of the LSCV is to find a smoothing parameter to balance the estimated biases and variances, so that the average squared errors of the smoothing estimators can be minimized. Extending the same argument to global smoothing estimation method, Huang, Wu and Zhou (2002) suggests that the same LSCV procedure can be modified as follows to select the smoothing parameter K for $\widehat{\mu}_B(t)$.

Leave-One-Subject-Out Cross-Validation:

(a) *Let $\widehat{\gamma}^{(-i)}$ be the estimator of γ obtained by minimizing $\ell_w(\gamma)$ of (4.3) using the data with the measurements of the ith subject deleted, and let $\widehat{\mu}_B^{(-i)}(t)$ be the estimator defined in (4.4) with $\widehat{\gamma}$ replaced by $\widehat{\gamma}^{(-i)}$.*

(b) *The LSCV score for K is defined by*

$$LSCV_\mu(K) = \sum_{i=1}^{n} \sum_{j=1}^{n_i} \left\{ w_i \left[Y_{ij} - \widehat{\mu}_B^{(-i)}(t_{ij}) \right]^2 \right\}. \tag{4.7}$$

The cross-validated smoothing parameter K_{LSCV} is the minimizer of $LSCV_\mu(K)$, provided that (4.7) can be uniquely minimized. □

2. Heuristic Justification of Cross-Validation

The above LSCV procedure for the selection of K can be justified as in Section 3.3 for the LSCV bandwidth choices of the kernel and local polynomial estimators. Specifically, there are two main reasons for using this LSCV procedure. First, deletion of the entire measurements of the subject one at a time preserves the correlation in the data. Second, this approach does not require us to model the intra-subject correlation structures of the data.

For an intuitive justification of K_{LSCV}, we consider the following average squared error

$$ASE_\mu(K) = \sum_{i=1}^{n} \sum_{j=1}^{n_i} \left\{ w_i \left[\mu(t_{ij}) - \widehat{\mu}_B(t_{ij}) \right]^2 \right\} \tag{4.8}$$

and the decomposition

$$LSCV_\mu(K) = \sum_{i=1}^{n} \sum_{j=1}^{n_i} \left\{ w_i \left[Y_{ij} - \mu(t_{ij}) \right]^2 \right\}$$

$$+2 \sum_{i=1}^{n} \sum_{j=1}^{n_i} \left\{ w_i \left[Y_{ij} - \mu\left(t_{ij}\right) \right] \left[\mu\left(t_{ij}\right) - \widehat{\mu}_B^{(-i)}\left(t_{ij}\right) \right] \right\}$$

$$+ \sum_{i=1}^{n} \sum_{j=1}^{n_i} \left\{ w_i \left[\mu\left(t_{ij}\right) - \widehat{\mu}_B^{(-i)}\left(t_{ij}\right) \right]^2 \right\}. \tag{4.9}$$

The first term at the right side of (4.9) does not depend on the smoothing parameter K. Because of the definition of $\widehat{\mu}_B^{(-i)}(t)$ and the fact that the subjects are independent, the expectation of the second term is zero. Thus, by minimizing $LSCV_\mu(K)$, K_{LSCV} approximately minimizes the third term at the right side of (4.9), which is an approximation of $ASE_\mu(K)$ in (4.8).

4.2   Bootstrap Inference Procedures

Statistical inferences for $\mu(t)$ based on $\widehat{\mu}_B(t)$ of (4.4), including pointwise confidence intervals, simultaneous confidence bands and hypothesis testing, are usually constructed through a *resampling-subject bootstrap* procedure similar to the ones discussed in Section 3.4. Although in principle the asymptotic distributions of the basis approximation estimator $\widehat{\mu}_B(t)$ can be used to construct approximate inference procedures for $\mu(t)$, such approximate inference procedures depend on the particular asymptotic assumptions and the unknown correlation structures, and may not be appropriate for a given longitudinal sample. The resampling-subject bootstrap procedure, on the other hand, rely on estimating the variability of the estimators based on the intra-subject correlation structures of the subjects, hence, are more appropriate to the specific longitudinal design of the given sample.

4.2.1   Pointwise Confidence Intervals

The bootstrap pointwise confidence intervals can be constructed by substituting the kernel or local polynomial estimators of Chapter 3.4 with the basis approximation estimator $\widehat{\mu}_B(t)$ of (4.4). The specific steps can be briefly described in the following.

Approximate Bootstrap Pointwise Confidence Intervals:
(a) **Bootstrap Samples and Estimators.** *Let* $\left\{ \left(Y_{ij}^*, t_{ij}^*\right) : 1 \leq i \leq n; 1 \leq j \leq n_i \right\}$ *be a bootstrap sample obtained as in Step (a) of Section 3.4.1. Compute the estimators* $\widehat{\gamma}^{boot} = \left(\widehat{\gamma}_1^{boot}, \ldots, \widehat{\gamma}_K^{boot}\right)^T$ *and* $\widehat{\mu}_B^{boot}(t)$ *based on (4.6) with the basis functions* $\left\{ B_k(t) : k = 1, \ldots, K \right\}$ *and the available bootstrap sample. With $B > 1$ independent replications, we obtain B bootstrap samples with their corresponding estimators* $\widehat{\gamma}^{boot}$ *and* $\widehat{\mu}_B^{boot}(t)$.

(b) **Percentile Bootstrap Intervals.** *A pointwise* $\left[100 \times (1 - \alpha)\right]\%$ *confidence interval for $\mu(t)$ based on the percentiles of the bootstrap samples is*

$$\left(L_{\alpha/2}\left[\widehat{\mu}_B^{boot}(t)\right], U_{\alpha/2}\left[\widehat{\mu}_B^{boot}(t)\right] \right), \tag{4.10}$$

where $L_{\alpha/2}[\widehat{\mu}_B^{boot}(t)]$ and $U_{\alpha/2}[\widehat{\mu}_B^{boot}(t)]$ are the $[100 \times (\alpha/2)]$th and $[100 \times (1-\alpha/2)]$th percentiles of the bootstrap estimators of $\mu(t)$.

(c) **Normal Approximated Bootstrap Intervals.** *If the distribution of* $\widehat{\mu}_B^{boot}(t)$ *can be approximated by a normal distribution, a pointwise* $[100 \times (1-\alpha)]\%$ *confidence interval for* $E[\widehat{\mu}(t)]$ *based on normal approximation is*

$$\widehat{\mu}_B(t) \pm z_{1-\alpha/2} \widehat{s}_\mu^{boot}(t), \tag{4.11}$$

where $z_{1-\alpha/2}$ *is the* $[100 \times (1-\alpha/2)]$th *percentile of the standard normal distribution and* $\widehat{s}_\mu^{boot}(t)$ *is the sample standard deviation of* $\widehat{\mu}_B^{boot}(t)$ *computed from the* B *bootstrap estimators* $\widehat{\mu}_B^{boot}(t)$ *at time point* t. □

Strictly speaking, because the bias of $\widehat{\mu}_B(t)$ has not been adjusted, the above approximate confidence intervals (4.10) and (4.11) are not precisely the confidence intervals for $\mu(t)$, because they may not have the nominal coverage probability of $[100 \times (1-\alpha)]\%$ unless it is appropriate to ignore the bias of $\widehat{\mu}_B(t)$. In theory, one may either estimate the bias or make it negligible by selecting a relatively larger K in the computation of $\widehat{\mu}_B(t)$. However, in practice, it is difficult to estimate the bias of a basis approximation estimator. Consequently, we treat $E[\widehat{\mu}_B(t)]$ as the function of interest and the estimable part of $\mu(t)$. This is a reasonable approach since $E[\widehat{\mu}_B(t)]$, as a good approximation of $\mu(t)$, is expected to capture the main feature of $\mu(t)$. A similar argument in the context of kernel smoothing can be found in Hart (1997, Section 3.5). Thus, for practical purposes, the intervals given in (4.10) and (4.11) are viewed as appropriate approximate confidence intervals for $\mu(t)$.

4.2.2 Simultaneous Confidence Bands

The above pointwise confidence intervals can be extended through the same approach described in Section 3.4 to construct simultaneous confidence bands for $\mu(t)$ over a given interval $[a, b]$. Using the Bonferroni adjustment to (4.10) or (4.11), we partition $[a, b]$ into $M+1$ equally spaced grid points $a = \xi_1 < \cdots < \xi_{M+1} = b$ for some integer $M \geq 1$, and construct the approximate $[100 \times (1-\alpha)]\%$ simultaneous confidence intervals

$$\left(L_{\alpha/[2(M+1)]}[\widehat{\mu}_B^{boot}(\xi_r)], \, U_{\alpha/[2(M+1)]}[\widehat{\mu}_B^{boot}(\xi_r)] \right) \tag{4.12}$$

or

$$\left(\widehat{\mu}_B(\xi_r) \pm z_{1-\alpha/[2(M+1)]} \widehat{s}_B^{boot}(\xi_r) \right), \tag{4.13}$$

respectively. Using (4.12) or (4.13), the comparable approximate $[100 \times (1-\alpha)]\%$ for the linear interpolation $\mu^{(l)}(t)$ as defined in (3.25) is $(l_\alpha^{(l)}(t), u_\alpha^{(l)}(t))$, which satisfies (3.26),

$$l_\alpha^{(l)}(t) = M\left(\frac{\xi_{r+1}-t}{b-a}\right) l_\alpha(\xi_r) + M\left(\frac{t-\xi_r}{b-a}\right) l_\alpha(\xi_{r+1}) \tag{4.14}$$

and

$$u_\alpha^{(I)}(t) = M\left(\frac{\xi_{r+1} - t}{b - a}\right) u_\alpha(\xi_r) + M\left(\frac{t - \xi_r}{b - a}\right) u_\alpha(\xi_{r+1}), \qquad (4.15)$$

where $l_\alpha(\xi_r)$ and $u_\alpha(\xi_r)$ are the lower and upper bounds, respectively, given in (4.12) or (4.13).

Taking the view that $E\left[\widehat{\mu}_B(t)\right]$ is the *estimable part* of $\mu(t)$, we assume that the smoothness conditions of (3.29) or (3.30) are equivalent to

$$\sup_{t \in [a,b]} \left|\left\{E\left[\widehat{\mu}_B(t)\right]\right\}'\right| \leq c_1, \qquad \text{for a known constant } c_1 > 0, \qquad (4.16)$$

or

$$\sup_{t \in [a,b]} \left|\left\{E\left[\widehat{\mu}_B(t)\right]\right\}''\right| \leq c_2, \qquad \text{for a known constant } c_2 > 0, \qquad (4.17)$$

respectively. Then, adjusting the simultaneous confidence bands for the linear interpolation of $E\left[\widehat{\mu}_B(t)\right]$, the approximate $[100 \times (1 - \alpha)]\%$ simultaneous confidence bands for $E\left[\widehat{\mu}_B(t)\right]$, hence $\mu(t)$, can be given by

$$\left(l_\alpha^{(I)}(t) - 2c_1 M\left[\frac{(\xi_{r+1} - t)(t - \xi_r)}{b - a}\right], u_\alpha^{(I)}(t) + 2c_1 M\left[\frac{(\xi_{r+1} - t)(t - \xi_r)}{b - a}\right]\right)$$
$$(4.18)$$

or

$$\left(l_\alpha^{(I)}(t) - (c_2/2)(\xi_{r+1} - t)(t - \xi_r), u_\alpha^{(I)}(t) + (c_2/2)(\xi_{r+1} - t)(t - \xi_r)\right), \quad (4.19)$$

when (4.16) or (4.17) holds, where $l_\alpha^{(I)}(t)$ and $u_\alpha^{(I)}(t)$ are given in (4.14) and (4.15), respectively.

The practical issues of improving the simultaneous confidence intervals (4.12) and (4.13) and selecting the number and location of the grid points have been discussed in Section 3.4. We omit this discussion here to avoid redundancy.

4.2.3 Hypothesis Testing

Some practical questions for the evaluation of $\mu(t) = E[Y(t)|t]$ are whether $E[Y(t)|t]$ is time-varying or belongs to a pre-specified sub-model. These questions can be evaluated under the current framework of *extended linear models* by a class of goodness-of-fit tests.

1. Testing Time-Varying based on Residual Sum of Squares

This test is constructed by comparing the weighted residual sum of squares from weighted least squares fits under the following null and alternative hypotheses,

$$\begin{cases} H_0: & \mu(t) = \gamma_1 \text{ for all } t \in \mathscr{T} \text{ and some unknown constant } \gamma_1; \\ H_1: & \mu(t) \text{ is time-varying.} \end{cases} \qquad (4.20)$$

Under the null hypothesis H_0, we can estimate $\mu(t)$ by $\widehat{\gamma}_1$ which minimizes $\ell_w(\gamma)$ of (4.3) with $B_1(t) = 1$ and $B_2(t) = \cdots = B_K(t) = 0$. Then, the weighted residual sum of squares under H_0 is

$$RSS_0(\widehat{\gamma}_1) = \sum_{i=1}^{n} \sum_{j=1}^{n_i} w_i \left(Y_{ij} - \widehat{\gamma}_1 \right)^2. \tag{4.21}$$

Under the alternative H_1, $\mu(t)$ is estimated by the basis approximation estimator $\widehat{\mu}_B(t)$ given in (4.4), so that the corresponding weighted residual sum of squares is

$$RSS_1(\widehat{\mu}_B) = \sum_{i=1}^{n} \sum_{j=1}^{n_i} w_i \left[Y_{ij} - \sum_{k=1}^{K} \widehat{\gamma}_k B_k(t_{ij}) \right]^2. \tag{4.22}$$

The difference between $RSS_0(\widehat{\gamma}_1)$ and $RSS_1(\widehat{\mu}_B)$ can be used to test whether there is sufficient evidence to accept or reject the null hypothesis H_0 in (4.20). If the null hypothesis H_0 holds, we expect that $RSS_0(\widehat{\gamma}_1)$ and $RSS_1(\widehat{\mu}_B)$ are close to each other. On the other hand, if H_1 holds, we expect that $RSS_0(\widehat{\gamma}_1)$ and $RSS_1(\widehat{\mu}_B)$ are apart from each other. A natural goodness-of-fit test statistic for (4.20) is

$$T_n(\widehat{\gamma}_1, \widehat{\mu}_B) = \frac{RSS_0(\widehat{\gamma}_1) - RSS_1(\widehat{\mu}_B)}{RSS_1(\widehat{\mu}_B)}. \tag{4.23}$$

The null hypothesis H_0 is rejected if the value of $T_n(\widehat{\gamma}_1, \widehat{\mu}_B)$ is larger than an appropriate critical value.

Theoretical justification of using $T_n(\widehat{\gamma}_1, \widehat{\mu}_B)$ is provided in Theorem 4.3 of Section 4.4. This theorem indicates that, under some mild regularity conditions which are satisfied in most practical situations, if the null hypothesis H_0 of (4.20) holds, $T_n(\widehat{\gamma}_1, \widehat{\mu}_B)$ converges to zero in probability as n tends to infinity. On the other hand, if the alternative H_1 of (4.20) holds, then $T_n(\widehat{\gamma}_1, \widehat{\mu}_B)$ is larger than a constant for sufficiently large n.

2. Resampling-Subject Bootstrap Critical Values

The theoretical justification of the test statistic in (4.23) motivates the use of a resampling-subject bootstrap test procedure that rejects H_0 when $T_n(\widehat{\gamma}_1, \widehat{\mu}_B)$ is larger than an appropriate critical value. This critical value can be computed based on the following resampling-subject bootstrap procedure under the null hypothesis H_0 of (4.20). Let

$$\widehat{\varepsilon}_{ij} = Y_{ij} - \sum_{k=1}^{K} \widehat{\gamma}_k B_k(t_{ij}) \tag{4.24}$$

be the residuals of (4.4). Based on $\{\widehat{\varepsilon}_{ij} : i = 1,\ldots,n; j = 1,\ldots,n_i\}$, we define

$$\{Y_{ij}^p = \widehat{\gamma}_1 + \widehat{\varepsilon}_{ij} : i = 1,\ldots,n; j = 1,\ldots,n_i\}, \tag{4.25}$$

to be a set of pseudo-responses under the null hypothesis H_0 of (4.20).

The following resampling-subject bootstrap procedure can be used to evaluate the distribution of $T_n(\gamma_1, \widehat{\mu}_B)$ under the null hypothesis H_0 and compute the level-α rejection region and the p-values of the test statistic $T_n(\widehat{\gamma}_1, \widehat{\mu}_B)$ for (4.20).

Resampling-Subject Bootstrap Testing Procedure:

(a) *Resample n subjects with replacement from $\{(Y_{ij}^p, t_{ij}) : i = 1, \ldots, n; j = 1, \ldots, n_i\}$ to obtain the bootstrap sample $\{(Y_{ij}^{p*}, t_{ij}^*) : i = 1, \ldots, n; j = 1, \ldots, n_i^*\}$.*

(b) *Repeat the above sampling procedure B times, so that B independent resampling-subject bootstrap samples are obtained.*

(c) *From each bootstrap sample, calculate the test statistic $T_n^*(\widehat{\gamma}_1^{boot}, \widehat{\mu}_B^{boot})$ using the method (4.23) and compute the empirical distribution of $T_n^*(\widehat{\gamma}_1^{boot}, \widehat{\mu}_B^{boot})$ based on the B independent bootstrap samples.*

(d) *Reject the null hypothesis H_0 at the significance level α when the observed test statistic $T_n(\widehat{\gamma}_1, \widehat{\mu}_B)$ is greater than or equal to the $[100 \times (1 - \alpha)]$th percentile of the empirical distribution of $T_n^*(\widehat{\gamma}_1^{boot}, \widehat{\mu}_B^{boot})$. The p-value of the test is the empirical probability of "$T_n^*(\widehat{\gamma}_1^{boot}, \widehat{\mu}_B^{boot}) \geq T_n(\widehat{\gamma}_1, \widehat{\mu}_B)$".* $\qquad\square$

3. Testing Sub-models based on Residual Sum of Squares

The above residual sum of squares testing procedure can be modified in a straightforward way to test other null hypotheses and alternatives. For example, a set of hypotheses, which are more general than the ones given in (4.20) and may include testing a linear model for $\mu(t)$, is

$$
\begin{cases}
H_0: & \mu(t) = \sum_{k=1}^{K_0} \gamma_k B_k(t), \\
& \text{for a given } 1 < K_0 < K \text{ and } \gamma_k \neq 0 \text{ for some } 1 \leq k \leq K_0; \\
H_1: & \mu(t) = \sum_{k=1}^{K} \gamma_k B_k(t), \\
& \text{with } \gamma_k \neq 0 \text{ for some } K_0 \leq k \leq K.
\end{cases}
\tag{4.26}
$$

A simple special case of (4.26) is $K \geq 3$, $K_0 = 2$ and $B_k(t) = t^{k-1}$ being a polynomial basis, so that the null hypothesis H_0 is that $\mu(t)$ is a simple linear function of t, i.e., $\mu(t) = \gamma_1 + \gamma_2 t$, and the alternative H_1 is that $\mu(t)$ is a polynomial of t with degree 2 or higher. By extending the residual sum of squares in (4.21) and (4.22) to RSS_0 and RSS_1 under H_0 and H_1, respectively, of (4.26), Theorem 4.3 of Section 4.4 can be adapted easily to the general situation of (4.26). The residual sum of squares test procedure given in (4.23), (4.24), (4.25) and the resampling-subject bootstrap steps (a) to (d) can be analogously adapted to compute test statistic $T_n(\cdot, \cdot)$ and its level-α rejection regions and p-values under (4.26).

We note that, in principle, similar goodness-of-fit test procedures, including the test statistic $T_n(\cdot, \cdot)$ of (4.23) and the resampling-subject bootstrap procedure, can also be developed using other estimation methods, such as the least squares kernel and local polynomial estimators. But, asymptotic properties of the test statistics $T_n(\cdot, \cdot)$ with other smoothing methods have not been systematically investigated in the literature. Given that the focus of this book is mainly on the nonparametric estimation methods, our coverage of hypothesis testing for nonparametric models is limited to basis approximation methods only.

4.3   R Implementation

4.3.1   The HSCT Data

The HSCT data has been described in Sections 1.2 and 3.5. As an alternative to the least squares kernel and local polynomial smoothing methods described in Chapter 3, we illustrate here how to apply the basis approximation method to estimate the nonparametric time-trend of the granulocyte recovery in patients undergone hematopoietic stem cell transplantation. To obtain a smoothing estimate of the time-trend, different basis systems may be used. We use B-splines (i.e., polynomial splines) here as an example to show that the smoothness of the fitted curve depends on the choices of the degree of the polynomial and the number of knots for the splines.

In R, the function bs(x, knots = , degree =) generates the B-spline basis matrix for a polynomial spline. The B-splines are computationally more efficient compared to the truncated polynomial splines. Different degrees for the polynomial splines may be used, such as the linear (degree = 1), quadratic (degree = 2) or cubic B-splines (degree = 3). The cubic B-spline basis is the default choice in bs(). Since the cubic B-splines are continuous and have continuous first and second derivatives at the knots, we usually do not need to use a spline with degrees higher than three to obtain a continuous smoothing estimator. With sufficient numbers of knots, the cubic B-splines can approximate most functions arbitrarily well. Typical choices of knots within the data range (or the internal breakpoints) are to use mean/median for one knot, and quantiles for multiple knots. As discussed in Section 4.1.2, LSCV is used to select the number of equally spaced knots. Moreover, if linear constraints at the boundaries are required, the natural cubic splines may be used, for which ns(x, knots=) in R generates its B-spline basis matrix.

We use the following commands to fit a cubic spline with two fixed knots by the least squares method described in Section 4.1. The spline fits with different degrees or knots may be specified similarly. By default the bs() function does not include the intercept in the basis matrix because an intercept term is automatically included in the model formula for most of the regression functions in R. The R functions are

```
> library(splines)
> data(HSCT)
```

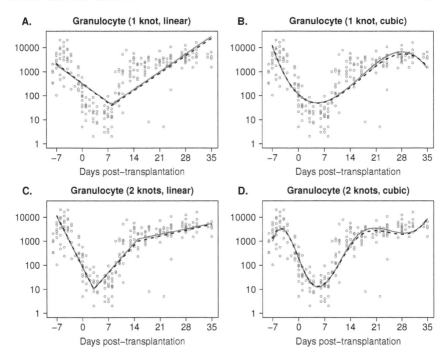

Figure 4.1 *The scatter plots and smoothing estimates of the granulocyte counts vs. time since stem cell transplantation. (A) 1 interior knot, linear; (B) 1 interior knot, cubic; (C) 2 interior knots, linear; and (D) 2 interior knots, cubic polynomial splines are used. Knots are chosen at median (day 8) for 1 knot and 33% and 67% percentiles (days 3 and 15) for 2 knots. The solid and dashed lines represent the spline estimates using the measurement uniform weight and the subject uniform weight, respectively.*

```
> attach(HSCT)
> Granu.log <- log10(Granu)
> KN2 <- quantile(Days, c(.33, .66))
> bs.Days <- bs(Days, knots=KN2, degree=3)

# Obtain coefficients for the spline basis, subject uniform
  weight
> Spline.fit <- lm(Granu.log ~ bs.Days, weights=1/ni)

# Obtain fitted estimates for a given x
> New.Days <- bs(-7:35, knots=KN2, degree=3)
> Spline.Est <- cbind(1, New.Days) %*% coef(Spline.fit)
```

Figure 4.1 displays the estimated smoothing curves using the linear or cubic B-splines (polynomial splines) with one knot (at median, day 8) or two knots (at 33% and 67% percentiles, days 3 and 15), respectively. In each plot, the smoothing curves based on the measurement uniform weight and the subject uniform weight are very similar. It is easy to visualize that the cubic B-spline fit with 2 knots in Figure 4.1(D) captures the entire nonlinear time-trend more adequately, compared to the linear spline fit and the smoothing estimates with only one knot in Figure 4.1(A)-(C). The granulocytes of patients are shown to decrease to the lowest levels following the conditioning regimen and gradually recover back to the pre-transplant level because of the engraftment of the donor stem cells and hematopoietic reconstitution post-transplantation.

4.3.2 The BMACS CD4 Data

The BMACS CD4 data has been described in Sections 1.2, 2.4 and 3.5. Using this dataset, we illustrate here how to select the number of knots with the LSCV procedure of Section 4.1.2 and obtain the pointwise and simultaneous confidence intervals based on the resampling-subject bootstrap procedure of Section 4.2.

Figure 4.2 shows the CD4 cell percentages at the study visits since HIV-infection for the 283 HIV infected men in the dataset. We apply spline fit to the CD4 data using the cubic B-splines with 1 interior knot (at 3 years) and 5 interior knots (at 1, 2, 3, 4 and 5 years), respectively. Both fits are based on equally spaced knots.

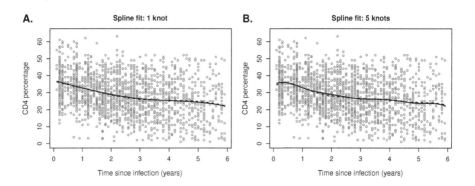

Figure 4.2 *Scatter plots and B-spline fitted curves of CD4 percentage vs. time since infection. The equally spaced knots for the cubic spline estimates are used with (A) 1 knot (at 3 years) and (B) 5 knots (at 1, 2, 3, 4 and 5 years). The solid and dashed lines represent the spline estimates using the measurement uniform weight and the subject uniform weight, respectively.*

The following `spline.fit()` functions are used to generate the fitted values with a given number of equally spaced knots, degree, and weight:

```
> library(npmlda)
```

```
> attach(BMACS)
> newX  <- seq(min(Time), max(Time),  by=0.1)
> fit5  <- spline.fit(newX, Time, CD4, nKnots=5, Degree=3)
> fit5W <- spline.fit(newX, Time, CD4, nKnots=5, Degree=3,
                      Wt=1/ni)
```

Figure 4.3(A) shows the fitted cubic spline curves with different numbers of knots ranging from 1 to 20 with measurement uniform weight. By the LSCV method in Section 4.1.2, the spline fit with 5 interior knots has the smallest LSCV score as shown in Figure 4.3(B). It also suggests that the complexity and curvature of fits are considerably increased with large numbers of knots, compared to much smoother estimates with few numbers of knots. We will discuss how to use "smoothing splines" in the next chapter to avoid knot selection and to penalize the roughness and curvature of the fitted curves.

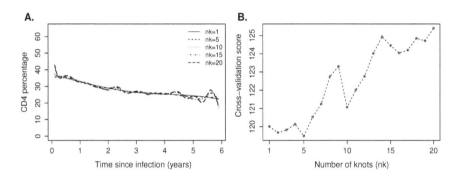

Figure 4.3 *B-spline smoothing fits and LSCV score of the CD4 percentage data. (A) Cubic spline fits of CD4 percentage vs. time since infection in years with various number of equally spaced knots. (B) LSCV scores vs. number of interior knots for selecting the number of knots.*

Figure 4.4 displays the bootstrap inference described in Section 4.2. Similar R code in Section 3.5 can be applied to generate the resampling-subject bootstrap samples with the smoothing method replaced by the cubic B-spline fit with five equally spaced knots as suggested by the LSCV procedure in each of the 1000 bootstrap samples. Both the 95% pointwise confidence intervals in (4.11) and the 95% simultaneous confidence band in (4.14) and (4.15) for the linear interpolation $\mu^{(l)}(t)$ with the Bonferroni adjustment are shown. The results based on (4.18) and (4.19) are similar (data not shown). For the linear interpolation simultaneous confidence band, $M = 59$ is used to cover all the distinct design time points from 0.1 to 5.9. Despite the conservativeness of the Bonferroni adjustment, Figure 4.4(B) still shows a clear indication that the mean CD4 percentage generally declines over time since HIV infection.

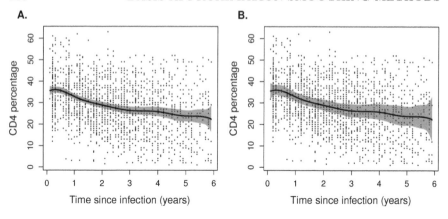

Figure 4.4 *CD4 percentage data and the cubic spline fit with 5 equally spaced knots, with (A) 95% pointwise confidence interval and (B) 95% simultaneous confidence band for the linear interpolation at 59 equally spaced time points.*

4.4   Asymptotic Properties

We derive in this section the asymptotic properties, including consistency and convergence rates, for the general basis approximation estimators $\widehat{\mu}_B(t)$ of (4.4) as well as the special case of B-spline estimators. Unlike the least squares kernel and local polynomial estimators of Chapter 3, we do not have the explicit expressions for the asymptotic biases of the basis approximation estimators $\widehat{\mu}_B(t)$, because the biases depend on the specific assumptions of the functional spaces containing $\mu(t)$. However, for the special case of B-splines, the upper bounds of the asymptotic biases and the explicit expressions for the asymptotic variances of $\widehat{\mu}_B(t)$ can be established. These asymptotic properties are used to establish the consistency of the test statistics in Section 4.4.5.

4.4.1   Conditional Biases and Variances

Let $\mathbf{t} = \left(t_1^T, t_2^T, \ldots, t_n^T\right)^T$ be the $(N \times 1)$ vector of all the observed time points. Conditioning on $\mathbf{t}$, it directly follows from the expression of $\widehat{\gamma}$ in (4.6) that the expectation of $\widehat{\gamma}$, denoted by $\widetilde{\gamma}$ is

$$\widetilde{\gamma} = E\left(\widehat{\gamma}\,\middle|\,\mathbf{t}\right) = \left\{\sum_{i=1}^{n}\left[B(t_i)^T W_i B(t_i)\right]\right\}^{-1}\left\{\sum_{i=1}^{n}\left[B(t_i)^T W_i E(Y_i|\mathbf{t})\right]\right\}, \qquad (4.27)$$

where $E\left(Y_i|\mathbf{t}\right) = \left(\mu\left(t_{i1}\right), \ldots, \mu\left(t_{in_i}\right)\right)^T$. Using the expression of $\widehat{\mu}_B(t)$ in (4.4), the bias of $\widehat{\mu}_B(t)$ given $\mathbf{t}$ is

$$B\left[\widehat{\mu}_B(t)\,\middle|\,\mathbf{t}\right] = E\left[\widehat{\mu}_B(t) - \mu(t)\,\middle|\,\mathbf{t}\right] = B(t)^T E\left(\widehat{\gamma}\,\middle|\,\mathbf{t}\right) - \mu(t). \qquad (4.28)$$

If $\mu(t)$ belongs to the linear space spanned by $\{B_1(t), \ldots, B_K(t)\}$ for a given K, we have that $\mu(t)$ belongs to the linear model $\mu(t) = B(t)^T \gamma$ for some unknown parameter vector γ, so that $E(\widehat{\gamma}|\mathbf{t}) = \gamma$ and the conditional bias is zero, i.e., $E\left[\widehat{\mu}_B(t) - \mu(t)|\mathbf{t}\right] = 0$. In general, when $\mu(t)$ does not necessarily belong to a linear model, the conditional bias $E\left[\widehat{\mu}_B(t) - \mu(t)|\mathbf{t}\right]$ does not have to be zero. But, by choosing a sufficiently large K when n is sufficiently large, the right side of (4.1) can be a good approximation of $\mu(t)$, so that asymptotically the conditional bias $B\left[\widehat{\mu}_B(t)|\mathbf{t}\right]$ tends to zero as n tends to infinity.

Let $C_\varepsilon(t_{ij}, t_{ij'}) = Cov\left[\varepsilon_i(t_{ij}), \varepsilon_i(t_{ij'})\right]$ be the covariance between $\varepsilon_i(t_{ij})$ and $\varepsilon_i(t_{ij'})$ for the ith subject at time points $(t_{ij}, t_{ij'})$. The variance-covariance matrix of Y_i given $\mathbf{t}$ is

$$V_i = Cov(Y_i|\mathbf{t}) = \begin{pmatrix} C_\varepsilon(t_{i1}, t_{i1}) & \cdots & C_\varepsilon(t_{i1}, t_{in_i}) \\ \vdots & \vdots & \vdots \\ C_\varepsilon(t_{in_i}, t_{i1}) & \cdots & C_\varepsilon(t_{in_i}, t_{in_i}) \end{pmatrix}. \tag{4.29}$$

Direct calculation from the expression of $\widehat{\gamma}$ in (4.6) then shows that the variance-covariance matrix of $\widehat{\gamma}$ given $\mathbf{t}$ is

$$\begin{aligned} Cov(\widehat{\gamma}|\mathbf{t}) &= \left\{\sum_{i=1}^{n} \left[B(t_i)^T W_i B(t_i)\right]\right\}^{-1} \left\{\sum_{i=1}^{n} \left[B(t_i)^T W_i V_i W_i B(t_i)\right]\right\} \\ &\times \left\{\sum_{i=1}^{n} \left[B(t_i)^T W_i B(t_i)\right]\right\}^{-1}. \end{aligned} \tag{4.30}$$

Substituting $Cov(\widehat{\gamma}|\mathbf{t})$ into (4.4), the variance of $\widehat{\mu}_B(t)$ is

$$V\left[\widehat{\mu}_B(t)|\mathbf{t}\right] = B(t)^T Cov(\widehat{\gamma}|\mathbf{t}) B(t). \tag{4.31}$$

When $\mu(t)$ belongs to the linear model spanned by $\{B_1(t), \ldots, B_K(t)\}$ and the $\varepsilon_i(t_{ij})$ are from mean zero Gaussian process, the conditional bias of the estimator $\widehat{\mu}_B(t)$ is zero and the conditional variance (4.31) can be used to construct statistical inferences, such as confidence intervals and hypothesis tests, based on $\widehat{\mu}_B(t)$. In such cases, the intra-subject covariances $C_\varepsilon(t_{ij}, t_{ij'})$ are unknown and have to be estimated in practice. When $\mu(t)$ belongs to a linear model but the distribution of $\varepsilon_i(t)$ is unknown, statistical inferences for $\mu(t)$ can be constructed based on asymptotically approximate distributions of $\widehat{\mu}_B(t)$, provided that the intra-subject covariances $C_\varepsilon(t_{ij}, t_{ij'})$ can be estimated. For the general case that $\mu(t)$ does not necessarily belong to a known linear model, statistical properties and inferences can be investigated through the asymptotic properties of $\widehat{\mu}_B(t)$ when n is sufficiently large.

4.4.2   Consistency of Basis Approximation Estimators

We establish here the consistency and convergence rates of $\widehat{\mu}_B(t)$ for any basis systems which may be used to approximate $\mu(t)$. More specific asymptotic representations, such as asymptotic expressions of biases, variances and mean squared errors, may be established when specific basis systems are given. In particular, the asymptotic properties for the B-spline, i.e., polynomial splines, estimators are discussed in Section 4.4.3. But, asymptotic properties similar to the ones developed for B-spline have not been systematically investigated in the literature, which warrants further research.

1. Asymptotic Assumptions

The following technical assumptions are made throughout this section.

(a) *The observation time points follow a random design in the sense that* $\{t_{ij} : j = 1, \ldots, n_i; i = 1, \ldots, n\}$ *are chosen independently from an unknown distribution function* $F(\cdot)$ *with a density function* $f(\cdot)$ *on the finite interval* $\mathscr{T}$. *The density function* $f(t)$ *is uniformly bounded away from 0 and infinity, i.e., there are constants* $M_1 > 0$ *and* $M_2 > 0$ *such that* $M_1 \leq f(t) \leq M_2$ *for all* $t \in \mathscr{T}$.

(b) *There is a positive constant* M_3 *such that* $E\left[\varepsilon(t)^2\right] \leq M_3$ *for all* $t \in \mathscr{T}$. $\square$

These assumptions correspond to the conditions given in Huang, Wu and Zhou (2002, Section 3.2, *Assumptions 1* and *2*), in which the basis approximation method is established for the more general *time-varying coefficient models* to be discussed in Chapters 6 to 9. It is reasonable to expect that Assumptions (a) and (b) are easily satisfied in most real applications.

2. Distance Measure and Definition of Consistency

We first introduce a distance measure to assess the performance of a smoothing estimator. Let

$$\left\| a(\cdot) \right\|_{L_2} = \left\{ \int_{\mathscr{T}} a^2(t)\, dt \right\}^{1/2}$$

be the L_2-norm of any square integrable real-valued function $a(t)$ on $\mathscr{T}$. We can then define the integrated squared error (ISE) of $\widehat{\mu}_B(t)$ on $t \in \mathscr{T}$ by

$$ISE\left(\widehat{\mu}_B\right) = \left\| \widehat{\mu}_B(\cdot) - \mu(\cdot) \right\|_{L_2}^2 = \int_{\mathscr{T}} \left[\widehat{\mu}_B(t) - \mu(t)\right]^2 dt. \tag{4.32}$$

A basis approximation estimator $\widehat{\mu}_B(\cdot)$ is defined to be a *consistent* smoothing estimator for $\mu(\cdot)$ if, as $n \to \infty$,

$$ISE\left(\widehat{\mu}_B\right) \to 0 \quad \text{in probability.}$$

Since $\mu(t)$ is approximated by functions in a linear space spanned by $\{B_1(t), B_2(t), \ldots\}$, the asymptotic derivations of $ISE\left(\widehat{\mu}_B\right)$ depend on the L_∞-norm between $\mu(t)$ and the chosen linear space. Let $\mathscr{G}$ be the linear space spanned by $\{B_1(t), \ldots, B_K(t)\}$, i.e.,

$$\mathscr{G} = Span\{B_1(t), \ldots, B_K(t)\}.$$

We define the L_∞-norm between $\mu(\cdot)$ and $\mathscr{G}$ to be

$$D(\mu, \mathscr{G}) = \inf_{g \in \mathscr{G}} \sup_{t \in \mathscr{T}} |\mu(t) - g(t)|. \tag{4.33}$$

The asymptotic properties of $ISE(\widehat{\mu}_B)$ depend on the quantity

$$A_n = \sup_{g \in \mathscr{G}, \|g(\cdot)\|_{L_2} \neq 0} \left[\frac{\sup_{t \in \mathscr{T}} |g(t)|}{\|g(\cdot)\|_{L_2}} \right]. \tag{4.34}$$

Examples of A_n for some commonly used basis systems, such as polynomials, splines and trigonometric bases, can be found in Huang (1998).

3. Consistency and Convergence Rates

Using the above distance measures and the quantity A_n in (4.34), the next theorem shows the consistency and the convergence rates of $\widehat{\mu}_B(\cdot)$, which is approximated by the basis functions $\{B_1(t), \ldots, B_K(t)\}$ where K may or may not tend to infinity as n tends to infinity.

Theorem 4.1. *If Assumptions (a) and (b) are satisfied,* $\lim_{n \to \infty} D(\mu, \mathscr{G}) = 0$ *and*

$$\lim_{n \to \infty} \left\{ A_n^2 K \max \left[\max_{1 \leq i \leq n} (n_i w_i), \sum_{i=1}^{n} n_i^2 w_i^2 \right] \right\} = 0, \tag{4.35}$$

then $\widehat{\mu}_B(\cdot)$ *uniquely exists with probability tending to one and is a consistent estimator of* $\mu(\cdot)$. *In addition, with* $a_n = O_p(b_n)$ *denoting the fact that* a_n/b_n *converging to a non-zero constant in probability as* $b_n \to \infty$, *the following convergence conclusions hold:*

(a) $\left\| \widehat{\mu}_B(\cdot) - E\left[\widehat{\mu}_B(\cdot)|\mathbf{t}\right] \right\|_{L_2}^2 = O_p(K\sum_{i=1}^{n} n_i^2 w_i^2)$;

(b) $\left\| E\left[\widehat{\mu}_B(\cdot)|\mathbf{t}\right] - \mu(\cdot) \right\|_{L_2} = O_p[D(\mu, \mathscr{G})]$;

(c) $ISE(\widehat{\mu}_B) = O_p[K \sum_{i=1}^{n} (n_i^2 w_i^2) + D^2(\mu, \mathscr{G})]$.   ∎

Since the above theorem gives the consistency of $\widehat{\mu}_B(\cdot)$ for general basis systems, including polynomials, splines and trigonometric bases, the convergence rates established in Theorem 4.1(a) to (c) may be improved when a particular type of basis is used.

Proof of Theorem 4.1:

We assume, without loss of generality, that $\{B_k(t) : k = 1, \ldots, K\}$ is an orthonormal basis for the linear space $\mathscr{G}$ with inner product $\langle f_1(\cdot), f_2(\cdot) \rangle = \int_{\mathscr{T}} f_1(t) f_2(t) \, dt$. Then, for any $g \in \mathscr{G}$, there is an unique representation $g(t) = \sum_{k=1}^{K} \gamma_k B_k(t)$, so that the L_2-norm of $g(t)$ is $\|g(\cdot)\|_{L_2} = \left(\sum_{k=1}^{K} \gamma_k^2\right)^{1/2}$. Following the notation of Huang (1998, p. 246), we write $a_n \asymp b_n$ if both a_n and b_n are positive and a_n/b_n and b_n/a_n are bounded for all n.

Let T be the random variable of time with distribution $F(\cdot)$ and density $f(\cdot)$. The proof is then derived from the following three technical lemmas.

Lemma 4.1. *If the assumption (4.35) is satisfied, then*

$$\sup_{g \in \mathcal{G}} \left| \frac{\sum_{i=1}^{n} \sum_{j=1}^{n_i} w_i [g(t_{ij})]^2}{E[g(T)]^2} - 1 \right| = o_p(1), \tag{4.36}$$

where $o_p(1)$ denotes converging to zero in probability. ■

Proof of Lemma 4.1:
The lemma can be proved using arguments similar to those in the proof of Lemma 10 of Huang (1998). Details are omitted. ■

Lemma 4.2. *Suppose that the assumption (4.35) holds and $B(\mathbf{t}) = \left(B^T(t_1), \cdots, B^T(t_n)\right)^T$ is the $(N \times K)$, $N = \sum_{i=1}^{n} n_i$, matrix with its $(n_i \times K)$ component matrix $B(t_i) = \left(B(t_{i1}), \ldots, B(t_{n_i})\right)^T$ defined in (4.5). There is an interval $[M_1^*, M_2^*]$ with $0 < M_1^* < M_2^*$ such that, as $n \to \infty$,*

$$P\left\{\text{all the eigenvalues of } \left[B^T(\mathbf{t}) W B(\mathbf{t})\right] \text{ are in } [M_1^*, M_2^*]\right\} \to 1, \tag{4.37}$$

where W is the block diagonal matrix with diagonal blocks $W_1, \ldots, W_n$ and $W_i = \operatorname{diag}(w_i, \ldots, w_i)$. Then, with probability tending to 1,

$$B^T(\mathbf{t}) W B(\mathbf{t}) = \sum_{i=1}^{n} \left[B^T(t_i) W_i B(t_i)\right]$$

is invertible and $\widehat{\mu}_B(\cdot)$ exists uniquely. ■

Proof of Lemma 4.2:
By Lemma 4.1, the following equations hold with probability tending to one as $n \to \infty$:

$$\gamma^T B^T(\mathbf{t}) W B(\mathbf{t}) \gamma = \sum_{i=1}^{n} \sum_{j=1}^{n_i} \left[w_i g(t_{ij})\right]^2 \asymp E[g'(T)],$$

where $g(t) = \sum_{k=1}^{K} \gamma_k B_k(t)$ and $\gamma = (\gamma_1, \ldots, \gamma_K)^T$. Using conditional expectations and Assumptions (a) and (b), we observe that

$$E[g^2(T)] = \int_{\mathcal{J}} g^2(t) f_T(t) \, dt \asymp \int_{\mathcal{J}} g^2(t) \, dt = \|g(\cdot)\|_{L_2}^2$$

holds uniformly for all $g(\cdot) \in \mathcal{G}$. Thus, with probability tending to one as $n \to \infty$, we have

$$\gamma^T B^T(\mathbf{t}) W B(\mathbf{t}) \gamma \asymp \gamma^T \gamma$$

holds uniformly for all γ, so that the conclusion (4.37) follows. Consequently, $B^T(\mathbf{t})\,W\,B(\mathbf{t}) = \sum_{i=1}^{n}\left[B^T(t_i)\,W_i\,B(t_i)\right]$ is invertible and $\widehat{\mu}_B(\cdot)$ uniquely exists. ∎

Lemma 4.3. *If the assumption (4.35) holds, then*

$$\left\|\widehat{\mu}_B(\cdot) - E\left[\widehat{\mu}_B(\cdot)\big|\mathbf{t}\right]\right\|_{L_2}^2 = O_p\left(K\sum_{i=1}^{n}\left[n_i^2\,w_i^2\right]\right) \tag{4.38}$$

and

$$\left\|E\left[\widehat{\mu}_B(\cdot)\big|\mathbf{t}\right] - \mu(\cdot)\right\|_{L_2} = O_p\left[D(\mu,\mathscr{G})\right], \tag{4.39}$$

which give the convergence rates of the above two terms. ∎

Proof of Lemma 4.3:
Since $\{B_k(t): k=1,2,\ldots,\}$ are assumed to be orthonormal, it follows from direct calculations that

$$\left\|\widehat{\mu}_B(\cdot) - E\left[\widehat{\mu}_B(\cdot)\big|\mathbf{t}\right]\right\|_{L_2}^2 = \sum_{k=1}^{K}\left|\widehat{\gamma}_k - E\left[\widehat{\gamma}_k\big|\mathbf{t}\right]\right|^2, \tag{4.40}$$

and

$$\begin{aligned}
\widehat{\gamma} - E\left[\widehat{\gamma}\big|\mathbf{t}\right] &= \left[\sum_{i=1}^{n}B^T(t_i)\,W_i\,B(t_i)\right]^{-1}\left[\sum_{i=1}^{n}B^T(t_i)\,W_i\,\varepsilon_i\right]\\
&= \left[B^T(\mathbf{t})\,W\,B(\mathbf{t})\right]^{-1}B^T(\mathbf{t})\,W\,\varepsilon.
\end{aligned} \tag{4.41}$$

By Lemma 4.2, we have that, with sufficiently large n and probability tending to 1,

$$\left|\left[B^T(\mathbf{t})\,W\,B(\mathbf{t})\right]^{-1}B^T(\mathbf{t})\,W\,\varepsilon\right|^2 \asymp \varepsilon^T\,W\,B(\mathbf{t})\,B^T(\mathbf{t})\,W\,\varepsilon. \tag{4.42}$$

Using the Cauchy–Schwarz inequality and Assumptions (a) and (b), we have that

$$E\left[\left|B^T(t_i)\,W_i\,\varepsilon_i\right|^2\right] = E\left\{\sum_{k=1}^{K}w_i^2\left[\sum_{j=1}^{n_i}B_k(t_{ij})\,\varepsilon_{ij}\right]^2\right\} = O\left(K\,n_i^2\,w_i^2\right). \tag{4.43}$$

Consequently, it follows from (4.42) and (4.43) that

$$\begin{aligned}
E\left[\varepsilon^T\,W\,B(\mathbf{t})\,B^T(\mathbf{t})\,W\,\varepsilon\right] &= \sum_{i=1}^{n}E\left[\varepsilon_i^T\,W_i\,B(t_i)\,B^T(t_i)\,W_i\,\varepsilon_i^T\right]\\
&= O\left(K\sum_{i=1}^{n}\left[n_i^2\,w_i^2\right]\right).
\end{aligned} \tag{4.44}$$

The Markov inequality then implies that

$$\left|\left[B^T(\mathbf{t})\,W\,B(\mathbf{t})\right]^{-1}B^T(\mathbf{t})\,W\,\varepsilon\right|^2 = O_P\left(K\sum_{i=1}^{n}\left[n_i^2\,w_i^2\right]\right). \tag{4.45}$$

The conclusion of (4.38) then follows (4.40) to (4.45).

To prove (4.39), we consider $g^*(\cdot) \in \mathcal{G}$ with $\sup_{t \in \mathcal{T}} |g^*(t) - \mu(t)| = D(\mu, \mathcal{G})$. Since

$$\left| E\left[\hat{\mu}_B(t)\big|\mathbf{t}\right] - \mu(t) \right| \leq \left| E\left[\hat{\mu}_B(t)\big|\mathbf{t}\right] - g^*(t) \right| + \left| g^*(t) - \mu(t) \right|,$$

it suffices to show that

$$\left\| E\left[\hat{\mu}_B(\cdot)\big|\mathbf{t}\right] - g^*(\cdot) \right\|_{L_2} = O_p\left[D(\mu, \mathcal{G})\right].$$

Since $g^*(\cdot) \in \mathcal{G}$, there is a γ^* such that $g^*(t) = B^T(t)\gamma^*$. Note that $E\left[\hat{\mu}_B(t)\big|\mathbf{t}\right] = B^T(t) E\left[\hat{\gamma}\big|\mathbf{t}\right]$. It follows from Lemma 4.2 that

$$\left\| E\left[\hat{\mu}_B(\cdot)\big|\mathbf{t}\right] - g^*(\cdot) \right\|_{L_2}^2 = \sum_{k=1}^{K} \left| E\left[\hat{\gamma}_k\big|\mathbf{t}\right] - \gamma_k^* \right|^2 \tag{4.46}$$

$$\asymp \left\{ E\left[\hat{\gamma}\big|\mathbf{t}\right] - \gamma^* \right\}^T \left[\sum_{i=1}^{n} B^T(t_i) W_i B(t_i) \right] \left\{ E\left[\hat{\gamma}\big|\mathbf{t}\right] - \gamma^* \right\}.$$

Since $\sum_{i=1}^{n} B^T(t_i) W_i \left\{ E(Y_i|\mathbf{t}) - B^T(t_i) E\left[\hat{\gamma}\big|\mathbf{t}\right] \right\} = 0$, we have that

$$\sum_{i=1}^{n} w_i \left| B^T(t_i) E\left[\hat{\gamma}\big|\mathbf{t}\right] - B^T(t_i)\gamma^* \right|^2 \leq \sum_{i=1}^{n} w_i \left| E\left[Y_i\big|\mathbf{t}\right] - B^T(t_i)\gamma^* \right|^2 \tag{4.47}$$

and, by (4.43),

$$\left| \mu(t_{ij}) - B^T(t_{ij})\gamma^* \right| = O\left[D(\mu, \mathcal{G})\right]. \tag{4.48}$$

Thus, it follows from (4.47) and (4.48) that

$$\left\{ E\left[\hat{\gamma}\big|\mathbf{t}\right] - \gamma^* \right\}^T \left[\sum_{i=1}^{n} B^T(t_i) W_i B(t_i) \right] \left\{ E\left[\hat{\gamma}\big|\mathbf{t}\right] - \gamma^* \right\} \leq \sum_{i=1}^{n} \sum_{j=1}^{n_i} w_i D^2(\mu, \mathcal{G})$$

$$= D^2(\mu, \mathcal{G}). \tag{4.49}$$

The assertion of (4.39) then follows from the computations in (4.46) to (4.49). This completes the proof of Lemma 4.3. ∎

The conclusions in Theorem 4.1 are then a direct consequence of Lemma 4.3 and the triangle inequality. ∎

4.4.3   Consistency of B-Spline Estimators

As a direct extension of the polynomial approaches in linear models, the B-spline estimators are a popular choice of basis approximations in biomedical studies. Theoretical justifications for the B-spline estimators deserve special attention in practice. Most applications of the global smoothing methods presented in this book are also based on the B-spline estimators.

The next theorem gives the convergence rates for the B-spline estimators.

In this theorem, we assume that the $\mathscr{G}$ is a space spanned by a set of B-spline basis functions on $\mathscr{T}$ with a fixed degree and the knots have bounded mesh ratio, that is, the ratios of the differences between consecutive knots are bounded away from zero and infinity uniformly in n.

Theorem 4.2. *Suppose that $\widehat{\mu}_B(t)$ is defined in (4.4) with a B-spline basis. If the assumptions of Theorem 4.1 are satisfied, then, the following equalities hold when n tends to infinity:*

(a) $\left\| \widehat{\mu}_B(\cdot) - E\left[\widehat{\mu}_B(\cdot)\big|\mathbf{t}\right] \right\|_{L_2}^2 = O_p\left\{ \sum_{i=1}^n n_i^2 w_i^2 \left[(K/n_i) + 1 \right] \right\};$

(b) $\left\| E\left[\widehat{\mu}_B(\cdot)\big|\mathbf{t}\right] - \mu(\cdot) \right\|_{L_2} = O_p\left[D(\mu, \mathscr{G}) \right];$

(c) $ISE\left(\widehat{\mu}_B\right) = O_p\left\{ \sum_{i=1}^n n_i^2 w_i^2 \left[(K/n_i) + 1 \right] + D^2(\mu, \mathscr{G}) \right\}.$ ∎

Proof of Theorem 4.2:

This theorem can be proved along the same lines as Theorem 4.1, but we need to use the special properties of the B-spline functions. Let

$$B_k(t) = K^{1/2} N_k(t), \ k = 1, \dots, K,$$

where $\{N_k(t) : k = 1, \dots, K\}$ are the B-splines as defined in de Boor (1978, Chapter IX). These B-splines $N_k(t)$, $k = 1, \dots, K$, are non-negative functions satisfying

$$\begin{cases} \sum_{k=1}^K N_k(t) = 1, & \text{for } t \in \mathscr{T}, \\ \int_{\mathscr{T}} N_k(t)\, dt \leq c/K, & \text{for some constant } c, \\ (c_1/K) \sum_{k=1}^K \gamma_k^2 \leq \int_{\mathscr{T}} \left[\sum_{k=1}^K \gamma_k N_k(t) \right]^2 dt \leq (c_2/K) \sum_{k=1}^K \gamma_k^2, \\ \qquad \text{for } \gamma_k \in R \text{ and } k = 1, \dots, K, \end{cases} \tag{4.50}$$

where c_1 and c_2 are positive constants. When the properties of B-splines in (4.50) are used, we get

$$E\left[\left| B(t_i)^T W_i \varepsilon_i \right|^2\right] = E\left\{ \sum_{k=1}^K w_i^2 \left[\sum_{j=1}^{n_i} B_k(t_{ij})\, \varepsilon_{ij} \right]^2 \right\} \leq w_i^2 \left[n_i + \frac{n_i^2 - n_i}{K} \right] K. \tag{4.51}$$

Using (4.51), the rest of the proof is similar to that of Theorem 4.1 and thus is omitted. ∎

4.4.4 Convergence Rates

We observe a few useful implications from Theorems 4.1 and 4.2.

1. Effects of Weight Choices

Different choices of the weight function w_i lead to different convergence rates of the estimators. For the general situation in Theorem 4.1, we have

(a) $\sum_{i=1}^{n} \left(K n_i^2 w_i^2 \right) = K/n$, when $w_i = 1/(n n_i)$;

(b) $\sum_{i=1}^{n} \left(K n_i^2 w_i^2 \right) = K \sum_{i=1}^{n} \left(n_i^2/N^2 \right)$, when $w_i = 1/N$.

As shown in Hoover et al. (1998), $\lim_{n \to \infty} \sum_{i=1}^{n} \left(n_i^2/N^2 \right) = 0$ if and only if $\lim_{n \to \infty} \max_{1 \le i \le n} \left(n_i/N \right) = 0$. Thus, as with the kernel and local polynomial smoothing methods of Chapter 3, the $w_i = 1/N$ weight may lead to inconsistent estimators $\widehat{\mu}_B(\cdot)$. On the other hand, $w_i = 1/(n n_i)$ leads to consistent $\widehat{\mu}_B(\cdot)$ for all choices of n_i.

2. Effects of Smoothness Conditions

When the specific smoothness conditions are given, more precise convergence rates can be derived by determining the size of $D(\mu, \mathscr{G})$, which gives the discrepancy between $\mu(\cdot)$ and the linear space $\mathscr{G}$. For example, when $\mu(t)$ has bounded second derivatives and $\mathscr{G}$ is a space of cubic splines with K interior knots on $\mathscr{T}$, we have $D(\mu, \mathscr{G}) = O(K^{-2})$ (Schumaker, 1981, Theorem 6.27) and, by Theorem 4.1, $ISE(\widehat{\mu}_B) = O_p(K/n + K^{-4})$. For the special choice of $K = O(n^{1/5})$, this reduces to $ISE(\widehat{\mu}_B) = O_p(n^{-4/5})$, which is the optimal convergence rate for nonparametric regression with the cross-sectional i.i.d. data under the same smoothness conditions (e.g., Stone, 1982).

4.4.5   Consistency of Goodness-of-Fit Test

We now show the asymptotic properties of the test statistic $T_n(\widehat{\gamma}_1, \widehat{\mu}_B)$. These asymptotic properties demonstrate that, when n is sufficiently large, the value of $T_n(\widehat{\gamma}_1, \widehat{\mu}_B)$ tends to zero when the null hypothesis H_0 of (4.20) holds, and the value of $T_n(\widehat{\gamma}_1, \widehat{\mu}_B)$ tends to some constant larger than zero when a specific alternative holds. Thus, at least theoretically, $T_n(\widehat{\gamma}_1, \widehat{\mu}_B)$ is an appropriate statistic for testing the null and alternative hypotheses in (4.20).

Theorem 4.3. *Suppose that the conditions of Theorem 4.1 are satisfied,* $\inf_{t \in \mathscr{T}} \sigma^2(t) > 0$, $\sup_{t \in \mathscr{T}} E\left[\varepsilon^4(t) \right] < \infty$, *and*

$$T_n(\widehat{\gamma}_1, \widehat{\mu}_B) = \frac{RSS_0(\widehat{\gamma}_1) - RSS_1(\widehat{\mu}_B)}{RSS_1(\widehat{\mu}_B)}.$$

The following conclusions hold:

(a) *Under H_0 of (4.20), $T_n(\widehat{\gamma}_1, \widehat{\mu}_B) \to 0$ in probability as $n \to \infty$.*

(b) *If, as stated in H_1 of (4.20), $\inf_{c \in R} \| \gamma_k - c \|_{L_2} > 0$ for some $k = 2, \ldots, K$, so that $\mu(t)$ is not a constant on $t \in \mathscr{T}$, then there exists a constant $\delta > 0$ such that, with probability tending to one, $T_n(\widehat{\gamma}_1, \widehat{\mu}_B) > \delta$.* ∎

Proof of Theorem 4.3:

Using direct calculation and the conclusions of Lemma 4.1, it can be shown that, with probability tending to one as $n \to \infty$,

$$
\begin{aligned}
RSS_0(\widehat{\gamma}_1) - RSS_1(\widehat{\mu}_B) &= \sum_{i=1}^{n} \sum_{j=1}^{n_i} w_i \left\{ [Y_{ij} - \widehat{\gamma}_1]^2 - [Y_{ij} - \widehat{\mu}_B]^2 \right\} \\
&= \sum_{i=1}^{n} \sum_{j=1}^{n_i} w_i \left[\widehat{\mu}_B(t_{ij}) - \widehat{\beta}^0(t) \} \right]^2 \\
&\asymp \left\| \widehat{\mu}_B(\cdot) - \widehat{\gamma}_1 \right\|_{L_2}^2,
\end{aligned}
\tag{4.52}
$$

where the second equality holds because the basis functions

$$
\{ B_1(t) = 1, B_2(t), \ldots, B_K(t) \}
$$

are assumed to be orthonormal. Under H_0, $\mu(t) = \gamma_1$ is a constant, so that

$$
\left\| \widehat{\mu}_B(\cdot) - \widehat{\gamma}_1 \right\|_{L_2} \le \left\| \widehat{\mu}_B(\cdot) - \gamma_1 \right\|_{L_2} + \left\| \widehat{\gamma}_1 - \gamma_1 \right\|_{L_2} \to 0,
\tag{4.53}
$$

in probability as $n \to \infty$. So that, by (4.52) and (4.53), we have that, under the null hypothesis H_0,

$$
RSS_0(\widehat{\gamma}_1) - RSS_1(\widehat{\mu}_B) \to 0,
\tag{4.54}
$$

in probability as $n \to \infty$.

On the other hand, because

$$
\left\| \widehat{\mu}_B(\cdot) - \widehat{\gamma}_1 \right\|_{L_2} \ge \left\| \widehat{\gamma}_1 - \mu(\cdot) \right\|_{L_2} - \left\| \widehat{\mu}_B(\cdot) - \mu(\cdot) \right\|_{L_2},
$$

we have that, when $\inf_{c \in R} \| \gamma_k - c \|_{L_2} > 0$ for some $k = 2, \ldots, K$, there is a $\delta^* > $ so that $\sum_{k=1}^{K} \inf_{c \in R} \| \gamma_k - c \|_{L_2} > \delta^*$, and consequently, as $n \to \infty$,

$$
\begin{aligned}
\left\| \widehat{\mu}_B(\cdot) - \widehat{\gamma}_1 \right\|_{L^2} &\ge \left\| \widehat{\gamma}_1 - \gamma_1 \right\|_{L_2} + \sum_{k=2}^{K} \| \gamma_k \|_{L_2} - o_p(1) \\
&\ge \sum_{k=2}^{K} \inf_{c \in R} \| \gamma_k - c \|_{L_2} - o_p(1) \\
&\ge \delta^* - o_p(1).
\end{aligned}
\tag{4.55}
$$

It then follows from (4.52) and (4.55) that, when $\inf_{c \in R} \| \gamma_k - c \|_{L_2} > 0$ for some $k = 2, \ldots, K$, with probability tending to one as $n \to \infty$,

$$
RSS_0(\widehat{\gamma}_1) - RSS_1(\widehat{\mu}_B) > \delta^*.
\tag{4.56}
$$

It remains to show that, with probability tending to one as $n \to \infty$, $RSS_1(\widehat{\mu}_B)$

is bounded away from zero and infinity. By the definition of $RSS_1(\widehat{\mu}_B)$, we have that

$$
RSS_1(\widehat{\mu}_B) = \sum_{i=1}^{n} \sum_{j=1}^{n_i} w_i \Big\{ Y_{ij} - \mu(t_{ij}) + \Big[\mu(t_{ij}) - E\big[\widehat{\mu}_B(t_{ij})\big|\mathbf{t}\big] \Big]
$$
$$
+ \Big[E\big[\widehat{\mu}_B(t_{ij})\big|\mathbf{t}\big] - \widehat{\mu}_B(t_{ij}) \Big] \Big\}^2, \tag{4.57}
$$

and, it follows from the proof of Theorem 4.1 that

$$
\begin{cases}
\sum_{i=1}^{n} \sum_{j=1}^{n_i} w_i \Big\{ \mu(t_{ij}) - E\big[\widehat{\mu}_B(t_{ij})\big|\mathbf{t}\big] \Big\}^2 = o_p(1), \\
\sum_{i=1}^{n} \sum_{j=1}^{n_i} w_i \Big\{ E\big[\widehat{\mu}_B(t_{ij})\big|\mathbf{t}\big] - \widehat{\mu}_B(t_{ij}) \Big\}^2 = o_p(1).
\end{cases} \tag{4.58}
$$

Thus, it suffices to show that, with probability tending to one as $n \to \infty$,

$$
\sum_{i=1}^{n} \sum_{j=1}^{n_i} w_i \big[Y_{ij} - \mu(t_{ij}) \big]^2 = \sum_{i=1}^{n} \sum_{j=1}^{n_i} w_i \varepsilon_i^2(t_{ij})
$$

is bounded away from zero and infinity. By $\sup_{t \in \mathscr{T}} E\big[\varepsilon^4(t)\big] < \infty$, there is a constant $c > 0$ such that, as $n \to \infty$,

$$
Var\Big[\sum_{i=1}^{n} \sum_{j=1}^{n_i} w_i \varepsilon_i^2(t_{ij}) \Big] \le \sum_{i=1}^{n} \Big\{ w_i^2 n_i \sum_{j=1}^{n_i} E\big[\varepsilon_i^4(t_{ij})\big] \Big\} \le \sum_{i=1}^{n} \big[n_i^2 w_i^2 c \big] \to 0. \tag{4.59}
$$

The Chebyshev inequality then implies that, by (4.59), as $n \to \infty$,

$$
\sum_{i=1}^{n} \sum_{j=1}^{n_i} w_i \varepsilon_i^2(t_{ij}) - E\Big[\sum_{i=1}^{n} \sum_{j=1}^{n_i} w_i \varepsilon_i^2(t_{ij}) \Big] \to 0, \qquad \text{in probability.} \tag{4.60}
$$

Since $\sum_{i=1}^{n}(n_i w_i) = 1$ and $E\big[\varepsilon_i^2(t_{ij})\big]$ is bounded away from zero and infinity, the result of Theorem 4.3(a) follows from (4.54) and (4.57) to (4.60), and the result of Theorem 4.3(b), i.e., $T_n(\widehat{\gamma}_1, \widehat{\mu}_B) > \delta$ for some $\delta > 0$ with probability tending to one as $n \to \infty$, follows from (4.56) to (4.60). ∎

4.5   Remarks and Literature Notes

The methods presented in this chapter are direct extensions of the basis approximation estimation and inference methods to the longitudinal data. In principle, the effects of the intra-subject correlations should be captured by the weight choices of w_i of (4.3). In practice, however, the optimal choices of w_i are unknown because the structures of the intra-subject correlations are unknown, so that w_i are often chosen subjectively. Consequently, for the longitudinal data, the statistical properties of the basis approximation estimator $\widehat{\mu}_B(t)$ of (4.4) depend on the choices of the basis functions as well as the choices of w_i. Although both the measurement uniform weight $w_i = 1/N$ and the subject uniform weight $w_i = 1/(nn_i)$ are common subjective weight choices in practice, the subject uniform weight has the attractive property that it leads to consistent estimator $\widehat{\mu}_B(t)$ under all choices of n_i when n tends to infinity.

Compared with the asymptotic properties of the local smoothing estimators of Chapter 3, we see from Theorems 4.1 and 4.2 that the asymptotic bias of a basis approximation estimator does not have an explicit expression in general, because it depends on how well the unknown function $\mu(t)$ is approximated using the *extended linear model* formed by the linear space spanned by the chosen basis functions. By adopting an intuitive connection with the well-known linear models, the basis approximation approach has the advantage of having simple interpretations similar to that of the linear models. Thus, hypothesis testing of a linear model can be naturally interpreted by testing a sub-family within the chosen family of extended linear models.

The methods of Sections 4.1 and 4.2 and the asymptotic derivations of Section 4.4 are adopted from the special case of Huang, Wu and Zhou (2002, 2004) without the inclusion of covariates other than time. These methods are extensions of the results in Stone et al. (1997), Huang (1998, 2001, 2003) to the longitudinal data.

Chapter 5

Penalized Smoothing Spline Methods

We introduce in this chapter the computational aspects and theoretical derivations of the penalized smoothing spline estimators for the mean function $\mu(t) = E[Y(t)|t]$ of (3.1) based on the sample $\{(Y_{ij}, t_{ij}) : i = 1, \ldots, n; j = 1, \ldots, n_i\}$. Extensions of the methods of this chapter to the time-varying coefficient models are presented in Chapter 9. The penalized smoothing spline methods have natural connections with both the local smoothing methods of Chapter 3 and the global basis approximation smoothing methods of Chapter 4. On one hand, through an approximation via the Green's function, the penalized smoothing spline estimators can be approximated by some *equivalent kernel* estimators. On the other hand, since a penalized smoothing spline estimator is obtained through a penalized least squares criterion, it is in fact an estimator based on the natural cubic splines with knots at the observed time points.

5.1 Estimation Procedures

Theory and methods of the penalized smoothing splines with cross-sectional i.i.d. data have been extensively studied in the literature. Summaries of the results with cross-sectional i.i.d. data can be found in Wahba (1975, 1990), Green and Silverman (1994) and Eubank (1999). Extensions of the penalized smoothing splines to the longitudinal data have been investigated by Rice and Silverman (1991), Hoover et al. (1998), Lin and Zhang (1999), and Chiang, Rice and Wu (2001), among others. The theory and methods presented in this chapter are a special case of the longitudinal data extension developed in Hoover et al. (1998) and Chiang, Rice and Wu (2001).

5.1.1 Penalized Least Squares Criteria

Suppose that the support of the design time points is contained in a compact set $[a, b]$ and $\mu(t)$ is twice differentiable for all $t \in [a, b]$. We can obtain a penalized least squares estimator $\widehat{\mu}_\lambda(t; w)$ of $\mu(t)$ by minimizing

$$J_w(\mu; \lambda) = \sum_{i=1}^{n} \sum_{i=1}^{n_i} \left\{ w_i \left[Y_{ij} - \mu(t_{ij}) \right]^2 \right\} + \lambda \int_a^b \left[\mu''(s) \right]^2 ds, \qquad (5.1)$$

where λ is a non-negative smoothing parameter, $\{w_i : i = 1, \ldots, n\}$ are non-negative weights, and $w = (w_1, \ldots, w_n)^T$ is the $(n \times 1)$ vector of the weights. We refer to (5.1) as the score function of the penalized least squares criterion.

5.1.2 Penalized Smoothing Spline Estimator

The minimizer $\widehat{\mu}_\lambda(t; w)$ of (5.1) is a cubic spline and a linear statistic of $\{Y_{ij} : i = 1, \ldots, n; j = 1, \ldots, n_i\}$. To see the linearity of $\widehat{\mu}_\lambda(t; w)$, we define $\mathscr{H}_{[a,b]}$ to be the set of compactly supported functions such that

$$\mathscr{H}_{[a,b]} = \left\{ g(\cdot) : g \text{ and } g' \text{ absolutely continuous on } [a, b], \int_a^b \left[g''(s) \right]^2 ds < \infty \right\}.$$

Setting the Gateaux derivative of $J_w(\mu; \lambda)$ to zero, $\widehat{\mu}(t; w)$ uniquely minimizes (5.1) if and only if it satisfies the normal equation

$$\sum_{i=1}^n \sum_{j=1}^{n_i} \left\{ w_i \left[Y_{ij} - \widehat{\mu}_\lambda \left(t_{ij}; w \right) \right] g(t_{ij}) \right\} = \lambda \int_a^b \widehat{\mu}''(s; w) g''(s) \, ds, \qquad (5.2)$$

for all g in a dense subset of $\mathscr{H}_{[a,b]}$. The same argument as in Wahba (1975) shows that there exists a symmetric function $S_\lambda(t, s)$, such that

$$S_\lambda(t, s) \in \mathscr{H}_{[a,b]}, \text{ when either } t \text{ or } s \text{ is fixed},$$

and $\widehat{\mu}_\lambda(t; w)$ is a natural cubic spline estimator with knots at the observed time points given by

$$\widehat{\mu}_\lambda(t; w) = \sum_{i=1}^n \sum_{j=1}^{n_i} \left[w_i S_\lambda \left(t, t_{ij} \right) Y_{ij} \right], \qquad (5.3)$$

which is referred to as the *penalized smoothing spline estimator*. The right side of (5.3) suggests that $\widehat{\mu}_\lambda(t; w)$ is a linear statistic of $\{Y_{ij} : i = 1, \ldots, n; j = 1, \ldots, n_i\}$ with weight functions $w_i S_\lambda(t, t_{ij})$. The explicit expression of $S_\lambda(t, s)$ is unknown. For the theoretical development of $\widehat{\mu}_\lambda(t; w)$, we approximate $S_\lambda(t, s)$ by an equivalent kernel function whose explicit expression can be derived or approximated.

As in the local and global smoothing estimators of Chapters 3 and 4, usual choices of w_i may include the measurement uniform weight $w_i^{**} = 1/N$ and the subject uniform weight $w_i^* = 1/(nn_i)$. Different choices of w generally lead to different finite sample and asymptotic properties for $\widehat{\mu}_\lambda(t; w)$. Ideally the optimal choice of w may depend on the correlation structures of the data. But, because the correlation structures are often unknown and may be difficult to estimate, we do not have a uniformly optimal choice of w. In practice, $w_i^{**} = 1/N$ and $w_i^* = 1/(nn_i)$ generally give satisfactory estimators.

5.1.3   Cross-Validation Smoothing Parameters

The smoothing parameter λ in (5.1) and (5.2), which determines the amount of roughness penalty, is the crucial term affecting the appropriateness of $\widehat{\mu}_\lambda(t; w)$. Adequate smoothing parameters for $\widehat{\mu}_\lambda(t; w)$ may depend on the structures of the possible intra-correlations. However, because the correlation structures are unknown as is often the case in practice, we naturally return to the useful approach of leave-one-subject-out cross-validation (LSCV) established in Chapters 3 and 4. This approach is carried out with the following steps.

Leave-One-Subject-Out Cross-Validation:

(a) *Compute the leave-one-subject-out smoothing spline estimator $\widehat{\mu}_\lambda^{(-i)}(t; w)$ from (5.3) using the remaining data with all the observations of the ith subject deleted. The ith subject's predicted outcome at time t_{ij} is $\widehat{\mu}_\lambda^{(-i)}(t_{ij}; w)$.*

(b) *Define the LSCV score of $\widehat{\mu}_\lambda(t; w)$ by*

$$LSCV(\lambda; w) = \sum_{i=1}^{n} \sum_{j=1}^{n_i} w_i \left[Y_{ij} - \widehat{\mu}_\lambda^{(-i)}(t_{ij}; w) \right]^2. \qquad (5.4)$$

If (5.4) can be uniquely minimized by λ_{LSCV} over all the positive values of $\lambda > 0$, the cross-validated smoothing parameter λ_{LSCV} is then the minimizer of $LSCV(\lambda; w)$. $\qquad\square$

Theoretical properties of λ_{LSCV} have not been systematically established. For a heuristic justification, it can be shown by the same arguments as Sections 3.3 and 4.1 that λ_{LSCV} approximately minimizes an average prediction error of $\widehat{\mu}_\lambda(t; w)$.

5.1.4   Bootstrap Pointwise Confidence Intervals

Similar to the local and global smoothing estimators of Chapters 3 and 4, statistical inferences based on $\widehat{\mu}_\lambda(t; w)$ are possibly influenced by the correlation structures of the data. In the absence of a known correlation structure, we return to the resampling-subject bootstrap procedure used in Chapters 3 and 4. The approximate bootstrap pointwise confidence intervals for $\mu(t)$ based on $\widehat{\mu}_\lambda(t; w)$ can then be computed using the following procedure.

Approximate Bootstrap Pointwise Confidence Intervals:

(a) **Computing Bootstrap Estimators.** *Generate B independent bootstrap samples using the resampling-subject bootstrap procedure of Section 3.4, and compute the penalized smoothing spline estimators*

$$\mathscr{B}_\mu^B(t; \lambda, w) = \left\{ \widehat{\mu}_\lambda^{boot,1}(t; w), \ldots, \widehat{\mu}_\lambda^{boot,B}(t; w) \right\} \qquad (5.5)$$

based on (5.3) and the corresponding bootstrap samples.

(b) Approximate Bootstrap Intervals. *Compute the percentiles and the estimated standard errors as in Section 3.4 based on the B bootstrap estimators in (5.5). The approximate $\left[100 \times (1-\alpha)\right]$th percentile bootstrap pointwise confidence interval for $\mu(t)$ is given by*

$$\left(L_{(\alpha/2)}(t), U_{(\alpha/2)}(t)\right), \tag{5.6}$$

where $L_{(\alpha/2)}(t)$ and $U_{(\alpha/2)}(t)$ are the $\left[100 \times (\alpha/2)\right]$th and $\left[100 \times (1-\alpha/2)\right]$th percentiles of $\mathscr{B}_{\mu}^{B}(t; \lambda, w)$, respectively. The normal approximated bootstrap pointwise confidence interval for $\mu(t)$ is

$$\widehat{\mu}_{\lambda}(t; w) \pm z_{(1-\alpha/2)} \times \widehat{se}_{B}\left(t; \widehat{\mu}_{\lambda}\right), \tag{5.7}$$

where $\widehat{\mu}_{\lambda}(t; w)$ is the penalized smoothing estimator computed from the original data, $z_{(1-\alpha/2)}$ is the $\left[100 \times (1-\alpha/2)\right]$th percentile of the standard normal distribution and $\widehat{se}_{B}\left(t; \widehat{\mu}_{\lambda}\right)$ is the estimated standard error of $\widehat{\mu}_{\lambda}(t; w)$ from the B bootstrap estimators $\mathscr{B}_{\mu}^{B}(t; \lambda, w)$. □

As discussed in Sections 3.4 and 4.2, (5.6) and (5.7) are only approximate confidence intervals because they ignore the biases of the estimator $\widehat{\mu}_{\lambda}(t; w)$. Bias adjustment for (5.6) and (5.7) generally do not work well in practice, because it is difficult to estimate the bias of $\widehat{\mu}_{\lambda}(t; w)$. The simulation results in the literature, e.g., Chiang, Rice and Wu (2001), suggest that both (5.6) and (5.7) have acceptable empirical coverage probabilities, and can be used as good approximate pointwise confidence intervals, although the theoretical properties of these intervals have not been systematically developed.

The approximate simultaneous confidence bands for $\mu(t)$ over an interval $t \in [a, b]$ can be established by applying the same procedure as in Sections 3.4 and 4.2 to the approximate pointwise intervals (5.6) or (5.7). Since this procedure is self-evident and can be straightforwardly adapted to the current situation, we omit its details in this chapter.

5.2   R Implementation

5.2.1   The HSCT Data

Following the examples of Sections 3.5 and 4.3, we illustrate here how to use the penalized smoothing spline estimator to estimate the mean time-trend of the HSCT data. In comparison to the local smoothing methods in Chapter 3 or the basis approximation smoothing methods in Chapter 4, there is no need to select the bandwidth or the number and location of the knots to control the smoothness of the estimated curve. For a smoothing spline estimator, a roughness penalty term is used to control the excess curvature of the smoothing estimate as described in Section 5.1. The minimizer of (5.1) is a natural cubic spline with knots located at the distinct design time points. However,

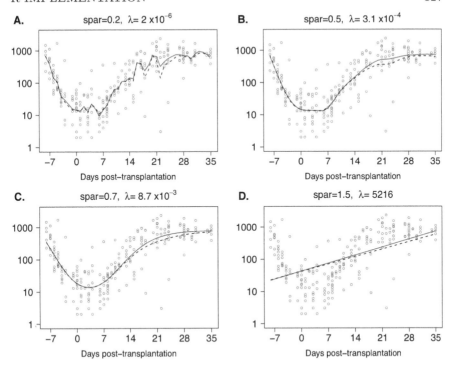

Figure 5.1 *The fluctuations and smoothing spline estimates of the lymphocyte counts relative to the time of stem cell transplantation. The solid and dashed lines represent the estimates using the measurement uniform weight and the subject uniform weight, respectively. (A)-(D) The smoothing parameter spar=0.2, 0.5, 0.7 and 1.5, and the corresponding λ for measurement uniform weight are shown, respectively.*

the selection of the smoothing parameter λ is important to determine the goodness-of-fit and curvature of penalized smoothing splines.

In R, the function smooth.spline() can be used to fit a cubic smoothing spline, which uses a smoothing parameter argument spar to control the smoothness, instead of λ. The usual smoothing parameter λ in the penalized criterion is a monotone function of spar, and the value of λ is given in the output of the smooth.spline fit for a specified or estimated spar. See help(smooth.spline) for details of the arguments for this function and the relationship between spar and λ. The following R code is used to fit the lymphocytes count in the HSCT data:

```
> attach(HSCT)
> plot(Days, LYM.log, xlab="Days post-transplantation", ylab="")
> smfit<-smooth.spline(Days, LYM.log,spar=0.7,cv=NA)
> smfit.w<-smooth.spline(Days, LYM.log, spar=0.7, cv=NA, w=1/ni)
```

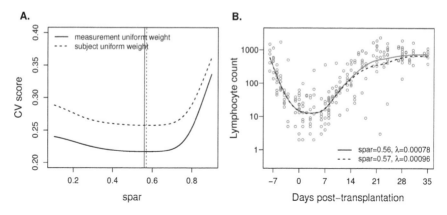

Figure 5.2 *Lymphocyte counts data. (A) Use of cross-validation to select smoothing parameters. (B) Smoothing spline fits with cross-validated smoothing parameters. The solid and dashed lines represent the estimates using the measurement uniform weight and the subject uniform weight, respectively.*

```
> lines(predict(smfit, -8:35), col ="gray40", lwd=1.5)
> lines(predict(smfit.w, -8:35), lwd=1.5, lty=2)
```

Figure 5.1 shows the lymphocyte measurements following the conditioning regimen and the allogeneic hematopoietic stem cell transplantation. We can see the different amount of smoothing for the estimated curves associated with the values of smoothing parameter λ. A very small λ as in Figure 5.1(A) gives little roughness penalty and may result in an undersmoothed fit. On the other hand, too large a λ as in Figure 5.1(D) gives excess roughness penalty and results in a linear regression fit, without allowing for any curvature. The plots in Figure 5.1(B)-(C) give the visually appealing trade-off between fitness and smoothness of the estimated curves, which adequately capture the nonlinear time-trend of the lymphocyte counts.

In practice, we can choose the smoothing parameter `spar` (or λ) subjectively by visually examining the fitted mean curve to the scatter plots of the data. Alternatively, the smoothing parameter may be selected automatically by the LSCV procedure discussed in Section 5.1.3. Figure 5.2(A) shows the LSCV scores of (5.4) plotted against a range of `spar` values with the two choices of weights $w_i^* = 1/(nn_i)$ and $w_i^{**} = 1/N$. The smoothing spline estimators shown in Figure 5.2(B) with `spar=0.56` and `spar=0.57` minimize the corresponding LSCV score functions, respectively. Note that the smoothing splines estimators in Figures 5.1 and 5.2 based on the measurement and subject uniform weights are similar, except in the region with some unusually low lymphocyte counts. Importantly, they both show that the HSCT is associated with two phases of change, lymphocytopenia and lymphocyte recovery, in the transplant recipients. First, lymphocytes reach the lowest concentration af-

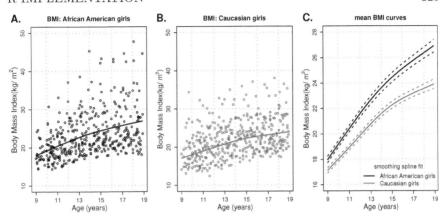

Figure 5.3 *The NGHS body mass index (BMI) data. (A) Smoothing spline estimate of mean curves for African American girls. (B) Smoothing spline estimate of mean curves for Caucasian girls. The BMI values for a sample of 50 girls are plotted for both (A) and (B). (C) The smoothing spline estimates with 95% bootstrap pointwise confidence interval.*

ter conditioning, then followed by a gradual recovery during the first month post-transplant with the achievement of the donor lymphocyte engraftment.

5.2.2   The NGHS BMI Data

The NGHS data has been described in Section 1.2. One aim of this multi-center longitudinal study is to examine the differences in childhood cardiovascular risk factors, such as overweight and obesity, between African American and Caucasian girls during adolescence. The study enrolled 1213 African American girls and 1166 Caucasian girls, who were followed up from ages 9 or 10 years to 18 or 19 years. The body mass index (BMI, weight in *kg* divided by height in m^2) is calculated from the ten annual measurements of height and weight. With very high retention rate throughout the study, the median number of follow-up visits for the individual girls is 9 with an interquartile range of 8 to 10. As adults who were overweight during childhood are more likely to have greater risk of cardiovascular disease, it is important to track the longitudinal change of BMI from childhood into adulthood. We illustrate here that smoothing splines can be used to provide flexible nonparametric estimates of the mean growth curves of BMI and to examine the racial difference in the NGHS girls.

Figures 5.3(A)-(B) show the estimated mean curves of BMI over time, i.e., age, for the study participants stratified by the two racial groups. To illustrate that the estimated curves adequately capture the overall time-trend of the mean BMI values, the BMI values for a randomly selected subset of 50 girls from each race group are plotted along with the estimated mean curves. The penalized smoothing splines are fitted to the BMI data separately for the African American girls and the Caucasian girls using the measurement uniform weight and the cross-validated $\lambda_{LSCV} = 2.32$ and $\lambda_{LSCV} = 1.76$, respectively. The curve estimates based on the subject uniform weight yield almost identical results since most of subjects have similar numbers of measurements. The approximate 95% percentile bootstrap pointwise confidence intervals (CIs) for the estimated curves are displayed in Figure 5.3(C), which are obtained from 1000 resampling-subject bootstrap samples as described in Section 5.1.4. These smoothing estimates and 95% CIs suggest that the mean BMI levels increase over time from 9 to 19 years of age for both racial groups and the rate of increase in BMI is greater in African American girls than that in Caucasian girls. Notably, there is already a significant racial difference in the mean BMI since early adolescence (age 9 to 10 years) and this difference increases significantly at late adolescence and young adulthood. These findings have important implications in the design of long-term pediatric studies and in developing guidelines for the primary prevention of atherosclerosis cardiovascular disease beginning in childhood.

5.3   Asymptotic Properties

We present in this section the asymptotic properties of the penalized smoothing spline estimator $\widehat{\mu}_\lambda(t; w)$ of (5.3) with the measurement uniform weight $w^{**} = (1/N, \ldots, 1/N)^T$. Without loss of generality, we assume that $a = 0$ and $b = 1$. Extension to general $[a, b]$ can be obtained using the affine transformation $u = (t - a)/(b - a)$ for $t \in [a, b]$. Asymptotic properties for the estimators with other weight functions, e.g., the subject uniform weight $w_i^* = 1/(n n_i)$, can be established analogously, so they are omitted from the presentation.

5.3.1   Assumptions and Equivalent Kernel Function

1. Asymptotic Assumptions

   We assume the following technical conditions, which are mainly imposed for mathematical simplicity and may be modified if necessary, for $\widehat{\mu}_\lambda(t; w^{**})$ throughout this chapter:

(a) *The design time points $\{t_{ij} : i = 1, \ldots, n; j = 1, \ldots, n_i\}$ are nonrandom and satisfy*

$$D_N = \sup_{t \in [0, 1]} \left| F_N(t) - F(t) \right| \to 0, \qquad as\ n \to \infty,$$

for some distribution function $F(\cdot)$ with strictly positive density $f(\cdot)$ on $[0, 1]$, where $F_N(t) = N^{-1} \sum_{i=1}^n \sum_{j=1}^{n_i} 1_{[t_{ij} \leq t]}$ and $1_{[t_{ij} \leq t]}$ is the indicator function such that $1_{[t_{ij} \leq t]} = 1$ if $t_{ij} \leq t$, and $1_{[t_{ij} \leq t]} = 0$ if $t_{ij} > t$. The density $f(\cdot)$ is three times differentiable and uniformly continuous on $[0, 1]$. The rth derivative $f^{(r)}(t)$ of $f(t)$ satisfies $f^{(r)}(0) = f^{(r)}(1) = 0$ for $r = 1, 2$.

(b) *The mean curve $\mu(t)$ is four times differentiable and satisfies the boundary conditions $\mu^{(r)}(0) = \mu^{(r)}(1) = 0$ for $r = 2, 3$. The fourth derivatives $\mu^{(4)}(t)$*

is Lipschitz continuous in the sense that $\left|\mu^{(4)}(s_1) - \mu^{(4)}(s_2)\right| \leq c_1 \left|s_1 - s_2\right|^{c_2}$ for all $s_1, s_2 \in [0, 1]$ and some positive constants c_1 and c_2.

(c) There exists a positive constant $\delta > 0$ such that $E\left(|\varepsilon(t)|^{2+\delta}\right) < \infty$.

(d) The smoothing parameter λ is nonrandom and satisfy $\lambda \to 0$, $N\lambda^{1/4} \to \infty$ and $\lambda^{-5/4} D_N \to 0$ as $n \to \infty$.

(e) Define $\sigma^2(t) = E\left[\varepsilon^2(t)\right]$ and $\rho_\varepsilon(t) = \lim_{t' \to t} E\left[\varepsilon(t)\varepsilon(t')\right]$. Both $\sigma^2(t)$ and $\rho_\varepsilon(t)$ are continuous at t. $\square$

These assumptions are sufficiently general and should be satisfied in most applications. A major distinction between the current longitudinal data and the classical cross-sectional i.i.d. data is the additional term $\rho_\varepsilon(t)$. As in Chapters 3 and 4, $\sigma^2(t)$ may not equal $\rho_\varepsilon(t)$ in general, and strict inequality between $\sigma^2(t)$ and $\rho_\varepsilon(t)$ appears, when $\varepsilon(t)$ is the sum of a stationary process of t and an independent measurement error. Because in most applications $\sigma^2(t)$ and $\rho_\varepsilon(t)$ are unknown, we do not require further specific structures for $\sigma^2(t)$ and $\rho_\varepsilon(t)$, except for their continuity in Assumption (e). When $\{t_{ij} : i = 1, \ldots, n; j = 1, \ldots, n_i\}$ are from random designs, we would need to require almost sure convergence of D_N to 0, as suggested in Nychka (1995, Section 2).

2. Equivalent Kernel Function

Because the $S_\lambda(t, s)$ of (5.3) does not have an explicit expression, we would like to approximate it by an explicit equivalent kernel function. Substituting $S_\lambda(t, s)$ with the equivalent kernel, the asymptotic properties of $\hat{\mu}(t; w^{**})$ can be established through the equivalent kernel function (e.g., Brauer and Nohel, 1973). For smoothing spline estimators with cross-sectional i.i.d. data, an equivalent kernel is usually obtained by approximating the Green's function of a differential equation. Motivations and heuristic justifications of considering an equivalent kernel through a differential equation have been discussed extensively in the literature, for example, Silverman (1986), Messer (1991), Messer and Goldstein (1993) and Nychka (1995).

Under the current context, we apply the same rationale established for the smoothing spline estimators with cross-sectional i.i.d. data to the estimator (5.3) and consider the following fourth-order differential equation

$$\lambda g^{(4)}(t) + f(t)g(t) = f(t)\mu(t), \quad t \in [0, 1], \tag{5.8}$$

with $g^{(v)}(0) = g^{(v)}(1) = 0$ for $v = 2, 3$. Let $G_\lambda(t, s)$ be the Green's function associated with (5.8). Then, any solution $g(t)$ of (5.8) satisfies

$$g(t) = \int_0^1 G_\lambda(t, s)\mu(s)f(s)\,ds.$$

Let $\gamma = \int_0^1 f^{1/4}(s)\,ds$ and $\tau(t) = \gamma^{-1} \int_0^t f^{1/4}(s)\,ds$. We define

$$H_\lambda(t,s) = H_{\lambda/\gamma^4}^U \left[\tau(t), \tau(s)\right] \tau^{(1)}(s) f^{-1}(s) \tag{5.9}$$

to be the equivalent kernel of $S_\lambda(t,s)$, where

$$H_\lambda^U(t,s) = \frac{\lambda^{-1/4}}{2\sqrt{2}} \left[\sin\left(\frac{\lambda^{-1/4}}{\sqrt{2}}|t-s|\right) + \cos\left(\frac{\lambda^{-1/4}}{\sqrt{2}}|t-s|\right)\right]$$

$$\times \exp\left(-\frac{\lambda^{-1/4}}{\sqrt{2}}|t-s|\right). \tag{5.10}$$

It is straightforward to verify from (5.9) that $H_\lambda(t,s)$ reduces to $H_\lambda^U(t,s)$ when $f(\cdot)$ is the uniform density. Substituting $S_\lambda(t, t_{ij})$ in (5.3) with $H_\lambda(t, t_{ij})$, the equivalent kernel estimator of $\mu(t)$ with the measurement uniform weight $w^{**} = (1/N, \ldots, 1/N)^T$ is

$$\widetilde{\mu}_\lambda(t; w^{**}) = \frac{1}{N} \sum_{i=1}^n \sum_{j=1}^{n_i} \left[H_\lambda(t, t_{ij}) Y_{ij}\right]. \tag{5.11}$$

The next lemma shows that $H_\lambda(t,s)$ is the dominating term of the Green's function $G_\lambda(t,s)$, which in turn approximates $S_\lambda(t,s)$.

Lemma 5.1. *Assume that Assumptions (a) and (d) are satisfied. When n is sufficiently large, there are positive constants α_1, α_2, κ_1 and κ_2 so that*

$$\left|G_\lambda(t,s) - H_\lambda(t,s)\right| \le \kappa_1 \exp\left(-\alpha_1 \lambda^{-1/4}|t-s|\right), \tag{5.12}$$

$$\left|\frac{\partial^\nu G_\lambda(t,s)}{\partial t^\nu}\right| \le \kappa_1 \lambda^{-(\nu+1)/4} \exp\left(-\alpha_2 \lambda^{-1/4}|t-s|\right), \tag{5.13}$$

$$\left|S_\lambda(t,s) - G_\lambda(t,s)\right| \le \kappa_2 \lambda^{-1/2} D_N \exp\left(-\alpha_1 \lambda^{-1/4}|t-s|\right), \tag{5.14}$$

$$\left|\frac{\partial^\nu S_\lambda(t,s)}{\partial t^\nu}\right| \le \kappa_2 \lambda^{-(\nu+1)/4} D_N \exp\left(-\alpha_2 \lambda^{-1/4}|t-s|\right) \tag{5.15}$$

hold uniformly for $t \in [0,1]$, $s \in [0,1]$ and $0 \le \nu \le 3$. ∎

Proof of Lemma 5.1 is given in Section 5.3.4.

It is worthwhile to note that $H_\lambda(t,s)$ of (5.9) is not the only equivalent kernel that could be considered, and there are other possible choices, such as the equivalent kernels suggested by Messer (1991) and Messer and Goldstein (1993). However, the theoretical derivation of this chapter is based on Chiang, Rice and Wu (2001), which relies on $H_\lambda(t,s)$ to approximate $S_\lambda(t,s)$.

5.3.2 Asymptotic Distributions, Risk and Inferences

We now summarize the main theoretical results of this chapter. Derivations of these results are deferred to Section 5.3.4.

1. Asymptotic Distributions

Recall by Assumption (e) that the variance of $Y(t)$ is $\sigma^2(t)$ and the covariance of $Y(t_1)$ and $Y(t_2)$ at two time points $t_1 \neq t_2$ is $\rho_\varepsilon(t_1, t_2)$. When $t_k \to t$ for $k = 1, 2$, we note that

$$\rho_\varepsilon(t_1, t_2) \to \rho_\varepsilon(t) \neq \sigma^2(t).$$

The next theorem shows that asymptotically $\hat{\mu}_\lambda(t; w^{**})$ of (5.3) has a normal distribution when N is sufficiently large.

Theorem 5.1. *Suppose that Assumptions (a) through (e) are satisfied, t is an interior point of $[0, 1]$, and there are constants $\lambda_0 \geq 0$ and $a_0 \geq 0$ such that $\lim_{n \to \infty} N^{1/2} \lambda^{9/8} = \lambda_0$, $\lim_{n \to \infty} N^{-1} \left(\sum_{i=1}^n n_i^2 \right) \lambda^{1/4} = a_0$ and $\lim_{n \to \infty} N n^{-9/8} = 0$. Then, as $n \to \infty$, $\hat{\mu}(t; w^{**})$ is asymptotically normal in the sense that*

$$\left(N \lambda^{1/4} \right)^{1/2} \left[\hat{\mu}(t; w^{**}) - \mu(t) \right] \to N\left(\lambda_0 b(t), \sigma_\mu^2(t) \right) \quad \text{in distribution,} \quad (5.16)$$

where

$$b(t) = -f^{-1}(t) \mu^{(4)}(t) \tag{5.17}$$

and

$$\sigma_\mu(t) = \left[\frac{1}{4\sqrt{2}} f^{-3/4}(t) \sigma^2(t) + a_0 \rho_\varepsilon(t) \right]^{1/2}. \tag{5.18}$$

The conclusions in (5.16) to (5.18) imply that, in general, the asymptotic distributions of $\hat{\beta}(t; w^{**})$ are affected by n, n_i and the intra-subject correlations of the data. ∎

Proof of Theorem 5.1 is given in Section 5.3.4.

A direct implication of Theorem 5.1 is that, by (5.18), the correlations of the data may only affect the asymptotic variance term $\sigma_\mu^2(t)$ if $a_0 > 0$, which holds if $\sum_{i=1}^n n_i^2$ tends to infinity sufficiently fast. Since, by Assumption (d), we have $\lim_{n \to \infty} \lambda = 0$, it follows from the condition $\lim_{n \to \infty} N^{-1} \left(\sum_{i=1}^n n_i^2 \right) \lambda^{1/4} = a_0$ that, for the special case that n_i are bounded, i.e., $n_i \leq m$ for all $i = 1, \ldots, n$ and some constant $m > 0$, the intra-subject correlation $\rho_\varepsilon(t)$ does not play a role in the asymptotic distribution of $\left(N \lambda^{1/4} \right)^{1/2} \left[\hat{\mu}(t; w^{**}) - \mu(t) \right]$, because

$$\lim_{n \to \infty} N^{-1} \left(\sum_{i=1}^n n_i^2 \right) \lambda^{1/4} \leq m^2 \lim_{n \to \infty} \lambda = 0.$$

When n_i, $i = 1, \ldots, n$, are bounded, the probability that there are at least two data points from the same subject in a shrinking neighborhood is zero.

2. Asymptotic Mean Squared Errors

Risks of spline estimators are usually measured by their asymptotic mean squared errors. We consider the mean squared error (MSE) given by

$$MSE\left[\widehat{\mu}(t; w^{**})\right] = E\left\{\left[\widehat{\mu}(t; w^{**}) - \mu(t)\right]^2\right\}. \tag{5.19}$$

The next theorem gives the asymptotic expression of $MSE\left[\widehat{\mu}(t; w^{**})\right]$.

Theorem 5.2. *Suppose that Assumptions (a) to (e) are satisfied and t is an interior point of* $[0, 1]$. *When n is sufficiently large, the MSE of (5.19) has the following asymptotic expression*

$$MSE\left[\widehat{\mu}(t; w^{**})\right] = \lambda^2 b^2(t) + V(t) + o_p\left[N^{-1}\lambda^{-1/4} + \sum_{i=1}^{n}(n_i/N)^2\right]$$
$$+ O_p\left(n^{-1/2}\lambda\right) + O_p\left(n^{-1}\right) + o_p\left(\lambda^2\right), \tag{5.20}$$

where b(t) is defined in (5.17) and

$$V(t) = \frac{1}{4\sqrt{2}}N^{-1}\lambda^{-1/4}f^{-3/4}(t)\sigma^2(t) + \left[\sum_{i=1}^{n}(n_i/N)^2\right]\rho(t). \tag{5.21}$$

Furthermore, it follows from (5.21) that the asymptotic variance V(t) is not affected by the covariance function $\rho(t)$ *if* $\lim_{n \to \infty}\left[\sum_{i=1}^{n}(n_i/N)^2\right] = 0$, *which holds if and only if* $\lim_{n \to \infty}\max_{1 \leq i \leq n}(n_i/N) = 0$. ∎

Proof of Theorem 5.2 is given in Section 5.3.4.

Since the assumptions in Theorems 5.1 and 5.2 are quite general, the above asymptotic properties provide theoretical justifications for the penalized smoothing spline estimators to be used in most practical situations.

3. Remarks on Asymptotic Properties

A number of interesting special cases of Theorems 5.1 and 5.2 can be derived under some specific but practical settings. These special cases lead to different asymptotic properties of $\widehat{\mu}(t; w^{**})$. The following remarks illustrate some of these useful special cases.

(a) Consistency and Convergence Rates:

Theorem 5.2 does not require any further rate condition on λ other than Assumption (d) and allows for any choice of nonrandom n_i. Thus, under the conditions of Theorem 5.2, $\widehat{\mu}(t; w^{**})$ is a consistent estimator of $\mu(t)$ in the sense that $MSE\left[\widehat{\mu}(t; w^{**})\right] \to 0$ in probability as $n \to \infty$.

By (5.21), the rate of $V(t)$ tending to zero depends on n, n_i, $i = 1, \ldots, n$, λ and the intra-subject covariance $\rho(t)$. If $\lambda^{-1/4}N^{-1}$ converges to zero in a rate slower than $\sum_{i=1}^{n}(n_i/N)^2$, then the second term of the right side of (5.21) becomes negligible, so that the effect of the intra-subject covariance $\rho(t)$ disappears from the asymptotic representation of $MSE\left[\hat{\mu}(t;w^{**})\right]$. This occurs, when the n_i are bounded, which is a case of practical interest. However, in general, the contribution of the intra-subject covariance $\rho(t)$ is not negligible. If $n_i \to \infty$ sufficiently fast as $n \to \infty$, which leads to the so-called *dense longitudinal data*, then the second term of the right side of (5.21) may not be ignored from $V(t)$. This occurs, for example, when $n_i = n^{\alpha}$ for some $\alpha > 0$.

(b) Random Design Time Points and Other Weight Choices:

The derivations of Theorem 5.2, which relies on nonrandom design time points and measurement uniform weight, w^{**}, can be extended to random design time points and other weight choices. Suppose that the design time points t_{ij} are independent identically distributed with distribution function $F(\cdot)$ and density $f(\cdot)$. For the measurement uniform weight w^{**}, we would require, as a modification of Assumption (d), the almost sure convergence of $\lambda^{-5/4}D_N$ to 0 as $n \to \infty$ and consider the same equivalent kernel estimator as defined in (5.9). For the subject uniform weight $w^* = \left(w_1^*, \ldots, w_n^*\right)^T$ with $w_i^* = 1/(nn_i)$, we would replace F_N and D_N in Assumption (a) by

$$F_N^*(t) = \sum_{i=1}^{n}\sum_{j=1}^{n_i}(nn_i)^{-1}1_{[t_{ij}\leq t]} \quad \text{and} \quad D_N^* = \sup_{t \in [0,1]}\left|F_N^*(t) - F(t)\right|,$$

respectively, and, under the almost sure convergence of $\lambda^{-5/4}D_N^*$ to 0 as $n \to \infty$, consider the equivalent kernel estimator

$$\tilde{\mu}(t;w^*) = \sum_{i=1}^{n}\sum_{j=1}^{n_i}\left[(nn_i)^{-1}H_\lambda(t,t_{ij})Y_{ij}\right].$$

The asymptotic distributions and the asymptotic conditional mean squared errors of these equivalent kernel estimators can be derived explicitly. However, as noted by Nychka (1995, Section 7), because the exponential bound of (5.9) may not be sharp enough to establish the asymptotic equivalence between the smoothing spline and the equivalent kernel estimators, further research is needed to develop the improved error bounds under these situations.

Suppose that the time design points t_{ij} are nonrandom, $w^* = \left(w_1^*, \ldots, w_n^*\right)$ with $w_i^* = 1/(nn_i)$ is used and Assumption (a) holds for $F_N^*(t)$ and D_N^*. By Lemma 5.1, we can show that the variance of $\hat{\mu}(t;w^*)$ can be approximated by

$$\sum_{i=1}^{n}\sum_{j=1}^{n_i}\left[\left(\frac{1}{nn_i}\right)^2 G_\lambda^2(t;t_{ij})\,Var(Y_{ij})\right]$$

$$+ \sum_{(i_1,j_1)\neq(i_2,j_2)}\left[\left(\frac{1}{n^2 n_{i_1}n_{i_2}}\right)G_\lambda(t,t_{i_1j_1})G_\lambda(t,t_{i_2j_2})\,Cov(Y_{i_1j_1},Y_{i_2j_2})\right].$$

Unfortunately, the above two summations cannot be easily approximated by some straightforward integrals without further assumptions on n_i. Similarly, we do not have an explicit asymptotic risk representation for $\widehat{\mu}(t; w)$ with a general weight w.

4. Asymptotically Approximate Confidence Intervals

The asymptotic distribution of Theorem 5.1 is potentially useful for making approximate inferences for $\mu(t)$ based on $\widehat{\mu}(t; w^{**})$. In particular, if n, n_i, $i = 1, \ldots, n$, and λ satisfy the conditions stated in Theorem 5.1 and there are consistent estimators $\{\widehat{b}(t), \widehat{\sigma}_\mu(t)\}$ of $\{b(t), \sigma_\mu(t)\}$, then an approximate $[100 \times (1 - \alpha)]\%$ confidence interval for $\mu(t)$ can be given by

$$\left[\widehat{\mu}(t; w^{**}) - \lambda\, \widehat{b}(t)\right] \pm z_{1-\alpha/2} N^{-1/2} \lambda^{-1/8}\, \widehat{\sigma}_\mu(t), \tag{5.22}$$

where $0 < \alpha < 1$, z_p is the pth quantile of the standard normal distribution $N(0, 1)$. In theory, it is possible to construct the consistent estimators $\widehat{b}(t)$ and $\widehat{\sigma}_\mu(t)$ by substituting the unknown quantities of (5.17) and (5.18) with their consistent estimators. But, in practice, $b_r(t)$ is difficult to estimate because, by (5.17), it depends on the fourth derivative of $\mu(t)$. One possible approach to circumvent the difficulty of estimating $b(t)$ is to select a small smoothing parameter λ so that the asymptotic bias $b(t)$ is negligible. For the estimation of $\sigma_\mu(t)$, one approach is to construct adequate smoothing estimators for the variance and covariance processes $\sigma^2(t)$ and $\rho_\varepsilon(t)$, respectively. But a practical smoothing spline estimator for $\sigma_\mu(t)$ is not yet available and requires further research. When the bias adjustment term $\lambda\, \widehat{b}(t)$ is ignored and $\sigma_\mu(t)$ is estimated by the resampling-subject bootstrap procedure, the confidence interval of (5.22) is the same as the one given in (5.7).

5.3.3   Green's Function for Uniform Density

We now give a brief discussion of the Green's function for the differential equation (5.8) with the uniform density on $[0, 1]$. The uniform density is an important and useful special case, because the Green's function associated with the uniform density gives an important linkage between the Green's function $G_\lambda(t, s)$ associated with the differential equation (5.8) and the equivalent kernel function $H_\lambda(t, s)$ in (5.9). The dominating term of $G_\lambda(t, s)$ can be used to establish Lemma 5.1.

Using direct calculation, it can be shown that, for the uniform density $f(t) = 1_{[0,1]}(t)$, the Green's function $G_\lambda^U(t, s)$ of (5.8) for $t \neq s$ is the solution of

$$\lambda \frac{\partial^4}{\partial t^4} G_\lambda^U(t, s) + G_\lambda^U(t, s) = 0, \tag{5.23}$$

subject to the following conditions:

(a) $G_\lambda^U(t, s) = G_\lambda^U(s, t) = G_\lambda^U(1-t, 1-s);$

(b) $(\partial^v/\partial t^v)G_\lambda^U(0,t) = (\partial^v/\partial t^v)G_\lambda^U(1,t) = 0$ for $v = 2,3$;

(c) $(\partial^v/\partial t^v)G_\lambda^U(t,s)\big|_{s=t^-} - (\partial^v/\partial t^v)G_\lambda^U(t,s)\big|_{s=t^+} = 0$ for $v = 0,1,2$;

(d) $(\partial^3/\partial t^3)G_\lambda^U(t,s)\big|_{s=t^-} - (\partial^3/\partial t^3)G_\lambda^U(t,s)\big|_{s=t^+} = \lambda^{-1}$.

The following lemma gives a crucial technical result, which shows that the equivalent kernel function $H_\lambda^U(t,s)$ of (5.10) is the dominating term of the Green's function $G_\lambda^U(t,s)$.

Lemma 5.2. *Suppose that $G_\lambda^U(t,s)$ is the Green's function of the differential equation (5.23) with the uniform density $f(t) = 1_{[0,1]}(t)$. When $\lambda \to 0$, the solution $G_\lambda^U(t,s)$ is given by*

$$G_\lambda^U(t,s) = H_\lambda^U(t,s)\left\{1 + O\left[\exp\left(-\lambda^{-1/4}/\sqrt{2}\right)\right]\right\}, \qquad (5.24)$$

where $H_\lambda^U(t,s)$ is defined in (5.10). ∎

Proof of Lemma 5.2:
Because the proof involves tedious algebraic calculations, only the main steps are sketched here, while some straightforward and tedious details are left out. By the well-known result in differential equations, for example, Brauer and Nohel (1973), a general solution $G_\lambda^U(t,s)$ of (5.23) can be expressed as

$$G_\lambda^U(t,s) = \sum_{l=1,3,5,7}\left\{\left[C_{jl}\sin\left(\lambda^{-1/4}\xi_l(t,s)/\sqrt{2}\right)\right.\right.$$
$$\left.\left. + C_{j,l+1}\cos\left(\lambda^{-1/4}\xi_{l+1}(t,s)/\sqrt{2}\right)\right] \times \exp\left(\lambda^{-1/4}\zeta_l(t,s)/\sqrt{2}\right)\right\},$$

where $j = 1$ or 2, when $t \le s$ or $t > s$,

$$\xi_1(t,s) = \zeta_1(t,s) = t - s, \quad \xi_3(t,s) = \zeta_3(t,s) = t + s,$$

$$\xi_5(t,s) = -\zeta_5(t,s) = t - s \ \text{ and } \ \xi_7(t,s) = -\zeta_7(t,s) = t + s.$$

Now, the objective is to evaluate the relationships among the C_{jl}'s. By $G_\lambda^U(t,s) = G_\lambda^U(s,t)$ in the condition (a) of (5.23), we can obtain that

$$C_{11} = -C_{25}, \ C_{12} = C_{26}, \ C_{13} = C_{23}, \ C_{14} = C_{24},$$

$$C_{15} = -C_{21}, \ C_{16} = C_{22}, \ C_{17} = C_{27} \ \text{ and } \ C_{18} = C_{28}.$$

Furthermore, by $G_\lambda^U(t,s) = G_\lambda^U(1-t,1-s)$, i.e., the condition (a) of (5.23), it can be shown that

$$C_{13} = \left[-\cos\left(\sqrt{2}\lambda^{-1/4}\right)C_{17} + \sin\left(\sqrt{2}\lambda^{-1/4}\right)C_{18}\right]\exp\left(-\sqrt{2}\lambda^{-1/4}\right) \quad (5.25)$$

and

$$C_{14} = \left[\sin\left(\sqrt{2}\lambda^{-1/4}\right)C_{17} + \cos\left(\sqrt{2}\lambda^{-1/4}\right)C_{18}\right]\exp\left(-\sqrt{2}\lambda^{-1/4}\right). \quad (5.26)$$

Now we denote

$$\lambda_1^* = 2^{-1/2}\lambda^{-1/4}, \ \lambda_2^* = 2^{-1/2}\lambda^{-1/4} - (\pi/4) \ \text{ and } \ \lambda_3^* = 2^{-1/2}\lambda^{-1/4} + (\pi/4).$$

Taking derivatives of $G_\lambda^U(t,s)$ with respect to t, we can derive from the condition (b) of (5.23) that

$$C_{11} + C_{13} - C_{15} - C_{17} = \sum_{j=1}^{4}(-1)^{j+1}C_{1j} + \sum_{j=5}^{8}C_{1j} = 0, \quad (5.27)$$

$$\begin{aligned}
\left[\cos\left(\lambda_1^*\right)\left(C_{11} - C_{17}\right) + \sin\left(\lambda_1^*\right)\left(C_{12} + C_{18}\right)\right]\exp\left(-2\lambda_1^*\right) & \\
+\cos\left(\lambda_1^*\right)\left(C_{13} - C_{15}\right) - \sin\left(\lambda_1^*\right)\left(C_{14} + C_{16}\right) &= 0 \quad (5.28)
\end{aligned}$$

and

$$\begin{aligned}
\left[\sin\left(\lambda_3^*\right)\left(C_{11} - C_{17}\right) + \sin\left(\lambda_2^*\right)\left(C_{12} + C_{18}\right)\right]\exp\left(-2\lambda_1^*\right) & \\
+\sin\left(\lambda_2^*\right)\left(C_{13} - C_{15}\right) - \sin\left(\lambda_3^*\right)\left(C_{14} + C_{16}\right) &= 0. \quad (5.29)
\end{aligned}$$

From the conditions (c) and (d) of (5.23), we get

$$C_{11} + C_{12} + C_{15} - C_{16} = 0 \quad \text{and} \quad C_{11} - C_{12} + C_{15} + C_{16} = -\lambda_1^*. \quad (5.30)$$

Using (5.27) and (5.28), we can express C_{15} through C_{18} as linear combinations of C_{1j}, $j = 1, \dots, 4$, and get

$$C_{15} = -C_{11} - 2^{-3/2}\lambda^{-1/4}, \ C_{16} = C_{12} - 2^{-3/2}\lambda^{-1/4},$$

$$C_{17} = 2C_{11} + C_{13} + 2^{-3/2}\lambda^{-1/4} \ \text{ and } \ C_{18} = -2\left(C_{11} + C_{13}\right) + C_{14} + 2^{-3/2}\lambda^{-1/4}.$$

Substituting C_{15} through C_{18} with their corresponding linear combinations of C_{1j}, $j = 1, \dots, 4$, it can be derived from (5.25), (5.26), (5.27) and (5.28) that

$$\begin{aligned}
2\exp\left(-2\lambda_1^*\right)C_{11} + \left[\exp\left(-2\lambda_1^*\right) + \cos\left(2\lambda_1^*\right)\right]C_{13} & \\
-\sin\left(2\lambda_1^*\right)C_{14} + 2^{-1}\lambda_1^*\exp\left(-2\lambda_1^*\right) &= 0, \quad (5.31)
\end{aligned}$$

$$\begin{aligned}
2\exp\left(-2\lambda_1^*\right)C_{11} + \left[2\exp\left(-2\lambda_1^*\right) + \sin\left(2\lambda_1^*\right)\right]C_{13} & \\
-\left[\exp\left(-2\lambda_1^*\right) - \cos\left(2\lambda_1^*\right)\right]C_{14} - 2^{-1}\lambda_1^*\exp\left(-2\lambda_1^*\right) &= 0, \quad (5.32)
\end{aligned}$$

$$\begin{aligned}
\left\{\cos\left(\lambda_1^*\right) - \left[\cos\left(\lambda_1^*\right) + 2\sin\left(\lambda_1^*\right)\right]\exp\left(-2\lambda_1^*\right)\right\}\left(C_{11} + C_{13}\right) & \\
-\sin\left(\lambda_1^*\right)\left[1 - \exp\left(-2\lambda_1^*\right)\right]\left(C_{12} + C_{14}\right) + 2^{-1}\lambda_1^*\cos\left(\lambda_1^*\right)\left[1 - \exp\left(-2\lambda_1^*\right)\right] & \\
-2^{-1}\lambda_1^*\sin\left(\lambda_1^*\right)\left[1 + \exp\left(-2\lambda_1^*\right)\right] &= 0 \quad (5.33)
\end{aligned}$$

and

$$\begin{aligned}
\left\{-\sin\left(\lambda_2^*\right) + \left[\sin\left(\lambda_3^*\right) + 2\sin\left(\lambda_2^*\right)\right]\exp\left(-2\lambda_1^*\right)\right\}\left(C_{11} + C_{13}\right) & \\
-\left[\sin\left(\lambda_3^*\right) + \sin\left(\lambda_2^*\right)\exp\left(-2\lambda_1^*\right)\right]\left(C_{12} + C_{14}\right) & \\
-2^{-1}\lambda_1^*\left[\sin\left(\lambda_2^*\right) - \sin\left(\lambda_3^*\right)\right]\left[1 + \exp\left(-2\lambda_1^*\right)\right] &= 0. \quad (5.34)
\end{aligned}$$

Suppose first that $\lambda \neq 2^{-2}[(k+2^{-1})\pi]^{-4}$ and $\lambda \neq 2^{-2}(k\pi)^{-4}$ for any positive integer k. When $\lambda \to 0$, it can be derived from equations (5.31) to (5.34) that

$$C_{1l} = (-1)^l \left[\lambda^{-1/4}/(2\sqrt{2})\right]\left\{1+O\left[\exp\left(-\lambda^{-1/4}/\sqrt{2}\right)\right]\right\}, \quad l=1,2, \quad (5.35)$$

and

$$C_{1l} = O\left[\lambda^{-1/4}\exp\left(-\lambda^{-1/4}/\sqrt{2}\right)\right], \quad l=3,4. \quad (5.36)$$

Finally, C_{15} through C_{18} can be directly calculated by using (5.35) and (5.36), so that

$$C_{1l} = O\left[\lambda^{-1/4}\exp\left(-\lambda^{-1/4}/\sqrt{2}\right)\right], \quad l=5,6, \quad (5.37)$$

$$C_{17} = -\left[\lambda^{-1/4}/(2\sqrt{2})\right]\left\{1+O\left[\exp\left(-\lambda^{-1/4}/\sqrt{2}\right)\right]\right\} \quad (5.38)$$

and

$$C_{18} = \left[3\lambda^{-1/4}/(2\sqrt{2})\right]\left\{1+O\left[\exp\left(-\lambda^{-1/4}/\sqrt{2}\right)\right]\right\}. \quad (5.39)$$

Then (5.24) is obtained by substituting (5.35) through (5.39) into the general expression of $G_\lambda^U(t,s)$.

When $\lambda = 2^{-2}[(k+2^{-1})\pi]^{-4}$ or $2^{-2}(k\pi)^{-4}$, the same argument as above shows that the coefficients in (5.35) through (5.39) also hold. This completes the proof. ∎

5.3.4   Theoretical Derivations

We now give the technical derivations for the proofs of Lemma 5.1, Theorems 5.1 and 5.2.

1. Derivations for Lemma 5.1

Proof of Lemma 5.1:

A key step for the proof is to establish the relationship between the Green's function for uniform density $G_\lambda^U(t,s)$ in (5.23) and the general Green's function $G_\lambda(t,s)$ for (5.8). For this purpose, we first consider a transformation $Q(t,s)$ such that

$$Q[\tau(t), \tau(s)]\,\tau^{(1)}(s) = G_\lambda(t,s)f(s), \quad (5.40)$$

where $\tau(t) = \gamma^{-1}\int_0^t f^{1/4}(s)\,ds$ and $\gamma = \int_0^1 f^{1/4}(s)\,ds$ are defined in (5.9). Now, define

$$q(u) = \int_0^1 Q(u,v)\beta\left[\tau^{-1}(v)\right]dv,$$

$$\phi_1(t) = \left\{6\left[\tau^{(1)}(t)\right]^2\tau^{(2)}(t)\right\}f^{-1}(t),$$

$$\phi_2(t) = \left\{3\left[\tau^{(2)}(t)\right]^2 + 4\tau^{(1)}(t)\,\tau^{(3)}(t)\right\}f^{-1}(t),$$

$$\phi_3(t) = \tau^{(4)}(t)f^{-1}(t).$$

By the definition of γ and $\tau(t)$, it can be verified by straightforward calculation that $g(t) = q[\tau(t)]$ and $q(u)$ is the solution of the following fourth-order differential equation

$$\left[(\lambda/\gamma^4) q^{(4)}(u) + q(u) \right] + \lambda \sum_{l=1}^{3} \phi_l(u) q^{(4-l)}(u) = \mu\left[\tau^{-1}(u) \right], \qquad (5.41)$$

subject to the boundary conditions that $q^{(\nu)}(0) = q^{(\nu)}(1) = 0$ for $\nu = 2, 3$.

To simplify the notation of (5.41), let $\mathscr{D}$ and $\mathscr{I}$ be the operators for differentiation and identity, and let $\mathscr{M}_\phi$ be the multiplication operator $\mathscr{M}_\phi g = \phi \cdot g$, so that, (5.41) can be expressed as

$$(\mathscr{I} + \mathscr{A})\mathscr{L}q(u) = \mu\left[\tau^{-1}(u) \right], \qquad (5.42)$$

where $\mathscr{L}$ and $\mathscr{A}$ are the composite operators defined by

$$\mathscr{L} = \left[(\lambda/\gamma^4)\mathscr{D}^4 + \mathscr{I} \right] \quad \text{and} \quad \mathscr{A} = \lambda\left(\sum_{l=1}^{3} \mathscr{M}_\phi \mathscr{D}^{4-l} \right)\mathscr{L}^{-1}.$$

Let $A^\nu(u, v)$ be the kernel associated with the integral operator $\mathscr{A}^\nu$. We can verify by the induction argument in the proof of (A.1) of Nychka (1995) that, when n is large, there are constants $\alpha_0 > 0$ and $\kappa_0 > 0$ such that

$$\left| A^\nu(u, v) \right| \leq \kappa_0 W^\nu \exp\left(-\alpha_0 \lambda^{-1/4} |u - v| \right), \qquad \nu \geq 1,$$

where W is some positive constant such that $W < 1$. Because $\left| A^\nu(u, v) \right| < 1$ for sufficiently small λ, the integral operator $\mathscr{L}^{-1}(\mathscr{I} + \mathscr{A})^{-1}$ has the expansion

$$\mathscr{L}^{-1}(\mathscr{I} + \mathscr{A})^{-1} = \mathscr{L}^{-1}\left[\mathscr{I} + \sum_{\nu=1}^{\infty} (-\mathscr{A})^\nu \right]. \qquad (5.43)$$

Thus, by interchanging the integration and summation signs, (5.43) implies that

$$Q(u, v) = G^U_{\lambda/\gamma^4}(u, v) + \sum_{\nu=1}^{\infty} (-1)^\nu \int_0^1 G^U_{\lambda/\gamma^4}(u, s) A^\nu(s, v)\, ds. \qquad (5.44)$$

Applying Lemma 4.2 of Nychka (1995) and Lemma A with $u = \tau(t)$ and $v = \tau(s)$ to (5.44), there are positive constants α_0^*, α_0^{**}, κ_0^* and κ_0^{**}, such that, uniformly for $t, s \in [0, 1]$,

$$\left| Q(u, v) - G^U_{\lambda/\gamma^4}(u, v) \right|$$

$$\leq \left| \sum_{\nu=1}^{\infty} (-1)^\nu \int_0^1 G^U_{\lambda/\gamma^4}(u, s) A^\nu_r(s, v)\, ds \right|$$

$$\leq \kappa_0^* \lambda^{-1/4} \left(\sum_{\nu=1}^{\infty} W^{\nu} \right) \int_0^1 \exp\left(-\frac{\lambda^{1/4}}{\sqrt{2}} |u-s| - \alpha_0 \lambda^{-1/4} |s-v| \right) ds$$

$$\leq \kappa_0^{**} \exp\left(-\alpha_0^* \lambda^{-1/4} |u-v| \right)$$

$$\leq \kappa_0^{**} \exp\left[-\alpha_0^* \lambda^{-1/4} |t-s| \inf_{s \leq u \leq t} |\tau^{(1)}(u)| \right]$$

$$\leq \kappa_0^{**} \exp\left(-\alpha_0^{**} \lambda^{-1/4} |t-s| \right). \tag{5.45}$$

From (5.40), we also have that

$$Q(u,v) - G_{\lambda/\gamma^4}^U(u,v) = \frac{f(s)}{\tau^{(1)}(s)} \left\{ G_{\lambda}(t,s) - G_{\lambda/\gamma^4}^U[\tau(t),\tau(s)] \frac{\tau^{(1)}(s)}{f(s)} \right\}. \tag{5.46}$$

Then, equation (5.12) is a direct consequence of Lemma 5.2, equations (5.45), (5.46) and (5.9). The exponential bounds of (5.13) can be obtained using the same method.

For the proofs of equations (5.14) and (5.15), we can show from equations (5.2) and (5.3) that

$$\int_0^1 S_{\lambda}(t_{ij},s) g(s) dF_N(s) + \lambda \int_0^1 \frac{\partial^2}{\partial s^2} S_{\lambda}(t_{ij},s) g^{(2)}(s) ds = g(t_{ij}). \tag{5.47}$$

Now, let $\mathscr{R}$ be the integral operator such that

$$\mathscr{R}[g(\cdot)](t) = \int_0^1 G_{\lambda}(t,s) g(s) d(F-F_N)(s).$$

By equations (5.12) and (5.13) and the induction argument in the proof of Nychka (1995), there are positive constants κ_1^*, κ_1^{**} and α_1, such that, uniformly for $t, s \in [0,1]$ and $0 \leq \mu \leq 3$,

$$\left| \frac{\partial^{\mu}}{\partial t^{\mu}} \mathscr{R}^{\nu}[G_{\lambda}(\cdot,s)](t) \right|$$

$$\leq \kappa_1^* \left(\kappa_1^{**} D_N \lambda^{-1/4} \right)^{\nu} \lambda^{-(\mu+1)/4} \exp\left[-\alpha_1 \lambda^{-1/4} |t-s| \right]. \tag{5.48}$$

In addition, by Lemma 3.1 of Nychka (1995), a solution of (5.47) satisfies

$$S_{\lambda}(t,t_{ij}) = G_{\lambda}(t,t_{ij}) + \mathscr{R}[S_{\lambda}(\cdot,t_{ij})](t)$$

and, when n is sufficiently large,

$$S_{\lambda}(t,t_{ij}) = G_{\lambda}(t,t_{ij}) + \sum_{\nu=1}^{\infty} \mathscr{R}^{\nu}[G_{\lambda}(\cdot,t_{ij})](t). \tag{5.49}$$

Taking $\kappa_2 \geq \left[\kappa_1^* \kappa_1^{**} / (1 - \kappa_1^{**} D_N \lambda^{-1/4}) \right]$, we can derive from equations (5.48) and (5.49) and condition (d) of Section 5.3.1 that, uniformly for $t, s \in [0, 1]$,

$$
\begin{aligned}
\left| S_\lambda(t, s) - G_\lambda(t, s) \right| &\leq \sum_{v=1}^{\infty} \left| \mathscr{R}^v [G_\lambda(\cdot, s)](t) \right| \\
&\leq \kappa_1^* \lambda^{-1/4} \left(\frac{\kappa_1^{**} D_N \lambda^{-1/4}}{1 - \kappa_1^{**} D_N \lambda^{-1/4}} \right) \exp\left(-\alpha_1 \lambda^{-1/4} |t - s| \right) \\
&\leq \kappa_2 \lambda^{-1/2} D_N \exp\left(-\alpha_1 \lambda^{-1/4} |t - s| \right).
\end{aligned}
$$

This completes the proof of equation (5.14). Again, equation (5.15) can be shown by similar derivations, so the details are not repeated. ∎

2. Three Technical Lemmas

We now present three technical lemmas. The results of these lemmas are used in the proofs of Theorems 5.1 and 5.2. Recall that the outcome process $Y(t)$ has variance $\sigma^2(t)$ at time point t and covariance $\rho_\varepsilon(t, s)$ at time points $t \neq s$, and the limit of $\rho_\varepsilon(t, s)$ is denoted by $\rho_\varepsilon(t) = \lim_{s \to t} \rho_\varepsilon(t, s)$.

Under the Green's function $G_\lambda(t, s)$ of Lemma 5.1, the next lemma gives the dominating terms of the integrals

$$
\int_0^1 G_\lambda^2(t, s) \sigma^2(s) f(s) \, ds \quad \text{and} \quad \int_0^1 G_\lambda(t, s) \rho_\varepsilon(t, s) f(s) \, ds
$$

as λ tends to zero.

Lemma 5.3. *If Assumptions (a) and (d) are satisfied, then, when λ is sufficiently small,*

$$
\int_0^1 G_\lambda^2(t, s) \sigma^2(s) f(s) \, ds = \frac{1}{4\sqrt{2}} f^{-3/4}(t) \lambda^{-1/4} \sigma^2(t) \left[1 + o(1) \right] \tag{5.50}
$$

and

$$
\int_0^1 G_\lambda(t, s) \rho_\varepsilon(t, s) f(s) \, ds = \rho_\varepsilon(t) \left[1 + o(1) \right] \tag{5.51}
$$

hold for all $t \in [\tau, 1 - \tau]$ with some $\tau > 0$. ∎

Proof of Lemma 5.3:

By Lemma 5.1, we can show, using the properties of double exponential distributions and straightforward algebra, that, for some positive constants κ, α and c, as $\lambda \to 0$,

$$
\begin{aligned}
&\left| \int_0^1 \left[G_\lambda^2(t, s) - H_\lambda^2(t, s) \right] \sigma^2(s) f(s) \, ds \right| \\
&\leq \int_0^1 \left| G_\lambda(t, s) - H_\lambda(t, s) \right| \left\{ \left| G_\lambda(t, s) \right| + \left| H_\lambda(t, s) \right| \right\} \sigma^2(s) f(s) \, ds
\end{aligned}
$$

$$\leq \int_0^1 \kappa^2 \lambda_r^{-1/4} \exp\left(-\alpha\lambda^{-1/4}|t-s|\right)\sigma^2(s)f(s)\,ds$$
$$\to c\sigma^2(t)f(t). \tag{5.52}$$

Similarly, denoting $u = \tau(t)$ and $v = \tau(s)$, we can show from (5.10) and the properties of double exponential distributions that, for λ sufficiently small,

$$\int_0^1 H_\lambda^2(t,s)\sigma^2(s)f(s)\,ds = \int_0^1 \left[H_{\lambda/\gamma^4}^U(u,v)\right]^2 \sigma^2\left[\tau^{-1}(v)\right]\left\{\frac{f^{1/4}[\tau^{-1}(v)]}{\gamma f[\tau^{-1}(v)]}\right\}dv$$
$$= \frac{1}{4\sqrt{2}}f^{-3/4}(t)\lambda^{-1/4}\sigma^2(t)\left[1+o(1)\right]. \tag{5.53}$$

Thus, (5.50) follows from (5.52) and (5.53). Then (5.51) can be shown by similar calculations. ∎

The following lemma establishes a useful connection between the solution $g(t)$ of the differential equation (5.8) and the mean curve of interest $\mu(t)$.

Lemma 5.4. *If the mean curve $\mu(t)$ satisfies Assumption (b) and $g(t)$ is a solution of (5.8), then $g^{(4)}(t) \to \mu^{(4)}(t)$ uniformly for $t \in [0, 1]$ as $\lambda \to 0$.* ∎

Proof of Lemma 5.4:
This lemma is a special case of Lemma 6.1 of Nychka (1995). We skip the tedious details here to avoid repetition. ∎

Finally, the following lemma establishes the asymptotic normality of the equivalent kernel estimator $\tilde{\mu}(t; w^{**})$ defined in (5.11).

Lemma 5.5. *Suppose that Assumptions (a) to (e) are satisfied, t is an interior point of $[0, 1]$, and there are positive constants λ_0 and a_0 such that, as $n \to \infty$,*

$$N^{1/2}\lambda^{9/8} \to \lambda_0, \quad N^{-1}\left(\sum_{i=1}^n n_i^2\right)\lambda^{1/4} \to a_0 \quad and \quad Nn^{-9/8} \to 0.$$

Then $\tilde{\mu}(t; w^{**})$ is asymptotically normal in the sense that, as $n \to \infty$,

$$\left(N\lambda^{1/4}\right)^{1/2}\left[\tilde{\mu}(t; \mathbf{w}_0) - \mu(t)\right] \to N\left(\lambda_0 b(t), \sigma_\mu^2(t)\right) \quad in\ distribution, \tag{5.54}$$

where $b(t)$ is defined in (5.17) and $\sigma_\mu(t)$ is defined in (5.18). ∎

Proof of Lemma 5.5:
By Assumptions (a) and (d), equations (5.9) and (5.10) and Lemma 5.4,

we have that

$$E\left[\tilde{\mu}(t, w^{**})\right] - \mu(t)$$

$$= \int_0^1 G_\lambda(t, s)\, \mu(s)\, f(s)\, ds - \mu(t) + \int_0^1 \left[H_\lambda(t, s) - G_\lambda(t, s)\right] \mu(s)\, f(s)\, ds$$

$$+ \int_0^1 H_\lambda(t, s)\, \mu(s)\, d\left[F_N(s) - F(s)\right]$$

$$= -\lambda\, f^{-1}(t)\, g^{(4)}(t) \left[1 + o(\lambda)\right]$$

$$= -\lambda\, b(t) \left[1 + o(\lambda)\right]. \tag{5.55}$$

To compute the variance of $\tilde{\mu}(t; w^{**})$, we consider

$$Var\left[\tilde{\mu}(t; w^{**})\right] = V_1 + V_2 + V_3,$$

where

$$V_1 = N^{-2} \sum_{i=1}^n \sum_{j=1}^{n_i} \left[H_\lambda^2(t, t_{ij})\, Var(Y_{ij})\right],$$

$$V_2 = \frac{1}{N^2} \sum_{i=1}^n \sum_{j_1 \neq j_2 = 1}^{n_i} \left[H_\lambda(t, t_{ij_1})\, H_\lambda(t, t_{ij_2})\, Cov(Y_{ij_1}, Y_{ij_2})\right]$$

and, because the subjects are independent,

$$V_3 = \frac{1}{N^2} \sum_{i_1 \neq i_2 = 1}^n \sum_{j_1, j_2} \left[H_\lambda(t, t_{i_1 j_1})\, H_\lambda(t, t_{i_2 j_2})\, Cov(Y_{i_1 j_1}, Y_{i_2 j_2})\right] = 0.$$

Because $Var(Y_{ij}) = \sigma^2(t_{ij})$, we have that, by Assumption (a) and equation (5.53),

$$V_1 = \frac{1}{4\sqrt{2}}\, f^{-3/4}(t)\, N^{-1}\, \lambda^{-1/4}\, \sigma^2(t) \left[1 + o(1)\right].$$

Similar to the derivation in (5.52), because $Cov(Y_{ij_1}, Y_{ij_2}) = \rho_\varepsilon(t_{ij_1}, t_{ij_2})$, it is straightforward to compute that

$$V_2 = \left[\sum_{i=1}^n \left(\frac{n_i}{N}\right)^2 - \frac{1}{N}\right]$$

$$\times \iint H_\lambda(t, s_1)\, H_\lambda(t, s_2)\, \rho_\varepsilon(s_1, s_2)\, f(s_1)\, f(s_2)\, ds_1\, ds_2 \left[1 + o(1)\right]$$

$$= \left[\sum_{i=1}^n \left(\frac{n_i}{N}\right)^2 - \frac{1}{N}\right] \rho_\varepsilon(t) \left[1 + o(1)\right].$$

The above equations and (5.18) imply that

$$Var\left[\tilde{\mu}(t; w^{**})\right] = N^{-1}\, \lambda^{-1/4}\, \sigma_\mu^2(t) \left[1 + o(1)\right].$$

Finally, it can verified from Assumption (c) and equations (5.9) and (5.10) that $\widetilde{\mu}(t; w^{**})$ satisfies Lindeberg's condition for double arrays of random variables. The results of the lemma follow from equation (5.55) and the central limit theorem for double arrays (e.g., Serfling, 1980, Section 1.9.3). ∎

3. Proofs of Theorems 5.1 and 5.2

Given the technical results of Lemma 5.1 through Lemma 5.5, the asymptotic properties of the penalized spline estimator $\widehat{\mu}(t; w^{**})$ described in Theorems 5.1 and 5.2 can be derived through straightforward calculations. We now sketch the main steps for these proofs.

Proof of Theorem 5.1:
By Assumptions (a), (c) and (d) and Lemma 5.1, we have that, when n is sufficiently large,

$$\widehat{\mu}(t; w^{**}) - \mu(t) = \frac{1}{N} \sum_{i=1}^{n} \sum_{j=1}^{n_i} \left\{ \left[S_\lambda(t, t_{ij}) - G_\lambda(t, t_{ij}) \right] Y_{ij} \right\} + o_p \left[N^{-1/2} \lambda^{-1/8} \right].$$

Then the asymptotic normality result of (5.16) is a direct consequence of Lemma 5.5 and the above equation. ∎

Proof of Theorem 5.2:
Using the variance-bias squared decomposition for (5.19), we have that

$$MSE\left[\widehat{\mu}(t; w^{**})\right] = \left\{ E\left[\widehat{\mu}(t; w^{**})\right] - \mu(t) \right\}^2 + Var\left[\widehat{\mu}(t; w^{**})\right]. \tag{5.56}$$

Because $Y_{i_1 j_1}$ and $Y_{i_2 j_2}$ are independent when $i_1 \neq i_2$,

$$Var\left[\widehat{\mu}(t; w^{**})\right] = V_1^* + V_2^*,$$

where

$$V_1^* = \frac{1}{N^2} \sum_{i=1}^{n} \sum_{j=1}^{n_i} \left[S_\lambda^2(t, t_{ij}) \, Var(Y_{ij}) \right]$$

and

$$V_2^* = \frac{1}{N^2} \sum_{i=1}^{n} \sum_{j_1 \neq j_2 = 1}^{n_i} \left[S_\lambda(t, t_{ij_1}) \, S_\lambda(t, t_{ij_2}) \, Cov(Y_{ij_1}, Y_{ij_2}) \right].$$

Using Lemma 5.1 and the derivation of (5.51), we can show that, for sufficiently large n,

$$Var\left[\widehat{\mu}(t; w^{**})\right] = \frac{1}{4\sqrt{2}} N^{-1} \lambda^{-1/4} f^{-3/4}(t) \sigma_\varepsilon^2(t) \left[1 + o_p(1)\right]$$

$$+ \left[\sum_{i=1}^{n} \left(\frac{n_i}{N}\right)^2 - \frac{1}{N} \right] \rho_\varepsilon(t) \left[1 + o_p(1)\right]. \tag{5.57}$$

For the bias term of (5.56), we consider that, for sufficiently large n,

$$E\left[\widehat{\mu}(t; w^{**})\right] - \mu(t) = \frac{1}{N} \sum_{i=1}^{n} \sum_{j=1}^{n_i} \left[S_\lambda\left(t, t_{ij}\right) \mu\left(t_{ij}\right)\right] - \mu(t). \qquad (5.58)$$

Then, by Lemma 5.5,

$$\left\{\frac{1}{N} \sum_{i=1}^{n} \sum_{j=1}^{n_i} \left[S_\lambda\left(t, t_{ij}\right) \mu\left(t_{ij}\right) - \mu(t)\right]\right\}^2 = \lambda^2 b^2(t) \left[1 + o_p(1)\right]. \qquad (5.59)$$

The conclusion in equation (5.20) of Theorem 5.2 is then a direct consequence of equations (5.56) to (5.59). ■

5.4   Remarks and Literature Notes

This chapter presents a number of results for the estimation of the conditional mean curve $\mu(t) = E[Y(t)|t]$ based on the penalized smoothing splines with longitudinal sample $\{(Y_{ij}, t_{ij}) : i = 1, \ldots, n; j = 1, \ldots, n_i\}$. The theoretical results demonstrate that, although a penalized smoothing spline estimator is obtained through a "global smoothing method," it has natural connections with a "local smoothing method" because it is asymptotically equivalent to an equivalent kernel estimator. These theoretical implications have been corroborated by the applications to the HSCT and NGHS studies in Section 5.2, since the numerical results obtained by the penalized smoothing spline estimators are similar to the results obtained by the local smoothing method in Chapter 3 or the global smoothing method in Chapter 4.

The theoretical derivations of this chapter depend on different techniques from the estimators in Chapters 4 and 5. As seen from the proofs of Theorems 5.1 and 5.2, the crucial step for establishing the asymptotic equivalence between the penalized smoothing spline estimator (5.3) and the equivalent kernel estimator (5.11) relies on obtaining the approximate Green's function for the differential equation (5.8). Consequently, the asymptotic properties for the penalized smoothing spline estimators are only established on a case-by-case basis using the equivalent kernel approach. For general settings, explicit forms of the Green's functions of such differential equations may not be readily available. Further research is needed to develop alternative approaches for establishing the asymptotic properties of the penalized smoothing spline estimators with longitudinal samples.

The results of this chapter are mainly adopted from Hoover et al. (1998) and Chiang, Rice and Wu (2001). Theoretical derivations rely heavily on the techniques described in Nychka (1995). Earlier results for the penalized smoothing spline estimators in nonparametric regression with cross-sectional i.i.d. data have been described in Silverman (1986), Eubank (1999), Wahba (1975, 1990), Rice and Silverman (1991), Messer (1991), Messer and Goldstein (1993) and Green and Silverman (1994). These are only a small fraction of the publications in this area.

Part III

Time-Varying Coefficient Models

Chapter 6

Smoothing with Time-Invariant Covariates

The estimation and inference methods of Chapters 3 to 5 are mainly concerned with evaluating the mean time curve $E[Y(t)] = \mu(t)$ without incorporating the effects of potential covariates. In most practical situations, the scientific interests of a longitudinal study are often focused on evaluating the effects of time t and a set of covariates $X^{(l)}(t)$, $l = 1, \ldots, k$, which may or may not depend on t, on the chosen time dependent outcome variable $Y(t)$. The objective of this chapter is to present a series of methods for modeling and estimating the effects of a set of time-invariant covariates on a real-valued longitudinal outcome variable. Methods for the general case involving time-dependent covariates are presented in Chapters 7 to 9.

6.1   Data Structure and Model Formulation

6.1.1   Data Structure

We assume throughout this chapter that, for each given t, $Y(t)$ is a real-valued, continuous and time-dependent variable, and there is a set of $k \geq 1$ covariates $\{X^{(1)}, \ldots, X^{(k)}\}$ which do not change with time t, so that the covariates are given by a time-invariant $(k+1) \times 1$ column vector $\mathbf{X} = \left(1, X^{(1)}, \ldots, X^{(k)}\right)^T$. The observations for $\left(Y(t), t, \mathbf{X}^T\right)$ are given by $\left\{\left(Y_{ij}, t_{ij}, \mathbf{X}_i^T\right) : i = 1, \ldots, n; j = 1, \ldots, n_i\right\}$. At the jth measurement time t_{ij} of the ith subject, the ith subject's observed covariates and outcome are $\mathbf{X}_i = \left(1, X_i^{(1)}, \ldots, X_i^{(k)}\right)^T$ and Y_{ij}, respectively, where $\mathbf{X}_i$ have the same values at all the time points $\{t_{ij} : j = 1, \ldots, n_i\}$. Since the subjects are assumed to be independent, the measurements $\left\{\left(Y_{ij}, t_{ij}, \mathbf{X}_i^T\right) : i = 1, \ldots, n; j = 1, \ldots, n_i\right\}$ are independent between different subjects, but are possibly correlated within the same subject. That is, $\left(Y_{i_1 j_1}, t_{i_1 j_1}, \mathbf{X}_{i_1}^T\right)$ and $\left(Y_{i_2 j_2}, t_{i_2 j_2}, \mathbf{X}_{i_2}^T\right)$ are independent for any $i_1 \neq i_2$ and all (j_1, j_2). On the other hand, $\left(Y_{ij_1}, t_{ij_1}, \mathbf{X}_i^T\right)$ and $\left(Y_{ij_2}, t_{ij_2}, \mathbf{X}_i^T\right)$ are possibly correlated for any $j_1 \neq j_2$ and all $1 \leq i \leq n$.

6.1.2   The Time-Varying Coefficient Model

Although the parametric and semiparametric models summarized in Chapter 2 can be used to evaluate the relationship between $Y(t)$ and $\{t, \mathbf{X}\}$, they are only useful when this relationship belongs to a known parametric or semiparametric family. When there are no justifiable parametric or semiparametric models available for the data, the resulting statistical inferences and conclusions based on a misspecified model could be misleading. On the other hand, the estimation and inference methods presented in Chapters 3 to 5 are only appropriate for the sample $\{(Y_{ij}, t_{ij}) : i = 1, \ldots, n; j = 1, \ldots, n_i\}$ without covariates other than time. When covariates other than time are also involved, unstructured nonparametric estimation of $E[Y(t)|t, \mathbf{X}]$ may require multivariate smoothing estimators, which could be numerically unstable and difficult to interpret in practice.

A promising alternative to the methods of Chapters 2 to 5 is to consider regression models that are more flexible than the classical parametric or semiparametric models and also have specific structures which can be easily interpreted in real applications. Hence, this leads to the term "structured nonparametric model." As a special case of nonparametric models with linear structures, the varying-coefficient models have been studied by Hastie and Tibshirani (1993) as an extension of the classical linear marginal models by allowing the linear coefficients to be nonparametric curves of another variable. For the analysis of $(Y(t), t, \mathbf{X}^T)$, Hoover et al. (1998) proposed to model the conditional means of $Y(t)$ given $\{t, \mathbf{X}\}$ by the time-varying coefficient model

$$Y(t) = \mathbf{X}^T \boldsymbol{\beta}(t) + \varepsilon(t), \tag{6.1}$$

where $\mathbf{X} = (1, X^{(1)}, \ldots, X^{(k)})^T$, $\{X^{(l)} : l = 1, \ldots, k\}$ are time-invariant covariates, $\{\beta_l(t) : l = 0, \ldots, k\}$ are smooth coefficient curves which are functions of time t, $\boldsymbol{\beta}(t) = (\beta_0(t), \ldots, \beta_k(t))^T$, $\varepsilon(t)$ is a mean zero stochastic process for the error term, and $\mathbf{X}$ and $\varepsilon(t)$ are independent.

The model (6.1) has simple and natural interpretations in real applications, because, when a time point t is fixed, the expression of (6.1) is a multivariate linear marginal model with the continuous outcome variable $Y(t)$ and covariate vector $\mathbf{X}$. Thus, interpretations for the classical multivariate linear models can be simply extended to the time-varying coefficient model (6.1) when t is fixed. When the time t changes, the coefficients in $\boldsymbol{\beta}(t)$ also change with t, so that $(Y(t), t, \mathbf{X}^T)$ follows a multivariate linear model with different coefficients $\{\beta_l(t) : l = 0, \ldots, k\}$ at different time points. Depending on the scientific nature of the variables, it is usually reasonable in most biological applications to assume that $\{\beta_l(t) : l = 0, \ldots, k\}$ satisfy some smoothness conditions. These smoothness assumptions ensure that the effects of the covariates $\mathbf{X}$ on the outcome variable $Y(t)$ do not change dramatically at any two adjacent time points $t_1 \neq t_2$.

6.1.3   A Useful Component-wise Representation

Since in most longitudinal studies, the subjects are randomly selected, it is reasonable to assume that the observed covariates $\mathbf{X}_i$ are random and the $(k+1) \times (k+1)$ matrix $E(\mathbf{X}\mathbf{X}^T)$ is nonsingular, so that $E(\mathbf{X}\mathbf{X}^T)$ has the unique inverse

$$E(\mathbf{X}\mathbf{X}^T)^{-1} = E_{\mathbf{X}\mathbf{X}^T}^{-1}.$$

Multiplying both sides of (6.1) with $\mathbf{X}$ and taking expectation, $\beta(t)$ can be expressed as

$$\beta(t) = \left(E_{\mathbf{X}\mathbf{X}^T}^{-1}\right) E\left[\mathbf{X}Y(t)\right]. \tag{6.2}$$

Let $e_{r+1,l+1}$ be the $(r+1, l+1)$th element of $E_{\mathbf{X}\mathbf{X}^T}^{-1}$. Then (6.2) shows that, for $r = 0, \ldots, k$,

$$\beta_r(t) = E\left\{\left[\sum_{l=0}^{k} e_{r+1,l+1} X^{(l)}\right] Y(t)\right\}. \tag{6.3}$$

The equation (6.3) gives the expression of each component of $\beta(t)$ based on the model (6.2) as the expectation of a function of the outcome variable $Y(t)$ and the covariate matrix $\mathbf{X}$.

Since $E(\mathbf{X}\mathbf{X}^T)$ is time-invariant, a simple estimator of $E(\mathbf{X}\mathbf{X}^T)$ is the sample mean

$$\widehat{E}_{\mathbf{X}\mathbf{X}^T} = n^{-1} \sum_{i=1}^{n} (\mathbf{X}_i \mathbf{X}_i^T). \tag{6.4}$$

If $\widehat{E}_{\mathbf{X}\mathbf{X}^T}$ is invertible, then a natural estimator of $E_{\mathbf{X}\mathbf{X}^T}^{-1}$ is $\left(\widehat{E}_{\mathbf{X}\mathbf{X}^T}\right)^{-1}$, so that $e_{r+1,l+1}$ of (6.3) can be estimated by $\widehat{e}_{r+1,l+1}$, where

$$\widehat{e}_{r+1,l+1} = \text{ the } (r+1, l+1)\text{th element of } \left(\widehat{E}_{\mathbf{X}\mathbf{X}^T}\right)^{-1}. \tag{6.5}$$

Substituting $e_{r+1,l+1}$ of $\left(\sum_{l=0}^{k} e_{r+1,l+1} X^{(l)}\right) Y(t)$ with $\widehat{e}_{rl}$, nonparametric smoothing estimators of $\beta_r(t)$ can be constructed by applying the univariate smoothing methods of Chapters 3 to 5 to the component-wise conditional expectation of $\left(\sum_{l=0}^{k} \widehat{e}_{r+1,l+1} X^{(l)}\right) Y(t)$ given t. Specifically, if $\widehat{\beta}_r(t)$, $r = 0, \ldots, k$, are smoothing estimators of $\beta_r(t)$ in (6.3), then the corresponding component-wise smoothing estimator of $\beta(t)$ is

$$\widehat{\beta}(t) = \left(\widehat{\beta}_0(t), \widehat{\beta}_1(t), \ldots, \widehat{\beta}_k(t)\right)^T. \tag{6.6}$$

Note that, because the component-wise smoothing estimator $\widehat{\beta}(t)$ relies on calculating the inverse of $\widehat{E}_{\mathbf{X}\mathbf{X}^T}$, it may be numerically unstable when $\widehat{E}_{\mathbf{X}\mathbf{X}^T}$ is nearly singular.

Another intuitive method for the estimation of $\beta(t)$ is to first obtain the estimators $\widehat{E}_{\mathbf{X}\mathbf{X}^T}^{-1}$ and $\widetilde{E}\left[\mathbf{X}Y(t)\right]$ for $E_{\mathbf{X}\mathbf{X}^T}^{-1}$ and $E\left[\mathbf{X}Y(t)\right]$, respectively, and then substitute $E_{\mathbf{XX}^T}^{-1}$ and $E[\mathbf{X}Y(t)]$ in (6.2) with $\widehat{E}_{\mathbf{XX}^T}^{-1}$ and $\widetilde{E}[\mathbf{X}Y(t)]$, so that $\beta(t)$ is estimated by

$$\widetilde{\beta}(t) = \widehat{E}_{\mathbf{XX}^T}^{-1} \widetilde{E}[\mathbf{X}Y(t)]. \tag{6.7}$$

Although $E[\mathbf{X}Y(t)]$ can be estimated by $\widetilde{E}[\mathbf{X}Y(t)]$ with different smoothness within each of its components, (6.7) suggests that the components of $\widetilde{\beta}(t)$ are estimated by the linear combinations of the components of $\widehat{E}_{\mathbf{XX}^T}^{-1}$ and $\widetilde{E}[\mathbf{X}Y(t)]$. Thus, the difference between $\widehat{\beta}(t)$ and $\widetilde{\beta}(t)$ is the result of estimating the different components of the right-side terms of (6.2).

The subtle difference between $\widehat{\beta}(t)$ and $\widetilde{\beta}(t)$ can be seen by considering the following special case of (6.2). Suppose that $\mathbf{X} = (1, X)^T$, $E(X) = 1$ and $Var(X) = 1$. Then, equations (6.1) and (6.2) give that $\beta(t) = (\beta_0(t), \beta_1(t))^T$,

$$\mathbf{XX}^T = \begin{pmatrix} 1 & X \\ X & X^2 \end{pmatrix} \quad \text{and} \quad E[\mathbf{X}Y(t)] = \begin{pmatrix} \beta_0(t) + \beta_1(t) \\ \beta_0(t) + 2\beta_1(t) \end{pmatrix}.$$

The first component of $\widetilde{E}[\mathbf{X}Y(t)]$ is a consistent estimator of $[\beta_0(t) + \beta_1(t)]$. The second component of $\widetilde{E}[\mathbf{X}Y(t)]$ is a consistent estimator of $[\beta_0(t) + 2\beta_1(t)]$. Thus, $\widetilde{\beta}(t)$ is constructed by a linear combination of the consistent smoothing estimators of $[\beta_0(t) + \beta_1(t)]$ and $[\beta_0(t) + 2\beta_1(t)]$ with random weights that depend on $\mathbf{X}_i$. When $\beta_0(t)$ and $\beta_1(t)$ satisfy different smoothness conditions, larger mean squared errors may arise from estimating $[\beta_0(t) + \beta_1(t)]$ and $[\beta_0(t) + 2\beta_1(t)]$ than estimating $\beta_0(t)$ and $\beta_1(t)$ separately. Thus, $\widetilde{\beta}(t)$ of (6.7) is in general less desirable than $\widehat{\beta}(t)$ of (6.6). Similar phenomena evidently hold for the general covariate vector $\mathbf{X}$ with $k \geq 1$.

6.2   Component-wise Kernel Estimators

Using (6.3) and the sample mean estimators $\widehat{E}_{\mathbf{XX}^T}$ and $\widehat{e}_{r+1,l+1}$ given in (6.4) and (6.5), we can estimate the coefficient curves $\beta(t) = (\beta_0(t), \ldots, \beta_k(t))^T$ by the kernel smoothing method. This method, which estimates each of the components in $\beta(t)$ by a kernel estimator of $E[(\sum_{l=0}^{k} \widehat{e}_{r+1,l+1} X^{(l)}) Y(t)]$, is motivated by the intuition that, by estimating each component of $\beta(t)$ separately, different smoothing needs of the coefficient curves $\beta_r(t)$, $r = 0, \ldots, k$, can be adapted by using different bandwidths.

6.2.1   Construction of Estimators through Least Squares

A component-wise kernel smoothing estimator of $\beta_r(t)$ can be constructed by extending the univariate kernel approach of Section 3.1 to the mean curve of (6.3) with a local least squares criterion based on (6.4), (6.5) and the longitudinal observations. By (6.5), we can substitute $e_{r+1,l+1}$ of (6.3) with $\widehat{e}_{r+1,l+1}$

and approximate $\left(\sum_{l=0}^{k} e_{r+1,l+1} X_i^{(l)}\right) Y_{ij}$ by a pseudo-observation

$$Y_{ijr}^* = \left(\sum_{l=0}^{k} \widehat{e}_{r+1,l+1} X_i^{(l)}\right) Y_{ij}.$$

Here Y_{ijr}^* can be viewed as an "observed value" for $\beta_r(t_{ij})$. Then, the kernel smoothing method of Section 3.1, such as equations (3.5) and (3.6), can be applied to the pseudo-sample

$$\mathscr{Y}_r = \left\{ Y_{ijr}^* : i = 1, \ldots, n; \, j = 1, \ldots, n_i \right\} \quad \text{with } 0 \le r \le k, \tag{6.8}$$

so that $E\left[\left(\sum_{l=0}^{k} \widehat{e}_{r+1,l+1} X^{(l)}\right) Y(t)\right]$ can be estimated by minimizing the component-specific local score function

$$L_{r,K_r}(t; h_r, w) = \sum_{i=1}^{n} \sum_{j=1}^{n_i} \left\{ w_i \left[Y_{ijr}^* - b_r(t) \right]^2 K_r\left(\frac{t - t_{ij}}{h_r}\right) \right\} \tag{6.9}$$

with respect to $b_r(t)$, where $K_r(\cdot)$ is a kernel function, $h_r > 0$ is a bandwidth, $w = (w_1, \ldots, w_n)^T$ and w_i, $i = 1, \ldots, n$, are the weights for the subjects. The minimizer of (6.9) leads to the kernel estimator $\widehat{\beta}_{r,K_r}(t; h_r, w)$ of $\beta_r(t)$, which is a linear statistic of Y_{ijr}^*, such that

$$\widehat{\beta}_{r,K_r}(t; h_r, w) = \frac{\sum_{i=1}^{n} \sum_{j=1}^{n_i} w_i Y_{ijr}^* K_r\left[(t - t_{ij})/h_r\right]}{\sum_{i=1}^{n} \sum_{j=1}^{n_i} w_i K_r\left[(t - t_{ij})/h_r\right]} \quad \text{for all } 0 \le r \le k. \tag{6.10}$$

The component-wise kernel estimator of $\beta(t) = (\beta_0(t), \ldots, \beta_k(t))^T$ is

$$\widehat{\beta}_{\mathbf{K}}(t; \mathbf{h}, w) = \left(\widehat{\beta}_{0,K_0}(t; h_0, w), \ldots, \widehat{\beta}_{k,K_k}(t; h_k, w) \right)^T, \tag{6.11}$$

where $\mathbf{K}(\cdot) = \{K_0(\cdot), \ldots, K_k(\cdot)\}$ is the collection of kernel functions and $\mathbf{h} = (h_0, \ldots, h_k)^T$ is the vector of bandwidths.

Similar to the kernel estimators in Section 3.1, the "subject uniform weight" $w_i^* = 1/(n n_i)$ and the "measurement uniform weight" $w_i^{**} = 1/N$ are the two commonly used weight choices in practice. When the subject uniform weight $w_i^* = 1/(n n_i)$, $w^* = (w_1^*, \ldots, w_n^*)^T$, is used, the component-specific local score function is

$$L_{r,K_r}(t; h_r, w^*) = \sum_{i=1}^{n} \sum_{j=1}^{n_i} \left\{ \left(\frac{1}{n n_i}\right) \left[Y_{ijr}^* - b_r(t) \right]^2 K_r\left(\frac{t - t_{ij}}{h_r}\right) \right\}. \tag{6.12}$$

Minimizing $L_{r,K_r}(t; h_r, w^*)$ with respect to $b_r(t)$, the kernel estimator of $\beta_r(t)$ is

$$\widehat{\beta}_{r,K_r}(t; h_r, w^*) = \frac{\sum_{i=1}^{n} n_i^{-1} \sum_{j=1}^{n_i} \left\{ Y_{ijr}^* K_r\left[(t - t_{ij})/h_r\right] \right\}}{\sum_{i=1}^{n} n_i^{-1} \sum_{j=1}^{n_i} K_r\left[(t - t_{ij})/h_r\right]}, \tag{6.13}$$

and the kernel estimator $\widehat{\beta}_{\mathbf{K}}(t; \mathbf{h}, w^*)$ is obtained by (6.11).

When the measurement uniform weight $w_i^{**} = 1/N$, $w^{**} = (w_1^{**}, \ldots, w_n^{**})^T$, is used, the component specific local score function is

$$L_{r, K_r}(t; h_r, w^{**}) = \sum_{i=1}^{n} \sum_{j=1}^{n_i} \left\{ \frac{1}{N} \left[Y_{ijr}^* - b_r(t) \right]^2 K_r \left(\frac{t - t_{ij}}{h_r} \right) \right\}. \qquad (6.14)$$

Minimizing $L_{r, K_r}(t; h_r, w^{**})$ with respect to $b_r(t)$, the kernel estimator of $\beta_r(t)$ is

$$\widehat{\beta}_{r, K_r}(t; h_r, w^{**}) = \frac{\sum_{i=1}^{n} \sum_{j=1}^{n_i} \left[Y_{ijr}^* K_r \left((t - t_{ij}) / h_r \right) \right]}{\sum_{i=1}^{n} \sum_{j=1}^{n_i} K_r \left[(t - t_{ij}) / h_r \right]}, \qquad (6.15)$$

and the kernel estimator $\widehat{\beta}_{\mathbf{K}}(t; \mathbf{h}, w^{**})$ is again obtained by (6.11).

Similar to the univariate kernel smoothing estimators of Chapter 3, under-smoothing or over-smoothing of the resulting estimator $\widehat{\beta}_{\mathbf{K}}(t; \mathbf{h}, w)$ is mainly caused by unsuitable bandwidth choices, while the effect of the kernel functions is rarely influential. Usual choices of kernel functions, such as the standard Gaussian kernel, the Epanechnikov kernel and other probability density functions, normally give satisfactory results. Since the component-wise kernel estimators in (6.11), (6.13) and (6.15) rely heavily on the time-invariant nature of $\mathbf{X}$ and the sample mean estimator of $E(\mathbf{X}\mathbf{X}^T)$, different smoothing needs of $\beta_r(t)$, $r = 0, \ldots, k$, can be adjusted by selecting appropriate bandwidths h_r.

The choices of weighting schemes may also have profound influences on the adequacy of the estimators. Ideally, it may be theoretically beneficial if $w_i^* = 1/(n n_i)$ in (6.13) or $w_i^{**} = N^{-1}$ in (6.15) could be replaced by non-negative weights w_i, $i = 1, \ldots, n$, which depend on the intra-subject correlations of the data. However, without knowing the structures of the intra-correlations, the natural choices $w_i^* = 1/(n n_i)$ and $w_i^{**} = N^{-1}$ appear to be reasonable in practice. It can be seen from the asymptotic results of Section 6.6 that neither $\widehat{\beta}_{r, K_r}(t; h_r, w^*)$ nor $\widehat{\beta}_{r, K_r}(t; h_r, w^{**})$ asymptotically dominates the other uniformly for all possible situations. The simulation study of Section 6.5 suggests that the "subject uniform weight" $w_i^* = 1/(n n_i)$ is practically preferable to the "measurement uniform weight" $w_i^{**} = 1/N$, since $\widehat{\beta}_{r, K_r}(t; h_r, w^*)$ provides better fits than $\widehat{\beta}_{r, K_r}(t; h_r, w^{**})$ in many realistic situations. Because $\widehat{\beta}_{r, K_r}(t; h_r, w^{**})$ assigns the uniform weight N^{-1} to all the measurement points, it is more influenced by those subjects with large numbers of repeated measurements.

6.2.2 Cross-Validation Bandwidth Choices

Since the smoothing method is applied one at a time to the univariate components of $\beta(t) = (\beta_0(t), \ldots, \beta_k(t))^T$, the bandwidths may be selected subjectively by examining the plots of the fitted curves and the pseudo-observations $\mathscr{Y}_r$ of (6.8). But finding automatic bandwidths suggested by the data is still of both theoretical and practical interest. The bandwidth vector $\mathbf{h}$ of (6.10) can be selected by the two cross-validation approaches described below.

1. Component-wise Cross-Validation Bandwidths

Given that $\widehat{\beta}_{r,K_r}(t; h_r, w)$ is simply a univariate kernel estimator for the pseudo-observations $\mathscr{Y}_r$ for any $0 \leq r \leq k$, the leave-one-subject-out cross-validation (LSCV) method of Section 3.3 can be extended to $\mathscr{Y}_r$ to select the component-wise bandwidth h_r. Let $\mathscr{Y}_r^{(-i)}$ be the pseudo-observations as defined in (6.8) but with the ith subject's pseudo-observations $\{Y_{ijr}^* : j = 1, \ldots, n_i\}$ deleted, so that,

$$\mathscr{Y}_r^{(-i)} = \left\{ Y_{i^* jr}^* = \left(\sum_{l=0}^{k} \widehat{e}_{r+1,l+1} X_{i^*}^{(l)} \right) Y_{i^* jr} : 1 \leq i^* \leq n; i^* \neq i \right\}. \quad (6.16)$$

Note that, by equations (6.4) and (6.5), the ith subject's covariate vector $\mathbf{X}_i$ is still used in the computation of $\widehat{e}_{r+1,l+1}$. Ideally, in order to totally remove the influence of the ith subject in $\mathscr{Y}_r^{(-i)}$, it may be tempting to replace the $\widehat{e}_{r+1,l+1}$ in $Y_{i^* jr}^*$ with $\widehat{e}_{r+1,l+1}^{(-i)}$, which is computed with the ith subject's covariate vector $\mathbf{X}_i$ removed. However, when the sample size n is sufficiently large, the values of $\widehat{e}_{r+1,l+1}$ and $\widehat{e}_{r+1,l+1}^{(-i)}$ are approximately the same. Thus, for computational simplicity, $\widehat{e}_{r+1,l+1}$ is still used in the definition of $\mathscr{Y}_r^{(-i)}$.

Let $\widehat{\beta}_{r,K_r}^{(-i)}(t; h_r, w)$ be the kernel estimator of (6.13) computed based on the pseudo sample $\mathscr{Y}_r^{(-i)}$. Following the approach of Section 3.3, the LSCV score of $\widehat{\beta}_{r,K_r}(t; h_r, w)$ based on Y_{ijr}^* and $\widehat{\beta}_{r,K_r}^{(-i)}(t; h_r, w)$ can be defined by

$$LSCV_{K_r}(h_r, w) = \sum_{i=1}^{n} \sum_{j=1}^{n_i} w_i \left[Y_{ijr}^* - \widehat{\beta}_{r,K_r}^{(-i)}(t_{ij}; h_r, w) \right]^2. \quad (6.17)$$

The component-wise LSCV bandwidth $h_{r,lscv}$ is the minimizer of $LSCV_{K_r}(h_r, w)$ provided that it can be uniquely minimized over h_r. The use of $h_{r,lscv}$ can be heuristically justified using the similar decomposition as in equation (3.17), such that, by (6.17),

$LSCV_{K_r}(h_r, w)$
$$= \sum_{i=1}^{n} \sum_{j=1}^{n_i} w_i \left[Y_{ijr}^* - \beta_r(t_{ij}) \right]^2 + \sum_{i=1}^{n} \sum_{j=1}^{n_i} w_i \left[\beta_r(t_{ij}) - \widehat{\beta}_{K_r}^{(-i)}(t_{ij}; h_r, w) \right]^2$$
$$+ 2 \sum_{i=1}^{n} \sum_{j=1}^{n_i} w_i \left[Y_{ijr}^* - \beta_r(t_{ij}) \right] \left[\beta_r(t_{ij}) - \widehat{\beta}_{K_r}^{(-i)}(t_{ij}; h_r, w) \right]. \quad (6.18)$$

The same heuristic arguments for the cross-validation score of (3.17) suggest that, when n is large, the third term of the right side of (6.18) is approximately zero, and the second term is approximately the average squared error (ASE)

$$ASE\left[\widehat{\beta}_{r,K_r}(\cdot; h_r, w)\right] = \sum_{i=1}^{n} \sum_{j=1}^{n_i} w_i \left[\beta_r(t_{ij}) - \widehat{\beta}_{r,K_r}(t_{ij}; h_r, w) \right]^2,$$

so that, the component-wise cross-validated bandwidth $h_{r,cv}$ approximately minimizes the average squared error $ASE\left[\widehat{\beta}_{r,K_r}(\cdot;h_r,w)\right]$.

2. Cross-Validation Bandwidth Vector

Another procedure for selecting the data-driven bandwidths, which is suggested by Wu and Chiang (2000), is to compute a bandwidth vector **h** for

$$\widehat{\beta}_K(t;\mathbf{h},w) = \left(\widehat{\beta}_{0,K_0}(t;h_0,w),\ldots,\widehat{\beta}_{k,K_k}(t;h_k,w)\right)^T,$$

which minimizes the squared distance between the predicted and observed values of $Y(t)$ based on the model (6.1). Let $\widehat{\beta}_{\mathbf{K}}^{(-i)}(t;\mathbf{h},w)$ be a kernel estimator of $\beta(t) = (\beta_0(t),\ldots,\beta_k(t))^T$ computed using the pseudo-sample

$$\left\{(Y_{i^*j},t_{i^*j},\mathbf{X}_{i^*}) : i^* \neq i; j=1,\ldots,n_{i^*}\right\}, \tag{6.19}$$

which is the longitudinal sample with all the observations of the ith subject deleted. Define

$$LSCV_\mathbf{K}(\mathbf{h};w) = \sum_{i=1}^{n}\sum_{j=1}^{n_i} w_i\left[Y_{ij} - \mathbf{X}_i^T\,\widehat{\beta}_{\mathbf{K}}^{(-i)}(t_{ij};\mathbf{h},w)\right]^2 \tag{6.20}$$

to be the LSCV score for $\mathbf{h} = (h_0,\ldots,h_k)^T$. The cross-validated bandwidth vector $\mathbf{h}_{LSCV} = (h_{0,LSCV},\ldots,h_{k,LSCV})^T$ is then defined to be the minimizer of $LSCV_\mathbf{K}(\mathbf{h};w)$, provided that $LSCV_\mathbf{K}(\mathbf{h};w)$ can be uniquely minimized.

Similar justifications as in Section 3.3 can also be used to evaluate the adequacy of the LSCV criterion in (6.20). In this case,

$$\begin{aligned} &LSCV_\mathbf{K}(\mathbf{h};w) \\ &= \sum_{i=1}^{n}\sum_{j=1}^{n_i} w_i\left[Y_{ij} - \mathbf{X}_i^T\,\beta(t_{ij})\right]^2 + \sum_{i=1}^{n}\sum_{j=1}^{n_i} w_i\left\{\mathbf{X}_i^T\left[\beta(t_{ij}) - \widehat{\beta}_\mathbf{K}^{(-i)}(t_{ij};\mathbf{h},w)\right]\right\}^2 \\ &\quad + 2\sum_{i=1}^{n}\sum_{j=1}^{n_i} w_i\left[Y_{ij} - \mathbf{X}_i^T\,\beta(t_{ij})\right]\left\{\mathbf{X}_i^T\left[\beta(t_{ij}) - \widehat{\beta}_\mathbf{K}^{(-i)}(t_{ij};\mathbf{h},w)\right]\right\}. \end{aligned} \tag{6.21}$$

The first term of the right side of (6.21) does not depend on the bandwidths, while, because of the definition of $\beta^{(-i)}(t;\mathbf{h},w)$, the third term of (6.21) is approximately zero. Denote by $ASE\left[\widehat{\beta}_\mathbf{K}(\cdot;\mathbf{h},w)\right]$ the average squared error of $\mathbf{X}_i^T\,\widehat{\beta}_\mathbf{K}(t_{ij};\mathbf{h},w)$, i.e.,

$$ASE\left[\widehat{\beta}_\mathbf{K}(\cdot;\mathbf{h},w)\right] = \sum_{i=1}^{n}\sum_{j=1}^{n_i} w_i\left\{\mathbf{X}_i^T\left[\beta(t_{ij}) - \widehat{\beta}_\mathbf{K}(t_{ij};\mathbf{h},w)\right]\right\}^2.$$

The expectation of the second term of the right side of (6.21) is actually the expectation of $ASE\left[\widehat{\beta}_{\mathbf{K}}^{(-i)}(\cdot; \mathbf{h}, w)\right]$, which approximates the expectation of $ASE\left[\widehat{\beta}_{\mathbf{K}}(\cdot; \mathbf{h}, w)\right]$ when n is large. Thus, $\mathbf{h}_{LSCV}$ is justifiable because it approximately minimizes the average squared error $ASE\left[\widehat{\beta}_{\mathbf{K}}(\cdot; \mathbf{h}, w)\right]$.

3. A Combined Cross-Validation Approach

When the dimensionality of $\mathbf{X}_i$ is high, i.e., $(k+1)$ is large, the search for $\mathbf{h}_{LSCV}$ could be computationally intensive and numerically infeasible. The amount of computation can escalate dramatically when the dimensionality $(k+1)$ is 3 or larger. In practice, it is usually easy to find a suitable range of the bandwidths by examining the plots of the fitted curves. Within a given range of $\mathbf{h} = (h_0, \ldots, h_k)$, the value of $\mathbf{h}_{LSCV}$ can be approximated by computing $LSCV_{\mathbf{K}}(h; w)$ through a series of $\mathbf{h} = (h_0, \ldots, h_k)$ choices. Given that the component-wise cross-validated bandwidth $h_{r, lscv}$ is computed based on a univariate minimization of the LSCV score (6.17), a computationally feasible approach is to combine the above two cross-validation procedures. In this combined approach, the component-wise bandwidths $(h_{0, lscv}, \ldots, h_{k, lscv})^T$ from (6.17) are used as the initial values for $\mathbf{h}$, and $\mathbf{h}_{LSCV}$ can be computed by (6.20) through a grid search around the nearby values of $(h_{0, lscv}, \ldots, h_{k, lscv})^T$. This combined cross-validation method for searching $\mathbf{h}_{LSCV}$ could be computationally faster than a global search of $\mathbf{h}_{LSCV}$ based on the minimization of $LSCV_{\mathbf{K}}(\mathbf{h}; w)$ in (6.20) alone.

6.3   Component-wise Penalized Smoothing Splines

Another smoothing method for the estimation of $\beta(t) = \left(\beta_0(t), \ldots, \beta_k(t)\right)^T$ is the roughness penalty approach based on the quantities of (6.3), (6.4) and (6.5), which extends the method of Chapter 5 to the model (6.1). This approach leads to a class of penalized smoothing spline estimators for $\beta(t)$. We describe here the estimators developed in Chiang, Rice and Wu (2001).

6.3.1   Estimators by Component-wise Roughness Penalty

Suppose that the design time points are contained in a compact set $[a, b]$ and $\beta_r(t)$ are twice differentiable for all $t \in [a, b]$. Extending the score function (5.1) to the pseudo-sample $\mathscr{Y}_r$ of (6.8), a roughness penalized least squares estimator, also known as penalized smoothing spline estimator, $\widehat{\beta}_{r, RP}(t; \lambda_r, w)$ of $\beta_r(t)$ for any $0 \leq r \leq k$ is obtained by minimizing the following score function, which is referred to as the penalized least squares criterion,

$$J_w\left(\beta_r; \lambda_r, RP\right) = \sum_{i=1}^{n} \sum_{i=1}^{n_i} \left\{ w_i \left[Y_{ijr}^* - \beta_r(t_{ij})\right]^2 \right\} + \lambda_r RP\left[\beta_r(\cdot)\right], \qquad (6.22)$$

where λ_r is a non-negative smoothing parameter, $w = (w_1, \ldots, w_n)^T$ with w_i being non-negative weights, and $RP\left[\beta_r(\cdot)\right]$ is a roughness penalizing function of $\beta_r(t)$ measuring the roughness or smoothness of the curve $\beta_r(t)$. In practice, $RP[\beta_r(\cdot)]$ is unknown in advance, so that the choice of $RP[\beta_r(\cdot)]$ is often subjective.

By penalizing the integrated squares of the second derivatives of $\beta_r(t)$, a penalized smoothing spline estimator $\widehat{\beta}_{r,J}(t; \lambda_r, w)$ of $\beta_r(t)$, $0 \leq r \leq k$, can be obtained by minimizing

$$J_w\left(\beta_r; \lambda_r, \beta_r''\right) = \sum_{i=1}^{n} \sum_{i=1}^{n_i} \left\{ w_i \left[Y_{ijr}^* - \beta_r(t_{ij})\right]^2 \right\} + \lambda_r \int_a^b \left[\beta_r''(s)\right]^2 ds, \qquad (6.23)$$

where λ_r, w and w_i are defined in equation (6.22). Let $\lambda = \left(\lambda_0, \dots, \lambda_k\right)^T$ be the vector of smoothing parameters. The penalized least squares estimator of $\beta(t) = \left(\beta_0(t), \dots, \beta_k(t)\right)^T$ based on (6.23) is

$$\widehat{\beta}_J\left(t; \lambda, w\right) = \left(\widehat{\beta}_{0,J}(t; \lambda_0, w), \dots, \widehat{\beta}_{k,J}(t; \lambda_k, w)\right)^T. \qquad (6.24)$$

Similar to the component-wise kernel estimators of Section 6.2 or the penalized least squares estimators of Chapter 5, usual choices for w_i include $w_i^* = 1/(n n_i)$ and $w_i^{**} = 1/N$. It follows from (6.23) that, when $\beta_r''(s)$ is given, a larger λ_r is associated with a larger penalty term $\lambda_r \int_a^b \left[\beta_r''(s)\right]^2 ds$, which leads to an over-smoothed penalized least squares estimator $\widehat{\beta}_{r,S}(t; \lambda_r, w)$. On the other hand, a smaller λ_r gives a smaller penalty term $\lambda_r \int_a^b \left[\beta_r''(s)\right]^2 ds$, which leads to a under-smoothed penalized least squares estimator $\widehat{\beta}_{r,S}(t; \lambda_r, w)$.

The minimizer $\widehat{\beta}_{r,J}(t; \lambda_r, w)$ of (6.23) is a cubic spline and a linear statistic of Y_{ijr}^*. This can be seen by considering the set of compactly supported functions

$$\mathscr{H}_{[a,b]} = \Big\{ g(\cdot): \quad g \text{ and } g' \text{ are absolutely}$$
$$\text{continuous on } [a, b], \text{ and } \int_a^b \left[g''(s)\right]^2 ds < \infty \Big\}.$$

Setting the Gateaux derivative of $J_w\left(\beta_r; \lambda_r, \beta''\right)$ to zero, $\widehat{\beta}_{r,J}(t; \lambda_r, w)$ uniquely minimizes (6.23) if and only if it satisfies the normal equation

$$\sum_{i=1}^{n} \sum_{j=1}^{n_i} \left\{ w_i \left[Y_{ijr}^* - \widehat{\beta}_{r,J}(t; \lambda_r, w)\right] g(t_{ij}) \right\} = \lambda_r \int_a^b \widehat{\beta}_{r,J}''(t; \lambda_r, w) \, g''(s) \, ds, \qquad (6.25)$$

for all g in a dense subset of $\mathscr{H}_{[a,b]}$. The same argument as in Wahba (1975) then shows that there is a symmetric function $S_{\lambda_r}(t, s)$, which belongs to $\mathscr{H}_{[a,b]}$ when either t or s is fixed, so that $\widehat{\beta}_{r,J}(t; \lambda_r, w)$ is a natural cubic spline estimator given by

$$\widehat{\beta}_{r,J}(t; \lambda_r, w) = \sum_{i=1}^{n} \sum_{j=1}^{n_i} \left[w_i S_{\lambda_r}(t, t_{ij}) Y_{ijr}^* \right]. \qquad (6.26)$$

As discussed in Section 5.1, the explicit expression of $S_{\lambda_r}(t, s)$ is unknown, and the theoretical properties of $\widehat{\beta}_{r,J}(t; \lambda_r, w)$ can be derived by approximating $S_{\lambda_r}(t, s)$ with an equivalent kernel function, which has an explicit expression.

6.3.2 Estimators by Combined Roughness Penalty

In addition to the component-wise roughness penalized approach based on the expression (6.3), an alternative method, which is described in Hoover et al. (1998), is to use the expression (6.1) directly and minimize

$$
J_w^*(\beta; \lambda, \beta'') = \sum_{i=1}^{n} \sum_{j=1}^{n_i} w_i \left\{ Y_{ij} - \left[\sum_{l=0}^{k} X_i^{(l)} \beta_l(t_{ij}) \right] \right\}^2 + \sum_{l=0}^{k} \lambda_l \int_a^b \left[\beta_l''(s) \right]^2 ds \quad (6.27)
$$

with respect to $\beta(t) = (\beta_0(t), \ldots, \beta_k(t))^T$, where λ_l, $l = 0, \ldots, k$, are non-negative smoothing parameters and $\lambda = (\lambda_0, \ldots, \lambda_k)^T$. The minimizer of (6.27),

$$
\widehat{\beta}_{J^*}(t; \lambda, w) = \left(\widehat{\beta}_{0, J^*}(t; \lambda, w), \ldots, \widehat{\beta}_{k, J^*}(t; \lambda, w) \right)^T, \quad (6.28)
$$

is then a smoothing spline estimator of $\beta(t)$ with $\widehat{\beta}_{l, J^*}(t; \lambda, w)$ as the component estimator of $\beta_l(t)$.

The penalized score functions, $J_w(\beta_r; \lambda_r, \beta_r'')$ of (6.23) and $J_w^*(\beta; \lambda, \beta'')$ of (6.27), use different squared errors and penalty terms. Thus, the estimators $\widehat{\beta}_J(t; \lambda, w)$ and $\widehat{\beta}_{J^*}(t; \lambda, w)$ given in (6.24) and (6.28), respectively, may not have the same numerical values. Computationally, $J_w(\beta_r; \lambda_r, \beta_r'')$ is minimized with respect to $\beta_r(t)$ only, but minimizing $J_w^*(\beta; \lambda, \beta'')$ requires solving a linear system which involves all the components of $\beta(t)$ simultaneously. Consequently, the computation involved in minimizing $J_w(\beta_r; \lambda_r, \beta_r'')$ is much simpler than that involved in minimizing $J_w^*(\beta; \lambda, \beta'')$. Because $\widehat{\beta}_r(t; \lambda_r, w)$ has a simple linear expression, its asymptotic properties can be developed by methods similar to that with cross-sectional i.i.d. data. Theoretical properties of the spline estimators obtained by minimizing $J_w^*(\beta; \lambda, \beta'')$ have not been developed. Thus, the smoothing spline estimator $\widehat{\beta}_J(t; \lambda, w)$ of (6.24) has the advantage of being computationally simple with known asymptotic properties over the smoothing spline estimator $\widehat{\beta}_{J^*}(t; \lambda, w)$ of (6.28). On the other hand, because $\widehat{\beta}_J(t; \lambda, w)$ relies on the component-wise expression (6.3), it cannot be applied to situations involving time-dependent covariates, while $\widehat{\beta}_{J^*}(t; \lambda, w)$ may be generally applied to situations with any covariates.

6.3.3 Cross-Validation Smoothing Parameters

The choice of smoothing parameter λ_r controls the size of the penalizing term of $J_w(\beta_r; \lambda_r, \beta_r'')$ and plays the key role for determining the appropriateness of $\widehat{\beta}_J(t; \lambda_r, w)$. Unlike the kernel-based local smoothing method of Section 6.2, which only uses the local observations with time points t_{ij} in a neighborhood around t, the smoothing spline estimator $\widehat{\beta}_{r,J}(t; \lambda_r, w)$ is in principle a global smoothing estimator, so that, an ideal choice of λ_r may depend on the structures of the intra-subject correlations. But, since the correlation structures of the data are often completely unknown and $\widehat{\beta}_{r,J}(t; \lambda_r, w)$ is asymptotically equivalent to an equivalent kernel estimator (see Section 6.7), the cross-validation procedure of Section 6.2.2 can be similarly used for selecting the smoothing parameters $\lambda = (\lambda_0, \dots, \lambda_k)^T$.

1. Component-wise Cross-Validated Parameters

When the pseudo-observations $\mathscr{Y}_r^{(-i)}$ in (6.16) are used, the component-wise cross-validation smoothing parameter λ_r for any $0 \leq r \leq k$ can be selected by minimizing the cross-validation score

$$LSCV_{r,J}(\lambda_r, w) = \sum_{i=1}^{n} \sum_{j=1}^{n_i} w_i \left[Y_{ijr}^* - \widehat{\beta}_{r,J}^{(-i)}(t_{ij}; \lambda_r, w) \right]^2 \qquad (6.29)$$

with respect to λ_r, where $\widehat{\beta}_{r,J}^{(-i)}(t_{ij}; \lambda_r, w)$ is the penalized smoothing spline estimator computed using the pseudo-sample $\mathscr{Y}_r^{(-i)}$ in (6.16) and the penalized least squares function (6.23). If $LSCV_{r,J}(\lambda_r, w)$ can be uniquely minimized, the minimizer $\lambda_{r, lscv}$ of $LSCV_{r,J}(\lambda_r, w)$ is the component-wise leave-one-subject-out cross-validated (LSCV) smoothing parameter. The use of $\lambda_{r, lscv}$ can be heuristically justified by the same derivations as in (6.18) and (6.19) that it approximately minimizes the average squared error of the penalized spline estimator $\widehat{\beta}_{r,J}(t; \lambda_r, w)$.

2. Cross-Validated Parameter Vectors

When the squared error of $\widehat{\beta}_J(t; \lambda, w)$ based on the original model (6.1) is considered, Chiang, Rice and Wu (2001) suggests to select the smoothing parameters by the LSCV procedure which minimizes the cross-validation score function

$$LSCV_J(\lambda, w) = \sum_{i=1}^{n} \sum_{j=1}^{n_i} \left\{ w_i \left[Y_{ij} - \sum_{l=0}^{k} X_i^{(l)} \widehat{\beta}_{l,J}^{(-i)}(t_{ij}; \lambda_l, w) \right]^2 \right\}, \qquad (6.30)$$

where $\widehat{\beta}_{r,J}^{(-i)}(t; \lambda_r, w)$ is the smoothing spline estimator computed from (6.24) using the sample (6.19), which is the remaining data with all the observations of the ith subject deleted. The cross-validation smoothing parameters $\lambda_{LSCV} = (\lambda_{0, LSCV}, \dots, \lambda_{k, LSCV})^T$ are defined to be the minimizer of $LSCV_J(\lambda, w)$, provided that $LSCV_J(\lambda, w)$ can be uniquely minimized with respect to $\lambda = (\lambda_0, \dots, \lambda_k)^T$. Using the similar decomposition as in (6.21), the cross-validated smoothing parameters $\lambda_{LSCV} = (\lambda_{0, LSCV}, \dots, \lambda_{k, LSCV})^T$ approximately minimizes the average squared error of $\mathbf{X}_i^T \widehat{\beta}_j^{(-i)}(t; \lambda, w)$ as a predictor of Y_{ij}.

3. Combined Cross-Validated Parameters

Since the computation of λ_{LSCV} could be intensive for a large k when $LSCV_J(\lambda, w)$ is minimized directly with respect to λ, the search for λ_{LSCV} can be simplified by combining the univariate optimization algorithm of separately minimizing $LSCV_{r,J}(\lambda_r, w)$ of (6.29) for each $0 \le r \le k$ with the algorithm of minimizing $LSCV_J(\lambda, w)$ of (6.30). In this combined approach, the first step is to compute the component-wise cross-validated smoothing parameters $\{\lambda_{0,lscv}, \ldots, \lambda_{k,lscv}\}$ based on (6.29) as the initial values, and the second step is to compute the combined cross-validated smoothing parameter vector which minimizes $LSCV_J(\lambda, w)$ in (6.30) over a grid value around $\{\lambda_{0,lscv}, \ldots, \lambda_{k,lscv}\}$. This combined cross-validation procedure, which uses $\{\lambda_{0,lscv}, \ldots, \lambda_{k,lscv}\}$ as the initial values, may speed up the computation for the approximate values of $\lambda_{LSCV} = (\lambda_{0,LSCV}, \ldots, \lambda_{k,LSCV})^T$.

6.4   Bootstrap Confidence Intervals

Since the numbers of repeated measurements n_i, $i = 1, \ldots, n$, are allowed to be different and the possible intra-subject correlations of the data are often completely unknown, the corresponding asymptotic distributions of the estimators may involve bias and correlation terms which are difficult to estimate. Statistical inferences based on the asymptotic distributions of the smoothing estimators may be difficult to implement in practice. The confidence intervals described in this section follow the same framework of resampling-subject bootstrap described in Chapters 3 to 5. When the context is clear, we denote by $\widehat{\beta}_r(t)$ any estimator of $\beta_r(t)$ given in Sections 6.2 and 6.3.

Approximate Bootstrap Pointwise Confidence Intervals:

(a) **Computing Bootstrap Estimators.** *Generate B independent bootstrap samples using the resampling-subject bootstrap procedure of Section 3.4.1 and compute the B bootstrap estimators* $\{\widehat{\beta}_{r,1}^b(t), \ldots, \widehat{\beta}_{r,B}^b(t)\}$ *of* $\beta_r(t)$.

(b) **Approximate Bootstrap Confidence Interval.** *Let* $L_{r,\alpha/2}^b(t)$ *and* $U_{r,\alpha/2}^b(t)$ *be the* $[100 \times (\alpha/2)]$th *and* $[100 \times (1 - \alpha/2)]$th, *i.e., lower and upper* $[100 \times (\alpha/2)]$th, *percentiles, respectively, calculated based on the above B bootstrap estimators. The approximate* $[100 \times (1 - \alpha)]\%$ *bootstrap confidence interval for* $\beta_r(t)$ *is given by*

$$\left(L_{r,\alpha/2}^b(t), \ U_{r,\alpha/2}^b(t) \right). \tag{6.31}$$

The normal approximated bootstrap confidence interval for $\beta_r(t)$ *is*

$$\widehat{\beta}_r(t) \pm z_{1-\alpha/2} \times \widehat{se}\left(t; \widehat{\beta}_r^b\right), \tag{6.32}$$

where $\widehat{se}(t; \widehat{\beta}_r^b)$ is the estimated standard deviation of $\widehat{\beta}_r(t)$ from the B bootstrap estimators $\{\widehat{\beta}_{r,1}^b(t), \dots, \widehat{\beta}_{r,B}^b(t)\}$, such that,

$$\widehat{se}\left(t; \widehat{\beta}_r^b\right) = \left\{\frac{1}{B-1} \sum_{s=1}^{B} \left[\widehat{\beta}_{r,s}^b(t) - \frac{1}{B}\sum_{l=1}^{B} \widehat{\beta}_{r,l}^b(t)\right]^2\right\}^{1/2}, \qquad (6.33)$$

and $z_{1-\alpha/2}$ is the $\left[100 \times (1-\alpha/2)\right]$th percentile of the standard normal distribution. □

Similar to the procedures of Sections 4.2 and 5.1, the bootstrap confidence intervals described above ignore the biases of the smoothing estimators. For datasets with large n, the biases of the smoothing estimators are small, so that ignoring the biases does not have a significant impact on the coverage probabilities of the confidence intervals given in (6.31), (6.32) and (6.33).

6.5   R Implementation

6.5.1   The BMACS CD4 Data

The BMACS CD4 data has been described in Section 1.2. In Section 2.4.1, we evaluated certain covariate effects on the post-infection CD4 percentage using a linear mixed-effects model where the covariate effects are assumed to be constant with time t. In Sections 3.5.2 and 4.3.2, we used the unstructured local smoothing method and spline-based smoothing method to estimate the mean time curve of CD4 percentage after HIV infection without considering other baseline covariates. Here, we use this dataset to illustrate how to fit a flexible structured nonparametric model (6.1), in which the model coefficients $\beta(t)$ are allowed to vary with time t.

We consider evaluating the effects of three time-invariant covariates, pre-HIV infection CD4 percentage, cigarette smoking, and age at HIV infection on the mean CD4 percentage after HIV infection using the model (6.1). Let t_{ij} be the time (in years) of the jth measurement for the ith individual after HIV infection, and Y_{ij} be the ith individual's CD4 percentage at time t_{ij} post-infection. For the covariates, let $X^{(1)}$ be the pre-infection CD4 percentage, $X^{(2)}$ be the individual's cigarettes smoking status (1 indicates a smoker, 0 indicates a nonsmoker), and $X^{(3)}$ be the ith individual's age at HIV infection. To obtain a better interpretation of our results, the covariates $X^{(1)}$ and $X^{(3)}$ are centered by subtracting their corresponding sample averages from the individual values. The observed centered covariates for the ith subject are

$$\begin{cases} X_i^{(1)} &= i\text{th subject's pre-infection CD4 percentage} \\ & \quad - \text{ sample mean of pre-infection CD4 percentage,} \\ X_i^{(3)} &= i\text{th subject's age at HIV infection} \\ & \quad - \text{ sample mean of age at HIV infection.} \end{cases} \qquad (6.34)$$

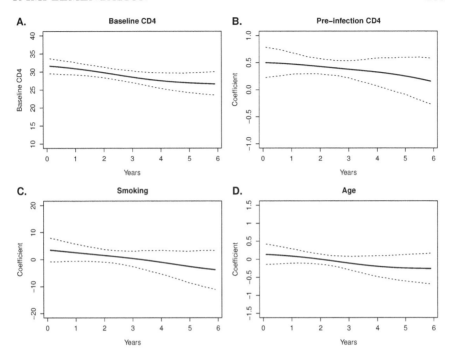

Figure 6.1 *Covariate effects on post-infection CD4 percentage using kernel esti-mators. The solid curves in (A)-(D) show the kernel estimators $\widehat{\beta}_{r,K_r}(t; h_r, w)$, $r = 0, 1, 2, 3$, respectively, based on (6.13) using a standard Gaussian kernel and $h_r = 1.5$, $r = 0, 1, 2, 3$. The dashed curves indicate the corresponding 95% pointwise bootstrap confidence intervals.*

Consequently, the time-varying coefficient model is

$$Y_{ij} = \beta_0(t_{ij}) + \beta_1(t_{ij}) X_i^{(1)} + \beta_2(t_{ij}) X_i^{(2)} + \beta_3(t_{ij}) X_i^{(3)} + \varepsilon_{ij}, \qquad (6.35)$$

where ε_{ij} is the mean zero error term at time t_{ij}. Based on (6.34), (6.35) and the computed sample means of pre-infection CD4 percentage and age at HIV infec-tion, the baseline CD4 percentage curve $\beta_0(t)$ represents the mean time curve of CD4 percentage for a nonsmoker with the sample average pre-infection CD4 of 42.9% and the average HIV infection age of 34.2 years, and $\beta_i(t)$, $i = 1, 2, 3$, represent the effects of pre-infection CD4 percentage, cigarette smoking, and age at infection, respectively, on the post-infection CD4 percentage.

Once we have the pseudo-samples for the baseline values and the three covariates of interest, we can estimate the baseline mean curve and each of the three covariate effects by the component-wise kernel smoothing method and penalized smoothing splines in Sections 6.2 and 6.3. The following R commands are used to compute the quantities in equations (6.4) and (6.5) and generate the pseudo-samples as in (6.8):

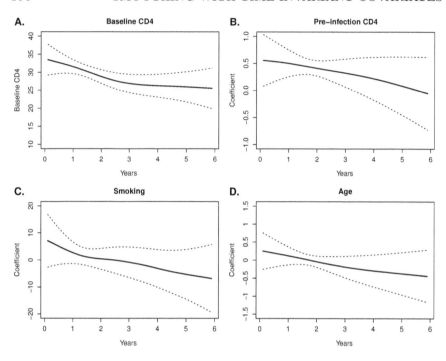

Figure 6.2 *Covariate effects on post-infection CD4 percentage using smoothing spline estimators. The solid curves in (A)-(D) show the smoothing spline estimators $\widehat{\beta}_{r,J}(t; \lambda_r, w)$, $r = 0, 1, 2, 3$, respectively, based on (6.23) using the smoothing parameters $\lambda_{LSCV} = (0.84, 4.4, 1.2, 5.2)^T$ for $r = 0, 1, 2, 3$. The dashed curves indicate the corresponding 95% pointwise bootstrap confidence intervals.*

```
# Obtain the sample size, N=283 for BMACS
> N <- length(unique(BMACS$ID))
# Obtain the baseline covariate:
# first observation per subject ID
> Xi <- do.call("rbind", as.list(by(BMACS[,c("preCD4C", "Smoke",
                  "ageC")], BMACS$ID, head,  n=1)))
# Formula (6.4-6.5)
> Xi <- as.matrix(cbind(1,Xi))
> EXX <- Reduce("+",
                  lapply(1:N, function(i) Xi[i,] %*% t(Xi[i,])))
> EstInv <- solve(EXX/N)
# Obtain the four pseudo longitudinal samples
> eX <- data.frame(IDD=1:N,
                  eX1 = apply(t(t(Xi)*EstInv[1,]), 1, sum),
                  eX2 = apply(t(t(Xi)*EstInv[2,]), 1, sum),
                  eX3 = apply(t(t(Xi)*EstInv[3,]), 1, sum),
                  eX4 = apply(t(t(Xi)*EstInv[4,]), 1, sum))
```

```
> BMACS <- merge(BMACS, eX, by="IDD")
> BMACS$PseudoY <- with(BMACS, cbind(eX1, eX2, eX3, eX4)*CD4)
```

Figure 6.1 shows the estimated $\beta_r(t)$, $r = 0, 1, 2, 3$, based on (6.13) using the standard Gaussian kernel and a subjective bandwidth choice of $h_r = 1.5$, $r = 0, 1, 2, 3$, by examining the estimated coefficient curves. The dashed lines present the corresponding bootstrap percentile 95% pointwise confidence intervals as in (6.31) for 59 equally spaced time points between 0.1 and 5.9 years. We can also choose the bandwidth vectors based on cross-validation as described in Section 6.2.2. With a quick grid search in the range of 0.5 to 3 for each component of the bandwidth vector, the bandwidth $\mathbf{h} = (0.5, 2.5, 3.0, 2.0)^T$ has the minimum LSCV score in (6.17). The kernel estimates with cross-validated bandwidths are similar compared to the coefficient curves with subjective bandwidths in Figure 6.1.

Figure 6.2 shows the estimated $\beta_r(t)$, $r = 0, 1, 2, 3$, based on (6.24) using the penalized smoothing splines (6.23) with the subject uniform measurement weight and the cross-validated smoothing parameters $\lambda_{LSCV} = (0.84, 4.4, 1.2, 5.2)^T$. For the ease of computation, a series of grid values for each λ_r in a small window around the cross-validated $\lambda_{r,LSCV}$ (selected only for the rth coefficient curve) are used for $r = 0, 1, 2, 3$, instead of searching over a 4-dimensional vector space in a wide range. The dashed lines represent the corresponding bootstrap 95% pointwise confidence intervals as in (6.32) for time points between 0.1 and 5.9 years.

Both Figures 6.1 and 6.2 show that the mean baseline CD4 percentage decreases quickly after HIV infection but the rate of declining is slowing down about 4 years post-infection. Consistent with the results from fitting a linear mixed-effects model in Section 2.4.1, neither smoking nor age at infection show a significant effect on the post-infection CD4 percentage with wide 95% confidence intervals covering the null effect. However, the pre-infection CD4 percentage seems to be positively associated with the post-infection CD4 percentage and its coefficient may be time-varying with a larger effect at the beginning of HIV infection.

6.5.2   A Simulation Study

We demonstrate here the performance of the estimation methods in Sections 6.2 and 6.3 through a simulation study. The simulation design has the data structure similar to the BMACS CD4 example. The simulated data are generated based on the time-varying coefficient model (6.1) with time-invariant covariates, $\mathbf{X} = \left(X^{(1)}, X^{(2)}\right)^T$, where $X^{(1)}$ is a binary random variable having the Bernoulli distribution with $p = 0.5$, $X^{(2)}$ is a continuous random variable having the normal distribution with mean 0 and standard deviation 4, and $X^{(1)}$ and $X^{(2)}$ are independent. The three nonlinear time-varying coeffi-

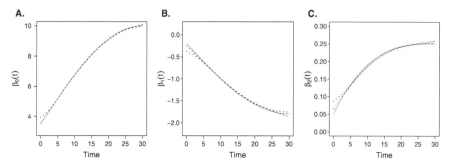

Figure 6.3 *The solid curves show the true values of $\beta_0(t)$, $\beta_1(t)$ and $\beta_2(t)$, respectively. The corresponding dotted and dashed curves are the averages of the estimated curves over 1000 simulated samples based on the kernel smoothing estimates with the standard Gaussian kernel and the penalized smoothing spline estimates using the cross-validated bandwidths and smoothing parameters, respectively.*

cients $\beta_r(t)$, $r = 0, 1, 2$, are given by

$$
\begin{cases}
\beta_0(t) &= 3.5 + 6.5 \sin\left(t\,\pi/60\right), \\
\beta_1(t) &= -0.2 - 1.6 \cos\left[(t-30)\,\pi/60\right], \\
\beta_2(t) &= 0.25 - 0.0074\left[(30-t)/10\right]^3.
\end{cases}
\tag{6.36}
$$

Taking into account the common scenario in epidemiological studies that the subjects may randomly miss some scheduled visits, the time points t_{ij} can be generated from a set of pre-specified values. As shown in the real data examples in Section 1.2.2, the number of repeated measurements for each subject often range from 1 to 20, we consider here the "unbalanced" design in the sense that the time points t_{ij} may not be the same for all the subjects. Thus, in each simulated sample with $n = 400$ subjects, we assume that the subjects are scheduled to be observed at 31 equally spaced time design points $\{0, 1, \ldots, 30\}$. However, at each given time point, a subject has 60% probability to be randomly missing. This leads to unequal numbers of repeated measurements n_i for individual subjects. In addition, the random errors ε_{ij} are generated from the Gaussian process with zero mean and covariance function

$$
Cov\left(\varepsilon_{i_1 j_1}, \varepsilon_{i_2 j_2}\right) =
\begin{cases}
0.0625 \exp\left(-\left|t_{i_1 j_1} - t_{i_2 j_2}\right|\right), & \text{if } i_1 = i_2; \\
0, & \text{if } i_1 \neq i_2.
\end{cases}
\tag{6.37}
$$

The time-dependent responses Y_{ij} are obtained by substituting t_{ij}, $\mathbf{X}_i$, ε_{ij} with the correlations (6.37), and the coefficient curves $\beta_r(t)$ of (6.36) into (6.1).

For each simulated sample $\left\{\left(Y_{ij}, t_{ij}, \mathbf{X}_i^T\right) : i = 1, \ldots, 400; j = 1, \ldots, n_i\right\}$, we estimate the coefficient curves, $\beta_r(t)$, $r = 0, 1, 2$, by applying the estimation procedures to obtain the kernel smoothing estimators in (6.13) with the standard Gaussian kernel and the penalized smoothing spline estimators in (6.23) using the cross-validated bandwidths and smoothing parameters, respectively. We repeat the simulation 1000 times. Figure 6.3 shows the true curves of $\beta_0(t)$, $\beta_1(t)$, $\beta_2(t)$, and the averages of the estimated curves over all the simulated samples. These simulation results demonstrate that both estimation approaches provide reasonably good estimators at least for the interior time points of the time-varying coefficients. But, compared to the penalized smoothing spline estimators, the kernel estimators have slightly larger bias near the boundary.

6.6 Asymptotic Properties for Kernel Estimators

We derive in this section the asymptotic representations of the mean squared errors and the mean integrated squared errors of $\widehat{\beta}_{r,K_r}(t; h_r, w)$ given in (6.10). Special cases for the $w_i^* = 1/(n n_i)$ and $w_i^{**} = 1/N$ weights are direct consequences of the general asymptotic results for $\widehat{\beta}_{r,K_r}(t; h_r, w)$.

6.6.1 Mean Squared Errors

Similar to the asymptotic setup of Section 3.6, we specify for mathematical simplicity that the time points $\{t_{ij} : i = 1, \ldots, n; j = 1, \ldots, n_i\}$, are randomly selected from a cumulative distribution function $F(\cdot)$ with density $f(\cdot)$. But n_i, $i = 1, \ldots, n$, are assumed to be nonrandom. This corresponds to random designs in regression analysis. Designs with nonrandom t_{ij} can be viewed as a special case with $F(\cdot)$ having point mass at the given values of t_{ij}.

Because, by the expression of (6.11), $\widehat{\beta}_{\mathbf{K}}(t; \mathbf{h}, w)$ is a R^{k+1}-valued vector, its distance from $\beta(t) = \left(\beta_0(t), \ldots, \beta_k(t)\right)^T$ can be measured by different indices. Suppose that the statistical objective is to evaluate the adequacy of one component $\widehat{\beta}_{r,K_r}(t; h_r, w)$ at a time point t. A natural risk index for $\widehat{\beta}_{r,K_r}(t; h_r, w)$ at time point t is the mean squared error defined by

$$MSE^* \left[\widehat{\beta}_{r,K_r}(t; h_r, w) \right] = E \left\{ \left[\widehat{\beta}_{r,K_r}(t; h_r, w) - \beta_r(t) \right]^2 \right\}. \tag{6.38}$$

For the risk of the vector estimator $\widehat{\beta}_{\mathbf{K}}(t; \mathbf{h}, w)$ at time point t, an obvious choice is the linear combination of the component-wise mean squared errors,

$$
\begin{aligned}
MSE^* \left[\widehat{\beta}_{\mathbf{K}}(t; \mathbf{h}, w) \right] &= E \left\{ \left[\widehat{\beta}_{\mathbf{K}}(t; \mathbf{h}, w) - \beta(t) \right]^T W \left[\widehat{\beta}_{\mathbf{K}}(t; \mathbf{h}, w) - \beta(t) \right] \right\} \\
&= \sum_{r=0}^{k} W_r E \left\{ \left[\widehat{\beta}_{r,K_r}(t; h_r, w) - \beta_r(t) \right]^2 \right\},
\end{aligned}
\tag{6.39}
$$

where W is a $(k+1) \times (k+1)$ diagonal matrix of weights with nonnegative elements $\{W_0, \ldots, W_k\}$.

Similar to the situations in kernel regression estimators discussed in Section 3.6.1, the conditional moments of the kernel estimators $\widehat{\beta}_{r,K_r}(t; h_r, w)$, hence, the right-side terms of (6.38) and (6.39), may not exist, so that modifications of the mean squared errors defined in (6.38) and (6.39) are used. It follows from (6.10) that $\widehat{\beta}_{r,K_r}(t; h_r, w)$ can be written as

$$\widehat{\beta}_{r,K_r}(t; h_r, w) = \left[\widehat{f}_{r,K_r}(t; h_r, w)\right]^{-1} \widehat{m}_{r,K_r}(t; h_r, w), \qquad (6.40)$$

where the right side terms are given by

$$\widehat{m}_{r,K_r}(t; h_r, w) = \sum_{i=1}^{n} \sum_{j=1}^{n_i} \left[\left(\frac{w_i}{h_r}\right) Y_{ijr}^* K_r\left(\frac{t - t_{ij}}{h_r}\right)\right] \qquad (6.41)$$

and

$$\widehat{f}_{r,K_r}(t; h_r, w) = \sum_{i=1}^{n} \sum_{j=1}^{n_i} \left[\left(\frac{w_i}{h_r}\right) K_r\left(\frac{t - t_{ij}}{h_r}\right)\right]. \qquad (6.42)$$

Similar to the derivation of (3.21), straightforward algebra using (6.40), (6.41) and (6.42) shows that

$$\left[1 - d_{r,K_r}(t; h_r, w)\right]\left[\widehat{\beta}_{r,K_r}(t; h_r, w) - \beta_r(t)\right]$$
$$= [f(t)]^{-1}\left[\widehat{m}_{r,K_r}(t; h_r, w) - \beta_r(t)\widehat{f}_{r,K_r}(t; h_r, w)\right], \qquad (6.43)$$

where $d_{r,K_r}(t; h_r, w) = 1 - [\widehat{f}_{r,K_r}(t; h_r, w)/f(t)]$. For any interior point t of the support of $f(\cdot)$, it can be shown by the same method used in (3.37) and (3.38) of Section 3.6 that $d_{r,K_r}(t; h_r, w) \to 0$ in probability as $n \to \infty$ and $h_r \to 0$. Then, applying direct algebra to (6.41), (6.42) and (6.43), the following approximation holds in probability when the sample size n is sufficiently large,

$$\left[1 + o_p(1)\right]\left[\widehat{\beta}_{r,K_r}(t; h_r, w) - \beta_r(t)\right] = f^{-1}(t)\widehat{R}_{r,K_r}(t; h_r, w), \qquad (6.44)$$

where $\widehat{R}_{r,K_r}(t; h_r, w) = \widehat{m}_{r,K_r}(t; h_r, w) - \beta_r(t)\widehat{f}_{r,K_r}(t; h_r, w)$.

Using the approximation give in (6.44), the local and global risks of $\widehat{\beta}_{r,K_r}(\cdot; h_r, w)$ can be defined by the modified mean squared error,

$$MSE\left[\widehat{\beta}_{r,K_r}(t; h_r, w)\right] = E\left\{\left[\widehat{R}_{r,K_r}(t; h_r, w)/f(t)\right]^2\right\} \qquad (6.45)$$

and the modified mean integrated squared error,

$$MISE\left[\widehat{\beta}_{r,K_r}(\cdot; h_r, w)\right] = \int MSE\left[\widehat{\beta}_{r,K_r}(s; h_r, w)\right]\pi(s)\,ds, \qquad (6.46)$$

respectively, where $\pi(s)$ is a known non-negative weight function supported by a compact subset in the interior of the support of $f(\cdot)$. As in (3.40), the assumption that $\pi(\cdot)$ has compact support within the support of $f(\cdot)$ is to remove the boundary effects of the kernel estimators. Based on the definitions of (6.45) and (6.46), the local and global risks of $\widehat{\beta}_{\mathbf{K}}(\cdot;\mathbf{h},w)$ are measured by

$$MSE\left[\widehat{\beta}_{\mathbf{K}}(t;\mathbf{h},w)\right] = \sum_{r=0}^{k} W_r\, MSE\left[\widehat{\beta}_{r,K_r}(t;h_r,w)\right] \qquad (6.47)$$

and

$$MISE\left[\widehat{\beta}_{\mathbf{K}}(\cdot;\mathbf{h},w)\right] = \sum_{r=0}^{k} W_r\, MISE\left[\widehat{\beta}_{r,K_r}(\cdot;h_r,w)\right], \qquad (6.48)$$

respectively, where $\{W_0,\ldots,W_k\}$ are the known non-negative constants defined in (6.39).

6.6.2 Asymptotic Assumptions

We make the following assumptions throughout this chapter for the asymptotic properties of $MSE\left[\widehat{\beta}_{\mathbf{K}}(t;\mathbf{h},w)\right]$ and $MISE\left[\widehat{\beta}_{\mathbf{K}}(\cdot;\mathbf{h},w)\right]$ defined in (6.47) and (6.48):

(a) *For all t on the real line, $f(t)$ is continuously differentiable and there are non-negative constants p_r, $r = 0,\ldots,k$, so that $\beta_r(t)$ are (p_r+2) times continuously differentiable with respect to t.*

(b) *For all $r,l = 0,\ldots,k$, $E[|X^{(r)}|^4]$ and the $(2+\delta)$th moments of $|\widehat{e}_{r+1,l+1}|$ are finite for some $\delta > 0$.*

(c) *The variance and covariance of the error process $\varepsilon(t)$ satisfy*

$$\sigma^2(t) = E\left[\varepsilon^2(t)\right] < \infty \quad and \quad \rho_\varepsilon(t) = \lim_{t'\to t} E\left[\varepsilon(t)\varepsilon(t')\right] < \infty.$$

Furthermore, $\sigma^2(t)$ and $\rho_\varepsilon(t)$ are continuous for all t on the real line.

(d) *The kernel function $K_r(\cdot)$ is a compactly supported (p_r+2)th order kernel which satisfies $\int u^j K_r(u)\,du = 0$ for all $1 \le j < p_r+2$, $\int K_r(u)\,du = 1$,*

$$M_{(p_r+2)}(K_r) = \int u^{p_r+2} K_r(u)\,du < \infty \quad and \quad R(K_r) = \int K_r^2(u)\,du < \infty.$$

(e) *The weight vector $w = (w_1,\ldots,w_n)^T$ satisfies $w_i \ge 0$ for all $1 \le i \le n$, $\sum_{i=1}^{n}(w_i n_i) = 1$, $\sum_{i=1}^{n}(w_i^2 n_i^2) = O(n^{-1})$ and $\sum_{i=1}^{n}(w_i^2 n_i) \to 0$ as $n \to \infty$.*

(f) *The bandwidth $h_r > 0$ satisfies $h_r \to 0$, $nh_r \to \infty$ and $\sum_{i=1}^{n}(w_i^2 n_i)/h_r \to 0$ as $n \to \infty$.* $\qquad\qquad\square$

Similar to the assumptions for the unstructured kernel estimation in Section 3.6, we have that $\sigma^2(t) \ge \rho_\varepsilon(t)$ in general, and the strict inequality between $\sigma^2(t)$ and $\rho_\varepsilon(t)$ happens when ε_{ij} includes an independent measurement error, e.g., ε_{ij} satisfies the white noise model $\varepsilon_{ij} = s(t_{ij}) + W_i$, where $s(t)$ is a mean zero Gaussian stationary process and W_i is an independent measurement error. Conditions such as the compact support of $K_r(\cdot)$ and the smoothness conditions of $f(t)$, $\beta_r(t)$, $\sigma^2(t)$ and $\rho_\varepsilon(t)$, are assumed for the simplicity of the derivations and may be relaxed in practice. Analogous asymptotic results may be derived when these conditions are modified or even weakened. For example, noncompact support kernels, such as the standard Gaussian kernels, are commonly used in practice and usually lead to satisfactory results.

6.6.3 Asymptotic Risk Representations

By (6.45), the mean squared error of $\widehat{\beta}_{r,K_r}(t; h_r, w)$ is defined through the second moment of $f^{-1}(t)\widehat{R}_{r,K_r}(t; h_r, w)$. The bias and variance of $f^{-1}(t)\widehat{R}_{r,K_r}(t; h_r, w)$ are

$$B\left[\widehat{\beta}_{r,K_r}(t; h_r, w)\right] = E\left[f^{-1}(t)\widehat{R}_{r,K_r}(t; h_r, w)\right] \qquad (6.49)$$

and

$$V\left[\widehat{\beta}_{r,K_r}(t; h_r, w)\right] = f^{-2}(t)\,Var\left[\widehat{R}_{r,K_r}(t; h_r, w)\right], \qquad (6.50)$$

respectively, and the mean squared error of (6.45) has the decomposition

$$MSE\left[\widehat{\beta}_{r,K_r}(t; h_r, w)\right] = B^2\left[\widehat{\beta}_{r,K_r}(t; h_r, w)\right] + V\left[\widehat{\beta}_{r,K_r}(t; h_r, w)\right]. \qquad (6.51)$$

An important fact for deriving the asymptotic expression of $B\left[\widehat{\beta}_{r,K_r}(t; h_r, w)\right]$ and $V\left[\widehat{\beta}_{r,K_r}(t; h_r, w)\right]$ is that

$$\widehat{e}_{r+1,l+1} = e_{r+1,l+1} + O_p(n^{-1/2}).$$

For simplicity, the derivations of $B\left[\widehat{\beta}_{r,K_r}(t; h_r, w)\right]$ and $V\left[\widehat{\beta}_{r,K_r}(t; h_r, w)\right]$ here only involve the $n^{1/2}$ convergence rate of $\widehat{e}_{r+1,l+1}$, but not the exact asymptotic expression of $\left(\widehat{e}_{r+1,l+1} - e_{r+1,l+1}\right)$.

The next theorem summarizes the asymptotic expressions of the bias, variance, mean squared errors and mean integrated squared errors of the component-wise kernel estimators of $\beta(t) = \left(\beta_0(t), \ldots, \beta_k(t)\right)^T$. Let

$$M_r^{(0)}(t) = \sum_{r_1=0}^{k}\sum_{r_2=0}^{k}\left\{\beta_{r_1}(t)\beta_{r_2}(t)E\left[X_{r_1}X_{r_2}\left(\sum_{l=0}^{k}e_{r+1,l+1}X_l\right)^2\right]\right\} - \beta_r^2(t),$$

$$M_r^{(1)}(t) = M_r^{(0)}(t) + \sigma^2(t)E\left[\left(\sum_{l=0}^{k}e_{r+1,l+1}X_l\right)^2\right],$$

$$M_r^{(2)}(t) = M_r^{(0)}(t) + \rho_\varepsilon(t)E\left[\left(\sum_{l=0}^{k}e_{r+1,l+1}X_l\right)^2\right],$$

$$Q_{1r}(t) = M_{(p_r+2)}(K_r)\left[\frac{\beta_r^{(p_r+2)}(t)}{(p_r+2)!} + \frac{\beta_r^{(p_r+1)}(t)f'(t)}{(p_r+1)!f(t)}\right],$$

$$Q_{2r}(t) = f^{-1}(t)R(K_r)M_r^{(1)}(t).$$

Theorem 6.1. *Suppose that t is in the interior of the support of $f(\cdot)$ and Assumptions (a) to (f) are satisfied. The following conclusions hold.*

(a) *When n is sufficiently large, the asymptotic bias and variance of $\widehat{\beta}_{r,K_r}(t; h_r, w)$ have the following expressions*

$$B\left[\widehat{\beta}_{r,K_r}(t; h_r, w)\right] = h_r^{p_r+2} Q_{1r}(t) + o\left(h_r^{p_r+2}\right) + O\left(n^{-1/2}\right) \qquad (6.52)$$

and

$$V\left[\widehat{\beta}_{r,K_r}(t; h_r, w)\right]$$

$$= \left\{h_r^{-1}\left[\sum_{i=1}^{n}\left(w_i^2 n_i\right)\right]Q_{2r}(t) + \left[\sum_{i=1}^{n}\left(w_i^2 n_i^2\right) - \sum_{i=1}^{n}\left(w_i^2 n_i\right)\right]M_r^{(2)}(t)\right\}[1 + o(1)]$$

$$+ B\left[\widehat{\beta}_{r,K_r}(t; h_r, w)\right]O\left(n^{-1/2}\right) + O\left(n^{-1}\right). \qquad (6.53)$$

(b) *If, in additional to Assumptions (a) to (f), h_r also satisfies*

$$n^{1/2} h_r^{p_r+2} \to \infty \quad \text{and} \quad n\sum_{i=1}^{n}\left(w_i^2 n_i\right)/h_r \to \infty \quad \text{as } n \to \infty,$$

the right side terms in (6.52) and (6.53) reduce to, respectively,

$$B\left[\widehat{\beta}_{r,K_r}(t; h_r, w)\right] = h_r^{p_r+2} Q_{1r}(t)\left[1 + o(1)\right],$$

and

$$V\left[\widehat{\beta}_{r,K_r}(t; h_r, w)\right] = h_r^{-1}\left[\sum_{i=1}^{n}\left(w_i^2 n_i\right)\right]Q_{2r}(t)\left[1 + o(1)\right] + O\left(n^{-1/2} h_r^{p_r+2}\right).$$

(c) *For sufficiently large n, the asymptotic expressions of $MSE\left[\widehat{\beta}_{r,K_r}(t; h_r, w)\right]$, $MISE\left[\widehat{\beta}_{r,K_r}(\cdot; h_r, w)\right]$, $MSE\left[\widehat{\beta}_K(t; \mathbf{h}, w)\right]$ and $MISE\left[\widehat{\beta}_K(\cdot; \mathbf{h}, w)\right]$ are obtained by substituting $B\left[\widehat{\beta}_{r,K_r}(t; h_r, w)\right]$ and $V\left[\widehat{\beta}_{r,K_r}(t; h_r, w)\right]$ in (6.51), (6.46), (6.47) and (6.48) with the right-side terms of (6.52) and (6.53), respectively.* ∎

Proof of Theorem 6.1 is given in Section 6.6.6.

The next theorem, which is a consequence of Theorem 6.1, shows the theoretically optimal bandwidth choices and the corresponding optimal mean squared errors of $\widehat{\beta}_{r,K_r}(t; h_r, w)$ when the numbers of repeated measurements $\{n_i : i = 1, \ldots, n\}$ are bounded. In practice, the asymptotic mean squared errors under the case of $\{n_i : i = 1, \ldots, n\}$ bounded and n tending to infinity may be used to approximate the finite sample mean squared errors when $\{n_i : i = 1, \ldots, n\}$ are small relative to n.

Theorem 6.2. *Suppose that the assumptions of Theorem 6.1 are satisfied, n_i are bounded, i.e., $n_i \leq c$ for some $c \geq 1$ and all $i = 1, \ldots, n$, and*

$$n^{-1} c_1 \leq \sum_{i=1}^{n} \left(w_i^2 \, n_i \right) \leq n^{-1} c_2 \quad \text{for some constants } c_1 > 0 \text{ and } c_2 > 0.$$

The following asymptotic results hold:

(a) *The optimal bandwidths $h_{r,opt}(t)$ and $h_{r,opt}$ minimizing $MSE\left[\widehat{\beta}_{r,K_r}(t; h_r, w)\right]$ and $MISE\left[\widehat{\beta}_{r,K_r}(\cdot; h_r, w)\right]$, respectively, for all $h_r > 0$, are given by*

$$h_{r,opt}(t) = \left[\sum_{i=1}^{n} \left(w_i^2 \, n_i \right) \right]^{1/(2p_r+5)} \left[\frac{Q_{2r}(t)}{2\,(p_r+2)\,Q_{1r}^2(t)} \right]^{1/(2p_r+5)} \tag{6.54}$$

and

$$h_{r,opt} = \left[\sum_{i=1}^{n} \left(w_i^2 \, n_i \right) \right]^{1/(2p_r+5)} \left\{ \frac{\int Q_{2r}(s)\,\pi(s)\,ds}{2\,(p_r+2)\,\left[\int Q_{1r}^2(s)\,\pi(s)\,ds\right]} \right\}^{1/(2p_r+5)}. \tag{6.55}$$

(b) *The asymptotically optimal MSE and MISE corresponding to $h_{r,opt}(t)$ and $h_{r,opt}$, respectively, are given by*

$$MSE\left[\widehat{\beta}_{r,K_r}\left(t; h_{r,opt}(t), w\right)\right]$$

$$= \left[\sum_{i=1}^{n} \left(w_i^2 \, n_i \right) \right]^{(2p_r+4)/(2p_r+5)} \left[Q_{2r}(t) \right]^{(2p_r+4)/(2p_r+5)} \left[Q_{1r}(t) \right]^{2/(2p_r+5)}$$

$$\times \left[\left(2\,p_r+4 \right)^{-(2p_r+4)/(2p_r+5)} + \left(2\,p_r+4 \right)^{1/(2p_r+5)} \right] \left[1 + o(1) \right] \tag{6.56}$$

and

$$MISE\left[\widehat{\beta}_{r,K_r}\left(\cdot; h_{r,opt}, w\right)\right]$$

$$= \left[\sum_{i=1}^{n} \left(w_i^2 \, n_i \right) \right]^{(2p_r+4)/(2p_r+5)} \left[\int Q_{2r}(s)\,\pi(s)\,ds \right]^{(2p_r+4)/(2p_r+5)}$$

$$\times \left[\int Q_{1r}^2(s)\,\pi(s)\,ds \right]^{1/(2p_r+5)} \tag{6.57}$$

$$\times \left[\left(2\,p_r+4 \right)^{-(2p_r+4)/(2p_r+5)} + \left(2\,p_r+4 \right)^{1/(2p_r+5)} \right] \left[1 + o(1) \right].$$

(c) *The $MSE\left[\widehat{\beta}_{\mathbf{K}}(t; \mathbf{h}_{opt}(t), w)\right]$ and $MISE\left[\widehat{\beta}_{\mathbf{K}}(\cdot; \mathbf{h}_{opt}, w)\right]$ corresponding to the theoretically optimal bandwidths*

$$\mathbf{h}_{opt}(t) = \left(h_{0,opt}(t), \ldots, h_{k,opt}(t) \right)^T \quad \text{and} \quad \mathbf{h}_{opt} = \left(h_{0,opt}, \ldots, h_{r,opt} \right)^T$$

are obtained by substituting $MSE\left[\widehat{\beta}_{r,K_r}(t; h_r, w)\right]$ and $MISE\left[\widehat{\beta}_{r,K_r}(\cdot; h_r, w)\right]$ in (6.47) and (6.48) with the right side of (6.56) and (6.57). ■

Proof of Theorem 6.2 is given in Section 6.6.6.

6.6.4   Remarks and Implications

The results of Theorems 6.1 and 6.2 lead to a few immediate consequences.

1. Consistency

The estimator $\widehat{\beta}_{r,K_r}(t; h_r, w)$ is asymptotically consistent (or simply consistent) for $\beta_r(t)$ if $MSE\big[\widehat{\beta}_{r,K_r}(t; h_r, w)\big] \to 0$ as $n \to \infty$. It immediately follows from Theorem 6.1 that $\widehat{\beta}_{r,K_r}(t; h_r, w)$ is consistent if and only if $h_r \to 0$ and $h_r^{-1}\sum_{i=1}^{n}\big(w_i^2 n_i\big) \to 0$ as $n \to \infty$. The intra-subject correlations of the data have no effect on the asymptotic expressions of the bias $B\big[\widehat{\beta}_{r,K_r}(t; h_r, w)\big]$. But, depending on the values of $\sum_{i=1}^{n}\big(w_i^2 n_i\big)$ when n is sufficiently large, the intra-subject correlations may influence the values of the asymptotic variance $V\big[\widehat{\beta}_{r,K_r}(t; h_r, w)\big]$.

2. Theoretically Optimal Bandwidths

It is important to note that, since $MSE\big[\widehat{\beta}_{r,K_r}(t; h_{r,opt}(t), w)\big]$ measures the risk of the estimator $\widehat{\beta}_{r,K_r}(t; h_{r,opt}(t), w)$ at a given time point t, the optimal bandwidth $h_{r,opt}(t)$ depends on t. In other words, the theoretically optimal bandwidths, which minimize the dominating term of (6.56), are possibly different at different time points. On the other hand, the theoretically optimal bandwidth $h_{r,opt}$, which minimizes the dominating term of (6.57) does not depend on any specific time point. Thus, depending on the objectives of the longitudinal analysis, the theoretically local optimal bandwidth $h_{r,opt}(t)$ and the theoretically global optimal bandwidth $h_{r,opt}$ are possibly different. In real applications, the statistical objective is usually to estimate the coefficient curve $\beta_r(t)$ for t over a range of interest, so that the global risk measure $MISE\big[\widehat{\beta}_{r,K_r}(\cdot; h_r, w)\big]$ and its corresponding optimal bandwidth $h_{r,opt}$ are usually more relevant.

3. Practical Bandwidths

The explicit expressions of the optimal bandwidths $h_{r,opt}(t)$ and $h_{r,opt}$ given in (6.54) and (6.55) are still not ready for practical use, because they depend on the derivatives of the unknown coefficient curves. Although, in principle, it is possible to estimate these unknown quantities and compute these bandwidths by plugging in the estimated quantities into the expressions of (6.54) and (6.55), these approaches do not generally work well in practice, because the derivatives of the coefficient curves are difficult to estimate. Thus, the expressions of (6.54) and (6.55) can only be used as a theoretical guideline for the convergence rates of $h_{r,opt}(t)$ and $h_{r,opt}$ as functions of n and $\{n_i : i = 1, \ldots, n\}$, while the exact values of $h_{r,opt}(t)$ and $h_{r,opt}$ are generally not available and difficult to estimate. The resampling subject cross-validation procedures of

Section 6.2.2 remain to be the most useful approach for selecting the bandwidths in practice.

6.6.5   Useful Special Cases

The results of Theorems 6.1 and 6.2 are derived for the general weight choices $w = (w_1, \ldots, w_n)^T$. Different expressions of MSE and MISE may be derived based on Theorems 6.1 and 6.2 when the specific form of w is provided. Since the subject uniform weight $w^* = (1/(nn_1), \ldots, 1/(nn_n))^T$ and the measurement uniform weight $w^{**} = (1/N, \ldots, 1/N)^T$ are the two most commonly used weight choices in practice, this section presents some direct consequences of Theorems 6.1 and 6.2 for these two weight choices.

1. Subject Uniform Weight

For the weight choice $w_i^* = 1/(nn_i)$, it follows that

$$\sum_{i=1}^n \left[(w_i^*)^2 n_i \right] = \sum_{i=1}^n (n^2 n_i)^{-1},$$

so that the following two corollaries are direct consequences of Theorems 6.1 and 6.2. First, the following corollary shows the convergence rates for the bias and variance terms of $\widehat{\beta}_{r, K_r}(t; h_r, w^*)$.

Corollary 6.1. *Suppose that t is an interior point of the support of $f(\cdot)$, the weight vector $w^* = (w_1^*, \ldots, w_n^*)^T$, $w_i^* = 1/(nn_i)$, is used, and Assumptions (a)-(d) and (f) are satisfied. When n is sufficiently large, $B\left[\widehat{\beta}_{r, K_r}(t; h_r, w^*)\right]$, $V\left[\widehat{\beta}_{r, K_r}(t; h_r, w^*)\right]$, $MSE\left[\widehat{\beta}_{r, K_r}(t; h_r, w^*)\right]$ and $MISE\left[\widehat{\beta}_{r, K_r}(t; h_r, w^*)\right]$ are given by the corresponding terms in Theorem 6.1 by substituting $\sum_{i=1}^n \left(w_i^2 n_i \right)$ with $\sum_{i=1}^n \left[1/(n^2 n_i) \right]$.* ∎

The next corollary shows that, under the theoretically optimal bandwidth choices, the optimal convergence rate for the MSE and MISE of $\widehat{\beta}_{r, K_r}(t; h_r, w^*)$ is $\left[\sum_{i=1}^n (n^2 n_i)^{-1} \right]^{-(2p_r+4)/(2p_r+5)}$.

Corollary 6.2. *Under the assumptions of Theorem 6.2, the pointwise optimal bandwidth $h_{r, opt}(t)$ and the global optimal bandwidth $h_{r, opt}$ for the weight $w_i^* = 1/(nn_i)$ are given by (6.54) and (6.55), respectively, with $\sum_{i=1}^n \left(w_i^2 n_i \right)$ substituted by $\sum_{i=1}^n (n^2 n_i)^{-1}$. The optimal $MSE\left[\widehat{\beta}_{r, K_r}(t; h_{r, opt}(t), w^*)\right]$ and $MISE\left[\widehat{\beta}_{r, K_r}(\cdot; h_{r, opt}, w^*)\right]$ are given by (6.56) and (6.57), respectively, with $\sum_{i=1}^n \left(w_i^2 n_i \right)$ substituted by $\sum_{i=1}^n (n^2 n_i)^{-1}$.* ∎

In Corollary 6.2, n_i, $i = 1, \ldots, n$, are assumed to be bounded, i.e., $n_i \leq$

c for some constant $c > 0$. Consequently, $n^{-1} c_1 \leq \sum_{i=1}^{n} \left(n^2 n_i\right)^{-1} \leq n^{-1} c_2$ for some constants $c_1 > 0$ and $c_2 > 0$. This implies that the best rate for both $MSE\left[\widehat{\beta}_{r,K_r}(t; h_{r,opt}(t), w^*)\right]$ and $MISE\left[\widehat{\beta}_{r,K_r}(\cdot; h_{r,opt}, w^*)\right]$ converging to zero is $n^{(2p_r+4)/(2p_r+5)}$.

2. Measurement Uniform Weight

For the weight choice $w_i^{**} = 1/N$, we have $\sum_{i=1}^{n} \left(w_i^{**}\right)^2 n_i = 1/N$. The following two corollaries are direct consequences of Theorems 6.1 and 6.2 by substituting $\sum_{i=1}^{n} \left(w_i^2 n_i\right)$ with $1/N$.

Corollary 6.3. *Suppose that t is an interior point of the support of $f(\cdot)$, the weight vector $w^{**} = \left(w_1^{**}, \ldots, w_n^{**}\right)^T$ with $w_i^{**} = 1/N$ is used, and Assumptions (a)-(d) and (f) are satisfied. When n is sufficiently large, $B\left[\widehat{\beta}_{r,K_r}(t; h_r, w^{**})\right]$, $V\left[\widehat{\beta}_{r,K_r}(t; h_r, w^{**})\right]$, $MSE\left[\widehat{\beta}_{r,K_r}(t; h_r, w^{**})\right]$ and $MISE\left[\widehat{\beta}_{r,K_r}(t; h_r, w^{**})\right]$ are given by the corresponding terms in Theorem 6.1 by substituting $\sum_{i=1}^{n} \left(w_i^2 n_i\right)$ with $1/N$.* ∎

As a consequence of Theorem 6.2, the next corollary shows that the optimal convergence rate for the MSE and MISE of $\widehat{\beta}_{r,K_r}(t; h_r, w^{**})$ is $N^{-(2p_r+4)/(2p_r+5)}$.

Corollary 6.4. *Under the assumptions of Theorem 6.2, the pointwise optimal bandwidth $h_{r,opt}(t)$ and the global optimal $h_{r,opt}$ for the weight $w_i^{**} = 1/N$ are given by (6.54) and (6.55), respectively, with $\sum_{i=1}^{n} \left(w_i^2 n_i\right)$ substituted by $1/N$. The optimal $MSE\left[\widehat{\beta}_{r,K_r}(t; h_{r,opt}(t), w^{**})\right]$ and $MISE\left[\widehat{\beta}_{r,K_r}(\cdot; h_{r,opt}, w^{**})\right]$ corresponding to $h_{r,opt}(t)$ and $h_{r,opt}$ are given by (6.46) and (6.47), respectively, with $\sum_{i=1}^{n} \left(w_i^2 n_i\right)$ substituted by $1/N$.* ∎

If the numbers of repeated measurements n_i, $i = 1, \ldots, n$, are bounded, then N/n is bounded, so that, by Corollary 6.4, the optimal rate for both $MSE\left[\widehat{\beta}_{r,K_r}(t; h_{r,opt}(t), w^{**})\right]$ and $MISE\left[\widehat{\beta}_{r,K_r}(\cdot; h_{r,opt}, w^{**})\right]$ converging to zero is $n^{(2p_r+4)/(2p_r+5)}$. More generally, the convergence rates of $MSE\left[\widehat{\beta}_{r,K_r}(t; h_r, w)\right]$ and $MISE\left[\widehat{\beta}_{r,K_r}(\cdot; h_r, w)\right]$ depend on whether and how n_i, $i = 0, \ldots, n$, converge to infinity relative to $n \to \infty$. In practice, it is usually unknown that whether or how n_i converge to infinity as $n \to \infty$, so that any bandwidth choices purely minimizing the asymptotic expressions of $MSE\left[\widehat{\beta}_{r,K_r}(t; h_r, w)\right]$ and $MISE\left[\widehat{\beta}_{r,K_r}(\cdot; h_r, w)\right]$ may not be preferable. In contrast, the cross-validation bandwidths of Section 6.2.2 and the bootstrap inference procedures of Section 6.4 only rely on the available data, which do not depend on the potentially unrealistic assumptions.

6.6.6   Theoretical Derivations

We provide here the proofs of Theorems 6.1 and 6.2.

1. Proof of Theorem 6.1:

    Following equations (6.1), (6.3), (6.4), (6.5) and (6.8), we have that

$$Y_{ijr}^* = \sum_{l_1=0}^{k} \left\{ X_i^{(l_1)} \left[\sum_{l_2=0}^{k} \widehat{e}_{r+1,l_2+1} X_i^{(l_2)} \right] \beta_{l_1}(t_{ij}) \right\} + \left[\sum_{l=0}^{k} \widehat{e}_{r+r,l+1} X_i^{(l)} \right] \varepsilon_{ij},$$

and the obvious identity

$$\beta_r(t) = E\left\{ \sum_{l_1=0}^{k} \left[X^{(l_1)} \left(\sum_{l_2=0}^{k} e_{r+1,l_2+1} X^{(l_2)} \right) \beta_{l_1}(t) \right] \right\} \tag{6.58}$$

holds for all $r = 0, \ldots, k$. By Assumption (b), the definition of $\widehat{e}_{r+r,l+1}$, equation (6.3) and the fact that

$$\widehat{e}_{r+r,l+1} = e_{r+r,l+1} + O_p(n^{-1/2})$$

imply that

$$
\begin{aligned}
E\left(Y_{ijr}^* | t_{ij} = s\right) &= \sum_{l_1=0}^{k} E\left\{ X_i^{(l_1)} \left[\sum_{l_2=0}^{k} \widehat{e}_{r+1,l_2+1} X_i^{(l_2)} \right] \right\} \beta_{l_1}(s) \\
&= \beta_r(s) + O(n^{-1/2}).
\end{aligned}
$$

It then follows from (6.36), (6.40), (6.45) and the change of variables that, when n is sufficiently large,

$$
\begin{aligned}
&B\left[\widehat{\beta}_{r,K_r}(t; h_r, \mathbf{w})\right] \\
&= \left[\frac{1}{h_r f(t)}\right] \sum_{i=1}^{n} \sum_{j=1}^{n_i} \int w_i \left[E\left(Y_{ijr}^* | t_{ij} = s\right) - \beta_r(t)\right] K_r\left(\frac{t-s}{h_r}\right) f(s)\, ds \\
&= f^{-1}(t) \int \left[\beta_r(t - h_r u) - \beta_r(t)\right] f(t - h_r u) K_r(u)\, du + O(n^{-1/2}).
\end{aligned}
$$

The asymptotic expression of (6.52) follows from Assumption (d) and the Taylor expansions of $\beta_r(t - h_r u)$ and $f(t - h_r u)$ at $\beta_r(t)$ and $f(t)$, respectively.

    To derive the asymptotic expression of $V\left[\widehat{\beta}_{r,K_r}(t; h_r, \mathbf{w})\right]$, it is useful to first consider the decomposition

$$\left[f^{-1}(t)\widehat{R}_{r,K_r}(t; h_r, w)\right]^2 = A_r^{(1)}(t) + A_r^{(2)}(t) + A_r^{(3)}(t),$$

where, with $Z_{ijr}(t) = w_i \left[Y_{ijr}^* - \beta_r(t)\right]$, the $A_r^{(l)}(t)$, $l = 1, 2, 3$, are defined by

$$
\begin{aligned}
A_r^{(1)}(t) &= f^{-2}(t) h_r^{-2} \sum_{i=1}^{n} \sum_{j=1}^{n_i} \left[Z_{ijr}^2(t) K_r^2\left(\frac{t - t_{ij}}{h_r}\right)\right], \\
A_r^{(2)}(t) &= f^{-2}(t) h_r^{-2} \sum_{i=1}^{n} \sum_{j_1 \neq j_2=1}^{n_i} \left[Z_{ij_1 r}(t) Z_{ij_2 r}(t) K_r\left(\frac{t - t_{ij_1}}{h_r}\right) K_r\left(\frac{t - t_{ij_2}}{h_r}\right)\right]
\end{aligned}
$$

and

$$A_r^{(3)}(t) = f^{-2}(t) h_r^{-2} \sum_{i_1 \neq i_2} \sum_{j_1, j_2=1}^{n_i} \left[Z_{i_1 j_1 r}(t) Z_{i_2 j_2 r}(t) K_r \left(\frac{t - t_{i_1 j_1}}{h_r} \right) K_r \left(\frac{t - t_{i_2 j_2}}{h_r} \right) \right].$$

Using (6.1), (6.2) and straightforward computation, it follows that

$$
\begin{aligned}
Z_{ijr}^2(t) &= w_i^2 \left[\xi_{ijr}(t) \right]^2 + 2 w_i^2 \xi_{ijr}(t) \left[\sum_{l=0}^{k} \widehat{e}_{r+1,l+1} X_i^{(l)} \right] \varepsilon_{ij} \\
&\quad + w_i^2 \left[\sum_{l=0}^{k} \widehat{e}_{r+1,l+1} X_i^{(l)} \right]^2 \varepsilon_{ij}^2,
\end{aligned}
$$

where

$$\xi_{ijr}(t) = \sum_{l_1=0}^{k} \left\{ X_i^{(l_1)} \left[\sum_{l_2=0}^{k} \widehat{e}_{r+1,l_2+1} X_i^{(l_2)} \right] \beta_{l_1}(t_{ij}) \right\} - \beta_r(t).$$

Since $\mathbf{X}_i$ and ε_{ij} are independent, it follows from (6.58) and the definition of $M_r^{(0)}(t)$ that

$$E \left\{ \xi_{ijr}(t) \left[\sum_{l=0}^{k} \widehat{e}_{r+1,l+1} X_i^{(l)} \right] \varepsilon_i(t_{ij}) \Big| t_{ij} = s \right\} = 0,$$

$$
\begin{aligned}
&E \left\{ \left[\xi_{ijr}(t) \right]^2 \Big| t_{ij} = s \right\} \\
&= E \left\{ \left[\sum_{l_1=0}^{k} \left[X_i^{(l_1)} \left(\sum_{l_2=0}^{k} \widehat{e}_{r+1,l_2+1} X_i^{(l_2)} \right) \beta_{l_1}(t_{ij}) \right] - \beta_r(t) \right]^2 \Big| t_{ij} = s \right\} \\
&= E \left\{ \left[\sum_{l_1=0}^{k} \left[X_i^{(l_1)} \left(\sum_{l_2=0}^{k} e_{r l_2} X_i^{(l_2)} \right) \beta_{l_1}(t_{ij}) \right] \right]^2 \Big| t_{ij} = s \right\} \\
&\quad - 2 \beta_r(t) E \left\{ \sum_{l_1=0}^{k} \left[X_i^{(l_1)} \left(\sum_{l_2=0}^{k} e_{r+1,l_2+1} X_i^{(l_2)} \right) \beta_{l_1}(t_{ij}) \right] \Big| t_{ij} = s \right\} \\
&\quad + \beta_r^2(t) + o(1) \\
&= M_r^{(0)}(s) + \beta_r^2(s) - 2 \beta_r(t) \beta_r(s) + \beta_r^2(t) + o(1)
\end{aligned}
$$

and

$$E \left\{ \left[\sum_{l=0}^{k} \widehat{e}_{r+1,l+1} X_i^{(l)} \right]^2 \varepsilon_{ij}^2 \Big| t_{ij} = s \right\} = \sigma^2(s) E \left\{ \left[\sum_{l=0}^{k} e_{r+1,l+1} X^{(l)} \right]^2 \right\} [1 + o(1)].$$

The above equations lead to the expectations of the right-side terms of $Z_{ijr}^2(t)$ given $t_{ij} = s$.

To compute the expectations of $A_r^{(l)}(t)$, $l = 1, 2, 3$, it then follows that

$$E \left[A_r^{(1)}(t) \right]$$

$$= [h_r f(t)]^{-2} \sum_{i=1}^{n} \sum_{j=1}^{n_i} w_i^2 \int E\left[Z_{ijr}^2(t) | t_{ij} = s\right] K_r^2\left(\frac{t-s}{h_r}\right) f(s)\, ds \qquad (6.59)$$

$$= [h_r f(t)]^{-2} \sum_{i=1}^{n} \sum_{j=1}^{n_i} w_i^2 \left\{ \int \left\{ M_r^{(0)}(s) + \beta_r^2(s) - 2\beta_r(t)\beta_r(s) + \beta_r^2(t) \right.\right.$$

$$\left.\left. + \sigma^2(s) E\left[\left(\sum_{l=0}^{k} e_{r+1,l+1} X^{(l)}\right)^2\right] \right\} K_r^2\left(\frac{t-s}{h_r}\right) f(s)\, ds \left[1 + o(1)\right] \right\}$$

$$= f^{-2}(t) \sum_{i=1}^{n} \left\{ \left(w_i^2 n_i\right) h_r^{-1} \sum_{j=1}^{n_i} \left[M_r^{(1)}(t) R(K_r) f(t)\right] \left[1 + o(1)\right] \right\}$$

$$= h_r^{-1} \left[\sum_{i=1}^{n} \left(w_i^2 n_i\right)\right] f^{-1}(t) R(K_r) M_r^{(1)}(t) \left[1 + o(1)\right].$$

Using similar computations as those for $A_r^{(1)}(t)$, it can be shown that

$$E\left[A_r^{(2)}(t)\right] = \left[\sum_{i=1}^{n} \left(w_i^2 n_i^2\right) - \sum_{i=1}^{n} \left(w_i^2 n_i\right)\right] M_r^{(2)}(t) \left[1 + o(1)\right] \qquad (6.60)$$

and

$$E\left[A_r^{(3)}(t)\right] = \left\{ B\left[\widehat{\beta}_{r,K_r}(t; h_r, w)\right] + O\left(n^{-1/2}\right) \right\}^2. \qquad (6.61)$$

When n is sufficiently large, the asymptotic expression of (6.49) follows from equations (6.59), (6.60), (6.61) and

$$V\left[\widehat{\beta}_{r,K_r}(t; h_r, w)\right]$$

$$= \sum_{l=1}^{3} E\left[A_r^{(l)}(t)\right] - B^2\left[\widehat{\beta}_{r,K_r}(t; h_r, w)\right]$$

$$= E\left[A_r^{(1)}(t)\right] + E\left[A_r^{(2)}(t)\right] + B\left[\widehat{\beta}_{r,K_r}(t; h_r, w)\right] O\left(n^{-1/2}\right) + O\left(n^{-1}\right).$$

The rest of the results in Theorem 6.1 then follow as direct consequences of the above expectation and variance expressions. ∎

2. Proof of Theorem 6.2

Because $n_1, \ldots, n_n$ are assumed to be bounded below by a constant $c \geq 1$ and the assumption that $n^{-1} c_1 \leq \sum_{i=1}^{n} \left(w_i^2 n_i\right) \leq n^{-1} c_2$ holds, there is a constant $c_3 > 0$ such that

$$\left| n^{-1} - \sum_{i=1}^{n} \left(w_i^2 n_i\right) \right| \leq n^{-1} c_3. \qquad (6.62)$$

If $n^{1/(2p_r+5)} h_r \to 0$ as $n \to \infty$, it follows from (6.62) that, as $n \to \infty$,

$$n^{(2p_r+4)/(2p_r+5)} h_r^{2p_r+4} Q_{1r}^2(t) \to 0,$$

$$n^{(2p_r+4)/(2p_r+5)} h_r^{-1} \left[\sum_{i=1}^{n} (w_i^2 n_i) \right] Q_{2r}(t) \geq h_r^{-1} n^{-1/(2p_r+5)} c_3^{-1} Q_{2r}(t) \to \infty$$

and

$$n^{(2p_r+4)/(2p_r+5)} \left| \left[\frac{1}{n} - \sum_{i=1}^{n} (w_i^2 n_i) \right] M_r^{(2)}(t) \right| \leq n^{(2p_r+4)/(2p_r+5)} n^{-1} c_3^{-1} \left| M_r^{(2)}(t) \right| \to 0,$$

so that, Theorem 6.1 implies that, as $n \to \infty$,

$$n^{(2p_r+4)/(2p_r+5)} MSE \left[\widehat{\beta}_{r,K_r}(t; h_r, w) \right] \to \infty. \tag{6.63}$$

If $n^{1/(2p_r+5)} h_r \to \infty$ as $n \to \infty$, similar calculations as above then show that, as $n \to \infty$,

$$n^{(2p_r+4)/(2p_r+5)} h_r^{2p_r+4} Q_{1r}^2(t) \to \infty, \quad n^{(2p_r+4)/(2p_r+5)} h_r^{-1} \left[\sum_{i=1}^{n} (w_i^2 n_i) \right] Q_{2r}(t) \to 0$$

and

$$n^{(2p_r+4)/(2p_r+5)} \left| \left[n^{-1} - \sum_{i=1}^{n} (w_i^2 n_i) \right] M_r^{(2)}(t) \right| \to 0,$$

so that, (6.63) still holds.

It then suffices to consider the case that $h_r = n^{-1/(2p_r+5)} c_n$ for some c_n which does not converge to either 0 or ∞ when $n \to \infty$. Since, by (6.62) and the inequality

$$n^{(2p_r+4)/(2p_r+5)} \left[n^{-1} - \sum_{i=1}^{n} (w_i^2 n_i) \right] \leq n^{-1/(2p_r+5)} c_3^{-1} = o(1),$$

equations (6.47), (6.48) and (6.49) imply that

$$n^{(2p_r+4)/(2p_r+5)} MSE \left[\widehat{\beta}_{r,K_r}(t; h_r, w) \right]$$

$$= c_n^{2p_r+4} Q_{1r}^2(t) + c_n^{-1} n \left[\sum_{i=1}^{n} (w_i^2 n_i) \right] Q_{2r}(t) + o(1). \tag{6.64}$$

Setting the derivative of the right side of (6.64) to zero, $MSE \left[\widehat{\beta}_{r,K_r}(t; h_r, w) \right]$ is uniquely minimized by

$$c_n = n^{1/(2p_r+5)} \left[\sum_{i=1}^{n} (w_i^2 n_i) \right]^{1/(2p_r+5)} \left[\frac{Q_{2r}(t)}{(2p_r+4) Q_{1r}^2(t)} \right]^{1/(2p_r+5)},$$

which shows that the theoretically optimal bandwidth $h_{r,opt}(t)$ is given by (6.54). The asymptotically optimal $MSE \left[\widehat{\beta}_{r,K_r}(t; h_{r,opt}(t), w) \right]$ of (6.56) is obtained by substituting the bandwidth h_r of (6.48) and (6.49) with $h_{r,opt}(t)$ in (6.50).

Repeating similar derivations as above for $MISE \left[\widehat{\beta}_{r,K_r}(\cdot; h_r, w) \right]$, the expression of the asymptotically optimal bandwidth $h_{r,opt}$ can be obtained as (6.51) and the asymptotically optimal $MISE \left[\widehat{\beta}_{r,K_r}(\cdot; h_{r,opt}, w) \right]$ is given as (6.53). The asymptotically optimal $MSE \left[\widehat{\beta}_K(t; \mathbf{h}_{opt}(t), w) \right]$ and $MISE \left[\widehat{\beta}_K(\cdot; \mathbf{h}_{opt}, w) \right]$ are direct consequences of equations (6.43), (6.44), (6.52) and (6.53). ∎

6.7 Asymptotic Properties for Smoothing Splines

We present in this section the asymptotic properties of the component-wise penalized smoothing spline estimator $\widehat{\beta}_{r,J}(t; \lambda_r, w)$ of (6.26) in Section 6.3.1 with the measurement uniform weight $w^{**} = (1/N, \ldots, 1/N)^T$ and $t \in [a, b]$. Without loss of generality, we assume that $a = 0$ and $b = 1$. Extension to general $[a, b]$ can be obtained using the affine transformation $u = (t - a)/(b - a)$ for $t \in [a, b]$.

6.7.1 Assumptions and Equivalent Kernel Functions

1. Asymptotic Assumptions

We assume the following technical assumptions for $\widehat{\beta}_{r,J}(t; \lambda_r, w)$ throughout the section, which extend the assumptions of Section 5.3.1 to the time-varying coefficient model (6.1):

(a) *The time design points $\{t_{ij} : i = 1, \ldots, n; j = 1, \ldots, n_i\}$ satisfy Assumption (a) of Section 5.3.1 with the same $F(t)$, $f(t)$, $F_N(t)$, D_N and $f^{(v)}(t)$ for $v = 1, 2$.*

(b) *The coefficient curves $\beta_r(t)$, $r = 0, \ldots, k$, are four times differentiable and satisfy the boundary conditions $\beta_r^{(v)}(0) = \beta_r^{(v)}(1) = 0$ for $v = 2, 3$. The fourth derivatives $\beta_r^{(4)}(t)$, $r = 0, \ldots, k$, are Lipschitz continuous in the sense that*

$$\left| \beta_r^{(4)}(s_1) - \beta_r^{(4)}(s_2) \right| \leq c_{1r} \left| s_1 - s_2 \right|^{c_{2r}}$$

for all $s_1, s_2 \in [0, 1]$ and some positive constants c_{1r} and c_{2r}.

(c) *There exists a positive constant $\delta > 0$ such that $E\left[|\varepsilon(t)|^{2+\delta}\right] < \infty$ and $E\left(X_r^{4+\delta}\right) < \infty$ for all $r = 0, \ldots, k$.*

(d) *The smoothing parameters λ_r, $r = 0, \ldots, k$, satisfy Assumption (d) of Section 5.3.1. Specifically, $\lambda_r \to 0$, $N\lambda_r^{1/4} \to \infty$ and $\lambda_r^{-5/4} D_N \to 0$ as $n \to \infty$.*

(e) *Same as Assumption (e) of Section 5.3.1, we define $\sigma^2(t) = E\left[\varepsilon^2(t)\right]$ and $\rho_\varepsilon(t) = \lim_{t' \to t} E\left[\varepsilon(t)\varepsilon(t')\right]$. Both $\sigma^2(t)$ and $\rho_\varepsilon(t)$ are continuous, and $\sigma^2(t) \geq \rho_\varepsilon(t)$, where the strict inequality holds if $\varepsilon(t)$ includes an independent measurement error.*   □

2. Equivalent Kernel Estimator

We now extend the equivalent kernel framework of Section 5.3 to approximate the unknown weight function $S_{\lambda_r}(t, s)$ of (6.26). The goal is to derive an explicit equivalent kernel function which can be used in place of $S_{\lambda_r}(t, s)$, so that the asymptotic properties of $\widehat{\beta}_{r,J}(t; \lambda_r, w^{**})$ can be established through the equivalent kernel estimator. From the expressions of (6.3), (6.8) and (6.26), the analogous differential equation of (5.8) under the current context is

$$\lambda_r g_r^{(4)}(t) + f(t) g_r(t) = f(t) \beta_r(t), \quad t \in [0,1], \tag{6.65}$$

with $g_r^{(v)}(0) = g_r^{(v)}(1) = 0$ for $v = 2, 3$. Then, any solution $g_r(t)$ of (6.65) is associated with a Green's function $G_{\lambda_r}(t, s)$ which satisfies

$$g_r(t) = \int_0^1 G_{\lambda_r}(t, s) \beta_r(s) f(s) \, ds.$$

We can then define

$$H_{\lambda_r}(t, s) = H_{\lambda_r / \gamma^4}^U [\Gamma(t), \Gamma(s)] \Gamma^{(1)}(s) f^{-1}(s) \tag{6.66}$$

to be the equivalent kernel of $S_{\lambda_r}(t, s)$, where

$$\gamma = \int_0^1 f^{1/4}(s) \, ds, \quad \Gamma(t) = \gamma^{-1} \int_0^t f^{1/4}(s) \, ds$$

and $H_{\lambda_r}^U(t, s)$ is the equivalent kernel for the uniform density which has the expression

$$H_{\lambda_r}^U(t, s) = \left(\frac{\lambda_r^{-1/4}}{2\sqrt{2}} \right) \left[\sin\left(\frac{\lambda_r^{-1/4}}{\sqrt{2}} |t - s| \right) + \cos\left(\frac{\lambda_r^{-1/4}}{\sqrt{2}} |t - s| \right) \right]$$
$$\times \exp\left(-\frac{\lambda_r^{-1/4}}{\sqrt{2}} |t - s| \right). \tag{6.67}$$

Substituting $S_{\lambda_r}(t, t_{ij})$ in (6.26) by the equivalent kernel $H_{\lambda_r}(t, t_{ij})$, our equivalent kernel estimator of $\beta_r(t)$ with the uniform weight $w_i^{**} = 1/N$ is

$$\tilde{\beta}_r(t; w^{**}) = \frac{1}{N} \sum_{i=1}^n \sum_{j=1}^{n_i} \left[H_{\lambda_r}(t, t_{ij}) Y_{ijr}^* \right]. \tag{6.68}$$

The next lemma shows that $H_{\lambda_r}(t, s)$ is the dominating term of $G_{\lambda_r}(t, s)$ and can be used to approximate $S_{\lambda_r}(t, s)$.

Lemma 6.1. *Assume that Assumptions (a) and (d) are satisfied. When n is sufficiently large, there are positive constants α_1, α_2, κ_1 and κ_2 so that*

$$\left| G_{\lambda_r}(t, s) - H_{\lambda_r}(t, s) \right| \leq \kappa_1 \exp\left(-\alpha_1 \lambda_r^{-1/4} |t - s| \right), \tag{6.69}$$

$$\left| \frac{\partial^v}{\partial t^v} G_{\lambda_r}(t, s) \right| \leq \kappa_1 \lambda_r^{-(v+1)/4} \exp\left(-\alpha_2 \lambda_r^{-1/4} |t - s| \right), \tag{6.70}$$

$$\left| S_{\lambda_r}(t, s) - G_{\lambda_r}(t, s) \right| \leq \kappa_2 \lambda_r^{-1/2} D_N \exp\left(-\alpha_1 \lambda_r^{-1/4} |t - s| \right) \tag{6.71}$$

$$\left| \frac{\partial^v}{\partial t^v} S_{\lambda_r}(t, s) \right| \leq \kappa_2 \lambda_r^{-(v+1)/4} D_N \exp\left(-\alpha_2 \lambda_r^{-1/4} |t - s| \right) \tag{6.72}$$

hold uniformly for $t, s \in [0, 1]$ and $0 \leq v \leq 3$. ∎

Proof of Lemma 6.1 is given in Section 6.7.3.

As discussed in Section 5.3.1, $H_{\lambda_r}(t, s)$ is not the only equivalent kernel that could be considered. Our presentation of approximating $S_{\lambda_r}(t, s)$ with $H_{\lambda_r}(t, s)$ is based on the theoretical derivation of Chiang, Rice and Wu (2001).

6.7.2 Asymptotic Distributions and Mean Squared Errors

We now summarize the main results of this section. Derivations of the theoretical results are deferred to Section 6.7.3.

1. Asymptotic Distributions

We first introduce a number of quantities to be used in the asymptotic expressions. Define, for each $r = 0, \ldots, k$, $t \in [0, 1]$ and $t_1 \neq t_2 \in [0, 1]$

$$
\begin{cases}
U_r &= \sum_{l=0}^{k} (e_{r+1, l+1} X^{(l)}), \\
M_r^{(0)}(t_1, t_2) &= \sum_{r_1, r_2=0}^{k} \left[\beta_{r_1}(t_1) \beta_{r_2}(t_2) E\left(X^{(r_1)} X^{(r_2)} U_r^2 \right) \right] - \beta_r(t_1) \beta_r(t_2), \\
M_r^{(1)}(t) &= M_r^{(0)}(t) + \sigma^2(t) e_{rr}, \\
M_r^{(2)}(t_1, t_2) &= M_r^{(0)}(t_1, t_2) + \rho_\varepsilon(t_1, t_2) e_{rr},
\end{cases}
$$

$M_r^{(0)}(t) = M_r^{(0)}(t, t)$ and $M_r^{(2)}(t) = M_r^{(2)}(t, t)$. The following theorem shows that asymptotically $\widehat{\beta}_{r,J}(t; \lambda_r, w^{**})$ has a normal distribution when N is sufficiently large.

Theorem 6.1. *Suppose that Assumptions (a) through (e) are satisfied, t is an interior point of $[0, 1]$, and there are constants $\lambda_{r,0} \geq 0$ and $a_0 \geq 0$, such that $\lim_{n \to \infty} N^{1/2} \lambda_r^{9/8} = \lambda_{r,0}$, $\lim_{n \to \infty} \left[N^{-1} \left(\sum_{i=1}^{n} n_i^2 \right) \lambda_r^{1/4} \right] = a_0$ and $\lim_{n \to \infty} N n^{-9/8} = 0$. Then, as $n \to \infty$, $\widehat{\beta}_{r,J}(t; \lambda_r, w^{**})$ is asymptotically normal in the sense that*

$$
\left(N \lambda_r^{1/4} \right)^{1/2} \left[\widehat{\beta}_{r,J}(t; \lambda_r, w^{**}) - \beta_r(t) \right] \to N\left(\lambda_{r,0} b_r(t), \tau_r^2(t) \right), \tag{6.73}
$$

in distribution, where

$$
b_r(t) = -f^{-1}(t) \beta_r^{(4)}(t) \tag{6.74}
$$

and

$$
\tau_r(t) = \left[\left(\frac{1}{4\sqrt{2}} \right) f^{-3/4}(t) M_r^{(1)}(t) + a_0 M_r^{(2)}(t) \right]^{1/2}. \tag{6.75}
$$

In general, the asymptotic distributions of $\widehat{\beta}_{r,J}(t; \lambda_r, w^{**})$ are affected by n, n_i and the intra-subject correlations. ∎

The correlations affect the asymptotic variance term $\tau_r^2(t)$ if a_0 is strictly positive, that is, $\sum_{i=1}^n n_i^2$ tends to infinity sufficiently fast. In the interesting case that n_i are bounded, the probability that there are at least two data points from the same subject in a shrinking neighborhood tends to be zero, hence, the intra-subject correlation does not play a role for the asymptotic distributions.

2. Asymptotic Mean Squared Errors

Unlike the simple regression model without covariates in Chapter 5, the risks of the penalized smoothing spline estimator $\widehat{\beta}_{r,J}(t; \lambda_r, w^{**})$ cannot be directly measured by their asymptotic mean squared errors (MSE). This is because Y_{ijr}^* involves the inverse of the estimator of $E(\mathbf{X}\mathbf{X}^T)$, so that the first and second moments, hence the MSE, of $\widehat{\beta}_{r,J}(t; \lambda_r, w^{**})$ may not exist. An alternative measure of the risks that has been used in the literature is the MSE conditioning on the observed covariates (e.g., Fan, 1992; Ruppert and Wand, 1994; and Fan and Gijbels, 1996). Denoting by

$$\mathscr{X}_n = \{\mathbf{X}_1, \ldots, \mathbf{X}_n\}$$

the set of observed covariates, we measure the risk of $\widehat{\beta}_{r,J}(t; \lambda_r, w^{**})$ by the following MSE conditioning on $\mathscr{X}_n$

$$MSE\left[\widehat{\beta}_{r,J}(t; \lambda_r, w^{**}) \middle| \mathscr{X}_n\right] = E\left\{\left[\widehat{\beta}_{r,J}(t; \lambda_r, w^{**}) - \beta_r(t)\right]^2 \middle| \mathscr{X}_n\right\}. \tag{6.76}$$

More generally, we measure the risk of $\widehat{\beta}_J(t; \lambda, w^{**})$ by

$$MSE_{\mathbf{p}}\left[\widehat{\beta}_J(t; \lambda_r, w^{**}) \middle| \mathscr{X}_n\right] = \sum_{r=0}^k \left\{p_r MSE\left[\widehat{\beta}_{r,J}(t; \lambda_r, w^{**}) \middle| \mathscr{X}_n\right]\right\}, \tag{6.77}$$

where $\mathbf{p} = (p_0, \ldots, p_k)^T$, $p_r \geq 0$, are known weights. The next theorem gives the asymptotic representation of the MSE in (6.76).

Theorem 6.2. *Suppose that Assumptions (a) through (e) are satisfied and t is an interior point of $[0, 1]$. When n is sufficiently large,*

$$MSE\left[\widehat{\beta}_{r,J}(t; \lambda_r, w^{**}) \middle| \mathscr{X}_n\right] = \lambda_r^2 b_r^2(t) + V_r(t) \tag{6.78}$$
$$+ o_p\left[N^{-1}\lambda_r^{-1/4} + \sum_{i=1}^n \left(\frac{n_i}{N}\right)^2\right]$$
$$+ O_p\left(n^{-1/2}\lambda_r\right) + O_p(n^{-1}) + o_p(\lambda_r^2),$$

where $b_r(t)$ is defined in (6.70) and

$$V_r(t) = \left(\frac{1}{4\sqrt{2}}\right) N^{-1}\lambda_r^{-1/4} f^{-3/4}(t) M_r^{(1)}(t) + \left[\sum_{i=1}^n \left(\frac{n_i}{N}\right)^2\right] M_r^{(2)}(t). \tag{6.79}$$

Furthermore, $\lim_{n\to\infty} V_r(t) = 0$ *if and only if* $\lim_{n\to\infty} \max_{1\le i\le n}(n_i/N) = 0$. ■

Proof of Theorem 6.2 is given in Section 6.7.3.

3. Remarks on Asymptotic Properties

The above asymptotic results can lead to a number of interesting special cases and practical implications.

Consistency and Convergence Rates:

Similar to the situations described in Section 5.3.2, we note that, unlike Theorem 6.1, Theorem 6.2 does not require any further convergence rate condition on λ_r other than Assumption (d) and allows for any choice of non-random n_i. Thus, under the conditions of Theorem 6.2, $\widehat{\beta}_{r,J}(t;\lambda_r,w^{**})$ is a consistent estimator of $\beta_r(t)$ in the sense that

$$MSE\left[\widehat{\beta}_{r,J}(t;\lambda_r,w^{**})\,\Big|\,\mathcal{X}_n\right] \to 0 \quad \text{in probability, as } n\to\infty.$$

The rate of $V_r(t) \to 0$ depends on n, n_i, $i=1,\ldots,n$, λ_r and the intra-subject correlations. If $\lambda_r^{-1/4}N^{-1}$ converges to zero in a rate slower than $\sum_{i=1}^{n}(n_i/N)^2$, then the second term of the right side of (6.79) becomes negligible, so that the effect of the intra-subject correlations disappears from the asymptotic representation of $MSE\left[\widehat{\beta}_{r,J}(t;\lambda_r,w^{**})\,\big|\,\mathcal{X}_n\right]$. As a special case of practical interest, this occurs when the n_i are bounded. In general, the contributions of the intra-subject correlations are not negligible. If, under the situation of *dense longitudinal data*, $n_i \to \infty$ sufficiently fast as $n\to\infty$, the second term of the right side of (6.79) may not be ignored from $V_r(t)$. This occurs, for example, when $n_i = n^\alpha$ for some $\alpha > 0$.

Random Design Time Points and Other Weight Choices:

Similar to the situations in Section 5.3.2, for the purpose of extending the derivations of Theorem 6.2 to random designs and other weight choices, we assume that t_{ij} are independent identically distributed with distribution function $F(\cdot)$ and density $f(\cdot)$. For the measurement uniform weight $w_i^{**} = 1/N$, we require the almost sure convergence $\lambda_r^{-5/4}D_N \to 0$ as $n\to\infty$ and consider the equivalent kernel estimator (6.68). For the subject uniform weight $w_i^* = 1/(nn_i)$, we replace $F_N(t)$ and D_N in Assumption (a) by

$$F_N^*(t) = \sum_{i=1}^{n}\sum_{j=1}^{n_i}(nn_i)^{-1}1_{[t_{ij}\le t]} \quad \text{and} \quad D_N^* = \sup_{t\in[0,1]}\left|F_N^*(t) - F(t)\right|,$$

respectively, and, under the almost sure convergence $\lambda_r^{-5/4}D_N^* \to 0$ as $n\to\infty$, consider the equivalent kernel estimator

$$\widetilde{\beta}_{r,J}(t; \lambda_r, w^*) = \sum_{i=1}^{n} \sum_{j=1}^{n_i} \left[(n\,n_i)^{-1} H_{\lambda_r}(t, t_{ij}) Y_{ijr}^* \right].$$

Further investigation is still needed to develop the explicit asymptotic distributions and the asymptotic conditional MSEs of these equivalent kernel estimators.

When the time design points t_{ij} are nonrandom, the subject measurement weight $w_i^* = 1/(n\,n_i)$ is used and Assumption (a) holds for $F_N^*(t)$ and D_N^*. Using the approach of Lemma 6.1, we can show that the variance of $\widehat{\beta}_{r,J}(t; \lambda_r, w^*)$ conditioning on $\mathcal{X}_n$ can be approximated by

$$\sum_{i=1}^{n} \sum_{j=1}^{n_i} \left[\left(\frac{1}{n\,n_i}\right)^2 G_{\lambda_r}^2(t, t_{ij}) \, Var\left(Y_{ijr}^* \big| \mathbf{X}_i\right) \right]$$

$$+ \sum_{(i_1, j_1) \neq (i_2, j_2)} \left[\left(\frac{1}{n^2 n_{i_1} n_{i_2}}\right) G_{\lambda_r}(t, t_{i_1 j_1}) G_{\lambda_r}(t, t_{i_2 j_2}) \right.$$

$$\left. \times Cov\left(Y_{i_1 j_1 r}^*, Y_{i_2 j_2 r}^* \big| \mathbf{X}_{i_1}, \mathbf{X}_{i_2}\right) \right],$$

but it is difficult to approximate the two summations above by some straightforward integrals without further assumptions on n_i. For the same reason, we do not have an explicit asymptotic risk representation for $\widehat{\beta}_{r,J}(t; \lambda_r, w)$ with a general weight vector w.

6.7.3 Theoretical Derivations

We outline here the main derivations used in the theoretical results of this section. Since many of the derivations are straightforward extensions of the methods presented in Section 5.3, we skip some tedious details to avoid repetition and only refer to the relevant steps in Section 5.3.

1. Green's Function for Uniform Density

The derivations here are analogous to their counterparts in Section 5.3.3. These results establish an important linkage between the Green's function $G_{\lambda_r}(t, s)$ and the equivalent kernel $H_{\lambda_r}(t, s)$. For the special case of the uniform density $f(t) = 1_{[0,1]}(t)$, the following equations and lemma are the same as equation (5.23) and Lemma 5.2 by substituting their λ with λ_r. The Green's function $G_{\lambda_r}^U(t, s)$ of (6.65) is the solution of

$$\lambda_r \frac{\partial^4}{\partial t^4} G_{\lambda_r}^U(t, s) + G_{\lambda_r}^U(t, s) = 0, \qquad \text{for } t \neq s, \tag{6.80}$$

subject to the following conditions:

(a) $G_{\lambda_r}^U(t, s) = G_{\lambda_r}^U(s, t) = G_{\lambda_r}^U(1 - t, 1 - s);$

(b) $(\partial^\nu / \partial t^\nu) \, G^U_{\lambda_r} (0, t) = (\partial^\nu / \partial t^\nu) \, G^U_{\lambda_r} (1, t) = 0$ for $\nu = 2, 3;$

(c) $(\partial^\nu / \partial t^\nu) \, G^U_{\lambda_r} (t, s) \big|_{s=t^-} - (\partial^\nu / \partial t^\nu) \, G^U_{\lambda_r} (t, s) \big|_{s=t^+} = 0$ for $\nu = 0, 1, 2;$

(d) $(\partial^3 / \partial t^3) \, G^U_{\lambda_r} (t, s) \big|_{s=t^-} - (\partial^3 / \partial t^3) \, G^U_{\lambda_r} (t, s) \big|_{s=t^+} = \lambda_r^{-1}.$

Lemma 6.2. *Suppose that $G^U_{\lambda_r} (t, s)$ is the Green's function of the differential equation (6.65) with $f(t) = 1_{[0, 1]}(t)$. When $\lambda_r \to 0$, the solution $G^U_{\lambda_r} (t, s)$ of (6.80) is given by*

$$G^U_{\lambda_r} (t, s) = H^U_{\lambda_r} (t, s) \left\{ 1 + O \left[\exp \left(- \lambda_r^{-1/4} / \sqrt{2} \right) \right] \right\}, \qquad (6.81)$$

where $H^U_{\lambda_r} (t, s)$ is defined in (6.67). ∎

Proof of Lemma 6.2 follows exactly the same steps in the proof of Lemma 5.2 by substituting λ with λ_r; hence, it is omitted.

2. Green's Function for General Cases

Based on the explicit expression of the Green's function $G^U_{\lambda_r} (t, s)$ for the uniform density $f(t) = 1_{[0, 1]}$, we can derive the explicit expression of the equivalent kernel function $H_{\lambda_r} (t, s)$ by establishing the relationship between $G^U_{\lambda_r} (t, s)$ and the Green's function $G_{\lambda_r} (t, s)$ for a general continuous density function $f(t)$. This relationship is established through the inequalities (6.69) to (6.72) of Lemma 6.1.

Proof of Lemma 6.1:
The inequalities (6.69) to (6.72) can be established using the same steps as the proof of Lemma 5.1 in Section 5.3.4. All the derivations in the proof of Lemma 5.1 remain valid by substituting $\{\lambda, \mu(t)\}$ with $\{\lambda_r, \beta_r(t)\}$. We omit the derivations here to avoid repetition. ∎

3. Three Technical Lemmas

The following three technical lemmas are analogous to Lemmas 5.3, 5.4 and 5.5. The proofs of these lemmas are not exactly the same as the ones in Chapter 5, although they share similarities.

Lemma 6.3. *Assume that Assumptions (a) and (d) are satisfied, then, when λ_r is sufficiently small,*

$$\int_0^1 G^2_{\lambda_r} (t, s) \, M_r^{(1)}(s) \, f(s) \, ds = \left(\frac{1}{4\sqrt{2}} \right) f^{-3/4}(t) \, \lambda_r^{-1/4} \, M_r^{(1)}(t) \, [1 + o(1)] \qquad (6.82)$$

and

$$\int_0^1 G_{\lambda_r} (t, s) \, M_r^{(2)} (t, s) \, f(s) \, ds = M_r^{(2)}(t) \, [1 + o(1)] \qquad (6.83)$$

hold for all $t \in [\tau, 1 - \tau]$ with some $\tau > 0$. ∎

Proof of Lemma 6.3:

By Lemma 6.1, the properties of double exponential distributions and straightforward algebra, we can show that, for some positive constants κ, α and c,

$$
\left| \int_0^1 \left[G_{\lambda_r}^2(t, s) - H_{\lambda_r}^2(t, s) \right] M_r^{(1)}(s) f(s) \, ds \right|
$$

$$
\leq \int_0^1 \left| G_{\lambda_r}(t, s) - H_{\lambda_r}(t, s) \right| \left[\left| G_{\lambda_r}(t, s) \right| + \left| H_{\lambda_r}(t, s) \right| \right] \left| M_r^{(1)}(s) \right| f(s) \, ds
$$

$$
\leq \int_0^1 \kappa^2 \lambda_r^{-1/4} \exp\left(-\alpha \lambda_r^{-1/4} |t - s| \right) \left| M_r^{(1)}(s) \right| f(s) \, ds
$$

$$
\rightarrow \ c \left| M_r^{(1)}(t) \right| f(t), \qquad \text{as } \lambda_r \to 0. \tag{6.84}
$$

Similarly, denoting $u = \Gamma(t)$ and $v = \Gamma(s)$, we can show from (6.67) and the properties of double exponential distributions that, for λ_r sufficiently small,

$$
\int_0^1 H_{\lambda_r}^2(t, s) M_r^{(1)}(s) f(s) \, ds
$$

$$
= \int_0^1 \left[H_{\lambda_r/\gamma^4}^U(u, v) \right]^2 M_r^{(1)} \left[\Gamma^{-1}(v) \right] \left\{ \frac{f^{1/4} [\Gamma^{-1}(v)]}{\gamma f [\Gamma^{-1}(v)]} \right\} dv
$$

$$
= \frac{1}{4\sqrt{2}} f^{-3/4}(t) \lambda_r^{-1/4} M_r^{(1)}(t) \left[1 + o(1) \right]. \tag{6.85}
$$

Thus, (6.82) follows from (6.84) and (6.85), and (6.83) can be shown by similar calculations. ∎

Lemma 6.4. *Assume that $\beta_r(t)$ satisfies Assumption (a) and $g_r(t)$ is a solution of (6.61), then $g_r^{(4)}(t) \to \beta_r^{(4)}(t)$ uniformly for $t \in [0, 1]$ as $\lambda_r \to 0$.* ∎

Proof of Lemma 6.4:

This lemma is a special case of Lemma 6.1 of Nychka (1995) and is the same as Lemma 5.4 by substituting $\mu(t)$ with $\beta_r(t)$. ∎

Lemma 6.5. *Consider the pseudo-equivalent kernel estimator*

$$
\widetilde{\beta}_r^*(t; w^{**}) = N^{-1} \sum_{i=1}^n \sum_{j=1}^{n_i} \left\{ H_{\lambda_r}(t; t_{ij}) \left[\sum_{l=0}^k \left(e_{r+1, l+1} X_i^{(l)} \right) Y_{ij} \right] \right\}.
$$

If the assumptions in Theorem 6.1 are satisfied, then $\widetilde{\beta}_r^*(t; w^{**})$ is asymptotically normal in the sense that (6.73) holds with $\widehat{\beta}_{r, J}(t; \lambda_r, w^{**})$ replaced by $\widetilde{\beta}_r^*(t; w^{**})$. ∎

Proof of Lemma 6.5:

Define $U_{ir} = \sum_{l=0}^{k}\left(e_{r+1,l+1}X_i^{(l)}\right)$ and $Z_{ijr} = U_{ir}Y_{ij}$. It then follows from (6.1) that

$$E\left(Z_{ijr}\right) = E\left[U_{ir}\mathbf{X}_i^T\beta\left(t_{ij}\right)\right] = \beta_r\left(t_{ij}\right).$$

By Assumptions (a) and (d), (6.65), (6.66), (6.69), (6.74) and Lemma 6.4, we have

$$E\left[\widetilde{\beta}_r^*\left(t;w^{**}\right)\right] - \beta_r(t)$$

$$= \int_0^1 G_{\lambda_r}(t,s)\,\beta_r(s)\,f(s)\,ds - \beta_r(t) + \int_0^1\left[H_{\lambda_r}(t,s) - G_{\lambda_r}(t,s)\right]\beta_r(s)\,f(s)\,ds$$

$$+ \int_0^1 H_{\lambda_r}(t,s)\,\beta_r(s)\,d\left[F_N(s) - F(s)\right]$$

$$= -\lambda_r\,f^{-1}(t)\,g_r^{(4)}(t)\left[1+o\left(\lambda_r\right)\right]$$

$$= -\lambda_r\,b_r(t)\left[1+o\left(\lambda_r\right)\right]. \tag{6.86}$$

For the variance of $\widetilde{\beta}_r^*\left(t;w^{**}\right)$, we consider

$$Var\left[\widetilde{\beta}_r^*\left(t;w^{**}\right)\right] = V_I + V_{II} + V_{III},$$

where

$$V_I = \frac{1}{N^2}\sum_{i=1}^{n}\sum_{j=1}^{n_i}\left[H_{\lambda_r}^2\left(t,t_{ij}\right)Var\left(Z_{ijr}\right)\right],$$

$$V_{II} = \frac{1}{N^2}\sum_{i=1}^{n}\sum_{j_1\neq j_2}\left[H_{\lambda_r}\left(t,t_{ij_1}\right)H_{\lambda_r}\left(t,t_{ij_2}\right)Cov\left(Z_{ij_1r},Z_{ij_2r}\right)\right]$$

and, because the subjects are independent,

$$V_{III} = \frac{1}{N^2}\sum_{i_1\neq i_2}\sum_{j_1,j_2}\left[H_{\lambda_r}\left(t,t_{i_1j_1}\right)H_{\lambda_r}\left(t,t_{i_2j_2}\right)Cov\left(Z_{i_1j_1r},Z_{i_2j_2r}\right)\right] = 0.$$

Because U_{ir} and ε_{ij} are independent, we have

$$Var\left(Z_{ijr}\right) = Var\left\{U_{ir}\left[\mathbf{X}_i^T\beta\left(t_{ij}\right)\right]\right\} + Var\left(U_{ir}\varepsilon_{ij}\right) = M_r^{(1)}\left(t_{ij}\right),$$

hence, by Assumption (a) and (6.85),

$$V_I = \frac{1}{4\sqrt{2}}f^{-3/4}(t)N^{-1}\lambda_r^{-1/4}M_r^{(1)}(t)\left[1+o(1)\right].$$

Similar to the derivation in (6.84), because

$$Cov\left(Z_{ij_1r},Z_{ij_2r}\right)$$

$$= Cov\left\{U_{ir}\left[\mathbf{X}_i^T\beta\left(t_{ij_1}\right)\right],U_{ir}\left[\mathbf{X}_i^T\beta\left(t_{ij_2}\right)\right]\right\} + Cov\left(U_{ir}\varepsilon_{ij_1},U_{ir}\varepsilon_{ij_2}\right)$$

$$= M_r^{(2)}\left(t_{ij_1},t_{ij_2}\right),$$

it is straightforward to compute that

$$
\begin{aligned}
V_{II} &= \left[\sum_{i=1}^{n}\left(\frac{n_i}{N}\right)^2 - \frac{1}{N}\right] \\
&\quad \times \iint H_{\lambda_r}(t, s_1)\, H_{\lambda_r}(t, s_2)\, M_r^{(2)}(s_1, s_2)\, f(s_1)\, f(s_2)\, ds_1\, ds_2 \\
&\quad \times [1 + o(1)] \\
&= \left[\sum_{i=1}^{n}\left(\frac{n_i}{N}\right)^2 - \frac{1}{N}\right] M_r^{(2)}(t)\, [1 + o(1)].
\end{aligned}
$$

The above equations and (6.75) imply that

$$
Var\left[\widetilde{\beta}_r^*(t; w^{**})\right] = N^{-1}\lambda_r^{-1/4}\,\tau_r^2(t)\,[1 + o(1)].
$$

Finally, we can check from Assumption (d), (6.66) and (6.67) that $\widetilde{\beta}_r^*(t; w^{**})$ satisfies the Lindeberg's condition for double arrays of random variables. The lemma follows from (6.86) and the central limit theorem for double arrays (e.g., Serfling, 1980, Section 1.9.3). ∎

4. Proofs of Main Theorems

Equipped with the above technical results, we can now derive the results of Theorems 6.1 and 6.2.

Proof of Theorem 6.1:
By $U_{ir} = \sum_{l=0}^{k}\left(e_{r+1,l+1}X_i^{(l)}\right)$ and $\widehat{U}_{ir} = \sum_{l=0}^{k}\left(\widehat{e}_{r+1,l+1}X_i^{(l)}\right)$, Assumptions (a), (c) and (d) and Lemma 6.1, we have that, when n is sufficiently large,

$$
\widetilde{\beta}_r(t; w^{**}) - \widetilde{\beta}_r^*(t; w^{**}) = \frac{1}{N}\sum_{i=1}^{n}\sum_{j=1}^{n_i}\left[H_{\lambda_r}(t, t_{ij})\left(\widehat{U}_{ir} - U_{ir}\right)Y_{ij}\right] = O_p\left(n^{-1/2}\right)
$$

and

$$
\begin{aligned}
\widehat{\beta}_r(t; w^{**}) - \widetilde{\beta}_r(t; w^{**}) &= \frac{1}{N}\sum_{i=1}^{n}\sum_{j=1}^{n_i}\left\{\left[S_{\lambda_r}(t, t_{ij}) - G_{\lambda_r}(t, t_{ij})\right]\widehat{U}_{ir}Y_{ij}\right\} \\
&= o_p\left(N^{-1/2}\lambda_r^{-1/8}\right).
\end{aligned}
$$

Then (6.73) follows from Lemma 6.5 and the above equalities. ∎

Proof of Theorem 6.2:
Using the variance-bias squared decomposition, we have

$$
\begin{aligned}
MSE\left[\widehat{\beta}_{r,J}(t; \lambda_r, w^{**})\,\big|\,\mathscr{X}_n\right] &= \left\{E\left[\widehat{\beta}_{r,J}(t; \lambda_r, w^{**})\,\big|\,\mathscr{X}_n\right] - \beta_r(t)\right\}^2 \\
&\quad + Var\left[\widehat{\beta}_{r,J}(t; \lambda_r, w^{**})\,\big|\,\mathscr{X}_n\right], \quad\quad (6.87)
\end{aligned}
$$

where, because $Y_{i_1 j_1}$ and $Y_{i_2 j_2}$ are independent when $i_1 \neq i_2$, we have

$$Var\left[\widehat{\beta}_{r,J}(t; \lambda_r, w^{**}) \,\middle|\, \mathcal{X}_n\right] = V_I^* + V_{II}^*$$

with

$$\begin{cases} V_I^* = N^{-2} \sum_{i=1}^n \sum_{j=1}^{n_i} \left[S_{\lambda_r}^2(t, t_{ij}) \, \widehat{U}_{ir}^2 \, Var(Y_{ij}) \right], \\ V_{II}^* = N^{-2} \sum_{i=1}^n \sum_{j_1 \neq j_2} \left[S_{\lambda_r}(t, t_{ij_1}) \, S_{\lambda_r}(t, t_{ij_2}) \, \widehat{U}_{ir}^2 \, Cov(Y_{ij_1}, Y_{ij_2}) \right]. \end{cases}$$

Using Lemma 6.1 and the derivation of (6.83), we can show that, for sufficiently large n,

$$\begin{aligned} Var\left[\widehat{\beta}_{r,J}(t; \lambda_r, w^{**}) \,\middle|\, \mathcal{X}_n\right] &= \frac{1}{4\sqrt{2}} N^{-1} \lambda_r^{-1/4} f^{-3/4}(t) \sigma^2(t) \, err \left[1 + o_p(1)\right] \\ &+ \left[\sum_{i=1}^n \left(\frac{n_i}{N} \right)^2 - \frac{1}{N} \right] \rho_{\varepsilon}(t) \, err \left[1 + o_p(1)\right]. \quad (6.88) \end{aligned}$$

For the conditional bias term of (6.87), we consider that, when n is sufficiently large,

$$\begin{aligned} &E\left[\widehat{\beta}_{r,J}(t; \lambda_r, w^{**}) \,\middle|\, \mathcal{X}_n\right] - \beta_r(t) \\ &= \frac{1}{N} \sum_{i=1}^n \sum_{j=1}^{n_i} \left\{ S_{\lambda_r}(t, t_{ij}) \left[U_{ir} \mathbf{X}_i^T \beta(t_{ij}) - E\left[U_{ir} \mathbf{X}_i^T \beta(t_{ij}) \right] \right] \right\} \quad (6.89) \\ &+ E\left\{ \frac{1}{N} \sum_{i=1}^n \sum_{j=1}^{n_i} \left[S_{\lambda_r}(t, t_{ij}) U_{ir} \mathbf{X}_i^T \beta(t_{ij}) \right] \right\} - \beta_r(t) + O_p(n^{-1/2}). \end{aligned}$$

By similar quadratic expansions as V_I^* and V_{II}^*, Lemma 6.1 and the weak law of large numbers, we can show that, when n is sufficiently large,

$$\begin{aligned} &\left\{ \frac{1}{N} \sum_{i=1}^n \sum_{j=1}^{n_i} \left\{ S_{\lambda_r}(t, t_{ij}) \left[U_{ir} \mathbf{X}_i^T \beta(t_{ij}) - E\left[U_{ir} \mathbf{X}_i^T \beta(t_{ij}) \right] \right] \right\} \right\}^2 \quad (6.90) \\ &= \left\{ \frac{1}{4\sqrt{2}} N^{-1} \lambda_r^{-1/4} f^{-3/4}(t) + \left[\sum_{i=1}^n \left(\frac{n_i}{N} \right)^2 - \frac{1}{N} \right] \right\} M_r^{(0)}(t) \left[1 + o_p(1)\right] \end{aligned}$$

and, furthermore, by Lemma 6.3,

$$\left\{ \frac{1}{N} \sum_{i=1}^n \sum_{j=1}^{n_i} \left[S_{\lambda_r}(t, t_{ij}) E\left[U_{ir} \mathbf{X}_i^T \beta(t_{ij}) \right] \right] - \beta_r(t) \right\}^2 = \lambda_r^2 b_r^2(t) \left[1 + o_p(1)\right]. \quad (6.91)$$

The conclusion of the theorem (6.78) is then a direct consequence of (6.87) through (6.91). ∎

6.8 Remarks and Literature Notes

The data structure and model presented in this chapter are the simplest case of the structured nonparametric models to be investigated in this book. Under this data structure, we assume that the covariates do not change with time, and only the outcome variable is time-dependent and repeatedly measured. Although the assumption of time-invariant covariates may be overly simplified in practice, it sets the stage for developing modeling strategies to meet the challenges with other more complex longitudinal data. The model that we have considered in this chapter is a special case of the "time-varying coefficient models." The appealing feature of this model is that the relationship between the time-invariant covariates and the outcome at a given time is described through a linear model, while the linear coefficients are nonparametric smooth functions of time. This modeling approach simultaneously retains the interpretable covariate effects and provides the flexibility by allowing the covariate effects to change with time. Useful variations of the varying coefficient models, such as the single-index varying coefficient model (Luo, Zhu and Zhu, 2016), are also widely used in various applications.

We summarize in this chapter a component-wise smoothing method for estimating the time-varying coefficients. Such a component-wise smoothing is possible because, based on the special feature of time-invariant covariates, the coefficient curves can be written as the conditional means of the products of the time-dependent outcome with a time-invariant matrix of the transformed covariates. The major advantage of this component-wise smoothing method is that different smoothing parameters can be used for different coefficient curves. This is particularly appealing when the coefficient curves satisfy different smoothing conditions, because the potentially different smoothing needs can be accommodated.

Through some straightforward modifications of the smoothing methods described in Chapters 3 to 5, both the local and global smoothing methods can be used for component-wise estimation of the coefficient curves. The results of this chapter are adopted from Wu and Chiang (2000), which describes the component-wise kernel estimation method, and Chiang, Rice and Wu (2001), which describes the component-wise smoothing spline estimation method. The asymptotic properties established in this chapter provide some useful theoretical justifications for the component-wise kernel and smoothing spline estimators. In order to focus on the estimation methods, we omit from the presentation a number of useful and interesting inference methods for the time-varying coefficient models, such as the confidence bands of Wu, Yu and Yuan (2000), the goodness-of-fit tests and empirical likelihood methods of Xue and Zhu (2007) and Xu and Zhu (2008, 2013).

Chapter 7

The One-Step Local Smoothing Methods

This chapter is concerned with the time-varying coefficient models with time-varying covariates. Unlike the special case in Chapter 6, where the covariates are time-invariant, the covariate effects described in this chapter are *instantaneous* in the sense that the model only describes the relationship between the covariates and the outcome at the same time. Despite this simplification, the time-varying coefficient models discussed in this chapter can be widely applied in practical longitudinal studies, because these models have the simple interpretation as being linear models at a specific given time point. The changing coefficient values at different time points allow the models to vary over time, which adds flexibility to the models. The coefficient curves of the models can be estimated by a number of different approaches, each with its own advantages and disadvantages. The estimation methods presented in this chapter are based on a one-step local least squares approach, which has the advantage of being conceptually simple. Additional local and global estimation methods for the same models and data structure are presented in Chapters 8 and 9.

7.1 Data Structure and Model Interpretations

7.1.1 Data Structure

In most longitudinal studies, both the outcome variables and the covariates are likely to be time-varying and repeatedly measured. Under this general setting, the random variables at each time point t are $(Y(t), t, \mathbf{X}^T(t))$, where $Y(t)$ is a real-valued outcome variable and $\mathbf{X}(t) = (1, X^{(1)}(t), \ldots, X^{(k)}(t))^T$ is the $(k+1)$ column covariate vector. The corresponding longitudinal observations for $(Y(t), t, \mathbf{X}^T(t))$ are given by $\{(Y_{ij}, t_{ij}, \mathbf{X}_{ij}^T) : i = 1, \ldots, n; j = 1, \ldots, n_i\}$, where the subjects are assumed to be independent, t_{ij} is the jth visit time of the ith subject, and Y_{ij} and $\mathbf{X}_{ij}$ are the ith subject's outcome and covariates observed at time t_{ij}, respectively. The statistical objective is to evaluate the effects of time t and the covariates $X^{(l)}(t)$, $l = 1, \ldots, k$, on the outcome $Y(t)$. Using the framework of conditional means, the effects of time t and covariates $\mathbf{X}(t)$ on $Y(t)$ can be described through $E[Y(t)|t, \mathbf{X}(t)]$. However, as discussed in Chapters 3 to 5, complete nonparametric estimation of $E[Y(t)|t, \mathbf{X}(t)]$ without assuming any modeling structures could be computationally infeasible and scientifically uninterpretable. In this chapter, we describe the formulation and interpretations of a class of popular structural nonparametric models for longitudinal data, the time-varying coefficient models, and present a number of smoothing estimation procedures of the coefficient curves based on local least squares.

7.1.2   Model Formulation

Using the linear model structure at each time point t, a simple and practical choice of the "structured nonparametric models" for the instantaneous relationship between $\mathbf{X}(t)$ and $Y(t)$ suggested by Hoover et al. (1998) is to use the time-varying coefficient model

$$Y(t) = \mathbf{X}^T(t)\,\beta(t) + \varepsilon(t), \tag{7.1}$$

where the coefficient curves $\beta(t) = \left(\beta_0(t), \dots, \beta_k(t)\right)^T$ describe the baseline time-trend $\beta_0(t)$ and the covariate effects $\{\beta_1(t), \dots, \beta_k(t)\}$ as functions of t, $\varepsilon(t)$ is a mean zero stochastic process with variance and covariance functions

$$\begin{cases} \sigma^2(t) & = & Var[\varepsilon(t)], \\ \rho(s,t) & = & Cov[\varepsilon(s), \varepsilon(t)], \quad \text{when } s \neq t, \end{cases}$$

and $\varepsilon(t)$ and $\mathbf{X}(t)$ are independent. Note that the covariance function $\rho(s,t)$ may not equal the variance function $\sigma^2(t)$ when $s = t$, which may happen, for example, when the error term $\varepsilon(t)$ is the sum of a mean zero stochastic process and an independent mean zero measurement error. In general, we can assume that

$$\rho(t,t) = \lim_{s \to t} \rho(s,t) \leq \sigma^2(t),$$

where the equality sign holds if there is no independent measurement error for $Y(t)$. To ensure the identifiability of $\beta(t)$, we assume that, when $\mathbf{X}(t)$ is random,

$$E\left[\mathbf{X}(t)\mathbf{X}^T(t)\right] = E_{\mathbf{X}\mathbf{X}^T}(t) \tag{7.2}$$

exists and invertible, so that its inverse $E_{\mathbf{X}\mathbf{X}^T}^{-1}(t)$ exists. The nonrandom covariates can be incorporated as a special case of (7.2).

Since the mean of $\varepsilon(t)$ is assumed to be zero, it is easily seen from (7.1) that the conditional mean of $Y(t)$ given $\mathbf{X}(t)$ is

$$E\left[Y(t)|t, \mathbf{X}(t)\right] = \mathbf{X}^T(t)\,\beta(t).$$

Furthermore, it follows from (7.1) and (7.2) that $\beta(t)$ uniquely minimizes the second moment of $\varepsilon(t)$, that is,

$$E\left\{\left[Y(t) - \mathbf{X}^T(t)\beta(t)\right]^2\right\} = \inf_{\text{all } b(\cdot)} E\left\{\left[Y(t) - \mathbf{X}^T(t)b(t)\right]^2\right\}, \tag{7.3}$$

and has the expression

$$\beta(t) = E_{\mathbf{XX}^T}^{-1}(t) E\left[\mathbf{X}(t) Y(t)\right]. \tag{7.4}$$

When all the covariates are time-invariant, i.e., $\mathbf{X}(t) = \mathbf{X}$, (7.4) reduces to the expression of $\beta(t)$ given in (6.2), and the estimation and inferences for $\beta(t)$ can be proceeded using the componentwise smoothing methods of Chapter 6. But, when some of the components in $\mathbf{X}(t)$ are time-dependent, the coefficient curves $\beta(t)$ given in (7.4) cannot be estimated by the methods of Chapter 6.

The objective of this chapter is to present a class of local least squares smoothing methods for the estimation of $\beta(t)$, which are described in Hoover et al. (1998), Wu, Chiang and Hoover (1998) and Wu, Yu and Chiang (2000). These estimation methods are motivated by the expressions of (7.3) and (7.4). Algorithms and R code for the estimation and inference procedures are presented in the applications to the NGHS and BMACS data.

7.1.3   Model Interpretations

The structural assumptions of (7.1) suggest that the time-varying coefficient model is, on one hand, a flexible nonparametric model and, on the other, a restricted linear model at each time point t. Interpretations of the model on time-trends and instantaneous covariate effects can be seen from the following four aspects.

1. Changing Covariate Effects

When t is fixed, (7.1) is a linear model with $\beta(t)$ being the coefficients describing the linear effects of $\mathbf{X}(t)$ on the mean of $Y(t)$. Because of the linear structure, the coefficients in $\beta(t)$ have the same interpretations as a usual linear model in the sense that they describe the instantaneous linear effects of the covariates $\mathbf{X}(t)$ on the outcome $Y(t)$ at any time point t. Due to the time-varying nature of $\beta(t)$, these linear effects may change with the time points. Thus, the model (7.1) is "local" in terms of the time point t with "structural restrictions" in terms of the covariates $\mathbf{X}(t)$. Since $X^{(0)}(t) = 1$ and the $\varepsilon(t)$ has zero mean, $\beta_0(t)$ is the mean value of $Y(t)$ when the values of $X^{(1)}(t), \ldots, X^{(k)}(t)$ are zero, which represents the baseline time-trend, while, for $1 \leq l \leq k$, $\beta_l(t)$ represents the average change of $Y(t)$ at time t caused by one unit change of $X^{(l)}(t)$, which describe the covariate effect of $X^{(l)}(t)$.

2. Local Linear Structure

Although the coefficients $\beta(t)$ change with t, the linear model structure is preserved by (7.1). It is certainly possible in some applications that the model structure also changes with t, such as, changing from a linear model to a nonlinear model as t changes. But, such generalizations are outside the framework of this book, which require additional research beyond the current literature.

3. Flexible Temporal Patterns

The model (7.1) is flexible in terms of its temporal patterns, because the co-efficient curves $\beta_l(t)$, $l = 0, \ldots, k$, are usually only subject to minimal smoothness assumptions as functions of t. The nonparametric assumptions for $\beta_l(t)$, $l = 0, \ldots, k$, reflect the desire to preserve the flexibility of the temporal patterns because, in real applications, it is usually unknown how the models, i.e., the coefficients, change with time. When more information is available, it is possible to consider parametric families for $\beta_l(t)$, $l = 0, \ldots, k$, so that, parametric sub-families of (7.1) may be investigated. Some sub-families of (7.1) lead to the parametric marginal models, such as the linear or nonlinear marginal models, which are already extensively studied in the literature.

4. Instantaneous Associations

The covariate effects on the outcome are "instantaneous" in the sense that (7.1) only includes the associations between $\mathbf{X}(t)$ and $Y(t)$ at the same time point t. The purpose of making the instantaneous association assumption on (7.1) is to reduce the computational complexity, so that the model can be parsimonious and practical at the same time. In many practical situations, the instantaneous association assumption can be reasonably justified because the current or most recent values of the covariates are most important on the outcome values. However, the model (7.1) does not allow "time lagging" or "cumulative" effects, which could be a potential concern in some longitudinal studies. Here the "time lagging" effects refer to the effects of the covariates or outcomes at earlier time points on the outcome at a later time point, and the "cumulative" effects refer to the influence of the cumulative or integrated values of the covariates over a time interval on the outcome at a later time point. Substantial further research is needed to better understand the potential approaches to adequately model and analyze "time lagging" and "cumulative" effects. The methods covered in this book do not sufficiently address these issues.

7.1.4 Remarks on Estimation Methods

Similar to other regression models, the model (7.1) and its nonparametric estimation methods presented in this chapter have their own advantages and disadvantages, which can be seen from the real applications and theoretical developments of this chapter. Details of these advantages and disadvantages are discussed in later sections. The following remarks give an overview of some important features of the model and estimation procedures of this chapter.

1. Impacts of Correlation Structures

Since the main objective is to estimate the time-varying effects of $\mathbf{X}(t)$ on the mean structure of $Y(t)$, our estimation methods to be presented in the next section do not use the potential correlations of $(Y_{ij}, t_{ij}, \mathbf{X}_{ij}^T)$ at two different visit times. This is mainly due to the fact that, in situations where the numbers of repeated measurements are not drastically different among different subjects, the asymptotic biases of our smoothing estimators of $\beta(t)$ are not affected by the correlation structures of the data (Section 7.7). The numbers of repeated measurements and the correlation structures may, in general, affect the asymptotic variances of our smoothing estimators of $\beta(t)$. As shown in Chapters 3 to 5, since the correlation structures of the data are often completely unknown in practice, developing a smoothing estimator based on the estimated correlations of the data is generally unrealistic. Thus, the smoothing estimators of $\beta(t)$ presented in this chapter are practical and have appropriate theoretical properties, although they may not have the smallest asymptotic variances. In some situations, the scientific objectives are best achieved by evaluating the relationships between the outcome values at different time points, so that the statistical interest is to evaluate the correlation structures conditioning on the covariates. Statistical models and estimation procedures related to the correlation structures of the data are discussed in Chapters 12 to 15.

2. Issues with Missing Data

In practice, the subjects may not be all observed at the same time points, so that the numbers of repeated measurements and the observation times are possibly different among subjects. In this sense, we do not assume that there are "missing data" in the longitudinal observations $\big\{ (Y_{ij}, t_{ij}, \mathbf{X}_{ij}^T) : i = 1, \ldots, n; j = 1, \ldots, n_i \big\}$. If, for some reason, a subject has missing observations at a scheduled visit, the *would be observations* of the subject at this specific visit time are equivalent to *data missing completely at random*, and consequently are ignored in the computation of the smoothing estimators. The estimation methods of this chapter, however, could be biased if some of the subjects have missing observations due to reasons other than "missing completely at random," such as not having observations because the subject's outcomes or covariates are "undesirable." Nonparametric regression models with data "not missing completely at random" may not be identifiable, and their estimation and inference methods require further investigation.

3. Outcome-Dependent Covariates

Similar to parametric models for data with time-varying covariates, an important assumption for the model (7.1) is that the values of the time-varying components in $\mathbf{X}(t)$ do not depend on the values of the outcome variable at time points prior to t. This situation does not appear in the model and estimation methods of Chapter 6, because the covariates there are all time-invariant. If $\mathbf{X}(t)$ is time-varying and "outcome-dependent" in the sense that the value of $\mathbf{X}(t)$ depends on the values of $Y(s)$ for some $s < t$, then the nonparametric smoothing estimators of $\beta(t)$ developed in this chapter are likely to be biased, because the relationships between $\mathbf{X}(t)$ and $Y(s)$ for $s < t$ are not included in (7.1). Useful models and estimation methods for several simple situations with outcome-dependent time-varying covariates are discussed in Chapter 10.

4. Componentwise Smoothness

Because $\beta(t)$ is a $(k+1)$ column vector of curves, it is possible that the different component curves in $\beta(t)$ satisfy different smoothness conditions, so that their estimators require different smoothing parameters. However, the componentwise estimation approach of Chapter 6 cannot be directly extended to the current model (7.1) because the covariates are time-varying. In Section 7.2, we introduce the popular local least squares approach suggested by Hoover et al. (1998) and Wu, Chiang and Hoover (1998), which relies on using one smoothing parameter to estimate all the component curves of $\beta(t)$. Thus this approach may only be used when all the component curves of $\beta(t)$ satisfy similar smoothness conditions. Because it depends on the intuitive idea of local linear fitting, this approach is computationally simple and can be used as an exploratory tool to gain some useful insight into the covariate effects.

To further refine the smoothing needs of different covariate curves, we introduce in Chapter 8 the two-step smoothing method proposed by Fan and Zhang (2000), which has the capability to select different smoothing parameters for different coefficient curves. This estimation approach depends on first obtaining some "raw estimates" from a preliminary parametric estimation procedure and then smoothing these "raw estimates" for each of the component curves. A further estimation method (discussed in Chapter 9) is based on the global fitting through basis approximation, which is also able to provide different smoothing needs for different coefficient curves. These different estimation approaches all have their advantages and disadvantages under different situations, and are all useful in practice.

5. Model Checking

An important question for modeling the relationship between $Y(t)$ and $\{t, \mathbf{X}(t)\}$ is whether the model (7.1) is appropriate for the dataset $\left\{ \left(Y_{ij}, t_{ij}, \mathbf{X}_{ij}^T \right) : i = 1, \ldots, n; j = 1, \ldots, n_i \right\}$. This question can be answered, at least partially, by a model checking procedure comparing the fitness of (7.1) with a more general alternative. Since model checking requires different statistical procedures, this topic is beyond the scope of this chapter, and some comparisons of the model (7.1) with a number of alternatives are deferred to Chapter 9. The focus of this chapter is based on the premise that the relationship between $Y(t)$ and $\mathbf{X}(t)$ is already appropriately described in (7.1).

7.2   Smoothing Based on Local Least Squares Criteria

7.2.1   General Formulation

The equation (7.3) suggests that $\beta(t)$ can be intuitively estimated by a local least squares method using the measurements observed within a neighborhood of t. Assume that, for each l and some integer $p \geq 0$, $\beta_l(t)$ is p times differentiable and its pth derivative is continuous. Approximating $\beta_l(t_{ij})$ by a pth order polynomial

$$\beta_l(t_{ij}) \approx \sum_{r=0}^{p} \left[b_{lr}(t) \left(t_{ij} - t \right)^r \right] \tag{7.5}$$

for all $l = 0, \ldots, k$, a local polynomial estimator of $\beta(t) = (\beta_0(t), \ldots, \beta_k(t))^T$ based on a kernel neighborhood is

$$\widehat{b}_0(t) = \left(\widehat{b}_{00}(t), \ldots, \widehat{b}_{k0}(t) \right)^T, \tag{7.6}$$

where $\left\{ \widehat{b}_{lr}(t) : l = 0, \ldots, k; r = 0, \ldots, p \right\}$ minimizes the local least squares score function

$$
\begin{aligned}
&L_{p,K}(t; h, w) \\
&= \sum_{i=1}^{n} \sum_{j=1}^{n_i} w_i \left\{ Y_{ij} - \sum_{l=0}^{k} \left\{ X_{ij}^{(l)} \left[\sum_{r=0}^{p} b_{lr}(t) \left(t_{ij} - t \right)^r \right] \right\} \right\}^2 K\left(\frac{t_{ij} - t}{h} \right),
\end{aligned} \tag{7.7}
$$

$w = (w_1, \ldots, w_n)^T$, w_i are the non-negative weights satisfying $\sum_{i=1}^{n} (w_i n_i) = 1$, $K(\cdot)$ is a kernel function, usually chosen to be a probability density function, and $h > 0$ is a bandwidth. For any $r = 1, \ldots, p$, $(r!)\widehat{b}_{lr}(t)$ is the local polynomial estimator of the rth derivative $\beta_l^{(r)}(t)$ of $\beta_l(t)$. In real applications, the choices of $\{p, h, K(\cdot), w\}$ in (7.7) lead to different smoothing estimators $\widehat{b}_0(t)$ of $\beta(t)$. Among these choices, the most influential one is the bandwidth h.

Notice that (7.7) is a naive local least-square criterion which attempts to use one bandwidth h for simultaneously computing all the local polynomial estimators $\widehat{b}_{l0}$ and $(r!)\widehat{b}_{lr}(t)$, $r = 1, \ldots, p$. Although (7.6) has the limitation of using a single bandwidth, the resulting estimators are still useful as a preliminary step for the estimation of $\beta(t)$. In particular, the naive local least squares criterion (7.7) may be modified by a few simple steps to obtain improved estimators that allow for componentwise specific bandwidths and other improvements over $\widehat{b}_0(t)$ given in (7.6). The following sections discuss some useful special cases of the estimators $\widehat{b}_0(t)$ and their modifications.

Similar to the smoothing estimators for longitudinal data in the previous chapters (e.g., Section 3.1), the subject uniform weight $w_i^* = 1/(n n_i)$ and the measurement uniform weight $w_i^{**} = 1/N$ are often used in practice. Theoretically, the choice of $w_i^{**} = 1/N$ may produce inconsistent estimators when some n_i are much larger than the others, while the estimators based on $w_i^* = 1/(n n_i)$ are always consistent regardless the choices of n_i. Details of these theoretical properties are discussed in Section 7.5.

7.2.2 Least Squares Kernel Estimators

The simplest special case of (7.6) is the least squares kernel estimator, also referred to as the local constant fit, which is obtained by minimizing the score function (7.7) with $p = 0$. Using the matrix representation

$$\mathbf{Y}_i = (Y_{i1}, \ldots, Y_{in_i})^T, \quad b_0(t) = (b_{00}(t), \ldots, b_{k0}(t))^T,$$

$$\mathbb{X}_i = \begin{pmatrix} 1 & X_{i1}^{(1)} & \cdots & X_{i1}^{(k)} \\ 1 & X_{i2}^{(1)} & \cdots & X_{i2}^{(k)} \\ \vdots & \vdots & \vdots & \vdots \\ 1 & X_{in_i}^{(1)} & \cdots & X_{in_i}^{(k)} \end{pmatrix} \quad \text{and} \quad \mathbb{K}_i(t) = \begin{pmatrix} K_{i1} & 0 & \cdots & 0 \\ 0 & K_{i2} & \cdots & 0 \\ \vdots & \vdots & \vdots & \vdots \\ 0 & \cdots & 0 & K_{in_i} \end{pmatrix}$$

with $K_{ij} = K\left[(t_{ij} - t)/h\right]$, the score function (7.7) with $p = 0$ reduces to

$$\begin{aligned} L_{0,K}(t; h, w) &= \sum_{i=1}^{n} \sum_{j=1}^{n_i} w_i \left\{ Y_{ij} - \sum_{l=0}^{k} \left[X_{ij}^{(l)} b_{l0}(t) \right] \right\}^2 K\left(\frac{t_{ij} - t}{h}\right), \\ &= \sum_{i=1}^{n} w_i \left[\mathbf{Y}_i - \mathbb{X}_i b_0(t) \right]^T \mathbb{K}_i(t) \left[\mathbf{Y}_i - \mathbb{X}_i b_0(t) \right]. \end{aligned} \tag{7.8}$$

If the matrix $\sum_{i=1}^{n} w_i \mathbb{X}_i^T \mathbb{K}_i(t) \mathbb{X}_i$ is invertible, i.e., $\left[\sum_{i=1}^{n} w_i \mathbb{X}_i^T \mathbb{K}_i(t) \mathbb{X}_i \right]^{-1}$ uniquely exists, then (7.8) can be uniquely minimized and its minimizer gives the following kernel estimator

$$\hat{\beta}_K^{LSK}(t; h, w) = \left[\sum_{i=1}^{n} w_i \mathbb{X}_i^T \mathbb{K}_i(t) \mathbb{X}_i \right]^{-1} \left[\sum_{i=1}^{n} w_i \mathbb{X}_i^T \mathbb{K}_i(t) \mathbf{Y}_i \right] \tag{7.9}$$

of $\beta(t) = (\beta_0(t), \ldots, \beta_k(t))^T$. For the special case of $k = 0$, i.e., the model incorporates no covariate other than time t, (7.1) reduces to the simple case (3.1) with $\beta_0(t) = \mu(t) = E[Y(t)|t]$, and (7.9) is the kernel estimator defined in (3.4).

In the expansion (7.5), the estimator $\hat{\beta}_K^{LSK}(t; h, w)$ uses only the first constant term, so that it is equivalent to a "local constant estimator." The main advantage of $\hat{\beta}_K^{LSK}(t; h, w)$ over its more general alternatives of (7.6) with $p \geq 1$ is its computational simplicity. A potential drawback for the kernel or local constant estimation approach, as demonstrated by Fan and Gijbels (1996) for the case of cross-sectional i.i.d. data, is its potential bias near the boundary of the time points. But, $\hat{\beta}_K^{LSK}(t; h, w)$ is still very competitive in most applications, when the main interest for the estimation $\beta(t)$ is not at t near the boundary of its support.

7.2.3 Least Squares Local Linear Estimators

An automatic procedure to reduce the potential boundary bias associated with the kernel estimators is to use a local polynomial estimator based on (7.5) and

(7.6) with $p \geq 1$. But, a high-order local polynomial fit, such as $p \geq 2$, can be impractical in some applications because it is computationally intensive. The simplest practical approach that provides immediate improvement on boundary bias over the kernel estimators (7.9) is to use a local linear fit that minimizes (7.7) with $p = 1$. Following equation (7.7), the local linear score function becomes

$$L_{1,K}(t; h, w) \tag{7.10}$$
$$= \sum_{i=1}^{n} \sum_{j=1}^{n_i} w_i \left\{ Y_{ij} - \sum_{l=0}^{k} \left\{ X_{ij}^{(l)} \left[b_{l0}(t) + b_{l1}(t) (t_{ij} - t) \right] \right\} \right\}^2 K\left(\frac{t_{ij} - t}{h} \right).$$

The minimizer of $L_{1,K}(t; h, w)$ can be derived by setting the derivatives of right side of (7.10) with respect to

$$\left\{ (b_{l0}(t), b_{l1}(t))^T : l = 0, \ldots, k \right\}$$

to zero, which leads to a set of normal equations. Assuming that the normal equations have a unique solution, the minimizer

$$\left\{ (\widehat{b}_{l0}(t), \widehat{b}_{l1}(t))^T : l = 0, \ldots, k \right\}$$

of (7.10) is the solution of the normal equations.

To simplify the notation, it is easier to formulate the minimizer of (7.10) with respect to $\{(b_{l0}(t), b_{l1}(t))^T : l = 0, \ldots, k\}$ using the following matrices and vectors:

$$\mathbb{N}_{lr} = \begin{pmatrix} \sum_{i=1}^{n} \sum_{j=1}^{n_i} w_i X_{ij}^{(l)} X_{ij}^{(r)} K_{ij} & \sum_{i=1}^{n} \sum_{j=1}^{n_i} w_i X_{ij}^{(l)} X_{ij}^{(r)} (t_{ij} - t) K_{ij} \\ \sum_{i=1}^{n} \sum_{j=1}^{n_i} w_i X_{ij}^{(l)} X_{ij}^{(r)} (t_{ij} - t) K_{ij} & \sum_{i=1}^{n} \sum_{j=1}^{n_i} w_i X_{ij}^{(l)} X_{ij}^{(r)} (t_{ij} - t)^2 K_{ij} \end{pmatrix},$$

$$\mathbb{M}_r = \left(\sum_{i=1}^{n} \sum_{j=1}^{n_i} w_i X_{ij}^{(r)} Y_{ij} K_{ij}, \sum_{i=1}^{n} \sum_{j=1}^{n_i} w_i X_{ij}^{(r)} (t_{ij} - t) Y_{ij} K_{ij} \right)^T,$$

$$\mathbb{N}_r = (\mathbb{N}_{0r}, \ldots, \mathbb{N}_{kr}), \quad \mathbb{N} = (\mathbb{N}_0^T, \ldots, \mathbb{N}_k^T)^T, \quad \mathbb{M} = (\mathbb{M}_0^T, \ldots, \mathbb{M}_k^T)^T,$$

$$b_l(t) = (b_{l0}(t), b_{l1}(t))^T \quad \text{and} \quad b(t) = (b_0(t), \ldots, b_k(t))^T$$

for $r = 0, \ldots, k$ and $l = 0, \ldots, k$, where $K_{ij} = K[(t_{ij} - t)/h]$ is given in (7.8). Setting the partial derivatives of $L_{1,K}(t; h, w)$ of (7.10) with respect to $b_{l0}(t)$ and $b_{l1}(t)$ to zero, the minimizer of (7.10), if exists, satisfies the following normal equation

$$\mathbb{N}\widehat{b}(t) = \mathbb{M}, \tag{7.11}$$

where $\widehat{b}(t) = (\widehat{b}_0(t), \ldots, \widehat{b}_k(t))^T$ and $\widehat{b}_l(t) = (\widehat{b}_{l0}(t), \widehat{b}_{l1}(t))^T$ for $l = 0, \ldots, k$. If the matrix $\mathbb{N}$ is invertible at t, i.e., $\mathbb{N}^{-1}$ exists, then the solution of (7.11) exists and is uniquely given by

$$\widehat{b}(t) = \mathbb{N}^{-1} \mathbb{M}. \tag{7.12}$$

For each $l = 0, \ldots, k$, the least squares local linear estimator $\widehat{\beta}_{l,K}^{LSL}(t; h, w)$ of $\beta_l(t)$ has the expression

$$\widehat{\beta}_{l,K}^{LSL}(t; h, w) = e_{2l+1}^T \, \widehat{b}(t), \qquad (7.13)$$

where e_q is the $2(k+1)$ column vector with 1 at its qth place and zero elsewhere.

Explicit expressions of the higher-order least squares local polynomial estimators can be, in principle, derived by setting the derivatives of the score function (7.7) with any $p \geq 2$ to zero. Although some results in the literature, such as Fan and Gijbels (1996), have shown that higher-order local polynomial estimators with $p \geq 2$ may have some theoretical advantages over the kernel estimators or local linear estimators, these higher-order local polynomial estimators are rarely used in practice, because they are more difficult to compute. The theoretical advantages of the higher-order local polynomial estimators over the kernel estimators or the local linear estimators depend on the smoothness assumptions of the coefficient curves $\beta(t)$, which are usually unknown in practice. Details of the general higher-order estimators based on (7.6) and (7.7) are not discussed in this book, since a local linear fitting is sufficiently satisfactory in almost all the biomedical studies that have appeared in the literature.

7.2.4   Smoothing with Centered Covariates

In some situations, modifications of the estimation methods in Sections 7.2.2 and 7.2.3 are needed in order to provide better scientific interpretations of the results. We discuss here a useful modification based on centered covariates.

1. Centered Covariates

The baseline coefficient curve $\beta_0(t)$ of the model (7.1) is generally interpreted as the mean value of $Y(t)$ at time t, when all the covariates $X^{(l)}(t)$, $l = 1, \ldots, k$, are zero. It is often the case that some of the covariates cannot have values at zero, so that the baseline coefficient curve $\beta_0(t)$ does not have a meaningful interpretation. Under such situations, the model (7.1) and the local least squares estimators (7.6) may not have useful scientific interpretations. This drawback has been noticed by Wu, Yu and Chiang (2000), which, as a remedy, has proposed to use a covariate centered modification of the model (7.1), so that the baseline coefficient can be interpreted as the conditional mean of $Y(t)$ when the centered covariates are set to zero. We present here the covariate centered time-varying coefficient model and the corresponding smoothing estimators studied by Wu, Yu and Chiang (2000).

For any $l = 1, \ldots, k$, let

$$\mu_{X^{(l)}}(t) = E\left[X^{(l)}(t)\right] \quad \text{and} \quad Z^{(l)}(t) = X^{(l)}(t) - \mu_{X^{(l)}}(t)$$

be the mean curve of $X^{(l)}(t)$ and the centered version of the time-varying covariate $X^{(l)}(t)$ $(Z^{(0)} = X^{(0)} \equiv 1)$, respectively. The covariate centered time-varying coefficient model is

$$Y(t) = \mathbf{Z}(t)^T \boldsymbol{\beta}^*(t) + \varepsilon(t), \tag{7.14}$$

where $\mathbf{Z}(t) = \left(1, Z^{(1)}(t), \ldots, Z^{(k)}(t)\right)^T$ is the vector of centered covariates and $\boldsymbol{\beta}^*(t) = \left(\beta_0^*(t), \beta_1(t), \ldots, \beta_k(t)\right)^T$ is the vector of coefficient curves. Compared with the original time-varying coefficient model (7.1), the relationship between the baseline coefficient curves $\beta_0(t)$ and $\beta_0^*(t)$ is given by

$$\beta_0^*(t) = \beta_0(t) + \sum_{l=1}^{k} \mu_{X^{(l)}}(t)\beta_l(t), \tag{7.15}$$

which represents the mean of $Y(t)$, when $X^{(l)}(t)$, $l = 1, \ldots, k$, have values at their means $\mu_{X^{(l)}}(t)$, i.e., $Z^{(l)}(t)$, $l = 1, \ldots, k$, are set to zero. Other coefficient curves of (7.1) and (7.14) remain the same. Thus, $\{\beta_1(t), \ldots, \beta_k(t)\}$ have the same interpretations in both (7.1) and (7.14).

2. Plug-In Estimation with Local Least Squares Criterion

A simple "plug-in" estimation approach based on the local least squares score function (7.7) is to first obtain the estimates of $\{\beta_0(t), \ldots, \beta_k(t)\}$ and $\{\mu_{X^{(1)}}(t), \ldots, \mu_{X^{(k)}}(t)\}$ and then estimate $\beta_0^*(t)$ by plugging in the corresponding curve estimates into (7.15). Under the framework of kernel estimators, if $X^{(l)}(t)$ is a time-dependent covariate, a centered version of $X_{ij}^{(l)}$ can be estimated by

$$\widehat{Z}_{ij}^{(l)} = X_{ij}^{(l)} - \widehat{\mu}_{X^{(l)}}(t_{ij}), \tag{7.16}$$

where, based on the kernel function $\kappa_l(\cdot)$ and bandwidth γ_l, $\widehat{\mu}_{X^{(l)}}(t_{ij})$ is the kernel estimator of $\mu_{X^{(l)}}(t)$ at $t = t_{ij}$ such that

$$\widehat{\mu}_{X^{(l)}}(t) = \frac{\sum_{i=1}^{n} \sum_{j=1}^{n_i} \left\{ w_i X_{ij}^{(l)} \kappa_l[(t - t_{ij})/\gamma_l] \right\}}{\sum_{i=1}^{n} \sum_{j=1}^{n_i} \left\{ w_i \kappa_l[(t - t_{ij})/\gamma_l] \right\}}. \tag{7.17}$$

For the special case of time-invariant covariates, covariate centering can be simply achieved by substituting $\widehat{\mu}_{X^{(l)}}(t)$ of (7.16) with the sample mean. Specifically, if $X^{(l)}(t) = X^{(l)}$ is time-invariant, then $X_{ij}^{(l)} = X_i^{(l)}$ for all $j = 1, \ldots, n_i$, and $Z_i^{(l)}$ can be estimated by $\widehat{Z}_i^{(l)} = X_i^{(l)} - \bar{X}^{(l)}$, where $\bar{X}^{(l)} = n^{-1} \sum_{i=1}^{n} X_i^{(l)}$ is the sample mean of $\{X_i^{(l)} : i = 1, \ldots, n\}$. If the kernel estimation method is used, Wu, Yu and Chiang (2000) suggest that a direct estimation approach based on (7.8), (7.15) and (7.17) is to estimate $\{\beta_0(t), \ldots, \beta_k(t)\}$ by

$$\left\{ \widehat{\beta}_0^{LSK}(t), \ldots, \widehat{\beta}_k^{LSK}(t) \right\},$$

which are the corresponding components of $\widehat{\beta}^{LSK}(t)$ given in (7.9), and then estimate $\beta_0^*(t)$ by

$$\widehat{\beta}_0^{*LSK}(t) = \widehat{\beta}_0^{LSK}(t) + \sum_{l=1}^{k} \widehat{\mu}_{X^{(l)}}(t)\, \widehat{\beta}_l^{LSK}(t), \tag{7.18}$$

so that the resulting estimator of $\beta^*(t) = (\beta_0^*(t), \beta_1(t), \ldots, \beta_k(t))^T$ is

$$\widehat{\beta}^{*LSK}(t) = \left(\widehat{\beta}_0^{*LSK}(t), \widehat{\beta}_1^{LSK}(t), \ldots, \widehat{\beta}_k^{LSK}(t)\right)^T. \tag{7.19}$$

We can see that the estimators in (7.18) and (7.19) depend on the kernel estimators of both $\{\beta_0(t), \ldots, \beta_k(t)\}$ and $\{\mu_{X^{(1)}}(t), \ldots, \mu_{X^{(k)}}(t)\}$.

3. Covariate-Centered Kernel Estimation

As an alternative to the direct least squares kernel estimators (7.19), a two-step covariate-centered kernel estimation method suggested by Wu, Yu and Chiang (2000) is to extend the local least squares score function (7.8) to the model (7.14). Let $\widehat{Z}_{ij} = \left(1, \widehat{Z}_{ij}^{(1)}, \ldots, \widehat{Z}_{ij}^{(k)}\right)^T$ be the estimator of the centered covariate vector $Z_{ij} = \left(1, Z_{ij}^{(1)}, \ldots, Z_{ij}^{(k)}\right)^T$ and

$$\widehat{Z}_i = \left(\widehat{Z}_{i1}, \ldots, \widehat{Z}_{in_i}\right)^T$$

be the estimator of the $n_i \times (k+1)$ centered covariate matrix. Substituting the $\mathbb{X}_i(t)$ of (7.8) by $\widehat{Z}_i$, the two-step covariate centered kernel estimator is obtained by minimizing

$$
\begin{aligned}
L_{0,K}^*(t; h, w) &= \sum_{i=1}^{n} \sum_{j=1}^{n_i} w_i \left\{ Y_{ij} - \sum_{l=0}^{k} \left[\widehat{Z}_{ij}^{(l)}\, b_{l0}(t)\right] \right\}^2 K\left(\frac{t_{ij} - t}{h}\right), \\
&= \sum_{i=1}^{n} w_i \left[\mathbf{Y}_i - \widehat{Z}_i\, b_0(t)\right]^T \mathbb{K}_i(t) \left[\mathbf{Y}_i - \widehat{Z}_i\, b_0(t)\right], \tag{7.20}
\end{aligned}
$$

with respect to $b_0(t) = (b_{00}(t), \ldots, b_{k0}(t))^T$. If $\left[\sum_{i=1}^{n} w_i \widehat{Z}_i^T \mathbb{K}_i(t) \widehat{Z}_i\right]$ is invertible and its inverse is $\left[\sum_{i=1}^{n} w_i \widehat{Z}_i^T \mathbb{K}_i(t) \widehat{Z}_i\right]^{-1}$, the two-step covariate centered kernel estimator of $\beta^*(t) = (\beta_0^*(t), \beta_1(t), \ldots, \beta_k(t))^T$ uniquely exists and is given by

$$
\begin{aligned}
\widetilde{\beta}_K^{*LSK}(t; h, w) &= \left(\widetilde{\beta}_{0,K}^{*LSK}(t; h, w), \widetilde{\beta}_{1,K}^{*LSK}(t; h, \mathbf{w}), \ldots, \widetilde{\beta}_{k,K}^{*LSK}(t; h, w)\right)^T \\
&= \left[\sum_{i=1}^{n} w_i \widehat{Z}_i^T \mathbb{K}_i(t) \widehat{Z}_i\right]^{-1} \left[\sum_{i=1}^{n} w_i \widehat{Z}_i^T \mathbb{K}_i(t) \mathbf{Y}_i\right], \tag{7.21}
\end{aligned}
$$

where $\mathbb{K}_i(t)$ and $\mathbf{Y}_i$ are defined as in (7.8).

4. Multiple Bandwidths and Kernels

The estimators mentioned above, both with and without covariate centering, rely on a single pair of bandwidth and kernel function $\{h, K(\cdot)\}$ to estimate all $k+1$ coefficient curves. In some practical situations, the coefficient curves $\{\beta_0(t), \beta_1(t), \ldots, \beta_k(t)\}$ or $\{\beta_0^*(t), \beta_1(t), \ldots, \beta_k(t)\}$ may belong to different smoothness families, so that an estimator which uses different pairs of bandwidth and kernel function $\{h, K(\cdot)\}$ for different components of $\beta(t)$ or $\beta^*(t)$ is generally preferred.

As an example of this approach based on the estimators given in (7.9) and (7.21), a straightforward modification of the above smoothing estimation procedures is to use a linear combination of the form

$$\widehat{\beta}_{\mathbf{K}}^{LSK}(t; \mathbf{h}, w) = \sum_{l=0}^{k} e_{l+1}^T \widehat{\beta}_{K_l}^{LSK}(t; h_l, w) \qquad (7.22)$$

for the estimation of $\beta(t) = \{\beta_0(t), \beta_1(t), \ldots, \beta_k(t)\}$ and

$$\widetilde{\beta}_{\mathbf{K}}^{*LSK}(t; \mathbf{h}, w) = \sum_{l=0}^{k} e_{l+1}^T \widetilde{\beta}_{K_l}^{*LSK}(t; h_l, w) \qquad (7.23)$$

for the estimation of $\beta^*(t) = \{\beta_0^*(t), \beta_1(t), \ldots, \beta_k(t)\}$, where, with kernel $K_l(\cdot)$ and bandwidth h_l for each $0 \le l \le k$, $\mathbf{K}(\cdot) = (K_0(\cdot), \ldots, K_k(\cdot))^T$, $\mathbf{h} = (h_0, \ldots, h_k)^T$, e_p is the $[(k+1) \times 1]$ vector with 1 at its pth place and zero elsewhere. In (7.22) and (7.23), $\widehat{\beta}_{\mathbf{K}}^{LSK}(t; \mathbf{h}, w)$ and $\widetilde{\beta}_{\mathbf{K}}^{*LSK}(t; \mathbf{h}, w)$ may use different pairs of bandwidth and kernel function to estimate different components of $\beta(t)$ or $\beta^*(t)$. As a general methodology, the idea of (7.22) and (7.23) may be applied to other smoothing estimators as well.

5. Some Additional Remarks

The large sample properties of both $\widehat{\beta}^{*LSK}(t)$ of (7.19) and $\widetilde{\beta}_K^{*LSK}(t; h, w)$ of (7.21) have been studied by Wu, Yu and Chiang (2000). These asymptotic results, which are presented in Section 7.5, suggest that both $\widehat{\beta}^{*LSK}(t)$ and $\widetilde{\beta}_K^{*LSK}(t; h, w)$ are useful in practice, and neither $\widehat{\beta}^{*LSK}(t)$ nor $\widetilde{\beta}_K^{*LSK}(t; h, w)$ is uniformly superior to the other. In particular, when all the covariates are time-invariant, $\widehat{\beta}^{*LSK}(t)$ and $\widetilde{\beta}_K^{*LSK}(t; h, w)$ are asymptotically equivalent. However, when $X^{(l)}(t)$ for $l \ge 1$ changes significantly with t, $\widetilde{\beta}_K^{*LSK}(t; h, w)$, which requires centering the time-varying $X^{(l)}(t)$ first, could be theoretically and practically superior to $\widehat{\beta}^{*LSK}(t)$. We discuss more details on the comparisons between these two kernel estimators in Section 7.5.

After a covariate is centered, the baseline coefficient curve of the model is changed, but the interpretations of the other coefficient curves, namely $\{\beta_1(t), \ldots, \beta_k(t)\}$, remain the same. Thus, the decision on whether a covariate should be centered or not primarily depends on the biological interpretations of the corresponding baseline coefficient curves $\beta_0(t)$ or $\beta_0^*(t)$. Clearly, smoothing methods other than the approaches in (7.18) and (7.21) may also be applied to the estimation of $\beta^*(t)$. But, because of the computational complications associated with covariate centering, statistical properties for local smoothing estimators other than (7.21) have not been systematically investigated in the literature.

7.2.5   Cross-Validation Bandwidth Choice

Following the approach of Chapter 6, we present here the leave-one-subject-out cross-validation (LSCV) procedure to select the bandwidths for the estimators (7.9), (7.13), (7.21), (7.22) and (7.23). Let $\mathbf{h} = (h_0, \ldots, h_k)^T$ be the bandwidths, $\widehat{\boldsymbol{\beta}}(t; \mathbf{h}, w)$ be a smoothing estimator of $\boldsymbol{\beta}(t) = (\beta_0(t), \ldots, \beta_k(t))^T$ or $(\beta_0^*(t), \beta_1(t), \ldots, \beta_k(t))^T$ based on $\mathbf{h}$, and $\widehat{\boldsymbol{\beta}}^{(-i)}(t; \mathbf{h}, w)$ be an estimator computed using the same method as $\widehat{\boldsymbol{\beta}}(t; \mathbf{h}, w)$ but with the ith subject's measurements deleted. The LSCV score for $\widehat{\boldsymbol{\beta}}(t; \mathbf{h}, w)$ is

$$LSCV(\mathbf{h}, w) = \sum_{i=1}^{n} \sum_{j=1}^{n_i} w_i \left[Y_{ij} - X_{ij}^T \widehat{\boldsymbol{\beta}}^{(-i)}(t; \mathbf{h}, w) \right]^2, \tag{7.24}$$

which measures the predictive error of $\widehat{\boldsymbol{\beta}}(t; \mathbf{h}, w)$ for the model (7.1). The LSCV bandwidth vector $\mathbf{h}_{LSCV}$ is the minimizer of the cross-validation score $LSCV(\mathbf{h}, w)$ provided that the right side of (7.24) can be uniquely minimized. An automatic search of the global minima of (7.24) usually requires a sophisticated optimization software. In practice, particularly when the dimensionality $(k+1)$ of $\mathbf{h}$ is high, it is often reasonable to use a bandwidth vector whose cross-validation score $LSCV(\mathbf{h}, w)$ is close to the global minima.

The reasons for using the LSCV score (7.24) are the same as the LSCV scores in Chapters 3 to 6. First, by deleting the subjects one at a time, the correlation structure of the remaining data and the measurements of the deleted subject is preserved. Second, since in real applications the correlation structures of the data are often completely unknown and difficult to estimate, the LSCV score (7.24) is a widely acceptable choice because it does not require any specific assumptions on the intra-subject correlations, hence, can be implemented in all practical situations. Third, when the number of subjects is sufficiently large, minimizing (7.24) leads to a bandwidth vector that approximately minimizes the following average squared error

$$ASE\left[\widehat{\boldsymbol{\beta}}(\cdot; \mathbf{h}, w)\right] = \sum_{i=1}^{n} \sum_{j=1}^{n_i} w_i \left\{ X_{ij}^T \left[\boldsymbol{\beta}(t_{ij}) - \widehat{\boldsymbol{\beta}}(t_{ij}; \mathbf{h}, w) \right] \right\}^2. \tag{7.25}$$

The last assertion follows heuristically from the decomposition

$$LSCV(\mathbf{h}, w) \quad = \quad \sum_{i=1}^{n} \sum_{j=1}^{n_i} \left\{ w_i \left[Y_{ij} - X_{ij}^T \boldsymbol{\beta}(t_{ij}) \right]^2 \right\}$$

$$+2\sum_{i=1}^{n}\sum_{j=1}^{n_i}\left\{w_i\left[Y_{ij}-X_{ij}^T\beta(t_{ij})\right]\right.$$

$$\left.\times\left[X_{ij}^T\left(\beta(t_{ij})-\widehat{\beta}^{(-i)}(t_{ij};\mathbf{h},w)\right)\right]^2\right\}$$

$$+\sum_{i=1}^{n}\sum_{j=1}^{n_i}\left\{w_i\left[X_{ij}^T\left(\beta(t_{ij})-\widehat{\beta}^{(-i)}(t_{ij};\mathbf{h},w)\right)\right]^2\right\}. \quad (7.26)$$

Here, (7.25) and the definition of $\widehat{\beta}^{(-i)}(t;\mathbf{h},w)$ imply that the third term at the right side of (7.26) is approximately the same as $ASE\left[\widehat{\beta}(\cdot;\mathbf{h},w)\right]$. Because the first term at the right side of (7.26) does not depend on the bandwidth and the second term is approximately zero, $\mathbf{h}_{LSCV}$ approximately minimizes the average squared error $ASE\left[\widehat{\beta}(\cdot;\mathbf{h},w)\right]$.

7.3 Pointwise and Simultaneous Confidence Bands

Since the asymptotic distributions for the smoothing estimators of $\beta(t)$ given in Section 7.2 have only been developed for a few special cases based on kernel smoothing, asymptotically approximate inferences for $\beta(t)$ are not generally available beyond these special cases. Even for the special cases where the asymptotic distributions are available, the corresponding approximate inferences require the "plug-in" approach by substituting the unknown quantities in the asymptotic biases and variances with their smoothing estimates. Thus, similar to the situations in Section 6.4, a more practical inference approach is to consider the resampling-subject bootstrap procedure. We describe here the bootstrap approximate pointwise and simultaneous inferences for $\beta(t)$.

7.3.1 Pointwise Confidence Intervals by Bootstrap

Let $\widehat{\beta}(t)=\left(\widehat{\beta}_0(t),\ldots,\widehat{\beta}_k(t)\right)^T$ be an estimator of $\beta(t)$ constructed based on any of the methods in Section 7.2, and $A=(a_0,\ldots,a_k)^T$ be a known $(k+1)$ column vector. Then,

$$A^T E\left[\widehat{\beta}(t)\right]=\sum_{l=0}^{k}a_l E\left[\widehat{\beta}_l(t)\right]$$

is a linear combination of the components of $E\left[\widehat{\beta}(t)\right]$. The resampling-subject bootstrap procedure constructs an approximate $\left[100\times(1-\alpha)\right]\%$ pointwise percentile interval for $A^T E\left[\widehat{\beta}(t)\right]$ using the following steps.

Approximate Bootstrap Pointwise Confidence Intervals:

(a) **Computing Bootstrap Estimators.** *Generate B independent bootstrap samples using the resampling-subject bootstrap procedure of Section 3.4.1 and compute the B bootstrap estimators $\left\{\widehat{\beta}_1^b(t),\ldots,\widehat{\beta}_B^b(t)\right\}$ of $\beta(t)$.*

(b) Approximate Bootstrap Confidence Intervals. *Calculate* $L^b_{A,\alpha/2}(t)$ *and* $U^b_{A,\alpha/2}(t)$, *the lower and upper* $\left[100 \times (\alpha/2)\right]$*th percentiles, respectively, of the B bootstrap estimators* $A^T \widehat{\beta}^b(t)$. *The approximate* $\left[100 \times (1-\alpha)\right]\%$ *bootstrap confidence interval for* $A^T E\left[\widehat{\beta}(t)\right]$ *based on percentiles is*

$$\left(L^b_{A,\alpha/2}(t), U^b_{A,\alpha/2}(t)\right). \tag{7.27}$$

The normal approximated bootstrap confidence interval for $A^T E\left[\widehat{\beta}(t)\right]$ *is*

$$A^T \widehat{\beta}(t) \pm z_{1-\alpha/2} \times \widehat{se}\left(t; A^T \widehat{\beta}^b\right), \tag{7.28}$$

where $\widehat{se}\left(t; A^T \widehat{\beta}^b\right)$ *is the estimated standard deviation of the B bootstrap estimates* $\{A^T \widehat{\beta}^b_1(t), \ldots, A^T \widehat{\beta}^b_B(t)\}$ *which is given by*

$$\widehat{se}\left(t; A^T \widehat{\beta}^b\right) = \left\{ \frac{1}{B-1} \sum_{s=1}^{B} \left[A^T \widehat{\beta}^b_s(t) - \frac{1}{B} \sum_{r=1}^{B} A^T \widehat{\beta}^b_r(t)\right]^2 \right\}^{1/2}, \tag{7.29}$$

and $z_{1-\alpha/2}$ *is the* $\left[100 \times (1-\alpha/2)\right]$*th percentile of the standard normal distribution.* □

The pointwise confidence intervals given in (7.27) and (7.28) are only approximate intervals for $A^T E\left[\widehat{\beta}(t)\right]$, because they do not contain bias corrections. When the asymptotic bias of $\widehat{\beta}(t)$ is small, the coverage probabilities of (7.27) and (7.28) containing $A^T \widehat{\beta}(t)$ are close to $[100 \times (1-\alpha)]\%$. Since the asymptotic bias of a smoothing estimator is often difficult to be accurately estimated in practice, the "plug-in" approach for correcting the unknown asymptotic bias may not lead to better coverage probabilities over the simple variability bands given in (7.27) and (7.28).

Pointwise confidence intervals for a single component of $\beta(t)$ can be constructed by selecting the corresponding component of A to be 1 and 0 elsewhere. In particular, the pointwise resampling-subject bootstrap confidence interval for $\beta_r(t)$, $0 \leq r \leq k$, can be computed by (7.27) or (7.28) with A being the $(k+1)$ column vector having 1 at its $(r+1)$th place and 0 elsewhere. Pointwise confidence intervals for the difference of two component curves of $\beta(t)$ can be similarly constructed by taking the corresponding elements of A to be 1 and -1 and 0 elsewhere. For example, the pointwise bootstrap confidence interval for $\left[\beta_{r_1}(t) - \beta_{r_2}(t)\right]$ can be computed by (7.27) or (7.28) with A being the $(k+1)$ column vector having 1 and -1 at its (r_1+1)th and (r_2+1)th places, respectively, and 0 elsewhere. Other special cases of (7.27) and (7.28) can be similarly constructed by choosing the appropriate vector A.

7.3.2 Simultaneous Confidence Bands

We now describe the "bridging-the-gap" simultaneous confidence bands for $A^T \beta(t) = \sum_{l=0}^{k} a_l \beta_l(t)$ over $t \in [a, b]$, where $[a, b]$ is any given interval in $\mathcal{T}$. The first step is to select a type of pointwise confidence intervals based on the estimator $\widehat{\beta}(t) = \left(\widehat{\beta}_0(t), \dots, \widehat{\beta}_k(t)\right)^T$ of Section 7.2. In general, let $l_{A,\alpha/2}(t)$ and $u_{A,\alpha/2}(t)$ be the lower and upper bounds, respectively, of a $\left[100 \times (1 - \alpha)\right]\%$ pointwise confidence interval for $A^T \beta(t)$ based on $\widehat{\beta}(t)$, so that the corresponding interval is given by

$$\left(l_{A,\alpha/2}(t), u_{A,\alpha/2}(t)\right). \tag{7.30}$$

A simple choice for (7.30) is to use the resampling-subject bootstrap confidence intervals given in (7.27) and (7.28).

The next step is to construct a set of confidence intervals which cover $A^T \beta(t)$ over a grid of time points in $[a, b] \in \mathcal{T}$ with at least $\left[100 \times (1 - \alpha)\right]\%$ of the coverage probability. Thus, we partition $[a, b]$ into M equally spaced intervals with grid points

$$a = \xi_1 < \cdots < \xi_{M+1} = b, \quad \text{such that } \xi_{j+1} - \xi_j = (b - a)/M \text{ for } j = 1, \dots, M.$$

The integer M is usually chosen subjectively with the intent that the resulting confidence bands are not too wide. Given $\{\xi_1, \cdots, \xi_{M+1}\}$, a set of approximate $\left[100 \times (1 - \alpha)\right]\%$ simultaneous confidence intervals for $A^T \beta(\xi_j)$, $j = 1, \dots, M+1$, is then the collection of intervals

$$\left\{\left(L_{A,\alpha/2}(\xi_j), U_{A,\alpha/2}(\xi_j)\right) : j = 1, \dots, M+1\right\}, \tag{7.31}$$

which satisfies the inequality

$$\lim_{n \to \infty} P\left[L_{A,\alpha/2}(\xi_j) \leq A^T \beta(\xi_j) \leq U_{A,\alpha/2}(\xi_j) \text{ for all } j = 1, \dots, M+1\right] \geq 1 - \alpha.$$

A simple choice of (7.31) is based on the Bonferroni adjustment of the pointwise confidence interval of (7.30), such that

$$\left(L_{A,\alpha/2}(\xi_j), U_{A,\alpha/2}(\xi_j)\right) = \left(l_{A,\alpha/[2(M+1)]}(\xi_j), u_{A,\alpha/[2(M+1)]}(\xi_j)\right). \tag{7.32}$$

To bridge the gaps between the grid points $\{\xi_j : j = 1, \dots, M+1\}$, we rely on some smoothness conditions for $A^T \beta(t)$. Given the values of $A^T \beta(\xi_j)$ and $A^T \beta(\xi_{j+1})$, the linear interpolation for $t \in [\xi_j, \xi_{j+1}]$ is

$$\left(A^T \beta\right)^{(I)}(t) = \left[\frac{M(\xi_{j+1} - t)}{b - a}\right]\left[A^T \beta(\xi_j)\right] + \left[\frac{M(t - \xi_j)}{b - a}\right]\left[A^T \beta(\xi_{j+1})\right]. \tag{7.33}$$

By (7.31) and (7.33), a simultaneous confidence band for the linear interpolation $\left(A^T \beta\right)^{(I)}(t)$ over $t \in [a, b]$ is

$$\left(L_{A,\alpha/2}^{(I)}(t), U_{A,\alpha/2}^{(I)}(t)\right), \tag{7.34}$$

where $L_{A,\alpha/2}^{(I)}(t)$ is the linear interpolation of $\{L_{A,\alpha/2}(\xi_j), L_{A,\alpha/2}(\xi_{j+1})\}$ and $U_{A,\alpha/2}^{(I)}(t)$ is the linear interpolation $\{U_{A,\alpha/2}(\xi_j), U_{A,\alpha/2}(\xi_{j+1})\}$. If the derivative of $A^T\beta(t)$ with respect to t is bounded, such that

$$\sup_{t\in[a,b]}\left|(A^T\beta)'(t)\right|\le c_1, \quad \text{for a known constant } c_1 > 0, \tag{7.35}$$

then it follows that

$$\left|A^T\beta(t) - (A^T\beta)^{(I)}(t)\right| \le 2c_1\left[\frac{M(\xi_{j+1}-t)(t-\xi_j)}{b-a}\right],$$

for all $t\in[\xi_j, \xi_{j+1}]$, and an approximate $[100\times(1-\alpha)]\%$ simultaneous confidence band for $A^T\beta(t)$ is

$$\left(L_{A,\alpha/2}^{(I)}(t) - 2c_1\left[\frac{M(\xi_{j+1}-t)(t-\xi_j)}{b-a}\right],\right.$$
$$\left.U_{A,\alpha/2}^{(I)}(t) + 2c_1\left[\frac{M(\xi_{j+1}-t)(t-\xi_j)}{b-a}\right]\right). \tag{7.36}$$

If the second derivative of $A^T\beta(t)$ with respect to t is bounded, such that

$$\sup_{t\in[a,b]}\left|(A^T\beta)''(t)\right|\le c_2, \quad \text{for a known constant } c_2 > 0, \tag{7.37}$$

then

$$\left|A^T\beta(t) - (A^T\beta)^{(I)}(t)\right| \le \frac{c_2}{2}\left[\frac{M(\xi_{j+1}-t)(t-\xi_j)}{b-a}\right],$$

for all $t\in[\xi_j, \xi_{j+1}]$, and an approximate $[100\times(1-\alpha)]\%$ simultaneous confidence band is given by

$$\left(L_{A,\alpha/2}^{(I)}(t) - \frac{c_2}{2}\left[\frac{M(\xi_{j+1}-t)(t-\xi_j)}{b-a}\right],\right.$$
$$\left.U_{A,\alpha/2}^{(I)}(t) + \frac{c_2}{2}\left[\frac{M(\xi_{j+1}-t)(t-\xi_j)}{b-a}\right]\right). \tag{7.38}$$

When $A^T\beta(t)$ satisfy smoothness conditions other than (7.35) and (7.37), simultaneous confidence bands for $A^T\beta(t)$ can be similarly constructed using the linear interpolation bands of (7.34) and the inequalities obtained from the smoothness conditions. The smoothness conditions defined through the derivatives as (7.35) and (7.37) are easy to interpret in real applications.

7.4   R Implementation

7.4.1   The NGHS BP Data

The NGHS data has been described in Section 1.2. In Section 5.2.2, we have analyzed the time-trends of body mass index (BMI) for the 1213 African

American girls and 1166 Caucasian girls, who were followed from ages 9 or 10 years to 18 or 19 years. Since the objective of this study is to investigate the patterns of cardiovascular risk factors, an important outcome variable is the subject's systolic blood pressure (SBP). Although the children's blood pressure generally increases with age and height, the longitudinal relationship between the time-trends of blood pressure and the covariates of interests may not follow a certain parametric family.

1. Estimation with the Time-Varying Coefficient Model

The structured nonparametric model (7.1) appears to be a natural option for an exploratory analysis. In this analysis, we evaluate the potential covariate effects of race, height and BMI on the time-trends of SBP using the time-varying coefficient model (7.1). For the ith NGHS participant, we denote by t_{ij} the age in years at the jth study visit, Y_{ij} the SBP level at t_{ij} during the study, ε_{ij} the error term at t_{ij}, $X_i^{(1)}$ the indicator variable for race ($X_i^{(1)} = 1$ for African American, $X_i^{(1)} = 0$ for Caucasian), $X_{ij}^{(2)}$ and $X_{ij}^{(3)}$ the height and BMI at t_{ij}. The model (7.1) can be written as

$$Y_{ij} = \beta_0(t_{ij}) + \beta_1(t_{ij}) X_i^{(1)} + \beta_2(t_{ij}) X_{ij}^{(2)} + \beta_3(t_{ij}) X_{ij}^{(3)} + \varepsilon_{ij}. \tag{7.39}$$

Instead of using the actual height and BMI values, we use in the above model the age-adjusted height and BMI percentiles for $X^{(2)}(t_{ij})$ and $X^{(3)}(t_{ij})$, respectively, which are computed by subtracting 50% from the individual's height and BMI percentiles obtained from the U.S. Center for Disease Control and Prevention (CDC) Growth Charts (https://www.cdc.gov/growthcharts/). Consequently, $\beta_0(t)$ represents the baseline SBP curve, i.e., the mean time curve of SBP for a Caucasian girl with a median height and a median BMI level, and $\beta_l(t)$, $l = 1, 2, 3$, represent the effects of race, height and BMI, respectively, on the individual's SBP during the adolescence years. This definition of $X^{(2)}(t_{ij})$ and $X^{(3)}(t_{ij})$ leads to meaningful clinical interpretations for the model (7.1). Hence, there is no need to consider further "covariate centering" as described in Section 7.2.4.

To obtain the local least squares smoothing estimates for the coefficient curves as described in Section 7.2, we need to select a kernel function $K(\cdot)$, an order of polynomial p and a bandwidth h. The bandwidth h is the most important parameter to determine the smoothness of the fitted curves. The resulting local polynomial estimators are the weighted least squares solution of (7.7), which may be obtained by simply applying the R function lm(). For a given t, we set the design matrix for lm() to include $\{\mathbb{X}^{(l)} : l = 0, 1, \ldots, k\}$, where $\mathbb{X}^{(l)}$ has p components,

$$\left\{ X_{ij}^{(l)}, X_{ij}^{(l)}(t_{ij} - t), \ldots, X_{ij}^{(l)}(t_{ij} - t)^p \right\},$$

and set the weight argument to include $w_i K[(t_{ij} - t)/h]$. Then we can take the $(2l+1)$th, $l = 0, 1, \ldots, k$, components of the coefficient estimates to obtain $\widehat{b}_0(t) = \left(\widehat{b}_{00}(t), \ldots, \widehat{b}_{k0}(t)\right)^T$ in (7.6). Alternatively, we may also use the algorithm in Section 7.2.3 to compute the matrices $\mathbb{N}$ and $\mathbb{M}$, and solve equation (7.11) to get the local linear estimators.

For the NGHS SBP data, we estimate the model coefficients $\{\beta_l(t) : l = 0, 1, 2, 3\}$ using a local linear fit with the Epanechnikov kernel and a subjective bandwidth $h = 3.5$ by using the following R commands:

```
> library(npmlda)
# Obtain a grid of time
> Age.grid <- seq(9, 19, by=0.5)
# Call function LocalLm.Beta() for local least squares fits for
# baseline and 3 covariate effects at  Age.grid
> Beta <- with(NGHS, LocalLm.Beta(Age.grid, AGE, X1=Black,
        X2=HTPCTc, X3=BMIPCTc, SBP, Bndwdth=3.5, Weight=1/ni))
```

Figure 7.1 displays the estimated coefficient curves for the covariates between 9 to 19 years of age for the NGHS girls and their 95% pointwise confidence intervals computed based on the resampling-subject bootstrap procedure of Section 7.3.1 with 1000 bootstrap replications. The subject uniform weight $w_i^* = 1/(nn_i)$ discussed in Section 7.2.1 is used. The fitted curves with the LSCV cross-validated bandwidth $h = 1.6$ are slightly undersmoothed but suggest very similar time-trends for the baseline and coefficient curves.

Remarkably, these coefficient curves in Figure 7.1 suggest very interesting baseline growth pattern and time-varying covariate effects on the time-varying SBP patterns of adolescent girls. As expected, the estimated baseline coefficient curve shows that the mean SBP for Caucasian girls with the median height and the median BMI increases with age during adolescent years. The effect of race, which changes with time, suggests that the African American girls tend to have higher mean SBP levels than the Caucasian girls. However, this racial difference is not statistically significant at younger ages but becomes more significant at later adolescent years. The lower panels of Figure 7.1 show the estimated coefficient curves for the two time-dependent covariates, height and BMI percentiles. The effect of height percentile on SBP is positive, suggesting that the SBP levels of adolescent girls tend to increase with height percentile and the influence of height percentile is larger for girls with younger ages and declines with age. Finally, the BMI percentile is also positively associated with SBP level, and its effect is larger at older ages compared to that at early adolescent years.

2. Estimation with the Linear Mixed-Effects Model

We see from Figure 7.1 that the above structured nonparametric results based on the model (7.39) suggest that the relationship between the mean of

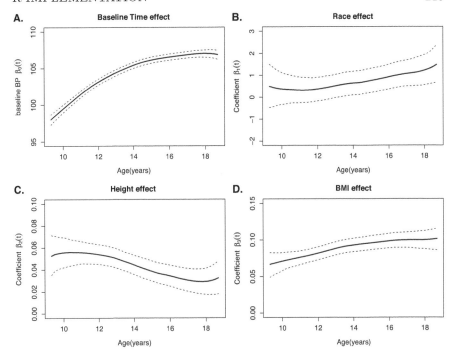

Figure 7.1 *Covariate effects on systolic blood pressures of NGHS girls using local linear estimators. The solid curves in (A)-(D) show the estimated coefficient curves* $\hat{\beta}_{l,K}^{LSL}(t; h, w^*)$, $l = 0, 1, 2, 3$, *respectively, based on (7.13) using a Epanechnikov kernel and a bandwidth* $h = 3.5$. *The dashed curves indicate the corresponding bootstrap 95% pointwise confidence intervals.*

SBP $Y(t)$ and the girls' age t, race $X^{(1)}$, height percentile $X^{(2)}(t)$ and BMI percentile $X^{(3)}(t)$ may be approximately described by a parametric linear mixed-effects model described in Section 2.1. We examine this possibility by fitting the data to the following second-order linear mixed-effects model

$$
\begin{aligned}
Y_{ij} &= \alpha_{0i} + \alpha_1 \times Age_{ij}^c + \alpha_2 \times X_i^{(1)} + \alpha_3 \times X_{ij}^{(2)} + \alpha_4 \times X_{ij}^{(3)} \\
&\quad + \alpha_5 \times Age_{ij}^c \times X_{ij}^{(1)} + \alpha_6 \times Age_{ij}^c \times X_{ij}^{(2)} \qquad (7.40) \\
&\quad + \alpha_7 \times Age_{ij}^c \times X_{ij}^{(3)} + e_{ij},
\end{aligned}
$$

where, for convenience and clinical interpretations, Age_{ij}^c is the girls' *centered* age computed by subtracting the starting age of 9 years from the girl's actual age at the jth visit, and, $X_i^{(1)}$, $X_{ij}^{(2)}$ and $X_{ij}^{(3)}$ are the girl's race, height percentile and BMI percentile as defined in (7.39). In (7.40), the time-varying covariate effects are described by the coefficients $\{\alpha_5, \alpha_6, \alpha_7\}$ for the interaction terms

$$
\left\{ Age_{ij}^c \times X_{ij}^{(1)}, \; Age_{ij}^c \times X_{ij}^{(2)}, \; Age_{ij}^c \times X_{ij}^{(3)} \right\}.
$$

The parameter estimates and their inferences of (7.40) can be computed using the following R code:

```
> NGHS.fit <- lme(SBP~AGEc+Race+Race:AGEc+HTPCTc+HTPCTc:AGEc
            + BMIPCTc + BMIPCTc:AGEc, random=~1|ID, data=NGHS)
> summary(NGHS.fit)

Linear mixed-effects model fit by REML

...
Fixed effects:
                  Value   Std.Error   DF    t-value p-value
(Intercept)   100.12814 0.21507240 16938 465.5555  0.0000
AGEc            0.87370 0.02517459 16938  34.7055  0.0000
Race            0.30588 0.30215527  2374   1.0123  0.3115
HTPCTc          0.04558 0.00478288 16938   9.5290  0.0000
BMIPCTc         0.08750 0.00439813 16938  19.8955  0.0000
AGEc:Race       0.07878 0.03540669 16938   2.2251  0.0261
AGEc:HTPCTc    -0.00131 0.00065129 16938  -2.0166  0.0438
AGEc:BMIPCTc    0.00252 0.00062072 16938   4.0556  0.0001
...
```

The output above shows that the baseline SBP increases with age because both $\widehat{\alpha}_0 = 100.128$ and $\widehat{\alpha}_1 = 0.874$ are statistically significantly larger than zero. There are significant positive associations of race, height percentile and BMI percentile with SBP over time. Furthermore, the significant interaction terms indicate that the covariate effects on SBP are varying with time. Hence, these results from the linear mixed-effects model (7.40) are consistent with those obtained from the time-varying coefficient model (7.39).

Since the cardiovascular risk factors track from childhood to adulthood, this example suggests that the time-varying coefficient models are useful approaches to explore the racial differences and correlates in blood pressures or other risk factors in this type of longitudinal studies. The findings could provide rationales for future interventions to reduce the excess cardiovascular mortality, as discussed in Daniels et al. (1998).

7.4.2   The BMACS CD4 Data

The BMACS CD4 data has been described in Section 1.2 and analyzed in several previous chapters. In this analysis, we consider using the local least squares smoothing method to estimate the effects of three time-invariant covariates, pre-HIV infection CD4 percentage, cigarette smoking, and age at HIV infection, on the mean CD4 percentage after the infection using the time-varying coefficient model (7.1). In comparison with the componentwise local method in Chapter 6, the estimators in Section 7.2 can be applied to more general models that also allow for time-dependent covariates.

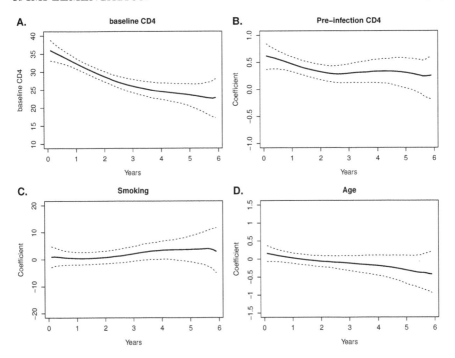

Figure 7.2 *Covariate effects on post-infection CD4 percentage using local linear estimators. The solid curves in (A)-(D) show the estimated coefficient curves $\widehat{\beta}_l(t)$, $l = 0, 1, 2, 3$, respectively, based on (7.6) using the Epanechnikov kernel and the cross-validated bandwidth $h = 1.8$. The dashed curves indicate the corresponding bootstrap 95% pointwise confidence intervals.*

Figure 7.2 displays the estimated coefficient curves from 0.1 to 5.9 years post-HIV infection for the BMACS CD4 data, where the continuous covariates of pre-infection CD4 percentage and age at HIV infection are first centered by subtracting their corresponding sample averages, 42.9% and 34.2 years, from the individual values, respectively. The four coefficients $\{\beta_l(t) : l = 0, ..., 3\}$ are obtained from a local linear estimator with the Epanechnikov kernel, subject uniform weight $w_i^* = 1/(nn_i)$, and the LSCV cross-validated bandwidth $h = 1.8$. Their 95% pointwise confidence intervals are computed using the resampling-subject bootstrap with 1000 bootstrap replications.

The results for the baseline and covariate effects from the local least squares estimates, shown in Figure 7.2, are very similar to the findings obtained from the componentwise methods in Figures 6.1 and 6.2. They all suggest that the mean baseline CD4 percentage decreases quickly after HIV infection, the pre-infection CD4 percentage is positively associated with post-infection CD4 percentage but this effect is declining with time since HIV infection. However, neither smoking nor age at infection has a significant association with the post-infection CD4 percentage.

7.5   Asymptotic Properties for Kernel Estimators

This section summarizes the asymptotic properties of the local least squares kernel estimators for the time-varying coefficient model (7.1). These smoothing estimators are representatives of the one-step smoothing methods for (7.1), and their asymptotic properties give an overview on how the smoothing method works in practice. By minimizing the local-squares score function (7.8), the expression of the least squares kernel estimator $\widehat{\beta}_K^{LSK}(t; h, w)$ of $\beta(t)$ in (7.1) is given by (7.9). The asymptotic risks of $\widehat{\beta}_K^{LSK}(t; h, w)$ are expressed using the mean squared errors.

The main results of this section have two interesting features, which distinguish the asymptotic risk of $\widehat{\beta}_K^{LSK}(t; h, w)$ from its counterpart of the Nadaraya-Watson kernel estimators in nonparametric regression with cross-sectional i.i.d. data. First, the asymptotic bias of the kernel estimator $\widehat{\beta}_K^{LSK}(t; h, w)$ is affected by the smoothness of the time-varying covariates $\mathbf{X}(t)$ as well as the smoothness of $\beta(t)$ and the underlying design density of t. Second, the asymptotic variance of $\widehat{\beta}_K^{LSK}(t; h, w)$ is influenced by the intra-subject correlation of the data and the numbers of repeated measurements n_i, as well as the variance of the error term $\varepsilon(t)$. Thus, as a consequence, the convergence rates of $\widehat{\beta}_K^{LSK}(t; h, w)$ in general depend on the number of subjects n and the numbers of repeated measurements n_i. In contrast, the convergence rates of the kernel estimators in nonparametric regression with cross-sectional i.i.d. data only depend on the sample size n.

7.5.1   Asymptotic Assumptions

The estimation methods of Section 7.2 can accommodate both fixed and random time points. We make the following technical assumptions for the asymptotic properties of $\widehat{\beta}_K^{LSK}(t; h, \mathbf{w})$ throughout this chapter:

(a) *The time points $\{t_{ij} : j = 1, \ldots, n_i; i = 1, \ldots, n\}$ are chosen independently according to some design distribution F_T and design density f_T.*

(b) *For all $i = 1, \ldots, n$, $j = 1, \ldots, n_i$, and $r, l = 0, \ldots, k$,*

$$\begin{cases} \sigma^2(t) & = & E[\varepsilon^2(t)], \\ \rho_\varepsilon(t) & = & \lim_{\delta \to 0} E[\varepsilon(t + \delta)\varepsilon(t)], \\ \xi_{lr}(t) & = & E(X_{ij}^{(l)} X_{ij}^{(r)} | t_{ij} = t). \end{cases} \qquad (7.41)$$

(c) *For all $l, r = 0, \ldots, k$, $\xi_{lr}(t)$ is Lipschitz continuous with order α_0, i.e.,*

$$|\xi_{lr}(s_1) - \xi_{lr}(s_2)| \le c_0 |s_1 - s_2|^{\alpha_0}$$

for s_1 and s_2 in the support of $f_T(t)$ and some $c_0 > 0$, and $\beta_l(t)$ and $f_T(t)$ are Lipschitz continuous, respectively, with orders $\alpha_1 > 0$ and $\alpha_2 > 0$.

(d) *The variance curve $\sigma^2(t)$ and the limiting covariance curve $\rho_\varepsilon(t)$ are continuous, and $E\left[\left(X_{ij}^{(l)}\right)^4\right] < \infty$.*

(e) *The kernel function $K(\cdot)$ is non-negative, has compact support on the real line, integrates to one and has finite second moment, that is, there are constants $c_1 < c_2$, such that,*

$$\begin{cases} K(u) > 0 \text{ if } c_1 \leq u \leq c_2, \ K(u) = 0 \text{ if } u < c_1 \text{ or } u > c_2, \\ \int K(u)\,du = 1 \quad \text{and} \quad \int K^2(u)\,du < \infty. \end{cases} \tag{7.42}$$

(f) *The bandwidth satisfies $h \to 0$ and $nh \to \infty$ as $n \to \infty$.* □

Similar to the assumptions in Section 6.6.2, $\rho_\varepsilon(t)$ in general does not necessarily equal $\sigma^2(t)$, and, when the error term $\varepsilon(t)$ includes an independent measurement error, $\sigma^2(t) > \rho_\varepsilon(t)$. These assumptions are comparable to the regularity conditions commonly used in nonparametric regression with cross-sectional i.i.d. data (e.g., Härdle, 1990), and are sufficiently general to be satisfied in many interesting practical situations. Theoretically, these assumptions can be further modified or weakened in various ways so that more desirable asymptotic properties of the kernel estimator $\widehat{\beta}_K^{LSK}(t; h, w)$ may be obtained.

7.5.2 Mean Squared Errors

The risk of $\widehat{\beta}_K^{LSK}(t; h, w)$ or any other smoothing estimator of $\beta(t)$ in Section 7.2 depends on the choice of loss functions. Let W_l, $l = 0, \dots, k$, be non-negative constants, and $\mathbf{W} = diag(W_0, \dots, W_k)$ be the $(k+1) \times (k+1)$ matrix with diagonal elements $\{W_0, \dots, W_k\}$ and zero elsewhere. The local mean squared error of $\widehat{\beta}^{LSK}(\cdot)$ with weights $\mathbf{W}$ at time t is

$$\begin{aligned} &MSE_{\mathbf{W}}^*\left[\widehat{\beta}_K^{LSK}(t; h, w)\right] \\ &= E\left\{ \left[\widehat{\beta}_K^{LSK}(t; h, w) - \beta(t)\right]^T \mathbf{W}\left[\widehat{\beta}_K^{LSK}(t; h, w) - \beta(t)\right] \right\}. \end{aligned} \tag{7.43}$$

Similar to the situation of Section 6.6, the mean squared error (7.43) may not exist in general. Since this minor technical problem does not have real implications about the practical value of an estimator, we use the following slightly modified version of (7.43) for $\widehat{\beta}_K^{LSK}(t; h, w)$.

1. *Approximation for $\left[\widehat{\beta}_K^{LSK}(t; h, w) - \beta(t)\right]$*

Let

$$\widehat{R}_w(t) = \sum_{i=1}^n (w_i/h) \left\{ \mathbb{X}_i^T \mathbb{K}_i(t) \left[\mathbf{Y}_i - \mathbb{X}_i \beta(t)\right] \right\}, \tag{7.44}$$

where $w = (w_1, \dots, w_n)^T$ is the weights used in (7.9). The objective is to express the risks of $\widehat{\beta}_K^{LSK}(t; h, w)$ through the second moment of $\widehat{R}_w(t)$. By straightforward algebra, which is given in the following, we have the approximation

$$
\begin{aligned}
D(t) &= [1 + o_p(1)] \left[\widehat{\beta}_K^{LSK}(t; h, w) - \beta(t) \right] \\
&= f_T^{-1}(t) E_{\mathbf{XX}^T}^{-1}(t) \widehat{R}_w(t),
\end{aligned}
\tag{7.45}
$$

where $f_T(t)$ is the density function of t_{ij} given in Assumption (a), $E_{\mathbf{XX}^T}(t) = E\left[\mathbf{X}(t) \mathbf{X}^T(t) \right]$. To derive (7.45), we first obtain from (7.9) that

$$
\begin{aligned}
&\widehat{\beta}_K^{LSK}(t; h, w) - \beta(t) \\
&= \left[\sum_{i=1}^n w_i \mathbb{X}_i^T \, \mathbb{K}_i(t) \, \mathbb{X}_i \right]^{-1} \left\{ \sum_{i=1}^n w_i \mathbb{X}_i^T \, \mathbb{K}_i(t) \left[\mathbf{Y}_i - \mathbb{X}_i \beta(t) \right] \right\} \\
&= \left[\sum_{i=1}^n (w_i/h) \mathbb{X}_i^T \, \mathbb{K}_i(t) \, \mathbb{X}_i \right]^{-1} \widehat{R}_w(t).
\end{aligned}
\tag{7.46}
$$

Then, multiplying $\sum_{i=1}^n (w_i/h) \mathbb{X}_i^T \, \mathbb{K}_i(t) \, \mathbb{X}_i$ to both sides of (7.46), we have

$$
\left[\sum_{i=1}^n (w_i/h) \mathbb{X}_i^T \, \mathbb{K}_i(t) \, \mathbb{X}_i \right] \left[\widehat{\beta}_K^{LSK}(t; h, w) - \beta(t) \right] = \widehat{R}_w(t).
\tag{7.47}
$$

By the definition of $\mathbb{K}_i(t)$ in Section 7.2.2, we have

$$
E\left[\sum_{i=1}^n (w_i/h) \mathbb{X}_i^T \, \mathbb{K}_i(t) \, \mathbb{X}_i \right] = \sum_{i=1}^n (w_i/h) E\left\{ \mathbb{X}_i^T \, E\left[\mathbb{K}_i(t) \right] \mathbb{X}_i \right\}.
\tag{7.48}
$$

To compute $E\left[\mathbb{K}_i(t) \right]$, we note that, by Assumptions (a) and (c),

$$
\begin{aligned}
E\left[K\left(\frac{t_{ij} - t}{h} \right) \Big/ h \right] &= \int \left[K\left(\frac{s - t}{h} \right) \Big/ h \right] f_T(s) \, ds \\
&= \int K(u) f_T(t - hu) \, du \\
&= f_T(t) \left[1 + o_p(1) \right].
\end{aligned}
\tag{7.49}
$$

It follows from (7.46), Assumptions (a) through (f), and $\sum_{i=1}^n (w_i n_i) = 1$ that

$$
E\left[\sum_{i=1}^n (w_i/h) \mathbb{X}_i^T \, \mathbb{K}_i(t) \, \mathbb{X}_i \right] = f_T(t) E_{\mathbf{XX}^T}(t) \left[1 + o_p(1) \right],
\tag{7.50}
$$

and, by the law of large numbers,

$$
\left[\sum_{i=1}^n (w_i/h) \mathbb{X}_i^T \, \mathbb{K}_i(t) \, \mathbb{X}_i \right] = f_T(t) E_{\mathbf{XX}^T}(t) \left[1 + o_p(1) \right].
\tag{7.51}
$$

The approximation of (7.45) then follows directly from equations (7.47) and (7.51).

2. Approximated Mean Squared Errors

To avoid the technical inconvenience that might arise because of the nonexistence of the mean squared errors, we describe the asymptotic risk of $\widehat{\beta}_K^{LSK}(t; h, w)$ through the mean squared error of $D(t)$ in (7.45), which is referred to as the modified mean squared error for $\widehat{\beta}_K^{LSK}(t; h, w)$,

$$MSE_{\mathbf{W}}^*\left[\widehat{\beta}_K^{LSK}(t; h, w)\right] = \sum_{l=0}^{k}\sum_{r=0}^{k}\left\{M_{lr}(t)\,E\left[\widehat{R}_{wl}(t)\,\widehat{R}_{wr}(t)\right]\right\}, \qquad (7.52)$$

where $\widehat{R}_{wl}(t)$ is the lth element of the $k+1$ column vector $\widehat{R}_w(t)$ and $M_{lr}(t)$ is the (l, r)th element of the $(k+1) \times (k+1)$ matrix

$$M(t) = f_T^{-2}(t)\left[E_{\mathbf{XX}^T}^{-1}(t)\right]^T \mathbf{W} E_{\mathbf{XX}^T}^{-1}(t).$$

Similar to the usual mean squared errors, $MSE_{\mathbf{W}}^*\left[\widehat{\beta}_K^{LSK}(t; h, w)\right]$ of (7.52) is formed by two components, the square of the expectation of $D(t)$ and the variance-covariance matrix of $D(t)$. The modified bias of $\widehat{\beta}_K^{LSK}(t; h, w)$ is defined to be the expectation of $D(t)$ given by

$$B^*\left[\widehat{\beta}_K^{LSK}(t; h, w)\right] = E\left[D(t)\right] = f_T^{-1}(t)\,E_{\mathbf{XX}^T}^{-1}(t)\,E\left[\widehat{R}_w(t)\right]. \qquad (7.53)$$

The modified variance-covariance matrix of $\widehat{\beta}^{LSK}(t)$ is defined to be the variance-covariance matrix of $D(t)$ given by

$$\begin{aligned}
&Cov^*\left[\widehat{\beta}_K^{LSK}(t; h, w)\right] \\
&= Cov\left[D(t)\right] \qquad\qquad\qquad\qquad\qquad\qquad\qquad (7.54)\\
&= \left[f_T^{-1}(t)\,E_{\mathbf{XX}^T}^{-1}(t)\right]Cov\left[\widehat{R}_w(t)\right]\left[f_T^{-1}(t)\,E_{\mathbf{XX}^T}^{-1}(t)\right]^T.
\end{aligned}$$

We similarly define the modified mean squared error of a component estimator of $\beta(t)$. For a given component $\widehat{\beta}_{K,l}^{LSK}(t; h, w)$ for some $0 \leq l \leq k$, its corresponding approximation component is $D_l(t)$, so that its modified bias and variance are given by

$$B^*\left[\widehat{\beta}_{K,l}^{LSK}(t; h, w)\right] = E\left[D_l(t)\right] \quad \text{and} \quad V^*\left[\widehat{\beta}_{K,l}^{LSK}(t; h, w)\right] = Var\left[D_l(t)\right], \quad (7.55)$$

respectively. Since the weight $\mathbf{W}$ defined in (7.43) is a diagonal matrix with diagonal elements $\{W_0, \ldots, W_k\}$, the modified mean squared error of $\widehat{\beta}_K^{LSK}(t; h, w)$ has the following variance-bias squared decomposition

$$\begin{aligned}
&MSE_{\mathbf{W}}^*\left[\widehat{\beta}_K^{LSK}(t; h, w)\right] \\
&= \sum_{l=0}^{k}W_l\left\{B^*\left[\widehat{\beta}_{K,l}^{LSK}(t; h, w)\right]\right\}^2 + \sum_{l=0}^{k}W_l\,V^*\left[\widehat{\beta}_{K,l}^{LSK}(t; h, w)\right]. \quad (7.56)
\end{aligned}$$

Taking $W_l = 1$ for any $0 \le l \le k$ and $W_r = 0$ for all $r \ne l$, the modified mean squared error for the component estimator $\widehat{\beta}^{LSK}_{K,l}(t; h, w)$ is given by

$$MSE^*\left[\widehat{\beta}^{LSK}_{K,l}(t; h, w)\right] = \left\{B^*\left[\widehat{\beta}^{LSK}_{K,l}(t; h, w)\right]\right\}^2 + V^*\left[\widehat{\beta}^{LSK}_{K,l}(t; h, w)\right]. \tag{7.57}$$

We derive in the next section the asymptotic representations of the bias and variance terms in the right side of (7.56) and (7.57).

7.5.3   Asymptotic Risk Representations

We present here only the asymptotic risks for the kernel estimator $\widehat{\beta}^{LSK}_K(t)$ of (7.9) with the $w_i^{**} = 1/N$ weight

$$\widehat{\beta}^{LSK}_K(t; h, w^{**}) = \left(\widehat{\beta}^{LSK}_{K,0}(t; h, w^{**}), \ldots, \widehat{\beta}^{LSK}_{K,k}(t; h, w^{**})\right)^T.$$

The asymptotic risks of $\widehat{\beta}^{LSK}_K(t; h, w)$ with weight $w \ne w^{**}$ can be derived as a straightforward extension from that of $\widehat{\beta}^{LSK}_K(t; h, w^{**})$.

1. Expressions of Asymptotic Mean Squared Errors

We first state and prove a general result for the mean squared risk of the kernel estimator $\widehat{\beta}^{LSK}_K(t; h, w^{**})$.

Theorem 7.1. *When the number of subjects n is large, t is an interior point of the support of f_T and Assumptions (a) through (f) are satisfied, the following conclusions hold:*

(a) *The asymptotic bias of $\widehat{\beta}^{LSK}_K(t; h, w^{**})$ is given by*

$$B^*\left[\widehat{\beta}^{LSK}_K(t; h, w^{**})\right] = f_T^{-1}(t) E^{-1}_{\mathbf{XX}^T}(t)\left(B_0(t), \ldots, B_k(t)\right)^T [1 + o(1)], \tag{7.58}$$

where, for $l = 0, \ldots, k$,

$$\begin{cases} B_l(t) = \sum_{r=0}^k \int \left[\beta_r(t - hu) - \beta_r(t)\right] \xi_{lr}(t - hu) f_T(t - hu) K(u)\, du, \\ B^*\left[\widehat{\beta}^{LSK}_K(t; h, w^{**})\right] = \left(B_0^*\left[\widehat{\beta}^{LSK}_K(t; h, w^{**})\right], \ldots, B_k^*\left[\widehat{\beta}^{LSK}_K(t; h, w^{**})\right]\right)^T. \end{cases}$$

(b) *The asymptotic variance term $V^*\left[\widehat{\beta}^{LSK}_K(t; h, w^{**})\right]$ is*

$$\sum_{l=0}^k W_l V^*\left[\widehat{\beta}^{LSK}_{K,l}(t; h, w^{**})\right] \tag{7.59}$$

$$= \left[\sum_{i=1}^n \left(\frac{n_i}{N}\right)^2\right] f_T^2(t) Z_1(t) + \frac{f_T(t)}{Nh}\left[\int K^2(u)\, du\right] Z_2(t) + o\left[\frac{1}{Nh} + \sum_{i=1}^n \left(\frac{n_i}{N}\right)^2\right],$$

where, with the matrix $M(t)$ defined in (7.52),

$$
\begin{cases}
Z_1(t) &= \rho_\varepsilon(t) \sum_{l_1=0}^{k} \sum_{l_2=0}^{k} \left[M_{l_1 l_2}(t)\, \xi_{l_1 l_2}(t) \right], \\
Z_2(t) &= \sigma^2(t) \sum_{l_1=0}^{k} \sum_{l_2=0}^{k} \left[M_{l_1 l_2}(t)\, \xi_{l_1 l_2}(t) \right].
\end{cases}
$$

(c) *The asymptotic mean squared error $MSE_W^* \left[\widehat{\beta}_K^{LSK}(t; h, w^{**}) \right]$, which is obtained by substituting (7.58) and (7.59) into (7.56), is given by*

$$
MSE_W^* \left[\widehat{\beta}_K^{LSK}(t; h, w^{**}) \right]
$$

$$
= \sum_{l=0}^{k} W_l B_l^* \left[\widehat{\beta}_K^{LSK}(t; h, w^{**}) \right] + \left[\sum_{i=1}^{n} \left(\frac{n_i}{N} \right)^2 \right] f_T^2(t)\, Z_1(t) \qquad (7.60)
$$

$$
+ \frac{f_T(t)}{Nh} \left[\int K^2(u)\, du \right] Z_2(t) + o \left[\frac{1}{Nh} + \sum_{i=1}^{n} \left(\frac{n_i}{N} \right)^2 \right].
$$

Furthermore, $MSE_W^* \left[\widehat{\beta}_K^{LSK}(t; h, w^{**}) \right] \to 0$ if and only if $\max_{1 \le i \le n} \left(n_i N^{-1} \right) \to 0$ as $n \to \infty$. ∎

Proof of Theorem 7.1 is given at the end of this section.

The general asymptotic bias expression of (7.58) leads to different special cases under different smoothness assumptions for $\xi_{lr}(t)$, $\beta_l(t)$, $f_T(t)$, $\sigma^2(t)$ and $\rho_\varepsilon(t)$. A commonly used smoothness condition in the literature is to assume that the nonparametric curves of interest are twice differentiable. Although it is usually impractical to validate whether the unknown curves in a practical situation are twice differentiable, asymptotic results developed under the twice differentiability assumption is often used as a general guideline to evaluate the appropriateness of the smoothing estimators. The next theorem gives the asymptotic mean squared error of $\widehat{\beta}_K^{LSK}(t; h, w^{**})$ under a special case of Assumptions (c) and (d).

Theorem 7.2. *Suppose that Assumptions (a) through (f) are satisfied, and, in addition, $\beta_l(t)$, $f_T(t)$ and $\xi_{lr}(t)$ are twice continuously differentiable. When n is sufficiently large, the modified bias of $\widehat{\beta}_K^{LSK}(t; h, w^{**})$ has the asymptotic expression*

$$
B^* \left[\widehat{\beta}_K^{LSK}(t; h, w^{**}) \right] = f_T^{-1}(t)\, E_{\mathbf{XX}^T}^{-1}(t)\, h^2 \left(b_0(t), \dots, b_k(t) \right)^T + o\left(h^2 \right), \qquad (7.61)
$$

where

$$
b_l(t) = \sum_{r=0}^{k} \sum_{a=0}^{1} \sum_{b=0}^{a} \left\{ \frac{\beta_r^{(2-a)}(t)\, \xi_{lr}^{(a-b)}(t)\, f_T^{(b)}(t)}{(2-a)!\,(a-b)!\,b!} \left[\int u^2 K(u)\, du \right] \right\},
$$

and the asymptotic repression for the modified variance $V^* \left[\widehat{\beta}_K^{LSK}(t; h, w^{**}) \right]$

is given in (7.59). The asymptotic representation of $MSE_{\mathbf{w}}^*[\widehat{\beta}_K^{LSK}(t; h, w^{**})]$, which is obtained by substituting the right-hand sides of (7.59) and (7.60) into (7.52), is given by

$$MSE_{\mathbf{w}}^*[\widehat{\beta}_K^{LSK}(t; h, w^{**})]$$

$$= h^4 f_T^{-2}(t) \sum_{l=0}^{k} W_l [b_l^*(t)]^2 + bigbl[\sum_{i=1}^{n} \left(\frac{n_i}{N}\right)^2] f_T^2(t) Z_1(t)$$

$$+ \frac{f_T(t)}{Nh} \left[\int K^2(u) \, du\right] Z_2(t) + o\left[h^4 + \frac{1}{Nh} + \sum_{i=1}^{n} \left(\frac{n_i}{N}\right)^2\right], \quad (7.62)$$

where $b_l^*(t)$ is the $(l+1)$th component of $E_{\mathbf{XX}^T}^{-1}(t) (b_0(t), \ldots, b_k(t))^T$. Further-more, $\widehat{\beta}_K^{LSK}(t; h, w^{**})$ is consistent, i.e., $MSE_{\mathbf{w}}^*[\widehat{\beta}_K^{LSK}(t; h, w^{**})] \to 0$ as $n \to \infty$, when $\sum_{i=1}^{n}(n_i N^{-1})^2 \to 0$, which holds if and only if $\max_{1 \leq i \leq n}(n_i N^{-1}) \to 0$ as $n \to \infty$. ∎

Proof of Theorem 7.2 is given at the end of this section.

2. Remarks on the Asymptotic Results

The theoretical results of Theorems 7.1 and 7.2 lead to a number of interesting special cases, which can guide the practical use of the least squares kernel estimator $\widehat{\beta}_K^{LSK}(t; h, w^{**})$. These asymptotic results have the following implications:

Consistency: In general, $\widehat{\beta}_K^{LSK}(t; h, w^{**})$ is not necessarily a consistent estimator of $\beta(t)$ in the sense that $MSE_{\mathbf{w}}^*[\widehat{\beta}_K^{LSK}(t; h, w^{**})]$ converges to 0 when N converges to infinity but the sizes of n_i, $i = 1, \ldots, n$ and n are unspecified. For example, if $n_i = m$ for all $i = 1, \ldots, n$ and m converges to infinity but n stays bounded, then, since $N^{-2}\sum_{i=1}^{n} n_i^2 = n^{-1}$ is bounded away from zero for sufficiently large N, $MSE_{\mathbf{w}}^*[\widehat{\beta}_K^{LSK}(t; h, w^{**})]$ does not converge to zero as N goes to infinity. □

Bounded Repeated Measurements: If n_i are bounded, i.e., $n_i \leq c$ for some integer $c \geq 1$ and all $i = 1, \ldots, n$, and n is sufficiently large, then the asymptotic variance term in (7.59) is dominated by the second term of the right side of (7.59). Then, if we minimize the dominating terms of the asymptotic $MSE_{\mathbf{w}}^*[\widehat{\beta}_K^{LSK}(t; h, w^{**})]$, the optimal bandwidth is $h_{opt} = O(N^{-1/5})$. Substituting h_{opt} into (7.60), the mean squared error $MSE_{\mathbf{w}}^*[\widehat{\beta}_K^{LSK}(t; h_{opt}, w^{**})]$ is of the order $N^{4/5}$. □

Effects of Correlations: It is seen from (7.60) that the asymptotic effects of the intra-subject correlations on $MSE_{\mathbf{w}}^*[\widehat{\beta}_K^{LSK}(t; h, w^{**})]$ are only included in $Z_1(t)$ in the variance term (7.59). Without this extra term, the asymptotic mean squared errors of $\widehat{\beta}_K^{LSK}(t; h, w^{**})$ would be the same as kernel estimators with cross-sectional i.i.d. data. The effects of the intra-subject correlations depends on the limiting values, $\rho_{\varepsilon}(t)$, of the covariances of $\varepsilon_i(t)$ and $\varepsilon_i(s)$ as $s \to t$. This is caused by the local averaging nature of kernel methods. Specifically, the estimators tend to ignore the measurements at time points t_{ij} which are outside a shrinking neighborhood of t. Since the bandwidths shrink to zero, any correlation between $\varepsilon_i(t)$ and $\varepsilon_i(s)$, $t \neq s$, is ignored when n is sufficiently large. This local nature makes the least squares kernel estimators useful under the current setting, since, in practice, we may only be aware of the presence of the intra-subject correlations but have very little knowledge about the specific correlation structures. By using a local smoothing method, we essentially choose to ignore the correlation structures. These asymptotic results provide some qualitative insight for the adequacy of the estimation procedures. □

3. Theoretical Derivations

We provide here the proofs of Theorems 7.1 and 7.2.

1. Proof of Theorem 7.1:

Following the approximation of (7.45), it suffices to study the asymptotic representations of $E\big[\widehat{R}_w(t)\big]$ and $E\big[\widehat{R}_{wl}(t)\widehat{R}_{wr}(t)\big]$ for $l, r = 0, \ldots, k$. Define

$$a_{ijl}(t) = \sum_{s=0}^{k} \left\{ X_{ij}^{(l)} X_{ij}^{(s)} \big[\beta_s(t_{ij}) - \beta_s(t)\big] \right\} + X_{ij}^{(l)} \varepsilon_i(t_{ij}).$$

It can be verified by direct computation from the definition of $\widehat{R}_w(t)$ in (7.46) that

$$\widehat{R}_{wl}(t) = \frac{1}{Nh} \sum_{i=1}^{n} \sum_{j=1}^{n_i} \left[a_{ijl}(t) K\left(\frac{t - t_{ij}}{h}\right) \right], \tag{7.63}$$

and, since $E\big[a_{ijl}(t)|t_{ij} = s\big] = \sum_{r=0}^{k} \left\{ \big[\beta_r(s) - \beta_r(t)\big] \xi_{lr}(s) \right\}$,

$$
\begin{aligned}
E\big[\widehat{R}_{wl}(t)\big] &= (Nh)^{-1} \sum_{i=1}^{n} \sum_{j=1}^{n_i} \int E\big[a_{ijl}(t)|t_{ij} = s\big] K\left(\frac{t - s}{h}\right) f_T(s) \, ds \\
&= B_l(t). \tag{7.64}
\end{aligned}
$$

Thus the asymptotic bias expression (7.58) follows from equations (7.53), (7.63) and (7.64).

To derive the asymptotic variance expression (7.59), we consider the following decomposition

$$\widehat{R}_{wl}(t)\widehat{R}_{wr}(t) = A_{lr1} + A_{lr2} + A_{lr3}$$

where

$$A_{lr1} = \left(\frac{1}{Nh}\right)^2 \sum_{i=1}^{n}\sum_{j=1}^{n_i} a_{ijl}(t)\, a_{ijr}(t)\, K^2\!\left(\frac{t-t_{ij}}{h}\right),$$

$$A_{lr2} = \left(\frac{1}{Nh}\right)^2 \sum_{i=1}^{n}\sum_{j\neq j'} a_{ijl}(t)\, a_{ij'r}(t)\, K\!\left(\frac{t-t_{ij}}{h}\right) K\!\left(\frac{t-t_{ij'}}{h}\right),$$

$$A_{lr3} = \left(\frac{1}{Nh}\right)^2 \sum_{i\neq i'}\sum_{j,j'} a_{ijl}(t)\, a_{i'j'r}(t)\, K\!\left(\frac{t-t_{ij}}{h}\right) K\!\left(\frac{t-t_{i'j'}}{h}\right).$$

By direct calculations with the change of variables, it is straightforward to verify that

$$\sum_{l=0}^{k}\sum_{r=0}^{k}\left[M_{lr}(t)\, E\left(A_{lr1}\right)\right] = \left(\frac{1}{Nh}\right) f_T(t)\left[\int K^2(u)\,du\right] Z_2(t) + o\!\left(N^{-1}h^{-1}\right).$$

Using the Cauchy-Schwarz inequality, we can show that

$$\sum_{l=0}^{k}\sum_{r=0}^{k}\left[M_{lr}(t)\, E\left(A_{lr2}\right)\right] = N^{-2}\left(\sum_{i=1}^{n}n_i^2 - N\right) f_T^2(t)\, Z_1(t) + o\!\left(N^{-2}\sum_{i=1}^{n}n_i^2\right).$$

Finally, we can verify directly that

$$\sum_{l=0}^{k} W_l\, V^*\left[\widehat{\beta}_{\mathbf{w}l}(t)\right] = \sum_{l=0}^{k}\sum_{r=0}^{k}\left[M_{lr}(t)\, E\left(A_{lr1}+A_{lr2}\right)\right],$$

which implies that the asymptotic expression (7.59) holds.

To show the last assertion in (c), we suppose $\sum_{l,r=0}^{k}\left[M_{lr}(t)\,\xi_{lr}(t)\right] > 0$. Then $MSE_{\mathbf{W}}^*\left[\widehat{\beta}_K^{LSK}(t;h,w^{**})\right] \to 0$ if and only if $N^{-2}\sum_{i=1}^{n}n_i^2 \to 0$ as $n \to \infty$. It is easy to see that $\sum_{i=1}^{n}\left(n_iN^{-1}\right)^2 \to 0$ implies $\max_{1\le i\le n}\left(n_iN^{-1}\right) \to 0$. It suffices to show that $\max_{1\le i\le n}\left(n_iN^{-1}\right) \to 0$ implies $\sum_{i=1}^{n}\left(n_iN^{-1}\right)^2 \to 0$. Assume now that $\max_{1\le i\le n}\left(n_iN^{-1}\right) \to 0$. Then, for any $\varepsilon > 0$, $\max_{1\le i\le n}\left(n_iN^{-1}\right) < (\varepsilon/2)$ for sufficiently large n. Let $1 = k_0 < k_1 < \cdots < k_m = n$ be positive integers such that $(\varepsilon/2) < \sum_{i=k_{l-1}}^{k_l}\left(n_iN^{-1}\right) < \varepsilon$ for $l = 1,\ldots,m-1$, and $\sum_{i=k_{m-1}}^{k_m}\left(n_iN^{-1}\right) < \varepsilon$. Then, for all $l = 1,\ldots,m$, $\sum_{i=k_{l-1}}^{k_l}\left(n_iN^{-1}\right)^2 < \varepsilon^2$. Since $N = \sum_{i=1}^{n}n_i$, we must have $m \le (2/\varepsilon)$, and, consequently, $\sum_{i=1}^{n}\left(n_iN^{-1}\right)^2 < 2\varepsilon$. Since ε can be arbitrarily small, this inequality implies that $\lim_{n\to\infty}\sum_{i=1}^{n}\left(n_iN^{-1}\right)^2 = 0$. This completes the proof of Theorem 7.1. ∎

2. Proof of Theorem 7.2:

Because the modified variance $V^*\left[\widehat{\beta}_K^{LSK}(t;h,w^{**})\right]$ is the same as (7.59) in Theorem 7.1, we only need to consider the asymptotic bias term $B^*\left[\widehat{\beta}_K^{LSK}(t;h,w^{**})\right]$ given in (7.58). The smoothness assumption of $\beta_l(t)$, $f_T(t)$

and $\xi_{lr}(t)$ in Assumption (c) and the compact support assumption of $K(u)$ in Assumption (e) and (7.42) suggest that the functions $\beta_r(t-hu)$, $\xi_{lr}(t-hu)$ and $f_T(t-hu)$ in the integral of $B_l(t)$ can be approximated by the following Taylor expansions

$$\beta_r(t-hu)-\beta_r(t) = -\beta_r'(t)hu+(1/2)\beta_r''(t)h^2u^2+r_1(h), \quad (7.65)$$
$$\xi_{lr}(t-hu)-\xi_{lr}(t) = -\xi_{lr}'(t)hu+(1/2)\xi_{lr}''(t)h^2u^2+r_2(h), \quad (7.66)$$
$$f_T(t-hu)-f_T(t) = -f_T'(t)hu+(1/2)f_T''(t)h^2u^2+r_3(h), \quad (7.67)$$

for $c_1 \leq u \leq c_2$ with the constants c_1 and c_2 given in Assumption (e), where $r_l(h)$, $l=1,2,3$, are functions of h such that $r_l(h)/h^2 \to 0$ as $h \to 0$. The right side of (7.61) is then obtained by substituting $\beta_r(t-hu)-\beta_r(t)$, $\xi_{lr}(t-hu)$ and $f_T(t-hu)$ of (7.58) with the corresponding terms at the right side of (7.65), (7.66) and (7.67). The asymptotic expression of the mean squared error $MSE_{\mathbf{W}}^*[\hat{\beta}_K^{LSK}(t;h,w^{**})]$ given at the right side of (7.62) is obtained by substituting $B^*[\hat{\beta}_K^{LSK}(t;h,w^{**})]$ and $V^*[\hat{\beta}_{Kl}^{LSK}(t;h,w^{**})]$ of (7.56) with the right side terms of (7.61) and (7.59), respectively. Since $h \to 0$ and $Nh \to \infty$ as $N \to \infty$, the last assertion of the theorem, that is $MSE_{\mathbf{W}}^*[\hat{\beta}_K^{LSK}(t;h,w^{**})] \to 0$ if and only if $\max_{1\leq i\leq n}(n_iN^{-1}) \to 0$ as $n \to \infty$ follows from Theorem 7.1. ∎

7.5.4   Asymptotic Distributions

We now derive the asymptotic distributions of the least squares kernel estimator $\hat{\beta}_K^{LSK}(t;h,w^{**})$ at a fixed time point $t_0 > 0$. As an alternative to the resampling-subject bootstrap confidence bands of Section 7.3, the asymptotically normal distributions derived in this section can be used to construct asymptotically approximate confidence intervals for $\beta(t_0)$. These intervals can then be used in conjunction with the method of Section 7.3.2 to construct simultaneous confidence bands for $\beta(t)$ with t inside a given interval (a,b).

1. Asymptotic Normality

In addition to Assumptions (a) through (f), we assume the stronger conditions that

$$\begin{cases} \beta_l(t),\ f_T(t),\ \xi_{rl}(t) \text{ have continuous second derivatives at } t_0 \\ \quad \text{for all } l,r=0,\dots,k; \\ E\left[|\varepsilon(t)|^{2+\delta}\right] \text{ and } E\left[|X_{ij}^{(l)}|^{4+\delta}\right] \text{ are finite for all } 1 \leq i \leq n, \\ \quad 1 \leq j \leq n_i,\ 0 \leq l \leq k,\ t \in \mathscr{S}(f_T) \text{ and some } \delta > 0; \\ h = N^{-1/5}h_0 \text{ for some constant } h_0 > 0; \\ \lim_{n\to\infty}N^{-6/5}\sum_{i=1}^n n_i^2 = \lambda \text{ for some } 0 \leq \lambda < \infty, \end{cases} \quad (7.68)$$

and denote, for all $l,r=0,\dots,k$,

$$\mu_1(K) = \int u^2 K(u)\,du, \quad \mu_2(K) = \int K^2(u)\,du,$$

$$b_l(t_0) = h_0^{3/2} \sum_{c=0}^{k} \left\{ \mu_1(K) \left[\beta_c'(t_0) \, \xi_{lc}'(t_0) \, f_T(t_0) + \beta_c'(t_0) \, \xi_{lc}(t_0) \, f_T'(t_0) \right. \right.$$

$$\left. \left. + (1/2) \beta_c''(t_0) \, \xi_{lc}(t_0) \, f_T(t_0) \right] \right\}, \quad \text{for } l = 0, \dots, k, \qquad (7.69)$$

$$\mathbf{B}(t_0) = f_T^{-1}(t_0) \, E_{\mathbf{XX}^T}^{-1}(t_0) \, (b_0(t_0), \dots, b_k(t_0))^T, \qquad (7.70)$$

$$D_{lr}(t_0) = \sigma^2(t_0) \, \xi_{lr}(t_0) \, f_T(t_0) \, \mu_2(K) + \lambda h_0 \, \rho_\varepsilon(t_0) \, \xi_{lr}(t_0) \, f_T^2(t_0), \qquad (7.71)$$

$$\mathbf{D}(t_0) = \begin{pmatrix} D_{00}(t_0) & D_{01}(t_0) & \cdots & D_{0k}(t_0) \\ \vdots & \vdots & \vdots & \vdots \\ D_{k0}(t_0) & D_{k1}(t_0) & \cdots & D_{kk}(t_0) \end{pmatrix},$$

$$\mathbf{D}^*(t_0) = f_T^{-2}(t_0) \, E_{\mathbf{XX}^T}^{-1}(t_0) \, \mathbf{D}(t_0) \, E_{\mathbf{XX}^T}^{-1}(t_0). \qquad (7.72)$$

The next theorem shows the asymptotic normality of $\widehat{\beta}_K^{LSK}(t; h, w^{**})$ at t_0.

Theorem 7.3. *Suppose that Assumptions (a) to (f) and (7.68) are satisfied. When n is sufficiently large, $\widehat{\beta}_K^{LSK}(t_0; h, w^{**})$ has asymptotically a multivariate normal distribution, such that*

$$(Nh)^{1/2} \left[\widehat{\beta}_K^{LSK}(t_0; h, w^{**}) - \beta(t_0) \right] \to N(\mathbf{B}(t_0), \mathbf{D}^*(t_0)) \qquad (7.73)$$

in distribution as $n \to \infty$, where $\mathbf{B}(t_0)$ and $\mathbf{D}^*(t_0)$ are defined in (7.70) and (7.72), respectively. ■

Proof of Theorem 7.3:
This theorem is a special case of Theorem 7.4 at the end of this section. ■

2. Remarks on Asymptotic Normality

A direct implication of Theorem 7.3 is that, to ensure good asymptotic properties of $\widehat{\beta}_K^{LSK}(t_0; h, w^{**})$, the numbers of repeated measurements $\{n_i : i = 1, \dots, n\}$ must be small relative to the overall sample size N. It was shown in Theorem 7.2 that $\widehat{\beta}_K^{LSK}(t_0; h, w^{**})$ is a consistent estimator of $\beta(t_0)$ if and only if $\sum_{i=1}^{n} n_i^2 = o(N^2)$, which is equivalent to $\max_{1 \le i \le n} (n_i/N) = o(1)$. Theorem 7.3 assumes a somewhat stronger condition, $\sum_{i=1}^{n} n_i^2 = O(N^{6/5})$, which ensures $\widehat{\beta}_K^{LSK}(t_0; h, w^{**})$ to have an attainable convergence rate of $N^{-2/5}$. If $\sum_{i=1}^{n} n_i^2$ converges to infinity faster than $N^{6/5}$, it can be shown with a slight modification of the proof of Theorem 7.3 that the attainable rate for $\widehat{\beta}_K^{LSK}(t_0; h, w^{**})$ is slower than $N^{-2/5}$.

Another important assumption in this section is that t_0 is an interior point of the support $\mathscr{S}(f_T)$. It is well known in cross-sectional i.i.d. data case that kernel estimators suffer from increased biases at the boundary of the design intervals. Methods for improving the theoretical and practical performance of kernel estimators have been extensively studied in the literature, such as Rice (1984), Hall and Wehrly (1991) and Müller (1993). Here, it is natural to expect $\widehat{\beta}_K^{LSK}(t_0; h, w^{**})$ to have a relatively larger bias when t_0 is near the boundary of $\mathscr{S}(f_T)$. The asymptotic properties of $\widehat{\beta}_K^{LSK}(t_0; h, w^{**})$ for t_0 near the boundary of $\mathscr{S}(f_T)$ have not been explicitly derived. Possible modifications to improve the boundary asymptotic properties of $\widehat{\beta}_K^{LSK}(t_0; h, w^{**})$ are still not well understood and warrant further investigation.

3. Theoretical Derivations

Before giving the proof of Theorem 7.3, we state and prove a technical lemma. These results are then used to prove a general asymptotic normality result of $\widehat{\beta}_K^{LSK}(t; h, w^{**})$ for t at a set of distinct time points within the support $\mathscr{S}(f_T)$. By (7.44) and the approximation (7.45), the asymptotic distributions of $\widehat{\beta}_K^{LSK}(t; h, w^{**})$ can be investigated through the asymptotic distributions of $\widehat{R}_{w^{**}}(t)$. Let $\mathbf{s} = (s_1, \ldots, s_J)$, $J \geq 1$, be a set of distinct interior points within the support $\mathscr{S}(f_T)$ and

$$\widehat{R}_{w^{**}}(\mathbf{s}) = \left(\widehat{R}_{w^{**}}^T(s_1), \ldots, \widehat{R}_{w^{**}}^T(s_J)\right)^T \tag{7.74}$$

It suffices to study the asymptotic distribution of $\widehat{R}_{w^{**}}(\mathbf{s})$. We first state and prove the technical lemma for $\widehat{R}_{w^{**}}(\mathbf{s})$.

Lemma 7.1. *Suppose that Assumptions (a) through (f) and (7.68) are satisfied, $\widehat{R}_{w^{**}}(\mathbf{s})$ is defined in (7.74), and $\xi_{lr}(s_1, s_2)$ and $\rho_\varepsilon(s_1, s_2)$ are continuous for any $s_1 \neq s_2$ in R^2. When n is sufficiently large, the asymptotically approximated bias and variance of $\widehat{R}_{w^{**}}(\mathbf{s})$ are given by*

$$\begin{cases} E\left[(Nh)^{1/2}\widehat{R}_{w^{**}}(\mathbf{s})\right] &= b(\mathbf{s}) + o_p(1), \\ Cov\left[(Nh)^{1/2}\widehat{R}_{w^{**}}(\mathbf{s})\right] &= \mathbf{D}(\mathbf{s}) + o_p(1), \end{cases} \tag{7.75}$$

respectively, where

$$b(\mathbf{s}) = \left(b_0(s_1), \ldots, b_k(s_1), \ldots, b_0(s_J), \ldots, b_k(s_J)\right)^T, \tag{7.76}$$

$$\mathbf{D}(\mathbf{s}) = \begin{pmatrix} \mathbf{D}(s_1, s_1) & \cdots & \mathbf{D}(s_1, s_J) \\ \vdots & \vdots & \vdots \\ \mathbf{D}(s_J, s_1) & \cdots & \mathbf{D}(s_J, s_J) \end{pmatrix},$$

$$\mathbf{D}(s_1, s_2) = \begin{pmatrix} D_{00}(s_1, s_2) & \cdots & D_{0k}(s_1, s_2) \\ \vdots & \vdots & \vdots \\ D_{k0}(s_1, s_2) & \cdots & D_{kk}(s_1, s_2) \end{pmatrix},$$

$$D_{lr}(s_1, s_2) = \begin{cases} \sigma^2(s_1)\, \xi_{lr}(s_1)\, f_T(s_1)\, \mu_2(K) \\ \quad + \lambda\, h_0\, \rho_\varepsilon(s_1)\, \xi_{lr}(s_1, s_2)\, f_T^2(s_1), & \text{if } s_1 = s_2, \text{ (7.77)} \\ \lambda\, h_0\, \rho_\varepsilon(s_1, s_2)\, \xi_{lr}(s_1, s_2)\, f_T(s_1)\, f_T(s_2), & \text{if } s_1 \neq s_2, \end{cases}$$

$$\xi_{lr}(s_1, s_2) = E\left[X_{ij_1}^{(l)} X_{ij_2}^{(r)} \big| t_{ij_1} = s_1, t_{ij_2} = s_2 \right]$$

and $b_l(s)$, $l = 0, \ldots, k$, defined in (7.69). ∎

Proof of Lemma 7.1:

We note first that the lth element of $\widehat{R}_{w^{**}}(s)$ at a single time point $s \in \mathscr{S}(f_T)$ is given by (7.63). For any $i = 1, \ldots, n$ and $l = 0, \ldots, k$, let

$$\psi_{il}(s) = \sum_{j=1}^{n_i} \left[a_{ijl}(s) K\left(\frac{s - t_{ij}}{h} \right) \right], \tag{7.78}$$

where $a_{ijl}(s)$ is defined in (7.63). Substituting the corresponding term in (7.63) by (7.78), $\widehat{R}_{w^{**}}(s)$ can be written as a sum of independent vectors

$$\widehat{R}_{w^{**}}(s) = (Nh)^{-1} \sum_{i=1}^{n} \mathbf{A}_i(s), \tag{7.79}$$

where $\mathbf{A}_i(s)$ is a $J(k+1)$ column vector such that

$$\mathbf{A}_i(s) = \left(\psi_{i0}(s_1), \ldots, \psi_{ik}(s_1), \ldots, \psi_{i0}(s_J), \ldots, \psi_{ik}(s_J) \right)^T.$$

Since the design time points are assumed to be independent, i.e., Assumption (a), direct calculation using the definition of $\xi_{lr}(t)$ in (7.41) and the change of variables show that

$$E\left[\psi_{il}(s) \right] = \sum_{j=1}^{n_i} \int E\left[a_{ijl}(s) \big| t_{ij} = v \right] K\left(\frac{s - v}{h} \right) f_T(v)\, dv$$

$$= n_i h \sum_{r=0}^{k} \int \left[\beta_r(s - hu) - \beta_r(s) \right] \xi_{lr}(s - hu)\, f_T(s - hu)\, K(u)\, du.$$

Then, by (7.79) and (7.68), and taking the Taylor expansions on the right side of the above equation, we have that

$$E\left[(Nh)^{1/2} \widehat{R}_{w^{**}}(s) \right] = b(s) + o(1).$$

To compute the covariance of $(Nh)^{1/2} \widehat{R}_{w^{**}}(s)$, we note first that

$$Cov\left[\widehat{R}_{w^{**}l}(s_1), \widehat{R}_{w^{**}r}(s_2) \right] = E\left[\widehat{R}_{w^{**}l}(s_1)\, \widehat{R}_{w^{**}r}(s_2) \right] - E\left[\widehat{R}_{w^{**}l}(s_1) \right] E\left[\widehat{R}_{w^{**}r}(s_2) \right],$$

and, by (7.79), it is sufficient to compute the right side terms of the following equation

$$E\left\{\left[(Nh)^{-1/2}\sum_{i=1}^{n}\psi_{il}(s_1)\right]\left[(Nh)^{-1/2}\sum_{i=1}^{n}\psi_{ir}(s_2)\right]\right\} \tag{7.80}$$

$$= (Nh)^{-1}\left\{\sum_{i=1}^{n}E\left[\psi_{il}(s_1)\psi_{ir}(s_2)\right]+\sum_{i_1\neq i_2}E\left[\psi_{i_1l}(s_1)\psi_{i_2r}(s_2)\right]\right\}.$$

For the first term of the right side of (7.80), we consider the further decomposition

$$\psi_{il}(s_1)\psi_{ir}(s_2) = \sum_{j=1}^{n_i}\left[a_{ijl}(s_1)a_{ijr}(s_2)K\left(\frac{s_1-t_{ij}}{h}\right)K\left(\frac{s_2-t_{ij}}{h}\right)\right] \tag{7.81}$$

$$+\sum_{j_1\neq j_2}\left[a_{ij_1l}(s_1)a_{ij_2r}(s_2)K\left(\frac{s_1-t_{ij_1}}{h}\right)K\left(\frac{s_2-t_{ij_2}}{h}\right)\right].$$

Using (7.68), the change of variables and the fact that $\varepsilon_i(\cdot)$ is a mean zero stochastic process independent of $\mathbf{X}_{ij}$, we can show by direct calculation that, as $n\to\infty$,

$$E\left[a_{ijl}(s_1)a_{ijr}(s_2)|t_{ij}=v\right]$$

$$= \sum_{c=0}^{k}\left\{\left[\beta_c(v)-\beta_c(s_1)\right]\left[\beta_c(v)-\beta_c(s_2)\right]E\left[X_{ij}^{(l)}\left(X_{ij}^{(c)}\right)^2X_{ij}^{(r)}\Big|t_{ij}=v\right]\right\}$$

$$+\sigma^2(v)E\left[X_{ij}^{(l)}X_{ij}^{(r)}|t_{ij}=v\right]$$

$$+\sum_{c_1\neq c_2}\left\{\left[\beta_{c_1}(v)-\beta_{c_1}(s_1)\right]\left[\beta_{c_2}(v)-\beta_{c_2}(s_2)\right]\right.$$

$$\left.\times E\left[X_{ij}^{(l)}X_{ij}^{(c_1)}X_{ij}^{(r)}X_{ij}^{(c_2)}|t_{ij}=v\right]\right\}$$

$$\to \sigma^2(s_c)\xi_{lr}(s_c),\quad\text{if }v\to s_c,\ c=1,2.$$

Then, it follows from the above equation that

$$E\left\{\sum_{j=1}^{n_i}\left[a_{ijl}(s_1)a_{ijr}(s_2)K\left(\frac{s_1-t_{ij}}{h}\right)K\left(\frac{s_2-t_{ij}}{h}\right)\right]\right\} \tag{7.82}$$

$$= \sum_{j=1}^{n_i}\int E\left[a_{ijl}(s_1)a_{ijr}(s_2)|t_{ij}=v\right]K\left(\frac{s_1-v}{h}\right)K\left(\frac{s_2-v}{h}\right)f_T(v)\,dv$$

$$= \begin{cases} n_ih\sigma^2(s_1)\xi_{lr}(s_1)f_T(s_1)\left[\int K^2(u)\,du\right]+o(n_ih), & \text{if }s_1=s_2, \\ o(n_ih), & \text{if }s_1\neq s_2. \end{cases}$$

Similarly, direct calculation then shows that, as $n\to\infty$, $v_1\to s_1$ and $v_2\to s_2$,

$$E\left[a_{ij_1l}(s_1)a_{ij_2r}(s_2)|t_{ij_1}=v_1,t_{ij_2}=v_2\right]$$

$$= \sum_{c=0}^{k} \left\{ \left[\beta_c(v_1) - \beta_c(s_1) \right] \left[\beta_c(v_2) - \beta_c(s_2) \right] \right.$$

$$\times E \left[X_{ij_1}^{(l)} X_{ij_1}^{(c)} X_{ij_2}^{(r)} X_{ij_2}^{(c)} \big| t_{ij_1} = v_1, t_{ij_2} = v_2 \right] \right\}$$

$$+ \rho_\varepsilon(v_1, v_2) E \left[X_{ij_1}^{(l)} X_{ij_2}^{(r)} \big| t_{ij_1} = v_1, t_{ij_2} = v_2 \right]$$

$$+ \sum_{c_1 \neq c_2} \left\{ \left[\beta_{c_1}(v_1) - \beta_{c_1}(s_1) \right] \left[\beta_{c_2}(v_2) - \beta_{c_2}(s_2) \right] \right.$$

$$\times E \left[X_{ij_1}^{(l)} X_{ij_1}^{(c_1)} X_{ij_2}^{(r)} X_{ij_2}^{(c_2)} \big| t_{ij_1} = v_1, t_{ij_2} = v_2 \right] \right\}$$

$$\rightarrow \begin{cases} \rho_\varepsilon(s_1, s_2) \, \xi_{lr}(s_1, s_2), & \text{if } s_1 \neq s_2, \\ \rho_\varepsilon(s_1) \, \xi_{lr}(s_1, s_1), & \text{if } s_1 = s_2, \end{cases}$$

and the expectation of the second term of the right side of (7.82) is

$$E \left\{ \sum_{j_1 \neq j_2} \left[a_{ij_1}^{(l)}(s_1) \, a_{ij_2}^{(r)}(s_2) K\left(\frac{s_1 - t_{ij_1}}{h} \right) K\left(\frac{s_2 - t_{ij_2}}{h} \right) \right] \right\} \qquad (7.83)$$

$$= \sum_{j_1 \neq j_2} \left\{ \int\!\!\int E \left[a_{ij_1}^{(l)}(s_1) \, a_{ij_2}^{(r)}(s_2) \big| t_{ij_1} = v_1, t_{ij_2} = v_2 \right] \right.$$

$$\left. \times K\left(\frac{s_1 - v_1}{h} \right) K\left(\frac{s_2 - v_2}{h} \right) f_T(v_1) f_T(v_2) \, dv_1 \, dv_2 \right\}$$

$$= \begin{cases} h^2 n_i \left(n_i - 1 \right) \rho_\varepsilon(s_1, s_2) f_T(s_1) f_T(s_2) \, \xi_{lr}(s_1, s_2) \\ \quad + o\left[h^2 n_i \left(n_i - 1 \right) \right], & \text{if } s_1 \neq s_2, \\ h^2 n_i \left(n_i - 1 \right) \rho_\varepsilon(s_1) f_T^2(s_1) \, \xi_{lr}(s_1, s_1) \\ \quad + o\left[h^2 n_i \left(n_i - 1 \right) \right], & \text{if } s_1 = s_2. \end{cases}$$

Combining (7.81), (7.82) and (7.83), it follows that, when n is sufficiently large,

$$(Nh)^{-1} \sum_{i=1}^{n} E \left[\psi_{il}(s_1) \, \psi_{ir}(s_2) \right] \qquad (7.84)$$

$$= \begin{cases} \sigma^2(s_1) \xi_{lr}(s_1) f_T(s_1) \left[\int K^2(u) \, du \right] + o\left[hN^{-1} \left(\sum_{i=1}^{n} n_i^2 - N \right) \right] \\ \quad + hN^{-1} \left(\sum_{i=1}^{n} n_i^2 - N \right) \rho_\varepsilon(s_1) \xi_{lr}(s_1, s_1) f_T^2(s_1), & \text{if } s_1 = s_2, \\ hN^{-1} \left(\sum_{i=1}^{n} n_i^2 - N \right) \rho_\varepsilon(s_1, s_2) \xi_{lr}(s_1, s_2) f_T(s_1) f_T(s_2) \\ \quad + o\left[hN^{-1} \left(\sum_{i=1}^{n} n_i^2 - N \right) \right], & \text{if } s_1 \neq s_2. \end{cases}$$

Since $h = N^{-1/5} h_0$ and $\lim_{n \to \infty} N^{-6/5} \sum_{i=1}^{n} n_i^2 = \lambda$, it is easy to see that, as $n \to \infty$,

$$hN^{-1} \left(\sum_{i=1}^{n} n_i^2 - N \right) = N^{-6/5} \left(\sum_{i=1}^{n} n_i^2 - N \right) h_0 \rightarrow \lambda \, h_0.$$

Next, we define

$$M_l(h, s) = \sum_{r=0}^{k} \int \left[\beta_r(s - hu) - \beta_r(s) \right] \xi_{lr}(s - hu) f_T(s - hu) K(u) \, du,$$

so that $E\left[\psi_{il}(s)\right]=n_i\,h\,M_l(h,s)$, and it directly follows that

$$(Nh)^{-1}\sum_{i_1\neq i_2=1}^{n}E\left[\psi_{i_1l}(s_1)\,\psi_{i_2r}(s_2)\right]$$

$$-E\left[(Nh)^{-1/2}\sum_{i=1}^{n}\psi_{il}(s_1)\right]E\left[(Nh)^{-1/2}\sum_{i=1}^{n}\psi_{ir}(s_2)\right]$$

$$=(Nh)^{-1}\sum_{i_1\neq i_2=1}^{n}E\left[\psi_{i_1l}(s_1)\right]E\left[\psi_{i_2r}(s_2)\right]$$

$$-(Nh)^{-1/2}\sum_{i=1}^{n}E\left[\psi_{il}(s_1)\right]\times(Nh)^{-1/2}\sum_{i=1}^{n}E\left[\psi_{ir}(s_2)\right]$$

$$=(Nh)^{-1}\sum_{i_1\neq i_2=1}^{n}\left[n_{i_1}\,h\,M_l(h,s_1)\,n_{i_2}\,h\,M_r(h,s_2)\right]$$

$$-(Nh)^{-1/2}\sum_{i=1}^{n}\left[n_i\,h\,M_l(h,s_1)\right]\times(nh)^{-1/2}\sum_{i=1}^{n}\left[n_i\,h\,M_r(h,s_2)\right]$$

$$=\frac{Nh^2}{N^2h}\left[\sum_{i=1}^{n}\left(n_i\sum_{i'\neq i=1}^{n}n_{i'}\right)-N^2\right]M_l(h,s_1)\,M_r(h,s_2)$$

$$=Nh\left[N^{-2}\sum_{i=1}^{n}\left(n_i\sum_{i'\neq i=1}^{n}n_{i'}\right)-1\right]M_l(h,s_1)\,M_r(h,s_2)$$

$$=(Nh)N^{-2}\left(\sum_{i=1}^{n}n_i^2\right)M_l(h,s_1)\,M_r(h,s_2)$$

$$=(Nh)N^{-4/5}M_l(h,s_1)\,M_r(h,s_2)\,\lambda\,o(1)\quad\text{(by Assumption (b))}$$

$$\to\quad 0,\tag{7.85}$$

since $M_l(h,s_1)\to0$, $M_r(h,s_2)\to0$ and, by Assumption (a),

$$N^{-4/5}M_l(h,s_1)\,M_r(h,s_2)=o(Nh).$$

Now, by (7.68) and the calculations given in (7.81), (7.84) and (7.85), we have shown that, for any interior points s_1 and s_2 in the support of f and $l,r=0,\ldots,k$,

$$Cov\left[(Nh)^{1/2}\widehat{R}_{w^{**}l}(s_1),(Nh)^{1/2}\widehat{R}_{w^{**}r}(s_2)\right]=D_{lr}(s_1,s_2)+o(1).$$

This implies the assertion of the lemma. ∎

The technical results of Lemma 7.1 lead to the following asymptotic normality result for $\widehat{\beta}_K^{LSK}\left(\mathbf{s},h,w^{**}\right)$, which covers the asymptotic normality result of Theorem 7.3 as a special case.

Theorem 7.4. *Under the assumptions of Lemma 7.1, $\widehat{\beta}_K^{LSK}\left(\mathbf{s};h,w^{**}\right)$ has* *asymptotically a multivariate normal distribution, such that, as $n \to \infty$,*

$$(Nh)^{1/2}\left[\widehat{\beta}_K^{LSK}(\mathbf{s}; h, w^{**}) - \beta(\mathbf{s})\right] \to N(\mathbf{B}(\mathbf{s}), \mathbf{D}^*(\mathbf{s})) \quad \text{in distribution,} \quad (7.86)$$

where, for $r_1, r_2 = 1, \ldots, J$,

$$
\begin{aligned}
\mathbf{B}(\mathbf{s}) &= \left(\mathbf{B}(s_1), \ldots, \mathbf{B}(s_J)\right)^T, \\
\mathbf{B}(s) &= \left[f_T(s)\right]^{-1} E_{\mathbf{X}\mathbf{X}^T}^{-1}(s)\left(b_0(s), \ldots, b_k(s)\right)^T, \\
\mathbf{D}^*(\mathbf{s}) &= \begin{pmatrix}
\mathbf{D}^*(s_1, s_1) & \cdots & \mathbf{D}^*(s_1, s_J) \\
\vdots & \vdots & \vdots \\
\mathbf{D}^*(s_J, s_1) & \cdots & \mathbf{D}^*(s_J, s_J)
\end{pmatrix}, \\
\mathbf{D}^*(s_{r_1}, s_{r_2}) &= f_T^{-1}(s_{r_1}) f_T^{-1}(s_{r_2}) E_{\mathbf{X}\mathbf{X}^T}^{-1}(s_{r_1}) \mathbf{D}(s_{r_1}, s_{r_2}) E_{\mathbf{X}\mathbf{X}^T}^{-1}(s_{r_2})
\end{aligned}
$$

and $b_l(s)$ and $\mathbf{D}(s_1, s_2)$ are defined in Lemma 7.1. ∎

Proof of Theorem 7.4:

By the assumptions of Lemma 7.1, we can directly verify that $\widehat{R}_{w^{**}}(\mathbf{s})$ satisfies the conditions of the Cramer-Wold theorem (cf. Theorem of Section 1.5.2, Serfling (1980)) and the Lindeberg's condition (cf. Theorem A of Section 1.9.2, Serfling (1980)). Thus,

$$(Nh)^{1/2}\widehat{R}_{w^{**}}(\mathbf{s}) \to N(b(\mathbf{s}), \mathbf{D}(\mathbf{s})) \quad \text{in distribution}$$

as $n \to \infty$. The theorem then follows from (7.46) and the limiting distribution of $(Nh)^{1/2}\widehat{R}_{w^{**}}(\mathbf{s})$. ∎

7.5.5 Asymptotic Pointwise Confidence Intervals

We establish a procedure for constructing approximate confidence intervals based on the asymptotic normal distributions established above. These inference procedures, which depend on the "plug-in" estimates of the derivatives in (7.68), may not be as practical as the bootstrap procedures described in Section 7.3. The objective here is to demonstrate that such an inference procedure is at least theoretically possible for the time-varying coefficient models (7.1) or (7.14), although their practical values appear to be somewhat limited.

1. Formulation of Confidence Intervals

The asymptotic normality of Theorem 7.3 suggests that we can construct an approximate $[100 \times (1 - \alpha)]$th pointwise confidence interval of $A^T\beta(t_0)$ for any known $(k+1)$ column vector A. The resulting lower and upper bounds, $L_\alpha\left[\widehat{\beta}_K^{LSK}(t_0; h, w^{**})\right]$ and $U_\alpha\left[\widehat{\beta}_K^{LSK}(t_0; h, w^{**})\right]$, respectively, of this asymptotically approximate confidence interval should satisfy

$$\lim_{n \to \infty} P\left\{L_\alpha\left[\widehat{\beta}_K^{LSK}(t_0; hw^{**})\right] \leq A^T\beta(t_0) \leq U_\alpha\left[\widehat{\beta}_K^{LSK}(t_0; h, w^{**})\right]\right\} = 1 - \alpha \quad (7.87)$$

and are given by

$$\left[A^T \widehat{\boldsymbol{\beta}}_K^{LSK}(t_0; h, w^{**}) - (Nh)^{-1/2} A^T \mathbf{B}(t_0) \right]$$
$$\pm Z_{\alpha/2} (Nh)^{-1/2} \left[A^T \mathbf{D}^*(t_0) A \right]^{1/2}, \tag{7.88}$$

where $Z_{\alpha/2}$ is the $\left[100 \times (1 - \alpha/2) \right]$ percentile of the standard normal distribution, $\mathbf{B}(t_0)$ is defined in (7.70) and $\mathbf{D}^*(t_0)$ is defined in (7.72). In particular, taking $A = (0, \ldots, 0, 1, 0, \ldots, 0)^T$ to be the vector with 1 at its $(l+1)$th place and 0 elsewhere, (7.88) gives the approximate $[100 \times (1 - \alpha)]\%$ confidence interval of $\beta_l(t_0)$ for any $0 \leq l \leq k$.

Because the bias and variance-covariance terms, $\mathbf{B}(t_0)$ and $\mathbf{D}^*(t_0)$, depend on the unknown functions, the lower and upper bounds given in (7.88) cannot be used directly in practice. If we have consistent estimators $\widehat{\mathbf{B}}(t_0)$ and $\widehat{\mathbf{D}}^*(t_0)$ for $\mathbf{B}(t_0)$ and $\mathbf{D}^*(t_0)$, respectively, then a "plug-in" asymptotically approximate $[100 \times (1 - \alpha)]\%$ confidence interval for $A^T \beta(t_0)$ is

$$\left(\widehat{L}_\alpha \left[\widehat{\boldsymbol{\beta}}_K^{LSK}(t_0; h, w^{**}) \right], \widehat{U}_\alpha \left[\widehat{\boldsymbol{\beta}}_K^{LSK}(t_0; h, w^{**}) \right] \right)$$

with the lower and upper bounds given by

$$\left[A^T \widehat{\boldsymbol{\beta}}_K^{LSK}(t_0; h, w^{**}) - (Nh)^{-1/2} A^T \widehat{\mathbf{B}}(t_0) \right]$$
$$\pm Z_{\alpha/2} (Nh)^{-1/2} \left[A^T \widehat{\mathbf{D}}^*(t_0) A \right]^{1/2}, \tag{7.89}$$

which is constructed by substituting $\mathbf{B}(t_0)$ and $\mathbf{D}^*(t_0)$ in (7.88) by $\widehat{\mathbf{B}}(t_0)$ and $\widehat{\mathbf{D}}^*(t_0)$. A class of kernel-type consistent estimators of $\mathbf{B}(t_0)$ and $\mathbf{D}^*(t_0)$ are given below.

2. Approximate Error Bars

In practice, it is usually difficult to estimate the asymptotic bias term because $\mathbf{B}(t_0)$ involves the derivatives of the unknown functions. A simple alternative to the approximate confidence interval (7.89) of $A^T \beta(t_0)$ is to ignore the bias term and use the "plug-in" approximate error bar

$$\left(\widehat{L}_\alpha^* \left[\widehat{\boldsymbol{\beta}}_K^{LSK}(t_0; h, w^{**}) \right], \widehat{U}_\alpha^* \left[\widehat{\boldsymbol{\beta}}_K^{LSK}(t_0; h, w^{**}) \right] \right)$$

with the lower and upper bounds given by

$$A^T \widehat{\boldsymbol{\beta}}_K^{LSK}(t_0; h, w^{**}) \pm z_{\alpha/2} (Nh)^{-1/2} \left[A^T \widehat{\mathbf{D}}^*(t_0) A \right]^{1/2}. \tag{7.90}$$

The asymptotic normality (7.73) suggest that, theoretically, when the bandwidth satisfies $h = o(N^{-1/5})$, the bias of $\widehat{\boldsymbol{\beta}}_K^{LSK}(t_0; h, w^{**})$ is negligible. Thus, the above error bar has approximately $[100 \times (1 - \alpha)]\%$ probability covering the true value of $A^T \beta(t_0)$ if a "small" bandwidth in the sense that $h = o(N^{-1/5})$ is

used. Although the bandwidth choice $h = o(N^{-1/5})$ leads to a slower conver-
gence rate for $\widehat{\beta}_K^{LSK}(t_0; h, w^{**})$ than the attainable rate of $N^{-2/5}$, the error bar
of (7.90) has the advantage of being computationally simple, since no estima-
tor of $\mathbf{B}(t_0)$ is needed. Thus, in practice, (7.90) can be used as an approximate
pointwise confidence interval.   □

3. Estimation of Asymptotic Bias and Variance

Following the definitions of (7.69) through (7.72), $\mathbf{B}(t_0)$ and $\mathbf{D}^*(t_0)$ depend
on the unknown functions $f_T(t_0)$, $f_T'(t_0)$, $\beta_r'(t_0)$, $\beta_r''(t_0)$, $\xi_{lr}(t_0)$, $\xi_{lr}'(t_0)$, $\rho_\varepsilon(t_0)$
and $\sigma^2(t_0)$. We present a class of intuitive kernel smoothers for the estimation
of these quantities, which can be used in (7.69) through (7.72) to construct
consistent estimators $\widehat{\mathbf{B}}(t_0)$ and $\widehat{\mathbf{D}}^*(t_0)$ for the approximate confidence interval
(7.89) or the approximate error bar (7.90). The construction of $\widehat{\mathbf{B}}(t_0)$ and
$\widehat{\mathbf{D}}^*(t_0)$ can be proceeded by the following four steps:

(1) **Estimation of $f_T(t_0)$ and $\xi_{lr}(t_0)$.** In addition to Assumption (e), we assume that the kernel functions $K(u)$ used here are twice continuously differentiable with respect to u. If $h_{(f,0)}$ and $h_{(\eta_{lr},0)}$ are bandwidths satisfying
$\lim_{n\to\infty} h_{(\cdot,0)} = 0$ and $\lim_{n\to\infty} N h_{(\cdot,0)} = \infty$, then $f_T(t_0)$ and $\xi_{lr}(t_0)$ can be estimated by

$$\widehat{f}_T\left(t_0; h_{(f,0)}\right) = \left[N h_{(f,0)}\right]^{-1} \sum_{i=1}^{n} \sum_{j=1}^{n_i} K\left(\frac{t_0 - t_{ij}}{h_{(f,0)}}\right) \qquad (7.91)$$

and

$$\begin{aligned}
\widehat{\xi}_{lr}\left(t_0; h_{(\xi_{lr},0)}\right) &= \left[N h_{(\eta_{lr},0)} \widehat{f}_T\left(t_0; h_{(\xi_{lr},0)}\right)\right]^{-1} \\
&\times \sum_{i=1}^{n} \sum_{j=1}^{n_i} \left[X_{ij}^{(l)} X_{ij}^{(r)} K\left(\frac{t_0 - t_{ij}}{h_{(\xi_{lr},0)}}\right)\right], \qquad (7.92)
\end{aligned}$$

respectively. Let $\widehat{E}_{\mathbf{XX}^T}(t_0)$ be the $(k+1) \times (k+1)$ matrix whose (l, r)th ele-
ment is $\widehat{\xi}_{lr}\left(t_0; h_{(\xi_{lr},0)}\right)$. Suppose that $\widehat{E}_{\mathbf{XX}^T}(t_0)$ is invertible. Then, $E_{\mathbf{XX}^T}(t_0)$
and $E_{\mathbf{XX}^T}^{-1}(t_0)$ can be estimated by $\widehat{E}_{\mathbf{XX}^T}(t_0)$ and $\widehat{E}_{\mathbf{XX}^T}^{-1}(t_0)$, respectively.

(2) **Estimation of Derivatives.** Let $f_T^{(d)}(t_0)$, $\beta_r^{(d)}(t_0)$ and $\xi_{lr}^{(d)}(t_0)$, $d = 1, 2$, be
the dth derivatives of $f_T(t_0)$, $\beta_r(t_0)$ and $\xi_{lr}(t_0)$, respectively. Let $h_{(f,d)}$, $h_{(\beta_r,d)}$
and $h_{(\xi_{lr},d)}$ be the corresponding bandwidths for the estimators of $f_T(t_0)$,
$\beta_r(t_0)$ and $\xi_{lr}(t_0)$, which satisfy $\lim_{n\to\infty} h_{(\cdot,d)} = 0$ and $\lim_{n\to\infty} N h_{(\cdot,d)}^{2d+1} = \infty$.
Then, following the kernel derivative estimators with cross-sectional i.i.d.
data, e.g., Härdle (1990, Chapter 3), $f_T^{(d)}(t_0)$, $\beta_r^{(d)}(t_0)$ and $\xi_{lr}^{(d)}(t_0)$ can be es-
timated by the corresponding dth derivatives $\widehat{f}_T^{(d)}\left(t_0; h_{(f,d)}\right)$, $\widehat{\beta}_r^{(d)}\left(t_0; h_{(\beta_r,d)}\right)$
and $\widehat{\xi}_{lr}^{(d)}\left(t_0; h_{(\xi_{lr},d)}\right)$ of $\widehat{f}_T\left(t_0; h_{(f,d)}\right)$, $\widehat{\beta}_r\left(t_0; h_{(\beta_r,d)}\right)$ and $\widehat{\xi}_{lr}\left(t_0; h_{(\eta_{lr},d)}\right)$ at time
t_0, where $\widehat{\beta}_r\left(t_0; h_{(\beta_r,d)}\right)$ is the rth component of $\widehat{\beta}_K^{LSK}(t_0; h, w^{**})$.

(3) **Estimation of Variances and Covariances.** For the estimation of variance $\sigma^2(t_0)$ and covariance $\rho_\varepsilon(t_0)$, we use smoothing estimators based on the residuals

$$\widehat{\varepsilon}_i(t_{ij}; h) = Y_{ij} - \mathbf{X}_{ij}^T \widehat{\boldsymbol{\beta}}_K^{LSK}(t_{ij}; h, w^{**}). \tag{7.93}$$

Let $h_{(\sigma)}$ be a bandwidth satisfying $\lim_{n\to\infty} h_{(\sigma)} = 0$ and $\lim_{n\to\infty} N h_{(\sigma)} = \infty$. The variance $\sigma^2(t_0)$ can be simply estimated by

$$\widehat{\sigma}^2(t_0; h_{(\sigma)}) = \left[\frac{1}{N h_{(\sigma)} \widehat{f}_T(t_0; h_{(\sigma)})}\right] \sum_{i=1}^{n} \sum_{j=1}^{n_i} \left[\widehat{\varepsilon}_i^2(t_{ij}; h) K\left(\frac{t_0 - t_{ij}}{h_{(\sigma)}}\right)\right]. \tag{7.94}$$

The estimation of $\rho_\varepsilon(t_0)$ is usually more difficult than the estimation of $\sigma^2(t_0)$. Since $\rho_\varepsilon(s_1, s_2) = E\left[\varepsilon_i(s_1)\varepsilon_i(s_2)\right]$ for $s_1 \neq s_2$ and $\rho_\varepsilon(t_0) = \lim_{s\to t_0} \rho_\varepsilon(s, t_0)$, we can estimate $\rho_\varepsilon(t_0)$ by smoothing $\left[\widehat{\varepsilon}_i(t_{ij_1}; h)\widehat{\varepsilon}_i(t_{ij_2}; h)\right]$ for $j_1 \neq j_2$ when n_i, $i = 1, \ldots, n$, are large, that is

$$\widehat{\rho}_\varepsilon(t_0; h_{(\rho)}) = \frac{\sum_{i=1}^{n} \sum_{j_1 \neq j_2 = 1}^{n_i} \left[\widehat{\varepsilon}_i(t_{ij_1}; h)\widehat{\varepsilon}_i(t_{ij_2}; h) K\left(\frac{t_0 - t_{ij_1}}{h_{(\rho)}}\right) K\left(\frac{t_0 - t_{ij_2}}{h_{(\rho)}}\right)\right]}{\sum_{i=1}^{n} \sum_{j_1 \neq j_2 = 1}^{n_i} \left[K\left(\frac{t_0 - t_{ij_1}}{h_{(\rho)}}\right) K\left(\frac{t_0 - t_{ij_2}}{h_{(\rho)}}\right)\right]}, \tag{7.95}$$

where $h_{(\rho)}$ satisfies $\lim_{n\to\infty} h_{(\rho)} = 0$ and $\lim_{n\to\infty} N h_{(\rho)} = \infty$.

(4) **Plug-in Bias and Variance Estimators.** Finally, we obtain the kernel estimator $\widehat{\mathbf{B}}(t_0)$ by substituting $f_T^{(d)}(t_0)$, $\boldsymbol{\beta}_c^{(d)}(t_0)$ and $\boldsymbol{\xi}_{lc}^{(d)}(t_0)$, $d = 0, 1, 2$, of (7.69) and (7.70) with $\widehat{f}^{(d)}(t_0; h_{(f,d)})$, $\widehat{\boldsymbol{\beta}}_c^{(d)}(t_0; h_{(\beta_c,d)})$ and $\widehat{\boldsymbol{\xi}}_{lc}^{(d)}(t_0; h_{(\xi_{lc},d)})$, and obtain $\widehat{\mathbf{D}}^*(t_0)$ by substituting $f_T(t_0)$, $\boldsymbol{\xi}_{lr}(t_0)$, $\sigma^2(t_0)$ and $\rho_\varepsilon(t_0)$ of (7.71) and (7.72) with $\widehat{f}_T(t_0; h_{(f,0)})$, $\widehat{\boldsymbol{\xi}}_{lr}(t_0; h_{(\xi_{lr},0)})$, $\widehat{\sigma}^2(t_0; h_{(\sigma)})$ and $\widehat{\rho}_\varepsilon(t_0; h_{(\rho)})$. □

An important requirement for the above plug-in estimation to work well is that the sample size n and the numbers of repeated measurements $\{n_1, \ldots, n_n\}$ must be large. For example, in the estimation of variances and covariances in step (3), since $\widehat{\rho}_\varepsilon(t_0; h_{(\rho)})$ is obtained by smoothing the adjacent residuals for each subject, it only works well asymptotically when the numbers of repeated measurements n_i, $i = 1, \ldots, n$, are large, so that one can actually smooth $\left[\widehat{\varepsilon}_i(t_{ij_1}; h)\widehat{\varepsilon}_i(t_{ij_2}; h)\right]$ in the vicinity of t_0. When there is no measurement error for $\{Y_{ij} : i = 1, \ldots, n; j = 1, \ldots, n_i\}$, that is, $\sigma^2(t_0) = \rho_\varepsilon(t_0)$, $\widehat{\sigma}^2(t_0; h_{(\sigma)})$ is practically a better estimator of $\rho_\varepsilon(t_0)$ than $\widehat{\rho}_\varepsilon(t_0; h_{(\rho)})$. But when $\{Y_{ij} : i = 1, \ldots, n; j = 1, \ldots, n_i\}$ are subject to measurement errors and $\{n_i : i = 1, \ldots, n\}$ are small, we have that $\sigma^2(t_0) > \rho_\varepsilon(t_0)$ and both $\widehat{\sigma}^2(t_0; h_{(\sigma)})$ and $\widehat{\rho}_\varepsilon(t_0; h_{(\rho)})$ are subject to large biases for the estimation of $\rho_\varepsilon(t_0)$. It can be seen from (7.94) and (7.95) that the adequacy of $\widehat{\sigma}^2(t_0; h_{(\sigma)})$ and $\widehat{\rho}_\varepsilon(t_0; h_{(\rho)})$ also depends on the bandwidth h of $\widehat{\boldsymbol{\beta}}_K^{LSK}(t; h, w^{**})$.

4. Consistency of Bias and Variance Estimators

An important requirement for the appropriateness of the asymptotically approximate confidence intervals (7.89) and error bars (7.90) is that the estimators $\widehat{\mathbf{B}}(t_0)$ and $\widehat{\mathbf{D}}^*(t_0)$ are consistent for $\mathbf{B}(t_0)$ and $\mathbf{D}^*(t_0)$. When $\widehat{\mathbf{B}}(t_0)$ and $\widehat{\mathbf{D}}^*(t_0)$ are chosen by the plug-in estimators, the consistency of $\widehat{\mathbf{B}}(t_0)$ and $\widehat{\mathbf{D}}^*(t_0)$ requires that the estimators of the unknown functions $f_T(t_0)$, $f_T'(t_0)$, $\beta_r'(t_0)$, $\beta_r''(t_0)$, $\xi_{lr}(t_0)$, $\xi_{lr}'(t_0)$, $\rho_\varepsilon(t_0)$ and $\sigma^2(t_0)$ are also consistent.

The next lemma shows the consistency of the kernel estimators for the unknown functions $f_T(t_0)$, $f_T'(t_0)$, $\beta_r'(t_0)$, $\beta_r''(t_0)$, $\xi_{lr}(t_0)$, $\xi_{lr}'(t_0)$, $\rho_\varepsilon(t_0)$ and $\sigma^2(t_0)$ given above. The asymptotic consistency of the plug-in bias and variance estimators $\widehat{\mathbf{B}}(t_0)$ and $\widehat{\mathbf{D}}^*(t_0)$ directly follows from the consistency of the kernel estimators of these components. This lemma suggests that, despite their practical drawbacks of computational complexity, the plug-in type asymptotically approximate pointwise confidence intervals of (7.89) and (7.90) can be indeed constructed in practice.

Lemma 7.2. *Suppose that Assumptions (a) through (f) and (7.68) are satisfied and $K(u)$ is continuously twice differentiable with a compact support on the real line. The following consistency results hold:*

(a) *If* $\lim_{n\to\infty} h_{(\cdot;d)} = 0$ *and* $\lim_{n\to\infty} N h_{(\cdot;d)}^{2d+1} = \infty$ *for* $d = 0, 1, 2$, *then*

$$
\begin{cases}
\widehat{f}_T^{(d)}(t_0; h_{(f,d)}) & \to & f_T^{(d)}(t_0), \\
\widehat{\beta}_r^{(d)}(t_0; h_{(\beta_r,d)}) & \to & \beta_r^{(d)}(t_0), \\
\widehat{\xi}_{lr}^{(d)}(t_0; h_{(\eta_{lr},d)}) & \to & \xi_{lr}(t_0), \quad l, r = 0, \dots, k,
\end{cases}
$$

in probability as $n \to \infty$.

(b) *If* $\lim_{n\to\infty} h_{(\cdot)} = 0$ *and* $\lim_{n\to\infty} N h_{(\cdot)} = \infty$, *then* $\widehat{\sigma}^2(t_0; h_{(\sigma)}) \to \sigma^2(t_0)$ *and* $\widehat{\rho}_\varepsilon(t_0; h_{(\rho)}) \to \rho_\varepsilon(t_0)$ *in probability as* $n \to \infty$. ∎

Proof of Lemma 7.2:

Because the derivations for all the estimators involve many tedious and repetitive computations, we only sketch the proofs for $\widehat{f}_T^{(1)}(t_0; h_{(f,1)})$, $\widehat{\xi}_{lr}(t_0; h_{(\eta_{lr},0)})$ and $\widehat{\rho}_\varepsilon(t_0; h_{(\rho)})$. The consistency of other estimators can be similarly derived with tedious but straightforward calculations.

(a) *Consistency of* $\widehat{f}_T^{(1)}(t_0; h_{(f,0)})$:

Since the subjects are independent, direct calculation, integration by parts and the change of variables show that, as $n \to \infty$,

$$
\left| E\left[\widehat{f}_T^{(1)}(t_0; h_{(f,1)})\right] - f_T^{(1)}(t_0) \right| = \left| \int K(u) \left[f_T^{(1)}(t_0 - h_{(f,1)}u) - f_T^{(1)}(t_0) \right] du \right| \to 0
$$

and

$$Var\left[\widehat{f}_T^{(1)}\left(t_0; h_{(f,1)}\right)\right] = N^{-1} h_{(f,1)}^{-3} f_T(t_0) \left\{ \int \left[K'(u)\right]^2 du \right\} + o\left(N^{-1} h_{(f,1)}^{-3}\right) \to 0.$$

Thus, the above limits imply that $\lim_{n\to\infty} E\{ [\widehat{f}_T^{(1)}\left(t_0; h_{(f,1)}\right) - f_T^{(1)}(t_0)]^2 \} = 0$. Consequently, $\widehat{f}_T^{(1)}\left(t_0; h_{(f,1)}\right) \to f_T^{(1)}(t_0)$ in probability as $n \to \infty$, which shows the consistency of $\widehat{f}_T^{(1)}\left(t_0; h_{(f,0)}\right)$.

(b) *Consistency of* $\widehat{\xi}_{lr}(t_0; h_{(\xi lr,0)})$:
First, let us denote

$$\widehat{v}_\xi\left(t_0; h_{(\eta lr,0)}\right) = \left(N h_{(\xi lr,0)}\right)^{-1} \sum_{i=1}^{n} \sum_{j=1}^{n_i} \left[X_{ij}^{(l)} X_{1j}^{(r)} K\left(\frac{t_0 - t_{ij}}{h_{(\eta lr,0)}}\right)\right].$$

Using the same method as in the proof of Lemma 7.1, it can be verified by direct calculations that $\widehat{f}_T\left(t_0; h_{(\xi lr,0)}\right) \to f(t_0)$ in probability as $n \to \infty$. Then, by the definition of $\widehat{\xi}_{lr}(t_0; h_{(\xi lr,0)})$ given in (7.92), it suffices to show that $\widehat{v}_\xi\left(t_0; h_{(\xi lr,0)}\right) \to \xi_{lr}(t_0) f_T(t_0)$ in probability as $n \to \infty$.
By direct calculations, it can be verified that, as $n \to \infty$,

$$\begin{aligned}
&E\left[\widehat{v}_\xi\left(t_0; h_{(\eta lr,0)}\right)\right] \\
&= \left(N h_{(\xi lr,0)}\right)^{-1} \sum_{i=1}^{n} \sum_{j=1}^{n_i} \left\{ \int E\left[X_{ij}^{(l)} X_{ij}^{(r)} \Big| t_{ij} = s\right] K\left(\frac{t_0 - s}{h_{(\xi lr,0)}}\right) f(s)\, ds \right\} \\
&\to \xi_{lr}(t_0) f(t_0)
\end{aligned}$$

and

$$\begin{aligned}
&Var\left[\widehat{v}_\eta\left(t_0, h_{(\xi lr,0)}\right)\right] \\
&= \left(N h_{(\xi lr,0)}\right)^{-2} \sum_{i=1}^{n} \sum_{j=1}^{n_i} \left\{ \int E\left[\left(X_{ij}^{(l)}\right)^2 \left(X_{ij}^{(r)}\right)^2 \Big| t_{ij} = s\right] K^2\left(\frac{t_0 - s}{h_{(\xi lr,0)}}\right) f(s)\, ds \right\} \\
&\quad + o\left(N^{-1} h_{(\xi lr,0)}^{-1}\right) \\
&\to 0.
\end{aligned}$$

The above two limits imply that $E\left[\widehat{v}_\xi\left(t_0; h_{(\xi lr,0)}\right) - \xi_{lr}(t_0) f(t_0)\right]^2 \to 0$ holds as $n \to \infty$, which further implies that $\widehat{v}_\xi\left(t_0; h_{(\xi lr,0)}\right) \to \xi_{lr}(t_0) f(t_0)$ in probability as $n \to \infty$. This shows the consistency of $\widehat{\xi}_{lr}(t_0; h_{(\xi lr,0)})$.

(c) *Consistency of* $\widehat{\rho}_\varepsilon(t_0; h_{(\rho)})$:
We first define

$$\widehat{v}_\rho\left(t_0; h_{(\rho)}\right) = \left(\sum_{i=1}^{n} n_i^2 - N\right)^{-1} h_{(\rho)}^{-2}$$

$$\times \sum_{i=1}^{n} \sum_{j_1 \neq j_2 = 1}^{n_i} \left[\widehat{\varepsilon}_i(t_{ij_1}, h) \, \widehat{\varepsilon}_i(t_{ij_2}, h) \, K\left(\frac{t_0 - t_{ij_1}}{h_{(\rho)}}\right) K\left(\frac{t_0 - t_{ij_2}}{h_{(\rho)}}\right) \right]$$

and

$$\widehat{V}_\rho^*(t_0; h_{(\rho)}) = \left(\sum_{i=1}^{n} n_i^2 - N\right)^{-1} h_{(\rho)}^{-2} \sum_{i=1}^{n} \sum_{j_1 \neq j_2 = 1}^{n_i} \left[K\left(\frac{t_0 - t_{ij_1}}{h_{(\rho)}}\right) K\left(\frac{t_0 - t_{ij_2}}{h_{(\rho)}}\right) \right].$$

By the definition of $\widehat{\rho}_\varepsilon(t_0; h_{(\rho)})$ in (7.95), it follows that

$$\widehat{\rho}_\varepsilon(t_0; h_{(\rho)}) = \widehat{V}_\rho(t_0; h_{(\rho)}) / \widehat{V}_\rho^*(t_0; h_{(\rho)}).$$

It then suffices to show that $\widehat{V}_\rho^*(t_0; h_{(\rho)}) \to f_T^2(t_0)$ and $\widehat{\rho}_\varepsilon(t_0; h_{(\rho)}) \to \rho_\varepsilon(t_0) f_T^2(t_0)$ in probability as $n \to \infty$.

For the consistency of $\widehat{V}_\rho^*(t_0; h_{(\rho)})$, we can show by direct calculations as those in the proof of Lemma 7.1 that, as $n \to \infty$,

$$E\left[\widehat{V}_\rho^*(t_0; h_{(\rho)})\right]$$

$$= \left(\sum_{i=1}^{n} n_i^2 - N\right)^{-1} h_{(\rho)}^{-2}$$

$$\times \left[\sum_{i=1}^{n} \sum_{j_1 \neq j_2 = 1}^{n_i} \int\int K\left(\frac{t_0 - s_1}{h_{(\rho)}}\right) K\left(\frac{t_0 - s_2}{h_{(\rho)}}\right) f_T(s_1) f_T(s_1) \, ds_1 \, ds_2\right]$$

$$\to f_T(t_0)$$

and

$$Cov\left[\widehat{V}_\rho^*(t_0; h_{(\rho)})\right] = \left(\sum_{i=1}^{n} n_i^2 - N\right)^{-1} h_{(\rho)}^{-2} \left[\int K^2(u) \, du\right]^2 f_T^2(t_0) + o(1) \to 0.$$

Thus, we have that, as $n \to \infty$, $E\left[\widehat{V}_\rho^*(t_0; h_{(\rho)}) - f_T^2(t_0)\right]^2 \to 0$, which implies that $\widehat{V}_\rho^*(t_0; h_{(\rho)}) \to f_T^2(t_0)$ in probability.

To show the consistency of $\widehat{V}_\rho(t_0; h_{(\rho)})$, we define the pseudo-residual

$$\widetilde{\varepsilon}_i(t_{ij}) = Y_{ij} - \mathbf{X}_{ij}^T \beta(t_{ij})$$

and

$$\widetilde{V}_\rho(t_0; h_{(\rho)}) = \left(\sum_{i=1}^{n} n_i^2 - N\right)^{-1} h_{(\rho)}^{-2}$$

$$\times \sum_{i=1}^{n} \sum_{j_1 \neq j_2 = 1}^{n_i} \left[\widetilde{\varepsilon}_i(t_{ij_1}) \, \widetilde{\varepsilon}_i(t_{ij_2}) \, K\left(\frac{t_0 - t_{ij_1}}{h_{(\rho)}}\right) K\left(\frac{t_0 - t_{ij_2}}{h_{(\rho)}}\right) \right].$$

By Theorem 7.1 and similar calculations in the consistency of $\widehat{V}_\rho^*(t_0; h_{(\rho)})$, we can verify that there are constants $a_1 > 0$ and $a_2 > 0$ such that, as $n \to \infty$,

$$\sup_{t_{ij_1} \in [t_0 - a_1, t_0 + a_1], \, t_{ij_2} \in [t_0 - a_2, t_0 + a_2]} \left| \widehat{\varepsilon}_i(t_{ij_1}; h) \, \widehat{\varepsilon}_i(t_{ij_2}; h) - \widetilde{\varepsilon}_i(t_{ij_1}) \, \widetilde{\varepsilon}_i(t_{ij_2}) \right| \to 0$$

in probability, and

$$\left(\sum_{i=1}^{n} n_i^2 - N\right)^{-1} h_{(\rho)}^{-2} \sum_{i=1}^{n} \sum_{j_1 \neq j_2} \left| K\left(\frac{t_0 - t_{ij_1}}{h_{(\rho)}}\right) K\left(\frac{t_0 - t_{ij_2}}{h_{(\rho)}}\right) \right|$$

is bounded in probability when n is sufficiently large. Thus,

$$\left| \widehat{v}_\rho\left(t_0; h_{(\rho)}\right) - \widetilde{v}_\rho\left(t_0; h_{(\rho)}\right) \right|$$

$$\leq \sup_{t_{ij_1} \in [t_0 - a_1, t_0 + a_1], \ t_{ij_2} \in [t_0 - a_2, t_0 + a_2]} \left| \widehat{\varepsilon}_i\left(t_{ij_1}; h\right) \widehat{\varepsilon}_i\left(t_{ij_2}; h\right) - \widetilde{\varepsilon}_i\left(t_{ij_1}\right) \widetilde{\varepsilon}_i\left(t_{ij_2}\right) \right|$$

$$\times \left(\sum_{i=1}^{n} n_i^2 - N\right)^2 h_{(\rho)}^2 \sum_{i=1}^{n} \sum_{j_1 \neq j_2 = 1}^{n_i} \left| K\left(\frac{t_0 - t_{ij_1}}{h_{(\rho)}}\right) K\left(\frac{t_0 - t_{ij_2}}{h_{(\rho)}}\right) \right|$$

$$\to \quad 0 \quad \text{in probability as } n \to \infty.$$

It suffices to show that $\widetilde{v}_\rho\left(t_0; h_{(\rho)}\right) \to \rho_\varepsilon(t_0) f_T^2(t_0)$ in probability as $n \to \infty$.

Similar calculations as in (b) shows that, as $n \to \infty$, $E\left[\widetilde{v}_\rho\left(t_0; h_{(\rho)}\right)\right] \to \rho_\varepsilon(t_0) f_T^2(t_0)$ and $Var\left[\widetilde{v}_\rho\left(t_0; h_{(\rho)}\right)\right] \to 0$. Thus, $\widetilde{v}_\rho\left(t_0; h_{(\rho)}\right) \to \rho_\varepsilon(t_0) f_T^2(t_0)$ in probability as $n \to \infty$. This completes the proof. ∎

7.6   Remarks and Literature Notes

The methods presented in this chapter focus on the time-varying coefficient model (7.1) using a series of local smoothing methods. This model has a wide range of applications in longitudinal studies. The main advantages of this model are: (i) its simple interpretation as a standard multiple linear model for the outcome variable $Y(t)$ and the covariate vector $\mathbf{X}(t)$ at each fixed time point t; (ii) its flexibility of allowing the coefficients $\beta(t)$ to be unknown curves of t, which leads to different linear models at different time points. Among different smoothing estimation methods, this chapter focuses on two main kernel-type smoothing methods, the one-step least squares kernel smoothing method and the covariate centered kernel smoothing method. In practice, both smoothing methods, with or without covariate centering, have their advantages and disadvantages. The theoretical developments, however, are only focused on the kernel smoothing method without covariate centering.

Appropriate confidence intervals of the smoothing estimators can be computed using the resampling-subject bootstrap method. These confidence intervals have been shown in the previous chapters to have appropriate coverage probabilities, and such a bootstrap approach has the potential to preserve the unknown correlation structures of the data. On the other hand, the asymptotically approximate confidence intervals rely on the asymptotic distributions of the smoothing estimators, which have to be developed on a case-by-case basis. Even if the asymptotic distributions of the smoothing estimators are available, the asymptotic biases and variances still have to be estimated in order to compute the lower and upper bounds of the approximated confidence intervals.

In many situations, however, it is not easy to obtain accurate estimates for the asymptotic biases and variances. Since the resampling-subject bootstrap confidence intervals are entirely data-driven and do not depend on the bias and variance estimates, it is practically a more convenient inference procedure than the asymptotically approximate inference methods.

Results for the least squares kernel smoothing method without covariate centering are based on Hoover et al. (1998) and Wu, Chiang and Hoover (1998). Methods for the kernel smoothing method with covariate centering are based on Wu, Yu and Chiang (2000).

Chapter 8

The Two-Step Local Smoothing Methods

We describe in this chapter a class of two-step local smoothing methods for the estimation of the coefficient curves $\beta(t)$ in the time-varying coefficient model (7.1). This class of estimators is based on the simple idea that the coefficient curves $\beta(t)$ can be first estimated by the least squares method at a sequence of isolated time points, and then these isolated estimates can be treated as pseudo-observations and smoothed over to produce the final smoothed estimator of $\beta(t)$. Compared with the one-step local smoothing methods in Chapter 7, this class of methods have two major advantages. First, the two-step smoothing methods of this chapter can naturally incorporate different bandwidths for different components of $\beta(t)$, which provides some additional flexibility for adjusting the possibly different smoothing needs in $\beta(t)$. Second, since the two-step smoothing methods only depend on the existing estimation methods, they are computationally simple, and their bandwidths can be easily selected by modifying the cross-validation procedures with the classical cross-sectional i.i.d. data. The idea of two-step smoothing can be generalized to other structured nonparametric models constructed from some local parametric or semiparametric models when t is fixed, such as the time-varying transformation models of Chapters 13 and 14, in which the two-step estimation approach is the only available option in the literature.

8.1 Overview and Justifications

The local least squares smoothing estimators of Chapter 7 may be impractical in many longitudinal settings, because it is often difficult to decide whether the bandwidth and kernel function (h_l, K_l) of the component curve estimator $\widehat{\beta}(t; K_l, h_l)$ are appropriate for $\beta_l(t)$. This is due to the lack of direct observations of $\beta_l(t)$, that is, the approximate shape of $\beta_l(t)$ cannot be directly detected from the data $\left\{ \left(Y_{ij}, t_{ij}, \mathbf{X}_{ij}^T \right)^T : i = 1, \ldots, n; j = 1, \ldots, n_i \right\}$. If we can construct some "pseudo-observations" of $\beta_l(t)$, then appropriate smoothing estimators can be constructed specifically for each $\beta_l(t)$, $l = 1, \ldots, k$, so that the final estimator of $\beta(t)$ has appropriate smoothness for each of its components.

This intuitive estimation method is described in Fan and Zhang (2000) as the "two-step estimation" for functional linear models.

1. The Two-Step Estimation Procedure

In order to describe the "raw estimates" step of Fan and Zhang's procedure, we first redefine the design time points. Let

$$\mathbf{T} = \{t_j : j = 1, \ldots, J\} \tag{8.1}$$

be the distinct time points among the time points $\{t_{ij} : i = 1, \ldots, n; j = 1, \ldots, n_i\}$ previously defined in Section 7.1.1. For each given time t_j in $\mathbf{T}$, let $\mathscr{S}_j$ be the collection of all the subjects observed at t_j, that is,

$$\mathscr{S}_j = \{i : 1 \leq i \leq n; t_{ij^*} = t_j \text{ for some } 1 \leq j^* \leq n_i\}, \tag{8.2}$$

and, m_j be the number of subjects observed at time design point t_j, i.e.,

$$m_j = \{\# \text{ of } i : i \in \mathscr{S}_j\}. \tag{8.3}$$

Then, the Fan and Zhang two-step estimation procedure for the time-varying coefficient model (7.1) is given as follows.

The Two-Step Estimation Procedure:

(a) **Raw Estimators.** *Compute the raw estimates $\widehat{\beta}_l^{RAW}(t_j)$ of $\beta_l(t)$ using the subjects in $\mathscr{S}_j$ for all the distinct design time points $\{t_j \in \mathbf{T} : j = 1, \ldots, J\}$ and $l = 0, \ldots, k$.*

(b) **Smoothing Estimators.** *Estimate each coefficient curve $\beta_l(t)$ for any t in the time range $\mathscr{T}$ by a kernel or local polynomial smoothing estimator based on $\widehat{\beta}_l^{RAW}(t_j)$ for all $t_j \in \mathbf{T}$, $j = 1, \ldots, J$.* □

Although Fan and Zhang (2000) uses the local polynomial estimators to illustrate their method, other smoothing methods such as splines may also be used. To ensure existence of the raw estimates in a real situation, the following time-point binning strategy is often used.

Time-Point Binning. Computation of the raw estimators $\widehat{\beta}_l^{RAW}(t_j)$ at t_j for any $1 \leq j \leq J$ requires m_j defined in (8.3) to be sufficiently large, so that there are enough subjects observed at t_j. In practice, particularly for a sparse longitudinal study, i.e., study with relatively small numbers of repeated measurements over time, the numbers of subjects with observations at some design time points in $\mathbf{T}$ given in (8.1) may be small. Consequently, the raw estimates at these design time points may not be computable. In such situations, a useful approach is to round off or group some of the adjacent design time points into small bins and compute the raw estimates within each bin. In biological studies, the time scale is often rounded off to an acceptable precision. Thus, "time-point binning" may have already been used at the data collection stage. Of course, "time-point binning" may only make practical sense when both the raw and smoothing estimators are interpretable. Typically, "time-point binning" is used when the number of design time points J is large and the t_j's are spread around the intended time range $\mathcal{T}$.

2. Justifications of the Two-Step Estimation Procedure

There are a number of practical advantages associated with the Fan and Zhang two-step estimation method:

(a) **Visualization of the Raw Estimates.** Since only the local observations at each of the "design time points" in $\mathbf{T}$ are used in the first step for computing the raw estimates, we can treat $\{\widehat{\beta}_l^{RAW}(t_j) : j = 1, \ldots, J\}$ as the "pseudo-observations" of $\beta_l(t)$ for t varying within its range $\mathcal{T}$. The plots and visualization of these "pseudo-observations" $\{\widehat{\beta}_l^{RAW}(t_j) : j = 1, \ldots, J\}$ can be used to guide the appropriate smoothness choices for the estimation of $\beta_l(t)$.

(b) **Componentwise Smoothing.** Since the raw estimators are usually not smooth in the sense that $\widehat{\beta}_l^{RAW}(t_j)$ and $\widehat{\beta}_l^{RAW}(t_{j+1})$ at two adjacent design time points t_j and t_{j+1} may vary significantly, the smoothing step is aimed at reducing the variability of the raw estimates, so that an appropriately smoothing estimator specifically for $\beta_l(t)$ can be constructed for all the time points t. This smoothing step is crucial since it allows us to pool information from neighboring time points to improve the raw estimators and leads to smoothing estimates for the underlying smooth coefficient functions for $t \in \mathcal{T}$. If we ignore the smoothing step and simply estimate $\beta_l(t)$ by linear interpolation of the raw estimates at the design time points $\mathbf{T}$, the resulting estimate of $\beta_l(t)$ could be spiky because of the variability of the raw estimates.

(c) **Computational Simplicity.** Both the raw and smoothing estimates can be easily computed using the existing computational methods and software packages. For a given design time point t_j, the model (7.1) is a linear model, so that the raw estimates $\{\widehat{\beta}_0^{RAW}(t_j), \ldots, \widehat{\beta}_k^{RAW}(t_j)\}$ can be computed by the standard statistical software packages for least squares estimation. The smoothing estimators of $\beta_l(t)$ based on the pseudo-observations $\{\widehat{\beta}_l^{RAW}(t_j) : j = 1, \ldots, J\}$ for any $0 \leq l \leq k$ can be easily computed using any existing smoothing technique and software package. In addition, the existing well-developed smoothing parameter selectors, such as the bandwidth selectors, can be easily adopted in the smoothing step.

(d) **Simple Interpretations of the Estimates.** A useful by-product of the scatter plots of the raw estimates is the intuitive interpretations of the coefficient functions. At each design time point t_j, the covariate effects are simply characterized by the linear model coefficients $\{\beta_0(t_j), \ldots, \beta_k(t_j)\}$, which are estimated by $\{\hat{\beta}_0^{RAW}(t_j), \ldots, \hat{\beta}_k^{RAW}(t_j)\}$. The coefficient functions $\{\beta_0(t), \ldots, \beta_k(t)\}$, which have values close to $\{\beta_0(t_j), \ldots, \beta_k(t_j)\}$ when t is close to t_j, are naturally estimated by borrowing information from the raw estimates at the design time points within some neighborhood of t.   □

8.2   Raw Estimators

For the first step of the two-step estimation procedure, we give the expressions of the least squares parameter estimators at a fixed time point in $\mathbf{T}$. These estimates are used as the pseudo-observations for the second step where the smoothing estimators of $\beta(t)$ are constructed.

8.2.1   General Expression and Properties

For a given t_j in the design time points $\mathbf{T}$, we can collect all the outcomes Y_{ij^*} and covariates $\mathbf{X}_{ij^*}$ from the subjects in $\mathscr{S}_j$, and denote by $\{\tilde{\mathbf{Y}}_j, \tilde{\mathbf{X}}_j\}$ the corresponding response vector and design matrix. Following the time-varying coefficient model (7.1), the observations $\{\tilde{\mathbf{Y}}_j, \tilde{\mathbf{X}}_j\}$ collected at time t_j follow the standard linear model

$$\tilde{\mathbf{Y}}_j = \tilde{\mathbf{X}}_j^T \beta(t_j) + \tilde{e}_j, \tag{8.4}$$

where $\tilde{e}_j$ is the vector of random error for the subjects in $\mathscr{S}_j$, and the distribution of any element of $\tilde{e}_j$ is the same as the distribution of $\varepsilon(t_j)$ in (7.1). It then follows from the definitions of the models (7.1) and (8.4) that

$$\begin{cases} E(\tilde{e}_j) & = & 0, \\ Cov(\tilde{e}_j) & = & \sigma^2(t_j)\mathbf{I}_{m_j}, \end{cases} \tag{8.5}$$

where $\sigma^2(t_j)$ is the variance of $\varepsilon(t_j)$ and $\mathbf{I}_{m_j}$ is the $(m_j \times m_j)$ identity matrix.

We note three issues related to the use of the local linear model (8.4):

(a) Since the measurements from the subjects not in $\mathscr{S}_j$ are not included in $\{\tilde{\mathbf{Y}}_j, \tilde{\mathbf{X}}_j\}$, this approach is equivalent to assuming that the subjects without measurements at t_j are missing completely at random.

(b) When the "time-point binning" approach is used, a subject's observation time t_{ij^*} is grouped into one of the design time points t_j, hence, is treated as t_j, if t_{ij^*} is not in $\mathbf{T}$ but stays within a small neighborhood of t_j. By using the model (8.4) with time-point binning, the differences between the subject's observations at t_{ij^*} and t_j are assumed to be small, and, therefore, ignored. Thus, the estimation results based on (8.4) and time-point binning may be slightly biased, but the size of bias can be reduced by using small time bins. In practice, it is preferable to use the time bins as small as possible provided that the raw estimates from (8.4) exist and there are meaningful interpretations for the raw and smoothing estimates. Time-point binning may affect the asymptotic properties of the smoothing estimators. But this issue has not been investigated in the literature.

(c) An implicit assumption of (8.4) is that each subject is observed at most once at a given design time point. This assumption, which eliminates the possibility of repeated measurements from the same subject at any $t_j \in \mathbf{T}$, is a practical one for most biomedical studies. If a subject has two or more observations at time points within a very small neighborhood of t_j, then it is reasonable to use the averages at these time points.   □

Suppose that the design matrix $\widetilde{\mathbf{X}}_j$ is of rank $(k+1)$. If we denote by $\left(\widetilde{\mathbf{X}}_j \widetilde{\mathbf{X}}_j^T\right)^{-1}$ the inverse of $\left(\widetilde{\mathbf{X}}_j \widetilde{\mathbf{X}}_j^T\right)$, the standard least squares estimator of $\beta(t_j)$ based on $\left\{\widetilde{\mathbf{Y}}_j, \widetilde{\mathbf{X}}_j\right\}$ is obtained by minimizing

$$L_j\left[\beta(t_j)\right] = \left[\widetilde{\mathbf{Y}}_j - \widetilde{\mathbf{X}}_j^T \beta(t_j)\right]^T \left[\widetilde{\mathbf{Y}}_j - \widetilde{\mathbf{X}}_j^T \beta(t_j)\right] \tag{8.6}$$

with respect to $\beta(t_j)$, and has the expression

$$\widehat{\beta}^{RAW}(t_j) = \left(\widetilde{\mathbf{X}}_j \widetilde{\mathbf{X}}_j^T\right)^{-1} \widetilde{\mathbf{X}}_j \widetilde{\mathbf{Y}}_j. \tag{8.7}$$

In standard linear regression models, the biases and variances of the least squares estimators are usually computed by conditioning on the observed co-variates or assuming that the covariates are fixed and nonrandom. For the sample $\left\{\left(Y_{ij}, t_{ij}, \mathbf{X}_{ij}^T\right)^T : i = 1, \ldots, n; j = 1, \ldots, n_i\right\}$, let

$$\mathscr{D} = \left\{\left(t_j, \mathbf{X}_{ij^*}^T\right) : j = 1, \ldots, J; i = 1, \ldots, n; j^* = 1, \ldots, n_i\right\} \tag{8.8}$$

be the observed time design points and covariates. Direct calculation based on (8.5), (8.6) and (8.7) shows that, conditioning on $\mathscr{D}$, the mean and covariance matrix of $\widehat{\beta}^{RAW}(t_j)$ are

$$\begin{cases} E\left[\widehat{\beta}^{RAW}(t_j) \big| \mathscr{D}\right] &= \beta(t_j), \\ Cov\left[\widehat{\beta}^{RAW}(t_j) \big| \mathscr{D}\right] &= \sigma^2(t_j)\left(\widetilde{\mathbf{X}}_j \widetilde{\mathbf{X}}_j^T\right)^{-1}. \end{cases} \tag{8.9}$$

The first equation of (8.9) indicates that, under the approximation of time-point binning, $\widehat{\beta}^{RAW}(t_j)$ is approximately an unbiased estimator of $\beta(t_j)$. Since each subject is assumed to be observed at most once at t_j, the second equation of (8.9) shows that the covariance matrix of $\widehat{\beta}^{RAW}(t_j)$ depends on the number of subjects m_j in $\mathscr{S}_j$. The variation of the raw estimate $\widehat{\beta}^{RAW}(t_j)$ is small if m_j is large. Since m_j may not increase with n, the variation of $\widehat{\beta}^{RAW}(t_j)$ is not necessarily smaller when the sample size n increases.

8.2.2 Component Expressions and Properties

Components of the raw estimator $\widehat{\beta}^{RAW}(t_j)$ can be expressed explicitly. For any $l = 0, 1, \ldots, k$, let $\widehat{\beta}_l^{RAW}(t_j)$ be the $(l+1)$th component of $\widehat{\beta}^{RAW}(t_j)$. Then, by (8.7), $\widehat{\beta}_l^{RAW}(t_j)$ has the expression

$$\widehat{\beta}_l^{RAW}(t_j) = e_{l+1,k+1}^T \left(\widetilde{\mathbf{X}}_j \widetilde{\mathbf{X}}_j^T\right)^{-1} \widetilde{\mathbf{X}}_j \widetilde{\mathbf{Y}}_j, \tag{8.10}$$

where $e_{l,k+1}$ is a $(k+1)$-dimensional unit vector with 1 at its lth entry and 0 elsewhere. By the conditional mean and the conditional covariance matrix of $\widehat{\beta}^{RAW}(t_j)$ given in (8.9), the mean and variance of $\widehat{\beta}_l^{RAW}(t_j)$ conditioning on the covariate set $\mathscr{D}$ are

$$\begin{cases} E\left[\widehat{\beta}_l^{RAW}(t_j)\big|\mathscr{D}\right] &= \beta_l(t_j), \\ Var\left[\widehat{\beta}_l^{RAW}(t_j)\big|\mathscr{D}\right] &= e_{l+1,k+1}^T \sigma^2(t_j) \left(\widetilde{\mathbf{X}}_j \widetilde{\mathbf{X}}_j^T\right)^{-1} e_{l+1,k}. \end{cases} \tag{8.11}$$

In particular, the variance of $\widehat{\beta}_l^{RAW}(t_j)$ conditioning on $\mathscr{D}$ is the $(l+1)$th diagonal element of $\sigma^2(t_j) \left(\widetilde{\mathbf{X}}_j \widetilde{\mathbf{X}}_j^T\right)^{-1}$.

The correlation structure of the data affects the covariance of $\widehat{\beta}_l^{RAW}(\cdot)$ at different design time points. Let $\{t_j, t_{j^*} : j \neq j^*\}$ be any design time points in $\mathbf{T}$ of (8.1). The covariance of $\widehat{\beta}_l^{RAW}(t_j)$ and $\widehat{\beta}_l^{RAW}(t_{j^*})$ conditioning on $\mathscr{D}$ is

$$\begin{aligned} &Cov\left[\widehat{\beta}_l^{RAW}(t_j), \widehat{\beta}_l^{RAW}(t_{j^*})\big|\mathscr{D}\right] \\ &= \rho(t_j, t_{j^*}) e_{l+1,k+1}^T \left(\widetilde{\mathbf{X}}_j \widetilde{\mathbf{X}}_j^T\right)^{-1} \widetilde{\mathbf{X}}_j \mathbf{M}_{jj^*} \widetilde{\mathbf{X}}_j^T \left(\widetilde{\mathbf{X}}_j \widetilde{\mathbf{X}}_j^T\right)^{-1} e_{l+1,k+1}, \end{aligned} \tag{8.12}$$

where $\rho(s, t)$ is the covariance function defined in (7.1) and $\mathbf{M}_{jj^*}$ is a matrix with entries 0 or 1, such that, its (a, b)th entry is 1, if the ath entry of $\widetilde{\mathbf{Y}}_j$ and the bth entry of $\widetilde{\mathbf{Y}}_{j^*}$ are from the same subject, and 0 otherwise. Since some subjects may have observations at only one of the time points t_j or t_{j^*} for $j \neq j^*$, the purpose of $\mathbf{M}_{jj^*}$ is to make sure that only the subjects who have been observed at both t_j and t_{j^*} are used for computing the conditional covariance $Cov\left[\widehat{\beta}_l^{RAW}(t_j), \widehat{\beta}_l^{RAW}(t_{j^*})\big|\mathscr{D}\right]$. When $j = j^*$, $\mathbf{M}_{jj}$ is an identity matrix.

8.2.3 Variance and Covariance Estimators

The variances and covariances of the raw estimate $\widehat{\beta}^{RAW}(t_j)$ given in (8.9), (8.11) and (8.12) depend on the unknown variance $\sigma^2(t_j)$ and covariance $\rho(t_j, t_{j^*})$ for $t_j \neq t_{j^*}$. Following the well-known residual-based approaches in linear models, these two quantities can be easily estimated based on the residuals from the least squares fit of the model (8.4). To compute the residuals of $\widehat{\beta}^{RAW}(t_j)$ in (8.7) from the local model (8.4), let $\mathbf{I}_{m_j}$ be the $m_j \times m_j$ identity matrix and

$$\mathbf{P}_j = \widetilde{\mathbf{X}}_j^T \left(\widetilde{\mathbf{X}}_j \widetilde{\mathbf{X}}_j^T\right)^{-1} \widetilde{\mathbf{X}}_j, \tag{8.13}$$

so that $\widetilde{\mathbf{X}}_j^T \widehat{\boldsymbol{\beta}}^{RAW}(t_j) = \mathbf{P}_j \widetilde{\mathbf{Y}}_j$. It follows from (8.4) that the residuals from the least squares fit is

$$\widehat{\widetilde{e}}_j = \left(\mathbf{I}_{m_j} - \mathbf{P}_j\right)\widetilde{\mathbf{Y}}_j. \tag{8.14}$$

Let $\mathrm{tr}\left(\widehat{\widetilde{e}}_j \widehat{\widetilde{e}}_{j^*}^T\right)$ be the trace of $\widehat{\widetilde{e}}_j \widehat{\widetilde{e}}_{j^*}^T$, i.e., the sum of the diagonal elements of $\widehat{\widetilde{e}}_j \widehat{\widetilde{e}}_{j^*}^T$, for any $j, j^* = 1, \ldots, J$. Conditioning on the covariate set $\mathscr{D}$, the expectation of $\mathrm{tr}\left(\widehat{\widetilde{e}}_j \widehat{\widetilde{e}}_{j^*}^T\right)$ is

$$E\left[\mathrm{tr}\left(\widehat{\widetilde{e}}_j \widehat{\widetilde{e}}_{j^*}^T\right)\Big|\mathscr{D}\right] = \mathrm{tr}\left[\left(\mathbf{I}_{m_j} - \mathbf{P}_j\right)\mathbf{M}_{jj^*}^T \left(\mathbf{I}_{m_j} - \mathbf{P}_j\right)^T\right]\gamma(t_j, t_{j^*}), \tag{8.15}$$

where $\gamma(t_j, t_{j^*}) = \sigma^2(t_j)$ if $j = j^*$, and $\gamma(t_j, t_{j^*}) = \rho(t_j, t_{j^*})$ if $j \neq j^*$. If the trace $\mathrm{tr}\left[\left(\mathbf{I}_{m_j} - \mathbf{P}_j\right)\mathbf{M}_{jj^*}^T \left(\mathbf{I}_{m_j} - \mathbf{P}_j\right)^T\right] \neq 0$, it follows from (8.15) that

$$\gamma(t_j, t_{j^*}) = E\left[\mathrm{tr}\left(\widehat{\widetilde{e}}_j \widehat{\widetilde{e}}_{j^*}^T\right)\Big|\mathscr{D}\right] \Big/ \mathrm{tr}\left[\left(\mathbf{I}_{m_j} - \mathbf{P}_j\right)\mathbf{M}_{jj^*}^T \left(\mathbf{I}_{m_j} - \mathbf{P}_j\right)^T\right].$$

Substituting $E\left[\mathrm{tr}\left(\widehat{\widetilde{e}}_j \widehat{\widetilde{e}}_{j^*}^T\right)\Big|\mathscr{D}\right]$ of the above equation with the observed trace of residuals $\mathrm{tr}\left(\widehat{\widetilde{e}}_j \widehat{\widetilde{e}}_{j^*}^T\right)$, a natural estimator of $\gamma(t_j, t_{j^*})$ is

$$\widehat{\gamma}(t_j, t_{j^*}) = \mathrm{tr}\left(\widehat{\widetilde{e}}_j \widehat{\widetilde{e}}_{j^*}^T\right) \Big/ \mathrm{tr}\left[\left(\mathbf{I}_{m_j} - \mathbf{P}_j\right)\mathbf{M}_{jj^*}^T \left(\mathbf{I}_{m_j} - \mathbf{P}_j\right)^T\right]. \tag{8.16}$$

In particular, when $j = j^*$, estimation of the variance $\sigma^2(t_j)$ requires the local sample size to be larger than the number of parameters in the model (8.4), i.e., $m_j > k+1$. The specific estimators of $\sigma^2(t_j)$ and $\rho(t_j, t_{j^*})$ are

$$\begin{cases} \widehat{\sigma}^2(t_j) = \widehat{\gamma}(t_j, t_j) = \mathrm{tr}\left(\widehat{\widetilde{e}}_j \widehat{\widetilde{e}}_{j^*}^T\right) \Big/ (m_j - k - 1), & \text{if } j = j^*, \\ \widehat{\rho}(t_j, t_{j^*}) = \widehat{\gamma}(t_j, t_{j^*}), & \text{if } j \neq j^*. \end{cases} \tag{8.17}$$

The estimators of

$$Cov\left[\widehat{\boldsymbol{\beta}}^{RAW}(t_j)\big|\mathscr{D}\right], \quad Var\left[\widehat{\beta}_l^{RAW}(t_j)\big|\mathscr{D}\right] \quad \text{and} \quad Cov\left[\widehat{\beta}_l^{RAW}(t_j), \widehat{\beta}_l^{RAW}(t_{j^*})\big|\mathscr{D}\right]$$

are obtained by substituting $\sigma^2(t_j)$ and $\rho(t_j, t_{j^*})$ in (8.9), (8.11) and (8.12) with their corresponding estimators $\widehat{\sigma}^2(t_j)$ and $\widehat{\rho}(t_j, t_{j^*})$ in (8.17). These variance and covariance estimators are useful for deriving the statistical properties of the final smoothing estimators of $\beta(t)$.

8.3   Refining the Raw Estimates by Smoothing

When the local sample size m_j is sufficient large at the design time point t_j, the raw estimates described above are asymptotically unbiased and have the same theoretical properties as with the standard linear models. These estimates are not suitable for practical use and need to be refined by a smoothing step.

8.3.1   Rationales for Refining by Smoothing

There are four reasons to refine the raw estimates through an appropriate smoothing procedure:

(a) Since the raw estimates are obtained at each of the design time points $\{t_j : j = 1, \ldots, J\}$, they are generally not smooth from one time point to the next. In addition to the temporal trends across the time points, the variations of the raw estimates between different time points are in part caused by the random errors.

(b) The raw estimates are generally not efficient for any design time point t_j, because the information across different time points is ignored and only the subjects observed within a neighborhood of t_j are used. Given that the observations from a nearby time point t_{j^*} may provide useful information about the value of $\beta(t_j)$, more efficient estimators of $\beta(t_j)$ may be obtained by utilizing the observations from neighboring design time points.

(c) It is possible that there are not enough subjects observed at some design time points, so that the raw estimates at these time points are missing. In such cases, it is natural to use the raw estimates from the neighboring design time points to estimate the coefficient curves at the time points with missing raw estimates.

(d) Since the statistical objective is to estimate the coefficient curves $\beta(t)$, raw estimates from the design time points do not give direct estimates of $\beta(t)$ when $t \in \mathcal{T}$ is not a design time point, i.e., $t \notin \mathbf{T}$. A smoothing step using the raw estimates near t is a natural approach to construct an estimator of $\beta(t)$. □

8.3.2   The Smoothing Estimation Step

Following Chapters 3 to 5, several smoothing methods can be applied to the raw estimates to obtain a smoothing estimator $\widehat{\beta}_l(t)$ of $\beta_l(t)$, $0 \leq l \leq k$, for all t within the time range $\mathcal{T}$. Using $\{\widehat{\beta}_l^{RAW}(t_j) : 0 \leq l \leq k; j = 1, \ldots, J\}$ as the pseudo-observations, the smoothing estimators are linear functions of these raw estimates.

1. Expression of the Smoothing Estimators

Suppose that $\beta_l(t)$ is $(p+1)$-times differentiable. A linear estimator of the qth derivative $\beta_l^{(q)}(t)$ of $\beta_l(t)$ for any $0 \leq q < p+1$ is given by the form

$$\widehat{\beta}_l^{(q)}(t) = \sum_{j=1}^{J} w_{l,q}(t_j, t) \, \widehat{\beta}_l^{RAW}(t_j), \qquad (8.18)$$

where $w_{l,q}(t_j, t)$ are the weights constructed by various smoothing methods, such as, splines and local polynomials. Setting $q = 0$, the smoothing estimator of $\beta_l(t)$ from (8.18) is

$$\widehat{\beta}_l(t) = \widehat{\beta}_l^{(0)}(t) = \sum_{j=1}^{J} w_{l,0}(t_j, t)\, \widehat{\beta}_l^{RAW}(t_j). \tag{8.19}$$

In most real applications, we are concerned with the estimation of $\beta_l(t)$, rather than its derivatives. Only in rare situations, such as estimating the changes or curvatures of $\beta_l(t)$, the derivative estimates $\widehat{\beta}_l^{(q)}(t)$ are required. For practical purposes, we focus on the smoothing estimators of $\beta_l(t)$, although the derivative estimates of (8.18) are presented as a theoretical possibility.

2. The Two-Step Local Polynomial Estimators

For a specific estimator, the choice of weights $w_{l,q}(t_j, t)$ in (8.18) has to be selected. The only two-step smoothing method that has been systematically studied in the literature is the local polynomial fitting by Fan and Zhang (2000). For any $t_j \in \mathbf{T}$ and $t \in \mathscr{T}$, let $\{h, K(\cdot)\}$ be the bandwidth and kernel function, $K_h(t) = K(t/h)/h$, and

$$\begin{cases} C_j &= \left(1, t_j - t, \ldots, (t_j - t)^p\right)^T, \; j = 1, \ldots, J, \\ \mathbf{C} &= \left(C_1, C_2, \ldots, C_J\right)^T, \\ W_j &= K_h(t_j - t), \; \mathbf{W} = \mathrm{diag}\left(W_1, \ldots, W_J\right). \end{cases} \tag{8.20}$$

The pth order local polynomial estimator of $\beta_l^{(q)}(t)$, $0 \le q < p+1$, is derived from (8.18) with the weights

$$w_{l,q}(t_j, t) = q!\, e_{q+1,p+1}^T \left(\mathbf{C}^T \mathbf{W} \mathbf{C}\right)^{-1} C_j W_j, \tag{8.21}$$

and has the expression

$$\widehat{\beta}_l^{(q)}(t) = \sum_{j=1}^{J} \left\{ \left[q!\, e_{q+1,p+1}^T \left(\mathbf{C}^T \mathbf{W} \mathbf{C}\right)^{-1} C_j W_j\right] \widehat{\beta}_l^{RAW}(t_j) \right\}. \tag{8.22}$$

The pth order local polynomial estimator of $\beta_l(t)$ is $\widehat{\beta}_l^{(0)}(t)$.

Taking $p = 1$ and $C_j = \left(1, t_j - t\right)^T$ in (8.21), the local linear estimator of $\beta_l(t)$ is derived from (8.19) with the weights

$$w_{l,0}(t_j, t) = e_{1,2}^T \left(\mathbf{C}^T \mathbf{W} \mathbf{C}\right)^{-1} C_j W_j, \tag{8.23}$$

and has the expression

$$\widehat{\beta}_l^L(t) = \widehat{\beta}_l^{(0)}(t) = \sum_{j=1}^{J} \left\{ \left[e_{1,2}^T \left(\mathbf{C}^T \mathbf{W} \mathbf{C}\right)^{-1} C_j W_j\right] \widehat{\beta}_l^{RAW}(t_j) \right\}. \tag{8.24}$$

Although it is possible to use higher-order local polynomial estimators with $p \geq 2$, the local linear estimators (8.24) are most commonly used in practice. This is mostly due to the fact that the additional structural complicity of higher-order local polynomials often does not translate into better fits with real data.

The local polynomial estimators (8.22) and (8.24) are obtained by treating $\{\widehat{\beta}_l^{RAW}(t_j) : 0 \leq l \leq k; j = 1, \ldots, J\}$ as the pseudo-observations and minimizing

$$L_{l,p}(t) = \sum_{j=1}^{J} \left\{ \widehat{\beta}_l^{RAW}(t_j) - \sum_{q=0}^{p} \left[b_{lq}(t)(t_j - t)^q \right] \right\}^2 K\left(\frac{t_j - t}{h}\right) \tag{8.25}$$

with respect to $b_{lq}(t)$, where different bandwidth h and kernel $K(\cdot)$ may be used for each $\beta_l(t)$. If $\widehat{b}_{lq}(t)$ for $q = 0, \ldots, p$ uniquely minimize (8.25), then $\widehat{b}_{l0}(t) = \widehat{\beta}_l^{(0)}(t)$ is the two-step pth order local polynomial estimator of $\beta_l(t)$ given in (8.22), and, for $q \geq 1$, $(q!)\widehat{b}_{lq}(t) = \widehat{\beta}_l^{(q)}(t)$ is the local polynomial estimator of the qth derivative of $\beta_l(t)$ with respect to t.

3. The Two-Step Kernel Estimators

The special case of local constant fitting leads to the kernel smoothing estimator of $\beta_l(t)$, which is obtained by minimizing

$$L_l(t) = \sum_{j=1}^{J} \left[\widehat{\beta}_l^{RAW}(t_j) - b_l(t) \right]^2 K\left(\frac{t_j - t}{h}\right) \tag{8.26}$$

with respect to $b_l(t)$. The resulting two-step kernel smoothing estimator of $\beta_l(t)$ is

$$\widehat{\beta}_l^K(t) = \sum_{j=1}^{J} \left[\widehat{\beta}_l^{RAW}(t_j) K\left(\frac{t_j - t}{h}\right) \right] \bigg/ \sum_{j=1}^{J} K\left(\frac{t_j - t}{h}\right), \tag{8.27}$$

which corresponds to (8.25) with $p = 0$.

4. Some Remarks on the Smoothing Step

A major advantage of using the two-step smoothing estimators in (8.18), (8.19), (8.22), (8.24) and (8.27) is their computational simplicity, because the existing computing software for cross-sectional i.i.d. data can be readily used. The raw estimates can be computed by the standard software for linear models. The smoothing estimates at the second step can be computed using the standard smoothing software for nonparametric regression. The entire procedure does not depend on the correlation structures of the observations over time. This is possible because the statistical objective is to estimate the unknown coefficient curves at any time point $t \in \mathscr{T}$, while the correlation structures of (7.1) at different time points are not estimated.

8.3.3 Bandwidth Choices

Component curve bandwidths can be selected using the usual cross-validation method in nonparametric regression by deleting the raw estimates one design time point at a time. The main idea is to find a reasonable bandwidth such that the resulting smoothing estimator $\widehat{\beta}_l^{(0)}(t)$ has appropriate smoothness for the pseudo-observations $\{\widehat{\beta}_l^{RAW}(t_j) : 0 \le l \le k; j = 1, \ldots, J\}$.

Let $\widehat{\beta}_l^{(0, -j)}(t; h_l)$ be any estimator of $\beta_l(t)$ given in (8.22), (8.24) or (8.27) computed using the bandwidth h_l kernel $K_l(\cdot)$ and the pseudo-observations of raw estimates with the one at the time design point t_j deleted, that is, $\{\widehat{\beta}_l^{RAW}(t_{j^*}) : 0 \le l \le k; j^* \ne j; 1 \le j^* \le J\}$. The leave-one-time-point-out cross-validation (LTCV) score is defined by

$$LTCV_l(h_l) = \sum_{j=1}^{J} \left[\widehat{\beta}_l^{RAW}(t_j) - \widehat{\beta}_l^{(0, -j)}(t_j; h_l)\right]^2. \tag{8.28}$$

The cross-validated bandwidth $h_{l, LTCV}$ is the minimizer of $LTCV_l(h_l)$ with respect to h_l, provided that the unique minimizer of (8.28) exists. Heuristically, $LTCV_l(h_l)$ measures the predictive error of $\widehat{\beta}_l^{(0, -j)}(t_j; h_l)$ for $\widehat{\beta}_l^{RAW}(t_j)$.

In contrast to the leave-one-subject-out cross-validation (LSCV) of Section 7.2.5 for the one-step local smoothing estimators, the LTCV (8.28) is specific for each component curve $\beta_l(t)$ and can provide different smoothing needs for different $0 \le l \le k$. The LTCV bandwidth vector from (8.28) is $\mathbf{h}_{LTCV} = \left(h_{0, LTCV}, \ldots, h_{k, LTCV}\right)^T$. However, because the raw estimates $\widehat{\beta}_l^{RAW}(t_j)$ and $\widehat{\beta}_l^{RAW}(t_{j^*})$ for any $j \ne j^*$ are possibly correlated, the approximate relationship linking the ASE in (7.25) with the LSCV in (7.26) may not necessarily hold for the LTCV score (8.28). Practical properties of the cross-validated bandwidths $\mathbf{h}_{LTCV}$ have to be investigated through simulation studies.

8.4 Pointwise and Simultaneous Confidence Bands

Approximate inferences for $\beta(t)$ based on the asymptotic distributions under general assumptions of $\{n, n_i : i = 1, \ldots, n\}$ and $\{m_j : j = 1, \ldots, J\}$ are still not available for the smoothing estimators of Section 8.3. Similar to Section 7.3, a practical inference approach is to consider the resampling subject bootstrap. We describe the bootstrap pointwise confidence intervals and the simultaneous confidence bands for $\beta(t)$ constructed using the similar procedures in Section 7.3

8.4.1 Pointwise Confidence Intervals by Bootstrap

Let $\widehat{\beta}(t) = \left(\widehat{\beta}_0(t), \ldots, \widehat{\beta}_k(t)\right)^T$ be an estimator of $\beta(t)$ constructed using any of the estimation methods of Section 8.3. Let $A = \left(a_0, \ldots, a_k\right)^T$ be a known $(k+1)$ column vector, so that $A^T E\left[\widehat{\beta}(t)\right] = \sum_{l=0}^{k} a_l E\left[\widehat{\beta}_l(t)\right]$ is a linear combination of the components of $E\left[\widehat{\beta}(t)\right]$. The resampling-subject bootstrap procedure for an approximate $\left[100 \times (1-\alpha)\right]\%$ pointwise percentile interval has the following steps.

Approximate Bootstrap Pointwise Confidence Intervals:

(a) **Computing Bootstrap Estimators.** *Generate B independent boot-strap samples as in Section 7.3 and compute B bootstrap estimators* $\{A^T\widehat{\beta}_1^b(t), \ldots, A^T\widehat{\beta}_B^b(t)\}$.

(b) **Approximate Bootstrap Intervals.** *Calculate* $l_{A,\alpha/2}^b(t)$ *and* $u_{A,\alpha/2}^b(t)$, *the lower and upper* $[100 \times (\alpha/2)]th$ *percentiles, respectively, of the B bootstrap estimators. The approximate* $[100 \times (1-\alpha)]\%$ *percentile bootstrap confidence interval for* $A^T\beta(t)$ *is*

$$\left(l_{A,\alpha/2}^b(t), u_{A,\alpha/2}^b(t)\right). \tag{8.29}$$

The normal approximated bootstrap confidence interval for $A^T\beta(t)$ *is*

$$A^T\widehat{\beta}(t) \pm z_{1-\alpha/2} \times \widehat{se}\left(t; A^T\widehat{\beta}^b\right), \tag{8.30}$$

where $\widehat{se}\left(t; A^T\widehat{\beta}^b\right)$ *is the sample standard deviation of the B bootstrap estimates, which is given by*

$$\widehat{se}\left(t; A^T\widehat{\beta}^b\right) = \left\{\frac{1}{B-1}\sum_{s=1}^{B}\left[A^T\widehat{\beta}_s^b(t) - \frac{1}{B}\sum_{r=1}^{B}A^T\widehat{\beta}_r^b(t)\right]^2\right\}^{1/2}, \tag{8.31}$$

and $z_{1-\alpha/2}$ *is the* $[100 \times (1-\alpha/2)]th$ *percentile of the standard normal distribution.*   □

Since the smoothing estimator $\widehat{\beta}(t)$ has a small asymptotic bias for $\beta(t)$, the above procedure only gives approximate pointwise confidence intervals for $A^T\beta(t)$, which, strictly speaking, is an approximate variability band based on $A^T\widehat{\beta}(t)$. A strict pointwise confidence interval would require an unbiased estimator for the asymptotic bias of $\widehat{\beta}(t)$. The intervals (8.29) and (8.30) do not contain the required bias correction. When small bandwidths are used, the asymptotic bias of $\widehat{\beta}(t)$ is small, so that the coverage probabilities that the intervals (8.29) and (8.30) containing $A^T\beta(t)$ are close to $\left[100 \times (1-\alpha)\right]\%$. The asymptotic bias of $\widehat{\beta}(t)$ is often difficult to be accurately estimated. As a result, the "plug-in" bias correction of substituting the unknown asymptotic bias with their estimates may not improve the coverage probability over the simple variability bands of (8.29) and (8.30).

Pointwise confidence intervals for $\beta_r(t)$, $0 \leq r \leq k$, can be computed using (8.29) and (8.30) with A being the $(k+1)$ column vector having 1 at its $(r+1)$th place and 0 elsewhere. Pointwise confidence intervals for the difference of two component curves of $\beta(t)$ can be similarly constructed by taking the corresponding elements of A to be 1 and -1 and 0 elsewhere. For example, the pointwise bootstrap confidence interval for $[\beta_{r_1}(t) - \beta_{r_2}(t)]$ can be computed by (8.29) and (8.30) with A being the $(k+1)$ column vector having 1 and -1 at its $(r_1 + 1)$th and $(r_2 + 1)$th places and 0 elsewhere. Other special cases of (8.29) and (8.30) can be similarly constructed with appropriate choice of A.

8.4.2 Simultaneous Confidence Bands

The "bridging-the-gap" simultaneous confidence bands for $A^T\beta(t)$ over $t \in [a, b]$ can be similarly constructed as in Section 7.3.2. In this case, we let

$$\left(l_{A,\alpha/2}(t), u_{A,\alpha/2}(t) \right) \tag{8.32}$$

be the $[100 \times (1 - \alpha)]\%$ pointwise confidence interval in (8.29) or (8.30), and partition $[a, b]$ into M equally spaced intervals with grid points $a = \xi_1 < \cdots < \xi_{M+1} = b$, such that $\xi_{j+1} - \xi_j = (b - a)/M$ for $j = 1, \ldots, M$. A set of approximate $[100 \times (1 - \alpha)]\%$ simultaneous confidence intervals for $\{A^T\beta(\xi_j) : j = 1, \ldots, M+1\}$ is the intervals

$$\left\{ \left(L_{A,\alpha/2}(\xi_j), U_{A,\alpha/2}(\xi_j) \right) : j = 1, \ldots, M+1 \right\}, \tag{8.33}$$

such that

$$\lim_{n \to \infty} P\left\{ L_{A,\alpha/2}(\xi_j) \leq A^T\beta(\xi_j) \leq U_{A,\alpha/2}(\xi_j) \text{ for all } j = 1, \ldots, M+1 \right\} \geq 1 - \alpha.$$

The special case of (8.33) based on the Bonferroni adjustment is

$$\left(L_{A,\alpha/2}(\xi_j), U_{A,\alpha/2}(\xi_j) \right) = \left(l_{A,\alpha/[2(M+1)]}(\xi_j), u_{A,\alpha/[2(M+1)]}(\xi_j) \right). \tag{8.34}$$

Based on (8.33) and the interpolation

$$(A^T\beta)^{(I)}(t) = \left[\frac{M(\xi_{j+1} - t)}{b - a} \right] [A^T\beta(\xi_j)] + \left[\frac{M(t - \xi_j)}{b - a} \right] [A^T\beta(\xi_{j+1})], \tag{8.35}$$

a simultaneous confidence band for $(A^T\beta)^{(I)}(t)$ over $t \in [a, b]$ is

$$\left(L_{A,\alpha/2}^{(I)}(t), U_{A,\alpha/2}^{(I)}(t) \right), \tag{8.36}$$

where $L_{A,\alpha/2}^{(I)}(t)$ and $U_{A,\alpha/2}^{(I)}(t)$ are the corresponding linear interpolations of lower and upper bounds.

Using the same derivations as in (7.35) to (7.38), we obtain the following two types of simultaneous confidence bands. If

$$\sup_{t \in [a,b]} \left| (A^T\beta)'(t) \right| \leq c_1 \text{ for a known constant } c_1 > 0, \tag{8.37}$$

an approximate $[100 \times (1 - \alpha)]\%$ simultaneous confidence band for $A^T \beta(t)$ is

$$\left(L_{A,\alpha/2}^{(l)}(t) - 2c_1 \left[\frac{M(\xi_{j+1}-t)(t-\xi_j)}{b-a} \right], \right.$$
$$\left. U_{A,\alpha/2}^{(l)}(t) + 2c_1 \left[\frac{M(\xi_{j+1}-t)(t-\xi_j)}{b-a} \right] \right). \tag{8.38}$$

If the second derivative of $A^T \beta(t)$ is bounded,

$$\sup_{t \in [a,b]} \left| (A^T \beta)''(t) \right| \le c_2 \quad \text{for a known constant } c_2 > 0, \tag{8.39}$$

an approximate $[100 \times (1 - \alpha)]\%$ simultaneous confidence band is

$$\left(L_{A,\alpha/2}^{(l)}(t) - \frac{c_2}{2} \left[\frac{M(\xi_{j+1}-t)(t-\xi_j)}{b-a} \right], \right.$$
$$\left. U_{A,\alpha/2}^{(l)}(t) + \frac{c_2}{2} \left[\frac{M(\xi_{j+1}-t)(t-\xi_j)}{b-a} \right] \right). \tag{8.40}$$

8.5 R Implementation

8.5.1 The NGHS BP Data

The NGHS data has been described in Section 1.2. In Chapter 7, we used a time-varying coefficient model and the one-step local smoothing method to evaluate the effect of race, height and body mass index (BMI) on the systolic blood pressure (SBP). We illustrate here the two-step smoothing method for the coefficient curves $\beta(t)$ in the same varying coefficient model

$$Y_{ij} = \beta_0(t_{ij}) + \beta_1(t_{ij}) X^{(1)} + \beta_2(t_{ij}) X^{(2)}(t_{ij}) + \beta_3(t_{ij}) X^{(3)}(t_{ij}) + \varepsilon_{ij}, \tag{8.41}$$

where, for the ith NGHS girl, t_{ij} denotes her age (in years) at the jth study visit, Y_{ij} and ε_{ij} are her SBP level and random measurement error at age t_{ij}, $X^{(1)}$ denotes her race (1 indicates African American, 0 indicates Caucasian), and $X^{(2)}(t_{ij})$ and $X^{(3)}(t_{ij})$ are the girl's age-adjusted height percentile and BMI percentile at age t_{ij}. In this model, $\beta_0(t)$ represents the baseline SBP curve, i.e., the mean time curve of SBP for Caucasian girls with the median height and the median BMI, and $\beta_l(t)$, $l = 1, 2, 3$, represent the effects of race, height and BMI, respectively, on the SBP during the adolescence years.

   To apply the two-step smoothing estimation procedure of Sections 8.2 and 8.3, we first obtain the set of design time points. The original observed visit times t_{ij} spread from 9 to 19 years for the 2379 NGHS girls and many visit times only have 1 to 2 observations. To use the time-point binning of Section 8.1, we round off the adjacent study visit times to the nearest tenth, so that the resulting set of design time points consists of 100 small time bins, $\mathbf{T} = \{t_1 = 9.0, t_2 = 9.1, \ldots, t_{100} = 18.9\}$. The number of observations at time t_j, $j = 1, \ldots, 100$, range from 77 to 277 with no more than one observation per subject at each time t_j, since each participant was only scheduled to have one visit per year.

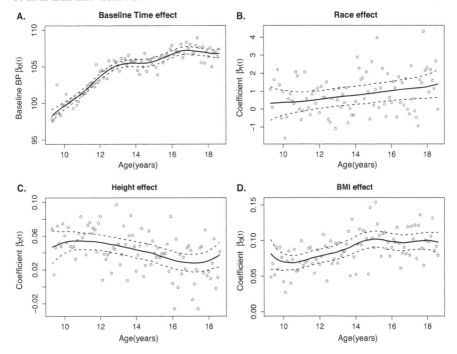

Figure 8.1 *Covariate effects on systolic blood pressure of NGHS girls. The scatter plots and the solid curves in (A)-(D) show the raw estimates, the two-step local linear estimates for the coefficient curves $\widehat{\beta}_l(t)$, $l = 0, 1, 2, 3$, respectively, with the cross-validated bandwidths. The dashed curves indicate the corresponding 95% pointwise bootstrap confidence intervals.*

As discussed in Sections 8.1 and 8.2, an advantage of the two-step approach is the computational simplicity. We can apply the standard statistical procedures to compute the raw estimates in the first step and then obtain the final smoothing estimates over time using any smoothing method described in Section 8.3. For this NGHS example, at each distinct design time t_j, the raw estimates $\beta_l(t_j)$, $l = 0, 1, 2, 3$ for the model (8.41) are the coefficients of a standard linear regression model. Figure 8.1 shows these raw estimates for $\hat{\beta}_l^{RAW}(t_j)$, $l = 0, 1, 2, 3$ computed using the lm() in R and the two-step local linear smoothing estimates using the Epanechnikov kernel and the cross-validated bandwidths.

Unlike the one-step procedure in Chapter 7, we can observe the general time-trend based on the scatter plots of the raw estimates and apply different amounts of smoothing or use different bandwidths for the four components of covariate coefficients. Following Section 8.2.3, we use the leave-one-time-point-out cross-validation (LTCV) for each covariate effect and select the bandwidth $\mathbf{h}_{LTCV} = (1.3, 4.0, 2.4, 1.5)^T$ for the coefficient curves of the baseline, race, height and BMI, respectively. For statistical inference, we compute the 95% pointwise confidence intervals based on the resampling-subject boot-strap of Section 8.4 with 1000 replications. To ease the computational burden, the cross-validated bandwidths obtained from the original sample were used for the bootstrap standard error bands.

Compared to the one-step smoothing estimates in Chapter 7, the coefficient curves in Figure 8.1 estimated by the two-step local smoothing method suggest similar findings for these covariate effects on the longitudinal SBP levels of the NGHS girls. Perhaps most importantly, the results obtained from both methods demonstrate that these covariate effects are not constant over time. The baseline coefficient curve shows that the mean SBP for the Caucasian girls with the median height and the median BMI increases with the girls' age. The coefficient curves for race and BMI percentile are all positive and gradually increase with age, while the coefficient curve for height percentile indicates that there is a positive association between height and SBP and the height effect decreases with age.

We note that, for a nonparametric time-varying coefficient model, one may use either the one-step or the two-step local smoothing methods to estimate the coefficient curves. These two estimation methods are generally expected to yield similar results. In practice, for long-term longitudinal cohort studies, such as the NGHS, the results from these estimation approaches would provide useful insight into the covariate effects on certain disease risk factors, which could lead to significant implications for tracking these risk factors over time and developing future intervention strategies to reduce the disease risks.

8.6   Remark on the Asymptotic Properties

Asymptotic properties, such as consistency and the asymptotic bias and vari-ance expressions, for the two-step local polynomial estimator $\widehat{\beta}_l^{(q)}(t)$ of (8.22) have been presented in Fan and Zhang (2000). Thus, we focus in this chap-ter on the practical issues of the two-step estimation method, and omit the presentation of its asymptotic properties. Detailed asymptotic derivations are referred to in Fan and Zhang (2000). Technically, the theoretical derivations of these asymptotic properties are similar to the two-step smoothing estimators to be presented in Chapters 13 and 14 for the time-varying transformation models. In these chapters, as well as the derivations of Fan and Zhang (2000), the asymptotic properties are established by first deriving the bias, variance and covariance expressions of the raw estimators and then computing the asymptotic expressions using the fact that the smoothing estimators are lin-ear statistics of the corresponding raw estimates. This technical approach is repeatedly used in Sections 13.5.4 and 14.5.4.

8.7   Remarks and Literature Notes

This chapter presents a two-step smoothing procedure for estimating the co-efficient curves of the time-varying coefficient model (7.1). The main idea of a two-step method is to first fit a linear model at the grid time design points $\mathbf{T}$ so that the raw coefficient estimates can be obtained at the distinct time points of $\mathbf{T}$, and then smooth these raw coefficient estimates over the time range. Compared with the one-step estimation methods for the same model in Chapter 7, a two-step method is intuitive and has the advantages of being computationally simple and having the flexibility of allowing different smoothing parameters for different coefficient curves in (7.1).

Although our presentation is focused on the kernel-type local polynomial estimators, it is clear that the smoothing estimators of the two-step procedure may also be computed using other smoothing methods, such as the B-splines and the penalized smoothing splines. However, statistical properties for the two-step estimation procedures of (7.1) based on splines have not been investigated in the literature, which remains to be a worthy topic of further statistical research. In practice, both the one-step and two-step smoothing procedures should be considered, as each of these procedures has its own advantages and disadvantages. For example, the two-step smoothing procedure on one hand has the extra flexibility than the one-step smoothing procedure, because it allows different bandwidths for different components of $\beta(t)$. On the other hand, the appropriateness of the two-step smoothing estimators depends on the time-point binning choices and the raw estimators. The two-step smoothing procedure can be essentially viewed as a favorable alternative to the one-step smoothing procedure.

Computationally, both the one-step and the two-step smoothing methods can be easily implemented in practice. Appropriate confidence intervals of the smoothing estimators can be computed using the resampling-subject bootstrap procedure. Although theoretical properties of the bootstrap confidence intervals still need further investigation, these confidence intervals have been shown in the literature to have appropriate coverage probabilities. Results of Sections 8.2, 8.3 and 8.4 are based on Fan and Zhang (2000).

Chapter 9

Global Smoothing Methods

We describe in this chapter a class of global smoothing methods based on basis approximations for estimating the coefficient curves of the time-varying coefficient model (7.1). These basis approximation smoothing methods are natural extensions of the methods in Chapter 4 to models with time-varying covariates. Compared with the local smoothing methods in Chapters 7 and 8, the basis approximation methods can be viewed as a flexible extension of the classical parametric regression models described in Section 2.1 to nonparametric models, which have also been referred to in the literature as extended linear models, for example, Stone et al. (1997).

9.1 Basis Approximation Model and Interpretations

Intuitively, a basis approximation intends to express an unknown coefficient function through a linear combination of some basis functions, and the linear coefficients are parameters which determine the shape of the coefficient function. Thus, on one hand, when the number of basis functions is fixed, the unknown coefficient curve is approximated by a parametric model. On the other hand, the model has the flexibility to accommodate the specific shape of the coefficient function by increasing the number of basis functions. This natural link between a linear model and a nonparametric regression model makes the basis approximation smoothing method a useful alternative to the local smoothing methods of Chapters 7 and 8 in real applications.

9.1.1 Data Structure and Model Formulation

Following the general setup of Chapter 7, including the data structure and the time-varying coefficient models, we consider the stochastic processes $\{(Y(t), t, \mathbf{X}^T(t)) : t \in \mathscr{T}\}$, where $Y(t)$ is a real-valued response process, $\mathbf{X}(t) = (X^{(0)}(t), \ldots, X^{(k)}(t))^T$ is a R^{k+1}-valued covariate process, and $\mathscr{T}$ is the range of time points. The realizations of these processes for each subject are obtained at a set of distinct and possibly irregularly spaced time points in the fixed interval $\mathscr{T}$. The statistical interest is focused on evaluating the mean effects of t and $\mathbf{X}(t)$ on the response $Y(t)$. A longitudinal sample of $\{(Y(t), t, \mathbf{X}^T(t)) : t \in \mathscr{T}\}$ from n randomly selected subjects is denoted by $\left\{\left(Y_{ij}, t_{ij}, \mathbf{X}_{ij}^T\right) : i = 1, \ldots, n; j = 1, \ldots, n_i\right\}$, where t_{ij} is the time of the jth measurement for the ith subject, n_i is the number of repeated measurements of the ith subject, and $Y_{ij} = Y_i(t_{ij})$ and $\mathbf{X}_{ij} = \left(X_{ij}^{(0)}, \ldots, X_{ij}^{(k)}\right)^T$ are the ith subject's observed outcome and covariates at time t_{ij}. The total number of observations in the sample is $N = \sum_{i=1}^{n} n_i$. As in regression models where a baseline effect is desired, the baseline for $\mathbf{X}(t)$ is set by $X^{(0)}(t) = X_{ij}^{(0)} = 1$. The time-varying coefficient model (7.1) for the longitudinal sample $\left\{\left(Y_{ij}, t_{ij}, \mathbf{X}_{ij}^T\right) : i = 1, \ldots, n; j = 1, \ldots, n_i\right\}$ is

$$Y_{ij} = \mathbf{X}_{ij}^T \boldsymbol{\beta}(t_{ij}) + \varepsilon_{ij}, \tag{9.1}$$

where, for all $t \in \mathcal{T}$, $\boldsymbol{\beta}(t) = \left(\beta_0(t), \ldots, \beta_k(t)\right)^T$ are smooth functions of t, $\varepsilon_{ij} = \varepsilon_i(t_{ij})$ is a realization of the zero-mean stochastic process $\varepsilon(t)$ as defined in (7.1), and $\mathbf{X}_{ij}$ and $\varepsilon_i = \left(\varepsilon_{i1}, \ldots, \varepsilon_{in_i}\right)^T$ are independent.

9.1.2   Basis Approximation

For local smoothing methods, the curves in the coefficient vector $\boldsymbol{\beta}(t) = \left(\beta_0(t), \ldots, \beta_k(t)\right)^T$ are estimated either by the one-step local smoothing method in Chapter 7 or the two-step local smoothing method in Chapter 8. For the basis approximation approach, we approximate each $\beta_l(t)$ in $\boldsymbol{\beta}(t)$ by a basis function expansion

$$\beta_l(t) \approx \sum_{s=0}^{K_l} \left[\gamma_{ls}^* B_{ls}(t)\right], \quad t \in \mathcal{T}, \tag{9.2}$$

where $\left\{B_{ls}(t) : s = 1, \ldots, K_l\right\}$ is a set of basis functions, such as polynomial bases, Fourier bases or B-splines (polynomial splines). Since the basis functions

$$\left\{B_{ls}(t) : l = 0, \ldots, k; s = 1, \ldots, K_l\right\},$$

are known, when the approximation sign in (9.2) is replaced by the equality sign, $\beta_l(t) = \sum_{s=0}^{K_l} \left[\gamma_{ls}^* B_{ls}(t)\right]$ for all $l = 0, \ldots, k$ lead to a multiple linear model in (9.1). But, in general, when $\beta_l(t)$ for some $0 \le l \le k$ are unknown functions, (9.2) only holds for an approximate sign "$\approx$" when the number of basis functions K_l is sufficiently large, so that the resulting model (9.1) is a structured nonparametric model, which is only approximated by a multiple linear model.

9.1.3   Remarks on Estimation Methods

The estimation methods of this chapter are concerned with the general structured nonparametric model (9.1) when (9.2) holds for some K_l, $l = 0, \ldots, k$, which may increase when the sample size n increases. This estimation relies on a simple one-step "global" procedure, in which the estimator of $\beta_l(t)$ for all $t \in \mathcal{T}$ is determined by the estimators of the coefficients γ_{ls}^*, and no binning of data as described in Chapter 8 is needed when the observations are sparse at distinct time points. Different amounts of smoothing can be used for different individual coefficient curves. The coefficients in the basis expansion, γ_{ls}^*, are estimated by least squares. The numbers of basis terms, $\{K_l : l = 0, \ldots, k\}$, play the role of smoothing parameters, and are selected using a leave-one-subject-out cross-validation procedure. By taking $B_{l1}(t) = 1$ to be the basis in the expansion (9.2) if $\beta_l(t)$ is known to be a constant, a basis approximation method suggests a non-iterative solution to the partially linear model discussed in Section 2.3.

In the main results of this chapter, we first describe the estimation methods based on both the general basis approximations and the special case of B-splines. Because of its popularity in applications, we describe the real data example and computing procedures exclusively for the B-spline method. But, in order to demonstrate the feasibility of basis approximations as a general global smoothing method, the asymptotic properties and theoretical derivations include both general basis choices and B-splines. Our goal is to show that, despite the popularity of B-splines in the literature, other basis functions should also be considered if they are deemed appropriate for a particular situation.

9.2   Estimation Method

9.2.1   Approximate Least Squares

We assume that the approximation (9.2) holds for a set of basis functions and constants $\{B_{ls}(t), \gamma_{ls}^* : l = 0, \ldots, k; s = 1, \ldots, K_l\}$. The basis approximation of the time-varying coefficient model (9.1) is given by

$$\begin{cases} Y_{ij} \approx Y_{ij}^* \\ Y_{ij}^* = \sum_{l=0}^{k} \sum_{s=1}^{K_l} \left[X_{ij}^{(l)} \gamma_{ls}^* B_{ls}(t_{ij}) \right] + \varepsilon_{ij}. \end{cases} \tag{9.3}$$

In (9.3), Y_{ij}^* is a pseudo-observation which represents the ith subject's "observation" at time t_{ij} if $\sum_{l=0}^{k} \sum_{s=1}^{K_l} \left[X_{ij}^{(l)} \gamma_{ls}^* B_{ls}(t_{ij}) \right]$ is the true mean of the process at time t_{ij}. Since Y_{ij}^* is not actually observed but approximates the real observed outcome Y_{ij}, we can estimate the coefficients γ_{ls}^* of (9.3) by minimizing the approximate square error

$$\ell(\gamma) = \sum_{i=1}^{n} \sum_{j=1}^{n_i} \left\{ w_i \left[Y_{ij} - \sum_{l=0}^{k} \sum_{s=1}^{K_l} X_{ij}^{(l)} B_{ls}(t_{ij}) \gamma_{ls} \right]^2 \right\} \tag{9.4}$$

with respect to the γ_{ls}, where w_i is a known non-negative weight satisfying $\sum_{i=1}^{n} (n_i w_i) = 1$ for the ith subject, $\gamma = (\gamma_0^T, \ldots, \gamma_k^T)^T$ and $\gamma_l = (\gamma_{l1}, \ldots, \gamma_{lK_l})^T$. We define $\widehat{\gamma} = (\widehat{\gamma}_0^T, \ldots, \widehat{\gamma}_k^T)^T$ with $\widehat{\gamma}_l = (\widehat{\gamma}_{l1}, \ldots, \widehat{\gamma}_{lK_l})^T$ for $l = 0, \ldots, k$ to be the approximate least squares estimator of $\gamma = (\gamma_0^T, \ldots, \gamma_k^T)^T$ if (9.4) is uniquely minimized by $\widehat{\gamma}$. Here $\widehat{\gamma}$ is referred to as an approximate least squares estimator because Y_{ij} is used in (9.4) instead of the unobserved pseudo-observation Y_{ij}^*. Using (9.2) and $\widehat{\gamma}_l$, it is natural to estimate $\beta_l(t)$ by

$$\widehat{\beta}_l(t) = \sum_{s=1}^{K_l} \left[\widehat{\gamma}_{ls} B_{ls}(t)\right], \tag{9.5}$$

which we refer to as the *least squares basis approximation estimator* of $\beta_l(t)$ based on $\{B_{ls}(t), \gamma_{ls}^* : l = 0, \ldots, k; s = 1, \ldots, K_l\}$.

To give the explicit expressions for $\widehat{\gamma}$ and $\widehat{\beta}_l(t)$, we define the following matrices from the basis functions, for $i = 1, \ldots, n$ and $j = 1, \ldots, n_i$,

$$\begin{cases} \mathbf{B}(t) = \begin{pmatrix} B_{01}(t) & \cdots & B_{0K_0}(t) & 0 & \cdots & 0 & 0 & \cdots & 0 \\ & \vdots & & & \vdots & & & \vdots & \\ 0 & \cdots & 0 & 0 & \cdots & 0 & B_{k1}(t) & \cdots & B_{kK_k}(t) \end{pmatrix}, \\ \mathbf{U}_i(t_{ij}) = \left[X_{ij}^T \mathbf{B}(t_{ij})\right]^T, \\ \mathbf{U}_i = \left(\mathbf{U}_i(t_{i1}), \ldots, \mathbf{U}_i(t_{in_i})\right)^T, \\ Y_i = \left(Y_{i1}, \ldots, Y_{in_i}\right)^T. \end{cases} \tag{9.6}$$

Then (9.4) under the matrix representation (9.6) is

$$\ell(\gamma) = \sum_{i=1}^{n} \left[(Y_i - \mathbf{U}_i\gamma)^T \mathbf{W}_i (Y_i - \mathbf{U}_i\gamma)\right], \tag{9.7}$$

where $\mathbf{W}_i = diag(w_i, \ldots, w_i)$ is the $n_i \times n_i$ diagonal matrix with w_i as its diagonal elements and 0 elsewhere. If $\sum_{i=1}^{n} (\mathbf{U}_i^T \mathbf{W}_i \mathbf{U}_i)$ is invertible, the approximate least squares estimator $\widehat{\gamma}$ is uniquely defined by

$$\widehat{\gamma} = \left[\sum_{i=1}^{n} (\mathbf{U}_i^T \mathbf{W}_i \mathbf{U}_i)\right]^{-1} \left[\sum_{i=1}^{n} (\mathbf{U}_i^T \mathbf{W}_i Y_i)\right]. \tag{9.8}$$

The least squares basis approximation estimator of $\beta(t)$ under the matrix representation (9.6) is

$$\widehat{\beta}(t) = \left(\widehat{\beta}_0(t), \ldots, \widehat{\beta}_k(t)\right)^T = \mathbf{B}(t)\widehat{\gamma}. \tag{9.9}$$

If the equality sign holds for (9.2), i.e., $Y_{ij} = Y_{ij}^*$ holds in (9.3), $\widehat{\beta}(t)$ reduces to the least squares estimator of $\beta(t)$ from a multiple linear model. This implies that the statistical properties of (9.9) depend on the functional space which contains the true vector of coefficient curves $\beta(t)$. The next section gives some further discussion of the issues related to the assumptions of functional spaces and basis choices.

9.2.2   Remarks on Basis and Weight Choices

We provide a few remarks on linear functional spaces for approximating $\beta_l(t)$, the choices of basis functions and the weights w_i used in (9.4).

1. Linear Functional Spaces

Let $\mathbb{G}_l$ be the linear functional space spanned by the basis functions $\{B_{l1}(t), \ldots, B_{lK_l}(t)\}$. The real coefficient curve $\beta_l(t)$ may or may not belong to $\mathbb{G}_l$. If $\beta_l(t)$ belongs to $\mathbb{G}_l$, then (9.2) holds with an equality sign, and $\beta_l(t)$ belongs to a multiple linear model. More generally, $\beta_l(t)$ does not necessarily belong to $\mathbb{G}_l$, so that, for a real study, we could only assume that (9.2) holds with sufficiently large K_l. Fortunately, in many biomedical studies, it is reasonable to assume that $\beta_l(t)$, $l = 0, \ldots, k$, are sufficiently smooth functions (see Section 9.5 for formal assumptions), so that (9.2) holds for a number of popular basis choices provided that K_l is allowed to increase as the sample size n increases. Once the basis $\{B_{l1}(t), \ldots, B_{lK_l}(t)\}$ is chosen, $\mathbb{G}_l$ and the least squares basis approximation estimator $\widehat{\beta}_l(t)$ are uniquely determined. On the other hand, different sets of basis functions can be used to span the same linear functional space $\mathbb{G}_l$, and thus give the same estimator $\widehat{\beta}_l(t)$, although the corresponding least squares estimator $\widehat{\gamma}$ may be different. For example, both the B-spline basis and the truncated power basis can be used to span a space of spline functions.

Any basis system for function approximation can be used. The Fourier basis may be desirable when the underlying functions exhibit periodicity, and polynomials are familiar choices which can provide good approximations to smooth functions. However, these bases may not be sensitive enough to exhibit certain local features without using a large K_l. In this respect, polynomial splines such as B-splines are often desirable. Ideally, a basis should be chosen to achieve an excellent approximation using a relatively small value of K_l. Some general guidance on basis choices are described in Ramsay and Silverman (2005, Section 3.2.2). We demonstrate the practical example of this chapter using B-splines because they can exhibit local features and provide stable numerical solutions (de Boor, 1978, Ch. II).

2. Weight Choices

The choice of w_i in (9.4) may have a significant influence on the theoretical and practical properties of $\widehat{\gamma}$ and $\widehat{\beta}(t)$. As in Section 4.1, the choice $w_i^* = 1/(nn_i)$ corresponds to the subject uniform weight, while $w_i^{**} = 1/N$ corresponds to the measurement uniform weight. It is conceivable that an ideal choice of w_i may also depend on the intra-subject correlation structures of the data $\{(Y_{ij}, t_{ij}, \mathbf{X}_{ij}^T) : i = 1, \ldots, n; j = 1, \ldots, n_i\}$. However, because the actual correlation structures are often completely unknown in practice, the subject uniform weight $w_i^* = 1/(nn_i)$ may be appropriate if the numbers of repeated measurements vary significantly for different subjects, while the measurement uniform weight $w_i^{**} = 1/N$ appears to be a practical choice if n_i, $i = 1, \ldots, n$, are relatively similar. Some theoretical implications of the choices for w_i are discussed later in Section 9.5.

9.2.3 Least Squares B-Spline Estimators

As discussed in Chapter 4 for evaluating the time-trend of $Y(t)$ without covariates, polynomial splines, or equivalently B-splines, are piecewise polynomials with the polynomial pieces jointing together smoothly at a set of interior knot points. A polynomial spline of degree $d \geq 0$ on $\mathscr{T}$ with knot sequence $\xi_0 < \xi_1 < \ldots < \xi_{M+1}$, where ξ_0 and ξ_{M+1} are the two end points of the interval $\mathscr{T}$, is a function that is a polynomial of degree d on each of the intervals $[\xi_m, \xi_{m+1})$, $0 \leq m \leq M - 1$, and $[\xi_M, \xi_{M+1}]$, and globally has continuous $d - 1$ continuous derivatives for $d \geq 1$. A piecewise constant function, linear spline, quadratic spline and cubic spline corresponds to $d = 0, 1, 2$ and 3, respectively. The collection of spline functions of a particular degree and knot sequence form a linear space. Good references for spline functions include de Boor (1978) and Schumaker (1981).

1. B-Splines Least Squares Criteria

Suppose that each $\beta_l(t)$, $l = 0, \ldots, k$, can be approximated by a set of B-spline basis functions $\mathscr{B}_l^{(S)}(t) = \{B_{ls}^{(S)}(t) : t \in \mathscr{T}; s = 1, \ldots, K_l\}$, which spans the linear space

$$\mathbb{G}_l^{(S)} = \left\{ \text{ linear function space spanned by } \mathscr{B}_l^{(S)}(t) \right\}. \tag{9.10}$$

Once $\mathscr{B}_l^{(S)}(t)$ is selected with a fixed K_l, $\mathbb{G}_l^{(S)}$ is a linear space with a fixed degree and knot sequence. We then have the B-spline basis approximation

$$\beta_l(t) \approx \sum_{s=1}^{K_l} \left[\gamma_{lk}^* B_{ls}^{(S)}(t) \right]. \tag{9.11}$$

Because of its good numerical properties, the B-spline basis functions are used for the applications of this chapter. The approximation sign in (9.11) is replaced by a strict equality sign with a fixed and known K_l, when $\beta_l(t)$ belongs to the linear space $\mathbb{G}_l^{(S)}$. For the general case that $\beta_l(t)$ may not belong to $\mathbb{G}_l$, it is natural to allow K_l to increase with the sample size n, so that a more accurate approximation in (9.11) is obtained when the sample size increases.

Following (9.1), (9.3) and (9.11), we have the B-spline approximated time-varying coefficient model

$$\begin{cases} Y_{ij} \approx Y_{ij}^{(S)}, \\ Y_{ij}^{(S)} = \sum_{l=0}^{k} \sum_{s=1}^{K_l} \left[X_{ij}^{(l)} \gamma_{ls}^* B_{ls}^{(S)}(t_{ij}) \right] + \varepsilon_{ij}, \end{cases} \tag{9.12}$$

where $Y_{ij}^{(S)}$ is the pseudo-observation with conditional mean given by the B-spline linear model $\sum_{l=0}^{k} \sum_{s=1}^{K_l} \left[X_{ij}^{(l)} \gamma_{ls}^* B_{ls}^{(S)}(t_{ij}) \right]$. We can then estimate the B-spline coefficients γ_{lk}^*, hence $\beta_l(t)$, based on (9.12) with any given K_l by minimizing

$$\ell_S(\gamma) = \sum_{i=1}^{n} \sum_{j=1}^{n_i} \left\{ w_i \left[Y_{ij} - \sum_{l=0}^{k} \sum_{s=1}^{K_l} X_{ij}^{(l)} B_{lk}^{(S)}(t_{ij}) \gamma_{lk} \right]^2 \right\} \tag{9.13}$$

with respect to γ_{lk}, where w_i are the non-negative weights defined in (9.4), $\gamma = \left(\gamma_0^T, \ldots, \gamma_k^T \right)^T$ and $\gamma_l = \left(\gamma_{l1}, \ldots, \gamma_{lK_l} \right)^T$.

2. Matrix Expression of B-Spline Estimators

For the expressions of (9.13) and the corresponding estimators, we define the following matrices from the B-spline basis, for $i = 1, \ldots, n$ and $j = 1, \ldots, n_i$,

$$\begin{cases} \mathbf{B}^{(S)}(t) = \begin{pmatrix} B_{01}^{(S)}(t) & \cdots & B_{0K_0}^{(S)}(t) & 0 & \cdots & 0 & 0 & \cdots & 0 \\ & \vdots & & & \vdots & & & \vdots & \\ 0 & \cdots & 0 & 0 & \cdots & 0 & B_{k1}^{(S)}(t) & \cdots & B_{kK_k}^{(S)}(t) \end{pmatrix}, \\ \mathbf{U}_i^{(S)}(t_{ij}) = \left[X_{ij}^T \mathbf{B}^{(S)}(t_{ij}) \right]^T, \\ \mathbf{U}_i^{(S)} = \left(\mathbf{U}_i^{(S)}(t_{i1}), \ldots, \mathbf{U}_i^{(S)}(t_{in_i}) \right)^T, \\ Y_i = \left(Y_{i1}, \ldots, Y_{in_i} \right)^T. \end{cases}$$

$$\tag{9.14}$$

Similar to (9.7), $\ell_S(\gamma)$ of (9.13) can be expressed using the matrix representation

$$\ell_S(\gamma) = \sum_{i=1}^{n} \left[\left(Y_i - \mathbf{U}_i^{(S)} \gamma \right)^T \mathbf{W}_i \left(Y_i - \mathbf{U}_i^{(S)} \gamma \right) \right], \tag{9.15}$$

where $\mathbf{W}_i = diag(w_i, \ldots, w_i)$ is the $n_i \times n_i$ diagonal matrix with w_i as its diagonal elements and 0 elsewhere.

Since we are considering the special case of polynomial spline basis functions, the matrix $\sum_{i=1}^{n} \left[(\mathbf{U}_i^{(S)})^T \mathbf{W}_i \mathbf{U}_i^{(S)} \right]$ is invertible under some mild conditions. A simple set of conditions for $\sum_{i=1}^{n} \left[(\mathbf{U}_i^{(S)})^T \mathbf{W}_i \mathbf{U}_i^{(S)} \right]$ to be invertible is given in Lemma 9.3 of Section 9.5. Then, under the conditions in Lemma 9.3, $\ell_S(\gamma)$ has a unique minimizer

$$\begin{aligned} \hat{\gamma}^{(S)} &= \left((\hat{\gamma}_0^{(S)})^T, \ldots, (\hat{\gamma}_k^{(S)})^T \right)^T \\ &= \left\{ \sum_{i=1}^{n} \left[(\mathbf{U}_i^{(S)})^T \mathbf{W}_i \mathbf{U}_i^{(S)} \right] \right\}^{-1} \left\{ \sum_{i=1}^{n} \left[(\mathbf{U}_i^{(S)})^T \mathbf{W}_i Y_i \right] \right\}, \end{aligned} \tag{9.16}$$

where, for $l = 0, \ldots, k$, $\hat{\gamma}_l^{(S)} = \left(\hat{\gamma}_{l1}^{(S)}, \ldots, \hat{\gamma}_{lK_l}^{(S)} \right)^T$. The polynomial spline estimator of $\beta(t)$ is

$$\widehat{\beta}^{(S)}(t) = \left(\widehat{\beta}_0^{(S)}(t), \ldots, \widehat{\beta}_k^{(S)}(t)\right)^T = \mathbf{B}^{(S)}(t)\,\widehat{\gamma}^{(S)}, \tag{9.17}$$

where, in particular, the polynomial spline estimator of the component curve $\beta_l(t)$ is

$$\widehat{\beta}_l^{(S)}(t) = \sum_{s=1}^{K_l}\left[\widehat{\gamma}_{ls}^{(S)}\, B_{ls}^{(S)}(t)\right]. \tag{9.18}$$

We note that, for an arbitrary smooth curve $\beta_l(t)$, $\widehat{\beta}_l^{(S)}(t)$ should be on average getting closer to $\beta_l(t)$ when the number of basis terms, K_l, increases. If the B-spline functions $B_{ls}^{(S)}(t)$ are polynomials with a fixed degree d, increasing K_l implies the number of knots needs to be increased and the locations of the knots are also changed accordingly. Thus, when d is fixed, the number and location of the knots determine the smoothing parameter K_l. In most applications, it is common to use linear ($d = 1$) or quadratic ($d = 2$) splines with equally spaced knots, i.e., all adjacent knots have equal distance between them, which leaves the number of knots to be the only factor determining the smoothing parameter K_l. In practice, we can select the smoothing parameters subjectively by examining the fitness of $\widehat{\beta}_l^{(S)}(t)$ graphically under different knot numbers. But, unlike simple regression models, we do not have observations for a direct scatter plot of $\beta_l(t)$, so that subjective choices of knot numbers through graphical examinations may not be justifiable in practice. A more appealing approach is to use the data-driven smoothing parameter choice based on the leave-one-subject-out cross-validation, which is discussed in detail in the next section.

9.2.4   Cross-Validation Smoothing Parameters

Similar to the smoothing parameter selection method of Section 4.1, the leave-one-subject-out cross-validation (LSCV) can be extended to the basis approximation estimators $\widehat{\beta}(t)$ for selecting the smoothing parameters $\{K_l : l = 0, \ldots, k\}$. The purpose of this cross-validation is to preserve the correlation structures of the data, which are often completely unknown in practice. The similar cross-validation has been used in the local smoothing methods of Chapter 7 for the time-varying coefficient models.

1. LSCV for Basis Estimators

Let $\widehat{\gamma}^{(-i)}$ be the least squares estimator given in (9.8) computed from the remaining sample after deleting all the observations of the ith subject. Substituting $\widehat{\gamma}^{(-i)}$ into (9.9), we obtain the least squares basis approximation estimator

$$\widehat{\beta}^{(-i)}(t) = \mathbf{B}(t)\,\widehat{\gamma}^{(-i)}. \tag{9.19}$$

We define the LSCV score for $K = (K_0, \ldots, K_k)^T$ based on $\widehat{\beta}^{(-i)}(t)$ to be

$$LSCV(K) = \sum_{i=1}^{n} \sum_{j=1}^{n_i} \left\{ w_i \left[Y_{ij} - X_{ij}^T \widehat{\beta}^{(-i)}(t_{ij}) \right]^2 \right\}, \qquad (9.20)$$

and the LSCV smoothing parameters $K_{LSCV} = (K_{0,LSCV}, \ldots, K_{k,LSCV})^T$ to be the minimizer of $LSCV(K)$ provided that (9.20) can be uniquely minimized with respect to K.

There are two advantages of using the LSCV (9.20) for the model (9.1) and the global smoothing estimator (9.9). First, as in the local smoothing for the time-varying coefficient models, this approach does not require modeling the intra-subject correlation structures of the data. In contrast to the local smoothing methods where only the observations around the time point t are used in computing the estimates, the global smoothing estimator of (9.9) uses all the observations throughout the time range $\mathcal{T}$, so that the LSCV method (9.20) selects the smoothing parameters based on all the observations. If the intra-subject correlation structured are modeled through w_i, then the LSCV of (9.20) selects the smoothing parameters with the correlations of the data taken into consideration. In most practical situations, it is difficult to model the correlation structures appropriately. Second, since the smoothing parameters K_l are allowed to be different for different coefficient curves $l = 0, \ldots, k$, the LSCV method (9.20) leads to possibly different cross-validated smoothing parameters $K_{l,LSCV}$. Thus, when different smoothing is required for different coefficient curves $\beta_l(t)$, the LSCV method (9.20) has the flexibility of providing appropriate smoothing for each component curve. This is in contrast to the one-step local smoothing methods of Chapter 7, where only one smoothing parameter, i.e., the bandwidth, is selected through the LSCV method.

An intuitive justification of K_{LSCV} based on (9.20) can be provided using the same rationale as in Section 7.2. Under the global smoothing setting, we consider the average squared error of $\widehat{\beta}(t_{ij})$ defined by

$$ASE(K) = \sum_{i=1}^{n} \sum_{j=1}^{n_i} \left\{ w_i \left[X_{ij}^T \left[\beta(t_{ij}) - \widehat{\beta}(t_{ij}) \right] \right]^2 \right\} \qquad (9.21)$$

and the decomposition

$$\begin{aligned} LSCV(K) &= \sum_{i=1}^{n} \sum_{j=1}^{n_i} \left\{ w_i \left[Y_{ij} - X_{ij}^T \beta(t_{ij}) \right]^2 \right\} \\ &+ 2 \sum_{i=1}^{n} \sum_{j=1}^{n_i} \left\{ w_i \left[Y_{ij} - X_{ij}^T \beta(t_{ij}) \right] \left[X_{ij}^T \beta(t_{ij}) - X_{ij}^T \widehat{\beta}^{(-i)}(t_{ij}) \right] \right\} \\ &+ \sum_{i=1}^{n} \sum_{j=1}^{n_i} \left\{ w_i \left[X_{ij}^T \left[\beta(t_{ij}) - \widehat{\beta}^{(-i)}(t_{ij}) \right] \right]^2 \right\}. \end{aligned} \qquad (9.22)$$

The first term on the right side of (9.22) does not depend on the smoothing parameters $K = (K_0, \ldots, K_k)^T$, and, because of the definition of $\widehat{\beta}^{(-i)}(t)$

in (9.19), the expectation of the second term is zero. Thus, by minimizing $LSCV(K)$, K_{LSCV} approximately minimizes the third term on the right side of (9.22) which approximates the average squared error $ASE(K)$ of (9.21).

2. LSCV for B-Spline Estimators

Because of computation complexity involved in the smoothing parameters, it is often impractical to automatically select all three smoothing components in B-spline smoothing: the degrees of spline, the number of knots and the locations of knot choices. Since the number of knots is the most influential factor affecting the smoothness of a B-spline approximation (Stone et al., 1997). A computationally simpler choice, which has been used in Chapter 4 and proposed by Rice and Wu (2001) and Huang, Wu and Zhou (2002, 2004), is to use splines with equally-spaced knots and fixed degrees and select only the number of knots based on the data. Since the knots are equally spaced within the time range $\mathscr{T}$ of interest, if $\mathscr{T} = [T_0, T_1]$ is a closed interval with lower and upper endpoints T_0 and T_1, then T_0 and T_1 are the two boundary knots, and, by (9.10), K_l is the dimension of $\mathbb{G}_l^{(S)}$ and is related to the number M_l of interior knots through

$$K_l = M_l + 1 + d, \tag{9.23}$$

where d is the degree of the spline.

For the LSCV based on the B-spline estimator $\widehat{\beta}^{(S)}(t)$ given in (9.17), we define $\widehat{\gamma}^{(-i,S)}$ to be the least squares B-spline estimator of (9.16) computed using the remaining sample after deleting all the observations of the ith subject, and

$$\widehat{\beta}^{(-i,S)}(t) = \mathbf{B}^{(S)}(t)\,\widehat{\gamma}^{(-i,S)} \tag{9.24}$$

to be the corresponding B-spline estimator of $\beta(t)$ based on $\widehat{\gamma}^{(-i,S)}$. The LSCV score for the B-spline estimator (9.17) is

$$LSCV^{(S)}(K) = \sum_{i=1}^{n} \sum_{j=1}^{n_i} \left\{ w_i \left[Y_{ij} - X_{ij}^T \widehat{\beta}^{(-i,S)}(t_{ij}) \right]^2 \right\}. \tag{9.25}$$

If the right side of (9.25) can be uniquely minimized with respect to $K = (K_0, \dots, K_k)^T$, the LSCV smoothing parameter vector

$$K_{LSCV}^{(S)} = \left(K_{0,LSCV}^{(S)}, \dots, K_{k,LSCV}^{(S)} \right)^T$$

is the minimizer of (9.25). It is seen from (9.23) that, when the degree of spline d is fixed, the LSCV smoothing parameter $K_{l,LSCV}^{(S)}$ is uniquely determined by the number of internal knots M_l for splines with equally spaced knots.

Although our attention in B-splines is restricted to splines with equally spaced knots, this choice of knots is computationally simple and works well for the real data example of Section 9.4. In general, however, it might be worthwhile to investigate using the data to also select the knot locations. B-splines without any specified knot numbers and locations are referred to as free-knot splines. There has been considerable work on free-knot splines with cross-sectional i.i.d. data, for example, Stone et al. (1997), Hansen and Kooperberg (2002) and Stone and Huang (2003). Extension of the methodology and theory of free-knot splines to the time-varying coefficient model (9.1) has not been systematically investigated in the literature.

3. M-fold LSCV

When the number of subjects n is large, calculating the LSCV score (9.20), or (9.25) for the case of B-splines, can be computationally intensive. In such situations, we can reduce the computational cost by using the leave-M-subjects-out cross-validation (M-fold LSCV), in which we split the subjects into M roughly equal-sized groups and computed the basis approximation estimators using the remaining sample after deleting all the observations from each of the M subject groups.

Let $M[i]$ be the group containing the ith subject and denote by $\widehat{\beta}^{-M[i]}(t)$ the estimate of $\beta(t)$ with the observations of the $M[i]$th group of the subjects removed. Then the M-fold LSCV score is

$$LMSCV(K) = \sum_{i=1}^{n} \sum_{j=1}^{n_i} \left\{ w_i \left[Y_{ij} - X_{ij}^T \widehat{\beta}^{(-M[i])}(t_{ij}) \right]^2 \right\}. \qquad (9.26)$$

If the right side of (9.26) can be uniquely minimized with respect to $K = (K_0, \ldots, K_k)^T$, the M-fold LSCV smoothing parameter vector

$$K_{LMSCV} = \left(K_{0,LMSCV}, \ldots, K_{k,LMSCV} \right)^T$$

is the minimizer of $LMSCV(K)$.

The standard LSCV of (9.20) is the special case of (9.26) with $M = 1$. The only reason of using the M-fold LSCV (9.26) with $M > 1$ is to save computing time by calculating the cross-validation scores one group at a time, instead of one subject at a time. For this reason, the choice of M depends on the available computing power and the desired computing time. Within the desirable computing time, it is preferable to choose M as small as possible.

9.2.5   *Conditional Biases and Variances*

Statistical properties of the basis approximation estimators can be first derived by evaluating their biases and variances conditioning on the observed time points and covariates. These conditional biases and variances can then be used to compute the conditional mean squared errors. Asymptotic expressions of the conditional biases, variances and mean squared errors of the least squares estimators (9.8) and (9.9) are presented in Section 9.5.

1. Conditional Bias

We first derive the general expressions of the conditional biases of (9.8) and (9.9). The conditional biases of the B-spline estimators (9.16) and (9.18) are just special cases with the spline basis functions $\mathscr{B}_l^{(S)}(t)$ in (9.10). Let

$$\mathscr{X} = \left\{ \mathbf{X}_{ij}, t_{ij} : i = 1, \ldots, n; j = 1, \ldots, n_i \right\} \tag{9.27}$$

be the set of the observed covariates and time points of the longitudinal sample. Using (9.8), the conditional expectation of $\widehat{\gamma}$ given $\mathscr{X}$ is

$$\begin{cases} \widetilde{\gamma} = E\left(\widehat{\gamma} \mid \mathscr{X}\right) = \left[\sum_{i=1}^{n} \left(\mathbf{U}_i^T \mathbf{W}_i \mathbf{U}_i\right)\right]^{-1} \left[\sum_{i=1}^{n} \left(\mathbf{U}_i^T \mathbf{W}_i \widetilde{Y}_i\right)\right], \\ \widetilde{Y}_i = E\left(Y_i \mid \mathscr{X}\right) = \left(\widetilde{Y}_{i1}, \ldots, \widetilde{Y}_{in_i}\right)^T, \\ \widetilde{Y}_{ij} = \mathbf{X}_{ij}^T \beta\left(t_{ij}\right). \end{cases} \tag{9.28}$$

It follows from (9.9) and (9.28) that the conditional bias of $\widehat{\beta}(t)$ given $\mathscr{X}$ is

$$Bias\left[\widehat{\beta}(t) \mid \mathscr{X}\right] = E\left[\widehat{\beta}(t) - \beta(t) \mid \mathscr{X}\right] = \mathbf{B}(t)\widetilde{\gamma} - \beta(t), \tag{9.29}$$

and, for $l = 0, \ldots, k$, the conditional bias of $\widehat{\beta}_l(t)$ given $\mathscr{X}$ is

$$Bias\left[\widehat{\beta}_l(t) \mid \mathscr{X}\right] = E\left[\widehat{\beta}_l(t) - \beta_l(t) \mid \mathscr{X}\right] = e_{l+1}^T \left[\mathbf{B}(t)\widetilde{\gamma} - \beta(t)\right], \tag{9.30}$$

where e_{l+1} is the $[(k+1) \times 1]$ column vector with 1 as its $(l+1)$th element and zero elsewhere.

If, for all $l = 0, \ldots, k$, $\beta_l(t)$ belongs to $\mathbb{G}_l$, the linear function space spanned by $\left\{B_{l1}(t), \ldots, B_{lK_l}(t)\right\}$, we can write $\beta(t) = \mathbf{B}(t)\gamma^*$ for some γ^*, so that $\widetilde{\gamma} = \gamma^*$ and the conditional bias $E\left[\widehat{\beta}(t) - \beta(t) \mid \mathscr{X}\right]$ is 0. In general, the conditional biases may not be zero, but it is possible to make the conditional biases asymptotically negligible as n tends to infinity by choosing large values of $K_0, \ldots, K_k$.

2. Conditional Variance and Covariance

To derive the conditional variances of the estimators given $\mathscr{X}$, we first note from (9.1) that the conditional variance-covariance matrix of Y_i by

$$\begin{cases} \mathbf{V}_i = Cov\left(Y_i \mid \mathscr{X}\right) = \begin{pmatrix} \rho\left(t_{i1}, t_{i1}\right) & \cdots & \rho\left(t_{i1}, t_{in_i}\right) \\ \vdots & \vdots & \vdots \\ \rho\left(t_{in_i}, t_{i1}\right) & \cdots & \rho\left(t_{in_i}, t_{in_i}\right) \end{pmatrix}, \\ \rho\left(t_{ij}, t_{ij'}\right) = Cov\left(\varepsilon_{ij}, \varepsilon_{ij'}\right) \quad \text{and} \quad \rho\left(t_{ij}, t_{ij}\right) = \sigma^2\left(t_{ij}\right), \end{cases} \tag{9.31}$$

where $\sigma^2\left(t_{ij}\right)$ and $\rho\left(t_{ij}, t_{ij'}\right)$ are the variance and covariance of ε_{ij} defined in

(7.1). Using (9.31) and the expression of $\widehat{\gamma}$ in (9.8), the conditional variance-covariance matrix of $\widehat{\gamma}$ is

$$Cov(\widehat{\gamma}|\mathcal{X}) = \left[\sum_{i=1}^{n}(\mathbf{U}_i^T\mathbf{W}_i\mathbf{U}_i)\right]^{-1}\left[\sum_{i=1}^{n}(\mathbf{U}_i^T\mathbf{W}_i\mathbf{V}_i\mathbf{W}_i\mathbf{U}_i)\right]$$
$$\times\left[\sum_{i=1}^{n}(\mathbf{U}_i^T\mathbf{W}_i\mathbf{U}_i)\right]^{-1}. \qquad (9.32)$$

It then follows from the (9.32) and the expression of $\widehat{\beta}(t)$ in (9.9) that

$$\begin{cases} Cov\left[\widehat{\beta}(t)\Big|\mathcal{X}\right] = \mathbf{B}(t)\,Cov(\widehat{\gamma}|\mathcal{X})\,\mathbf{B}^T(t), \\ Var\left[\widehat{\beta}_l(t)\Big|\mathcal{X}\right] = e_{l+1}^T\,Cov\left[\widehat{\beta}(t)\Big|\mathcal{X}\right]e_{l+1}, \quad l=0,\ldots,k, \end{cases} \qquad (9.33)$$

where e_{l+1} is the same $[(k+1)\times 1]$ column vector defined in (9.30).

If the random errors ε_{ij} of (9.1) are from a known Gaussian process and the conditional biases of the estimators are negligible, the above conditional variance-covariance matrices (9.32) and (9.33) can be used for statistical inferences. However, the intra-subject correlation structure $\rho(t_{ij}, t_{ij'})$ is often unknown in practice and needs to be estimated. Without the normality assumption on ε_{ij}, we can construct the asymptotically approximated inference procedures for the smoothing estimators, if the asymptotic distributions of the smoothing estimators can be derived.

9.2.6 Estimation of Variance and Covariance Structures

We present here a tensor product spline method for the estimation of the covariance function $\rho(t, s)$ for $t, s \in \mathcal{T}$ and variance function $\sigma^2(t) = Var[\varepsilon(t)]$. Since, by (9.33), the conditional variance of $\widehat{\beta}_l(t)$ is determined by the covariance structure of the error process $\varepsilon(t)$ of (9.1), a crucial step for the estimation of the conditional variance of $\widehat{\beta}_l(t)$ is to estimate the variance-covariance matrix $\mathbf{V}_i$ of (9.31).

1. Spline Estimation of Covariances

Nonparametric estimation for the covariance structures of longitudinal data has been investigated in the literature. For example, Diggle and Verbyla (1998) suggests a local smoothing method to estimate the covariance structures, Li et al. (2009) studies influence diagnostics for outliers, and Li et al. (2012) presents a testing method based on variance components. However, estimation methods based on local smoothing could be computationally intensive under the current context, since the estimated covariance function needs to be evaluated at each distinct pair of observation times. On the other hand, the tensor product spline estimators are much more computationally feasible.

The B-spline approach depends on approximating the covariance function

$$\rho(t, s) = Cov\left[\varepsilon(s), \varepsilon(t)\right]$$

of (7.1) by a tensor product spline on $\mathcal{T} \times \mathcal{T}$, such that,

$$\rho(t, s) \approx \sum_{l_1=1}^{K} \sum_{l_2=1}^{K} \left[u_{l_1 l_2} B_{l_1}(t) B_{l_2}(s)\right], \quad t, s \in \mathcal{T} \text{ and } t \neq s, \qquad (9.34)$$

where $\{B_l(t) : l = 1, \ldots, K\}$ is a spline basis on $\mathcal{T}$ with a fixed knot sequence. The above approximation is only required to hold when $t \neq s$, since, in most practical longitudinal settings, the correlation function $\rho(t, s)$ is not necessarily continuous at $t = s$ in the sense that

$$\lim_{s \to t} \rho(t, s) < \sigma^2(t).$$

Furthermore, we note that $E\left(\varepsilon_{ij}\varepsilon_{ij'}\right) = \rho\left(t_{ij}, t_{ij'}\right)$ for $j \neq j'$ and $\rho(t, s) = \rho(s, t)$.

To motivate the tensor product spline estimation method, we consider first that, if the errors $\{\varepsilon_{ij} : i = 1, \ldots, n; j = 1, \ldots, n_i\}$ are observed, then $\rho(t, s)$, $t \neq s$, could be estimated by finding $\{u_{l_1 l_2} : u_{l_1 l_2} = u_{l_2 l_1}; l_1, l_2 = 1, \ldots, K\}$ which minimize

$$\sum_{i=1}^{n} \sum_{j,j'=1, j<j'}^{n_i} \left\{\varepsilon_{ij}\varepsilon_{ij'} - \sum_{l_1=1}^{K} \sum_{l_2=1}^{K} \left[u_{l_1 l_2} B_{l_1}\left(t_{ij}\right) B_{l_2}\left(t_{ij'}\right)\right]\right\}^2. \qquad (9.35)$$

Since ε_{ij} are not really observed, we can estimate ε_{ij} by the residuals

$$\widehat{\varepsilon}_{ij} = Y_{ij} - \mathbf{X}_{ij}^T \widehat{\boldsymbol{\beta}}\left(t_{ij}\right), \qquad (9.36)$$

where, in principle, $\widehat{\boldsymbol{\beta}}\left(t_{ij}\right)$ may be chosen as any appropriate basis approximation estimator of $\boldsymbol{\beta}(t)$.

To be consistent with the spline basis function of (9.34), a simple and intuitive choice is to use the B-spline estimator $\widehat{\boldsymbol{\beta}}^{(S)}(t)$ in (9.36). Replacing $\left(\varepsilon_{ij}\varepsilon_{ij'}\right)$ with $\left(\widehat{\varepsilon}_{ij}\widehat{\varepsilon}_{ij'}\right)$ in (9.35), the residual least squares estimator $\widehat{u}_{l_1 l_2}$ of $u_{l_1 l_2}$ minimizes

$$\sum_{i=1}^{n} \sum_{j,j'=1, j<j'}^{n_i} \left\{\widehat{\varepsilon}_{ij}\widehat{\varepsilon}_{ij'} - \sum_{l_1=1}^{K} \sum_{l_2=1}^{K} \left[u_{l_1 l_2} B_{l_1}\left(t_{ij}\right) B_{l_2}\left(t_{ij'}\right)\right]\right\}^2 \qquad (9.37)$$

with respect to $u_{l_1 l_2}$, assuming that (9.37) can be uniquely minimized. Based on $\widehat{u}_{l_1 l_2}$, the tensor product spline estimator of $\rho(t, s)$ is

$$\widehat{\rho}(t, s) = \sum_{l_1=1}^{K} \sum_{l_2=1}^{K} \left[\widehat{u}_{l_1 l_2} B_{l_1}(t) B_{l_2}(s)\right] \qquad (9.38)$$

for all $t, s \in \mathcal{T}$ and $t \neq s$. $\qquad\qquad\qquad\qquad\qquad\qquad\qquad\qquad\qquad\qquad\qquad\square$

2. Spline Estimation of Variances

For the estimation of the variance $\sigma^2(t)$ of $\varepsilon(t)$ defined in (7.1), we use the B-spline approximation

$$\sigma^2(t) \approx \sum_{l=1}^{K} v_l B_l(t) \tag{9.39}$$

based on the B-spline basis functions $\{B_l(t) : l = 1, \ldots, K\}$. Using the residuals $\widehat{\varepsilon}_{ij}$ given in (9.36), the least squares estimator $\widehat{v}_l$ of v_l is obtained by minimizing the square risk

$$\sum_{i=1}^{n} \sum_{j=1}^{n_i} \left[\widehat{\varepsilon}_{ij}^2 - \sum_{l=1}^{K} v_l B_l\left(t_{ij}\right) \right]^2, \tag{9.40}$$

provided that (9.40) can be uniquely minimized with respect to v_l. Substituting $\widehat{v}_l$ into (9.39), we obtain

$$\widehat{\sigma}^2(t) = \sum_{l=1}^{K} \widehat{v}_l B_l(t) \tag{9.41}$$

as the B-spline estimator of $\sigma^2(t)$ based on $\{B_l(t) : l = 1, \ldots, K\}$ and the residuals $\widehat{\varepsilon}_{ij}$.

3. Estimation of the Variances and Covariances

Replacing the $\rho\left(t_{ij}, t_{ij'}\right)$ and $\sigma^2\left(t_{ij}\right)$ of (9.31) by their tensor product spline estimators $\widehat{\rho}\left(t_{ij}, t_{ij'}\right)$ of (9.38) and $\widehat{\sigma}^2\left(t_{ij}\right)$ of (9.41), we can estimate $\mathbf{V}_i$ by

$$\widehat{\mathbf{V}}_i = \widehat{Cov}\left(Y_i \middle| \mathcal{X}\right) \tag{9.42}$$

$$= \begin{pmatrix} \widehat{\sigma}^2\left(t_{i1}\right) & \widehat{\rho}\left(t_{i1}, t_{i2}\right) & \cdots & \widehat{\rho}\left(t_{i1}, t_{i(n_i-1)}\right) & \widehat{\rho}\left(t_{i1}, t_{in_i}\right) \\ \widehat{\rho}\left(t_{i2}, t_{i1}\right) & \widehat{\sigma}^2\left(t_{i2}, t_{i2}\right) & \cdots & \widehat{\rho}\left(t_{i2}, t_{i(n_i-1)}\right) & \widehat{\rho}\left(t_{i2}, t_{in_i}\right) \\ \vdots & \vdots & \vdots & \vdots & \vdots \\ \widehat{\rho}\left(t_{in_i}, t_{i1}\right) & \widehat{\rho}\left(t_{in_i}, t_{i2}\right) & \cdots & \widehat{\rho}\left(t_{in_i}, t_{i(n_i-1)}\right) & \widehat{\sigma}^2\left(t_{in_i}\right) \end{pmatrix}.$$

Consequently, replacing $\mathbf{V}_i$ of (9.32) with $\widehat{\mathbf{V}}_i$, the tensor product spline estimator of $Cov\left(\widehat{\gamma} \middle| \mathcal{X}\right)$ is

$$\widehat{Cov}\left(\widehat{\gamma} \middle| \mathcal{X}\right) = \left[\sum_{i=1}^{n} \left(\mathbf{U}_i^T \mathbf{W}_i \mathbf{U}_i\right) \right]^{-1} \left[\sum_{i=1}^{n} \left(\mathbf{U}_i^T \mathbf{W}_i \widehat{\mathbf{V}}_i \mathbf{W}_i \mathbf{U}_i\right) \right]$$

$$\times \left[\sum_{i=1}^{n} \left(\mathbf{U}_i^T \mathbf{W}_i \mathbf{U}_i\right) \right]^{-1}, \tag{9.43}$$

and, by (9.33), the estimators of $Cov\left[\widehat{\beta}(t) \middle| \mathcal{X}\right]$ and $Var\left[\widehat{\beta}_l(t) \middle| \mathcal{X}\right]$ are

$$\begin{cases} \widehat{Cov}\left[\widehat{\beta}(t) \middle| \mathcal{X}\right] = \mathbf{B}(t) \widehat{Cov}\left(\widehat{\gamma} \middle| \mathcal{X}\right) \mathbf{B}^T(t), \\ \widehat{Var}\left[\widehat{\beta}_l(t) \middle| \mathcal{X}\right] = e_{l+1}^T \widehat{Cov}\left[\widehat{\beta}(t) \middle| \mathcal{X}\right] e_{l+1}, \quad l = 0, \ldots, k. \end{cases} \tag{9.44}$$

The estimation of $\rho(t,s)$ and $\sigma^2(t)$ relies on choosing the appropriate spline spaces. In practice, we can use equally spaced knot sequences and select the number of knots either subjectively or through the cross-validation procedures described in Section 9.2.4. However, such data-driven choices are often computationally intensive. In most applications, the number of knots between 5 and 10 often gives satisfactory results. The spline estimators of the covariance and variance functions given above need not be positive definite for a given finite sample, although, because of their consistency, they are asymptotically positive definite. So far, there is no satisfactory solution to the problem of constructing a nonparametric covariance or variance function estimator that is positive definite under finite longitudinal samples. How to impose the finite sample positive definiteness constraint to the current spline estimator is an important problem that deserves further investigation.

9.3   Resampling-Subject Bootstrap Inferences

Statistical inferences for the basis approximation estimators $\widehat{\gamma}$ and $\widehat{\beta}(t)$ can be constructed using the same resampling subject bootstrap methods as in Section 7.3. In this section, we present the resampling subject bootstrap methods for the construction of (i) pointwise confidence intervals for $\beta(t_0)$ at a time point $t_0 \in \mathscr{T}$, (ii) a class of simultaneous confidence bands for $\beta(t)$ for all $t \in [a, b] \subset \mathscr{T}$, and (iii) a simple method for testing the null hypothesis that $\beta_l(t)$ is a constant for all $t \in \mathscr{T}$.

9.3.1   Pointwise Confidence Intervals

Based on the estimation procedure of Section 9.2, let

$$\begin{cases} \widehat{\gamma} = \left(\widehat{\gamma}_0^T, \ldots, \widehat{\gamma}_k^T \right)^T \text{ with } \widehat{\gamma}_l = \left(\widehat{\gamma}_{l1}, \ldots, \widehat{\gamma}_{lK_l} \right)^T \\ \widehat{\beta}(t) = \left(\widehat{\beta}_0(t), \ldots, \widehat{\beta}_k(t) \right)^T \end{cases}$$

be the basis approximation estimators of γ^* and $\beta(t) = \left(\beta_0(t), \ldots, \beta_k(t) \right)^T$ defined in (9.8) and (9.9), respectively, based on a given set of basis functions $\left\{ B_{ls}(t) : s = 1, \ldots, K_l; l = 0, \ldots, k \right\}$. Let $A = \left(a_0, \ldots, a_k \right)^T$ be a known $(k+1)$ column vector. For any given $0 \leq \alpha \leq 1$ and time point $t \in \mathscr{T}$, our objective is to construct an approximate $\left[100 \times (1 - \alpha) \right]$th confidence interval for $A^T E \left[\beta(t) \right]$ based on $A^T E \left[\widehat{\beta}(t) \right]$. In particular, the $\left[100 \times (1 - \alpha) \right]$th confidence interval for $E \left[\beta_l(t) \right]$ can be constructed by selecting A with $a_l = 1$ and $a_s = 0$ for all $s \neq l$. By selecting the polynomial spline basis functions $\left\{ \mathscr{B}_l^{(S)}(t), \ldots, \mathscr{B}_k^{(S)}(t) \right\}$ of (9.10), our procedure leads to an approximate $\left[100 \times (1 - \alpha) \right]$th confidence interval for $A^T E \left[\beta(t) \right]$ based on $\widehat{\beta}^{(S)}(t)$ of (9.17). Using similar procedures as in Section 7.3, the resampling subject bootstrap pointwise confidence intervals for $A^T E \left[\beta(t) \right]$ can be constructed by the following steps.

Approximate Bootstrap Pointwise Confidence Intervals:

(a) **Computing Bootstrap Estimators.** *Generate B independent bootstrap samples using the resampling-subject bootstrap procedure of Section 3.4.1 and compute the B bootstrap estimators* $\{\widehat{\gamma}^b, \widehat{\beta}^b(t) : b = 1, \ldots, B\}$.

(b) **Approximate Percentile Bootstrap Confidence Intervals.** *Calculate* $l_{A,\alpha/2}(t)$ *and* $u_{A,\alpha/2}(t)$, *the lower and upper* $[100 \times (1 - \alpha/2)]$*th percentiles, respectively, of the B bootstrap estimators* $\{A^T\widehat{\beta}^b(t) : b = 1, \ldots, B\}$ *computed above. The approximate* $[100 \times (1 - \alpha/2)]$*th pointwise bootstrap confidence interval for* $A^T\beta(t)$ *is*

$$\left(l_{A,\alpha/2}(t), u_{A,\alpha/2}(t)\right). \tag{9.45}$$

In particular, the approximate $[100 \times (1 - \alpha/2)]$*th pointwise bootstrap confidence interval for* $\beta_l(t)$, $0 \leq l \leq k$, *is*

$$\left(l_{l,\alpha/2}(t), u_{l,\alpha/2}(t)\right), \tag{9.46}$$

where $l_{l,\alpha/2}(t)$ *and* $u_{l,\alpha/2}(t)$ *are the lower and upper* $[100 \times (1 - \alpha/2)]$*th percentiles, respectively, of the B bootstrap estimators* $\{\widehat{\beta}_l^b(t) : b = 1, \ldots, B\}$.

(c) **Normal Approximated Bootstrap Confidence Intervals.** *Using normal approximation of the critical values, the normal approximated bootstrap confidence interval is given by*

$$A^T\widehat{\beta}(t) \pm z_{1-\alpha/2}\,\widehat{se}\left[A^T\widehat{\beta}^b(t)\right], \tag{9.47}$$

where z_p *is the* $[100 \times p]$*th percentile of the standard normal distribution and* $\widehat{se}\left[A^T\widehat{\beta}^b(t)\right]$ *is the sample standard error of* $\widehat{\beta}^b(t)$ *computed from the B bootstrap estimators* $\{A^T\widehat{\beta}^b(t) : b = 1, \ldots, B\}$. *The normal approximated confidence interval for* $\beta_l(t)$, $0 \leq l \leq k$, *is*

$$\widehat{\beta}_l(t) \pm z_{1-\alpha/2}\,\widehat{se}\left[\widehat{\beta}_l^b(t)\right], \tag{9.48}$$

with $\widehat{se}\left[\widehat{\beta}^b(t)\right]$ *being the sample standard error of* $\widehat{\beta}_l^b(t)$ *computed from the B bootstrap estimators* $\{\widehat{\beta}_l^b(t) : b = 1, \ldots, B\}$. □

We note that, since neither the percentile intervals in step (b) nor the normal approximated error bars in step (c) adjust for the bias of $\widehat{\beta}(t)$, these intervals may not lead to proper pointwise confidence intervals for $A^T\beta(t)$, unless the bias of $\widehat{\beta}(t)$ is negligible relative to its variance. In the local smoothing estimation methods of Chapter 7, one potential approach is to consider a "plug-in" approach, which adjusts the locations of the intervals using an estimated bias of $\widehat{\beta}(t)$. But, the explicit expressions for the basis approximation estimators of $\beta(t)$ depend on the appropriateness of the chosen basis functions and are generally not available (see Section 9.5). Even for a specifically given basis choice, such as the B-splines, the bias term of $\widehat{\beta}(t)$ is difficult to derive explicitly, because it depends on the "closeness" between the true coefficient curves $\beta(t)$ and the linear functional space $\mathbb{G}_l$ spanned by $\{B_{l1}(t), \ldots, B_{lK_l}(t)\}$. Thus, unlike the local smoothing methods of Chapter 7, the "plug-in" approach for bias correction is not applicable for the basis approximation estimators $\widehat{\beta}(t)$.

Because it is not exactly known how well the basis expansion approximates the true coefficient curves $\beta(t)$, $E\left[\widehat{\beta}(t)\right]$ is in essence the estimable part of $\beta(t)$. Within the framework of basis approximations, we can only treat $E\left[\widehat{\beta}(t)\right]$ as the parameter of interest and simply use the intervals in steps (b) and (c) as the approximate confidence intervals for $A^T\beta(t)$. This is a sensible approach since $E\left[\widehat{\beta}(t)\right]$, as a good approximation of $\beta(t)$, is expected to capture the main feature of $\beta(t)$. To obtain proper coverage probabilities for the approximate intervals (9.45) and (9.47), we may make the bias of the basis approximation estimator $\widehat{\beta}(t)$ negligible by selecting a set of large $\{K_0, \ldots, K_k\}$ in the computation of $\widehat{\beta}(t)$.

9.3.2 Simultaneous Confidence Bands

The same "bridging-the-gap" approach of Section 7.3.2 can also be used to extend the above pointwise confidence intervals to simultaneous confidence bands for $A^T E\left[\beta(t)\right]$ over a given sub-interval $[a, b]$ of $\mathscr{T}$. This approach has three main steps.

1. Confidence Bands on Partitioned Time Points

For the first step, we partition $[a, b]$ into $M+1$ equally spaced grid points $a = \xi_1 < \cdots < \xi_{M+1} = b$ for some integer $M \geq 1$, and construct a set of approximate $\left[100 \times (1-\alpha)\right]$ percent simultaneous confidence intervals

$$\left(L_{A,\alpha}(\xi_r), U_{A,\alpha}(\xi_r)\right) \qquad (9.49)$$

for $\{A^T E\left[\beta(\xi_r)\right] : r = 1, \ldots, M+r\}$, such that

$$\lim_{n \to \infty} P\left\{L_{A,\alpha}(\xi_r) \leq A^T E\left[\widehat{\beta}(\xi_r)\right] \leq U_{A,\alpha}(\xi_r), \quad \text{for all } r = 1, \ldots, M+1\right\}$$
$$\geq \quad 1 - \alpha. \qquad (9.50)$$

The simple choice based on Bonferroni adjustment of the bootstrap pointwise intervals (9.45) or (9.47) gives

$$\left(L_{A,\alpha}(\xi_r), U_{A,\alpha}(\xi_r)\right) = \left(l_{A,\alpha}(\xi_r), u_{A,\alpha}(\xi_r)\right) \qquad (9.51)$$

or

$$\left(L_{A,\alpha}(\xi_r), U_{A,\alpha}(\xi_r)\right) = A^T\widehat{\beta}(\xi_r) \pm z_{1-\alpha/[2(M+1)]}\, \widehat{se}\left[A^T\widehat{\beta}^b(\xi_r)\right]. \qquad (9.52)$$

For refinements, one may use the inclusion-exclusion identities to calculate $\left(L_{A,\alpha}(\xi_r), U_{A,\alpha}(\xi_r)\right)$ with more accurate coverage probabilities. These refinements, however, involve more extensive computations, and may not be always practical for large longitudinal studies. As in Section 7.3.2, the integer M is chosen subjectively, since the optimal choices of M under the current situation are not available.

2. Linear Interpolated Confidence Bands

We now construct the simultaneous confidence bands for the linear interpolations of $\{A^T E[\widehat{\beta}(\xi_r)] : r = 1, \dots, M+1\}$ at any time point $t \in [a, b]$. Let $E^{(l)}[\widehat{\beta}_l(t)]$ be the linear interpolation of $E[\widehat{\beta}_l(\xi_r)]$ and $E[\widehat{\beta}_l(\xi_{r+1})]$ for $\xi_r \le t \le \xi_{r+1}$ and any $0 \le l \le k$, such that

$$E^{(l)}[\widehat{\beta}_l(t)] = M\left(\frac{\xi_{r+1}-t}{b-a}\right) E[\widehat{\beta}_l(\xi_r)] + M\left(\frac{t-\xi_r}{b-a}\right) E[\widehat{\beta}_l(\xi_{r+1})]. \qquad (9.53)$$

The linear interpolation of $A^T E[\widehat{\beta}(\xi_r)]$ and $A^T E[\widehat{\beta}(\xi_{r+1})]$ for $\xi_r \le t \le \xi_{r+1}$ is

$$A^T E^{(l)}[\widehat{\beta}(t)] = A^T \left(E^{(l)}[\widehat{\beta}_0(t)], \dots, E^{(l)}[\widehat{\beta}_k(t)] \right)^T, \qquad (9.54)$$

where $E^{(l)}[\widehat{\beta}(t)] = \left(E^{(l)}[\widehat{\beta}_0(t)], \dots, E^{(l)}[\widehat{\beta}_k(t)] \right)^T$. Let $\left(L^{(l)}_{A,\alpha}(t), U^{(l)}_{A,\alpha}(t) \right)$ be the linear interpolation of $\left(L_{A,\alpha}(\xi_r), U_{A,\alpha}(\xi_r) \right)$ and $\left(L_{A,\alpha}(\xi_r), U_{A,\alpha}(\xi_r) \right)$ for any $\xi_r \le t \le \xi_{r+1}$ defined by

$$\begin{cases} L^{(l)}_{A,\alpha}(t) = M\left(\frac{\xi_{r+1}-t}{b-a}\right) L_{A,\alpha}(\xi_r) + M\left(\frac{t-\xi_r}{b-a}\right) L_{A,\alpha}(\xi_{r+1}), \\ U^{(l)}_{A,\alpha}(t) = M\left(\frac{\xi_{r+1}-t}{b-a}\right) U_{A,\alpha}(\xi_r) + M\left(\frac{t-\xi_r}{b-a}\right) U_{A,\alpha}(\xi_{r+1}). \end{cases} \qquad (9.55)$$

It follows from (9.49), (9.53), (9.54) and (9.55) that

$$\lim_{n\to\infty} P\left\{ L^{(l)}_{A,\alpha}(t) \le A^T E^{(l)}[\widehat{\beta}(t)] \le U^{(l)}_{A,\alpha}(t), \text{ for all } t \in [a, b] \right\} \ge 1 - \alpha, \qquad (9.56)$$

and, consequently, $\left(L^{(l)}_{A,\alpha}(t), U^{(l)}_{A,\alpha}(t) \right)$ is an approximate $[100 \times (1 - \alpha)]$th confidence band for $A^T E^{(l)}[\widehat{\beta}(t)]$.

3. Bridging-the-Gap Confidence Bands

To bridge the gap between the real estimators $A^T E[\widehat{\beta}(t)]$ and the linear interpretations $A^T E^{(l)}[\widehat{\beta}(t)]$ of (9.54), we construct the bands for $A^T E[\widehat{\beta}(t)]$ based on the following two smoothness assumptions:

(a) Bounded First Derivatives. If the derivative of $A^T E\big[\widehat{\beta}(t)\big]$ with respect to t satisfies

$$\sup_{t\in[a,b]}\left|\left\{A^T E\big[\widehat{\beta}(t)\big]\right\}'\right|\leq c_1, \quad \text{for a known constant } c_1 > 0, \tag{9.57}$$

then it follows from direct calculations that

$$\left|A^T E\big[\widehat{\beta}(t)\big] - A^T E^{(l)}\big[\widehat{\beta}(t)\big]\right|\leq 2c_1\left[\frac{M\,(\xi_{r+1}-t)\,(t-\xi_r)}{b-a}\right] \tag{9.58}$$

for all $t \in \big[\xi_r, \xi_{r+1}\big]$, so that

$$\left(L_{A,\alpha}^{(l)}(t) - 2c_1\left[\frac{M\,(\xi_{r+1}-t)\,(t-\xi_r)}{b-a}\right],\; U_{A,\alpha}^{(l)}(t) + 2c_1\left[\frac{M\,(\xi_{r+1}-t)\,(t-\xi_r)}{b-a}\right]\right). \tag{9.59}$$

is the approximate $\big[100 \times (1-\alpha)\big]$ confidence band for $A^T E\big[\widehat{\beta}(t)\big]$.

(b) Bounded Second Derivatives. If the second derivative of $A^T E\big[\widehat{\beta}(t)\big]$ with respect to t satisfies

$$\sup_{t\in[a,b]}\left|\left\{A^T E\big[\widehat{\beta}(t)\big]\right\}''\right|\leq c_2, \quad \text{for a known constant } c_2 > 0, \tag{9.60}$$

then it follows from direct calculations that

$$\left|A^T E\big[\widehat{\beta}(t)\big] - A^T E^{(l)}\big[\widehat{\beta}(t)\big]\right|\leq \frac{c_2}{2}\left[\frac{M\,(\xi_{r+1}-t)\,(t-\xi_r)}{b-a}\right] \tag{9.61}$$

for all $t \in \big[\xi_r, \xi_{r+1}\big]$, so that

$$\left(L_{A,\alpha}^{(l)}(t) - \frac{c_2}{2}\left[\frac{M\,(\xi_{r+1}-t)\,(t-\xi_r)}{b-a}\right],\; U_{A,\alpha}^{(l)}(t) + \frac{c_2}{2}\left[\frac{M\,(\xi_{r+1}-t)\,(t-\xi_r)}{b-a}\right]\right). \tag{9.62}$$

is the approximate $\big[100 \times (1-\alpha)\big]$ confidence band for $A^T E\big[\widehat{\beta}(t)\big]$.

4. Remarks on Practical Concerns

As discussed in the local smoothing estimators of $\beta(t)$ in Section 7.3.2, smoothing conditions other than (9.57) and (9.60) can also be considered using the similar "bridging-the-gap" inequalities (9.58) and (9.61). We use the smoothing conditions (9.57) and (9.60) because they are intuitive and easy to interpret in practice. For (9.57), we require that the slope of $A^T E\big[\widehat{\beta}(t)\big]$ is bounded above by a known constant c_1. For (9.60), we require that the curvature of $A^T E\big[\widehat{\beta}(t)\big]$ is bounded above by a known constant c_2. In biomedical studies, the upper bounds of the slopes and curvatures can be determined by the specific biological mechanism or scientific nature of the studies.

We note also that the simultaneous confidence bands of (9.59) and (9.62) are for $A^T E\left[\widehat{\beta}(t)\right]$, which are different from $A^T \beta(t)$ because the bias term of $\widehat{\beta}(t)$ is ignored. As discussed in Section 9.3.1, $E\left[\widehat{\beta}(t)\right]$ already captures the main features of $\beta(t)$, and it is difficult to estimate the bias of $\widehat{\beta}(t)$ without knowing how close the function space spanned by the basis functions is to the true curves $\beta(t)$. Simulation results in the literature, such as Huang, Wu and Zhou (2002, 2004), suggest that both (9.59) and (9.62) can be used as approximate simultaneous bands for $A^T \beta(t)$ and they give appropriate coverage probabilities for most practical situations.

9.3.3   Hypothesis Testing for Constant Coefficients

Because of its nature as an "extended linear model," the basis approximation (9.2) allows a simple bootstrap goodness-of-fit procedure for testing whether the coefficient curves are time-varying, i.e., $\beta_l(t) = \beta_l^0$ for some constants β_l^0 with $l = 0, \ldots, k$. This test procedure gives a simple and useful tool because we are often interested in knowing whether the effects of a set of covariates are time-varying. Using the weighted least squares basis approximations (9.8) and (9.9), a goodness-of-fit test statistic can be constructed based on comparing the weighted residual sum of squares under both the null and the alternative hypotheses.

1. Testing Time-Varying Covariate Effects

For clarity, we illustrate the main idea of this goodness-of-fit test procedure using the example of testing

$$\begin{cases} H_0 : \beta_l(t) = \beta_l^0 \text{ for all } t \in \mathcal{T} \text{ and all } 1 \le l \le k; \\ H_1 : \beta_l(t) \text{ are time-varying at least for some } 1 \le l \le k, \end{cases} \tag{9.63}$$

where β_l^0 are unknown constants. The null hypothesis H_0 in (9.63) suggests that none of the coefficient functions, except the baseline curve, is time-varying, and the alternative H_1 is that at least some of the covariates have time-varying coefficients. The testing procedure presented for (9.63) can be modified for testing other null and alternative hypotheses. Since the baseline curve $\beta_0(t)$ is allowed to be time-varying under both the null hypothesis H_0 and the alternative H_1 in (9.63), we can, by (9.2), use the following approximation

$$\begin{cases} \beta_0(t) \approx \sum_{s=1}^{K_0} \left[\gamma_{0s}^* B_{0s}(t)\right], & \text{under both } H_0 \text{ and } H_1; \\ \beta_l(t) = \beta_l^0, & \text{for all } 1 \le l \le k \text{ under } H_0; \\ \beta_l(t) \approx \sum_{s=1}^{K_l} \left[\gamma_{ls}^* B_{ls}(t)\right], & \text{for all } 1 \le l \le k \text{ under } H_1. \end{cases} \tag{9.64}$$

Using (9.64), we can estimate the coefficient curves and compute the residual sum of squares for the fitted models under both the null hypothesis H_0 and the alternative H_1.

The test statistics for (9.63) can be derived by comparing these residual sum of squares. Under the null hypothesis H_0, (9.64) suggests that we set the basis functions $B_{ls}(t) = 1$ for all $l = 1, \ldots, k$ and $K_l = 1$. By (9.8) and (9.9), the least squares basis approximation estimators of $\beta_0(t)$ and $\{\beta_1^0, \ldots, \beta_k^0\}$ are given by

$$\begin{cases} \widehat{\beta}_0(t) = \sum_{s=1}^{K_0} \left[\widehat{\gamma}_{0s}^0 B_{0s}(t) \right], & \text{under } H_0; \\ \widehat{\beta}_l(t) = \widehat{\beta}_l^0, & \text{for all } 1 \le l \le k \text{ under } H_0. \end{cases} \tag{9.65}$$

The weighted residual sum of squares under H_0 computed using the estimators in (9.65) is

$$RSS_0 = \sum_{i=1}^{n} \sum_{j=1}^{n_i} w_i \left\{ Y_{ij} - \sum_{s=1}^{K_0} \left[X_{ij}^{(0)} B_{0s}(t_{ij}) \, \widehat{\gamma}_{0s}^0 \right] - \sum_{l=1}^{k} \sum_{s=1}^{K_l} \left[X_{ij}^{(l)} \widehat{\beta}_l^0 \right] \right\}^2. \tag{9.66}$$

On the other hand, under the general alternative H_1 and (9.64), we can directly apply (9.8) and (9.9), and obtain the estimators

$$\widehat{\beta}_l(t) = \sum_{s=1}^{K_l} \left[\widehat{\gamma}_{ls} B_{ls}(t) \right] \quad \text{for all } 0 \le l \le k \text{ under } H_1, \tag{9.67}$$

and obtain the corresponding weighted residual sum of squares

$$RSS_1 = \sum_{i=1}^{n} \sum_{j=1}^{n_i} w_i \left\{ Y_{ij} - \sum_{l=0}^{k} \sum_{s=1}^{K_l} \left[X_{ij}^{(l)} B_{ls}(t_{ij}) \, \widehat{\gamma}_{lk} \right] \right\}^2. \tag{9.68}$$

Since the model under H_0 is a special case of the model under H_1, it follows that $RSS_0 \ge RSS_1$. The models under H_0 and H_1 are similar if $[RSS_0 - RSS_1]$ is small. By measuring the relative size of $[RSS_0 - RSS_1]$ with RSS_1, we define the goodness-of-fit test statistic for the hypotheses in (9.63) to be

$$T_n = \frac{RSS_0 - RSS_1}{RSS_1}, \tag{9.69}$$

and we can reject the null hypothesis H_0 of (9.63) if T_n is larger than an appropriate critical value. In Section 9.5, we show that T_n is asymptotically consistent in the sense that, if H_0 holds, T_n converges to zero in probability, and, if H_1 holds, T_n is larger than a positive constant with probability tending to one.

2. Resampling-Subject Bootstrap Critical Values

We compute the critical value for rejecting H_0 using a resampling-subject bootstrap procedure similar to the one described in Section 4.2.3. Since the observed outcomes $\{Y_{ij} : i = 1, \ldots, n; j = 1, \ldots, n_i\}$ are not necessary from the model under H_0, we need to construct some pseudo-outcome observations based on our available estimators and residuals, so that the distribution of T_n under H_0 can be estimated.

Based on the estimators under the alternative H_1 in (9.65), the residuals of the model (9.1) are

$$\widehat{\varepsilon}_{ij} = Y_{ij} - \sum_{l=0}^{k} \sum_{s=1}^{K_l} \left[X_{ij}^{(l)} B_{ls}\left(t_{ij}\right) \widehat{\gamma}_{lk} \right] \tag{9.70}$$

for $i = 1, \ldots, n$ and $j = 1, \ldots, n_i$. Using the estimators under the null hypothesis H_0 in (9.65) and the residuals in (9.70), the pseudo-outcome values under the model in H_0 are given by $\{Y_{ij}^p : i = 1, \ldots, n; j = 1, \ldots, n_i\}$ with

$$Y_{ij}^p = \sum_{s=1}^{K_0} \left[X_{ij}^{(0)} B_{0s}\left(t_{ij}\right) \widehat{\gamma}_{0s}^0 \right] + \sum_{l=1}^{k} \sum_{s=1}^{K_l} \left[X_{ij}^{(l)} \widehat{\beta}_l^0 \right] + \widehat{\varepsilon}_{ij}. \tag{9.71}$$

The distributions of T_n under the null hypothesis H_0 can then be evaluated from a set of bootstrap samples of $\{Y_{ij}^p : i = 1, \ldots, n; j = 1, \ldots, n_i\}$. The following steps describe this resampling-subject bootstrap procedure for computing the null distribution of T_n and the p-value of the test for (9.63).

Resampling-Subject Bootstrap Testing Procedure:

(a) *Resample n subjects with replacement from $\{(Y_{ij}^p, t_{ij}, \mathbf{X}_{ij}) : i = 1, \ldots, n; j = 1, \ldots, n_i\}$ to obtain the bootstrap sample $\{(Y_{ij}^{p,b}, t_{ij}^b, \mathbf{X}_{ij}^b) : i = 1, \ldots, n; j = 1, \ldots, n_i\}$.*

(b) *Repeat the above bootstrap sampling procedure B times, so that B independent resampling-subject bootstrap samples are obtained.*

(c) *From each bootstrap sample, calculate the test statistic T_n^b using (9.66), (9.68) and (9.69), and derive the empirical distribution of $\{T_n^b : b = 1, \ldots, B\}$ based on the B independent bootstrap samples.*

(d) *Reject the null hypothesis H_0 at the significance level α when the observed test statistic T_n is greater than or equal to the $[100 \times (1 - \alpha)]$th percentile of the empirical distribution of T_n^b. The p-value of the test is the empirical probability of $(T_n^b \geq T_n)$.* □

For ease of computation, the same basis system can be used for each $\beta_l(t)$ in the calculations of T_n^b and T_n in the above resampling bootstrap testing procedure. Theoretically, it is possible to use different basis approximations for different bootstrap replications. But, changing basis functions is likely to significantly increase the computational complexity.

The above testing procedure can be modified in a straightforward way to test other null and alternative hypotheses. For example, instead of the hypotheses in (9.63), we may want to test the null hypothesis that a subset of the coefficient functions are constants versus the general alternative that the coefficient functions are time-varying. To do this, we need to proceed with the following steps: (a) modifying (9.64) to establish the appropriate basis approximations under both the null and alternative hypotheses, (b) deriving the corresponding estimators and weighted residual sum of squares as in (9.65) through (9.68), (c) computing the test statistic T_n defined in (9.69) and its distribution under the null hypothesis using the above resampling-subject bootstrap testing procedure, and (d) computing the p-value or critical value of the test.

9.4 R Implementation with the NGHS BP Data

9.4.1 Estimation by B-Splines

The NGHS Blood Pressure (BP) data has been analyzed in Sections 7.4.1 and 8.4 with the time-varying coefficient model (7.1) and local smoothing methods. We illustrate here the global smoothing method by B-splines for estimating the coefficient curves with the same dataset. As in Chapters 7 and 8, we denote that, for the ith NGHS girl, t_{ij} is the age in years at the jth study visit, Y_{ij} is the systolic blood pressure (SBP) level at age t_{ij}, $X_i^{(1)}$ is the indicator for race (1 if African American, 0 if Caucasian), $X_{ij}^{(2)}$ and $X_{ij}^{(3)}$ are the girl's age-adjusted height and body mass index (BMI) percentiles, respectively, at age t_{ij}, which are centered by subtracting 50% from the girl's actual height and BMI percentiles. Same as (7.39), the time-varying coefficient model (7.1) for these variables is

$$Y_{ij} = \beta_0(t_{ij}) + \beta_1(t_{ij}) X_i^{(1)} + \beta_2(t_{ij}) X_{ij}^{(2)} + \beta_3(t_{ij}) X_{ij}^{(3)} + \varepsilon_{ij}, \qquad (9.72)$$

where ε_{ij} is the error term for the ith girl at age t_{ij}. In this model, $\beta_0(t)$ represents the baseline SBP curve, i.e., the mean time curve of SBP for a Caucasian girl with the median height and the median BMI level in the population, and $\beta_l(t)$, $l = 1, 2, 3$, represent the effects of race, height and BMI, respectively, on the SBP level during the adolescence years.

In contrast to the one-step local methods of Chapter 7, the global smoothing method can provide different amounts of smoothing for the coefficient estimates by using different knots for $\beta_l(t)$, $l = 0, 1, 2, 3$. Unlike the two-step local smoothing methods of Chapter 8, the global smoothing method based on basis approximations does not require data binning when the observations are sparse at distinct observation time points.

In our application to this dataset, we use the weighted least squares estimation procedure described in (9.13) to (9.18) with cubic B-splines, equally spaced knots and the subject uniform weight $w_i^* = 1/(n_i n)$. Based on the cross-validation procedure (9.25), the numbers of internal knots for $\widehat{\beta}_l(t)$, $l = 0, 1, 2, 3$, are chosen to be $K_{LSCV} = (4, 0, 0, 0)^T$, which suggests that $\{\widehat{\beta}_1(t), \widehat{\beta}_2(t), \widehat{\beta}_3(t)\}$ were actually estimated using cubic polynomials. We note that different choices of the degrees or knot locations for polynomial splines or other types of basis functions may also be used similarly.

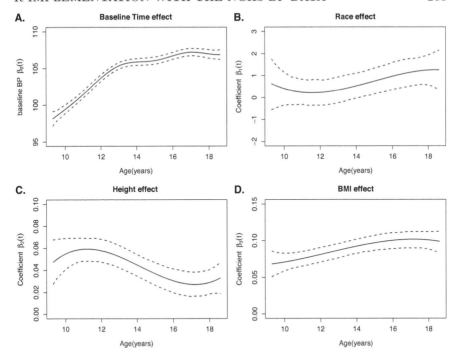

Figure 9.1 *Covariate effects on systolic blood pressure of NGHS girls. The solid curves in (A)-(D) show the cubic B-spline estimates for the coefficient curves $\widehat{\beta}_l(t)$, $l = 0, 1, 2, 3$, respectively. The dashed curves indicate the corresponding 95% pointwise confidence intervals computed from the bootstrap percentile procedure (9.46).*

The following R commands are used to obtain the weighted least squares solution for the B-spline estimated coefficients $\widehat{\gamma}$ from (9.13) using the standard R regression procedure:

```
# set up knots
> nknots <- c(4,0,0,0)+2
> KN0 <- seq(from=9, to=19, length=nknots[1])[-c(1,nknots[1])]
> KN1 <- seq(from=9, to=19, length=nknots[2])[-c(1,nknots[2])]
> KN2 <- seq(from=9, to=19, length=nknots[3])[-c(1,nknots[3])]
> KN3 <- seq(from=9, to=19, length=nknots[4])[-c(1,nknots[4])]

# generate cubic B-spline regression basis
> Bs.age  <- bs(NGHS$AGE, knots=KN0, intercept=T)
> Bs.race <- bs(NGHS$AGE, knots=KN1, intercept=T)* NGHS$Black
> Bs.HT   <- bs(NGHS$AGE, knots=KN2, intercept=T)* NGHS$HTPCTc
> Bs.BMI  <- bs(NGHS$AGE, knots=KN3, intercept=T)* NGHS$BMIPCTc
```

```
# Fit a weight least squares models to get the coefficient
# estimates
> fit.WLS <- lm(SBP ~ 0 + Bs.age + Bs.race + Bs.HT + Bs.BMI,
                weights=1/ni, data=NGHS)
> summary(fit.WLS)

...

Coefficients:
            Estimate Std. Error t value Pr(>|t|)
Bs.age1     97.675065   0.651817 149.850  < 2e-16 ***
Bs.age2     98.762326   0.511900 192.933  < 2e-16 ***
Bs.age3    101.184166   0.444347 227.714  < 2e-16 ***
Bs.age4    106.366454   0.363764 292.405  < 2e-16 ***
Bs.age5    105.505418   0.388155 271.813  < 2e-16 ***
Bs.age6    107.951453   0.477944 225.867  < 2e-16 ***
Bs.age7    106.571675   0.501244 212.614  < 2e-16 ***
Bs.age8    107.006195   0.531819 201.208  < 2e-16 ***
Bs.race1    0.760543   0.586170   1.297    0.194
Bs.race2   -0.794169   1.085292  -0.732    0.464
Bs.race3    1.517382   1.075168   1.411    0.158
Bs.race4    1.224773   0.508564   2.408    0.016 *
Bs.HT1      0.043056   0.011007   3.912 9.20e-05 ***
Bs.HT2      0.096770   0.019829   4.880 1.07e-06 ***
Bs.HT3     -0.004087   0.019150  -0.213    0.831
Bs.HT4      0.037284   0.008931   4.175 3.00e-05 ***
Bs.BMI1     0.067379   0.010274   6.558 5.59e-11 ***
Bs.BMI2     0.076266   0.018980   4.018 5.89e-05 ***
Bs.BMI3     0.115021   0.018910   6.082 1.21e-09 ***
Bs.BMI4     0.096801   0.008589  11.270  < 2e-16 ***
...
```

The coefficient curves $\widehat{\beta}_l(t)$ are computed from the cubic B-spline coefficient estimates (9.18) over a grid of time points using the following R commands:

```
> tgrid <- seq(from=9, to=19, by=0.1)
> Bs0  <- bs(tgrid, knots=KN0, intercept=T)
> Bs1  <- bs(tgrid, knots=KN1, intercept=T)
> Bs2  <- bs(tgrid, knots=KN2, intercept=T)
> Bs3  <- bs(tgrid, knots=KN3, intercept=T)
> nn <- cumsum(c(ncol(Bs0), ncol(Bs1), ncol(Bs2), ncol(Bs3)))
> Beta0 <- Bs0 %*% fit.WLS$coef[1:nn[1]]
> Beta1 <- Bs1 %*% fit.WLS$coef[(nn[1]+1):nn[2]]
> Beta2 <- Bs2 %*% fit.WLS$coef[(nn[2]+1):nn[3]]
> Beta3 <- Bs3 %*% fit.WLS$coef[(nn[3]+1):nn[4]]
```

Figure 9.1 shows the estimated coefficients curves of the baseline SBP time-trend, and the time-varying effects of race, height and BMI. Although all four curves are relatively smooth, the coefficient curves for race, height and BMI have relative lower curvatures compared to the baseline SBP curve. This suggests that the global cubic polynomial is quite reasonable for the estimation of these covariate effects, while the extra curvature of the baseline SBP curve justifies the four internal knots selected by the cross-validation procedure. The approximate 95% pointwise confidence intervals were obtained using the approximate percentile bootstrap procedure (9.46) with 1000 replications and the same knot choice obtained from the original B-spline estimators. The findings in Figure 9.1 for the coefficient curves are similar to the results obtained from the local smoothing estimates in Chapters 7 and 8.

9.4.2 Testing Constant Coefficients

The estimated curves in Figure 9.1 seem to suggest that the coefficient curves $\{\beta_l(t) : l = 0, 1, 2, 3\}$ of (9.72) are indeed time-varying. To verify this finding, we apply the goodness-of-fit test procedure of Section 9.3.3 to test the hypotheses that some of the coefficient curves are not time-varying or identically zero.

We consider the following eight tests of the null hypotheses for the coefficient curves versus the general alternatives that the corresponding null hypotheses are not true and all the coefficients are allowed to be time-varying as in the model (9.72).

Null Hypotheses:

(a) H_0: $\beta_0(t) = \beta_0$ with unknown β_0 and $\beta_l(t) = 0$, $l = 1, 2, 3$. The baseline SBP curve is an unknown constant, and the effects of race, height and BMI are identically zero across the age range.

(b) H_0: $\beta_l(t) = 0$, $l = 1, 2, 3$. The baseline SBP curve may be time-varying, but the effects of race, height and BMI are identically zero across the age range.

(c) H_0: $\beta_1(t) = 0$. The baseline SBP curve and the effects of height and BMI may be time-varying, but the effect of race is identically zero across the age range.

(d) H_0: $\beta_2(t) = 0$. The baseline SBP curve and the effects of race and BMI may be time-varying, but the effect of height is identically zero across the age range.

(e) H_0: $\beta_3(t) = 0$. The baseline SBP curve and the effects of race and height may be time-varying, but the effect of BMI is identically zero across the age range.

(f) H_0: $\beta_1(t) = \beta_1$ with unknown β_1. The baseline SBP curve and the effects of

Table 9.1 *The null and alternative hypotheses, observed test statistics T_n computed from the NGHS BP data and the empirical p-values obtained from the resampling-subject bootstrap procedure.*

Null hypothesis H_1: H_0 is not true	Test Statistic T_n	Empirical P-value
(a) H_0: $\beta_0(t) = \beta_0$, unknown β_0 & $\beta_l(t) = 0$, $l = 1, 2, 3$	0.262	<0.0002
(b) H_0: $\beta_l(t) = 0$, $l = 1, 2, 3$	0.136	<0.0002
(c) H_0: $\beta_1(t) = 0$	0.0019	0.206
(d) H_0: $\beta_2(t) = 0$	0.0212	<0.0002
(e) H_0: $\beta_3(t) = 0$	0.0867	<0.0002
(f) H_0: $\beta_1(t) = \beta_1$, unknown β_1	0.00049	0.233
(g) H_0: $\beta_2(t) = \beta_2$, unknown β_2	0.00151	0.0064
(h) H_0: $\beta_3(t) = \beta_3$, unknown β_3	0.00139	<0.0002

height and BMI may be time-varying, but the effect of race is time-invariant across the age range.

(g) H_0: $\beta_2(t) = \beta_2$ with unknown β_2. The baseline SBP curve and the effects of race and BMI may be time-varying, but the effect of height is time-invariant across the age range.

(h) H_0: $\beta_3(t) = \beta_3$ with unknown β_3. The baseline SBP curve and the effects of race and height may be time-varying, but the effect of BMI is time-invariant across the age range.

To apply the testing procedure of Section 9.3.3 to the above tests, we compute the weighted residual sum of squares RSS_0 under each of the specific null hypotheses (a) through (h), and the weighted residual sum of squares RSS_1 under the general alternative that the corresponding null hypothesis is not true. Here, the RSS_0 and RSS_1 are not computed exactly from (9.66) and (9.68) but should be the modified versions of (9.66) and (9.68) by taking the specific null and alternative hypotheses in (a) through (h) into account. To compute the critical values of the test statistics T_n as defined in (9.69), we apply the resampling-subject bootstrap procedure in Section 9.3.3 with $B = 5000$ replications within each bootstrap run.

Table 9.1 summarizes the observed test statistics T_n computed from the original NGHS BP dataset and their corresponding empirical p-values obtained from the resampling-subject bootstrap procedure. As expected and consistent with the results in Figure 9.1, we observe that the baseline SBP curve and the covariate effects of height and BMI are all significantly time-varying with very small empirical p-values. However, there is insufficient evidence for a significant time-varying effect of race after including the time-varying baseline SBP curve and covariate effects of height and BMI into the model.

9.5 Asymptotic Properties

The asymptotic properties of this section are discussed separately for the general basis approximation estimators and the B-spline approximation estimators. For estimators based on the general basis functions, we only establish the consistency and convergence rates of $\widehat{\beta}(t)$. Further details of the asymptotic properties, such as the asymptotic mean squared errors and the asymptotic distributions, depend on the choices of basis functions and may not be expressed under a general formulation. When the estimators are obtained from the B-spline basis approximations, their asymptotic mean squared errors and asymptotic distributions can be explicitly derived using the special properties of B-splines.

9.5.1 Integrated Squared Errors

We first establish a framework for the asymptotic consistency and convergence rates of the estimator $\widehat{\beta}(t)$ of (9.9) based on a set of known basis functions $\{B_{ls}(t) : s = 1, \ldots, K_l; l = 0, \ldots, k\}$. The forms of the basis functions are not specified as long as the approximations (9.2) can be used. Consequently, the asymptotic results of this section may be applied to the commonly used basis functions, such as polynomial bases, Fourier bases, Wavelet bases, and B-spline bases.

Unlike the local smoothing estimators of Chapters 7 and 8, the basis approximation estimator $\widehat{\beta}(t)$ depends on the linear function space

$$\begin{cases} \mathbb{G}_l = \{ \text{ linear function space spanned by } \mathscr{B}_l(t)\}, \\ \mathscr{B}_l = \{B_{ls}(t) : s = 1, \ldots, K_l\} \end{cases} \tag{9.73}$$

used to approximate $\beta_l(t)$. Thus, in order to assess the performance of $\widehat{\beta}(t)$, it is important to introduce a distance which is used to measure the closeness between $\widehat{\beta}(t)$ and $\beta(t)$. Let

$$\|a\|_{L_2} = \left[\int_{\mathscr{T}} a^2(t)\, dt \right]^{1/2} \tag{9.74}$$

be the L_2-norm (L_2-distance) of any square integrable real-valued function $a(t)$ on $\mathscr{T}$ and let

$$\|A\|_{L_2} = \left(\sum_{l=0}^{k} \|a_l\|_{L_2}^2 \right)^{1/2} \tag{9.75}$$

be the L_2-norm of any vector of real-valued functions $A(t) = \left(a_0(t), \ldots, a_k(t)\right)^T$ on $\mathscr{T}$. Based on the L_2-distances (9.74) and (9.75), we define the integrated squared error (ISE) of $\beta_l(t)$ as

$$ISE\left(\widehat{\beta}_l\right) = \left\|\widehat{\beta}_l - \beta_l\right\|_{L_2}^2 = \int_{\mathscr{T}} \left[\widehat{\beta}_l(t) - \beta_l(t)\right]^2 dt \tag{9.76}$$

and the integrated squared error of $\beta(t) = (\beta_0(t), \cdots, \beta_k(t))^T$ as

$$ISE\left(\widehat{\beta}\right) = \sum_{l=0}^{k} \left[ISE\left(\widehat{\beta}_l\right)\right]. \tag{9.77}$$

The above integrated squared errors lead to the following definition of asymptotic consistency and useful measure of functional approximation for the estimator $\widehat{\beta}(t)$ for $t \in \mathscr{T}$:

Consistency. The estimator $\widehat{\beta}(t)$ is asymptotically consistent for $\beta(t)$ for $t \in \mathscr{T}$ if $\lim_{n \to \infty} ISE\left(\widehat{\beta}\right) = 0$ holds in probability, or equivalently $\lim_{n \to \infty} ISE\left(\widehat{\beta}_l\right) = 0$ holds in probability for all $l = 0, \ldots, k$. $\square$

Bias-Variance Decomposition. Similar to the well-known strategy of bias-squared and variance decomposition, we can decompose the integrated squared error $ISE\left(\widehat{\beta}\right)$ into the sum of two components. First, let

$$\widetilde{\beta}(t) = \left(\widetilde{\beta}_0(t), \ldots, \widetilde{\beta}_k(t)\right)^T = E\left[\widehat{\beta}(t) \big| \mathscr{X}\right] \tag{9.78}$$

be the conditional expectation of the estimator $\widehat{\beta}(t)$ given the covariates $\mathscr{X}$, where

$$\mathscr{X} = \left\{ (t_{ij}, \mathbf{X}_{ij}^T)^T : i = 1, \ldots, n; j = 1, \ldots, n_i \right\}. \tag{9.79}$$

The Cauchy-Schwarz inequality and the definition of consistency given above imply that, $\widehat{\beta}_l(t)$ is an asymptotically consistent estimator of $\beta_l(t)$ for $t \in \mathscr{T}$, i.e., $\lim_{n \to \infty} ISE\left(\widehat{\beta}_l\right) = 0$ in probability, if and only if both $\left\|\widehat{\beta}_l - \widetilde{\beta}_l\right\|_{L_2}$ and $\left\|\widetilde{\beta}_l - \beta_l\right\|_{L_2}$ tend to zero in probability. Consequently, the consistency of $\widehat{\beta}(t)$ holds for $t \in \mathscr{T}$ if and only if $\left\|\widehat{\beta}_l - \widetilde{\beta}_l\right\|_{L_2}$ and $\left\|\widetilde{\beta}_l - \beta_l\right\|_{L_2}$ tend to zero in probability for all $l = 0, \ldots, k$. $\square$

Functional Approximation. Since we approximate $\beta_l(t)$ by functions in the linear space $\mathbb{G}_l$ defined in (9.73), the asymptotic derivations of $ISE\left(\widehat{\beta}\right)$ depend on some L_∞-distances between $\beta_l(t)$ and $\mathbb{G}_l$. Specifically, let

$$D\left(\beta_l, \mathbb{G}_l\right) = \inf_{g \in \mathbb{G}_l} \sup_{t \in \mathscr{T}} |\beta_l(t) - g(t)| \tag{9.80}$$

be the L_∞-distance between $\beta_l(t)$ and $\mathbb{G}_l$. Then, as will be seen in Section 9.5.3, the asymptotic properties of $ISE\left(\widehat{\beta}\right)$ depend on

$$\begin{cases} \rho_n &= \sum_{l=0}^{k} D\left(\beta_l, \mathbb{G}_l\right), \\ A_{n,l} &= \sup_{g \in \mathbb{G}_l, \|g\|_{L_2} \neq 0} \left(\sup_{t \in \mathscr{T}} |g(t)| / \|g\|_{L_2}\right), \\ A_n &= \max_{0 \leq l \leq k} A_{n,l}. \end{cases} \tag{9.81}$$

Examples of ρ_n and A_n for the commonly used basis functions, such as polynomials, splines and trigonometric bases, have been given in Huang (1998, Section 2.2). Intuitively, ρ_n describes the overall L_∞-distance between $\beta_l(t)$ and $\mathbb{G}_l$ for all $l = 0, \ldots, k$, while A_n describes the largest ratio of the L_∞-distance over the L_2-distance for all the functions in $\mathbb{G}_l$ and all $l = 0, \ldots, k$. The purpose of (9.81) is to provide an upper bound for the "distance" between the true coefficient curves $\beta(t)$ and the product space $\{\mathbb{G}_0 \times \cdots \times \mathbb{G}_k\}$, which contains the basis approximations of $\beta(t)$.   □

9.5.2 Asymptotic Assumptions

We make the following technical assumptions for the derivations of asymptotic consistency and convergence rates of the basis approximation estimator $\widehat{\beta}(t)$ of (9.9) under any known basis functions $\{B_{ls}(t) : s = 1, \ldots, K_l; l = 0, \ldots, k\}$:

(a) *The observation time points follow a random design in the sense that $\{t_{ij} : i = 1, \ldots, n; j = 1, \ldots, n_i\}$ are chosen independently from an unknown distribution $F_T(\cdot)$ with a density $f_T(\cdot)$ on the finite interval $\mathscr{T}$. The density function $f_T(t)$ is uniformly bounded away from 0 and infinity, i.e., there are positive constants M_1 and M_2 such that $M_1 \le f(t) \le M_2$ for all $t \in \mathscr{T}$.*

(b) *Let $E_{\mathbf{XX}^T}(t) = E\left[\mathbf{X}(t)\mathbf{X}^T(t)\right]$ and $\lambda_0(t) \le \ldots \le \lambda_k(t)$ be the eigenvalues of $E_{\mathbf{XX}^T}(t)$. Then, $\lambda_l(t)$ for all $l = 0, \ldots, k$ are uniformly bounded away from 0 and infinity for all $t \in \mathscr{T}$.*

(c) *The range of the covariates $X^{(l)}(t)$ is bounded in the sense that there is a positive constant M_3 such that $\left|X^{(l)}(t)\right| \le M_3$ for all $t \in \mathscr{T}$ and $l = 0, \ldots, k$.*

(d) *There is a positive constant M_4 such that the second moment of the error process $\varepsilon(t)$ is bounded by M_4, i.e., $E\left[\varepsilon(t)^2\right] \le M_4$, for all $t \in \mathscr{T}$.*   □

The above assumptions, with some minor exceptions, are essentially similar to the assumptions of Section 7.5.1 for the least squares kernel estimators. These assumptions can be easily verified in real applications. It is important to note that, because our objective is to establish the asymptotic consistency of a general basis approximation estimator $\widehat{\beta}(t)$, the above assumptions are sufficiently weak for this purpose. Some stronger assumptions will be used later in Section 9.5.3 to establish the asymptotic mean square errors and asymptotic distributions of the B-spline estimators of $\beta(t)$.

9.5.3 Convergence Rates for Integrated Squared Errors

1. General Basis Approximation Estimators

The next theorem, whose proof is given in Section 9.5.4, establishes the general results of asymptotic consistency and convergence rates for the least squares basis approximation estimator $\widehat{\beta}(t)$ of (9.9).

Theorem 9.1. *Let $K_n = \max_{0 \le l \le k} K_l$ be the maximum number of basis approximation terms with K_l for any $0 \le l \le k$ defined in (9.2), where K_n may or may not tend to infinity as n tends to infinity. If Assumptions (a) through (d) are satisfied, $\lim_{n \to \infty} \rho_n = 0$ and*

$$\lim_{n \to \infty} \left\{ A_n^2 K_n \max \left[\max_{1 \le i \le n} (n_i w_i), \sum_{i=1}^{n} (n_i^2 w_i^2) \right] \right\} = 0 \qquad (9.82)$$

for ρ_n and A_n defined in (9.81), then $\widehat{\beta}(t)$ uniquely exists with probability tending to one and is a consistent estimator of $\beta(t)$ for any $t \in \mathcal{T}$.

In addition, the convergence rates under the L_2-norms are given as follows:

(a) $\left\| \widehat{\beta} - \widetilde{\beta} \right\|_{L_2}^2 = O_p \left[K_n \sum_{i=1}^{n} (n_i^2 w_i^2) \right]$,

(b) $\left\| \widetilde{\beta} - \beta \right\|_{L_2} = O_p(\rho_n)$, *and*

(c) $ISE\left(\widehat{\beta}\right) = O_p \left[K_n \sum_{i=1}^{n} (n_i^2 w_i^2) + \rho_n^2 \right]$. ∎

Proof of Theorem 9.1 is given in Section 9.5.4.

Each of the convergence rates in Theorem 9.1(a)-(b) gives a specific component of the overall convergence rate of $IST\left(\widehat{\beta}\right)$. For Theorem 9.1(a), $\widetilde{\beta}(t)$, which is defined in (9.78) to be the conditional mean of $\widehat{\beta}(t)$ given the covariates, can be viewed as the "estimable part" of $\beta(t)$, and the convergence rate under the L_2-norm depends on K_n, the numbers of basis approximation terms, as well as $\sum_{i=1}^{n} (n_i^2 w_i^2)$. Here $\left\| \widehat{\beta} - \widetilde{\beta} \right\|_{L_2}^2$ represents the variance part of $ISE\left(\widehat{\beta}\right)$, which tends to zero slower when K_n increases. The component $\left\| \widetilde{\beta} - \beta \right\|_{L_2}$ in Theorem 9.1(b) is the L_2-distance between $\widetilde{\beta}(t)$ and the true coefficient curves $\beta(t)$. Since, by (9.81), ρ_n measures how well the linear space $\mathbb{G}_l$ approximates $\beta_l(t)$ for all $l = 0, \ldots, k$, it is not surprising that $\left\| \widetilde{\beta} - \beta \right\|_{L_2}$, which can be viewed as the bias part of $ISE\left(\widehat{\beta}\right)$, depends on ρ_n. In practice, it is impossible to know the value of ρ_n for a given set of function spaces $\{\mathbb{G}_0, \ldots, \mathbb{G}_k\}$, so that $\left\| \widetilde{\beta} - \beta \right\|_{L_2}$ is the un-estimable component of $ISE\left(\widehat{\beta}\right)$. This is in contrast to the local smoothing estimator results of Section 7.5, where the bias of a local smoothing estimator of $\beta(t)$ depends on the smoothness assumptions of the unknown coefficient curves $\beta(t)$, which can be estimated in practice. Since Theorem 9.1 is for $\widehat{\beta}(\cdot)$ with general basis choices, the convergence rate of $ISE\left(\widehat{\beta}\right)$ given Theorem 9.1(c) is not necessarily the optimal rate for a particular type of basis functions.

2. Optimal Convergence Rate for B-Spline Estimators

For the special case of B-splines, the next theorem gives the improved convergence rates for $\widehat{\beta}^{(S)}(t)$ of (9.18). In this theorem, we assume that each $\mathbb{G}_l$ is a space of polynomial splines on $\mathcal{T}$ with a fixed degree and the knots have bounded mesh ratio, that is, the ratios of the differences between consecutive knots are bounded away from zero and infinity uniformly in n.

Theorem 9.2. *Suppose that $\widehat{\beta}^{(S)}(t)$ is defined as in (9.18) with a B-spline basis. If the conditions of Theorem 9.1 are satisfied, then*

(a) $\left\| \widehat{\beta}^{(S)} - \widetilde{\beta}^{(S)} \right\|_{L_2}^2 = O_p \left\{ \sum_{i=1}^{n} n_i^2 \, w_i^2 \left[(K_n/n_i) + 1 \right] \right\}$,

(b) $\left\| \widetilde{\beta}^{(S)} - \beta \right\|_{L_2} = O_p(\rho_n)$, *and*

(c) $ISE\left(\widehat{\beta}^{(S)} \right) = O_p \left\{ \sum_{i=1}^{n} n_i^2 \, w_i^2 \left[(K_n/n_i) + 1 \right] + \rho_n^2 \right\}$.   ∎

Proof of Theorem 9.1 is given in Section 9.5.4.

Similar interpretations for the results of Theorem 9.1(a)-(c) can be extended to Theorem 9.2(a)-(c). In the following remarks, we illustrate a number of possible convergence rates under some special choices of w_i and basis functions.

Effect of Weight Choices. As shown in both Theorems 9.1 and 9.2, different choices of w_i generally lead to different convergence rates for the estimators. For the general basis approximation estimators $\widehat{\beta}(t)$ in Theorem 9.1, the convergence rates for the variance part $\|\widehat{\beta} - \widetilde{\beta}\|_{L_2}^2$ given in Theorem 9.1(a) are

$$
\begin{cases}
\sum_{i=1}^{n} \left(K_n \, n_i^2 \, w_i^2 \right) = K_n/n, & \text{when } w_i = 1/(n \, n_i); \\
\sum_{i=1}^{n} \left(K_n \, n_i^2 \, w_i^2 \right) = K_n \sum_{i=1}^{n} \left(n_i^2/N^2 \right), & \text{when } w_i = 1/N,
\end{cases}
\tag{9.83}
$$

and, by Theorem 9.1(c), the convergence rates for $ISE\left(\widehat{\beta} \right)$ are

$$
\begin{cases}
\sum_{i=1}^{n} \left(K_n \, n_i^2 \, w_i^2 \right) + \rho_n^2 = K_n/n + \rho_n^2, & \text{when } w_i = 1/(n \, n_i); \\
\sum_{i=1}^{n} \left(K_n \, n_i^2 \, w_i^2 \right) + \rho_n^2 = K_n \sum_{i=1}^{n} \left(n_i^2/N^2 \right) + \rho_n^2, & \text{when } w_i = 1/N.
\end{cases}
\tag{9.84}
$$

Similarly, for the B-spline approximation estimators $\widehat{\beta}^{(S)}(t)$ in Theorem 9.2, the convergence rates for the variance part $\|\widehat{\beta}^{(S)} - \widetilde{\beta}^{(S)}\|_{L_2}^2$ given in Theorem 9.2(a) are

$$
\begin{cases}
\sum_{i=1}^{n} n_i^2 \, w_i^2 \left[(K_n/n_i) + 1 \right] = K_n \sum_{i=1}^{n} \left(1/n_i n^2 \right) + K_n/n, \\
\qquad \text{when } w_i = 1/(n \, n_i); \\
\sum_{i=1}^{n} n_i^2 \, w_i^2 \left[(K_n/n_i) + 1 \right] = K_n/N + \sum_{i=1}^{n} \left(n_i^2/N^2 \right), \\
\qquad \text{when } w_i = 1/N,
\end{cases}
\tag{9.85}
$$

and, by Theorem 9.2(c), the convergence rates for $ISE(\widehat{\beta})$ are

$$
\begin{cases}
\sum_{i=1}^n n_i^2 w_i^2 \left[(K_n/n_i) + 1 \right] + \rho_n^2 = K_n \sum_{i=1}^n \left(1/n_i n^2 \right) + K_n/n + \rho_n^2, \\
\quad \text{when } w_i = 1/(nn_i); \\
\sum_{i=1}^n n_i^2 w_i^2 \left[(K_n/n_i) + 1 \right] + \rho_n^2 = K_n/N + \sum_{i=1}^n \left(n_i^2/N^2 \right) + \rho_n^2, \\
\quad \text{when } w_i = 1/N.
\end{cases}
\tag{9.86}
$$

As shown in Theorems 7.1 and 7.2 of Section 7.5, $\lim_{n\to\infty} \sum_{i=1}^n \left(n_i^2/N^2 \right) = 0$ if and only if $\lim_{n\to\infty} \max_{1\le i\le n} \left(n_i/N \right) = 0$. Thus, similar to the local smoothing methods of Chapter 7, the $w_i = 1/N$ weight may lead to inconsistent estimators $\widehat{\beta}(\cdot)$, while $w_i = 1/(nn_i)$ leads to consistent $\widehat{\beta}(\cdot)$ for all choices of n_i. □

Effects of Functional Approximations. When specific smoothness conditions for $\beta(t)$ are given, more precise convergence rates can be deduced by determining the size of $D(\beta_l, \mathbb{G}_l)$, the discrepancy between $\beta_l(t)$ and the linear space $\mathbb{G}_l$. For example, when $\beta_l(t)$ has bounded second derivatives and $\mathbb{G}_l$ is a space of cubic B-splines with K_n interior knots on $\mathscr{T}$, we have $D(\beta_l, \mathbb{G}_l) = O(K_n^{-2})$ (Schumaker, 1981, Theorem 6.27) and, by Theorem 9.1(c), $ISE(\widehat{\beta}) = O_p(K_n/n + K_n^{-4})$. For the special choice of $K_n = O(n^{1/5})$, this reduces to $ISE(\widehat{\beta}) = O_p(n^{-4/5})$, which is the optimal convergence rate for nonparametric regression with cross-sectional i.i.d. data under the same smoothness conditions (Stone, 1982). □

9.5.4 Theoretical Derivations

We provide a number of technical results, which are useful for the theoretical derivations of the main consistency results, and then give the proofs of Theorems 9.1 and 9.2. We assume, without loss of generality, that $\{ B_{ls}(t) : s = 1, \ldots, K_l \}$ is an orthonormal basis for the linear space $\mathbb{G}_l$ for any $l = 1, \ldots, k$ with inner product

$$
\langle f_1, f_2 \rangle = \int_{\mathscr{T}} f_1(t) f_2(t) \, dt.
$$

Then, for any $g_l \in \mathbb{G}_l$ there is an unique representation $g_l(t) = \sum_{s=1}^{K_l} \gamma_{ls} B_{ls}(t)$, so that the L_2-norm of $g(t) = (g_0(t), \ldots, g_k(t))^T$ is

$$
\|g\|_{L_2} = \left[\sum_{l=0}^k \|g_l\|_{L_2}^2 \right]^{1/2} = \left(\sum_{l=0}^k \sum_{s=1}^{K_l} \gamma_{ls}^2 \right)^{1/2}.
$$

Following the notation of Huang (1998, p. 246), we write $a_n \asymp b_n$ if both a_n and b_n are positive and a_n/b_n and b_n/a_n are bounded for all n. Let T be the random variable of time with distribution $F(t)$ and density $f(t)$ for any $t \in \mathscr{T}$.

1. Three Technical Lemmas

The following series of lemmas summarize the useful technical results for the proof of Theorem 9.1.

Lemma 9.1. *If the condition (9.82) in Theorem 9.1 is satisfied, then*

$$\sup_{g_l \in \mathbb{G}_l, l=0,\dots,k} \left| \frac{\sum_{i=1}^{n} \sum_{j=1}^{n_i} w_i \left[\sum_{l=0}^{k} X_{ij}^{(l)} g_l(t_{ij}) \right]^2}{E \left[\sum_{l=0}^{k} X^{(l)}(T) g_l(T) \right]^2} - 1 \right| = o_p(1) \tag{9.87}$$

holds, when n is sufficiently large.   ∎

Proof of Lemma 9.1:

The lemma can be proved using arguments similar to those in the proof of Lemma 10 of Huang (1998). Thus, we do not repeat the tedious details here, since all the derivations can be directly adapted by changing the notation used under the current context.   ∎

Lemma 9.2. *If the condition (9.82) in Theorem 9.1 holds, $\mathbf{U} = \left(\mathbf{U}_1^T, \cdots, \mathbf{U}_n^T \right)^T$ with $\mathbf{U}_i$ defined in (9.6) and $\mathbf{W}$ is the block diagonal matrix with diagonal matrices $\mathbf{W}_1, \dots, \mathbf{W}_n$ defined in (9.7), there is an interval $[M_1, M_2]$ with some positive constants endpoints $0 < M_1 < M_2$, such that, as $n \to \infty$,*

$$P\left\{ \text{all the eigenvalues of } \mathbf{U}^T \mathbf{W} \mathbf{U} \text{ fall in } [M_1, M_2] \right\} \to 1. \tag{9.88}$$

Then, with probability tending to 1, $\mathbf{U}^T \mathbf{W} \mathbf{U} = \sum_{i=1}^{n} \left(\mathbf{U}_i^T \mathbf{W}_i \mathbf{U}_i \right)$ is invertible with inverse matrix $\left(\mathbf{U}^T \mathbf{W} \mathbf{U} \right)^{-1}$, and $\widehat{\beta}(t)$ exists uniquely for all $t \in \mathcal{T}$.   ∎

Proof of Lemma 9.2:

By Lemma 9.1, the following equation holds with probability tending to one as $n \to \infty$,

$$\gamma^T \mathbf{U}^T \mathbf{W} \mathbf{U} \gamma = \sum_{i=1}^{n} \sum_{j=1}^{n_i} \left\{ w_i \left[\sum_{l=0}^{k} X_{ij}^{(l)} g_l(t_{ij}) \right]^2 \right\}$$

$$\asymp E\left\{ \left[\sum_{l=0}^{k} X^{(l)}(T) g_l(T) \right]^2 \right\}, \tag{9.89}$$

where $g_l(t) = \sum_{s=1}^{K_l} \left[\gamma_{ls} B_{ls}(t) \right]$ for $l = 0, \dots, k$, and γ is the vector with entries γ_{ls} for $s = 1, \dots, K_l$ and $l = 0, \dots, k$. Using conditional expectations and Assumptions (a) and (b), we have that

$$E\left\{ \left[\sum_{l=0}^{k} X^{(l)}(T) g_l(T) \right]^2 \right\} = \int_{\mathcal{T}} g^T(t) E_{\mathbf{X}\mathbf{X}^T}(t) g(t) f_T(t) dt$$

$$\asymp \int_{\mathcal{T}} g^T(t) g(t) \, dt \qquad (9.90)$$

$$= \sum_{l=0}^{k} \|g_l\|_{L_2}^2$$

holds uniformly for all $g_l \in \mathbb{G}_l$, $l = 0, \ldots, k$. Consequently, $\gamma^T \mathbf{U}^T \mathbf{W} \mathbf{U} \gamma \asymp \gamma^T \gamma$ holds uniformly for all γ. The conclusion of (9.88) follows. ∎

Lemma 9.3 *If the condition (9.82) holds, then*

$$\|\widehat{\beta} - \widetilde{\beta}\|_{L_2}^2 = O_p \left[K_n \sum_{i=1}^{n} \left(n_i^2 \, w_i^2 \right) \right] \qquad (9.91)$$

and

$$\left(\sum_{l=0}^{k} \|\widetilde{\beta}_l - \beta_l\|_{L_2}^2 \right)^{1/2} = O_p(\rho_n) \qquad (9.92)$$

hold, when n is sufficiently large. ∎

Proof of Lemma 9.3:
First, using direct calculation, we get

$$\begin{cases} \|\widehat{\beta} - \widetilde{\beta}\|_{L_2}^2 = \sum_{l=0}^{k} \sum_{s=1}^{K_l} |\widehat{\gamma}_{ls} - \widetilde{\gamma}_{ls}|^2, \\ \widehat{\gamma} - \widetilde{\gamma} = \left[\sum_{i=1}^{n} \left(\mathbf{U}_i^T \mathbf{W}_i \mathbf{U}_i \right) \right]^{-1} \sum_{i=1}^{n} \left(\mathbf{U}_i^T \mathbf{W}_i \varepsilon_i \right) = \left(\mathbf{U}^T \mathbf{W} \mathbf{U} \right)^{-1} \mathbf{U}^T \mathbf{W} \varepsilon \end{cases} \qquad (9.93)$$

and, by Lemma 9.2, with probability tending to 1 as $n \to \infty$,

$$\left| \left(\mathbf{U}^T \mathbf{W} \mathbf{U} \right)^{-1} \mathbf{U}^T \mathbf{W} \varepsilon \right|^2 \asymp \varepsilon^T \mathbf{W} \mathbf{U} \mathbf{U}^T \mathbf{W} \varepsilon. \qquad (9.94)$$

Using the Cauchy-Schwarz inequality and Assumptions (c) and (d), we have that

$$E \left(|\mathbf{U}_i^T \mathbf{W}_i \varepsilon_i|^2 \right) = E \left\{ \sum_{l=0}^{k} \sum_{s=1}^{K_l} w_i^2 \left[\sum_{j=1}^{n_i} X_{ij}^{(l)} B_{ls}(t_{ij}) \varepsilon_{ij} \right]^2 \right\} = O(K_n n_i^2 w_i^2). \qquad (9.95)$$

Consequently, it follows from (9.95) that

$$E \left(\varepsilon^T \mathbf{W} \mathbf{U} \mathbf{U}^T \mathbf{W} \varepsilon \right) = \sum_{i=1}^{n} E \left(\varepsilon_i^T \mathbf{W}_i \mathbf{U}_i \mathbf{U}_i^T \mathbf{W}_i \varepsilon_i^T \right) = O \left[K_n \sum_{i=1}^{n} \left(n_i^2 \, w_i^2 \right) \right]. \qquad (9.96)$$

The Markov inequality then implies that, by (9.96),

$$\left| \left(\mathbf{U}^T \mathbf{W} \mathbf{U} \right)^{-1} \mathbf{U}^T \mathbf{W} \varepsilon \right|^2 = O_P \left[K_n \sum_{i=1}^{n} \left(n_i^2 \, w_i^2 \right) \right]. \qquad (9.97)$$

The first assertion of the lemma (9.91) follows from (9.93) and (9.97).

To prove the second assertion (9.92), we consider the vector of functions $g^*(t) = \left(g_0^*(t), \ldots, g_k^*(t)\right)^T$ with $g_l^* \in \mathbb{G}_l$, such that

$$\sup_{t \in \mathscr{T}} \left|g_l^*(t) - \beta_l(t)\right| = D\left(\beta_l, \mathbb{G}_l\right). \tag{9.98}$$

Since, by the triangle inequality,

$$\left|\widetilde{\beta}_l(t) - \beta_l(t)\right| \leq \left|\widetilde{\beta}_l(t) - g_l^*(t)\right| + \left|g_l^*(t) - \beta_l(t)\right|, \tag{9.99}$$

it follows from (9.98) and (9.99) that it suffices to show that

$$\left(\sum_{l=0}^{k} \left\|\widetilde{\beta}_l - g_l^*\right\|_{L_2}^2\right)^{1/2} = O_p\left(\rho_n\right). \tag{9.100}$$

Since $g_l^* \in \mathbb{G}_l$, there is a $\gamma^* = \left((\gamma_0^*)^T, \ldots, (\gamma_k^*)^T\right)^T$ with $\gamma_l^* = \left(\gamma_{l1}^*, \ldots, \gamma_{lK_l}^*\right)^T$ such that $g^*(t) = B(t)\gamma^*$. In addition, we note that $\widetilde{\beta}(t) = B(t)\widetilde{\gamma}$. It follows from Lemma 9.2 that

$$
\begin{aligned}
\sum_{l=0}^{k} \left\|\widetilde{\beta}_l - g_l^*\right\|_{L_2}^2 &= \sum_{l=0}^{k}\sum_{s=1}^{K_l} \left|\widetilde{\gamma}_{ls} - \gamma_{ls}^*\right|^2 \\
&\asymp \left(\widetilde{\gamma} - \gamma^*\right)^T \left(\sum_{i=1}^{n} \mathbf{U}_i^T \mathbf{W}_i \mathbf{U}_i\right)\left(\widetilde{\gamma} - \gamma^*\right). \tag{9.101}
\end{aligned}
$$

Let $\widetilde{Y}_i = \left(\widetilde{Y}_{i1}, \ldots, \widetilde{Y}_{in_i}\right)$ with $\widetilde{Y}_{ij} = \mathbf{X}_{ij}^T \beta\left(t_{ij}\right)$ for $j = 1, \ldots, n_i$ and $i = 1, \ldots, n$. Since, by the definition of $\widetilde{\gamma}$,

$$\sum_{i=1}^{n} \left[\mathbf{U}_i^T \mathbf{W}_i \left(\widetilde{Y}_i - \mathbf{U}_i \widetilde{\gamma}\right)\right] = 0,$$

it follows from (9.98) and the definition of γ^* that

$$\sum_{i=1}^{n} w_i \left|\mathbf{U}_i \widetilde{\gamma} - \mathbf{U}_i \gamma^*\right|^2 \leq \sum_{i=1}^{n} w_i \left|\widetilde{Y}_i - \mathbf{U}_i \gamma^*\right|^2 \tag{9.102}$$

and, by Assumption (c),

$$\left|\mathbf{X}_{ij}^T \left[\beta\left(t_{ij}\right) - B\left(t_{ij}\right)\gamma^*\right]\right| = O\left(\rho_n\right). \tag{9.103}$$

Thus, combining (9.102) and (9.103), we have the inequality

$$\left(\widetilde{\gamma} - \gamma^*\right)^T \left(\sum_{i=1}^{n} \mathbf{U}_i^T \mathbf{W}_i \mathbf{U}_i\right)\left(\widetilde{\gamma} - \gamma^*\right) \leq \sum_{i=1}^{n}\sum_{j=1}^{n_i} \left(w_i \rho_n^2\right) = \rho_n^2. \tag{9.104}$$

The assertion of the lemma (9.92) then follows from (9.100) and (9.104). This completes the proof of Lemma 9.3. ∎

2. Proofs of Main Theorems

Proof of Theorem 9.1:

Theorem 9.1 is a direct consequence of Lemma 9.3 and the triangle inequality. ∎

Proof of Theorem 9.2:

This theorem can be proved along the same lines as Theorem 9.1. But, since in this case the basis functions are B-splines, we need to use the special properties of B-spline functions. For $l = 1, \ldots, k$ and $s = 1, \ldots, K_l$, we denote by $N_{ls}(t)$ the B-splines as defined in de Boor (1978, Ch. IX), and set $B_{ls}(t) = K_l^{1/2} N_{ls}(t)$, where the B-splines $N_{ls}(t)$ are non-negative functions satisfying

$$\begin{cases} \sum_{s=1}^{K_l} N_{ls}(t) = 1 & \text{for } t \in \mathcal{T}, \\ \int_{\mathcal{T}} N_{ls}(t)\, dt \leq M/K_l & \text{for some constant } M, \end{cases} \tag{9.105}$$

and, in addition, there are positive constants $M_1 > 0$ and $M_2 > 0$, such that

$$\frac{M_1}{K_l} \left(\sum_{s=1}^{K_l} \gamma_{ls}^2 \right) \leq \int_{\mathcal{T}} \left[\sum_{s=1}^{K_l} \gamma_{ls} N_{ls}(t) \right]^2 dt \leq \frac{M_2}{K_l} \left(\sum_{s=1}^{K_l} \gamma_{ls}^2 \right), \tag{9.106}$$

for $\gamma_{ls} \in R$ and $s = 1, \ldots, K_l$.

If we use the above properties (9.105) and (9.106) of B-splines, (9.95) can be strengthened to

$$\begin{aligned} E\left[|\mathbf{U}_i^T \mathbf{W}_i \boldsymbol{\varepsilon}_i|^2 \right] &= E\left\{ \sum_{l=0}^{k} \sum_{s=1}^{K_l} w_i^2 \left[\sum_{j=1}^{n_i} \mathbf{X}_{ij}^{(l)} B_{ls}(t_{ij})\, \varepsilon_{ij} \right]^2 \right\} \\ &\leq w_i^2 \sum_{l=0}^{k} \left\{ \left[n_i + (n_i^2 - n_i) \frac{1}{K_l} \right] K_l \right\}. \end{aligned} \tag{9.107}$$

The rest of the proof is proceeded the same as that of Theorem 9.1, and thus is omitted to avoid repetition. ∎

9.5.5 Consistent Hypothesis Tests

The concept of asymptotic consistency can be analogously extended to the procedures of hypothesis testing. As described in the previous section, the purpose of asymptotic consistency for a smoothing estimator is to ensure the good property that the estimator is getting close to the true coefficient curves $\beta(t)$ when the number of subjects n increases.

1. Consistency of Test Procedure

For hypothesis testing, we would like to have a test statistic which has the following property: When the number of subjects n is large, the value of

the test statistic is close to zero if the null hypothesis holds, and the value of
the test statistic is larger than a constant if the alternative hypothesis holds.
Thus, we have the following definition for the asymptotic consistency of a test
statistic.

Consistency of Test Statistic. A test statistic T_n for testing the null
hypothesis H_0 versus the alternative H_1 is asymptotic consistent if $T_n \to 0$
as $n \to \infty$ in probability when H_0 holds, and $T_n > \delta$ in probability for some
constant $\delta > 0$ when n is sufficiently large and H_1 holds.   □

2. Rejection Region of Consistent Test

If a test statistic T_n is asymptotically consistent (or simply consistent),
our decision of rejecting or accepting the null hypothesis H_0 can be made
by evaluating whether T_n is larger than a critical value. Especially, once an
appropriate critical value is obtained, our testing decision is:

$$\begin{cases} \text{Rejecting the null hypothesis } H_0 \text{ if } T_n \geq c; \\ \text{Accepting the null hypothesis } H_0 \text{ if } T_n < c. \end{cases} \qquad (9.108)$$

When the number of subject n is sufficiently large and T_n is consistent, this
testing procedure will ensure a small type I error, $P(T_n \geq c | H_0)$, and a large
power, $P(T_n \geq c | H_1)$. The next theorem shows that the test statistic T_n de-
fined by (9.68) and (9.69) is asymptotically consistent for testing the null and
alternative hypotheses H_0 and H_1, respectively, in (9.63). Consequently, the
decision rule given in (9.108) is appropriate.

Theorem 9.3. *Suppose that the assumptions of Theorem 9.1 are satisfied,*
$\inf_{t \in \mathscr{T}} \sigma^2(t) > 0$, $\sup_{t \in \mathscr{T}} E\left[\varepsilon^4(t)\right] < \infty$, *and* $T_n = \left(RSS_0 - RSS_1\right)/RSS_1$ *as defined
in (9.69), where the residual sum of squares RSS_0 and RSS_1 are defined in
(9.68). Under the null hypothesis H_0 of (9.63), $T_n \to 0$ in probability as $n \to \infty$.
Otherwise, if the alternative hypothesis H_1 of (9.63) holds and $\inf_{c \in R} \|\beta_l -
c\|_{L_2} > 0$ for some $l = 1, \ldots, k$, then there exists a positive constant $\delta > 0$ such
that $T_n > \delta$ with probability tending to one.*   ■

Proof of Theorem 9.3:
Under the null hypothesis H_0 of (9.63), we can write

$$\widehat{\beta}^0(t) = \left(\widehat{\beta}_0^0(t), \widehat{\beta}_1^0, \ldots, \widehat{\beta}_k^0\right)^T \text{ with } \widehat{\beta}_0^0(t) = \sum_{s=1}^{K_0} \left[\widehat{\gamma}_{0s}^0 B_{0s}(t)\right]. \qquad (9.109)$$

It can be shown by direct calculation and Lemma 9.1 that, with probability
tending to one as $n \to \infty$,

$$RSS_0 - RSS_1 = \sum_{i=1}^{n} \sum_{j=1}^{n_i} \left\{ w_i \left[\mathbf{X}_{ij}^T \left(\widehat{\beta}(t_{ij}) - \widehat{\beta}^0(t)\right)\right]^2 \right\} \asymp \|\widehat{\beta} - \widehat{\beta}^0\|_{L_2}^2. \qquad (9.110)$$

Then, under H_0, we have that, by the triangle inequality,

$$\left\|\widehat{\beta} - \widehat{\beta}^0\right\|_{L_2} \leq \left\|\widehat{\beta} - \beta\right\|_{L_2} + \left\|\widehat{\beta}^0 - \beta\right\|_{L_2} \to 0 \text{ in probability, as } n \to \infty. \quad (9.111)$$

It then follows from (9.110) and (9.111) that, under H_0,

$$RSS_0 - RSS_1 \to 0 \quad \text{in probability, as } n \to \infty. \quad (9.112)$$

On the other hand, under the alternative H_1, because

$$\left\|\widehat{\beta} - \widehat{\beta}^0\right\|_{L_2} \geq \left\|\widehat{\beta}^0 - \beta\right\|_{L_2} - \left\|\widehat{\beta} - \beta\right\|_{L_2}, \quad (9.113)$$

it follows from $\left\|\widehat{\beta} - \beta\right\|_{L_2} = o_p(1)$ and (9.113) that, as $n \to \infty$,

$$\left\|\widehat{\beta} - \widehat{\beta}^0\right\|_{L^2} \geq \sum_{l=1}^{k} \left\|\widehat{\beta}_l^0 - \beta_l\right\|_{L_2} - o_p(1) \geq \sum_{l=1}^{k} \inf_{c \in R} \left\|\beta_l - c\right\|_{L_2} - o_p(1). \quad (9.114)$$

The assumption that $\inf_{c \in R} \left\|\beta_l - c\right\|_{L_2} > 0$ for some $l = 1, \ldots, k$ imply that, by (9.114), there is a positive constant $\delta > 0$ such that, under H_1,

$$RSS_0 - RSS_1 > \delta \quad \text{in probability, as } n \to \infty. \quad (9.115)$$

It remains to show that, with probability tending to 1, RSS_1 is bounded away from zero and infinity. By the definition of RSS_1 in (9.68) and $\widetilde{\beta}(t) = E\left[\widehat{\beta}(t) \mid \mathscr{X}\right]$, we have that

$$RSS_1 = \sum_{i=1}^{n} \sum_{j=1}^{n_i} \left\{ Y_{ij} - \mathbf{X}_{ij}^T \beta(t_{ij}) + \mathbf{X}_{ij}^T \left[\beta(t_{ij}) - \widetilde{\beta}(t_{ij})\right] \right.$$

$$\left. + \mathbf{X}_{ij}^T \left[\widetilde{\beta}(t_{ij}) - \widehat{\beta}(t_{ij})\right] \right\}^2. \quad (9.116)$$

It follows from the proof of Theorem 9.1 that

$$\begin{cases} \sum_{i=1}^{n} \sum_{j=1}^{n_i} w_i \left\{ \mathbf{X}_{ij}^T \left[\beta(t_{ij}) - \widetilde{\beta}(t_{ij})\right] \right\}^2 = o_p(1), \\ \sum_{i=1}^{n} \sum_{j=1}^{n_i} w_i \left\{ \mathbf{X}_{ij}^T \left[\widetilde{\beta}(t_{ij}) - \widehat{\beta}(t_{ij})\right] \right\}^2 = o_p(1). \end{cases} \quad (9.117)$$

Thus, by (9.116) and (9.117), it suffices to show that, with probability tending to 1, $\sum_{i=1}^{n} \sum_{j=1}^{n_i} \left(w_i \varepsilon_{ij}^2 \right)$ is bounded away from zero and infinity. By the assumption that $\sup_{t \in \mathscr{T}} E\left[\varepsilon^4(t)\right] < \infty$, there is a constant $c > 0$ such that

$$Var\left\{ \sum_{i=1}^{n} \sum_{j=1}^{n_i} \left(w_i \varepsilon_{ij}^2 \right) \right\} \leq \sum_{i=1}^{n} \left\{ w_i^2 n_i \sum_{j=1}^{n_i} E\left(\varepsilon_{ij}^4\right) \right\} \leq \sum_{i=1}^{n} \left(n_i^2 w_i^2 c \right) \to 0. \quad (9.118)$$

The Chebyshev inequality then implies that, by (9.118),

$$\sum_{i=1}^{n} \sum_{j=1}^{n_i} \left(w_i \varepsilon_{ij}^2 \right) - E\left\{ \sum_{i=1}^{n} \sum_{j=1}^{n_i} \left(w_i \varepsilon_{ij}^2 \right) \right\} \to 0 \quad \text{in probability, as } n \to \infty. \quad (9.119)$$

Since $\sum_{i=1}^{n} \left(n_i w_i \right) = 1$ and $E\left(\varepsilon_{ij}^2\right)$ is bounded away from zero and infinity, the right side of (9.116) is bounded away from zero and infinity. The conclusions of the theorem follow from (9.112), (9.115) and (9.116). ∎

9.6 Remarks and Literature Notes

This chapter presents the global smoothing methods by basis approximation for the time-varying coefficient model (7.1) or equivalently (9.1). These methods are generalizations of the methods of Chapter 4 to the structured linear model with time-varying covariates and rely on approximating the unknown coefficient functions by some expansions of known basis functions. The basis approximated time-varying coefficient models are often referred to as the *extended linear models*, as they are equivalent to the classical linear models when the terms of basis expansions are fixed.

The methods in this chapter have two major differences from the local smoothing methods in Chapters 7 and 8. First, because the basis approximation approach has a natural connection to the parametric linear models, we have the added benefit of performing simple goodness-of-fit tests for the coefficient curves. On the other hand, due to the lack of an approximate parametric modeling structure, this type of simple test statistics cannot be directly constructed using the local smoothing methods in Chapters 7 and 8. Another difference between the global smoothing methods in this chapter and the local smoothing methods in Chapters 7 and 8 is the frameworks of their asymptotic properties. In the local smoothing methods, the asymptotic biases of the estimators are determined by the smoothness conditions, such as the derivatives, of the unknown coefficient curves. In the basis approximation smoothing methods, the asymptotic biases of the estimators are measured by the "closeness" of the basis approximated curves to the unknown true curves. Thus, unlike the local smoothing methods where the asymptotic biases of the estimators can be expressed explicitly, the asymptotic biases of the basis approximation estimators do not have explicit expressions. Despite the differences in their theoretical frameworks, both the global smoothing methods and the local smoothing methods are popular methods in practice, because they often lead to very similar results in real applications.

The materials in Sections 9.1 to 9.3 are adopted from Huang, Wu and Zhou (2002). The asymptotic properties and their proofs in Section 9.5 are based on Huang, Wu and Zhou (2002) and Huang, Wu and Zhou (2004). Interpretations and implications of basis approximations and the extended linear models with cross-sectional i.i.d. data have been discussed in Stone et al. (1997) and Huang (1998, 2001 and 2003). These interpretations and implications still hold under the current framework of time-varying coefficient models with longitudinal data. Some technical results of Section 9.5 are derived from the results of Huang (1998).

Part IV

Shared-Parameter and Mixed-Effects Models

Chapter 10

Models for Concomitant Interventions

The time-varying coefficient models discussed in Chapters 6 to 9 present a simple and flexible structured nonparametric approach for modeling the time-varying covariate effects on the outcome variable of interest. Despite their success in many longitudinal studies, these models have a number of restrictions which limit their practical values. For example, the time-varying coefficient models, as shown in (7.1), require that the linear coefficients to be functions of time only and the values of the time-dependent covariates do not depend on the values of the outcome variable at the previous time points. To broaden the scope of applications, we present generalizations of the time-varying coefficient models in three areas: (a) structured modeling for the effects outcome-adaptive covariates; (b) mixed-effects extension to incorporate subject-specific effects; (c) incorporating time-lagging effects using the conditional distributions. In this chapter, we consider the extension in (a) with concomitant interventions. The results of this chapter demonstrate that, because the initiation of concomitant interventions often depends on the outcome variable, the usual approach of mixed-effects models or nonparametric models may lead to biased inferences for the effects of a concomitant intervention. The extensions in (b) and (c) are discussed in Chapters 11 to 14.

10.1 Concomitant Interventions

10.1.1 Motivation for Outcome-Adaptive Covariate

It is well-known in the literature of longitudinal analysis that, when a covariate depends on the values of the outcome variable at the previous time points, the conditional-mean based regression methods in the previous chapters may lead to unsatisfactory results (e.g., Pepe and Anderson, 1994). This type of covariates, whose values may depend on the previous outcome values, are referred herein as "outcome-adaptive" covariates. In general, outcome-adaptive covariates are necessarily time-dependent. Examples of outcome-adaptive covariates in epidemiological studies or longitudinal clinical trials may include behavior variables, psychological risk factors, and medication use, among others, because, for a study participant, he or she may take actions and change the values of these variables over time depending on his or her health status observed at the previous time points.

1. Concomitant Intervention as Outcome-Adaptive Covariate

As a special case of the outcome-adaptive covariates, concomitant interventions are common in longitudinal clinical trials. It has been long recognized in the medical and statistical literature that, because of its outcome-adaptive nature, a concomitant intervention cannot be treated as a regular covariate in a mixed-effects model or a nonparametric regression model, and special consideration has to be given in the model to characterize the initiation and change of this variable. In this chapter, after a brief introduction of the data structure with a concomitant intervention, we describe the modeling and estimation approaches of Wu, Tian and Bang (2008) and Wu, Tian and Jiang (2011), and summarize the similarities and differences between these approaches.

For longitudinal clinical trials with randomly assigned study treatments, effects of the study treatments are modeled through a time-invariant categorical covariate vector, while other factors of interest, such as age, gender, ethnicity and disease risk factors, can be modeled through either time-invariant or time-dependent covariates. Since the study treatments are randomly assigned at the start of the study, these study treatments do not depend on the values of the outcome variables which are observed at different time points during the study. Because of randomization, the time-invariant covariates, such as age, gender, ethnicity and the baseline disease risk factors, are also expected to be balanced among patients receiving different study treatments, and do not depend on the later values of the outcome variables. However, some of the time-dependent covariates, particularly the covariates which can be controlled by the study participants and investigators, are possibly outcome-adaptive, because, due to ethical or logistical reasons, changes of these variables at a time point may depend on the values or time-trends of the outcome variable prior to that time point.

A typical scenario of outcome-adaptive covariates is the presence of concomitant interventions. Unlike the randomly assigned study treatments, concomitant interventions are not randomly assigned and often initiated to some of the study subjects when they exhibit less satisfactory trends in their health outcomes. The scenario of concomitant interventions bears some similarities to longitudinal studies with informative missing data. In the case of informative missing data, the study subjects with undesirable outcome time-trends tend to drop out from the study earlier than those with more desirable outcome time-trends. The only difference between concomitant intervention and informative drop-out is that, in studies with concomitant interventions, the outcomes of the study subjects continue to be observed after the start of the concomitant interventions, while in informative drop-out settings the study subjects who drop out of the study have no follow-up observations after the drop-out times.

In a clinical trial, study subjects who have taken a concomitant intervention in addition to their assigned study treatments may generally have different disease pathology from those who do not need the concomitant intervention. Thus, in addition to the primary objective of evaluating the study treatment effects, it is also important, perhaps as a secondary objective, to evaluate the effects of the additional concomitant intervention on the outcome variables of the study population.

2. Motivation from the ENRICHD Study

The Enhancing Recovery in Coronary Heart Disease Patients (ENRICHD) Study described in Section 1.2 and analyzed in Section 2.4.2 is a typical example which involves a concomitant intervention in addition to the randomly assigned treatment regimens. In this randomized clinical trial for evaluating the efficacy of a six-month cognitive behavior therapy (CBT) versus usual cardiovascular (UC) care, the Beck Depression Inventory (BDI) scores for patients in the CBT arm were repeatedly measured at weekly visits during the treatment and four yearly follow-up visits, while BDI scores for patients in the UC arm were only measured at baseline, the six-month visit and the yearly follow-up visits. By the study design (cf. ENRICHD, 2001), pharmacotherapy with antidepressants was allowed as a concomitant intervention in both the CBT and the UC arms, if a patient had high baseline BDI scores or nondecreasing BDI trends five weeks after enrollment or antidepressants were requested by the patient or the primary-care physicians. Although Taylor et al. (2005) reported that pharmacotherapy improved survival among 1834 depressed ENRICHD patients, their results, however, did not address the question of whether pharmacotherapy was beneficial for lowering the patients' depression severity.

10.1.2 Two Modeling Approaches

Although the issues of evaluating the concomitant intervention effects have been discussed in the literature of statistical methods for clinical trials (e.g., Cook and DeMets, 2008, Section 6.2), regression methods adjusting for the potential biases in statistical estimation and inferences with longitudinal data have just been suggested recently.

1. The Varying-Coefficient Mixed-Effects Model

Using the repeatedly measured BDI scores in a sub-sample of 91 patients who received pharmacotherapy as a concomitant intervention within the treatment period in the CBT arm of the ENRICHD study, Wu, Tian and Bang (2008) suggests that the naive mixed-effects models may give misleading results for the pharmacotherapy effects on the BDI trends over time, and a varying-coefficient mixed-effects model can be used to reduce the potential bias associated with the estimated pharmacotherapy effects. A main drawback of the approach of Wu, Tian and Bang (2008) is that the proposed regression model uses only part of the sample, i.e., the subjects who started the concomitant pharmacotherapy intervention during the study, while information provided by the subjects who did not start the pharmacotherapy during the study is ignored. This may lead to the loss of information.

2. The Shared-Parameter Change-Point Model

In order to adequately capture the concomitant intervention effects using all subjects of the sample, an alternative regression model based on shared parameters is proposed in Wu, Tian and Jiang (2011). Using the framework of the shared-parameter models in Follmann and Wu (1995), the approach of Wu, Tian and Jiang (2011) describes the covariate effects on the response variable through a change-point mixed-effects model, and incorporates the random coefficients and the intervention starting time (change-point time) through a series of joint distributions. Patients who have received a concomitant intervention at baseline or have not received any concomitant intervention during the study period are treated as censored. A likelihood-based method is established for statistical estimation and inferences, and its computation is implemented through a two-stage iteration procedure. Applying their procedures to the ENRICHD pharmacotherapy data, the results of Wu, Tian and Jiang (2011) suggest that their proposed method leads to adequate estimates when a concomitant intervention is present, while the naive mixed-effects model is likely misspecified under such situations.

10.1.3   Data Structure with a Single Intervention

For ease of presentation, we consider in this chapter the longitudinal samples with one possible concomitant intervention and only time-invariant covariates. The main reason of excluding the cases with time-dependent covariates is to distinguish the regular time-dependent covariates, whose values at a given time point do not depend on the values of the outcome variable at previous times, from the concomitant intervention, whose values may change based on the previous outcome values. For simplicity, our models and methods of this chapter are limited to a single possible concomitant intervention and the study subjects can only change from without concomitant intervention to with concomitant intervention. The more complicated cases with possibly multiple concomitant interventions, multi-levels of concomitant interventions (e.g., different intervention dosages) and other time-dependent covariates require more complicated modeling strategies. Although they are interesting and useful for real applications, these more complicated cases require substantial further research and are beyond the scope of this book.

Following the notation of Chapter 6, the longitudinal sample consists of the following components.

Longitudinal Sample with a Concomitant Intervention:

(a) *The study has a total of n randomly selected subjects within the time range $[\mathscr{T}_0, \mathscr{T}_1]$, where $\mathscr{T}_0$ and $\mathscr{T}_1$ are the beginning and ending times of the study.*

(b) *The ith subject has n_i visits and observations $\{t_{ij}, Y_{ij}, \mathbf{X}_i\}$ at the jth visit, where t_{ij}, the study time, is defined by the time elapsed from the beginning of the study to the jth visit, $\mathbf{X}_i = (1, X_i^{(1)}, \ldots, X_i^{(P)})^T$ is the R^{P+1}-valued time-invariant covariate vector, and Y_{ij} is the real-valued outcome variable.*

(c) *The study involves only one concomitant intervention. The ith subject can change from "without concomitant intervention" to "with concomitant intervention" only once during the study.*

(d) *For the ith subject, $1 \leq i \leq n$, the concomitant intervention starting time or change-point time is denoted by S_i, and the concomitant intervention indicator is defined by*

$$\lambda_{ij} = \begin{cases} 0, & \text{if } t_{ij} < S_i; \\ 1, & \text{if } t_{ij} \geq S_i. \end{cases} \tag{10.1}$$

(e) *Since not every subject has a change-point time during the study, the ith subject's change-point time is observed if $t_{i1} \leq S_i \leq t_{in_i}$. If $S_i < t_{i1}$ or $S_i > t_{in_i}$, the subject's change-point time is left or right censored, respectively. The indicator variable for censoring $\delta_i^{(c)}$ is defined by*

$$\delta_i^{(c)} = \begin{cases} 0, & \text{if } t_{i1} \leq S_i \leq t_{in_i}; \\ 1, & \text{if } S_i > t_{in_i}; \\ 2, & \text{if } S_i < t_{i1}. \end{cases} \tag{10.2}$$

The observed change-point times are

$$\begin{cases} \mathscr{S}^{(c)} &= \{\mathscr{S}_i^{(c)} = (S_i^{(c)}, \delta_i^{(c)}); i = 1, \ldots, n\}; \\ S_i^{(c)} &= S_i, \text{ if } \delta_i^{(c)} = 0; \\ S_i^{(c)} &= t_{in_i}, \text{ if } \delta_i^{(c)} = 1; \\ S_i^{(c)} &= t_{i1}, \text{ if } \delta_i^{(c)} = 2. \end{cases} \tag{10.3}$$

For subjects whose concomitant intervention time is observed during the study period, i.e., any subject $1 \leq i \leq n$ with $\delta_i^{(c)} = 0$, the time elapsed from the start of the concomitant intervention to the jth, $1 \leq j \leq n_i$, visit time is defined by

$$R_{ij} = \begin{cases} t_{ij} - S_i, & \text{if } \delta_i^{(c)} = 0; \\ 0, & \text{if } \delta_i^{(c)} = 1; \\ t_{ij}, & \text{if } \delta_i^{(c)} = 2. \end{cases} \tag{10.4}$$

(f) *The overall longitudinal sample with a concomitant intervention is*

$$\mathscr{D} = \left\{ \left(Y_{ij}, t_{ij}, \mathbf{X}_i^T, S_i^{(c)}, \delta_i^{(c)}, R_{ij}, \lambda_{ij} \right) : i = 1, \ldots, n; j = 1, \ldots, n_i \right\}, \quad (10.5)$$

where the subjects may or may not receive the concomitant intervention during the study.   □

In many practical situations, the above longitudinal sample (10.5) can be simplified, so that simpler statistical models for the effects of $\left\{ t_{ij}, \mathbf{X}_i^T, S_i^{(c)}, \delta_i^{(c)}, R_{ij}, \lambda_{ij} \right\}$ on Y_{ij} can be considered. On the other hand, when there are more than one concomitant interventions or a concomitant intervention has more than one change-point, the data structure in (10.5) should be generalized to account for multiple concomitant intervention indicators and multiple change-points. These possible special cases and generalizations are discussed in the following remarks.

Observed Concomitant Intervention for All Subjects: If we consider only the subjects in (10.5) who have the observed concomitant intervention during the study period, i.e., subjects with $\delta_i^{(c)} = 0$, and ignore those with "censored change-point times," i.e., subjects with $\delta_i^{(c)} = 1$ or 2, we obtain the longitudinal sample $\mathscr{D}_0$, which is a sub-sample of $\mathscr{D}$ given by

$$\mathscr{D}_0 = \left\{ \left(Y_{ij}, t_{ij}, \mathbf{X}_i^T, S_i, R_{ij}, \lambda_{ij} \right) : i = 1, \ldots, n^{(0)}; j = 1, \ldots, n_i; \delta_i^{(c)} = 0 \right\}. \quad (10.6)$$

Because $\mathscr{D}_0$ is a sub-sample of $\mathscr{D}$, the subject indices in $\mathscr{D}_0$ have been relabeled from the subject indices in $\mathscr{D}$, i.e., the ith subject in $\mathscr{D}_0$ is not necessarily the ith subject in $\mathscr{D}$, and $n^{(0)} \leq n$. The reason for considering $\mathscr{D}_0$ is that, since the change-point time S_i is observed for every subject in $\mathscr{D}_0$, it is possible to evaluate the concomitant intervention effect by comparing the time-trends of the outcome variable $Y(t)$ for t before and after the change-point time S_i. This is the approach used by Wu, Tian and Bang (2008). But, the potential drawback of using $\mathscr{D}_0$ is that the information provided by the subjects who have "censored change-point times," i.e., subjects with $\delta_i^{(c)} = 1$ or 2, is ignored in the analysis, which may lead to inefficient estimation of the concomitant intervention effect.   □

"Right-Censored" Intervention Starting Times: The longitudinal sample $\mathscr{D}$ in (10.5) includes also the subjects who have started the concomitant intervention before the study starting time $\mathscr{T}_0$, i.e., subjects with $\delta_i^{(c)} = 2$, which, borrowing from the term in the survival analysis literature, is referred herein as the "left-censored intervention starting time" or "left-censored intervention change-point time." This is intended to extract the potentially useful information from the study subjects who started the concomitant intervention before the study starting time and have continued the concomitant intervention throughout the study period. But, in most studies, the exact starting time of the concomitant intervention before the trial's starting time $\mathcal{T}_0$ is unknown, so that it is usually the intent of the study investigators to exclude the subjects who already started the concomitant intervention before the trial's starting time $\mathcal{T}_0$. As such, the longitudinal sample includes only the "right-censored intervention starting time" and is given by

$$\mathcal{D}_1 = \left\{ \left(Y_{ij}, t_{ij}, \mathbf{X}_i^T, S_i^{(c)}, \delta_i^{(c)}, R_{ij}, \lambda_{ij} \right) : \right.$$
$$\left. \delta_i^{(c)} = 0 \text{ or } 1; i = 1, \ldots, n; j = 1, \ldots, n_i \right\}. \quad (10.7)$$

The subjects in $\mathcal{D}_1$ who never start the concomitant intervention during the study period $[\mathcal{T}_0, \mathcal{T}_1]$ are viewed as having the "right-censored intervention starting times." These subjects may provide some useful information for modeling the time-trends of $Y(t)$ for $\mathcal{T}_0 < t < S_i$. The modeling and estimation methods of Wu, Tian and Jiang (2011) are intended to avoid the potential shortcomings of using only $\mathcal{D}_0$ in (10.6) by including the subjects with "right-censored intervention starting times." □

Multiple Concomitant Intervention Times: Generalization of the sample $\mathcal{D}$ in (10.5) to include multiple concomitant intervention times is also possible. In such cases, there are possibly multiple types of concomitant interventions available to the trial subjects and each subject may start and stop a concomitant intervention multiple times during the trial period $[\mathcal{T}_0, \mathcal{T}_1]$. It is also possible that the dosage of a given concomitant intervention is continuous, which can be changed during the trial period $[\mathcal{T}_0, \mathcal{T}_1]$. If we recognize that the decision of changing a concomitant intervention or changing the dosage of a concomitant intervention at a given time may depend on the trial subject's values of the outcome variable at the prior time points, then these concomitant interventions, their starting times and their dosages should be modeled differently from other covariates $\mathbf{X}_i$. But, appropriate statistical models for data beyond the scope of $\mathcal{D}$ in (10.5) are subject to various complexities in computation and interpretations, and have not been systematically investigated in the literature. Thus, this generalization, although practically interesting, is beyond the scope of this book. □

10.2   Naive Mixed-Effects Change-Point Models

Based on the data structures summarized in (10.5), a number of regression models can be considered as natural candidates to describe the time-trends between the outcome variable Y_{ij} and the covariates, including $\mathbf{X}_i$ as well as the presence and time length of the concomitant intervention which are denoted by $\{\lambda_{ij}, R_{ij}\}$. Because the starting time S_i of the concomitant intervention may depend on the values and time-trends of Y_{ij} at time points t_{ij} before S_i, some of the seemingly "natural" regression models may lead to biased results if the relationship between S_i and the time-trends of Y_{ij} for $t_{ij} < S_i$ is not properly taken into account. As an illustration of this crucial point, we demonstrate below in this section that some of the seemingly "natural" mixed-effects models may indeed lead to erroneous conclusions. As a remedy, we establish in Sections 10.4 and 10.5 two main modeling approaches to remove or reduce the potential biases in longitudinal models.

Since a concomitant intervention is not randomly assigned, it is understood in practice that the effects of a concomitant intervention cannot be properly evaluated by directly comparing the outcome values between the subjects who received the intervention and those who did not receive the intervention. This fact has been noted in the ENRICHD trial by a number of publications, such as, ENRICHD (2003) and Taylor et al. (2005). Thus, we use the time-trends of BDI scores among patients who used antidepressants and who did not use antidepressants in the ENRICHD trial (Section 10.1.1) as an illustrating example. In the discussions below, we outline the practical justifications and considerations for modeling the concomitant intervention effects through "mixed-effects change-point models."

10.2.1   Justifications for Change-Point Models

Given that the concomitant intervention is not randomly assigned, it is generally inappropriate to compare the summary statistics or time-trends of the outcome variable between patients with or without the concomitant intervention. As described in Section 10.1.1, since a patient in the ENRICHD trial could start using antidepressants as a concomitant intervention if he or she had high or nondecreasing depression severity measured by the BDI scores, it could be potentially misleading to compare the mean BDI scores at the six-month visit between the patients who used antidepressants and who did not use antidepressants during the period from baseline to six months. The reason is that the patients with the concomitant intervention, i.e., using antidepressants, could have higher mean BDI scores than patients without the concomitant intervention even if the concomitant intervention via antidepressants could indeed have the true effects of lowering depression severity through BDI scores. For the same reason, it may not be appropriate to compare the mean values of the BDI changes from baseline to six months between the patients who used antidepressants and those who did not use antidepressants during the six-month period, because it is possible that the patients who had antidepressants as the concomitant intervention could be those whose depression severity could not be lowered by treatments other than antidepressants.

The practical issues considered above suggest that it is generally inappropriate to compare the means of the outcome variable at the end of the study or the means of the changes of the outcome variable at two time points between patients with and without the concomitant intervention. It is reasonable, however, to compare the time-trends of the outcome variable before and after the concomitant intervention starting time through a "subject-specific change-point" regression model. Although a subject with unfavorable health status is more likely to start a concomitant intervention than a subject with acceptable health status, it is possible to use the available data to model and estimate the values and time-trends of the outcome variable before and after the concomitant intervention. We can then compare the subject-specific time-trends and trajectories of the outcome variable before and after the concomitant intervention. If, on average, the time-trends and trajectories of the outcome variable exhibit favorable health status after the concomitant intervention than before the concomitant intervention, then it is plausible to attribute the favorable change of health status to the positive "effect" of the concomitant intervention.

10.2.2 Model Formulation and Interpretation

To make the change-point approach more precise, let $\mu_0(t_{ij}, \mathbf{X}_i; \mathbf{a}_i)$ be the ith subject's trajectory before the concomitant intervention, which is parameterized by the subject-specific parameters $\mathbf{a}_i = (a_{i1}, \ldots, a_{id_0})^T$ for some $d_0 \geq 1$, and let $\mu_1(t_{ij}, \mathbf{X}_i, R_{ij}; \mathbf{b}_i)$ be the change of the trajectory after the concomitant intervention, which is parameterized by the subject-specific parameters $\mathbf{b}_i = (b_{i1}, \ldots, b_{id_1})^T$ for some $d_1 \geq 1$. Since $\mu_1(t_{ij}, \mathbf{X}_i, R_{ij}; \mathbf{b}_i)$ describes the changing trajectory of the ith subject after the concomitant intervention starting time (or change-point time) S_i, it may depend on the "intervention duration time" $R_{ij} = t_{ij} - S_i$ as well as t_{ij} and $\mathbf{X}_i$. The ith subject's outcome variable Y_{ij} at time point t_{ij} is then given by

$$Y_{ij} = \mu_0(t_{ij}, \mathbf{X}; \mathbf{a}_i) + \lambda_{ij}\,\mu_1(t_{ij}, \mathbf{X}_i, R_{ij}; \mathbf{b}_i) + \varepsilon_{ij}, \tag{10.8}$$

where ε_{ij} are mean zero random errors such that ε_{i_1,j_1} and ε_{i_2,j_2} are independent if $i_1 \neq i_2$, $Var(\varepsilon_{ij}) = \sigma^2(t_{ij})$ and $Cov(\varepsilon_{i,j_1}, \varepsilon_{i,j_2}) = \rho(t_{i,j_1}, t_{i,j_2})$. For the special case that ε_{ij} is simply a measurement error, the ε_{ij} for all $i = 1, \ldots, n$ and $j = 1, \ldots, n_i$ are independent random variables with mean zero and $Var(\varepsilon_{ij}) = \sigma^2(t_{ij})$. When the ith subject's jth visit time is before the concomitant intervention, i.e., $t_{ij} \leq S_i$, the subject's concomitant intervention indicator is $\lambda_{ij} = 0$, and the subject's trajectory of the outcome variable is $\mu_0(t_{ij}, \mathbf{X}_i; \mathbf{a}_i)$. Similarly, when t_{ij} is after the concomitant intervention time S_i, i.e., $t_{ij} > S_i$ and $\lambda_{ij} = 1$, the ith subject's trajectory of the outcome variable is

$$\mu_0(t_{ij}, \mathbf{X}_i; \mathbf{a}_i) + \mu_1(t_{ij}, \mathbf{X}_i, R_{ij}; \mathbf{b}_i).$$

The usual mixed-effects model framework, e.g., Verbeke and Molenberghs (2000) and Diggle et al. (2002), suggests that a naive mixed-effects model for evaluating the pre- and post-intervention trajectories can be established by imposing a joint distribution for the random parameters $\mathbf{a}_i$ and $\mathbf{b}_i$ to the model (10.8). Assuming that the subject-specific parameter $(\mathbf{a}_i^T, \mathbf{b}_i^T)$ has a known joint distribution $G(\cdot)$, (10.8) suggests that the naive mixed-effects change-point model for the concomitant intervention can be given by

$$\begin{cases} Y_{ij} = \mu_0(t_{ij}, \mathbf{X}_i; \mathbf{a}_i) + \lambda_{ij}\,\mu_1(t_{ij}, \mathbf{X}_i, R_{ij}; \mathbf{b}_i) + \varepsilon_{ij}, \\ (\mathbf{a}_i^T, \mathbf{b}_i^T)^T \sim \text{ joint distribution } G(\cdot), \\ E(\mathbf{a}_i^T, \mathbf{b}_i^T)^T = (\alpha^T, \beta^T)^T \text{ and } Cov(\mathbf{a}_i^T, \mathbf{b}_i^T)^T = \mathbf{V}, \\ Var(\varepsilon_{ij}) = \sigma^2(t_{ij}) \text{ and } Cov(\varepsilon_{i,j_1}, \varepsilon_{i,j_2}) = \rho(t_{i,j_1}, t_{i,j_2}), \end{cases} \tag{10.9}$$

where the mean (α^T, β^T) and covariance matrix $\mathbf{V}$ are unknown, $(\mathbf{a}_i^T, \mathbf{b}_i^T)$ and ε_{ij} are independent, and ε_{i_1, j_1} and ε_{i_2, j_2} are independent for any $i_1 \neq i_2$ and any j_1 and j_2. A mathematically convenient special case is to assume that $(\mathbf{a}_i^T, \mathbf{b}_i^T)$ has the multivariate normal distribution $N[(\alpha^T, \beta^T)^T, \mathbf{V}]$. Under (10.9), a positive (or negative) value for $\mu_1(t_{ij}, \mathbf{X}_i, R_{ij}; \mathbf{b}_i)$ would suggest that the concomitant intervention tends to increase (or decrease) the mean of Y_{ij} given $\{t_{ij}, \mathbf{X}_i, R_{ij}\}$.

There are several interesting special cases of the mixed-effects change-point model (10.9), which can be derived by setting special parametric models for $\mu_0(t_{ij}, \mathbf{X}_i; \mathbf{a}_i)$ and $\mu_1(t_{ij}, \mathbf{X}_i, R_{ij}; \mathbf{b}_i)$. For example, let $d_0 = d_1 = P+2$, so that $\mathbf{a}_i = (a_{i0}, \ldots, a_{i,P+2})^T$ and $\mathbf{b}_i = (b_{i0}, \ldots, b_{i,P+2})^T$. A linear model of (10.9) with normal subject-specific parameters is given by

$$\begin{cases} Y_{ij} = (\mathbf{X}_i^T, t_{ij})\,\mathbf{a}_i + \lambda_{ij}\,(\mathbf{X}_i^T, R_{ij})\,\mathbf{b}_i + \varepsilon_{ij}, \\ = a_{i0} + a_{i1} X_i^{(1)} + \cdots + a_{i,P+1} X_i^{(P)} + a_{i,P+2} t_{ij} \\ \qquad + \lambda_{ij}\big(b_{i0} + b_{i1} X_i^{(1)} + \cdots + b_{i,P+1} X_i^{(P)} + b_{i,P+2} R_{ij}\big) + \varepsilon_{ij}, \\ (\mathbf{a}_i^T, \mathbf{b}_i^T)^T \sim N[(\alpha^T, \beta^T)^T, \mathbf{V}], \\ E(\mathbf{a}_i^T, \mathbf{b}_i^T)^T = (\alpha^T, \beta^T)^T \text{ and } Cov(\mathbf{a}_i^T, \mathbf{b}_i^T)^T = \mathbf{V}, \\ Var(\varepsilon_{ij}) = \sigma^2(t_{ij}) \text{ and } Cov(\varepsilon_{i,j_1}, \varepsilon_{i,j_2}) = \rho(t_{i,j_1}, t_{i,j_2}), \end{cases} \tag{10.10}$$

where, with $\sigma^2(t_{ij})$ and $\rho(t_{i,j_1}, t_{i,j_2})$ as defined in (10.9), $\alpha = (\alpha_0, \alpha_1, \ldots, \alpha_{P+2})^T$ and $\beta = (\beta_0, \beta_1, \ldots, \beta_{P+2})^T$, which are the means of $\mathbf{a}_i$ and $\mathbf{b}_i$, respectively, are the population parameters of interest. The parameters $\{a_{i0}, a_{i1}, \ldots, a_{i,P+2}\}$ in (10.10) describe the linear effects of $\{1, X_i^{(1)}, \ldots, X_i^{(P)}, t_{ij}\}$ for the ith subject before starting the concomitant intervention, i.e., $\lambda_{ij} = 0$, in which the mean of Y_{ij} has the linear time-trend $a_{i,P+2} t_{ij}$. After the concomitant intervention, $\lambda_{ij} = 1$, the effects of the covariates $\{1, X_i^{(1)}, \ldots, X_i^{(P)}\}$ on the mean of Y_{ij} are described by

$$\{(a_{i0} + b_{i0}), (a_{i1} + b_{i1}), \ldots, (a_{i,P+1} + b_{i,P+1})\},$$

while the time-trend for the mean of Y_{ij} has the linear form

$$a_{i,P+2} t_{ij} + b_{i,P+2} R_{ij} = (a_{i,P+2} + b_{i,P+2}) t_{ij} - b_{i,P+2} S_i,$$

which depends on the time elapsed after the start of the concomitant intervention $R_{ij} = t_{ij} - S_i$.

Other linear or nonlinear special cases of (10.9) can be similarly developed as (10.10). In particular, the time-trends for the mean of Y_{ij} may not be necessarily linear or depend on R_{ij} as characterized in (10.10). The major implication of (10.9) and its special linear case (10.10) is that the concomitant intervention effect for each subject is described by the subject-specific parameter vector $\mathbf{b}_i$. Any $\mathbf{b}_i$ that gives $\mu_1(t_{ij}, \mathbf{X}_i, R_{ij}; \mathbf{b}_i) = 0$ would indicate that the concomitant intervention does not affect the values of the outcome variable Y_{ij} for the ith subject.

10.2.3 Biases of Naive Mixed-Effects Models

Although the mixed-effects change-point model in (10.9) seems natural and intuitive, it does not really distinguish the concomitant intervention from other covariates in a fundamental way, because it only involves the concomitant intervention through the time-dependent indicator variable λ_{ij}. A major consequence of this naive approach is that this model may lead to bias and erroneous statistical inferences when the initiation of the concomitant intervention is "self-selective" in the sense that it strongly depends on the values of the outcome variable of the subject. Because the model (10.9) does not really take the "self-selectiveness" of the concomitant intervention into account, Wu, Tian and Bang (2008) show that (10.9) can be a misspecified model even if $\mu_0(\cdot; \mathbf{a}_i)$, $\mu_1(\cdot; \mathbf{b}_i)$, ε_{ij} and $G(\cdot)$ are all correctly specified. Consequently, misleading conclusions may occur even under very simple situations where (10.9) appears to have natural interpretations.

A Synthetic Example of Concomitant Intervention

To illustrate the potential bias associated with the naive modeling approach of (10.9), we present a simple synthetic example which shows that (10.9) leads to erroneous conclusions, and then we show that some seemingly "not-so-intuitive" alternatives of (10.9) can be used to reduce the bias. This example motivates the more general approaches for modeling the concomitant intervention to be discussed in Sections 10.4 and 10.5.

1. Data Generation

Suppose that we have a longitudinal study with $n = 24$ independent subjects. For the ith subject, $1 \le i \le 24$, there are $n_i = 10$ repeated measurements at the set of integer time points $\{1, 2, \ldots, 10\}$, and the jth measurement time, $j = 1, \ldots, n_i$, $n_i = 10$, is given by $t_{ij} = j$. Following the data structure (10.6), let S_i be the change-point time for the ith subject changing from "without concomitant intervention" to "with concomitant intervention," and let λ_{ij} be the corresponding "concomitant intervention indicator" such that $\lambda_{ij} = 0$ if $S_i < t_{ij}$, and $\lambda_{ij} = 1$ if $S_i \geq t_{ij}$. Assume that the ith subject's outcome variable Y_{ij} observed at the jth measurement time $t_{ij} = j$ only depends on the concomitant intervention status λ_{ij} at time $t_{ij} = j$ and a mean zero measurement error ε_{ij}, so that the subject-specific model for Y_{ij} is

$$Y_{ij} = a_i + b_i \lambda_{ij} + \varepsilon_{ij}, \tag{10.11}$$

where a_i is the ith subject's mean of Y_{ij} for $t_{ij} = j$ before the concomitant intervention, b_i is the change of the ith subject's mean of Y_{ij} from "without concomitant intervention" to "with concomitant intervention," and the $\{\varepsilon_{ij} : i = 1, \ldots, 24; j = 1, \ldots, 10\}$ are i.i.d. with the $N(0, \sigma^2)$ distribution. The mean of Y_{ij} in the model (10.11) does not depend on $t_{ij} = j$.

Given the model (10.11), the next issue is to specify how the starting time of concomitant intervention S_i depends on the outcome variable at time points t_{ij} prior to S_i. For simplicity, let S_i depend on the outcome variable Y_{ij} through some specified values of (a_i, b_i) and (α, β) in the following manner:

Steps for Generating Synthetic Data:

(a) *Let $\alpha = 19.5$ and $\beta = -1.5$ be the means of $\{a_1, \ldots, a_{24}\}$ and $\{b_1, \ldots, b_{24}\}$, respectively.*

(b) *For the first 12 subjects, i.e., $1 \leq i \leq 12$, their change-point time is $S_i = 2$, and the values of their outcome variable Y_{ij} are generated using $a_i = 20$, $b_i = -1.0$ and $\varepsilon \sim N(0,3)$.*

(c) *For the remaining 12 subjects, i.e., $13 \leq i \leq 24$, their change-point time is $S_i = 8$, and Y_{ij} are generated using $a_i = 19$, $b_i = -2.0$ and $\varepsilon \sim N(0, 3)$.* □

In the form of (10.11), the true model of the data is

$$\begin{cases} Y_{ij} = 20 - \lambda_{ij} + \varepsilon_{ij}, \ \lambda_{ij} = \begin{cases} 0, & \text{if } 1 \leq j \leq 2, \\ 1, & \text{if } 3 \leq j \leq 10, \end{cases} \quad i = 1, \ldots, 12; \\ Y_{ij} = 19 - 2\lambda_{ij} + \varepsilon_{ij}, \ \lambda_{ij} = \begin{cases} 0, & \text{if } 1 \leq j \leq 8, \\ 1, & \text{if } 9 \leq j \leq 10, \end{cases} \quad i = 12, \ldots, 24, \end{cases} \tag{10.12}$$

where ε_{ij} are i.i.d with $N(0, 3)$. Our objective based on the observations generated from (10.12) is to estimate the population mean parameters $(\alpha, \beta) = (19.5, -1.5)$, where $\beta = -1.5$ represents the mean concomitant intervention effect on the outcome variable Y_{ij}.

2. Estimation with Naive Mixed-Effects Models

Suppose that we already know that, for all $1 \leq i \leq 24$, Y_{ij} depends on j only through λ_{ij} for all $j = 1, \ldots, 10$, but the distributions of the random intercepts

a_i and b_i are unknown. Then, a natural model for the current situation is

$$\begin{cases} Y_{ij} & = & a_i + b_i \lambda_{ij} + \varepsilon_{ij}, \\ (a_i, b_i)^T & \sim & \text{joint distribution } G(\cdot), \end{cases} \tag{10.13}$$

where, for all $1 \leq i \leq 24$ and $1 \leq j \leq 10$, ε_{ij} are i.i.d. mean zero measurement errors, $G(\cdot)$ is an unknown distribution function, $E(a_i) = \alpha$ and $E(b_i) = \beta$ are the fixed-effects parameters. Clearly, (10.13) is a special case of the naive mixed-effects change-point model (10.9). Given a sample

$$\left\{ (Y_{ij}, t_{ij}, S_i) : i = 1, \ldots, 24; j = 1, \ldots, 10 \right\} \tag{10.14}$$

generated from the unknown true model (10.12), we can estimate the mean and covariance parameters of (10.13) using a number of standard statistical analysis software packages, such as SAS and R. Among them, the mean concomitant intervention effect β is the primary parameter of interest. A negative value of β would suggest that the concomitant intervention could on average lower the value of the outcome variable $Y(t)$. Conversely, a positive β on average increases the value of $Y(t)$.

Since the exact distribution function $G(\cdot)$ is unknown, we use the common practice in the literature, e.g., the linear mixed-effects models in Section 2.1, and estimate the mean parameters $(\alpha, \beta)^T$ by assuming that $G(\cdot)$ is a multivariate normal distribution with a given correlation structure. Since the correlation structures of the data are generally unknown, three linear mixed-effects model (LME) estimation procedures in the R statistical package can be used to estimate $(\alpha, \beta)^T$:

- LME with working independent (LMEWI) correlation structure;
- LME with random intercept (LMERI);
- LME with random intercept and slope (LMERIS).

Similarly, parameter estimates can also be obtained by the generalized estimating equation (GEE) procedure with three correlation structures:

- GEE with working independent (GEEWI) correlation structure;
- GEE with exchangeable correlation (GEEEC) structure;
- GEE with unstructured correlation (GEEUC) structure.

Further details for the parameter estimations with linear mixed-effects models can be found in Verbeke and Molenberghs (2000).

3. Simulation Results of Naive Mixed-Effects Models

To examine whether the naive mixed-effects change-point model (10.13) can lead to appropriate estimates for the mean concomitant intervention effect β, we present here the results from a simulation study with 10,000 independent samples of (9.14). In this study, we compute the estimators $\widehat{\beta}$ and their standard errors from each of the simulated samples using each of the standard estimation procedures, namely LMEWI, LMERI, LMERIS, GEEWI, GEEEC and GEEUC, in SAS. Table 10.1 summarizes the averages of the estimators and their standard errors (SE) and the empirical coverage probabilities of the corresponding 95% confidence intervals covering the true parameter $\beta = -1.5$ computed from the 10,000 simulated samples and the naive mixed-effects model (10.13).

Table 10.1 *Averages and standard errors (SE) of the parameter estimates and the empirical coverage probabilities of the corresponding 95% confidence intervals (CI) covering the true parameter $\beta = -1.5$ computed from 10,000 simulated samples with the naive mixed-effects change-point model (9.13)*

Correlation Structure	Estimate	SE	Empirical Coverage Probability of 95% CI
LME			
Working Independence	-0.598	0.395	37.8%
Random Intercept	-0.731	0.403	51.0%
Random Intercept and Slope	-0.788	0.431	60.4%
GEE			
Working Independence	-0.598	0.385	36.2%
Exchangeable	-0.709	0.385	46.7%
Unstructured	-0.752	0.572	43.9%

The results of Table 10.1 suggest that all these LME and GEE procedures with different correlation structure assumptions give similar estimates for β, which are around -0.6 and -0.7 and far from the true value of $\beta = -1.5$. Since the standard errors for these estimates are comparable, it is not surprising that the 95% confidence intervals shown in Table 10.1 have low empirical coverage probabilities, which suggest that the model (10.13) leads to excessive biases and inadequate estimates for β.

4. An Alternative Using "Individual Fitting"

We now consider an ad hoc and seemingly *oversimplified* approach which can lead to better estimates of the mean concomitant intervention effect β than the LME and GEE results shown in Table 10.1. For the simple setup considered here, we use an "individual fitting" estimator of β based on the mixed-effects model (10.13), in which we ignore the standard estimation procedures in the literature and compute the mean changes of Y_{ij} before and after the concomitant intervention for each subject individually.

Since the time-trends of Y_{ij} do not depend on $t_{ij} = j$ when λ_{ij} is given, we can simply calculate the subject-specific differences for the mean values of Y_{ij} before and after the concomitant intervention, and then estimate the population parameter β by the sample mean of the differences, which lead to the "individual fitting" estimator of β given by

$$\widehat{\beta}_{ind} = \left(\frac{1}{n}\right) \sum_{i=1}^{n} \left[\frac{\sum_{j=1}^{n_i} \left(Y_{ij} \, 1_{[\delta_{ij}=1]}\right)}{\sum_{j=1}^{n_i} 1_{[\delta_{ij}=1]}} - \frac{\sum_{j=1}^{n_i} \left(Y_{ij} \, 1_{[\delta_{ij}=0]}\right)}{\sum_{j=1}^{n_i} 1_{[\delta_{ij}=0]}}\right]. \tag{10.15}$$

A simple application of (10.15) to our simulated samples leads to estimates of β very close to its true value of -1.5. This suggests that it is indeed possible to reduce the biases shown in Table 10.1 by considering the subject-specific model of (10.11), while the mixed-effects model (10.13) and the well-known likelihood and estimating equation based procedures are misspecified and may lead to significant estimation biases.

However, the "individual fitting" approach of (10.15) is generally not used in the literature due to two reasons. First, it is known in the literature that "individual fitting" estimation methods using the repeated measurements separately from each subject, such as (10.15), are generally less efficient than the well-known procedures, such as the maximum likelihood estimates, the restricted maximum likelihood estimates, and the generalized estimating equation estimates (e.g., Verbeke and Molenberghs, 2000; Diggle et al., 2002). Second, it is difficult to generalize the "individual fitting" approaches to regression models with more complicated terms and patterns than the simple cases exhibited in (10.11). Fitting data from the repeated measurements of each subject require the numbers of repeated measurements to be sufficiently large.

5. An Alternative Incorporating Concomitant Intervention Starting Time

Comparing the underlying mechanism for generating the data in (10.14) with the model (10.13), a potential flaw of using (10.13) is that the model does not incorporate the potential relationship between the change-point time S_i and the values of Y_{ij} before the change-point time, i.e., the values of Y_{ij} when $\delta_{ij} = 0$. Indeed, under the real data generating mechanism (10.12), subjects with subject-specific mean value of Y_{ij} at $j = 1$ to be 20 have change-point time at $S_i = 2$, while subjects with subject-specific mean value of Y_{ij} at $j = 1$ to be 19 have change-point time at $S_i = 8$. Although this fact is unknown at the estimation stage, the potential relationship between S_i and Y_{ij} is not incorporated in the model (10.13).

To see whether the potential bias for the estimation of β could be reduced by incorporating the change-point time S_i into the model, we consider a simple generation of (10.13) given by

$$\begin{cases} Y_{ij} = a_{0i} + a_{1i} S_i + b_i \lambda_{ij} + \varepsilon_{ij}, \\ (a_{0i}, a_{1i}, b_i)^T \sim N\left[(\alpha_0, \alpha_1, \beta)^T, \mathbf{V}\right], \end{cases} \tag{10.16}$$

where $(\alpha_0, \alpha_1, \beta)^T$ is the vector of mean parameters, $\mathbf{V}$ is the unknown covariance matrix for $(a_{0i}, a_{1i}, b_i)^T$ and ε_{ij} are the measurement errors as in (10.13).

We note that (10.16) is not at all the true model, since the model used to generate the data in (10.14) is in fact (10.12) and Y_{ij} does not depend on S_i through a simple linear model. We will give some justifications of considering (10.16) in Section 10.4, where it is described as a special case of the varying-coefficient mixed-effects models of Wu, Tian and Bang (2008). But for now, (10.16) simply suggests that, for subjects with different values of S_i, the mean values of Y_{ij} before the concomitant intervention are also different, and the relationship between Y_{ij} and S_i is approximated by a simple linear model.

The structures of $\mathbf{V}$ generally do not have major influences on the estimation of $(\alpha_0, \alpha_1, \beta)^T$, and many commonly used parametric structures may be used when implementing the estimation procedures (Diggle et al., 2002). It is important to note that the interpretation of β is the same in both (10.13) and (10.16). Although (10.16) is at best a rough approximation of the true underlying data generating mechanism, the main intent here is to evaluate whether the bias for the estimation of the concomitant intervention effect can be reduced by incorporating S_i into the model.

Table 10.2 *Averages and standard errors (SE) of the parameter estimates and the empirical coverage probabilities of the corresponding 95% confidence intervals (CI) covering the true parameter $\beta = -1.5$ computed from the simulated samples with the mixed-effects model (10.16)*

Correlation Structure	Estimate	SE	Empirical Coverage Probability of 95% CI
LME			
Working Independence	-1.498	0.485	94.7%
Random Intercept	-1.498	0.483	94.6%
Random Intercept and Slopes	-1.498	0.496	95.2%

Table 10.2 summarizes the averages of the estimators and their standard errors (SE), and the empirical coverage probabilities of the corresponding 95% confidence intervals covering the true parameter $\beta = -1.5$ computed from the 10,000 simulated samples, the mixed-effects model (10.16) and the same LME procedures as the ones used in Table 10.1. The mean estimates for β in Table 10.2 are very close to the true value of $\beta = -1.5$, with comparable standard errors in both Tables 10.1 and 10.2. The 95% confidence intervals in Table 10.2 have empirical coverage probabilities that are very close to the nominal level of 95% and much higher than the ones shown in Table 10.1. Clearly, (10.16) leads to much smaller bias for the estimation of β than (10.13).

10.3   General Structure for Shared Parameters

The data structure of Section 10.1.3 can be viewed as a special case of outcome-adaptive covariates, because the data involve only one concomitant intervention and each study subject has at most one change-point from "without con-comitant intervention" to "with concomitant intervention." This simple struc-ture of outcome adaptiveness, which has only one change-point time, suggests that a natural extension for the mixed-effects model (10.9) is to incorporate the initiation of the concomitant intervention into the regression model. Specifi-cally, the model should allow the intervention starting time S_i to be correlated with the subject-specific parameters $\{a_i, b_i\}$.

Let $\mu_0(\cdot; a_i)$ and $[\mu_0(\cdot; a_i) + \mu_1(\cdot; b_i)]$ be the subject-specific response curves before and after the start of the concomitant intervention, respectively, where $\mu_1(\cdot; b_i)$ is known as the concomitant intervention effect. A shared-parameter change-point model for the dataset $\{Y_{ij}, t_{ij}, X_i, S_i\}$ is

$$
\begin{cases}
Y_{ij} = \mu_0(t_{ij}, X_i; a_i) + \delta_{ij} \mu_1(t_{ij}, X_i, R_{ij}; b_i) + \varepsilon_{ij}, \\
(a_i^T, b_i^T, S_i)^T \sim \text{Joint Distribution},
\end{cases}
\tag{10.17}
$$

where $R_{ij} = T_{ij} - S_i$, ε_{ij} are mean zero errors with $Cov(\varepsilon_{ij_1}, \varepsilon_{ij_2}) = \sigma_{ij_1 j_2}$, $\varepsilon_{i_1 j_1}$ and $\varepsilon_{i_2 j_2}$ are independent if $i_1 \neq i_2$, S_i and $\{t_{ij}, X_i\}$ are independent conditioning on $\{a_i, b_i\}$, and $\{a_i, b_i\}$ and $\{t_{ij}, X_i\}$ are independent.

Using the matrix representation $Y_i = (Y_{i1}, \ldots, Y_{in_i})^T$ and $t_i = (t_{i1}, \ldots, t_{in_i})^T$, the joint likelihood function of $(Y_i^T, S_i)^T$ given $\{t_i, X_i\}$ based on (10.17) is

$$
f(Y_i, S_i | t_i, X_i) = \int f(Y_i | t_i, X_i, S_i, a_i, b_i) f(S_i | a_i, b_i) dH(a_i, b_i), \tag{10.18}
$$

where $f(\cdot | \cdot)$ denotes the conditional density and $H(\cdot, \cdot)$ is the joint distribu-tion function of $\{a_i, b_i\}$. The $f(S_i | a_i, b_i)$ term in the integration distinguishes (10.18) from the usual likelihood functions for the mixed-effects models, such as (2.8). The subject-specific parameters $\{a_i, b_i\}$ in (10.17) are associated with both the response curves of Y_{ij} and the distribution of S_i, which suggests the name of "shared-parameter model." The population parameters $E(a_i) = \alpha$ and $E(b_i) = \beta$ are also shared in both the upper and lower displays of (10.17).

The shared-parameter model was proposed in Follmann and Wu (1995) for the purpose of modeling the behaviors of informative missing data. Unlike Follmann and Wu (1995), the subjects are still being observed after the change-point time S_i in (10.17). The correlation between S_i and a_i suggests that the ith subject's change-point time is affected by the pre-intervention response curve $\mu_0(\cdot; a_i)$, and the correlation between S_i and b_i suggests that S_i may also influence the response curve $\mu_1(\cdot; b_i)$, which describes the intervention effects.

The main advantage of (10.17) is that it distinguishes the concomitant intervention from randomly assigned study treatments and other covariates by modeling the association between the change-point time S_i and the out-come variable Y_{ij}. However, at its current form, (10.17) is still too general to be used in real applications, because the likelihood based estimation and inferences cannot be easily derived without specific expressions of (10.18). In real applications, sub-classes of (10.17) are needed with specific expressions of

$\mu_0(\cdot; \mathbf{a}_i)$, $\mu_1(\cdot; \mathbf{b}_i)$ and the joint distributions of $\{\mathbf{a}_i, \mathbf{b}_i, S_i\}$. We focus on two different special cases of (10.17), each of which has its own advantages and disadvantages and is suitable for a specific data structure.

10.4   The Varying-Coefficient Mixed-Effects Models

As discussed above, direct application of (10.17) and (10.18) is usually limited by two issues: (a) the joint likelihood function (10.18) requires a known distribution function for $\{\mathbf{a}_i, \mathbf{b}_i, S_i\}$; (b) parameter estimation by maximizing the likelihood function (10.18) can be computationally intensive when the number of parameters is large. When all the subjects have observed change-point times within the study period, that is, $\mathcal{T}_0 < S_i < \mathcal{T}_1$ for all $1 \leq i \leq n$, we can extend the varying-coefficient approach of Chapters 6 to 9 to the change-point time S_i and consider a computationally simple regression model, which does not depend on the distribution function of S_i.

10.4.1   Model Formulation and Interpretation

1. General and Linear Model Expressions

When S_i are observed for all $1 \leq i \leq n$, we can consider the conditional distribution

$$f\big(\mathbf{Y}_i \big| S_i, \mathbf{t}_i, \mathbf{X}_i\big) = \int f\big(\mathbf{Y}_i \big| \mathbf{t}_i, \mathbf{X}_i, S_i, \mathbf{a}_i, \mathbf{b}_i\big) \, dG\big(\mathbf{a}_i, \mathbf{b}_i \big| S_i\big), \qquad (10.19)$$

and rewrite (10.17) as a varying-coefficient model using the conditional distribution of $\{\mathbf{a}_i, \mathbf{b}_i\}$ given S_i. Although $\mu_0(\cdot)$ and $\mu_1(\cdot)$ are allowed to take general parametric or nonparametric forms, this approach is illustrated here assuming that $\mu_0(\cdot)$ and $\mu_1(\cdot)$ are linear functions of the form

$$\mu_0\big(t_{ij}, \mathbf{X}_i; \mathbf{a}_i\big) = \mathbf{Z}_{ij}^T \mathbf{a}_i \quad \text{and} \quad \mu_1\big(t_{ij}, \mathbf{X}_i, S_i; \mathbf{b}_i\big) = \mathbf{W}_{ij}^T \mathbf{b}_i, \qquad (10.20)$$

where $\mathbf{Z}_{ij} = \big(Z_{ij0}, \ldots, Z_{ijD_1}\big)^T$ is generated by $\big\{ (t_{ij}, \mathbf{X}_i^T) : 1 \leq j \leq n_i; \delta_{ij} = 0 \big\}$ and $\mathbf{W}_{ij} = \big(W_{ij0}, \ldots, W_{ijD_2}\big)^T$ is generated by $\big\{ (t_{ij}, \mathbf{X}_i^T, S_i) : 1 \leq j \leq n_i; \delta_{ij} = 1 \big\}$.

If we view S_i as a possible confounder for the means of Y_{ij} in (10.17), we can in principle evaluate the concomitant intervention effect by comparing the mean trajectories of Y_{ij} before and after the concomitant intervention conditioning on a given S_i. Let

$$\alpha(S_i) = E(\mathbf{a}_i | S_i), \quad \beta(S_i) = E(\mathbf{b}_i | S_i), \quad \mathbf{a}_i^* = \mathbf{a}_i - \alpha(S_i) \quad \text{and} \quad \mathbf{b}_i^* = \mathbf{b}_i - \beta(S_i).$$

A varying-coefficient mixed-effects model for the data

$$\big\{ (Y_{ij}, t_{ij}, \mathbf{X}_i^T, S_i) : i = 1, \ldots, n; j = 1, \ldots, n_i \big\}$$

based on (10.17) and (10.20) is

$$\begin{cases} Y_{ij} = \mathbf{Z}_{ij}^T \big[\alpha(S_i) + \mathbf{a}_i^* \big] + \delta_{ij} \, \mathbf{W}_{ij}^T \big[\beta(S_i) + \mathbf{b}_i^* \big] + \varepsilon_{ij}, \\ \big(\mathbf{a}_i^{*T}, \mathbf{b}_i^{*T} \big)^T \big| S_i \sim G(\cdot | S_i) \end{cases} \qquad (10.21)$$

where, for $S_i = s$, $G(\cdot|s)$ is a distribution function with mean zero and covariance matrix $Cov\left[\left(\mathbf{a}_i^{*T}, \mathbf{b}_i^{*T}\right)^T | s\right] = \mathbf{C}(s)$. The population-mean parameters are unknown curves $\alpha(s)$ and $\beta(s)$, which, in this case, are both smooth functions of s. When $S_i = s$, the mean concomitant intervention effect is $\beta(s)$. The special choice of $\beta(s) = 0$ for all $s \in (\mathcal{T}_0, \mathcal{T}_1)$ implies that the concomitant intervention has no population-mean effect on the time-trend curve of Y_{ij}.

2. Interpretations and Remarks

Parametric vs. Nonparametric Models:
The models (10.17) and (10.21) characterize the concomitant intervention effects using different parameters, which lead to different types of models and estimation methods. In (10.17), the concomitant intervention effect is summarized in $\mu_1(\cdot; \mathbf{b}_i)$, so that the population parameter determining the concomitant intervention effects is $\beta = E(\mathbf{b}_i)$, although the estimation of β is possibly influenced by the observations of S_i. In (10.21), on the other hand, because of the definition of $\mathbf{b}_i^*$ as a *re-centered* subject-specific parameter, the statistical inference is based on $\beta(s)$, which is the concomitant intervention effects conditioning on the $S_i = s$. Thus, when $\alpha(s)$ and $\beta(s)$ are smoothing curves of s without further parametric assumptions, (10.21) is a structured nonparametric model, while (10.17) may be either a parametric or nonparametric model depending on the assumptions on $\mu_0(\cdot; \mathbf{a}_i)$, $\mu_1(\cdot; \mathbf{b}_i)$ and the joint distribution of $(\mathbf{a}_i^T, \mathbf{b}_i^T, S_i)$. In this section, the estimation of $\alpha(s)$ and $\beta(s)$ of (10.21) is proceeded using the basis approximations as in Chapter 9, and parameter estimation of the parametric versions of (10.17) is proceeded using the maximum likelihood method.   □

Advantage of Conditional Change-Point Approach:
The advantage of (10.21) over a naive mixed-effects model, such as (10.9), is that the statistical inference of the concomitant intervention effect of (10.21) is made separately for subjects with different values of concomitant intervention change-point time $S_i = s$. This is somewhat analogous to the *sub-group analysis* for subjects with different categorical values of S_i, and, as demonstrated in Section 10.2.3, may reduce the estimation bias associated with a naive mixed-effects model. This conditional approach is reasonable, because subjects with different concomitant intervention change-point times are likely to be different. The *change-point* coefficient $\beta(s)$ gives a precise description of the concomitant intervention effect for subjects with $S_i = s$.

The advantage of the conditional approach of (10.21) is also seen from the synthetic example of Section 10.2.3. In this example, the true concomitant intervention effects in (10.12) are different for individuals with $S_i = 2$ and individuals with $S_i = 8$, and the population mean concomitant intervention effect is $\beta = -1.5$. The estimators shown in Table 10.1, which are obtained from the model (10.13) without conditioning on the change-point time S_i, all have significant biases no matter what correlation structures are used. On the other hand, the estimators shown in Table 10.2, which rely on (10.16), are all nearly unbiased, although (10.16) is not necessarily close to the true model (10.12). Thus, the bias for the estimation of β can be reduced by models conditioning on S_i.   □

10.4.2   Special Cases of Conditional-Mean Effects

A number of interesting special cases of (10.21) may be considered in real applications by specifying the forms of $\alpha(s)$, $\beta(s)$ and $G(\cdot|S_i = s)$. An obvious choice for $G(\cdot|S_i = s)$ is the multivariate normal distribution with mean zero and covariance matrix $\mathbf{C} = Cov\big[(\mathbf{a}_i^{*T}, \mathbf{b}_i^{*T})\big|s\big]$, which, for simplicity, is assumed to be time-invariant. Extension to time-dependent covariances can be made by modeling $\mathbf{C}(s)$. Since the main objective is to estimate the population-mean effects of the concomitant intervention and the explicit forms of $G(\cdot|S_i)$ are often unknown, using appropriate models for $\alpha(s)$ and $\beta(s)$ is often more important than using a suitable model for $\mathbf{C}(s)$.

1. Parametric Models for Mean Effects

Linear models for $\alpha(s)$ and $\beta(s)$ of (10.21) can be expressed as

$$
\begin{cases}
\alpha(s; \gamma) = \big(\alpha_0(s; \gamma_0), \ldots, \alpha_{D_1}(s; \gamma_{D_1})\big)^T, & \alpha_d(s; \gamma) = \sum_{l=0}^{L_d} \gamma_{dl}\, \mathcal{T}_{dl}(s); \\
\beta(s; \tau) = \big(\beta_0(s; \tau_0), \ldots, \beta_{D_2}(s; \tau_{D_2})\big)^T, & \beta_d(s; \tau) = \sum_{m=0}^{M_d} \tau_{dm}\, \mathcal{T}_{dm}^{*}(s); \\
\gamma_d = \big(\gamma_{d0}, \ldots, \gamma_{dL_d}\big)^T, & \gamma = \big(\gamma_0, \ldots, \gamma_{D_1}\big)^T; \\
\tau_d = \big(\tau_{d0}, \ldots, \tau_{dM_d}\big)^T, & \tau = \big(\tau_0, \ldots, \tau_{D_2}\big)^T,
\end{cases}
$$

$$\tag{10.22}$$

where $\{L_d, M_d\}$ are fixed, and $\{\mathcal{T}_{dl}(s), \mathcal{T}_{dm}^{*}(s)\}$ are known transformations of s. The choice of $\mathcal{T}_{dl}(s) = s^l$ and $\mathcal{T}_{dm}^{*}(s) = s^m$ leads to the global polynomials for $\alpha_d(s; \gamma)$ and $\beta_d(s; \tau)$. The parameters γ and τ characterize the time-trends of Y_{ij}. In particular, the concomitant intervention effects are determined by τ. Other parametric models for $\alpha(s)$ and $\beta(s)$, such as, piece-wise linear models, piece-wise polynomials and other nonlinear models, are also possible choices. But, since these models are less often used compared with the linear models of (10.22), explicit forms of these models are not described in detail here.

2. Basis Approximations for Mean Effects

When the parametric forms of $\alpha(s)$ and $\beta(s)$ of (10.21) are unknown, nonparametric analysis can be performed by approximating $\alpha(s)$ and $\beta(s)$ with basis expansions. If

$$
\Big\{ \mathscr{B}_{d_1}(s) = \big(\mathscr{B}_{d_1 0}(s), \ldots, \mathscr{B}_{d_1 \mathscr{L}_{d_1}}(s)\big)^T : 0 \le d_1 \le D_1 \Big\}
$$

and

$$
\Big\{ \mathscr{B}_{d_2}^{*}(s) = \big(\mathscr{B}_{d_2 0}^{*}(s), \ldots, \mathscr{B}_{d_2 \mathscr{M}_{d_2}}^{*}(s)\big)^T : 0 \le d_2 \le D_2 \Big\}
$$

are two sets of pre-specified basis functions, their basis approximations for $\alpha(s)$ and $\beta(s)$ are given by

$$\alpha_d(s;\gamma) \approx \sum_{l=0}^{\mathscr{L}_d} \gamma_{dl}\, \mathscr{B}_{dl}(s) \quad \text{and} \quad \beta_d(s;\tau) \approx \sum_{m=0}^{\mathscr{M}_d} \tau_{dm}\, \mathscr{B}_{dm}^*(s), \qquad (10.23)$$

where $\mathscr{L}_d$ and $\mathscr{M}_d$ may tend to infinity as $n \to \infty$. Unlike (10.22), the dimensionality of $(\gamma_{d0}, \ldots, \gamma_{d\mathscr{L}_d})^T$ and $(\tau_{d0}, \ldots, \tau_{d\mathscr{M}_d})^T$ may tend to infinity, so that we may view (10.23) as a special case of the *infinite dimensional extended linear models*. The real quantities of interest in practice are the curves $\alpha_d(s;\gamma)$ and $\beta_d(s;\tau)$ instead of the coefficients γ_{dl} and τ_{dm}.

Common choices of basis functions include truncated polynomial bases, Fourier bases or B-splines. Currently, only B-splines with fixed knot sequences have been investigated for the model (10.21) in the literature (Wu, Tian and Bang, 2008). An alternative smoothing approach could be to approximate $\alpha(s)$ and $\beta(s)$ by smoothing splines as in Chapter 5. But nonparametric estimation and inference with smoothing splines in (10.21) have not been studied, since explicit expressions and statistical properties of smoothing spline estimators require different mathematical derivations from B-splines.

10.4.3   *Likelihood-Based Estimation*

When the conditional mean curves $\alpha(s)$ and $\beta(s)$ are specified by parametric forms and the distribution of ε_{ij} and the conditional distribution $G(\cdot|S_i)$ belong to known parametric families, parameter estimation of the model (10.21) can be obtained by a likelihood-based estimation procedure. If the conditional density function (10.18) has an explicit parametric expressions, the parameters can be in principle estimated by maximizing the log-likelihood $\sum_{i=1}^n \log f(\mathbf{Y}_i, S_i | \mathbf{t}_i, \mathbf{X}_i)$. But, since the distribution of S_i is usually unknown in practice, the approach suggested by Wu, Tian and Bang (2008) is to maximize the *partial likelihood* composed by the conditional joint densities given $\{S_i : i = 1, \ldots, n\}$.

1. *Estimators for Linear Models*

We demonstrate this likelihood-based estimation method using the linear conditional model formed by (10.21) and (10.22). When (10.21) and (10.22) are satisfied, $G(\cdot|s)$ is Gaussian with mean zero and covariance matrix $\mathbf{C}(S_i)$, and

$$\left(\varepsilon_{i1}, \ldots, \varepsilon_{in_i}\right)^T \sim N\left(0, \Gamma_i\right),$$

the model under consideration is

$$\begin{cases} Y_{ij} = \sum_{d=0}^{D_1}\sum_{l=0}^{L_d} Z_{ijd}\, \gamma_{dl}\, \mathscr{T}_{dl}(s) + \mathbf{Z}_{ij}^T \mathbf{a}_i^* \\ \qquad + \sum_{d=0}^{D_2}\sum_{m=0}^{M_d} \delta_{ij}\, W_{ijd}\, \tau_{dm}\, \mathscr{T}_{dm}^*(s) + \delta_{ij}\, \mathbf{W}_{ij}^T \mathbf{b}_i^* + \varepsilon_{ij}, \\ \left(\mathbf{a}_i^{*T}, \mathbf{b}_i^{*T}\right)^T \big| S_i \sim N\left(\mathbf{0}, \mathbf{C}(s)\right), \end{cases} \qquad (10.24)$$

where the parameters of interest γ and τ are defined in (10.22). Since the marginal distribution of S_i in the model (10.24) is unknown, we can estimate γ and τ by maximizing the following partial log-likelihood function,

$$L\big(\gamma, \tau \big| \mathbf{t}_i, \mathbf{X}_i, S_i\big) = \sum_{i=1}^{n} \log\bigg[\int f\big(\mathbf{Y}_i \big| \mathbf{t}_i, \mathbf{X}_i, S_i, \mathbf{a}_i, \mathbf{b}_i\big)\, dG\big(\mathbf{a}_i, \mathbf{b}_i \big| S_i\big)\bigg], \qquad (10.25)$$

which is the conditional log-likelihood function of $\{\mathbf{Y}_i : i = 1, \dots, n\}$ given $\{S_i, \mathbf{t}_i, \mathbf{X}_i : i = 1, \dots, n\}$,

To obtain the matrix representation of the maximum partial log-likelihood estimators of γ and τ, we define

$$\begin{cases} \mathscr{W}_i & = \quad \text{the matrix whose } j\text{th row is } \big(\mathbf{Z}_{ij}^T, \delta_{ij}\, \mathbf{W}_{ij}^T\big), \\[4pt] \mathscr{T}_d(s) & = \quad \big(\mathscr{T}_{d1}(s), \dots, \mathscr{T}_{dL_d}(s)\big)^T, \\[4pt] \mathscr{T}_d^*(s) & = \quad \big(\mathscr{T}_{d1}^*(s), \dots, \mathscr{T}_{dM_d}^*(s)\big)^T, \\[4pt] \mathscr{T}(s) & = \quad \mathrm{diag}\big\{\mathscr{T}_0^T(s), \dots, \mathscr{T}_{D_1}^T(s), \mathscr{T}_0^{*T}(s), \dots, \mathscr{T}_{D_2}^{*T}(s)\big\}, \\[4pt] \mathscr{T}(S_i) & = \quad \mathscr{T}_i, \\[4pt] \mathbf{V}_i & = \quad \text{the covariance matrix of} \\[4pt] & \qquad e_{ij} = \mathbf{Z}_{ij}^T \mathbf{a}_i^* + \delta_{ij}\, \mathbf{W}_{ij}^T \mathbf{b}_i^* + \varepsilon_{ij}, \quad 1 \le j \le n_i, \end{cases} \qquad (10.26)$$

where $diag(A_1, \dots, A_k)$ represents the block diagonal matrix with diagonal block matrices $\{A_1, \dots, A_k\}$.

Using the notation in (10.26), the matrix representation for (10.22) is

$$\begin{cases} \big(\boldsymbol{\alpha}^T(s; \gamma), \boldsymbol{\beta}^T(s; \tau)\big)^T = \mathscr{T}(s)\big(\gamma^T, \tau^T\big)^T, \\[4pt] \gamma_d = \big(\gamma_{d0}, \dots, \gamma_{dL_d}\big)^T, \quad \gamma = \big(\gamma_0^T, \dots, \gamma_{D_1}^T\big)^T, \\[4pt] \tau_d = \big(\tau_{d0}, \dots, \tau_{dM_d}\big)^T, \quad \tau = \big(\tau_0^T, \dots, \tau_{D_2}^T\big)^T. \end{cases} \qquad (10.27)$$

When $\mathbf{V}_i$ are known, maximizing (10.25) for the model (10.24) leads to the maximum likelihood estimator of $\big(\gamma^T, \tau^T\big)^T$ of the form

$$\begin{pmatrix} \widehat{\gamma}_{ML}(\mathscr{T}) \\ \widehat{\tau}_{ML}(\mathscr{T}) \end{pmatrix} = \bigg[\sum_{i=1}^{n} \big(\mathscr{W}_i \mathscr{T}_i\big)^T \mathbf{V}_i^{-1} \big(\mathscr{W}_i \mathscr{T}_i\big)\bigg]^{-1} \bigg[\sum_{i=1}^{n} \big(\mathscr{W}_i \mathscr{T}_i\big)^T \mathbf{V}_i^{-1} \mathbf{Y}_i\bigg] \qquad (10.28)$$

provided that $\sum_{i=1}^{n} \big[(\mathscr{W}_i \mathscr{T}_i)^T \mathbf{V}_i^{-1} (\mathscr{W}_i \mathscr{T}_i)\big]$ is nonsingular. When $\mathbf{V}_i$ are unknown but can be consistently estimated by a nonsingular estimator $\widehat{\mathbf{V}}_i$, we can estimate $\big(\gamma^T, \tau^T\big)^T$ by

$$\begin{pmatrix} \widetilde{\gamma}_{ML}(\mathscr{T}) \\ \widetilde{\tau}_{ML}(\mathscr{T}) \end{pmatrix} = \bigg[\sum_{i=1}^{n} \big(\mathscr{W}_i \mathscr{T}_i\big)^T \widehat{\mathbf{V}}_i^{-1} \big(\mathscr{W}_i \mathscr{T}_i\big)\bigg]^{-1} \bigg[\sum_{i=1}^{n} \big(\mathscr{W}_i \mathscr{T}_i\big)^T \widehat{\mathbf{V}}_i^{-1} \mathbf{Y}_i\bigg]. \qquad (10.29)$$

A class of covariance estimators $\widehat{\mathbf{V}}_i$ is derived in Section 10.4.5.

2. Estimators for Basis Approximated Models

When $\alpha_d(s; \gamma)$ and $\beta_d(s; \tau)$ are nonparametric curves which can be approximated by the right-side expansions of (10.23), an infinite dimensional extended linear model version of (10.24) is

$$
\begin{cases}
Y_{ij} \approx \sum_{d=0}^{D_1} \sum_{l=0}^{\mathcal{L}_d} Z_{ijd}\, \gamma_{dl}\, \mathscr{B}_{dl}(s) + \mathbf{Z}_{ij}^T \mathbf{a}_i^* \\
\qquad + \sum_{d=0}^{D_2} \sum_{m=0}^{\mathcal{M}_d} \delta_{ij} Z_{ijd}\, \tau_{dm}\, \mathscr{B}_{dm}^*(s) \\
\qquad + \delta_{ij} \mathbf{W}_{ij}^T \mathbf{b}_i^* + \varepsilon_{ij}, \\
\left(\mathbf{a}_i^{*T}, \mathbf{b}_i^{*T}\right)^T \big| S_i \sim N\big(\mathbf{0}, \mathbf{C}(s)\big).
\end{cases}
\tag{10.30}
$$

For any given $\mathcal{L}_d$ and $\mathcal{M}_d$, we substitute $\mathscr{T}_{dl}(s)$ and $\mathscr{T}_{dm}^*(s)$ in (10.26) with the basis functions $\mathscr{B}_{dl}(s)$ and $\mathscr{B}_{dm}^*(s)$, respectively, and get

$$
\begin{cases}
\mathscr{B}_d(s) &= \left(\mathscr{B}_{dl}(s), \ldots, \mathscr{B}_{d\mathcal{L}_d}(s)\right)^T, \\
\mathscr{B}_d^*(s) &= \left(\mathscr{B}_{d1}^*(s), \ldots, \mathscr{B}_{d\mathcal{M}_d}^*(s)\right)^T, \\
\mathscr{B}(s) &= diag\{\mathscr{B}_0^T(s), \ldots, \mathscr{B}_{D_1}^T(s), \mathscr{B}_0^{*T}(s), \ldots, \mathscr{B}_{D_2}^{*T}(s)\}, \\
\mathscr{B}(S_i) &= \mathscr{B}_i.
\end{cases}
\tag{10.31}
$$

The approximate maximum likelihood estimators $\widehat{\gamma}_{ML}(\mathscr{B})$ and $\widehat{\tau}_{ML}(\mathscr{B})$ for the case of known $\mathbf{V}_i$ are then given by

$$
\begin{pmatrix} \widehat{\gamma}_{ML}(\mathscr{B}) \\ \widehat{\tau}_{ML}(\mathscr{B}) \end{pmatrix} = \left[\sum_{i=1}^n (\mathscr{W}_i \mathscr{B}_i)^T \mathbf{V}_i^{-1} (\mathscr{W}_i \mathscr{B}_i)\right]^{-1} \left[\sum_{i=1}^n (\mathscr{W}_i \mathscr{B}_i)^T \mathbf{V}_i^{-1} \mathbf{Y}_i\right]
\tag{10.32}
$$

provided that $\sum_{i=1}^n \left[(\mathscr{W}_i \mathscr{B}_i)^T \mathbf{V}_i^{-1} (\mathscr{W}_i \mathscr{B}_i)\right]$ is nonsingular, where $\mathscr{W}_i$ and $\mathbf{V}_i$ are defined in (10.26). The likelihood-based nonparametric basis approximation estimators of $\left(\alpha^T(s), \beta^T(s)\right)^T$ under the basis approximations of (10.23) with known $\mathbf{V}_i$ is

$$
\left(\widehat{\alpha}_{ML}^T(s; \mathscr{B})), \widehat{\beta}_{ML}^T(s; \mathscr{B})\right)^T = \mathscr{B}(s) \left(\widehat{\gamma}_{ML}^T(\mathscr{B}), \widehat{\tau}_{ML}^T(\mathscr{B})\right)^T.
\tag{10.33}
$$

Using the same approach as in (10.29), when $\mathbf{V}_i$ is unknown but a consistent estimator $\widehat{\mathbf{V}}_i$ (Section 10.4.5) of $\mathbf{V}_i$ is available, the approximate maximum likelihood estimators $\widetilde{\gamma}_{ML}(\mathscr{B})$ and $\widetilde{\tau}_{ML}(\mathscr{B})$ are then given by

$$
\begin{pmatrix} \widetilde{\gamma}_{ML}(\mathscr{B}) \\ \widetilde{\tau}_{ML}(\mathscr{B}) \end{pmatrix} = \left[\sum_{i=1}^n (\mathscr{W}_i \mathscr{B}_i)^T \widehat{\mathbf{V}}_i^{-1} (\mathscr{W}_i \mathscr{B}_i)\right]^{-1} \left[\sum_{i=1}^n (\mathscr{W}_i \mathscr{B}_i)^T \widehat{\mathbf{V}}_i^{-1} \mathbf{Y}_i\right],
\tag{10.34}
$$

and the nonparametric basis approximation estimators of $\left(\alpha^T(s), \beta^T(s)\right)^T$ computed using (10.34) is

$$
\left(\widetilde{\alpha}_{ML}^T(s; \mathscr{B}), \widetilde{\beta}_{ML}^T(s; \mathscr{B})\right)^T = \mathscr{B}(s) \left(\widetilde{\gamma}_{ML}^T(\mathscr{B}), \widetilde{\tau}_{ML}^T(\mathscr{B})\right)^T.
\tag{10.35}
$$

provided that $\sum_{i=1}^n \left[(\mathscr{W}_i \mathscr{B}_i)^T \widehat{\mathbf{V}}_i^{-1} (\mathscr{W}_i \mathscr{B}_i)\right]$ is nonsingular.

10.4.4   Least Squares Estimation

Likelihood-based estimators of $\alpha(s)$ and $\beta(s)$ for (10.21) cannot be computed if the explicit forms of $G(\cdot|S_i)$ and the distribution of ε_{ij} are unknown. In this case, the special case (10.24) may not be satisfied. A practical approach is to directly estimate the coefficient curves $\alpha(s)$ and $\beta(s)$ in (10.21) using least squares.

1. Least Squares Estimators for Linear Models

When $\alpha(s) = \alpha(s; \gamma)$ and $\beta(s) = \beta(s; \tau)$ belong to some parametric families, which are determined by Euclidean space parameters γ and τ, respectively, a practical approach is to compute the weighted least squares estimators $\widehat{\gamma}_{LS}$ and $\widehat{\tau}_{LS}$ by minimizing the score function

$$\ell(\gamma, \tau) \;=\; \sum_{i=1}^{n} \left\{ \left[\mathbf{Y}_i - \left(\mathbf{Z}_i^T \, \alpha(S_i; \gamma) + (\delta \mathbf{W})_i^T \, \beta(S_i; \tau) \right) \right]^T \right.$$
$$\left. \times \Lambda_i \left[\mathbf{Y}_i - \left(\mathbf{Z}_i^T \, \alpha(S_i; \gamma) + (\delta \mathbf{W})_i^T \, \beta(S_i; \tau) \right) \right] \right\} \quad (10.36)$$

with respect to γ and τ, where

$$\mathbf{Z}_i = \left(\mathbf{Z}_{i1}, \ldots, \mathbf{Z}_{in_i} \right)^T, \quad (\delta \mathbf{W})_i = \left(\delta_{i1} \, \mathbf{W}_{i1}, \ldots, \delta_{in_i} \, \mathbf{W}_{in_i} \right)^T,$$

and Λ_i are pre-specified symmetric nonsingular $n_i \times n_i$ weight matrices. The main advantage of the weighted least squares estimators is that (10.36) does not depend on the explicit distribution functions of ε_{ij} and the conditional distribution function $G(\cdot|s)$, so that (10.36) is not limited to (10.24) or other special cases and can be applied to the general model (10.21). Statistical properties of the weighted least squares estimators $\widehat{\gamma}_{LS}$ and $\widehat{\tau}_{LS}$ depend on the chosen parametric families for $\alpha(s; \gamma)$ and $\beta(s; \tau)$. At the very minimum, the forms of $\alpha(s; \gamma)$ and $\beta(s; \tau)$ have to ensure the existence and uniqueness of the minimizer $(\widehat{\gamma}^T, \widehat{\tau}^T)$ of (10.36).

For the special case that $\alpha(s; \gamma)$ and $\beta(s; \tau)$ are given by the linear model (10.22), explicit expressions of the weighted least squares estimator of (γ^T, τ^T) uniquely exists and is given by

$$\left(\begin{array}{c} \widehat{\gamma}_{LS}(\mathscr{T}) \\ \widehat{\tau}_{LS}(\mathscr{T}) \end{array} \right) = \left[\sum_{i=1}^{n} \left(\mathscr{W}_i \, \mathscr{T}_i \right)^T \Lambda_i \left(\mathscr{W}_i \, \mathscr{T}_i \right) \right]^{-1} \left[\sum_{i=1}^{n} \left(\mathscr{W}_i \, \mathscr{T}_i \right)^T \Lambda_i \, \mathbf{Y}_i \right], \quad (10.37)$$

provided that $\sum_{i=1}^{n} [\mathscr{W}_i \, \mathscr{T}_i]^T \Lambda_i [\mathscr{W}_i \, \mathscr{T}_i]$ is nonsingular, where $\mathscr{W}_i$, $\mathscr{T}_i$ and $\mathscr{T}$ are defined in (10.26).

2. Least Squares Basis Approximation Estimators

For the case that $\alpha(s)$ and $\beta(s)$ are nonparametric functions but can be

approximated by the basis approximations (10.23), we have the infinite dimensional extended linear model, which is a basis approximation version of
(10.21) such that

$$
\begin{cases}
Y_{ij} \approx \sum_{d=0}^{D_1} \sum_{l=0}^{\mathscr{L}_d} Z_{ijd}\, \gamma_{dl}\, \mathscr{B}_{dl}(s) + \mathbf{Z}_{ij}^T \mathbf{a}_i^* \\
\qquad + \sum_{d=0}^{D_2} \sum_{m=0}^{\mathscr{M}_d} \delta_{ij} W_{ijd}\, \tau_{dm}\, \mathscr{B}_{dm}^*(s) + \delta_{ij}\, \mathbf{W}_{ij}^T \mathbf{b}_i^* + \varepsilon_{ij}, \\
\left(\mathbf{a}_i^{*T}, \mathbf{b}_i^{*T}\right)^T \big| S_i \ \sim \ G(\cdot | S_i).
\end{cases}
\tag{10.38}
$$

Substituting the $\alpha_d(s;\gamma)$ and $\beta_d(s;\tau)$ in (10.36) with the right side approximation terms of (10.23), the weighted least squares nonparametric estimator
$\left(\widetilde{\gamma}_{LS}^T(\mathscr{B}), \widetilde{\tau}_{LS}^T(\mathscr{B})\right)^T$ has the expression

$$
\begin{pmatrix} \widetilde{\gamma}_{LS}(\mathscr{B}) \\ \widetilde{\tau}_{LS}(\mathscr{B}) \end{pmatrix} = \left[\sum_{i=1}^{n} \left(\mathscr{W}_i\, \mathscr{B}_i\right)^T \Lambda_i \left(\mathscr{W}_i\, \mathscr{B}_i\right) \right]^{-1} \left[\sum_{i=1}^{n} \left(\mathscr{W}_i\, \mathscr{B}_i\right)^T \Lambda_i \mathbf{Y}_i \right], \tag{10.39}
$$

provided that $\sum_{i=1}^{n}[\mathscr{W}_i\, \mathscr{B}_i]^T \Lambda_i [\mathscr{W}_i\, \mathscr{B}_i]$ is nonsingular, where $\mathscr{W}_i$ is defined in
(10.26) and $\{\mathscr{B}_i, \mathscr{B}\}$ are defined in (10.31). Substituting the coefficients γ_{dl}
and τ_{dm} in (10.23) with the corresponding estimates given in (10.39), the basis
approximation estimators of the coefficient curves $\alpha(s;\gamma)$ and $\beta(s;\tau)$ are

$$
\left(\widetilde{\alpha}_{LS}^T(s;\mathscr{B}), \widetilde{\beta}_{LS}^T(s;\mathscr{B})\right)^T = \mathscr{B}(s) \left(\widetilde{\gamma}_{LS}^T(\mathscr{B}), \widetilde{\tau}_{LS}^T(\mathscr{B})\right)^T. \tag{10.40}
$$

In practice, a number of choices for the weight function Λ_i in (10.36) may
be considered. When $\Lambda_i = \mathbf{V}_i^{-1}$ and the distribution function of ε_{ij} and the
conditional distribution function $G(\cdot | s)$ are assumed to be normal, (10.37)
and (10.39) are the same as the maximum likelihood estimators or their basis
approximated versions. When $\mathbf{V}_i$ are unknown, as is often the case in practice,
subjective choices for Λ_i may be used. One potential *plug-in approach* is to
estimate $\mathbf{V}_i$ from the data and compute the estimates by substituting Λ_i with
the estimates of $\mathbf{V}_i^{-1}$. But, in practice, $\mathbf{V}_i$ is often difficult to estimate. The
question of whether such *plug-in* estimators have superior statistical proper-
ties than the estimators with subjective Λ_i choices has not been satisfactorily
answered in the literature, and requires further systematic investigation.

10.4.5 Estimation of the Covariances

1. Models for Covariances

We now discuss a number of ways to model and estimate the covariance
structure $\mathbf{V}_i$ defined in (10.26). By the definition of e_{ij} in (10.26), the (j_1, j_2)th
component of $\mathbf{V}_i$ is

$$
V_{i,j_1,j_2} = E\left[e_{ij_1} e_{ij_2}\right] = \rho_{i,j_1,j_2}(\mathbf{A}, \mathbf{B}, \mathbf{C}) + \sigma_{i,j_1,j_2}, \tag{10.41}
$$

where $\mathbf{A} = E\left(\mathbf{a}_i^* \mathbf{a}_i^{*T}\right)$, $\mathbf{B} = E\left(\mathbf{b}_i^* \mathbf{b}_i^{*T}\right)$, $\mathbf{C} = E\left(\mathbf{a}_i^* \mathbf{b}_i^{*T}\right)$, $\sigma_{i,j_1,j_2} = E\left(\varepsilon_{ij_1} \varepsilon_{ij_2}\right)$ and

$$
\begin{aligned}
\rho_{i,j_1,j_2}(\mathbf{A}, \mathbf{B}, \mathbf{C}) \;=\; & \mathbf{Z}_{ij_1}^T \mathbf{A} \mathbf{Z}_{ij_2} + \mathbf{Z}_{ij_1}^T \mathbf{C}\left(\delta_{ij_2} \mathbf{W}_{ij_2}\right) + \left(\delta_{ij_1} \mathbf{W}_{ij_1}^T\right) \mathbf{C} \mathbf{Z}_{ij_2} \\
& + \left(\delta_{ij_1} \mathbf{W}_{ij_1}^T\right) \mathbf{B} \left(\delta_{ij_2} \mathbf{W}_{ij_2}\right).
\end{aligned}
$$

For the special case that ε_{ij} are independent measurement errors, such that $\sigma_{i,j_1,j_2} = 0$ if $j_1 \neq j_2$ and σ^2 if $j_1 = j_2$, $\mathbf{V}_i$ follows the parametric model $\mathbf{V}_i(\mathbf{A}, \mathbf{B}, \mathbf{C}, \sigma^2)$ with

$$
V_{i,j_1,j_2} = \begin{cases} \rho_{i,j_1,j_2}(\mathbf{A}, \mathbf{B}, \mathbf{C}), & \text{if } j_1 \neq j_2; \\ \rho_{i,j,j}(\mathbf{A}, \mathbf{B}, \mathbf{C}) + \sigma^2, & \text{if } j_1 = j_2 = j. \end{cases} \tag{10.42}
$$

Other modeling structures for $\mathbf{V}_i$ can be constructed by incorporating parametric or nonparametric models of σ_{i,j_1,j_2} (e.g., Diggle et al., 2002).

For the general case of ε_{ij} having unknown correlation structures, σ_{i,j_1,j_2} is a nonparametric component in (10.41), hence, can be either directly estimated or approximated by a basis approximation. Under a different regression model, Diggle and Verbyla (1998) suggests a local smoothing method for estimating the covariance structures. However, local smoothing could be computationally intensive, since the covariance estimates have to be computed at all the distinct pairs of observation times. To ease the burden of computation, Huang, Wu and Zhou (2004) suggests that a consistent covariance estimator can be constructed by B-spline approximations.

Using the B-spline approach, we can approximate σ_{i,j_1,j_2} by

$$
\sigma_{i,j_1,j_2}(\mathbf{u}, \mathbf{v}) = \begin{cases} \sum_{k=1}^{K_1} \sum_{l=1}^{K_1} u_{kl} \, \mathscr{B}_k\left(t_{ij_1}\right) \mathscr{B}_l\left(t_{ij_2}\right), & \text{if } j_1 \neq j_2; \\ \sum_{k=1}^{K_2} v_k \, \mathscr{B}_k\left(t_{ij}\right), & \text{if } j_1 = j_2 = j, \end{cases} \tag{10.43}
$$

where $\{\mathscr{B}_k\}$ is a spline basis with a fixed knot sequence,

$$
\mathbf{u} = \left\{u_{kl} = u_{lk} : k, l = 1, \ldots, K_1\right\} \quad \text{and} \quad \mathbf{v} = \left\{v_k : k = 1, \ldots, K_2\right\}.
$$

Substituting the right-side expressions of $\sigma_{i,j_1,j_2}(\mathbf{u}, \mathbf{v})$ in (10.43) into (10.41), $\mathbf{V}_i$ is approximated by $\mathbf{V}_i(\mathbf{A}, \mathbf{B}, \mathbf{C}, \mathbf{u}, \mathbf{v})$, such that

$$
V_{i,j_1,j_2} = \begin{cases} \rho_{i,j_1,j_2}(\mathbf{A}, \mathbf{B}, \mathbf{C}) \\ \quad + \sum_{k=1}^{K_1} \sum_{l=1}^{K_1} u_{kl} \, \mathscr{B}_k\left(t_{ij_1}\right) \mathscr{B}_l\left(t_{ij_2}\right), & \text{if } j_1 \neq j_2; \\ \rho_{i,j_1,j_2}(\mathbf{A}, \mathbf{B}, \mathbf{C}) + \sum_{k=1}^{K_2} v_k \, \mathscr{B}_k\left(t_{ij}\right), & \text{if } j_1 = j_2 = j \end{cases} \tag{10.44}
$$

gives the basis approximated components defined in (10.41).

2. Least Squares Covariance Estimators

Once an approximate parametric model for $\mathbf{V}_i$ is established, such as

(10.42) and (10.44), the estimators of $\mathbf{V}_i$ can be computed by a least squares method. Let

$$\widehat{e}_{ij} = Y_{ij} - \left[\mathbf{Z}_{ij}^T \widehat{\alpha}(S_i) + \delta_{ij} \mathbf{W}_{ij}^T \widehat{\beta}(S_i)\right] \tag{10.45}$$

be the residual of Y_{ij} computed based on some consistent estimators $\widehat{\alpha}(s)$ and $\widehat{\beta}(s)$. If $V_{i,j_1,j_2} = \rho_{i,j_1,j_2}(\mathbf{A}, \mathbf{B}, \mathbf{C}) + \sigma_{i,j_1,j_2}(\mathbf{u}, \mathbf{v})$, we can estimate $\mathbf{V}_i$ by $\mathbf{V}_i(\widehat{\mathbf{A}}, \widehat{\mathbf{B}}, \widehat{C}, \widehat{\mathbf{u}}, \widehat{\mathbf{v}})$, where $\{\widehat{\mathbf{A}}, \widehat{\mathbf{B}}, \widehat{C}, \widehat{\mathbf{u}}, \widehat{\mathbf{v}}\}$ minimizes the score function

$$L(\mathbf{A}, \mathbf{B}, C, \mathbf{u}, \mathbf{v}) = \begin{cases} \sum_{i=1}^n \sum_{j_1,j_2=1, j_1<j_2}^{n_i} \left\{ \widehat{e}_{ij_1}\widehat{e}_{ij_2} - \left[\rho_{i,j_1,j_2}(\mathbf{A}, \mathbf{B}, \mathbf{C}) \right.\right. \\ \qquad\qquad \left.\left. + \sum_k \sum_l u_{kl}\, \mathscr{B}_k\left(t_{ij_1}\right) \mathscr{B}_l\left(t_{ij_2}\right)\right]\right\}^2, \\ \text{subject to } u_{kl} = u_{lk}, \text{ when } j_1 \neq j_2; \\ \sum_{i=1}^n \sum_{j=1}^{n_i} \left\{ \widehat{e}_{ij}^2 - \left[\rho_{i,j,j}(\mathbf{A}, \mathbf{B}, \mathbf{C}) + \sum_k v_k\, \mathscr{B}_k\left(t_{ij}\right)\right]\right\}^2, \\ \text{when } j_1 = j_2 = j. \end{cases}$$
$$\tag{10.46}$$

Since the estimator of $\mathbf{V}_i$ may also be used in the estimation procedures of Sections 10.4.3 and 10.4.4, we can adopt a two-step procedure by first obtaining the preliminary estimators $\widehat{\alpha}(s)$ and $\widehat{\beta}(s)$ using the estimation procedures of Sections 10.4.3 and 10.4.4 with $\mathbf{V}_i$ ignored, and then estimating $\mathbf{V}_i$ using

$$\widehat{\mathbf{V}}_i = \mathbf{V}_i(\widehat{\mathbf{A}}, \widehat{\mathbf{B}}, \widehat{C}, \widehat{\mathbf{u}}, \widehat{\mathbf{v}}).$$

The estimator $\widehat{\mathbf{V}}_i$ can then be used in the estimation procedures of Sections 10.4.3 and 10.4.4 to obtain the refined estimators of $\alpha(s)$ and $\beta(s)$.

Because (10.46) involves many parameters, minimizing the score function $L(\mathbf{A}, \mathbf{B}, C, \mathbf{u}, \mathbf{v})$ can be computationally intensive in practice. Fortunately, as discussed in Sections 10.4.3 and 10.4.4, the choice of $\mathbf{V}_i$ may only affect the variability of the estimators for the mean coefficient curves $\alpha(s)$ and $\beta(s)$ in the model (10.21). As discussed in Huang, Wu and Zhou (2004), $\mathbf{V}_i(\widehat{\mathbf{A}}, \widehat{\mathbf{B}}, \widehat{C}, \widehat{\mathbf{u}}, \widehat{\mathbf{v}})$ need not be positive definite for a finite sample, although, by consistency, it is asymptotically positive definite. The problem of imposing finite sample positive definiteness to the spline estimators of $\mathbf{V}_i$ deserves substantial further investigation. The adequacy of $\mathbf{V}_i(\widehat{\mathbf{A}}, \widehat{\mathbf{B}}, \widehat{C}, \widehat{\mathbf{u}}, \widehat{\mathbf{v}})$ depends on the choices of knots and the degrees of the splines. Although it is possible to develop data-driven knots using some cross-validation or generalized cross-validation procedures, statistical properties of such procedures are still unknown. Subjective knot choices, such as using a few equally spaced knots, often give satisfactory results in biomedical applications.

10.5   The Shared-Parameter Change-Point Models

As an alternative to the varying-coefficient models of the previous section, we summarize in this section a class of Shared-Parameter Change-Point Models, which, as a special case of the general model (10.17), can be used to evaluate the concomitant intervention effects with possibly censored change-point times in (10.5). This class of models is developed in Wu, Tian and Jiang (2011), which extends the shared-parameter model framework described in Follmann and Wu (1995) to the context with concomitant interventions.

10.5.1   Model Formulation and Justifications

Despite its advantages over the naive mixed-effects models in handling concomitant intervention, the varying-coefficient mixed effects model (10.21) has the main drawback of requiring the change-point times S_i to be observed for all the subjects. This is because the concomitant intervention effects on the outcome are described through the coefficient curves $\alpha(s)$ and $\beta(s)$ and the estimation methods of Section 10.4 require the observed values of $\{S_i : i = 1, \ldots, n\}$. For many situations, including the ENRICHD study (Section 1.2.2), some subjects may start the concomitant intervention at baseline, and some other subjects may not use the concomitant intervention during the study period at all, so that the structure fits into the general framework of (10.5). Regression models requiring the observed values of the change-point times for all the subjects, such as (10.21), can only be applied to a subset of the study subjects in (10.5), which may lead to biased conclusions or information loss.

In order to incorporate the censored change-point times $\{S_i^{(c)}, \delta_i^{(c)} : i = 1, \ldots, n\}$, we further specify the general shared-parameter change-point model (10.17) using a number of practical scenarios. When the concomitant intervention change-point time only depends on the pre-intervention trends $\mathbf{a}_i$ of the outcome variable, a useful special case of the shared-parameter change-point model (10.17) is

$$
\begin{cases}
Y_{ij} = \mu_0(t_{ij}, \mathbf{X}_i; \mathbf{a}_i) + \lambda_{ij}\mu_1(t_{ij}, \mathbf{X}_i, R_{ij}; \mathbf{b}_i) + \varepsilon_{ij}, \\
\mathbf{a}_i \sim F_a(\cdot), \quad S_i|\mathbf{a}_i \sim F_s(\cdot|\mathbf{a}_i), \quad \mathbf{b}_i|\mathbf{a}_i \sim F_b(\cdot|\mathbf{a}_i),
\end{cases}
\tag{10.47}
$$

where $F_a(\cdot)$ is the cumulative distribution function (CDF) of $\mathbf{a}_i$, $F_s(\cdot|\mathbf{a}_i)$ and $F_b(\cdot|\mathbf{a}_i)$ are the conditional CDFs of S_i and $\mathbf{b}_i$ given $\mathbf{a}_i$, $\mathbf{b}_i$ and S_i are independent given $\mathbf{a}_i$ and the random error term ε_{ij} are defined as in (10.17).

In contrast to the varying-coefficient model (10.21), where the conditional means of $\mathbf{a}_i$ and $\mathbf{b}_i$ given S_i are used, (10.47) specifies that both the change-point time S_i and the concomitant intervention effect $\mathbf{b}_i$ are affected by the pre-intervention time-trends $\mathbf{a}_i$, and these dependence relationships are described through the conditional distribution functions $F_s(\cdot|\mathbf{a}_i)$ and $F_b(\cdot|\mathbf{a}_i)$. By modeling the conditional distribution of S_i given $\mathbf{a}_i$, (10.47) allows S_i to be left- or right-censored. For simplicity, (10.47) assumes that $\mathbf{b}_i$ does not depend on S_i, although further generalizations may allow the distribution of $\mathbf{b}_i$ to depend on both S_i and $\mathbf{a}_i$ simultaneously. Further special cases of (10.47) can be derived by specifying the function forms of $\mu_0(\cdot; \mathbf{a}_i)$, $\mu_0(\cdot; \mathbf{b}_i)$, $F_s(\cdot|\mathbf{a}_i)$, $F_b(\cdot|\mathbf{a}_i)$ and $F_a(\cdot)$. These special cases lead to the models described below.

10.5.2 The Linear Shared-Parameter Change-Point Model

Further specifications of $F_a(\cdot)$, $F_s(\cdot|\mathbf{a}_i)$ and $F_b(\cdot|\mathbf{a}_i)$ in (10.47) may be considered in practice to balance the computational feasibility and flexibility of the model. A useful and mathematically tractable specification for (10.47) is to assume that S_i and $\mathbf{b}_i$ depend on $\mathbf{a}_i$ only through some linear functions of their conditional means. This approach leads to a linear shared-parameter change-point model of the form

$$
\begin{cases}
Y_{ij} &= \mu_0\left(t_{ij}, \mathbf{X}_i; \mathbf{a}_i\right) + \lambda_{ij}\,\mu_1\left(t_{ij}, \mathbf{X}_i, R_{ij}; \mathbf{b}_i\right) + \varepsilon_{ij}, \\
\mathbf{a}_i &= \alpha + e_i^{(a)}, \quad S_i = \gamma^T\left(1, \mathbf{a}_i^T\right)^T + e_i^{(s)}, \quad \mathbf{b}_i = \beta^T\left(1, \mathbf{a}_i^T\right)^T + e_i^{(b)}, \\
\alpha &= \left(\alpha_1, \ldots, \alpha_{d_0}\right)^T, \quad \alpha_d \in R, \\
\beta &= \left(\beta_1^T, \ldots, \beta_{d_1}^T\right)^T, \quad \beta_l = \left(\beta_{l0}, \ldots, \beta_{ld_0}\right)^T, \quad \beta_{ld} \in R, \\
\gamma &= \left(\gamma_0, \ldots, \gamma_{d_0}\right)^T, \quad \varepsilon_i = \left(\varepsilon_{i1}, \ldots, \varepsilon_{in_i}\right)^T,
\end{cases}
$$

$$(10.48)$$

where ε_i, $e_i^{(a)}$, $e_i^{(b)}$ and $e_i^{(s)}$ are independent mean zero random errors with covariance matrices $\mathbf{V}_y$, $\mathbf{V}_a$, $\mathbf{V}_b$ and σ_s^2, respectively. The unknown parameters in (10.48) are the mean components and the covariance structures

$$
\begin{cases}
\theta &= \left(\alpha^T, \beta_1^T, \ldots, \beta_{d_1}^T, \gamma^T\right)^T, \\
\mathbf{V} &= \left\{\mathbf{V}_y, \mathbf{V}_a, \mathbf{V}_b, \sigma_s^2\right\}.
\end{cases}
$$

$$(10.49)$$

The linearity in (10.48) refers only to the linear relationship between $\mathbf{a}_i$, S_i and $\mathbf{b}_i$. The functions $\mu_0(t_{ij}, \mathbf{X}_i; \mathbf{a}_i)$ and $\mu_1(t_{ij}, \mathbf{X}_i, R_{ij}; \mathbf{b}_i)$ may or may not have linear parametric forms. Since (10.48) involves many parameters, it is often motivated to use linear models for $\mu_0(t_{ij}, \mathbf{X}_i; \mathbf{a}_i)$ and $\mu_1(t_{ij}, \mathbf{X}_i, R_{ij}; \mathbf{b}_i)$ in practice to reduce the model complexity of (10.48), so that computationally feasible estimation procedures can be developed. If $\mu_0(\cdot; \mathbf{a}_i)$ and $\mu_1(\cdot; \mathbf{b}_i)$ are both linear functions, a fully linear model of (10.48) can be derived by writing

$$
\mu_0\left(t_{ij}, \mathbf{X}_i; \mathbf{a}_i\right) = \mathbf{a}_i^T\,\mathbf{Z}_{ij}^{(0)} \quad \text{and} \quad \mu_1\left(t_{ij}, \mathbf{X}_i, R_{ij}; \mathbf{b}_i\right) = \mathbf{b}_i^T\,\mathbf{Z}_{ij}^{(1)},
$$

where $\mathbf{Z}_{ij}^{(0)}$ is a $d_0 \times 1$ vector specified by $\{t_{ij}, \mathbf{X}_i\}$ and $\mathbf{Z}_{ij}^{(1)}$ is a $d_1 \times 1$ vector specified by $\{t_{ij}, S_i, \mathbf{X}_i\}$. This leads to the following fully linear shared-parameter change-point model

$$
\begin{cases}
Y_{ij} &= \mathbf{a}_i^T\,\mathbf{Z}_{ij}^{(0)} + \lambda_{ij}\,\mathbf{b}_i^T\,\mathbf{Z}_{ij}^{(1)} + \varepsilon_{ij}, \\
\mathbf{a}_i &= \alpha + e_i^{(a)}, \quad S_i = \gamma^T\left(1, \mathbf{a}_i^T\right)^T + e_i^{(s)}, \quad \mathbf{b}_i = \beta^T\left(1, \mathbf{a}_i^T\right)^T + e_i^{(b)}, \\
\alpha &= \left(\alpha_1, \ldots, \alpha_{d_0}\right)^T, \quad \alpha_d \in R, \\
\beta &= \left(\beta_1^T, \ldots, \beta_{d_1}^T\right)^T, \quad \beta_l = \left(\beta_{l0}, \ldots, \beta_{ld_0}\right)^T, \quad \beta_{ld} \in R, \\
\gamma &= \left(\gamma_0, \ldots, \gamma_{d_0}\right)^T, \quad \varepsilon_i = \left(\varepsilon_{i1}, \ldots, \varepsilon_{in_i}\right)^T,
\end{cases}
$$

$$(10.50)$$

where the population parameters of interest are the same as in (10.49).

The main advantage of (10.50) is its simplicity. For an intuitive interpretation, this model suggests that the outcome value of the ith subject at time point t_{ij} would have the subject-specific mean $\mathbf{a}_i^T \mathbf{Z}_{ij}^{(0)}$ if t_{ij} is before the concomitant intervention. If t_{ij} is after the concomitant intervention, the subject-specific mean outcome would be $\mathbf{a}_i^T \mathbf{Z}_{ij}^{(0)} + \mathbf{b}_i \mathbf{Z}_{ij}^{(1)}$ with $\mathbf{b}_i \mathbf{Z}_{ij}^{(1)}$ being the change of the subject's outcome due to the concomitant intervention. However, since the population-mean coefficients are α and β, the predicted outcome at time t_{ij} for a new subject with design vectors $\mathbf{Z}_{ij}^{(0)}$ and $\mathbf{Z}_{ij}^{(1)}$ would be $\alpha^T \mathbf{Z}_{ij}^{(0)}$ if t_{ij} is before the concomitant intervention, and $\alpha^T \mathbf{Z}_{ij}^{(0)} + \beta^T \mathbf{Z}_{ij}^{(1)}$ if t_{ij} is after the concomitant intervention. Given α and β, we would predict that the concomitant intervention would on average increase (or decrease) the predicted outcome for a new subject with design vectors $\mathbf{Z}_{ij}^{(0)}$ and $\mathbf{Z}_{ij}^{(1)}$ if $\beta^T \mathbf{Z}_{ij}^{(1)} > 0$ (or $\beta^T \mathbf{Z}_{ij}^{(1)} < 0$). Since not all subjects have change-point times S_i fully observed during the study, adequate estimation of α and β from the data (10.5) depends on the appropriateness of the models. We will see later in the numerical results of Section 10.7 that, although the linear relationships among $\mathbf{a}_i$, $\mathbf{b}_i$ and S_i may be subjective and "over-simplified," (10.50) still represents a remarkable improvement over the models without considering the effects of $\mathbf{a}_i$ on S_i and $\mathbf{b}_i$.

10.5.3 The Additive Shared-Parameter Change-Point Model

When the relationship between S_i and $\mathbf{a}_i$ in (10.47) is unknown, a nonparametric model for $\{S_i, \mathbf{a}_i\}$ is

$$S_i = \mu^{(s)}(\mathbf{a}_i) + \varepsilon_i^{(s)},$$

where $\mu^{(s)}(\mathbf{a}_i) = E(S_i|\mathbf{a}_i)$ is a smooth function of $\mathbf{a}_i$. Since unstructured estimation of $\mu^{(s)}(\mathbf{a}_i)$ could be difficult when $\mathbf{a}_i$ is a high-dimensional vector, a simple additive approach is to replace the relationship between S_i and $\mathbf{a}_i$ in (10.47) with

$$S_i = \sum_{d=0}^{d_0} \mu_d^{(s)}(a_{id}) + \varepsilon_i^{(s)}, \tag{10.51}$$

where $\mu_d^{(s)}(a_{id})$ are smooth functions of a_{id}. Based on (10.51), an additive shared-parameter model for (10.47) is

$$\begin{cases} Y_{ij} &= \mu_0(t_{ij}, \mathbf{X}_i; \mathbf{a}_i) + \lambda_{ij}\mu_1(t_{ij}, \mathbf{X}_i, R_{ij}; \mathbf{b}_i) + \varepsilon_{ij}, \\ \mathbf{a}_i &= \alpha + e_i^{(a)}, \ S_i = \sum_{d=0}^{d_0}\mu_d^{(s)}(a_{id}) + \varepsilon_i^{(s)}, \mathbf{b}_i = \beta^T\left(1, \mathbf{a}_i^T\right)^T + e_i^{(b)}, \end{cases} \tag{10.52}$$

where $\alpha = (\alpha_1, \ldots, \alpha_{d_0})^T$ and $\beta = (\beta_1, \ldots, \beta_{d_1})^T$ are defined as in (10.48). Further generalizations of (10.52) are theoretically possible but at the expense of computational complexity.

Additional special cases of (10.52) can be constructed by specifying the forms of $\mu_0(\cdot\,; \mathbf{a}_i)$ and $\mu_1(\cdot\,; \mathbf{b}_i)$. The additive relationship between S_i and $\mathbf{a}_i$ in (10.51) follows a nonparametric model, so that (10.52) is a semiparametric model if $\mu_0(\cdot\,; \mathbf{a}_i)$ and $\mu_1(\cdot\,; \mathbf{b}_i)$ are specified by parametric functions. Since (10.52) is a relatively complex model involving many parameters and nonparametric curves, it is of practical interest to consider some simple structures, such as the linear models, for $\mu_0(\cdot\,; \mathbf{a}_i)$ and $\mu_1(\cdot\,; \mathbf{b}_i)$, so that computationally feasible estimators of the parametric and nonparametric components can be obtained. For this reason, the practical applications of (10.48) and (10.52) in this chapter all assume linear models for $\mu_0(\cdot\,; \mathbf{a}_i)$ and $\mu_1(\cdot\,; \mathbf{b}_i)$. When linear models of the form $\mu_0(\cdot\,; \mathbf{a}_i) = \mathbf{a}_i^T \mathbf{Z}_{ij}^{(0)}$ and $\mu_1(\cdot\,; \mathbf{b}_i) = \mathbf{b}_i^T \mathbf{Z}_{ij}^{(1)}$ are used in (10.52), we obtain a semiparametric additive shared-parameter change-point model

$$
\begin{cases}
Y_{ij} &= \mathbf{a}_i^T \mathbf{Z}_{ij}^{(0)} + \lambda_{ij} \mathbf{b}_i^T \mathbf{Z}_{ij}^{(1)} + \varepsilon_{ij}, \\
\mathbf{a}_i &= \alpha + e_i^{(a)}, \ S_i = \sum_{d=0}^{d_0} \mu_d^{(s)}(a_{id}) + \varepsilon_i^{(s)}, \mathbf{b}_i = \beta^T \left(1, \mathbf{a}_i^T\right)^T + e_i^{(b)},
\end{cases}
$$
$$(10.53)$$

where $Z_{ij}^{(0)}$ and $\mathbf{Z}_{ij}^{(1)}$ are the design vectors defined in (10.50).

10.5.4   Likelihood-Based Estimation

For both models (10.48) and (10.52), the main objective is to estimate the mean parameters α and β, while other parameters, such as γ and the variance parameters, are generally not of primary interest. Because of the complexity of the models, the estimators of α and β are obtained through the maximum likelihood method for the parametric model (10.48) and the approximate maximum likelihood method for the semiparametric model (10.50). For both of these likelihood-based methods, the underlying marginal and conditional distributions are assumed to be known.

1. Maximum Likelihood Estimation

We first consider the case that (10.47) is a parametric model. If the distribution functions are explicitly specified with a known parametric form, the parameters in (10.47) can be estimated by the following maximum likelihood estimators (MLE). Denote by

$$
\mathbf{Y}_i = \left(Y_{i1}, \ldots, Y_{in_i}\right)^T, \ \mathbf{t}_i = \left(t_{i1}, \ldots, t_{in_i}\right)^T \ \text{and} \ \mathscr{D}_i = \left\{\mathbf{t}_i, \mathbf{X}_i\right\}.
$$

Let $f_y(\cdot)$, $f_b(\cdot)$, $f_s(\cdot)$ and $f_a(\cdot)$ be the density functions of Y_{ij}, $\mathbf{b}_i$, S_i and $\mathbf{a}_i$. The joint density of $\left\{\mathbf{b}_i, S_i, \mathbf{a}_i\right\}$ in (10.47) can be expressed as

$$
f\left(\mathbf{b}_i, S_i, \mathbf{a}_i\right) = f_b\left(\mathbf{b}_i \mid \mathbf{a}_i\right) f_s\left(S_i \mid \mathbf{a}_i\right) f_a\left(\mathbf{a}_i\right).
$$

The conditional density of $\left\{\mathbf{Y}_i, S_i\right\}$ given $\mathscr{D}_i = \left\{\mathbf{t}_i, \mathbf{X}_i\right\}$ can be derived by integrating over $\mathbf{a}_i$ and $\mathbf{b}_i$ and is given by

$$
f_{(y,s)}\left(\mathbf{Y}_i, S_i \mid \mathscr{D}_i\right) = \int\int f_y\left(\mathbf{Y}_i \mid \mathscr{D}_i, S_i, \mathbf{a}_i, \mathbf{b}_i\right) f_b\left(\mathbf{b}_i \mid \mathbf{a}_i\right)
$$

$$\times f_s\left(S_i\big|\mathbf{a}_i\right) f_a\left(\mathbf{a}_i\right) d\mathbf{a}_i\, d\mathbf{b}_i. \qquad (10.54)$$

Since the observed change-point time is the double censored version $\{S_i^{(c)}, \delta_i^{(c)}\}$, we may not be able to use the above conditional density function (10.54) directly in estimation.

Recall from (10.2) that $\delta_i^{(c)} = 1$ implies that the ith subject does not start the concomitant intervention at or before the last measurement time t_{in_i}, while $\delta_i^{(c)} = 2$ implies that the ith subject has already started the concomitant intervention before the first measurement time t_{i1}. In this case, we can consider the conditional density of $\mathscr{S}_i^{(c)} = \{S_i^{(c)}, \delta_i^{(c)}\}$ given $\mathbf{a}_i$,

$$f_s\left(S_i^{(c)}, \delta_i^{(c)}\big|\mathbf{a}_i\right) = \begin{cases} f_s\left(S_i\big|\mathbf{a}_i\right), & \text{if } \delta_i^{(c)} = 0, \\ 1 - F_s\left(t_{in_i}\big|\mathbf{a}_i\right), & \text{if } \delta_i^{(c)} = 1, \\ F_s\left(t_{i1}\big|\mathbf{a}_i\right), & \text{if } \delta_i^{(c)} = 2. \end{cases} \qquad (10.55)$$

Let $f_{(y,1)}(\cdot|\mathscr{D}_i, \mathbf{a}_i)$ and $f_{(y,2)}(\cdot|\mathscr{D}_i, \mathbf{a}_i, \mathbf{b}_i)$ be the densities of $\mathbf{Y}_i$ given $\{\mathscr{D}_i, \mathbf{a}_i, \delta_i^{(c)} = 1\}$ and $\{\mathscr{D}_i, \mathbf{a}_i, \mathbf{b}_i, \delta_i^{(c)} = 2\}$, respectively. Integrating out $\mathbf{a}_i$ and $\mathbf{b}_i$ using (10.55), the corresponding conditional densities of $\mathbf{Y}_i$ given $\mathscr{D}_i = \{\mathbf{t}_i, \mathbf{X}_i\}$ for subjects with $\delta_i^{(c)} = 1$ and $\delta_i^{(c)} = 2$ are

$$f_{(y,1)}\left(\mathbf{Y}_i\big|\mathscr{D}_i\right) = \int f_{(y,1)}\left(\mathbf{Y}_i\big|\mathscr{D}_i, \mathbf{a}_i\right)\left[1 - F_s\left(t_{in_i}\big|\mathbf{a}_i\right)\right] f_a\left(\mathbf{a}_i\right) d\mathbf{a}_i$$

and

$$f_{(y,2)}\left(\mathbf{Y}_i\big|\mathscr{D}_i\right) = \int\!\!\int f_{(y,2)}\left(\mathbf{Y}_i\big|\mathscr{D}_i, \mathbf{a}_i, \mathbf{b}_i\right) F_s\left(t_{i1}\big|\mathbf{a}_i\right) f_b\left(\mathbf{b}_i\big|\mathbf{a}_i\right) f_a\left(\mathbf{a}_i\right) d\mathbf{a}_i\, d\mathbf{b}_i.$$

The log-likelihood function of $\{\mathbf{Y}_i, \mathscr{S}_i^{(c)}\}$ conditioning on $\mathscr{D}_i$, $i = 1, \ldots, n$, and the MLE $\widehat{\phi}_{ML} = \{\widehat{\theta}_{ML}, \widehat{\mathbf{V}}_{ML}\}$ are given by

$$\begin{cases} L_c(\phi) & = \quad (1/n) \sum_{i:\,\delta_i^{(c)}=0} \log f_{(y,s)}\left(\mathbf{Y}_i, S_i\big|\mathscr{D}_i\right) \\ & \quad + (1/n) \sum_{l=1,2} \sum_{i:\,\delta_i^{(c)}=l} \log f_{(y,l)}\left(\mathbf{Y}_i\big|\mathscr{D}_i\right), \qquad (10.56) \\ L_c\left(\widehat{\phi}_{ML}\right) & = \quad \max_{\phi} L_c(\phi), \end{cases}$$

where $f_{(y,s)}(\cdot|\cdot)$ is given in (10.54), $\phi = \{\theta, \mathbf{V}\}$ is defined in (10.49), and the parametric family for $f_{(y,s)}(\cdot|\cdot)$ is denoted by $\{f_{(y,s)}(\cdot; \phi|\mathscr{D}_i); \phi\}$. We assume in (10.56) that the maximizer $\widehat{\phi}_{ML} = \{\widehat{\theta}_{ML}, \widehat{\mathbf{V}}_{ML}\}$ of $L_c(\phi)$ exists and is unique, and the log-likelihood function $L_c(\phi)$ satisfies the regularity conditions of the maximum likelihood estimators (cf. Serfling, 1980, Section 4.2).

2. Basis Approximation Estimation

Estimation for the additive shared-parameter change-point model (10.52) can be achieved through maximizing an approximate likelihood function for (10.52) by substituting $\mu_d^{(s)}(\cdot)$, $d = 0, \ldots, d_0$, with some expansions of basis functions. Under some mild smoothness conditions on $\mu_d^{(s)}(\cdot)$ (Section 4.4.2), we can consider a set of B-spline basis functions $\{B_1^{(d)}(\cdot), \ldots, B_{P_d}^{(d)}(\cdot)\}$, and approximate $\mu_d^{(s)}(\cdot)$ by the B-spline expansion

$$\mu_d^{(s)}(a_{id}) \approx \sum_{p=1}^{P_d} \gamma_p^{(d)} B_p^{(d)}(a_{id}) = \left(\gamma^{(d)}\right)^T \mathbf{B}^{(d)}(a_{id}) \tag{10.57}$$

for some $P_d \geq 1$, where

$$\mathbf{B}^{(d)}(a_{id}) = \left(B_1^{(d)}(a_{id}), \ldots, B_{P_d}^{(d)}(a_{id})\right)^T$$

and $\gamma^{(d)} = \left(\gamma_1^{(d)}, \ldots, \gamma_{P_d}^{(d)}\right)^T$ is a set of real-valued coefficients. Substituting (10.57) into (10.51), we have

$$S_i \approx \sum_{d=0}^{d_0} \left(\gamma^{(d)}\right)^T \mathbf{B}^{(d)}(a_{id}) + \varepsilon_i^{(s)}. \tag{10.58}$$

Substituting $\sum_{d=0}^{d_0} \mu_d^{(s)}(a_{id})$ of (10.52) with $\sum_{d=0}^{d_0} (\gamma^{(d)})^T \mathbf{B}^{(d)}(a_{id})$, the basis approximated additive shared-parameter change-point model is

$$\begin{cases} Y_{ij} &= \mu_0\left(t_{ij}, \mathbf{X}_i; \mathbf{a}_i\right) + \lambda_{ij} \mu_1\left(t_{ij}, \mathbf{X}_i, R_{ij}; \mathbf{b}_i\right) + \varepsilon_{ij}, \\ \mathbf{a}_i &= \alpha + e_i^{(a)}, \\ S_i &\approx \sum_{d=0}^{d_0} (\gamma^{(d)})^T \mathbf{B}^{(d)}(a_{id}) + \varepsilon_i^{(s)}, \\ \mathbf{b}_i &= \beta^T \left(1, \mathbf{a}_i^T\right)^T + e_i^{(b)}, \end{cases} \tag{10.59}$$

which is specified by the parameters

$$\begin{cases} \theta &= \left(\alpha^T, \beta_1^T, \ldots, \beta_{d_1}^T, \gamma^T\right)^T, \\ \gamma &= \left((\gamma^{(0)})^T, \ldots, (\gamma^{(d_0)})^T\right)^T, \\ \mathbf{V} &= \{\mathbf{V}_y, \mathbf{V}_a, \mathbf{V}_b, \sigma_s^2\}. \end{cases} \tag{10.60}$$

The objective for (10.59) is to estimate the mean parameters $\left(\alpha^T, \beta_1^T, \ldots, \beta_{d_1}^T\right)^T$ in (10.60) with γ and $\mathbf{V}$ taken as nuisance parameters. To do this, we need to write down the approximated log-likelihood function for (10.59), where the approximation refers to substituting S_i with the right-side term $\sum_{d=0}^{d_0} (\gamma^{(d)})^T \mathbf{B}^{(d)}(a_{id}) + \varepsilon_i^{(s)}$.

Let $f_s^*(\cdot; \gamma, \sigma_s | \mathbf{a}_i)$ be the conditional density of $\sum_{d=0}^{d_0} (\gamma^{(d)})^T \mathbf{B}^{(d)}(a_{id}) + \varepsilon_i^{(s)}$ given $\mathbf{a}_i$. If the distributions of ε_{ij}, $e_i^{(a)}$, $e_i^{(b)}$ and $e_i^{(s)}$, which all have zero means, are determined by the vector of variance parameters $\mathbf{V}$, and the density

$f_s(S_i|\mathbf{a}_i)$ can be approximated by $f_s^*(S_i; \gamma, \sigma_s|\mathbf{a}_i)$, the approximate maximum likelihood estimators $\widehat{\phi}_{AML} = \{\widehat{\theta}_{AML}, \widehat{\mathbf{V}}_{AML}\}$ of the parameters $\phi = \{\theta^T, \mathbf{V}^T\}^T$ of (10.60) can be obtained by maximizing the approximate log-likelihood function for $\{\mathbf{Y}_i, \mathscr{S}_i^{(c)}\}$ given $\mathscr{D}_i = \{\mathbf{t}_i, \mathbf{X}_i\}$, such that,

$$
\left\{
\begin{array}{rcl}
L_c^*(\phi) & = & (1/n)\sum_{i:\,\delta_i^{(c)}=0}\log f_{(y,s)}^*\left(\mathbf{Y}_i, S_i; \phi\,\middle|\,\mathscr{D}_i\right) \\[2mm]
 & & +(1/n)\sum_{l=1,2}\sum_{i:\,\delta_i^{(c)}=l}\log f_{(y,l)}^*\left(\mathbf{Y}_i; \phi\,\middle|\,\mathscr{D}_i\right), \qquad (10.61) \\[2mm]
L_c^*(\widehat{\phi}_{AML}) & = & \max_{\phi} L_c^*(\phi),
\end{array}
\right.
$$

where $f_{(y,s)}^*(\cdot|\mathscr{D}_i)$ and $f_{(y,k)}^*(\cdot|\mathscr{D}_i)$, $k = 1, 2$, are given in (10.56) with $f_s(S_i|\mathbf{a}_i)$ replaced by $f_s^*(S_i; \gamma, \sigma_s|\mathbf{a}_i)$. We assume in (10.61) that the maximizer $\widehat{\phi}_{AML} = \{\widehat{\theta}_{AML}, \widehat{\mathbf{V}}_{AML}\}$ of $L_c^*(\phi)$ exists and is unique, and, for any fixed P_d in (10.57), the approximate log-likelihood function $L_c^*(\phi)$ satisfies the regularity conditions of the MLEs (cf. Serfling, 1980, Section 4.2).

10.5.5   Gaussian Shared-Parameter Change-Point Models

To make the likelihood-based estimation methods of Section 10.5.4 more precise, we derive the expressions of the likelihood functions for the Gaussian linear shared-parameter change-point model (10.48) and the Gaussian additive shared-parameter change-point model (10.52). This Gaussian assumption refers that $F_a(\cdot)$, $F_s(\cdot|\mathbf{a}_i)$, $F_b(\cdot|\mathbf{a}_i)$ and the distribution of ε_{ij} are all Gaussian.

1. The Gaussian Linear Shared-Parameter Change-Point Model

The model (10.48) is a Gaussian linear shared-parameter change-point model if the error terms ε_i, $e_i^{(a)}$, $e_i^{(b)}$ and $e_i^{(s)}$ have mean zero multivariate normal distributions. As noted in Section 10.5.2, the linearity of (10.48) refers to the linear relationships of $\{S_i, \mathbf{a}_i\}$ and $\{\mathbf{b}_i, \mathbf{a}_i\}$, while $\mu_0(T_{i1}, \mathbf{X}_i; \mathbf{a}_i)$ and $\mu_1(T_{i1}, \mathbf{X}_i, R_{i1}; \mathbf{b}_i)$ could be either linear or nonlinear functions.

Let $\mathbf{V}_y$, $\mathbf{V}_a$, $\mathbf{V}_b$ and σ_s^2 be the corresponding covariance matrices of ε_i, $e_i^{(a)}$, $e_i^{(b)}$ and $e_i^{(s)}$,

$$
\mathbf{m}_i^{(0)} = \left(\mu_0(T_{i1}, \mathbf{X}_i; \mathbf{a}_i), \ldots, \mu_0(T_{in_i}, \mathbf{X}_i; \mathbf{a}_i)\right)^T
$$

and

$$
\mathbf{m}_i^{(1)} = \left(
\begin{array}{c}
\mu_0(T_{i1}, \mathbf{X}_i; \mathbf{a}_i) + \lambda_{i1}\,\mu_1(T_{i1}, \mathbf{X}_i, R_{i1}; \mathbf{b}_i) \\
\vdots \\
\mu_0(T_{in_i}, \mathbf{X}_i; \mathbf{a}_i) + \lambda_{in_i}\,\mu_1(T_{in_i}, \mathbf{X}_i, R_{in_i}; \mathbf{b}_i)
\end{array}
\right).
$$

The log-likelihood function (10.56) is determined by the mean structure

$$
\theta = \left(\alpha^T, \beta_1^T, \ldots, \beta_{d_1}^T, \gamma^T\right)^T
$$

and covariance structure $\mathbf{V}$ of (10.49). Direct computation using $f_{(y,s)}(\mathbf{Y}_i, S_i | \mathcal{D}_i)$, $f_{(y,1)}(\mathbf{Y}_i | \mathcal{D}_i)$, $f_{(y,2)}(\mathbf{Y}_i | \mathcal{D}_i)$ and the normality assumption for the distributions of the error terms ε_i, $e_i^{(a)}$, $e_i^{(b)}$ and $e_i^{(s)}$ shows that, for the ith subject, the summation terms involved in (10.56) can be expressed as

$$
\begin{aligned}
\log f_{(y,s)} & (\mathbf{Y}_i, S_i; \boldsymbol{\theta}, \mathbf{V} | \mathcal{D}_i) \\
= \quad & C_{(y,s)} - \frac{1}{2} \log \left(\sigma_s^2 \cdot |\mathbf{V}_y| \cdot |\mathbf{V}_b| \cdot |\mathbf{V}_a| \right) \\
& + \log \int \int \exp \Big\{ -\frac{1}{2} \Big[(\mathbf{Y}_i - \mathbf{m}_i^{(1)})^T \mathbf{V}_y^{-1} (\mathbf{Y}_i - \mathbf{m}_i^{(1)}) \\
& + (\mathbf{b}_i - \boldsymbol{\beta}^T (1, \mathbf{a}_i^T)^T)^T \mathbf{V}_b^{-1} (\mathbf{b}_i - \boldsymbol{\beta}^T (1, \mathbf{a}_i^T)^T) \\
& + \frac{(S_i - \boldsymbol{\gamma}^T (1, \mathbf{a}_i^T)^T)^2}{\sigma_s^2} \\
& + (\mathbf{a}_i - \boldsymbol{\alpha})^T \mathbf{V}_a^{-1} (\mathbf{a}_i - \boldsymbol{\alpha}) \Big] \Big\} \, d\mathbf{a}_i \, d\mathbf{b}_i,
\end{aligned}
\tag{10.62}
$$

$$
\begin{aligned}
\log f_{(y,1)} & (\mathbf{Y}_i; \boldsymbol{\theta}, \mathbf{V} | \mathcal{D}_i) \\
= \quad & C_{(y,1)} - \frac{1}{2} \log \left(\sigma_s^2 \cdot |\mathbf{V}_y| \cdot |\mathbf{V}_a| \right) \\
& + \log \int \int \exp \Big\{ -\frac{1}{2} \Big[(\mathbf{Y}_i - \mathbf{m}_i^{(0)})^T \mathbf{V}_y^{-1} (\mathbf{Y}_i - \mathbf{m}_i^{(0)}) \\
& + (\mathbf{a}_i - \boldsymbol{\alpha})^T \mathbf{V}_a^{-1} (\mathbf{a}_i - \boldsymbol{\alpha}) \Big] \Big\} \\
& \times \Big\{ \int_{T_{in_i}}^{\infty} \exp \Big[-\frac{(S - \boldsymbol{\gamma}^T (1, \mathbf{a}_i^T)^T)^2}{2\sigma_s^2} \Big] dS \Big\} \, d\mathbf{a}_i,
\end{aligned}
\tag{10.63}
$$

and

$$
\begin{aligned}
\log f_{(y,2)} & (\mathbf{Y}_i; \boldsymbol{\theta}, \mathbf{V} | \mathcal{D}_i) \\
= \quad & C_{(y,2)} - \frac{1}{2} \log \left(\sigma_s^2 \cdot |\mathbf{V}_y| \cdot |\mathbf{V}_b| \cdot |\mathbf{V}_a| \right) \\
& + \log \int \int \Big\{ \exp \Big[-\frac{1}{2} ((\mathbf{Y}_i - \mathbf{m}_i^{(1)})^T \mathbf{V}_y^{-1} (\mathbf{Y}_i - \mathbf{m}_i^{(1)}) \\
& + (\mathbf{b}_i - \boldsymbol{\beta}^T (1, \mathbf{a}_i^T)^T)^T \mathbf{V}_b^{-1} (\mathbf{b}_i - \boldsymbol{\beta}^T (1, \mathbf{a}_i^T)^T) \\
& + (\mathbf{a}_i - \boldsymbol{\alpha})^T \mathbf{V}_a^{-1} (\mathbf{a}_i - \boldsymbol{\alpha})) \Big] \\
& \times \int_{-\infty}^{T_{i1}} \exp \Big[-\frac{(S - \boldsymbol{\gamma}^T (1, \mathbf{a}_i^T)^T)^2}{2\sigma_s^2} \Big] dS \Big\} \, d\mathbf{a}_i \, d\mathbf{b}_i,
\end{aligned}
\tag{10.64}
$$

where $C_{(y,s)}$, $C_{(y,1)}$ and $C_{(y,2)}$ are the normalizing constants. The likelihood function for the Gaussian model of (10.48) is obtained by substituting the summation terms of (10.56) with the corresponding log-likelihood terms (10.62), (10.63) and (10.64).

2. The Gaussian Additive Shared-Parameter Change-Point Model

Similar to the above Gaussian linear model, the model (10.52) is a Gaussian additive shared-parameter change-point model if the error terms ε_i, $e_i^{(a)}$, $e_i^{(b)}$ and $e_i^{(s)}$ have mean zero multivariate normal distributions. Since $\mu_d^{(s)}(\cdot)$, $d = 0, \ldots, d_0$, are nonparametric functions, the parameters to be estimated from (10.52) are $\{\alpha, \beta, \mathbf{V} = \{\mathbf{V}_y, \mathbf{V}_a, \mathbf{V}_b, \sigma_s^2\}\}$, unless basis expansions are used to approximate $\mu_d^{(s)}(\cdot)$.

We first show the general expression of (10.61) without using basis approximations for $\mu_d^{(s)}(\cdot)$. Under the Gaussian assumption, the log-likelihood functions involved in (10.61) are given by

$$
\begin{aligned}
\log f_{(y,s)} & \left(\mathbf{Y}_i, S_i; \boldsymbol{\theta}, \mathbf{V} \mid \mathscr{D}_i\right) \\
= \ & C_{(y,s)} - \frac{1}{2} \log\left(\sigma_s^2 \cdot |\mathbf{V}_y| \cdot |\mathbf{V}_b| \cdot |\mathbf{V}_a|\right) \\
& + \log \int\!\!\int \exp\Bigg\{-\frac{1}{2}\Big[(\mathbf{Y}_i - \mathbf{m}_i^{(1)})^T \mathbf{V}_y^{-1}(\mathbf{Y}_i - \mathbf{m}_i^{(1)}) \\
& + \big(\mathbf{b}_i - \boldsymbol{\beta}^T(1, \mathbf{a}_i^T)^T\big)^T \mathbf{V}_b^{-1}\big(\mathbf{b}_i - \boldsymbol{\beta}^T(1, \mathbf{a}_i^T)^T\big) \\
& + \frac{\big(S_i - \sum_{d=0}^{d_0} \mu_d^{(s)}(a_{id})\big)^2}{\sigma_s^2} \\
& + (\mathbf{a}_i - \boldsymbol{\alpha})^T \mathbf{V}_a^{-1}(\mathbf{a}_i - \boldsymbol{\alpha})\Big]\Bigg\} d\mathbf{a}_i \, d\mathbf{b}_i,
\end{aligned}
\tag{10.65}
$$

$$
\begin{aligned}
\log f_{(y,1)} & \left(\mathbf{Y}_i; \boldsymbol{\theta}, \mathbf{V} \mid \mathscr{D}_i\right) \\
= \ & C_{(y,1)} - \frac{1}{2} \log\left(\sigma_s^2 \cdot |\mathbf{V}_y| \cdot |\mathbf{V}_a|\right) \\
& + \log \int\!\!\int \exp\Bigg\{-\frac{1}{2}\Big[(\mathbf{Y}_i - \mathbf{m}_i^{(0)})^T \mathbf{V}_y^{-1}(\mathbf{Y}_i - \mathbf{m}_i^{(0)}) \\
& + (\mathbf{a}_i - \boldsymbol{\alpha})^T \mathbf{V}_a^{-1}(\mathbf{a}_i - \boldsymbol{\alpha})\Big]\Bigg\} \\
& \times \left\{\int_{T_{in_i}}^{\infty} \exp\left[-\frac{\big(S - \sum_{d=0}^{d_0}\mu_d^{(s)}(a_{id})\big)^2}{2\sigma_s^2}\right] dS\right\} d\mathbf{a}_i,
\end{aligned}
\tag{10.66}
$$

and

$$
\begin{aligned}
\log f_{(y,2)} & \left(\mathbf{Y}_i; \boldsymbol{\theta}, \mathbf{V} \mid \mathscr{D}_i\right) \\
= \ & C_{(y,2)} - \frac{1}{2} \log\left(\sigma_s^2 \cdot |\mathbf{V}_y| \cdot |\mathbf{V}_b| \cdot |\mathbf{V}_a|\right) \\
& + \log \int\!\!\int \left\{\exp\left[-\frac{1}{2}\big((\mathbf{Y}_i - \mathbf{m}_i^{(1)})^T \mathbf{V}_y^{-1}(\mathbf{Y}_i - \mathbf{m}_i^{(1)})\right.\right.
\end{aligned}
$$

$$+\left(\mathbf{b}_i - \boldsymbol{\beta}^T\left(1, \mathbf{a}_i^T\right)^T\right)^T \mathbf{V}_b^{-1}\left(\mathbf{b}_i - \boldsymbol{\beta}^T\left(1, \mathbf{a}_i^T\right)^T\right)$$
$$+\left(\mathbf{a}_i - \boldsymbol{\alpha}\right)^T \mathbf{V}_a^{-1}\left(\mathbf{a}_i - \boldsymbol{\alpha}\right)\bigg]$$
$$\times \int_{-\infty}^{T_{i1}} \exp\left[-\frac{\left(S - \sum_{d=0}^{d_0}\mu_d^{(s)}\left(a_{id}\right)\right)^2}{2\sigma_s^2}\right] dS\bigg\} d\mathbf{a}_i\, d\mathbf{b}_i, \qquad (10.67)$$

where $C_{(y,s)}$, $C_{(y,1)}$ and $C_{(y,2)}$ are the normalizing constants. The likelihood function for the Gaussian model of (10.52) is obtained by substituting the summation terms of (10.56) with the corresponding log-likelihood terms (10.65), (10.66) and (10.67).

In order to numerically evaluate the log-likelihood functions obtained from (10.56), (10.65), (10.66) and (10.67), we have to consider specific expressions of $\mu_d^{(s)}(\cdot)$. When $\mu_d^{(s)}(a_{id})$ is approximated by the B-spline expansion (10.57), the approximate log-likelihood function under the Gaussian model is (10.61) with

$$\log f_{(y,s)}^*\left(\mathbf{Y}_i, S_i; \boldsymbol{\theta}, \mathbf{V}\big|\mathscr{D}_i\right)$$
$$= C_{(y,s)} - \frac{1}{2}\log\left(\sigma_s^2 \cdot |\mathbf{V}_y| \cdot |\mathbf{V}_b| \cdot |\mathbf{V}_a|\right)$$
$$+ \log\int\int \exp\bigg\{-\frac{1}{2}\Big[\left(\mathbf{Y}_i - \mathbf{m}_i^{(1)}\right)^T \mathbf{V}_y^{-1}\left(\mathbf{Y}_i - \mathbf{m}_i^{(1)}\right)$$
$$+ \left(\mathbf{b}_i - \boldsymbol{\beta}^T\left(1, \mathbf{a}_i^T\right)^T\right)^T \mathbf{V}_b^{-1}\left(\mathbf{b}_i - \boldsymbol{\beta}^T\left(1, \mathbf{a}_i^T\right)^T\right)$$
$$+ \frac{\left(S_i - \sum_{d=0}^{d_0}(\boldsymbol{\gamma}^{(d)})^T \mathbf{B}^{(d)}\left(a_{id}\right)\right)^2}{\sigma_s^2}$$
$$+ \left(\mathbf{a}_i - \boldsymbol{\alpha}\right)^T \mathbf{V}_a^{-1}\left(\mathbf{a}_i - \boldsymbol{\alpha}\right)\Big]\bigg\} d\mathbf{a}_i\, d\mathbf{b}_i, \qquad (10.68)$$

$$\log f_{(y,1)}^*\left(\mathbf{Y}_i; \boldsymbol{\theta}, \mathbf{V}\big|\mathscr{D}_i\right)$$
$$= C_{(y,1)} - \frac{1}{2}\log\left(\sigma_s^2 \cdot |\mathbf{V}_y| \cdot |\mathbf{V}_a|\right)$$
$$+ \log\int\int \exp\bigg\{-\frac{1}{2}\Big[\left(\mathbf{Y}_i - \mathbf{m}_i^{(0)}\right)^T \mathbf{V}_y^{-1}\left(\mathbf{Y}_i - \mathbf{m}_i^{(0)}\right)$$
$$+ \left(\mathbf{a}_i - \boldsymbol{\alpha}\right)^T \mathbf{V}_a^{-1}\left(\mathbf{a}_i - \boldsymbol{\alpha}\right)\Big]\bigg\}$$
$$\times \bigg\{\int_{T_{in_i}}^{\infty} \exp\left[-\frac{\left(S - \sum_{d=0}^{d_0}(\boldsymbol{\gamma}^{(d)})^T \mathbf{B}^{(d)}\left(a_{id}\right)\right)^2}{2\sigma_s^2}\right] dS\bigg\} d\mathbf{a}_i, \qquad (10.69)$$

and

$$\log f_{(y,2)}^*\left(\mathbf{Y}_i; \boldsymbol{\theta}, \mathbf{V}\big|\mathscr{D}_i\right)$$
$$= C_{(y,2)} - \frac{1}{2}\log\left(\sigma_s^2 \cdot |\mathbf{V}_y| \cdot |\mathbf{V}_b| \cdot |\mathbf{V}_a|\right)$$

$$+\log \int\int \left\{ \exp\left[-\frac{1}{2}\left(\mathbf{Y}_i - \mathbf{m}_i^{(1)}\right)^T \mathbf{V}_y^{-1}\left(\mathbf{Y}_i - \mathbf{m}_i^{(1)}\right) \right. \right.$$

$$+\left(\mathbf{b}_i - \boldsymbol{\beta}^T\left(1, \mathbf{a}_i^T\right)^T\right)^T \mathbf{V}_b^{-1}\left(\mathbf{b}_i - \boldsymbol{\beta}^T\left(1, \mathbf{a}_i^T\right)^T\right)$$

$$+\left(\mathbf{a}_i - \boldsymbol{\alpha}\right)^T \mathbf{V}_a^{-1}\left(\mathbf{a}_i - \boldsymbol{\alpha}\right)\Big)\bigg]$$

$$\left. \times \int_{-\infty}^{T_{i1}} \exp\left[-\frac{\left(S - \sum_{d=0}^{d_0}(\gamma^{(d)})^T \mathbf{B}^{(d)}\left(a_{id}\right)\right)^2}{2\sigma_s^2} \right] dS \right\} d\mathbf{a}_i \, d\mathbf{b}_i. \quad (10.70)$$

The approximate log-likelihood functions under the Gaussian additive shared-parameter change-point model (10.59) is given by substituting (10.68), (10.69) and (10.70) into (10.61). The parameters in (10.60) can be estimated by the approximate maximum likelihood estimators through a Newton-Raphson algorithm.

10.5.6　A Two-Stage Estimation Procedure

The log-likelihood functions (10.56) and (10.61) involve nonlinear terms of the parameters. A global maximization of (10.50) or (10.61) over $\boldsymbol{\theta}$ and $\mathbf{V}$ simultaneously could be computationally infeasible even under the Gaussian models specified in Section 10.5.5. In order to alleviate the computational burden, a suggestion by Wu, Tian and Jiang (2011) is to use the following two-stage estimation procedure, which combines restricted maximum likelihood estimation (REMLE) procedure with the Newton-Raphson algorithm:

Two-Stage Maximum Likelihood Algorithm:

(a) *Assume that* $\{\varepsilon_{ij}, \mathbf{a}_i, \mathbf{b}_i, S_i\}$ *of (10.48) or (10.52) are independent random variables with covariance matrices* $\mathbf{V} = \{\mathbf{V}_y, \mathbf{V}_a, \mathbf{V}_b, \sigma_s^2\}$, *that is, the naive mixed-effects change-point model (10.9) holds with* $\mathbf{a}_i$ *and* $\mathbf{b}_i$ *independent. Compute* $\widehat{\mathbf{V}}$ *of* $\mathbf{V}$ *using the REMLE procedure.*

(b) *Substitute* $\mathbf{V}$ *with* $\widehat{\mathbf{V}}$, *and maximize* $L_c(\boldsymbol{\theta}, \widehat{\mathbf{V}})$ *of (10.56) or* $L_c^*(\boldsymbol{\theta}, \widehat{\mathbf{V}})$ *of (10.61) with respect to* $\boldsymbol{\theta}$ *using the Newton-Raphson procedure. The maximizer* $\widehat{\boldsymbol{\theta}} = \arg\max_{\boldsymbol{\theta}} L_c(\boldsymbol{\theta}, \widehat{\mathbf{V}})$ *is the maximum likelihood estimator for* $\boldsymbol{\theta}$ *of (10.49). Similarly, the maximizer* $\widehat{\boldsymbol{\theta}} = \arg\max_{\boldsymbol{\theta}} L_c^*(\boldsymbol{\theta}, \widehat{\mathbf{V}})$ *is the approximate maximum likelihood estimator for* $\boldsymbol{\theta}$ *of (10.60).* □

From the expressions of $f_{(y,s)}(\mathbf{Y}_i, S_i | \mathscr{D}_i)$ and $f_{(y,k)}(\mathbf{Y}_i | \mathscr{D}_i)$ for $k = 1, 2$, it is easy to see that the Newton-Raphson algorithm for maximizing $L_c(\boldsymbol{\theta}, \widehat{\mathbf{V}})$ at stage (b) involves multidimensional integrations over the functions of $\mathbf{a}_i$, $\mathbf{b}_i$ and S_i with respect to the joint distributions of $\mathbf{a}_i$ and $\mathbf{b}_i$. All the necessary quantities involved in the Newton-Raphson algorithm, including the log-likelihood functions, and their gradients and Hessian matrices, can be computed using Monte Carlo simulations, in which case large Monte-Carlo samples are required to compute the gradient and the Hessian matrix in each iteration, so that a complete Newton-Raphson algorithm can be costly to implement. If a suitable initial estimator is available, computation of the algorithm can be significantly reduced by a "one-step" Newton-Raphson procedure (e.g., Bickel, 1975). In Wu, Tian and Jiang (2011), the authors suggest to use the estimators computed from the REMLE procedure as a natural candidate for the initial estimator $\widehat{\theta}_0$ and to compute the initial estimators of γ by fitting the regression model $S_i = \gamma^T \left(1, \widetilde{\mathbf{a}}_i^{pred}\right) + \varepsilon_i^{(s)}$ using the subjects with S_i observed, i.e., $\delta_i^{(c)} = 0$, where $\widetilde{\mathbf{a}}_i^{pred}$ is the predicted value for $\mathbf{a}_i$.

10.6   Confidence Intervals for Parameter Estimators

We now present the confidence intervals for the mean concomitant intervention effects estimated in Sections 10.4 and 10.5. The asymptotic confidence intervals can only be used for the linear shared-parameter change-point model (10.48) when the distributions of the error terms are known. When nonparametric components are present, the resampling-subject bootstrap procedures have to be used for all the models in Sections 10.4 and 10.5 due to the lack of asymptotic normality results.

10.6.1   Asymptotic Confidence Intervals

Approximated inferences for the linear shared-parameter change-point model (10.48) can be constructed using the asymptotic distribution of the maximum likelihood estimator $\widehat{\theta}$, when n is large and the model (10.48) follows a known parametric family. Under the usual regularity conditions for the maximum likelihood estimators (e.g., Serfling, 1980, Chapter 4), the asymptotic normality of the estimators implies that $\widehat{\theta}$ has approximately the multivariate normal distribution $N(\theta, \Sigma)$, where Σ is the asymptotic variance-covariance matrix. An approximate $\left[100 \times (1-\alpha)\right]$th confidence interval for a linear combination $\ell(\theta)$ of θ is

$$\ell\left(\widehat{\theta}\right) \pm Z_{\alpha/2} \widehat{\Sigma}^{1/2}, \tag{10.71}$$

where $\widehat{\Sigma}$ is the estimator of the asymptotic variance-covariance matrix of $\ell(\theta)$ and $Z_{\alpha/2}$ is the $\left[100 \times (1-\alpha/2)\right]$th percentile of the standard normal distribution.

It follows from (10.48) and (10.49) that $\widehat{\Sigma}$ can be constructed by a "plug-in" type method using the available estimators $\widehat{\theta}$ and $\widehat{\mathbf{V}} = \left\{\widehat{\mathbf{V}}_y, \widehat{\mathbf{V}}_a, \widehat{\mathbf{V}}_b, \widehat{\sigma}_s^2\right\}$, where $\widehat{\mathbf{V}}$ is the REMLE described in Section 10.5.6. However, a potential drawback of the "plug-in" approximate confidence interval (10.71) is its computational complexity, since in practice many parameters from the mean and variance-covariance structures have to appropriately estimated at the same time. For this reason, the following resampling-subject bootstrap procedure is often a more practical procedure than the "plug-in" approach based on (10.71).

10.6.2 Bootstrap Confidence Intervals

For the varying-coefficient mixed-effects model (10.21) and the additive shared-parameter change-point model (10.52), the unknown components include the finite dimensional parameters as well as the nonparametric coefficient curves. The asymptotic distributions of the corresponding least squares estimators and approximate maximum likelihood estimates have not been systematically investigated in the literature. As a practical alternative, the resampling-subject bootstrap procedure introduced in Section 3.4 can be used to compute the confidence intervals for the unknown finite dimensional parameters or the coefficient curves.

Let ξ be an unknown finite dimensional parameter or coefficient curve of the model (10.21) or (10.52), and $\widehat{\xi}$ be an estimator of ξ computed based on any of the appropriate estimation methods in Sections 10.4 and 10.5. Specific steps for constructing the bootstrap confidence intervals for ξ based on $\widehat{\xi}$ are given in the following.

Approximate Bootstrap Confidence Intervals:

(a) **Computing Bootstrap Estimators.** *Generate B independent bootstrap samples using the resampling-subject bootstrap procedure of Section 3.4.1 and compute the B bootstrap estimators* $\widehat{\xi}^{boot} = \{\widehat{\xi}^b : b = 1, \ldots, B\}$ *of* ξ.

(b) **Approximate Bootstrap Confidence Intervals.** *The* $[100 \times (1 - \alpha)]\%$ *bootstrap confidence interval for* ξ *based on percentiles is given by*

$$\left(L_{\alpha/2}^{boot}(\xi), U_{\alpha/2}^{boot}(\xi)\right), \qquad (10.72)$$

where $L_{\alpha/2}^{boot}(\xi)$ *and* $U_{\alpha/2}^{boot}(\xi)$ *are the corresponding lower and upper* $[100 \times (\alpha/2)]$ *th percentiles of the B bootstrap estimators* $\widehat{\xi}^{boot}$. *The* $[100 \times (1 - \alpha)]\%$ *bootstrap confidence interval for* ξ *based on the normal approximation is given by*

$$\widehat{\xi} \pm Z_{\alpha/2} SD\left(\widehat{\xi}^{boot}\right). \qquad (10.73)$$

where $\widehat{\xi}$ *is the estimator of* ξ *computed based on the original sample and* $SD\left(\widehat{\xi}^{boot}\right)$ *is the sample standard deviation of* $\widehat{\xi}^{boot}$. □

The above bootstrap approach has been used in Wu, Tian and Bang (2008) and Wu, Tian and Jiang (2011). The percentile bootstrap interval (10.72) is not necessarily symmetric about $\widehat{\xi}$. On the other hand, the normal approximated interval (10.73) may not be appropriate if the distribution of $\widehat{\xi}$ is not approximated by the normal distribution with mean zero and standard deviation $SD\left(\widehat{\xi}^{boot}\right)$. In the simulation results of Wu, Tian and Bang (2008) and Wu, Tian and Jiang (2011), both (10.72) and (10.73) lead to intervals with appropriate empirical coverage probabilities for any final dimensional parameters or coefficient curves of the models (10.21) and (10.52).

10.7   R Implementation to the ENRICHD Data

10.7.1   The Varying-Coefficient Mixed-Effects Models

A brief summary of the ENRICHD study has been described in Section 1.2. A preliminary analysis of the patterns of depression severity measured by the Beck Depression Inventory (BDI) score over time has been presented in Section 2.4.2. The objective here is to evaluate the additional effects of pharmacotherapy (i.e., the use of antidepressants) on the time-trends of BDI scores for patients who received pharmacotherapy during the six-month cognitive behavior therapy (CBT) treatment period. Pharmacotherapy with antidepressants is a concomitant intervention in this clinical trial, because the decision of using antidepressants and its starting time was made by the patients or their primary care physicians.

1. The ENRICHD BDI Data

Our dataset here includes 557 depressed patients (with a total of 7117 observations) in the CBT arm who had their status of antidepressant use (yes or no) and exact dates of the antidepressant starting time recorded and attended 5 or more treatment sessions during the CBT treatment period. In order to have a meaningful clinical interpretation, three types of patients were excluded from our analysis: (1) patients in the usual care arm due to the lack of BDI scores observed between the baseline and the end of six months, (2) patients whose pharmacotherapy status or starting dates of antidepressant use were not recorded, and (3) patients who had poor adherence to the required weekly treatment sessions, that is, attended less than 5 (approximately 20%) of the sessions. Because antidepressant use for each patient was individually monitored and recorded as accurate as possible by study psychiatrists, it is reasonable to assume that the missing records on the antidepressant starting dates were missing completely at random. Within our sample, 11 patients used antidepressants before baseline, 92 started antidepressant use during the treatment period, and 454 did not use antidepressants before and during the treatment period. The number of visits for these patients ranges from 5 to 36 and has a median of 12.

2. A Subset Analysis of the ENRICHD BDI Data

We first illustrate the application of the varying-coefficient mixed-effects models of Section 10.4 to a subset of the ENRICHD data, i.e., the 92 ENRICHD patients in the CBT arm who started antidepressant use during the treatment period. This subset of 92 patients contains a total of 1465 observations with clear records of the pharmacotherapy starting time. Among the 92 patients analyzed here, 45 of them started pharmacotherapy at baseline and the remaining 47 patients started pharmacotherapy between 10 and 172 days.

Following the data structure in Section 10.1.3, we denote by Y_{ij}, t_{ij}, S_i,

$R_{ij} = t_{ij} - S_i$ and $\lambda_{ij} = 1_{[t_{ij} \geq S_i]}$ the ith patient's BDI score, trial visit time, starting time of pharmacotherapy, time from initiation of pharmacotherapy, and pharmacotherapy indicator, respectively, at the jth visit. The time units are months in the analysis. A simple case of the naive mixed-effects change-point models (10.9) for evaluating the trends of BDI score over t_{ij} is the linear mixed-effects model

$$\begin{cases} Y_{ij} = a_{i0} + a_{i1}\,t_{ij} + b_{i0}\,\lambda_{ij} + b_{i1}\,\lambda_{ij}\,R_{ij} + \varepsilon_{ij}\,, \\ \left(a_{i0}, a_{i1}, b_{i0}, b_{i1}\right)^T \sim N\!\left((\alpha_0, \alpha_1, \beta_0, \beta_1)^T, \Sigma\right). \end{cases} \tag{10.74}$$

where ε_{ij} is an independent measurement error with mean zero and unknown variance σ^2, and $\left(a_{i0}, a_{i1}, b_{i0}, b_{i1}\right)^T$ and ε_{ij} are independent. When $\lambda_{ij} = 1$ and $R_{ij} = r$, the term $\beta_0 + \beta_1 r$ describes the mean pharmacotherapy effect at r months since the start of pharmacotherapy. As illustrated in Section 2.4.2, we can use the R lme to fit the naive linear mixed-effects change-point model (10.74) and obtain the parameter estimates from the fixed effects. The results are summarized in Table 10.3.

Table 10.3 *Parameter estimates, their standard errors (SE) and p-values for testing the null hypothesis that the corresponding parameter is zero obtained from the naive linear mixed-effects change-point model (10.74) based on the subsample of ENRICHD patients with observed pharmacotherapy starting time.*

Parameter	Estimate	SE	p-value
α_0	23.360	1.115	<0.0001
α_1	-0.610	0.478	0.2024
β_0	-3.582	1.000	0.0004
β_1	-1.547	0.516	0.0028

However, the above model (10.74) ignores the correlation between S_i and the pre-pharmacotherapy depression trends, which may lead to potential bias and erroneous conclusions. One source of the correlation between S_i and the pre-pharmacotherapy depression trends is the design of the ENRICHD study, since, by the study protocol, patients who had 25 or higher on baseline Hamilton Depression Rating Scale or showed less than 50% reduction in BDI scores after 5 weeks of CBT treatment were referred to study psychiatrists for consideration of pharmacotherapy. As demonstrated in Figure 10.1, those patients with higher BDI scores at baseline or undesirable BDI trends were more likely to take medication sooner.

To model the effects of pharmacotherapy start time S_i, we can use the

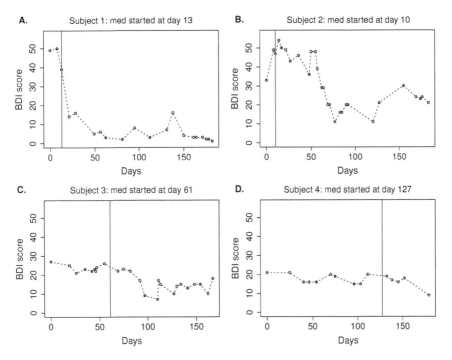

Figure 10.1 *The BDI observations of four patients randomly chosen from the EN-RICHD BDI dataset. The vertical line indicates the starting time of pharmacotherapy.*

following varying-coefficient mixed-effects (VCME) model

$$
\begin{cases}
Y_{ij} &= \alpha_0(S_i) + \alpha_1(S_i) t_{ij} + \beta_0 \lambda_{ij} + \beta_1 \lambda_{ij} R_{ij} + e_{ij}, \\
\alpha_0(S_i) &= \gamma_{00} + \gamma_{01} S_i, \\
\alpha_1(S_i) &= \gamma_{10} + \gamma_{11} S_i, \\
e_{ij} &= a_{i0}^* + a_{i1}^* t_{ij} + b_{i0}^* \lambda_{ij} + b_{i1}^* \lambda_{ij} R_{ij} + \varepsilon_{ij}.
\end{cases}
\tag{10.75}
$$

The mean pre-pharmacotherapy BDI trend in this VCME model (10.75) is associated with S_i through intercept $\alpha_0(S_i)$ and slope $\alpha_1(S_i)$. Similarly, the mean pharmacotherapy effect at r months after the start of pharmacotherapy is described by the term $\beta_0 + \beta_1 r$. A negative (positive) value for $\beta_0 + \beta_1 r$ corresponds to a beneficial (harmful) effect for reducing depression.

For simplicity, this VCME model (10.75) assumes that $\beta_0(S_i) = \beta_0$ and $\beta_1(S_i) = \beta_1$, so that the effects of pharmacotherapy only depend on how long the antidepressant has been used. Under this assumption, β_0 and β_1 have the same interpretations in both the naive linear mixed-effects change-point model (10.74) and the VCME model (10.75), although they differ in modeling the pre-pharmacotherapy time-trends of BDI scores.

The following R code is used to fit the VCME model (10.75) to the subset of

92 patients in the ENRICHD study with complete records of pharmacotherapy starting dates:

```
> library(nlme)
> data(BDIdata)
> BDIsub <-  subset(BDIdata, med.time >=0 & med.time < 200,)

# Recode the covariates and time variables in months
> BDIsub$Tijm <- BDIsub$time*12/365.25
> BDIsub$Sim <- BDIsub$med.time*12/365.25
> BDIsub$TijSim <- with(BDIsub,Tijm*Sim)
> BDIsub$Med <- with(BDIsub, ifelse(time-med.time>=0, 1,0))
> BDIsub$Rijm <- with(BDIsub, Med*(Tijm -Sim))

# Fit the VCME model
> VCME.fit <- lme(BDI ~ 1+Sim + Tijm + TijSim + Med + Rijm,
                  random=~Tijm + Med + Rijm|ID, data=BDIsub)
> summary(VCME.fit)

Linear mixed-effects model fit by REML
Data: BDIsub
...
Fixed effects: BDI ~ 1 + Sim + Tijm + TijSim + Med + Rijm
                Value Std.Error   DF   t-value p-value
(Intercept) 25.640517 1.4418923 1369 17.782546  0.0000
Sim         -1.377896 0.5901003   90 -2.335021  0.0218
Tijm        -0.237399 0.8153425 1369 -0.291165  0.7710
TijSim       0.069672 0.1728152 1369  0.403161  0.6869
Med         -4.489470 1.0506389 1369 -4.273086  0.0000
Rijm        -2.052332 0.7675756 1369 -2.673785  0.0076
...
Number of Observations: 1465
Number of Groups: 92
```

After running the above R commands for the VCME model (10.75), we summarize in Table 10.4 the estimates and their corresponding standard errors and p-values obtained by the R lme procedure based on unstructured covariance matrix and restricted maximum likelihood (REML). The negative estimates for (β_0, β_1) in Tables 10.3 and 10.4 suggest that the beneficial effect of pharmacotherapy for this patient population is detected under both models, when only the patients who had pharmacotherapy change-point time within the CBT period are included in the analysis. Remarkably, we also observe from Table 10.4 a slightly stronger depression lowering effect and a negative estimate of γ_{01}, which suggests a negative correlation between the antidepressant start time S_i and the baseline BDI scores. Thus, as demonstrated in the

Table 10.4 *Parameter estimates, their standard errors (SE) and p-values for test-ing the null hypothesis that the corresponding parameter is zero obtained from the varying-coefficient mixed-effects model (10.75) based on the sub-sample of ENRICHD patients with observed pharmacotherapy starting time.*

Parameter	Estimate	SE	p-value
γ_{00}	25.641	1.442	<0.0001
γ_{01}	-1.378	0.590	0.0218
γ_{10}	-0.237	0.815	0.771
γ_{11}	0.070	0.173	0.6869
β_0	-4.489	1.051	<0.0001
β_1	-2.052	0.768	0.0076

simulation study in Wu, Tian and Bang (2008), by allowing the random co-efficients to be correlated with S_i, the VCME model would lead to less biased estimates for the effects of the concomitant pharmacotherapy intervention.

10.7.2 Shared-Parameter Change-Point Models

The models discussed above can only be applied to the 92 patient subset of the ENRICHD BDI data, since they require the patient's change-point time S_i to be observed during the CBT treatment period. Thus, the estimated beneficial effects of antidepressants for lowering BDI scores obtained from these models ignore the information from those patients who did not start pharmacother-apy during the CBT treatment period. Using the shared-parameter change-point models, the analysis illustrated here is based on all the 557 patients in the ENRICHD BDI dataset, which includes 11 patients who used antide-pressants before baseline and 454 patients who did not use antidepressants during the CBT treatment period. Following the notation in Section 10.1.3, let Y_{ij}, t_{ij}, S_i, $R_{ij} = t_{ij} - S_i$ and $\lambda_{ij} = 1_{[t_{ij} \geq S_i]}$ be the ith patient's BDI score, trial visit time, starting time of pharmacotherapy, time from initiation of pharma-cotherapy, and pharmacotherapy indicator, respectively, at the jth visit. For all $1 \leq i \leq n$, the observed $(S_i^{(c)}, \delta_i^{(c)})$ is $(S_i^{(c)} = S_i, \delta_i^{(c)} = 0)$ if the ith patient used antidepressants within the CBT treatment period, $(S_i^{(c)} = t_{in_i}, \delta_i^{(c)} = 1)$ if the patient did not use antidepressant within the CBT treatment period, and $(S_i^{(c)} = t_{in_1}, \delta_i^{(c)} = 2)$ if the patient used antidepressants before baseline or at the start of the clinical trial.

1. A Linear Shared-Parameter Change-Point Model

A series of analyses described in Wu, Tian and Jiang (2011) suggest that a special case of the shared-parameter change-point model (10.47) can be used as a parsimonious approximation to the BDI time-trend for this study. When the linear functions $\mu_0(t_{ij}; \mathbf{a}_i) = a_{i0} + a_{i1} t_{ij}$ and $\mu_1(t_{ij}; \mathbf{b}_i) = b_{i0} + b_{i1} R_{ij}$ are used, the linear shared-parameter change-point model is

$$
\begin{cases}
Y_{ij} = a_{i0} + a_{i1}\, T_{ij} + \lambda_{ij}\left(b_{i0} + b_{i1}\, R_{ij}\right) + \varepsilon_{ij}, \\
\left(a_{i0}, a_{i1}, b_{i0}, b_{i1}\right)^T \sim N\left((\alpha_0, \alpha_1, \beta_0, \beta_1)^T, \Sigma\right),
\end{cases}
\tag{10.76}
$$

where $(\alpha_0, \alpha_1, \beta_0, \beta_1)^T$ and Σ are the unknown mean vector and the unstructured variance-covariance matrix for the multivariate normal distribution, and ε_{ij} are independent measurement errors as in (10.47) with the $N(0, \sigma^2)$ distribution. In (10.76), $\{a_{i0}, a_{i1}\}$ represent the intercept and slope of the ith subject's BDI trajectory before pharmacotherapy, and $\{b_{i0}, b_{i1}\}$ are the intercept and slope of the change of the subject's BDI trajectory after pharmacotherapy.

We fit the naive linear mixed-effects change-point model (10.76) to all the 557 patients in the ENRICHD BDI dataset, which includes the 465 patients with censored pharmacotherapy starting times. For patients who already used antidepressants before baseline, we set $S_i = 0$ and $R_{ij} = t_{ij}$. For patients who did not use antidepressants during the CBT treatment period, we have $\lambda_{ij} = 0$ for any t_{ij}, so that the BDI time-trend for these patients is described only by $\mu_0\left(t_{ij}; \mathbf{a}_i\right) = a_{i0} + a_{i1}\, t_{ij}$.

The following R commands are used to generate Table 10.5, which summarizes the results for fitting the naive linear mixed-effects change-point model for all the patients with observed or censored times of pharmacotherapy:

```
> library(nlme)
> data(BDIdata)
> nrow(BDIdata ) # No. of observations

[1] 7117

> length(unique(BDIdata$ID)) # No. of patients

[1] 557

# Recode the covariates and time variables in months
> BDIdata$Tijm <- BDIdata$time*12/365.25
> BDIdata$med.time[BDIdata$med.time <0] <- 0
> BDIdata$Med  <- with(BDIdata, ifelse(time-med.time>=0, 1,0))
> BDIdata$Rijm <- with(BDIdata, Med*(time-med.time)*12/365.25)
> Naive.LME    <- lme(BDI ~ Tijm + Med + Rijm, data=BDIdata,
                   random=~ Tijm + Med + Rijm|ID )
> summary(Naive.LME)
```

We note that the positive estimates of β_0 and β_1 in Table 10.5 contradict the results summarized in Tables 10.3 and 10.4, where an improvement of depression measured by BDI was seen for the patients who started pharmacotherapy during the CBT treatment period. If we interpret the positive estimates of β_0 and β_1 in Table 10.5 as an indication for the increase of patients' mean BDI scores (i.e., having worsening depression) after the start of

Table 10.5 *Parameter estimates, their standard errors (SE) and p-values for testing the null hypothesis that the corresponding parameter is zero obtained from the naive linear mixed-effects change-point model based on all the 557 patients in the ENRICHD BDI dataset.*

Parameter	Estimate	SE	p-value
α_0	14.453	0.312	<0.0001
α_1	-1.887	0.067	<0.0001
β_0	3.579	0.825	<0.0001
β_1	0.035	0.227	0.8758

pharmacotherapy, we would reach an erroneous conclusion. Since the "self-selectiveness" of pharmacotherapy is not considered in the model (10.74), the positive estimates of β_0 and β_1 under this model for the full ENRICHD BDI data do not reflect the real effect of pharmacotherapy on depression severity. We show in the next analysis that a proper approach is to incorporate the relationship between the pharmacotherapy starting time S_i and the pre-pharmacotherapy trends into the model. By doing so, the beneficial effects of pharmacotherapy can be demonstrated as in Tables 10.3 and 10.4.

2. A Shared-Parameter Change-Point Model

To account for the possible link between pharmacotherapy starting time and the BDI trend before the start of pharmacotherapy, a shared-parameter model that directly generalizes the model (10.76) is

$$
\begin{cases}
Y_{ij} = a_{i0} + a_{i1} T_{ij} + \lambda_{ij} \left(b_{i0} + b_{i1} R_{ij} \right) + \varepsilon_{ij}, \\
\left(a_{i0}, a_{i1} \right)^T = \left(\alpha_0, \alpha_1 \right)^T + \varepsilon_i^{(a)}, \\
S_i = \gamma_0 + \gamma_1 a_{i0} + \varepsilon_i^{(s)}, \\
\left(b_{i0}, b_{i1} \right)^T = \left(\beta_{00} + \beta_{01} a_{i0} + \beta_{02} a_{i1}, \right. \\
\qquad\qquad \left. \beta_{10} + \beta_{11} a_{i0} + \beta_{12} a_{i1} \right)^T + \varepsilon_i^{(b)},
\end{cases} \tag{10.77}
$$

where $\varepsilon_i^{(s)}$ and $\varepsilon_i^{(b)}$ are mean zero bivariate normal random vectors with unstructured covariance matrices $\Sigma^{(a)}$ and $\Sigma^{(b)}$, respectively, $\varepsilon_i^{(s)}$ is a mean zero normal random variable with variance σ_s^2, and $\varepsilon_i^{(s)}$, $\varepsilon_i^{(b)}$, and $\varepsilon_i^{(s)}$ are independent. It is interesting to see that in (10.77) we model the dependence between $\left(b_{i0}, b_{i1} \right)^T$ and $\left(a_{i0}, a_{i1} \right)^T$ through a linear model. The interpretations of the population-mean parameters in (10.77) are similar to their counterparts specified in the model (10.74).

We apply the algorithm of Section 10.5.5 to compute the maximum likelihood estimates for the Gaussian model (10.77) and obtain the standard errors and 95% confidence intervals for the estimators based on the resampling-subject bootstrap procedure of Section 10.6.2 with 1000 bootstrap replications. The results are summarized in the left panel of Table 10.6. After examining the 95% CIs for the parameter estimates obtained from (10.77), we set β_{01}, β_{02}, β_{10} and β_{11} of (10.77) to zero, repeat the estimation procedure for the sub-model of (10.77), and summarize the results in the right panel of Table 10.6.

The following R functions are used to compute the parameter estimates, their standard errors and 95% bootstrap confidence intervals:

```
> library(npmlda)
> Results.full <- SPM.full(data=BDIdata, n.iter=10, n.boot=1000)
> Results.sub <- SPM.sub(data=BDIdata, n.iter=10, n.boot=1000)
```

Table 10.6 *Parameter estimates, their standard errors (SE) and 95% bootstrap confidence intervals obtained from the linear shared-parameter change-point model (10.77) and its sub-model with β_{01}, β_{02}, β_{10} and β_{11} set to zero based on all the 557 patients in the ENRICHD BDI dataset.*

Parameter	Shared-Parameter Model			Shared-Parameter Sub-model		
	Est	SE	95% CI	Est	SE	95% CI
α_0	15.94	0.35	(15.24, 16.63)	15.84	0.35	(15.17, 16.55)
α_1	-2.03	0.07	(-2.18, -1.89)	-2.00	0.07	(-2.15, -1.86)
β_{00}	20.94	6.90	(7.56, 34.58)	-6.44	0.98	(-8.41, -4.61)
β_{01}	-0.62	0.33	(-1.29, 0.04)	—	—	—
β_{02}	3.52	1.51	(-0.06, 5.98)	—	—	—
β_{10}	-2.11	2.34	(-6.98, 2.24)	—	—	—
β_{11}	-0.06	0.09	(-0.24, 0.12)	—	—	—
β_{12}	-1.52	0.34	(-2.18, -0.87)	-0.23	0.08	(-0.38, -0.07)
γ_0	15.68	1.21	(13.63, 18.51)	15.71	1.29	(13.50, 18.51)
γ_1	-0.49	0.05	(-0.61, -0.40)	-0.50	0.06	(-0.62, -0.40)

Comparing the results in Tables 10.5 and 10.6, we observe that, under the naive mixed-effects change-point model and the shared-parameter change-point models, the estimates of α_1 are negative and mostly similar, which suggests that the mean BDI score for these patients tends to decrease over the trial time since the start of the CBT sessions. However, the parameters for describing the BDI score after pharmacotherapy are very different between the two types of modeling. For the naive mixed-effects change-point model summarized in Table 10.5, the estimate $\widehat{\beta}_0 = 3.579$ ($SE = 0.825$), while, under the shared parameter sub-model (10.77), the mean effect of pharmacotherapy on BDI is always negative, i.e., $\widehat{\beta}_{00} + \widehat{\beta}_{12}\,\widehat{\alpha}_1\,R_{ij} < 0$, since $\widehat{\beta}_{00} = -6.441$, $\widehat{\beta}_{12} = -0.232$, $\widehat{\alpha}_1 = -2.001$ and $0 \le R_{ij} \le 6$. Thus the results of the shared-parameter change-point model (10.77) suggest that pharmacotherapy has a beneficial effect on average for lowering a patient's depression score, i.e., reducing depression severity. We also note from Table 10.6 that the 95% CI for γ_1 obtained from the model (10.77) indicates that the starting time of pharmacotherapy S_i is significantly negatively correlated with the subject-specific baseline BDI values, therefore the naive mixed-effects change-point model (10.74) without incorporating this correlation is likely to be misspecified and would lead to biased estimate for the effect of the concomitant intervention for this dataset.

10.8   Consistency

Asymptotic properties for the maximum likelihood estimators of the parametric models (10.24) (Section 10.5.3) and (10.50) (Section 10.5.4) can be derived by direct applications of the general asymptotic theory of the maximum likelihood estimators, for example, Serfling (1980, Chapter 4). Because the focus is on structured nonparametric models, we only briefly comment in this section on the asymptotic derivations for the maximum likelihood estimators, and devote the main effort on the general results of asymptotic consistency for the B-spline approximation estimators (10.40) and (10.61).

10.8.1   The Varying-Coefficient Mixed-Effects Models

We establish here the rates of convergence, hence consistency, for the polynomial spline, i.e., B-spline, estimators $\left(\widetilde{\alpha}_{LS}^T(s;\mathscr{B}), \widetilde{\beta}_{LS}^T(s;\mathscr{B})\right)^T$ of (10.40) with weight $\Lambda_i = diag\left(1/n_i, \cdots, 1/n_i\right)$, where the basis functions $\mathscr{B}_{dl}(s)$ and $\mathscr{B}_{dm}^*(s)$ of (10.23) are chosen as the B-spline basis functions. Consistency for basis approximation estimators using other basis functions can in principle be derived using similar approaches in conjunction with the special properties of the chosen basis functions. But, asymptotic derivations for basis approximation estimators other than the B-splines have not been developed in the literature, hence, are out of scope for this chapter.

Following the definition of asymptotic consistency in Sections 4.4.2 and 9.5.1, we define an estimator $\widetilde{g}(s)$ of an unknown curve $g(s)$ on $s \in \mathscr{S}$ to be consistent if its L_2-norm satisfies

$$\left\|\widetilde{g} - g\right\|_{L_2} = \left\{\int_{s \in \mathscr{S}} \left[\widetilde{g}(s) - g(s)\right]^2 ds\right\}^{1/2} \to 0, \qquad (10.78)$$

as $n \to \infty$ in probability. From the unknown coefficient curves and finite dimensional parameters defined in (10.21) and (10.23), we denote by $\widetilde{\alpha}_{LS,d}(s;\mathscr{B})$ the B-spline estimator of $\alpha_d(s)$, $0 \le d \le D_1$, and by $\widetilde{\beta}_{LS,d}(s;\mathscr{B})$ the B-spline estimator of $\beta_d(s)$, $0 \le d \le D_2$.

The objective is to show that $\widetilde{\alpha}_{LS,d}(s;\mathscr{B})$ and $\widetilde{\beta}_{LS,d}(s;\mathscr{B})$ are asymptotically consistent estimators of $\alpha_d(s)$ and $\beta_d(s)$ for any d. We assume in this section the following assumptions for $\widetilde{\alpha}_{LS,d}(s;\mathscr{B})$ and $\widetilde{\beta}_{LS,d}(s;\mathscr{B})$.

Asymptotic Assumptions:

(a) *The concomitant intervention change-point times $\{S_1, \ldots, S_n\}$ are independent continuous random variables with unknown distribution function $F_S(\cdot)$ and density function $f_S(\cdot)$. The density function $f_S(\cdot)$ is uniformly bounded away from 0 and infinity on its support $[\mathcal{T}_0, \mathcal{T}_1]$.*

(b) *The conditional expectation $E\left[(\mathbf{Z}_{ij}^T, \delta_{ij} \mathbf{W}_{ij}^T)^T (\mathbf{Z}_{ij}^T, \delta_{ij} \mathbf{W}_{ij}^T) \mid S_i = s\right]$ has eigenvalues $\lambda_1(s) \leq \cdots \leq \lambda_D(s)$ with $D = D_1 + D_2 + 2$, which are uniformly bounded away from 0 and infinity for all $s \in [\mathcal{T}_0, \mathcal{T}_1]$. The elements of $\mathbf{Z}_{ij}$ and $\mathbf{W}_{ij}$ are also bounded.*

(c) *For the B-spline approximation model (10.38), let*

$$e_{ij} = \mathbf{Z}_{ij}^T \mathbf{a}_i^* + \delta_{ij} \mathbf{W}_{ij}^T \mathbf{b}_i^* + \varepsilon_{ij} \qquad (10.79)$$

be the random error terms of Y_{ij}. Conditioning on $S_i = s$, $E\left[e_{ij}^2 \mid S_i = s\right]$ are uniformly bounded for all $s \in [\mathcal{T}_0, \mathcal{T}_1]$.

(d) *For the B-spline approximations of (10.23), let*

$$\begin{cases} K_n = \max\left\{\max_{0 \leq d \leq D_1} \mathcal{L}_d, \max_{0 \leq d \leq D_2} \mathcal{M}_d\right\}, \\ L_\infty\left(\alpha_d, \mathcal{G}_d^{(1)}\right) = \inf_{g^{(1)} \in \mathcal{G}_d^{(1)}} \sup_{s \in \mathcal{S}} \left|\alpha_d(s) - g^{(1)}(s)\right|, \\ L_\infty\left(\beta_d, \mathcal{G}_d^{(2)}\right) = \inf_{g^{(2)} \in \mathcal{G}_d^{(2)}} \sup_{s \in \mathcal{S}} \left|\beta_d(s) - g^{(2)}(s)\right|, \end{cases} \qquad (10.80)$$

where $\mathcal{G}_d^{(1)}$ and $\mathcal{G}_d^{(2)}$ are the linear spaces spanned by $\mathcal{B}_d$ and $\mathcal{B}_d^*$, respectively. Then, the following limits

$$\frac{K_n \log K_n}{n} \to 0, \quad L_\infty\left(\alpha_d, \mathcal{G}_d^{(1)}\right) \to 0 \quad \text{and} \quad L_\infty\left(\beta_d, \mathcal{G}_d^{(2)}\right) \to 0. \qquad (10.81)$$

hold for all d as $n \to \infty$. □

The above assumptions are in fact similar to Assumptions (a)-(d) of Section 9.5.2 with a slight twist toward the estimators of (10.40). The next theorem summarizes the asymptotic consistency of the B-spline estimators $\left(\tilde{\alpha}_{LS}^T(s; \mathcal{B}), \tilde{\beta}_{LS}^T(s; \mathcal{B})\right)^T$ of (10.40) with weight $\Lambda_i = diag\left(1/n_i, \cdots, 1/n_i\right)$.

Theorem 10.1. *If Assumptions (a)-(d) above are satisfied, then the following conclusions hold:*

(a) *For $d = 0, \ldots, D_1$, $\tilde{\alpha}_{LS,d}(s; \mathcal{B})$ are consistent estimators of $\alpha_d(s)$.*

(b) *For K_n, $L_\infty\left(\alpha_d, \mathcal{G}_d^{(1)}\right)$ and $L_\infty\left(\beta_d, \mathcal{G}_d^{(2)}\right)$ defined in (10.80), we have*

$$\left\|\tilde{\alpha}_{LS,d} - \alpha_d\right\|_{L_2}^2 = O_P\left(\frac{1}{n} + \frac{K_n}{n^2} \sum_{i=1}^n \frac{1}{n_i} + \rho_n^2\right), \qquad (10.82)$$

where

$$\rho_n = \max\left\{\max_{0 \leq d \leq D_1} L_\infty\left(\alpha_d, \mathcal{G}_d^{(1)}\right), \max_{0 \leq d \leq D_2} L_\infty\left(\beta_d, \mathcal{G}_d^{(2)}\right)\right\}.$$

(c) *The conclusions in (a) and (b) also hold for* $\widetilde{\beta}_{LS,d}(s; \mathscr{B})$, $d = 0, \ldots, D_2$. ∎

The conclusions of Theorem 10.1(a) show the asymptotic consistency of $\widetilde{\alpha}_{LS,d}(s; \mathscr{B})$ and $\widetilde{\beta}_{LS,d}(s; \mathscr{B})$, while those of Theorem 10.1(b) give the convergence rates of these estimators. Since, in addition to n, the right side of (10.82) is also a function of K_n, $\sum_{i=1}^{n}(1/n_i)$ and ρ_n, we observe that the convergence rates of $\widetilde{\alpha}_{LS,d}(s; \mathscr{B})$ and $\widetilde{\beta}_{LS,d}(s; \mathscr{B})$ also depend on the numbers of repeated measurements as well as how well the functional linear spaces spanned by the B-spline bases approximate the unknown true curves $\alpha(s)$ and $\beta(s)$.

Proof of Theorem 10.1:
Since the result of Theorem 10.1(a) is a direct consequence of (10.82), we only prove (10.82). Let

$$
\begin{cases}
\overline{\gamma} &= \left(\overline{\gamma}_{00}, \ldots, \overline{\gamma}_{dl}, \ldots, \overline{\gamma}_{D_1, \mathscr{L}_{D_1}}\right)^T = E\left[\widetilde{\gamma}_{LS}(\mathscr{B})\right], \\
\overline{\alpha}_{LS,d}(s; \mathscr{B}) &= \sum_{l=0}^{\mathscr{L}_d} \overline{\gamma}_{dl} \mathscr{B}_{dl}(s).
\end{cases}
\tag{10.83}
$$

By the triangular inequality, it follows from (10.83) that

$$
\left|\widetilde{\alpha}_{LS,d}(s; \mathscr{B}) - \alpha_d(s)\right| \le \left|\widetilde{\alpha}_{LS,d}(s; \mathscr{B}) - \overline{\alpha}_{LS,d}(s)\right| + \left|\overline{\alpha}_{LS,d}(s; \mathscr{B}) - \alpha_d(s)\right|. \tag{10.84}
$$

For two sequences of positive numbers a_n and b_n, let $a_n \asymp b_n$ denote that both a_n/b_n and b_n/a_n are bounded for all n. It then follows from the properties of the B-splines, (9.105) and (9.106), that

$$
\left\|\widetilde{\alpha}_{LS,d} - \overline{\alpha}_{LS,d}\right\|_{L_2}^2 = \left\|\sum_{l=0}^{\mathscr{L}_d} \left(\widetilde{\gamma}_{LS,dl} - \overline{\gamma}_{dl}\right) \mathscr{B}_{dl}\right\|_{L_2}^2 \asymp K_n^{-1} \left|\widetilde{\gamma}_{LS,d} - \overline{\gamma}_d\right|^2, \tag{10.85}
$$

where $|\cdot|$ is the Euclidean norm. Let Δ_{LS} be the L_2-norm

$$
\Delta_{LS} = \left\|\left(\widetilde{\alpha}_{LS}^T, \widetilde{\beta}_{LS}^T\right)^T - \left(\overline{\alpha}_{LS}, \overline{\beta}_{LS}\right)^T\right\|_{L_2}. \tag{10.86}
$$

Applying (10.39), (10.40) and Lemma A.3 of Huang, Wu and Zhou (2004) to (10.86), we get

$$
\begin{aligned}
\Delta_{LS}^2 &= \left\|\left[\sum_{i=1}^{n} \left(\mathscr{W}_i \mathscr{B}_i\right)^T \Lambda_i \left(\mathscr{W}_i \mathscr{B}_i\right)\right]^{-1} \left[\sum_{i=1}^{n} \left(\mathscr{W}_i \mathscr{B}_i\right)^T \Lambda_i e_i\right]\right\|^2 \\
&\asymp K_n^2 n^{-2} \left\|\left[\sum_{i=1}^{n} \left(\mathscr{W}_i \mathscr{B}_i\right)^T \Lambda_i e_i\right]^T \left[\sum_{i=1}^{n} \left(\mathscr{W}_i \mathscr{B}_i\right)^T \Lambda_i e_i\right]\right\|,
\end{aligned}
\tag{10.87}
$$

where $e_i = (e_{i1}, \ldots, e_{in_i})^T$. The same argument as in the proof of Lemma A.4

of Huang, Wu and Zhou (2004) shows that

$$
\left[\sum_{i=1}^{n} \left(\mathscr{W}_i \mathscr{B}_i \right)^T \Lambda_i e_i \right]^T \left[\sum_{i=1}^{n} \left(\mathscr{W}_i \mathscr{B}_i \right)^T \Lambda_i e_i \right]
$$
$$
= O_P \left\{ \sum_{i=1}^{n} \left[n_i^{-1} + K_n^{-1} \left(1 - n_i^{-1} \right) \right] \right\}. \tag{10.88}
$$

Consequently, it follows from (10.39), (10.83) and (10.88) that

$$
\left| \widetilde{\gamma}_{LS, dl} - \overline{\gamma}_{LS, dl} \right| = O_P \left\{ K_n^2 n^{-2} \sum_{i=1}^{n} \left[n_i^{-1} + K_n^{-1} \left(1 - n_i^{-1} \right) \right] \right\}, \tag{10.89}
$$

and by (10.85) and (10.89),

$$
\left\| \widetilde{\alpha}_{LS, d} - \overline{\alpha}_{LS, d} \right\|_{L_2}^2 = O_P \left\{ K_n n^{-2} \sum_{i=1}^{n} \left[n_i^{-1} + K_n^{-1} \right] \right\}. \tag{10.90}
$$

On the other hand, let $g_d^{(1)}(s) \in \mathscr{G}_d^{(1)}$ be a linear element in $\mathscr{G}_d^{(1)}$ such that

$$
g_d^{(1)}(s) = \sum_{l=0}^{\mathscr{L}_d} \gamma_{dl}^* \mathscr{B}_{dl}(s)
$$

and the L_∞-norm $\left\| g_d^{(1)} - \alpha_d \right\|_\infty$ is smaller than or equal to ρ_n. Similarly, we define τ_{dm}^*, γ_d^*, τ_d^*, and

$$
g_d^{(2)}(s) = \sum_{l=0}^{\mathscr{L}_d} \gamma_{dl}^* \mathscr{B}_{dl}^*(s).
$$

The same argument as in (10.85) shows that

$$
\left\| \overline{\alpha}_{LS, d} - g_d^{(1)} \right\|_{L_2}^2 \asymp K_n^{-1} \left| \overline{\gamma}_d - \gamma_d^* \right|^2. \tag{10.91}
$$

The same argument as in the proof of Lemma A.7 of Huang, Wu and Zhou (2004) shows that

$$
\left| \overline{\gamma}_{LS} - \gamma^* \right|^2 \leq K_n \rho_n^2, \tag{10.92}
$$

Hence, combining (10.91) and (10.92), we get

$$
\left\| \overline{\alpha}_{LS, d} - g_d^{(1)} \right\|_{L_2}^2 = O_P \left(\rho_n^2 \right). \tag{10.93}
$$

It then follows from (10.93) and the definition of ρ_n that

$$
\left\| \overline{\alpha}_{LS, d} - \alpha_d \right\|_{L_2} = O_P(\rho_n). \tag{10.94}
$$

The assertion of (10.82) follows from (10.90) and (10.94). The conclusions for $\widehat{\beta}_{LS, d}(s)$ can be proved using the same method.   ∎

10.8.2 Maximum Likelihood Estimators

The shared-parameter change-point model (10.47) belongs to a parametric family when $F_a(\cdot)$, $F_s(\cdot|\cdot)$, $F_b(\cdot|\cdot)$ and the distribution function of ε_{ij} are determined by a finite dimensional parameter ϕ. In particular, the linear shared-parameter change-point model (10.48) is parametrized by $\phi = \{\theta, \mathbf{V}\}$ with θ and $\mathbf{V}$ defined in (10.49), if the covariance structure $\mathbf{V}$ is specified by a finite dimensional parameter, where $\theta = (\alpha^T, \beta_1^T, \ldots, \beta_{d_1}^T, \gamma^T)^T$ are the mean structure parameters of interest.

Even for the fully parametric model of (10.48), the asymptotic properties of the maximum likelihood estimators of based on (10.56) may depend on the assumptions for the numbers of repeated measurements. If the numbers of repeated measurements are bounded by a positive constant, that is, $n_i \leq m$ for some $m > 0$ and all $1 \leq i \leq n$, then it follows from the classical asymptotic theory for the MLEs (cf. Serfling, 1980, Section 4.2) that $\widehat{\phi}_{ML} = \{\widehat{\theta}_{ML}, \widehat{\mathbf{V}}_{ML}\}$ obtained by maximizing (10.56) is consistent and has asymptotically a multivariate normal distribution when $f_{y,s}(\cdot)$, $f_{y,1}(\cdot)$ and $f_{y,2}(\cdot)$ satisfy the regularity conditions for the maximum likelihood estimators, such as Conditions (R1), (R2) and (R3) of Serfling (1980, page 144). It is straightforward to check through some tedious calculations that the likelihood functions (10.56), (10.62), (10.63) and (10.64) of the Gaussian linear shared-parameter change-point model (10.48) satisfy Conditions (R1), (R2) and (R3) of Serfling (1980), so that its maximum likelihood estimators are consistent and have asymptotically normal distributions. Asymptotic inferences, such as asymptotic confidence intervals and testing procedures, of the maximum likelihood estimators can be derived using the methods in Serfling (1980, Sections 4.2 and 4.4).

If the numbers of repeated measurements n_i also tend to infinity as n tends to infinity, the existing asymptotic results for the maximum likelihood estimators with cross-sectional i.i.d. data may not apply. The asymptotic distributions of the maximum likelihood estimators will then depend on the rates of n_i tending to infinity as well as the structures of the intra-subject correlations. Asymptotic inferences for the maximum likelihood estimators for the model (10.48) under the general situations of $\lim_{n \to \infty} n_i \to \infty$ and different intra-subject correlation structures have not been systematically investigated in the literature and require substantial further development.

10.8.3 The Additive Shared-Parameter Models

We summarize here the asymptotic properties of the B-spline approximate maximum likelihood estimators $\widehat{\phi}_{AML} = \{\widehat{\theta}_{AML}, \widehat{\mathbf{V}}_{AML}\}$ of (10.61) using the B-spline basis functions $\mathbf{B}^{(d)}(\cdot)$ in (10.58).

1. Framework of Consistency

The general framework of asymptotic consistency and convergence rates for regression models with cross-sectional i.i.d. data is described in Stone (1994) and Stone et al. (1997). As an extension, Huang (2001) establishes a general asymptotic framework for the approximate maximum likelihood estimation with cross-sectional i.i.d. data and a number of nonparametric models, such as the partially linear models, generalized additive models, and functional analysis of variance models. Our theoretical development for $\hat{\phi}_{AML} = \{\hat{\theta}_{AML}, \hat{\mathbf{V}}_{AML}\}$ of (10.61) is based on the framework of Huang (2001).

General Framework:

(a) *The variance-covariance structure $\mathbf{V}$ in (10.60) is specified by a parameter ρ in a Euclidean space.*

(b) *For each $1 \leq d \leq d_0$, the nonparametric function $\mu_d^{(s)}(\cdot)$ of (10.57) is a function on the real line which belongs to a Hilbert space $\mathcal{H}_d$. Then, the set of nonparametric functions $\{\mu_1^{(s)}(\cdot), \ldots, \mu_{d_0}^{(s)}(\cdot)\}$ belongs to the tensor product Hilbert space*

$$\left\{\mu_1^{(s)}(\cdot), \ldots, \mu_{d_0}^{(s)}(\cdot)\right\} \in \mathcal{H}_1 \otimes \cdots \otimes \mathcal{H}_{d_0}.$$

(c) *Let $\mathcal{R}$ be the Euclidean space containing $\{\alpha, \beta, \rho\}$. Then*

$$\left\{\alpha, \beta, \rho, \mu_1^{(s)}(\cdot), \ldots, \mu_{d_0}^{(s)}(\cdot)\right\}$$

belongs to the model space

$$\left\{\alpha, \beta, \rho, \mu_1^{(s)}(\cdot), \ldots, \mu_{d_0}^{(s)}(\cdot)\right\} \in \mathcal{H} = \mathcal{R} \otimes \mathcal{H}_1 \otimes \cdots \otimes \mathcal{H}_{d_0}.$$

(d) *For a function $h \in \mathcal{H}$ with $\{\mathbf{Y}_i, \mathcal{S}_i^{(d)}; i = 1, \ldots, n\}$, the log-likelihood function is*

$$\begin{cases} L_c(h) &= L_c^{(0)}(h) + L_c^{(1)}(h) + L_c^{(2)}(h), \\ L_c^{(0)}(h) &= (1/n)\sum_{i:\delta_i^{(c)}=0} \log f_{(y,s)}\left(\mathbf{Y}_i, S_i; h | \mathcal{D}_i\right), \\ L_c^{(l)}(h) &= (1/n)\sum_{i:\delta_i^{(c)}=l} \log f_{(y,l)}\left(\mathbf{Y}_i; h | \mathcal{D}_i\right) \quad \text{for } l = 1, 2. \end{cases} \qquad (10.95)$$

(e) *The expected log-likelihood is $\Lambda_c(h) = E\{L_c(h)\}$, where the expectation is taken with respect to the true function $\eta \in \mathcal{H}$ defined to the maximizer of $\Lambda_c(h)$ in $\mathcal{H}$.*

(f) *The nonparametric function $\mu_d^{(s)}(\cdot)$ is approximated using (10.57) with a spline basis $\mathbf{B}^{(d)}(\cdot)$.*

(g) *The parameters are $\gamma^{(d)}$ which belong to a Euclidean space $\mathcal{G}_d$. The parameters in (10.60) belong to the estimation space*

$$\mathcal{G} = \mathcal{R} \otimes \mathcal{G}_1 \otimes \cdots \otimes \mathcal{G}_{d_0} \subset \mathcal{H},$$

which is a finite-dimensional space.

(h) *The log-likelihood function $L_c(h)$ and its expectation $\Lambda_c(h)$ are well defined on $\mathscr{H}$ and $\mathscr{G}$.*

(i) *In case that the true function is $\eta(\cdot)$, which may be outside of $\mathscr{H}$, the objective is to estimate*

$$\eta^* = \arg\max_{h \in \mathscr{H}} \Lambda_c(h),$$

defined to be the best approximation to η in $\mathscr{H}$.

(j) *Let $\overline{\eta} = \arg\max_{g \in \mathscr{G}} \Lambda_c(g)$ be the best approximation to η^* in $\mathscr{G}$. The approximate maximum likelihood estimate of η^* is $\widehat{\eta} = \arg\min_{g \in \mathscr{G}} L_c(g)$, and η^*, $\overline{\eta}$ and $\widehat{\eta}$ are characterized by the decomposition*

$$\widehat{\eta} - \eta^* = (\overline{\eta} - \eta^*) + (\widehat{\eta} - \overline{\eta}), \qquad (10.96)$$

where $\overline{\eta} - \eta^*$ and $\widehat{\eta} - \overline{\eta}$ are the approximation and estimation errors, respectively.

(k) *Since each $h \in \mathscr{H}$ is a function on a Euclidean space $\mathscr{U}$, we define $\|h\|_\infty = \sup_{u \in \mathscr{U}} |h(u)|$ and $\|h\|$ to be the L_∞ and L_2 norms of $h(\cdot)$, respectively, such that $\|h\| < \infty$ and $\|h\| \leq C_0 \|h\|_\infty$ for a positive constant C_0.* □

2. Asymptotic Assumptions

Let $N_n = \dim(\mathscr{G})$ be the dimensionality of $\mathscr{G}$, i.e., the number of elements in $\mathscr{G}$,

$$A_n = \sup_{g \in \mathscr{G}, \|g\| \neq 0} \frac{\|g\|_\infty}{\|g\|} \geq 1 \quad \text{and} \quad \rho_n = \inf_{g \in \mathscr{G}} \|g - \eta^*\|_\infty. \qquad (10.97)$$

We make the following assumptions for the B-spline approximate maximum likelihood estimators $\widehat{\phi}_{AML}$, which are almost identical to Conditions A.1 to A.4 of Huang (2001):

(a) *The numbers of repeated measurements are bounded by a positive constant, that is $n_i \leq m$ for some $m > 0$ and all $1 \leq i \leq n$. The best approximation η^* in $\mathscr{H}$ to η exists, and there is a positive constant K_0 such that $\|\eta^*\|_\infty \leq K_0$.*

(b) *For each pair of bounded functions $\{h_l(\cdot) \in \mathscr{H} : l = 1, 2\}$, $\Lambda_c[h_1 + \tau(h_2 - h_1)]$ is twice continuously differentiable with respect to τ for $0 \leq \tau \leq 1$. For any positive constant K, there are positive numbers M_1 and M_2 such that*

$$-M_1 \|h_2 - h_1\|^2 \leq \frac{d^2}{d\tau^2} \Lambda_c[h_1 + \tau(h_2 - h_1)] \leq -M_2 \|h_2 - h_1\|^2 \qquad (10.98)$$

for $0 \leq \tau \leq 1$ and all $h_1(\cdot) \in \mathscr{H}$, $h_2(\cdot) \in \mathscr{H}$ with $\|h_1\|_\infty \leq K$ and $\|h_2\|_\infty \leq K$.

(d) *There is a positive constant K_0 such that, for n sufficiently large, the best approximation $\overline{\eta}$ in $\mathscr{G}$ to η uniquely exists and $\|\overline{\eta}\|_\infty \leq K_0$.*

(e) *For any pair $\{g_l \in \mathscr{G} : l = 1, 2\}$, $L_c[g_1 + \tau(g_2 - g_1)]$ is twice continuously* *differentiable with respect to* $0 \leq \tau \leq 1$. *In addition,*

$$\sup_{g \in \mathcal{G}} \frac{|dL_c(\overline{\eta} + \tau g)/d\tau|_{\tau=0}|}{\|g\|} = O_P\left[\left(\frac{N_n}{n}\right)^{1/2}\right], \tag{10.99}$$

and, for any positive constant K, *there is a positive number* M *such that*

$$\frac{d^2}{d\tau^2} L_c\big[g_1 + \tau\left(g_2 - g_1\right)\big] \leq -M \|g_2 - g_1\|^2 \tag{10.100}$$

for $0 \leq \tau \leq 1$ *and any* $\{g_l \in \mathcal{G} : l = 1, 2\}$ *with* $\|g_1\|_\infty \leq K$ *and* $\|g_2\|_\infty \leq K$, *except on an event whose probability tends to zero as* $n \to \infty$. $\qquad\square$

For a few brief remarks of the above assumptions, the boundedness assumption for n_i in (a) ensures that the proofs of Huang (2001) can be directly extended to the longitudinal data setting. Assumption (b), i.e., (10.97), guarantees that $\Lambda_c(\cdot)$ is strictly concave on $\mathcal{H}$, so its unique maximizer over $\mathcal{H}$ exists. Assumptions (c) and (d) are comparable to the assumptions in Huang (2001). Assumptions (a) and (b) are used to quantify the errors of the approximation (10.57) using the spline basis functions. Assumptions (c) and (d) are used to derive the errors of estimation within the approximated linear spaces spanned by the spline basis function.

3. Main Asymptotic Results

The following theorems summarize the convergence rates of the approximation and estimation errors.

Theorem 10.2. (Approximation Error) *If Assumptions (a) and (b) are satisfied,* K_1 *is a positive constant such that* $K_1 > K_0$ *with* K_0 *given in Assumption (a) and* $\lim_{n \to \infty} A_n \rho_n = 0$, *then* $\overline{\eta}$ *exists uniquely,* $\|\overline{\eta}\|_\infty \leq K_1$ *for* n *sufficiently large, and* $\|\overline{\eta} - \eta^*\|^2 = O_p(\rho_n^2)$. $\qquad\blacksquare$

Proof of Theorem 10.2:
The derivations are identical to the proofs of Theorem A.1 of Huang (2001). Therefore, we do not repeat the derivations here and refer to the technical details in Huang (2001). $\qquad\blacksquare$

Theorem 10.3. (Estimation Error) *If Assumptions (c) and (d) hold,* $\lim_{n \to \infty} (A_n^2 N_n/n) = 0$, *and* K_1 *is a positive constant such that* $K_1 > K_0$ *with* K_0 *given in Assumption (c), then* $\widehat{\eta}$ *exists uniquely,* $\|\widehat{\eta}\|_\infty \leq K_1$, *except on an event whose probability tends to zero as* $n \to \infty$, *and* $\|\widehat{\eta} - \overline{\eta}\|^2 = O_p(N_n/n)$. $\qquad\blacksquare$

Proof of Theorem 10.3:
The derivations are identical to the proofs of Theorem A.2 of Huang (2001).

Therefore, we do not repeat the derivations here and refer to the technical details in Huang (2001). ∎

Combining Theorems 10.2 and 10.3, the following theorem summarizes the convergence rate for the B-spline approximate maximum likelihood estimator $\widehat{\eta}$.

Theorem 10.4. (Overall Error) *If Assumptions (a) to (d) are satisfied,* $\lim_{n\to\infty}\left(A_n\rho_n\right) = 0$ *and* $\lim_{n\to\infty}\left(A_n^2 N_n/n\right) = 0$, *then* $\overline{\eta}$ *and* $\widehat{\eta}$ *exist uniquely for* n *sufficiently large and* $\left\|\widehat{\eta} - \eta^*\right\|^2 = O_p\left(N_n/n + \rho_n^2\right)$. ∎

Proof of Theorem 10.4:
This theorem is a direct consequence of Theorems 10.2 and 10.3, so its proof is omitted, because it is identical to the ones in Huang (2001). ∎

4. Remarks of Asymptotic Results

We make the following remarks to further clarify some implications of Theorems 10.2, 10.3 and 10.4.

Practical Implications:
Although the above theorems provide a general framework for the asymptotic behaviors of the approximate maximum likelihood estimators, their applications in practice are limited because their asymptotic distributions and asymptotic variances have not been derived. Inference procedures based on the asymptotic distributions of B-spline estimators have been available only for certain nonparametric models, for example, Huang, Wu and Zhou (2004). Similar procedures, however, have been mostly lacking for likelihood-based spline methods. □

Strict Concavity Condition:
The strict concavity condition of (10.97) could be potentially too strong and difficult to verify in practical situations. Wu, Tian and Jiang (2011) use the Gaussian additive shared-parameter model as a natural choice in the simulation study. However, it is not clear that, under the model (10.59) and the approximate log-likelihood functions (10.61), (10.68) through (10.70), what conditions on $\mu_d^{(s)}(\cdot)$ are necessary and sufficient to satisfy the inequalities of (10.99). In the maximum likelihood estimation with parametric models, it is possible to encounter situations where the log-likelihood functions are not strictly concave but can be uniquely maximized over a pre-specified parameter space. An asymptotic theory that generalizes Assumption (b) to include non-strictly concave expected log-likelihood functions would yield useful insight on the asymptotic behaviors of the approximate maximum likelihood estimators. Therefore, further research along this direction is warranted. □

10.9   Remarks and Literature Notes

The main results of this chapter show that concomitant interventions, as a special case of the outcome-adaptive covariates, should not be treated as usual time-dependent covariates in naive mixed-effects models. For the simple case of having only one concomitant intervention with only one change-point from "without concomitant intervention" to "with concomitant intervention," a shared-parameter change-point model may be considered to reduce the estimation bias and correct the *self-selectiveness* of the concomitant intervention. The methods presented in this chapter have a narrow focus on a single concomitant intervention in a longitudinal clinical trial. Concomitant interventions may commonly appear in other settings, such as in an epidemiological study where study subjects may take antihypertensive medication during the study when their blood pressure levels either exhibit some undesirable trends or stay in an intolerable range. In the ENRICHD study (Section 10.7), pharmacotherapy as a concomitant was initiated under a vague guideline and a linear shared-parameter change-point model appears to be a reasonable choice. However, this model may not be suitable when the intervention selection mechanism is different, and in some situations the entire shared-parameter approach may have to be re-evaluated.

As a special case of the shared-parameter change-point models, a varying-coefficient mixed-effects model may be considered mainly because it has a simple and clear biological interpretation for the simple situation where there is only one concomitant intervention and the change-point time is observed for all subjects in the study. Compared with the shared-parameter change-point models, the least squares based estimation method for the varying-coefficient mixed-effects models does not require the known parametric forms of the distribution functions. The shared-parameter change-point models, on the other hand, may be applied to concomitant interventions with *double censored change-point time*, but their estimation requires computationally intensive maximum likelihood and approximate maximum likelihood algorithms.

The results of this chapter are mainly adopted from Wu, Tian and Bang (2008) and Wu, Tian and Jiang (2011). Technical derivations for B-spline estimators depend heavily on the results presented in Huang (2001, 2003) and Huang, Wu and Zhou (2004). Future research in this area may be pursued with several potentially worthy extensions. First, subjects in longitudinal studies may have single or multiple concomitant interventions which can be turned on or off at different time points. In such situations, more general shared-parameter change-point models may be needed to accommodate the possibility of multiple interventions and/or multiple change-points. Second, all shared-parameter change-point models studied in this section rely on linear functions to describe the time-trends before and after the intervention, but it is possible that linear response curves are inadequate for certain disease outcomes. Models with nonlinear response curves can be justified in practice and should be investigated. Third, the estimation approach of this section depends on the classical frequentist's framework for the B-spline methods. In a different context, Fahrmeir and Lang (2000) demonstrate a promising Bayesian inference procedure for generalized additive mixed models based on Markov random field priors. Similar Bayesian estimation and inference approaches for the models of this section may lead to computationally simpler estimation and inference procedures. Fourth, large sample properties, such as convergence rates and asymptotic distributions, of the maximum likelihood and approximate maximum likelihood estimators are still not well understood and should be systematically developed to provide theoretical justifications for these estimators. Finally, since it may not be always clear whether an intervention is a concomitant intervention, a model diagnostic method would be a valuable tool to be developed.

Chapter 11

Nonparametric Mixed-Effects Models

The classical linear mixed-effects models of Chapter 2 involve four components: the population-mean terms describing the average time trends and covariate effects; the subject-specific terms describing the individual deviations from the population-mean terms; the distributions of the subject-specific parameters; and the overall measurement errors. These classical models depend on the crucial assumption that both the population-mean and subject-specific terms follow the linear model structure. In practice, the parametric forms of the population mean and subject-specific terms are often unknown, and a misspecified parametric family may lead to erroneous conclusions. We present in this chapter a class of nonparametric mixed-effects models to describe the following aspects from the data: (a) the differences in population time-trends among different subgroups and covariates, (b) the population and individual covariate effects on the outcomes, (c) the population and individual changes over time, and (d) the percentile curves of the outcomes. By extending the population-mean and subject-specific terms to nonparametric functions of time, the models of this chapter are more flexible than the classical linear mixed-effects models.

11.1 Objectives of Nonparametric Mixed-Effects Models

The estimation methods described in Chapters 3 through 10 are all based on the so-called "marginal models", meaning that the models describe the time-trends and covariate effects for the outcome variable of the population being studied. A main feature of the marginal models in these chapters is that the correlation structures are completely unknown and unspecified. Consequently, the marginal models are not equipped to describe the covariate effects of an individual and the individual differences from the general population. In many longitudinal studies, the study objectives require both the population and subject-specific time-trends and covariate effects, so that a structured nonparametric model is needed to incorporate the population and subject-specific structures simultaneously.

As a nonparametric generalization of the classical parametric mixed-effects models, the methods discussed in this chapter are based on the assumption that the marginal and subject-specific functions for the mean time-trends and the covariate effects are smooth functions of time, which can be approximated by some expansions of a class of basis functions. Specifically, we focus on the B-spline, i.e., polynomial splines, bases because of their numerical stability. As a general principle, however, other bases, such as wavelet and Fourier bases, can be used as well.

The nonparametric methods then depend on the specified basis approximations and are focused on the following objectives:

(a) *Estimation of the population mean time-trends and covariate effects;*

(b) *Estimation of the covariance function without a requirement that it be stationary or belong to any particular parametric family;*

(c) *Estimation of the eigenfunctions and eigenvalues of the covariance function;*

(d) *Prediction of the individual outcome values;*

(e) *Estimation of functionals of individual time-trends, such as derivatives and location of extrema;*

(f) *Estimation of the effects of covariates on the shapes of individual time-trends;*

(g) *Exploration of the patterns of variability among the individual time-trends and identification of unusual ones.*

With the exception of the objective (a), none of the objectives in (b) through (g) above can be accomplished by the marginal models described in Chapters 3 to 10. We demonstrate in the real data application of Section 11.4 that adequate estimation and prediction results for the objectives (b) through (g) can lead to useful insights in a longitudinal study. We note that the asymptotic properties for the B-spline estimators of the population mean parameter curves of this chapter are the same as the ones discussed in Chapter 9. Thus, this chapter is primarily focused on the methods and applications of the nonparametric mixed-effects models.

11.2 Data Structure and Model Formulation

Following the general framework of Chapters 7 and 8, we consider the stochastic processes $\left\{ \left(Y(t), t, \mathbf{X}^T(t) \right)^T : t \in \mathscr{T} \right\}$ for a population of interest, where $Y(t)$ is a real-valued response process, $\mathbf{X}(t) = \left(X^{(1)}(t), \ldots, X^{(k)}(t) \right)^T$ is a R^k-valued covariate process, and $\mathscr{T}$ is the set of time points.

11.2.1 Data Structure

For n randomly selected subjects from the population, the subject-specific processes within the time range $t \in \mathscr{T}$ are denoted by

$$\left\{ \mathbf{Z}_i(t) = \left(Y_i(t), t, \mathbf{X}_i^T(t) \right)^T : \mathbf{X}_i(t) = \left(X_i^{(1)}(t), \ldots, X_i^{(k)}(t) \right)^T; t \in \mathscr{T}; i = 1, \ldots, n \right\}.$$

For each $i = 1, \ldots, n$, there are n_i observations of $\{\mathbf{Z}_i(t) : t \in \mathscr{T}\}$ obtained at a set of time design points $\mathbf{t}_i = (t_{i1}, \ldots, t_{in_i})^T$. The longitudinal sample is then denoted by

$$\mathscr{L} = \left\{ \mathbf{Z}_{ij} = (Y_{ij}, t_{ij}, \mathbf{X}_{ij}^T)^T : i = 1, \ldots, n; j = 1, \ldots, n_i \right\},$$

where $Y_{ij} = Y_i(t_{ij})$, $\mathbf{X}_{ij} = \mathbf{X}_i(t_{ij})$, $\mathbf{X}_{ij} = (X_{ij}^{(1)}, \ldots, X_{ij}^{(k)})^T$ and $X_{ij}^{(l)} = X_i^{(l)}(t_{ij})$ for $l = 1, \ldots, k$.

11.2.2   Mixed-Effects Models without Covariates

Without incorporating the covariate vector $\mathbf{X}(t)$, our objective is to evaluate the population mean of $Y(t)$ and the subject-specific time-trends of $Y_i(t)$ for $1 \leq i \leq n$ based on the observations $\{(Y_{ij}, t_{ij})^T : i = 1, \ldots, n; j = 1, \ldots, n_i\}$.

1. Model Formulation

As a direct generalization of the parametric mixed-effects models of Chapter 2, we consider the following nonparametric mixed-effects model for the time-trends of $Y_i(t)$,

$$Y_i(t) = \beta_0(t) + \beta_{0i}(t) + \varepsilon_i(t), \tag{11.1}$$

where $\beta_0(t) = E[Y_i(t)]$ is a smooth function of t representing the population mean time-trend of $Y(t)$, $\beta_{0i}(t)$, also a smooth function of t, represents the ith individual's departure from the population mean time-trend which satisfies $E[\beta_{0i}(t)] = 0$, and $\varepsilon_i(t)$ is the measurement error which satisfies

$$E[\varepsilon_i(t)] = 0, \quad Var[\varepsilon_i(t)] = \sigma^2(t) \quad \text{and} \quad Cov[\varepsilon_i(t_1), \varepsilon_i(t_2)] = 0 \quad \text{if } t_1 \neq t_2.$$

Applying (11.1) to the observations $\{(Y_{ij}, t_{ij})^T : i = 1, \ldots, n; j = 1, \ldots, n_i\}$, the model (11.1) is

$$Y_{ij} = \beta_0(t_{ij}) + \beta_{0i}(t_{ij}) + \varepsilon_{ij}, \tag{11.2}$$

where $\varepsilon_{ij} = \varepsilon_i(t_{ij})$ is the measurement error at time point t_{ij} which satisfies $E(\varepsilon_{ij}) = 0$, $Var(\varepsilon_{ij}) = \sigma^2$ and $Cov(\varepsilon_{ij_1}, \varepsilon_{ij_2}) = 0$ if $j_1 \neq j_2$.

Interpretations of $\beta_0(t)$ and $\beta_{0i}(t)$ are similar to their counterparts in parametric mixed-effects models of Chapter 2, except that, under the current context, $\beta_0(t)$ and $\beta_{0i}(t)$ are flexible curves of t which do not belong to any parametric families. It is common in the literature, for example, Shi, Weiss and Taylor (1996), Rice and Wu (2001), Guo (2002), Liang, Wu and Carroll (2003) and Wu and Liang (2004), to refer to $\beta_0(t)$ as the *population-mean coefficient curve*, $\beta_{0i}(t)$ the ith individual's *subject-specific deviation* and $\beta_0(t) + \beta_{0i}(t)$ the ith individual's *subject-specific outcome trajectory*. The estimated subject-specific outcome trajectory is used for the prediction of an individual's outcome time-trend.

In contrast to the parametric mixed-effects models, we assume that $\beta_0(t)$

and $\beta_{0i}(t)$ of (11.1) and (11.2) are some flexible and smooth functions of time t, which may not belong to known parametric families. In practice, however, we have to assume that $\beta_0(t)$ and $\beta_{0i}(t)$ satisfy some structural assumptions. These assumptions are generally more flexible than the classical parametric families and are imposed to ensure that $\beta_0(t)$ and $\beta_{0i}(t)$ are estimable.

2. B-spline Approximated Mixed-Effects Model

Using the ideas of basis approximation illustrated in Chapter 9, we assume here that $\beta_0(t)$ and $\beta_{0i}(t)$ can be approximated by the following B-spline basis expansions

$$\begin{cases} \beta_0(t) \approx \sum_{s=1}^{K_0} \left[\gamma_{0s} B_{0s}(t) \right] = \gamma_0^T B_0(t), \\ \beta_{0i}(t) \approx \sum_{s=1}^{K_0^*} \left[\gamma_{0si} B_{0s}(t) \right] = \gamma_{0i}^T B_0^*(t), \end{cases} \tag{11.3}$$

where $B_0(t) = \left(B_{01}(t), \ldots, B_{0K_0}(t) \right)^T$ and $B_0^*(t) = \left(B_{01}(t), \ldots, B_{0K_0^*}(t) \right)^T$ are B-spline basis functions with numbers of terms $K_0 \geq 1$ and $K_0^* \geq 1$, respectively,

$$\gamma_0 = \left(\gamma_{01}, \ldots, \gamma_{0K_0} \right)^T \quad \text{and} \quad \gamma_{0i} = \left(\gamma_{01i}, \ldots, \gamma_{0K_0^* i} \right)^T \tag{11.4}$$

are the corresponding coefficients, and $\{ \gamma_{0i} : i = 1, \ldots, n \}$ are assumed to be mutually uncorrelated random vectors. As noted in Chapter 9, many basis choices can be used to approximate $\beta_0(t)$ and $\beta_{0i}(t)$. We use B-splines in (11.3) because they exhibit local features and provide stable numerical solutions.

Substituting $\beta_0(t)$ and $\beta_{0i}(t)$ with the right-side terms of (11.3) and assuming that γ_{0i} and $\varepsilon_i(t)$ have normal distributions, the *B-spline approximated nonparametric mixed-effects model* (11.1) is given by

$$\begin{cases} Y_i(t) \approx Y_i^*(t), \\ Y_i^*(t) = \sum_{s=1}^{K_0} \left[\gamma_{0s} B_{0s}(t) \right] + \sum_{s=1}^{K_0^*} \left[\gamma_{0si} B_{0s}(t) \right] + \varepsilon_i(t) \\ \qquad = \gamma_0^T B_0(t) + \gamma_{0i}^T B_0^*(t) + \varepsilon_i(t), \\ \gamma_{0i} \sim N(\mathbf{0}, \mathbf{D}_0), \\ \varepsilon_i(t) \sim N(0, \sigma^2), \end{cases} \tag{11.5}$$

where $B_0(t)$, $B_0^*(t)$, γ_{0s}, γ_{0si} and $\varepsilon_i(t)$ are defined in (11.1), (11.3) and (11.4), $\sigma > 0$ is the unknown time-invariant standard deviation of the measurement error $\varepsilon_i(t)$, γ_{0i} and $\varepsilon_i(t)$ are independent, and $\mathbf{D}_0$ is the variance-covariance matrix of γ_{0i}. Substituting t with t_{ij}, (11.5) gives the B-spline approximated model for (11.2).

Since $Y_i^*(t)$ follows a linear mixed-effects model when K_0 and K_0^* are fixed and known, (11.5) can be viewed as an *extended linear* mixed-effects model. In model (11.5), both the population mean function $\beta_0(t)$ and the individual random effects $\beta_{0i}(t)$ are approximated by linear combinations of the B-spline basis. This allows the population mean time-trend $\beta_0(t)$ to be a general smooth function and each individual to have their own smooth time-trend

$\beta_0(t) + \beta_{0i}(t)$. By using subject-specific individual time-trends, this modeling approach does not force all subjects to have the same time-trends. In contrast, the parametric models for $\beta_0(t)$ and $\beta_{0i}(t)$, such as the linear or polynomial functions, force the time-trends of all the individuals to have similar patterns, which can be unrealistic in real applications.

The mean zero assumption for γ_{0i} and $\varepsilon_i(t)$ reflects the facts that γ_{0i} describes the individual's deviation from the population mean time-trends and $\varepsilon_i(t)$ is a measurement error. The normality assumption for γ_{0i} and $\varepsilon_i(t)$ is made for the purpose of computational convenience. Although other parametric distribution families may be assumed for γ_{0i} and $\varepsilon_i(t)$, the B-spline approximated mixed-effects models with non-normal distributions for γ_{0i} and $\varepsilon_i(t)$ have not been systematically studies, mostly due to computational complexity of the estimation procedures.

3. B-spline Approximation of Variance-Covariance Matrix

Using the B-spline approximated model (11.5), the covariance structure, including serial correlation, is modeled through the random coefficients γ_{0i}. The covariance of the random process $\{Y(t) : t \in \mathscr{T}\}$ at time points $t_1 \neq t_2$ can be approximated as

$$Cov[Y(t_1), Y(t_2)] = \sum_{s_1=1}^{K_0^*} \sum_{s_2=1}^{K_0^*} \left[d_{0s_1s_2} B_{0s_1}(t_1) B_{0s_2}(t_2) \right], \tag{11.6}$$

where, because $E(\gamma_{0si}) = 0$, $d_{0s_1s_2} = E(\gamma_{0s_1i} \times \gamma_{0s_2i})$ is the (s_1, s_2)th element of $\mathbf{D}_0$. The expression at the right side of (11.6) shows that, when measurement error $\varepsilon_i(t)$ exists, the variance of $Y(t)$ at time t is

$$Var[Y(t)] = \sum_{s_1=1}^{K_0^*} \sum_{s_2=1}^{K_0^*} \left[d_{0s_1s_2} B_{0s_1}(t) B_{0s_2}(t) \right] + \sigma^2. \tag{11.7}$$

Combining (11.6) and (11.7), we can write the variance-covariance of $Y(t_1)$ and $Y(t_2)$ at $\{t_1, t_2\}$ as

$$Cov[Y(t_1), Y(t_2)] = \sum_{s_1=1}^{K_0^*} \sum_{s_2=1}^{K_0^*} \left[d_{0s_1s_2} B_{0s_1}(t_1) B_{0s_2}(t_2) \right] + \sigma^2 1_{[t_1=t_2]}, \tag{11.8}$$

where $1_{[t_1=t_2]}$ is the indicator function with value 1 if $t_1 = t_2$, and 0 if $t_1 \neq t_2$. It then follows from (11.8) that, as $t_2 \to t_1$,

$$\lim_{t_2 \to t_1} Cov[Y(t_1), Y(t_2)] < Var[Y(t_1)],$$

which shows that the covariance process $Cov[Y(t), Y(t+\delta)]$ has an upward jump of size $\sigma^2 > 0$ at $\delta = 0$.

4. Matrix Representations

To rewrite (11.5) using the matrix expression, we define

$$
\begin{cases}
\mathbf{t}_i = \left(t_{i1}, \ldots, t_{in_i}\right)^T, & Y_i(\mathbf{t}_i) = \left(Y_{i1}, \ldots, Y_{in_i}\right)^T, \\
B_{0s}(\mathbf{t}_i) = \left(B_{0s}(t_{i1}), \ldots, B_{0s}(t_{in_i})\right)^T, & s = 1, \ldots, \max\left\{K_0, K_0^*\right\}, \\
B_0(\mathbf{t}_i) = \left(B_{01}(\mathbf{t}_i), \ldots, B_{0K_0}(\mathbf{t}_i)\right), & B_0^*(\mathbf{t}_i) = \left(B_{01}(\mathbf{t}_i), \ldots, B_{0K_0^*}(\mathbf{t}_i)\right), \\
\gamma_0 = \left(\gamma_{01}, \ldots, \gamma_{0K_0}\right)^T, & \gamma_{0i} = \left(\gamma_{01i}, \ldots, \gamma_{0K_0^*i}\right)^T, \\
\varepsilon_i(\mathbf{t}_i) = \left(\varepsilon_i(t_{i1}), \ldots, \varepsilon_i(t_{in_i})\right)^T, & i = 1, \ldots, n,
\end{cases}
$$

and then rewrite the B-spline approximated model (11.5) as

$$
\begin{cases}
Y_i(\mathbf{t}_i) \approx Y_i^*(\mathbf{t}_i), \\
Y_i^*(\mathbf{t}_i) = B_0(\mathbf{t}_i)\,\gamma_0 + B_0^*(\mathbf{t}_i)\,\gamma_{0i} + \varepsilon_i(\mathbf{t}_i), \\
\gamma_{0i} \sim N(\mathbf{0}, \mathbf{D}_0), \\
\varepsilon_i(\mathbf{t}_i) \sim N(\mathbf{0}, \sigma^2 I_{n_i}),
\end{cases}
\tag{11.9}
$$

where I_{n_i} is the $n_i \times n_i$ identity matrix and Γ_{0i} and $\varepsilon_i(\mathbf{t}_i)$ are independent. The variance-covariance matrix of $Y_i^*(\mathbf{t}_i)$ is

$$
\mathbf{V}_i = Cov\left[Y_i^*(\mathbf{t}_i)\right] = B_0^*(\mathbf{t}_i)\,\mathbf{D}_0\,B_0^*(\mathbf{t}_i)^T + \sigma^2 I_{n_i}.
\tag{11.10}
$$

The above matrix expressions are useful in Section 11.3 for estimating the coefficient curves and constructing the outcome trajectories.

11.2.3 Mixed-Effects Models with a Single Covariate

To illustrate the effect of incorporating covariates, we first present the simple case of modeling the stochastic processes $\left\{(Y(t), t, X(t))^T : t \in \mathscr{T}\right\}$ with a single continuous covariate $\mathbf{X}(t) = X(t)$. The corresponding subject-specific curves and observations are

$$
\begin{cases}
\left\{Z_i(t) = \left(Y_i(t), t, X_i(t)\right)^T : t \in \mathscr{T}; i = 1, \ldots, n\right\}, \\
\left\{Z_{ij} = \left(Y_{ij}, t_{ij}, X_{ij}\right)^T : i = 1, \ldots, n; j = 1, \ldots, n_i\right\},
\end{cases}
$$

where $Y_{ij} = Y_i(t_{ij})$ and $X_{ij} = X_i(t_{ij})$. Since the covariate process may also be measured with error, we consider first the covariate measured without error and then the covariate measured with error.

1. Models without Measurement Error

As the simplest flexible extension of the nonparametric mixed-effects model (11.1), we assume that, at any time point $t \in \mathscr{T}$, the ith subject's outcome process $Y_i(t)$ depends on its covariate process $X_i(t)$ through a linear model with population and subject-specific coefficients. This gives the following simplest mixed-effects varying-coefficient model, which was proposed by Liang, Wu and Carroll (2003),

$$Y_i(t) = \beta_0(t) + \left[\beta_1(t) + \beta_{1i}(t)\right] X_i(t) + \varepsilon_i(t), \tag{11.11}$$

where $\beta_0(t) = E\left[Y_i(t) \big| X_i(t) = 0\right]$ and $\beta_1(t)$ are smooth functions of t describing the population intercept and slope of $Y_i(t)$ at time t, $\beta_{1i}(t)$, also a smooth function of t, represents the ith individual's subject-specific deviation of the slope function, $\varepsilon_i(t)$ is the mean zero measurement error as in (11.1) which satisfies $Var\left[\varepsilon_i(t)\right] = \sigma^2$ and $Cov\left[\varepsilon_i(t_1), \varepsilon_i(t_2)\right] = 0$ for $t_1 \neq t_2$, and $X_i(t)$ and $\varepsilon_i(t)$ are independent.

We note that, in (11.11), there is no random intercept term, and the subject-specific deviation is only assumed for the coefficient term $\beta_{1i}(t)$. This assumption is made for the simple purpose of illustrating the effects of including a random intercept term to describe the individual covariate effects on the outcome process. In a real application, it is often appropriate to consider a more general model of the form

$$Y_i(t) = \beta_0(t) + \beta_{0i}(t) + \left[\beta_1(t) + \beta_{1i}(t)\right] X_i(t) + \varepsilon_i(t), \tag{11.12}$$

which has a subject-specific intercept curve $\beta_{0i}(t)$ satisfying $E\left[\beta_{0i}(t)\right] = 0$ in addition to $\beta_0(t)$, $\beta_1(t)$, $\beta_{1i}(t)$ and $\varepsilon_i(t)$ defined in (11.11).

To rewrite (11.11) and (11.12) under the observed data, let

$$
\begin{cases}
\mathbf{t}_i &= \left(t_{i1}, \ldots, t_{in_i}\right)^T, & Y_i(\mathbf{t}_i) &= \left(Y_{i1}, \ldots, Y_{in_i}\right)^T, \\
\beta_0(\mathbf{t}_i) &= \left(\beta_0(t_{i1}), \ldots, \beta_0(t_{in_i})\right)^T, & \beta_{0i}(\mathbf{t}_i) &= \left(\beta_{0i}(t_{i1}), \ldots, \beta_{0i}(t_{in_i})\right)^T, \\
\beta_1(\mathbf{t}_i) &= \left(\beta_1(t_{i1}), \ldots, \beta_1(t_{in_i})\right)^T, & \beta_{1i}(\mathbf{t}_i) &= \left(\beta_{1i}(t_{i1}), \ldots, \beta_{1i}(t_{in_i})\right)^T, \\
X_i(\mathbf{t}_i) &= \left(X_i(t_{i1}), \ldots, X_i(t_{in_i})\right)^T, & \varepsilon_i(\mathbf{t}_i) &= \left(\varepsilon_i(t_{i1}), \ldots, \varepsilon_i(t_{in_i})\right)^T.
\end{cases}
\tag{11.13}
$$

For the observations $\{Y_i(\mathbf{t}_i), \mathbf{t}_i, X_i(\mathbf{t}_i) : i = 1, \ldots, n\}$, (11.11) can be written as

$$Y_i(\mathbf{t}_i) = \beta_0(\mathbf{t}_i) + \left[\beta_1(\mathbf{t}_i) + \beta_{1i}(\mathbf{t}_i)\right] * X_i(\mathbf{t}_i) + \varepsilon_i(\mathbf{t}_i), \tag{11.14}$$

and (11.12) can be written as

$$Y_i(\mathbf{t}_i) = \beta_0(\mathbf{t}_i) + \beta_{0i}(\mathbf{t}_i) + \left[\beta_1(\mathbf{t}_i) + \beta_{1i}(\mathbf{t}_i)\right] * X_i(\mathbf{t}_i) + \varepsilon_i(\mathbf{t}_i), \tag{11.15}$$

where, for any $a(\mathbf{t}_i) = \left(a(t_{i1}), \ldots, a(t_{in_i})\right)^T$, $a(\mathbf{t}_i) * X_i(\mathbf{t}_i)$ is the vector of component-wise products, i.e.,

$$a(\mathbf{t}_i) * X_i(\mathbf{t}_i) = \left(a(t_{i1}) X_i(t_{i1}), \ldots, a(t_{in_i}) X_i(t_{in_i})\right)^T.$$

The expressions in (11.14) and (11.15) are useful for describing the estimation procedures.

2. Approximated Models without Covariate Measurement Error

Assuming that the coefficient curves of (11.11) and (11.12) can be approximated by some expansions of basis functions, we can derive the basis approximated models for (11.11) and (11.12). Using the B-spline approach of (11.3), we assume that the coefficient curves $\{\beta_0(t), \beta_{0i}(t), \beta_1(t), \beta_{1i}(t)\}$ can be approximated by the following B-spline basis expansions

$$
\begin{cases}
\beta_0(t) & \approx \sum_{s=1}^{K_0} \left[\gamma_{0s} B_{0s}(t) \right] = \gamma_0^T B_0(t), \\
\beta_{0i}(t) & \approx \sum_{s=1}^{K_0^*} \left[\gamma_{0si} B_{0s}(t) \right] = \gamma_{0i}^T B_0^*(t), \\
\beta_1(t) & \approx \sum_{s=1}^{K_1} \left[\gamma_{1s} B_{1s}(t) \right] = \gamma_1^T B_1(t), \\
\beta_{1i}(t) & \approx \sum_{s=1}^{K_1^*} \left[\gamma_{1si} B_{1s}(t) \right] = \gamma_{1i}^T B_1^*(t),
\end{cases}
\tag{11.16}
$$

where

$$
\begin{cases}
B_0(t) & = \left(B_{01}(t), \dots, B_{0K_0}(t) \right)^T, \quad B_0^*(t) = \left(B_{01}(t), \dots, B_{0K_0^*}(t) \right)^T, \\
B_1(t) & = \left(B_{11}(t), \dots, B_{1K_1}(t) \right)^T, \quad B_1^*(t) = \left(B_{11}(t), \dots, B_{1K_1^*}(t) \right)^T
\end{cases}
\tag{11.17}
$$

are B-spline basis functions with $K_0 \geq 1$, $K_0^* \geq 1$, $K_1 \geq 1$ and $K_1^* \geq 1$ numbers of terms, respectively,

$$
\begin{cases}
\gamma_0 & = \left(\gamma_{01}, \dots, \gamma_{0K_0} \right)^T, \quad \gamma_{0i} = \left(\gamma_{01i}, \dots, \gamma_{0K_0^* i} \right)^T, \\
\gamma_1 & = \left(\gamma_{11}, \dots, \gamma_{1K_1} \right)^T, \quad \gamma_{1i} = \left(\gamma_{11i}, \dots, \gamma_{1K_1^* i} \right)^T,
\end{cases}
\tag{11.18}
$$

are the corresponding coefficients for the intercept and slope curves, $\left(\gamma_{0i}, \gamma_{1i} \right)^T$ are mean zero random vectors with some covariance matrix Γ, which satisfies

$$
E\left(\gamma_{0i}^T, \gamma_{1i}^T \right)^T = 0 \quad \text{and} \quad Cov\left(\gamma_{0i}^T, \gamma_{1i}^T \right)^T = \Gamma.
\tag{11.19}
$$

Substituting the B-spline approximations at the right side terms of (11.16) to the model (11.11) and assuming normal distributions for the subject-specific effects and the measurement errors, the B-spline approximated nonparametric mixed-effects model for (11.11) is

$$
\begin{cases}
Y_i(t) & \approx Y_i^*(t), \\
Y_i^*(t) & = \sum_{s=1}^{K_0} \left[\gamma_{0s} B_{0s}(t) \right] + \sum_{s=1}^{K_1} \left[\gamma_{1s} B_{1s}(t) X_i(t) \right] \\
& \quad + \sum_{s=1}^{K_1^*} \left[\gamma_{1si} B_{1s}(t) X_i(t) \right] + \varepsilon_i(t) \\
& = \gamma_0^T B_0(t) + \left[\gamma_1^T B_1(t) + \gamma_{1i}^T B_1^*(t) \right] X_i(t) + \varepsilon_i(t), \\
\gamma_{1i} & \sim N(\mathbf{0}, \mathbf{D}_1), \\
\varepsilon_i(t) & \sim N(0, \sigma^2),
\end{cases}
\tag{11.20}
$$

where $B_0(t)$, $B_1(t)$, $B_1^*(t)$, γ_{0s}, γ_{1s}, γ_{1si} and $\varepsilon_i(t)$ are defined in (11.17), (11.18) and (11.19), $\sigma > 0$ is the unknown time-invariant standard deviation of the measurement error $\varepsilon_i(t)$, γ_{1i} and $\varepsilon_i(t)$ are independent, and $\mathbf{D}_1$ is the variance-covariance matrix of γ_{1i}. Similarly, the generalization of (11.20) with the inclusion of subject-specific intercept curve is then given by

$$
\begin{cases}
Y_i(t) \approx Y_i^*(t), \\
\begin{aligned}
Y_i^*(t) &= \sum_{s=1}^{K_0} \left[\gamma_{0s} B_{0s}(t) \right] + \sum_{s=1}^{K_0^*} \left[\gamma_{0si} B_{0s}(t) \right] \\
&\quad + \left\{ \sum_{s=1}^{K_1} \left[\gamma_{1s} B_{1s}(t) \right] + \sum_{s=1}^{K_1^*} \left[\gamma_{1si} B_{1s}(t) \right] \right\} X_i(t) \\
&\quad + \varepsilon_i(t) \\
&= \left[\gamma_0^T B_0(t) + \gamma_{0i}^T B_0^*(t) \right] + \left[\gamma_1^T B_1(t) + \gamma_{1i}^T B_1^*(t) \right] X_i(t) \\
&\quad + \varepsilon_i(t),
\end{aligned} \\
\left(\gamma_{0i}^T, \gamma_{1i}^T \right)^T \sim N(\mathbf{0}, \mathbf{D}_{01}), \\
\varepsilon_i(t) \sim N(0, \sigma^2),
\end{cases}
\tag{11.21}
$$

where $B_0(t)$, $B_0^*(t)$, $B_1(t)$, $B_1^*(t)$, γ_{0s}, γ_{0si}, γ_{1s}, γ_{1si} and $\varepsilon_i(t)$ are defined in (11.17), (11.18) and (11.19), $\sigma > 0$ is the unknown time-invariant standard deviation of the measurement error $\varepsilon_i(t)$, $(\gamma_{0i}^T, \gamma_{1i}^T)$ and $\varepsilon_i(t)$ are independent, and $\mathbf{D}_{01}$ is the variance-covariance matrix of $(\gamma_{0i}^T, \gamma_{1i}^T)$.

Comparing (11.20) with (11.21), we notice that, although (11.12) only involves one more subject-specific intercept curve than (11.11), their B-spline approximated mixed-effects models have significantly different model complexity. Specifically, (11.21) involves K_0^* more individual coefficients and a more complex covariance matrix $\mathbf{D}_{01}$ than (11.20). In practice, the additional model complexity may lead to computational complexity. Thus, it is generally advisable to start with a simpler model in real applications. For this reason, we illustrate the estimation methods by focusing on the simple B-spline approximated model (11.20) in the following sections.

Using the notation of (11.13) and

$$
\begin{cases}
B_{ls}(\mathbf{t}_i) = \left(B_{ls}(t_{i1}), \ldots, B_{ls}(t_{in_i}) \right)^T, \\
B_{ls}(\mathbf{t}_i) * X_i(\mathbf{t}_i) = \left(B_{ls}(t_{i1}) X_i(t_{i1}), \ldots, B_{ls}(t_{in_i}) X_i(t_{in_i}) \right)^T, \quad l = 0, 1,
\end{cases}
\tag{11.22}
$$

the expressions of (11.20) and (11.21) for the observations $\{ Y_i(\mathbf{t}_i), \mathbf{t}_i, X_i(\mathbf{t}_i) : i = 1, \ldots, n \}$ are

$$
\begin{cases}
Y_i(\mathbf{t}_i) \approx Y_i^*(\mathbf{t}_i), \\
\begin{aligned}
Y_i^*(\mathbf{t}_i) &= \sum_{s=1}^{K_0} \left[\gamma_{0s} B_{0s}(\mathbf{t}_i) \right] + \sum_{s=1}^{K_1} \left[\gamma_{1s} B_{1s}(\mathbf{t}_i) * X_i(\mathbf{t}_i) \right] \\
&\quad + \sum_{s=1}^{K_1^*} \left[\gamma_{1si} B_{1s}(\mathbf{t}_i) * X_i(\mathbf{t}_i) \right] + \varepsilon_i(\mathbf{t}_i)
\end{aligned} \\
\gamma_{1i} \sim N(\mathbf{0}, \mathbf{D}_1), \\
\varepsilon_i(t) \sim N(0, \sigma^2),
\end{cases}
\tag{11.23}
$$

and

$$
\left\{
\begin{aligned}
Y_i(\mathbf{t}_i) &\approx Y_i^*(\mathbf{t}_i), \\
Y_i^*(\mathbf{t}_i) &= \textstyle\sum_{s=1}^{K_0}\left[\gamma_{0s}B_{0s}(\mathbf{t}_i)\right] + \sum_{s=1}^{K_0^*}\left[\gamma_{0si}B_{0s}(\mathbf{t}_i)\right] \\
&\quad + \left\{\textstyle\sum_{s=1}^{K_1}\left[\gamma_{1s}B_{1s}(\mathbf{t}_i)\right] + \sum_{s=1}^{K_1^*}\left[\gamma_{1si}B_{1s}(\mathbf{t}_i)\right]\right\} * X_i(\mathbf{t}_i) \\
&\quad + \varepsilon_i(\mathbf{t}_i) \\
(\gamma_{0i}^T, \gamma_{1i}^T)^T &\sim N(\mathbf{0}, \mathbf{D}_{01}), \\
\varepsilon_i(t) &\sim N(0, \sigma^2).
\end{aligned}
\right.
\tag{11.24}
$$

The above expressions are useful for describing the estimation procedures for (11.11) and (11.12).

3. Models with Covariate Measurement Error

In some applications, the actual values of the covariate process $X_i(t)$ are not directly observed but are measured with errors. In this case, we have a measurement error model for $X_i(t)$ at time $t \in \mathscr{T}$ with the following form,

$$
W_i(t) = X_i(t) + u_i(t),
\tag{11.25}
$$

where $W_i(t)$ is the observable covariate value, $X_i(t)$ is the underlying true covariate value and $u_i(t)$ represents the measurement error of the covariate process. Similar to model (11.1), we assume in (11.25) that the measurement error process $u_i(t)$ satisfies

$$
E\left[u_i(t)\right] = 0,\ Var\left[u_i(t)\right] = \sigma_u^2(t)\ \text{ and }\ Cov\left[u_i(t_1), u_i(t_2)\right] = 0 \text{ if } t_1 \neq t_2. \tag{11.26}
$$

When the covariate $X_i(t)$ is measured with error, the varying-coefficient mixed-effects models (11.11) and (11.12) of Liang, Wu and Carroll (2003) become

$$
\left\{
\begin{aligned}
Y_i(t) &= \beta_0(t) + \left[\beta_1(t) + \beta_{1i}(t)\right]X_i(t) + \varepsilon_i(t), \\
W_i(t) &= X_i(t) + u_i(t)
\end{aligned}
\right.
\tag{11.27}
$$

and

$$
\left\{
\begin{aligned}
Y_i(t) &= \beta_0(t) + \beta_{0i}(t) + \left[\beta_1(t) + \beta_{1i}(t)\right]X_i(t) + \varepsilon_i(t), \\
W_i(t) &= X_i(t) + u_i(t),
\end{aligned}
\right.
\tag{11.28}
$$

respectively, with the coefficient curves defined in (11.11), (11.12), (11.25) and (11.26). Since $X_i(t)$ is measured with error, the observed sample for $\{Y_i(\mathbf{t}_i), \mathbf{t}_i, X_i(\mathbf{t}_i) : i = 1, \ldots, n\}$ is

$$
\left\{Y_i(\mathbf{t}_i), \mathbf{t}_i, W_i(\mathbf{t}_i) : W_i(\mathbf{t}_i) = \left(W_i(t_{i1}), \ldots, W_i(t_{in_i})\right)^T; i = 1, \ldots, n\right\}.
\tag{11.29}
$$

Based on (11.29), the models (11.27) and (11.28) become

$$\begin{cases} Y_i(\mathbf{t}_i) &=& \beta_0(\mathbf{t}_i) + [\beta_1(\mathbf{t}_i) + \beta_{1i}(\mathbf{t}_i)] X_i(\mathbf{t}_i) + \varepsilon_i(\mathbf{t}), \\ W_i(\mathbf{t}_i) &=& X_i(\mathbf{t}_i) + u_i(\mathbf{t}_i) \end{cases} \tag{11.30}$$

and

$$\begin{cases} Y_i(\mathbf{t}_i) &=& \beta_0(\mathbf{t}_i) + \beta_{0i}(\mathbf{t}_i) + [\beta_1(\mathbf{t}_i) + \beta_{1i}(\mathbf{t}_i)] X_i(\mathbf{t}_i) + \varepsilon_i(\mathbf{t}_i), \\ W_i(\mathbf{t}_i) &=& X_i(\mathbf{t}_i) + u_i(\mathbf{t}_i), \end{cases} \tag{11.31}$$

respectively, where $u_i(\mathbf{t}_i) = \left(u_i(t_{i1}), \ldots, u_i(t_{in_i})\right)^T$. $\qquad\qquad\square$

4. Approximated Models with Covariate Measurement Error

We use the same approach as (11.1) to write $X_i(t)$ as the sum of a fix-effect curve $\alpha(t)$ and a subject-specific curve $\alpha_i(t)$, such that,

$$X_i(t) = \alpha(t) + \alpha_i(t) \tag{11.32}$$

with $E\left[\alpha_i(t)\right] = 0$. Here, we approximate $\alpha(t)$ and $\alpha_i(t)$ by the B-spline basis expansions

$$\begin{cases} \alpha(t) &\approx& \sum_{s=0}^{p} \xi_s B_{xs}(t) &=& \xi^T B_x(t), \\ \alpha_i(t) &\approx& \sum_{s=0}^{q} \xi_{si} B_{xs}^*(t) &=& \xi_i^T B_x^*(t), \end{cases} \tag{11.33}$$

where $\xi = \left(\xi_0, \ldots, \xi_p\right)^T$ and $\xi_i = \left(\xi_{0i}, \ldots, \xi_{qi}\right)^T$ which satisfies $E\left(\xi_{is}\right) = 0$ for $s = 1, \ldots, q$ and $Cov\left(\xi_i\right) = \Sigma_\xi$. Using the B-spline approximation in (11.32) and (11.33), the B-spline approximated nonparametric mixed-effects model of (11.27) with covariate measurement error under normal distribution assumption is

$$\begin{cases} Y_i(t) &\approx& Y_i^*(t), \\ X_i(t) &\approx& X_i^*(t), \\ X_i^*(t) &=& \sum_{s=0}^{p} \xi_s B_{xs}(t) + \sum_{s=0}^{q} \xi_{si} B_{xs}^*(t) \\ &=& \xi^T B_x(t) + \xi_i^T B_x^*(t), \\ W_i(t) &\approx& X_i^*(t) + u_i(t), \\ Y_i^*(t) &=& \sum_{s=1}^{K_0} \left[\gamma_{0s} B_{0s}(t)\right] + \sum_{s=1}^{K_1} \left[\gamma_{1s} B_{1s}(t) X_i(t)\right] \\ && + \sum_{s=1}^{K_1^*} \left[\gamma_{1si} B_{1s}(t) X_i^*(t)\right] + \varepsilon_i(t) \\ &=& \gamma_0^T B_0(t) + \left[\gamma_1^T B_1(t) + \gamma_{1i}^T B_1^*(t)\right] X_i^*(t) + \varepsilon_i(t), \\ \xi_i &\sim& N\left(\mathbf{0}, \Sigma_\xi\right), \\ u_i(t) &\sim& N\left(0, \sigma_u^2(t)\right), \ Cov\left[u_i(t_1), u_i(t_2)\right] &=& 0 \text{ if } t_1 \neq t_2, \\ \gamma_{1i} &\sim& N\left(\mathbf{0}, \mathbf{D}_1\right), \\ \varepsilon_i(t) &\sim& N\left(0, \sigma^2\right), \end{cases} \tag{11.34}$$

and the B-spline approximated nonparametric mixed-effects model of (11.28) with covariate measurement under normal distribution assumption is

$$
\left\{
\begin{aligned}
Y_i(t) &\approx Y_i^*(t), \\
X_i(t) &\approx X_i^*(t), \\
X_i^*(t) &= \sum_{s=0}^{p} \xi_s B_{xs}(t) + \sum_{s=0}^{q} \xi_{si} B_{xs}^*(t) \\
&= \xi^T B_x(t) + \xi_i^T B_x^*(t), \\
W_i(t) &\approx X_i^*(t) + u_i(t), \\
Y_i^*(t) &= \sum_{s=1}^{K_0} \left[\gamma_{0s} B_{0s}(t) \right] + \sum_{s=1}^{K_0^*} \left[\gamma_{0si} B_{0s}(t) \right] \\
&\quad + \left\{ \sum_{s=1}^{K_1} \left[\gamma_{1s} B_{1s}(t) \right] + \sum_{s=1}^{K_1^*} \left[\gamma_{1si} B_{1s}(t) \right] \right\} X_i^*(t) \\
&\quad + \varepsilon_i(t) \\
&= \left[\gamma_0^T B_0(t) + \gamma_{0i}^T B_0^*(t) \right] + \left[\gamma_1^T B_1(t) + \gamma_{1i}^T B_1^*(t) \right] X_i^*(t) \\
&\quad + \varepsilon_i(t), \\
\xi_i &\sim N(\mathbf{0}, \Sigma_\xi), \\
u_i(t) &\sim N(0, \sigma_u^2(t)), \; Cov\left[u_i(t_1), u_i(t_2) \right] = 0 \;\text{ if } t_1 \neq t_2, \\
(\gamma_{0i}^T, \gamma_{1i}^T)^T &\sim N(\mathbf{0}, \mathbf{D}_{01}), \\
\varepsilon_i(t) &\sim N(0, \sigma^2),
\end{aligned}
\right.
\tag{11.35}
$$

where the corresponding parameters and coefficient curves are defined in (11.20), (11.21), (11.32) and (11.33). Substituting t with $\mathbf{t}_i$ in (11.34) and (11.35), we can obtain the B-spline approximated models for the observed data (11.29). To avoid repetition, the exact expressions of these models are omitted.

11.2.4 Extensions to Multiple Covariates

We now extend the models of Section 11.2.3 to the case with multiple covariates. This extension follows the approach for the univariate covariate case, as all the covariate effects are included as simple linear terms.

1. Models without Covariate Measurement Errors

With the presence of the covariate vector $\mathbf{X}(t)$, we have the additional objective of evaluating the population-average and subject-specific covariate effects based on the observations $\left\{ (Y_{ij}, t_{ij}, \mathbf{X}_{ij}^T)^T : i = 1, \ldots, n; j = 1, \ldots, n_i \right\}$. Extending the varying-coefficient mixed-effects model (11.12) to the case of multivariate covariates, we can incorporate the additional covariate effects by adding the additional population-mean and subject-specific terms to (11.12).

For the convenience of exposition, we decompose $\mathbf{X}(t)$ to

$$\begin{cases} \mathbf{X}(t) &= \left(\mathbf{X}^{(1)}(t)^T, \mathbf{X}^{(2)}(t)^T\right)^T, \\ \mathbf{X}^{(1)}(t) &= \left(X^{(1)}(t),\ldots,X^{(L)}(t)\right)^T, \\ \mathbf{X}^{(2)}(t) &= \left(X^{(L+1)}(t),\ldots,X^{(k)}(t)\right)^T \end{cases} \tag{11.36}$$

for some $L \leq k$, and express the varying-coefficient mixed-effects model as

$$\begin{aligned} Y_i(t) &= \beta_0(t) + \beta_{0i}(t) + \sum_{l=1}^{L} \left[\beta_l(t) + \beta_{li}(t)\right] X_i^{(l)}(t) \\ &\quad + \sum_{l=L+1}^{k} \beta_l(t) X_i^{(l)}(t) + \varepsilon_i(t). \end{aligned} \tag{11.37}$$

Under the observations $\left\{ \left(Y_{ij}, t_{ij}, \mathbf{X}_{ij}^T\right)^T : i = 1, \ldots, n; \; j = 1, \ldots, n_i\right\}$, (11.37) becomes

$$\begin{aligned} Y_{ij} &= \beta_0(t_{ij}) + \beta_{0i}(t_{ij}) + \sum_{l=1}^{L} \left[\beta_l(t_{ij}) + \beta_{li}(t_{ij})\right] X_{ij}^{(l)} \\ &\quad + \sum_{l=L+1}^{k} \beta_l(t_{ij}) X_{ij}^{(l)} + \varepsilon_i(t_{ij}). \end{aligned} \tag{11.38}$$

The second summation term, i.e., from $L+1$ to k, disappears when all the covariates have subject-specific effects, i.e., $L = k$. On the other hand, when $L = 0$, the summation from $l = 1$ to L disappears, which suggests that none of the coefficients have subject-specific effects. The coefficient curves in (11.37) have the same interpretations as in (11.12). Specifically, $\{\beta_0(t), \beta_{0i}(t)\}$ are the population-mean curve and the subject-specific deviation, $\{\beta_l(t) : l = 1, \ldots, k\}$ are the population-mean coefficient curves which represent the population-mean covariate effects of $\mathbf{X}_i(t)$, and $\{\beta_{li}(t) : l = 1, \ldots, L\}$ describe the subject-specific covariate effect deviations. □

2. Approximated Models without Covariate Measurement Errors

The B-spline approximation for the varying-coefficient mixed-effects model (10.38) can be constructed by substituting the following approximations

$$\begin{cases} \beta_l(t) &\approx \sum_{s=1}^{K_l} \gamma_{ls} B_{ls}(t) = \gamma_l^T B_l(t), \quad l = 0,\ldots,k, \\ \beta_{li}(t) &\approx \sum_{s=1}^{K_l^*} \gamma_{lsi} B_{ls}(t) = \gamma_{li}^T B_l^*(t), \quad l = 0,\ldots,L, \end{cases} \tag{11.39}$$

to (11.38), where, for $l = 0, \ldots, k$,

$$B_l(t) = \left(B_{l1}(t), \ldots, B_{lK_l}(t)\right)^T \text{ and } B_l^*(t) = \left(B_{l1}^*(t), \ldots, B_{lK_l^*}^*(t)\right)^T, \tag{11.40}$$

are B-spline basis functions, and

$$\gamma_l = \left(\gamma_{l1}, \ldots, \gamma_{lK_l}\right)^T \text{ and } \gamma_{li} = \left(\gamma_{l1i}, \ldots, \gamma_{lK_l^*}\right)^T \tag{11.41}$$

are the corresponding coefficients. Under the assumption of normal distributions, the B-spline approximated mixed-effects varying-coefficient model is given by

$$
\left\{
\begin{aligned}
Y_i(t) &\approx Y_i^*(t), \\
Y_i^*(t) &= \sum_{s=1}^{K_0} \left[\gamma_{0s} B_{0s}(t) \right] + \sum_{s=1}^{K_0^*} \left[\gamma_{0si} B_{0s}(t) \right] \\
&\quad + \sum_{l=1}^{L} \Big\{ \sum_{s=1}^{K_l} \left[\gamma_{ls} B_{ls}(t) \right] \\
&\qquad\qquad + \sum_{s=1}^{K_l^*} \left[\gamma_{lsi} B_{ls}(t) \right] \Big\} X_i^{(l)}(t) \\
&\quad + \sum_{l=L+1}^{k} \Big\{ \sum_{s=1}^{K_l} \gamma_{ls} B_{ls}(t) \Big\} X_i^{(l)}(t) + \varepsilon_i(t) \\
&= \left[\gamma_0^T B_0(t) + \gamma_{0i}^T B_0^*(t) \right] \\
&\quad + \sum_{l=1}^{L} \left[\gamma_l^T B_l(t) + \gamma_{li}^T B_l^*(t) \right] X_i^{(l)}(t) \\
&\quad + \sum_{l=L+1}^{k} \left[\gamma_l^T B_l(t) X_i^{(l)}(t) \right] + \varepsilon_i(t), \\
(\gamma_{0i}^T, \ldots, \gamma_{Li}^T)^T &\sim N(\mathbf{0}, \mathbf{D}_{0L}), \\
\varepsilon_i(t) &\sim N(0, \sigma^2),
\end{aligned}
\right.
\tag{11.42}
$$

where, for $l = 0, \ldots, K$, $B_l(t)$, $B_l^*(t)$, γ_{ls}, γ_{lsi}, $\varepsilon_i(t)$ and $\sigma > 0$ are defined in (11.17), (11.18) and (11.19), $(\gamma_{0i}^T, \ldots, \gamma_{Li}^T)$ and $\varepsilon_i(t)$ are independent, and $\mathbf{D}_{0L}$ is the variance-covariance matrix of $(\gamma_{0i}^T, \ldots, \gamma_{Li}^T)$.

3. Models with Covariate Measurement Errors

Extensions of the covariate measurement error models (11.28) and (11.35) can be constructed by combining the measurement error model (11.32) with (11.38) and (11.42). This extension depends on which covariates are measured with error. If the components in $\mathbf{X}^{(1)}(t)$ of (11.36) satisfy the measurement error model, we have the expression

$$
W_i^{(l)}(t) = X_i^{(l)}(t) + u_i^{(l)}(t), \quad l = 1, \ldots, L,
\tag{11.43}
$$

for $i = 1, \ldots, n$, where $X_i^{(l)}(t)$ is the ith subject's underlying value of $X^{(l)}(t)$, $W_i^{(l)}(t)$ is the ith subject's observed value for $X_i^{(l)}(t)$, and $u_i^{(l)}(t)$ is the corresponding measurement error which, similar to (11.26), satisfies

$$
\left\{
\begin{aligned}
E\left[u_i^{(l)}(t) \right] &= 0, \; Var\left[u_i^{(l)}(t) \right] = \sigma_{u^{(l)}}^2(t), \\
Cov\left[u_i^{(l)}(t_1), u_i^{(l)}(t_2) \right] &= 0 \; \text{ if } t_1 \neq t_2, \\
u_i^{(l_1)}(t) \text{ and } u_i^{(l_2)}(t) & \text{ are independent if } l_1 \neq l_2.
\end{aligned}
\right.
\tag{11.44}
$$

The observed data under the covariate measurement error model (11.43) are

$$
\left\{
\begin{aligned}
& \left\{ \left(Y_{ij}, t_{ij}, \mathbf{W}_{ij}^T, (\mathbf{X}_{ij}^{(2)})^T \right)^T : i = 1, \ldots, n; j = 1, \ldots, n_i \right\}, \\
& \mathbf{W}_{ij} = \left(W_{ij}^{(1)}, \ldots, W_{ij}^{(L)} \right)^T, \; W_{ij}^{(l)} = W_i^{(l)}(t_{ij}) \text{ for } l = 1, \ldots, L, \\
& \mathbf{X}_{ij}^{(2)} = \left(X_{ij}^{(L+1)}, \ldots, X_{ij}^{(k)} \right)^T, \; X_{ij}^{(l)} = X_i^{(l)}(t_{ij}) \text{ for } l = L+1, \ldots, k.
\end{aligned}
\right.
\tag{11.45}
$$

The extension of (11.37) with covariate measurement errors is

$$
\begin{cases}
Y_i(t) &= \beta_0(t) + \beta_{0i}(t) + \sum_{l=1}^{L} \left[\beta_l(t) + \beta_{li}(t) \right] X_i^{(l)}(t) \\
&\quad + \sum_{l=L+1}^{k} \beta_l(t) X_i^{(l)}(t) + \varepsilon_i(t), \\
W_i^{(l)}(t) &= X_i^{(l)}(t) + u_i^{(l)}(t), \ l = 1, \dots, L
\end{cases}
\tag{11.46}
$$

with the coefficient curves defined in (11.37), (11.43) and (11.44). Under the observations (11.45), (11.46) becomes

$$
\begin{cases}
Y_{ij} &= \beta_0(t_{ij}) + \beta_{0i}(t_{ij}) + \sum_{l=1}^{L} \left[\beta_l(t_{ij}) + \beta_{li}(t_{ij}) \right] X_{ij}^{(l)} \\
&\quad + \sum_{l=L+1}^{k} \beta_l(t_{ij}) X_{ij}^{(l)} + \varepsilon_i(t_{ij}), \\
W_{ij}^{(l)} &= X_{ij}^{(l)} + u_{ij}^{(l)}, \ l = 1, \dots, L,
\end{cases}
\tag{11.47}
$$

where $u_{ij}^{(l)} = u_i^{(l)}(t_{ij})$.

The B-spline approximated models for (11.46) and (11.47) can be constructed using the approach of (11.32) and (11.33). In this case, we first decompose $X_i^{(l)}(t)$ of (11.43) into the sum of a fixed effect curve $\alpha^{(l)}(t)$ and a subject-specific curve $\alpha_i^{(l)}(t)$, so that

$$
X_i^{(l)}(t) = \alpha^{(l)}(t) + \alpha_i^{(l)}(t), \ l = 1, \dots, L,
\tag{11.48}
$$

with $E\left[\alpha_i^{(l)}(t) \right] = 0$, and then approximate $\alpha^{(l)}(t)$ and $\alpha_i^{(l)}(t)$ by the following B-spline expansions

$$
\begin{cases}
\alpha^{(l)}(t) &\approx \sum_{s=0}^{p_l} \xi_s^{(l)} B_{xs}^{(l)}(t) = \left(\xi^{(l)} \right)^T B_x^{(l)}(t), \\
\alpha_i^{(l)}(t) &\approx \sum_{s=0}^{q_l} \xi_{si}^{(l)} B_{xs}^{*(l)}(t) = \left(\xi_i^{(l)} \right)^T B_x^{*(l)}(t),
\end{cases}
\tag{11.49}
$$

so that, for $l = 1, \dots, L$, $X_i^{(l)}(t)$ has the following approximation,

$$
\begin{cases}
X_i^{(l)}(t) &\approx X_i^{*(l)}(t), \\
X_i^{*(l)}(t) &= \sum_{s=0}^{p_l} \xi_s^{(l)} B_{xs}^{(l)}(t) + \sum_{s=0}^{q_l} \xi_{si}^{(l)} B_{xs}^{*(l)}(t) \\
&= \left(\xi^{(l)} \right)^T B_x^{(l)}(t) + \left(\xi_i^{(l)} \right)^T B_x^{*(l)}(t).
\end{cases}
\tag{11.50}
$$

Combining (11.42), (11.43) and (11.50), the B-spline approximated model for

(11.46) is given by

$$
\left\{
\begin{aligned}
Y_i(t) &\approx Y_i^*(t), \\
X_i^{(l)}(t) &\approx X_i^{*(l)}(t), \\
X_i^{*(l)}(t) &= \sum_{s=0}^{p_l} \xi_s^{(l)} B_{xs}^{(l)}(t) + \sum_{s=0}^{q_l} \xi_{si}^{(l)} B_{xs}^{*(l)}(t) \\
&= \left(\xi^{(l)}\right)^T B_x^{(l)}(t) + \left(\xi_i^{(l)}\right)^T B_x^{*(l)}(t), \\
W_i^{(l)}(t) &\approx X_i^{*(l)}(t) + u_i^{(l)}(t), \\
Y_i^*(t) &= \sum_{s=1}^{K_0} \left[\gamma_{0s} B_{0s}(t)\right] + \sum_{s=1}^{K_0^*} \left[\gamma_{0si} B_{0s}(t)\right] \\
&\quad + \sum_{l=1}^{L} \Big\{ \sum_{s=1}^{K_l} \left[\gamma_{ls} B_{ls}(t)\right] \\
&\qquad\qquad + \sum_{s=1}^{K_l^*} \left[\gamma_{lsi} B_{ls}(t)\right] \Big\} X_i^{(l)}(t) \\
&\quad + \sum_{l=L+1}^{k} \Big\{ \sum_{s=1}^{K_l} \gamma_{ls} B_{ls}(t) \Big\} X_i^{(l)}(t) \\
&\quad + \varepsilon_i(t) \\
&= \left[\gamma_0^T B_0(t) + \gamma_{0i}^T B_0^*(t)\right] \\
&\quad + \sum_{l=1}^{L} \left[\gamma_l^T B_l(t) + \gamma_{li}^T B_l^*(t)\right] X_i^{(l)}(t) \\
&\quad + \sum_{l=L+1}^{k} \left[\gamma_l^T B_l(t) X_i^{(l)}(t)\right] + \varepsilon_i(t), \\
u_i^{(l)}(t) &\sim N\big(0, \sigma_{u^{(l)}}^2(t)\big), \\
Cov\big[u_i^{(l)}(t_1), u_i^{(l)}(t_2)\big] &= 0 \ \text{if } t_1 \neq t_2, \\
\left(\gamma_{0i}^T, \ldots, \gamma_{Li}^T\right)^T &\sim N(\mathbf{0}, \mathbf{D}_{0L}), \\
\varepsilon_i(t) &\sim N(0, \sigma^2),
\end{aligned}
\right.
\tag{11.51}
$$

where the corresponding parameters and coefficient curves are defined in (10.41), (10.42) and (10.49).

11.3 Estimation and Prediction without Covariates

Using the B-spline approximated models of Section 10.2, the B-spline coefficients, hence, the population-mean and subject-specific coefficient curves, can be estimated using the method of least squares. These least squares estimates can be used to evaluate the population-mean time-trends of the outcome variable and covariate effects, as well as to predict the subject-specific outcome trajectories. We first consider the B-spline approximated nonparametric mixed-effects model (11.5) or its matrix representation (11.9). Since, when the expansion terms k_0 and k_0^* are fixed, the model (11.5) is a linear mixed-effects model, we can then estimate the population-mean coefficients γ_0 and predict the subject-specific coefficients γ_{0i} of (11.9) using the procedures described in Chapter 2.1 and the classical estimation and prediction results cited there, e.g., Laird and Ware (1982), Verbeke and Molenberghs (2000), Diggle et al. (2002) and Fitzmaurice et al. (2009). We consider two situations separately: the components of the covariance matrix (11.10) are known, and the covariance matrix (11.10) is unknown and has to be estimated from the data.

11.3.1   Estimation with Known Covariance Matrix

When the covariance structure is known, i.e., the components $\mathbf{D}_0$ and σ of (11.9) are known, and $\mathbf{V}_i$ of (11.10) is invertible with inverse

$$\mathbf{V}_i^{-1} = \left[B_0^*(\mathbf{t}_i) \, \mathbf{D}_0 \, B_0^*(\mathbf{t}_i)^T + \sigma^2 I_{n_i} \right]^{-1}, \tag{11.52}$$

the normality assumption of (11.9) suggests that the population-mean coefficients γ_0 can be estimated by

$$\widehat{\gamma_0} = \left[\sum_{i=1}^n B_0(\mathbf{t}_i)^T \, \mathbf{V}_i^{-1} \, B_0(\mathbf{t}_i) \right]^{-1} \left[\sum_{i=1}^n B_0(\mathbf{t}_i)^T \, \mathbf{V}_i^{-1} \, Y_i(\mathbf{t}_i) \right]. \tag{11.53}$$

Using the *best linear unbiased prediction* (BLUP) described in Laird and Ware (1982) and Robinson (1991), the BLUP estimate of the spline coefficients of the random effect for subject i is

$$\widehat{\gamma}_{0i} = \mathbf{D}_0 \, B_0^*(\mathbf{t}_i)^T \, \mathbf{V}_i^{-1} \left[Y_i(\mathbf{t}_i) - B_0(\mathbf{t}_i) \, \widehat{\gamma_0} \right]. \tag{11.54}$$

From the expressions of (11.53) and (11.54), both $\widehat{\gamma_0}$ and $\widehat{\gamma}_{0i}$ are linear estimators in the sense that they are linear functions of the observations $\{ Y_i(\mathbf{t}_i) : i = 1, \ldots, n \}$. When the B-spline model holds exactly, i.e., $Y_i(\mathbf{t}_i) = Y_i^*(\mathbf{t}_i)$, for any fixed k_0 and k_0^*, then the estimator $\widehat{\gamma_0}$ of (11.53) maximizes the likelihood based on the marginal distribution of the data and is also the minimum variance estimator. The expression for $\widehat{\gamma}_{0i}$ in (11.54) is of course not maximum likelihood but, as described in Laird and Ware (1982, Section 3.1), can be derived by an extension of the Gauss-Markov theorem to cover random effects.

Since $\widehat{\gamma_0}$ and $\widehat{\gamma}_{0i}$ are linear functions of $Y_i(\mathbf{t}_i)$, the approximate variances of $\widehat{\gamma_0}$ and $\widehat{\gamma}_{0i}$ can be given by

$$Var(\widehat{\gamma_0}) = \left[\sum_{i=1}^n B_0(\mathbf{t}_i)^T \, \mathbf{V}_i^{-1} \, B_0(\mathbf{t}_i) \right]^{-1} \tag{11.55}$$

and

$$\begin{aligned} Var(\widehat{\gamma}_{0i}) &= \mathbf{D}_0 \, B_0^*(\mathbf{t}_i)^T \left\{ \mathbf{V}_i^{-1} - \mathbf{V}_i^{-1} \, B_0(\mathbf{t}_i) \right. \\ &\quad \times \left. \left[\sum_{i=1}^n B_0(\mathbf{t}_i)^T \, \mathbf{V}_i^{-1} \, B_0(\mathbf{t}_i) \right]^{-1} B_0(\mathbf{t}_i)^T \, \mathbf{V}_i^{-1} \right\} B_0^*(\mathbf{t}_i) \, \mathbf{D}_0, \end{aligned} \tag{11.56}$$

respectively. These variances are useful to assess the errors of the estimation. However, because (11.56) ignores the variation of γ_{0i}, (10.55) cannot be used to assess the estimation error of $(\widehat{\gamma}_{0i} - \gamma_{0i})$. As suggested in Laird and Ware (1982), the variance of $(\widehat{\gamma}_{0i} - \gamma_{0i})$ can be approximated by

$$
\begin{aligned}
Var\big(\widehat{\gamma}_{0i} - \gamma_{0i}\big) \;=\; & \mathbf{D}_0 - \mathbf{D}_0 B_0^*(\mathbf{t}_i)^T \mathbf{V}_i^{-1} B_0^*(\mathbf{t}_i) \mathbf{D}_0 \\
& + \mathbf{D}_0 B_0^*(\mathbf{t}_i)^T \mathbf{V}_i^{-1} B_0(\mathbf{t}_i) \Big[\sum_{i=1}^n B_0(\mathbf{t}_i)^T \mathbf{V}_i^{-1} B_0(\mathbf{t}_i) \Big]^{-1} \\
& \times B_0(\mathbf{t}_i)^T \mathbf{V}_i^{-1} B_0^*(\mathbf{t}_i) \mathbf{D}_0.
\end{aligned}
\tag{11.57}
$$

These expressions of the approximate variances are approximations of the general formulas given by Harville (1974).

11.3.2 Estimation with Unknown Covariance Matrix

The components $\mathbf{D}_0$ and σ of (11.10) are usually unknown in practice, so that the covariance matrix $\mathbf{V}_i$, hence $\mathbf{V}_i^{-1}$, is unknown. A practical approach in such situations is to estimate $\mathbf{D}_0$, σ and $\mathbf{V}_i$ from the data, and compute the estimators $\widehat{\gamma}_0$ and $\widehat{\gamma}_{0i}$ using (11.53) and (11.54) by substituting $\mathbf{V}_i$ with its consistent estimator. Using this approach, if we have consistent estimators $\widehat{\mathbf{D}}_0$ and $\widehat{\sigma}$ for $\mathbf{D}_0$ and σ, respectively, we can obtain a consistent variance-covariance estimator of $\mathbf{V}_i$ by

$$
\widehat{\mathbf{V}}_i = B_0^*(\mathbf{t}_i) \widehat{\mathbf{D}}_0 B_0^*(\mathbf{t}_i)^T + \widehat{\sigma}^2 I_{n_i},
\tag{11.58}
$$

and, thus, estimate $\mathbf{V}_i^{-1}$ by $\widehat{\mathbf{V}}_i^{-1}$ provided that $\widehat{\mathbf{V}}_i$ is invertible. Substituting $\mathbf{D}_0$ and $\widehat{\mathbf{V}}_i$ of (11.53) and (11.54) with $\widehat{\mathbf{D}}_0$ and $\widehat{\mathbf{V}}_i$, respectively, our practical approximate estimator of γ_0 is

$$
\widehat{\gamma}_0\big(\widehat{\mathbf{D}}_0, \widehat{\sigma}\big) = \Big[\sum_{i=1}^n B_0(\mathbf{t}_i)^T \widehat{\mathbf{V}}_i^{-1} B_0(\mathbf{t}_i) \Big]^{-1} \Big[\sum_{i=1}^n B_0(\mathbf{t}_i)^T \widehat{\mathbf{V}}_i^{-1} Y_i(\mathbf{t}_i) \Big],
\tag{11.59}
$$

and the practical BLUP of γ_{0i} is

$$
\widehat{\gamma}_{0i}\big(\widehat{\mathbf{D}}_0, \widehat{\sigma}\big) = \widehat{\mathbf{D}}_0 B_0^*(\mathbf{t}_i)^T \widehat{\mathbf{V}}_i^{-1} \Big[Y_i(\mathbf{t}_i) - B_0(\mathbf{t}_i) \widehat{\gamma}_0\big(\widehat{\mathbf{D}}_0, \widehat{\sigma}\big) \Big].
\tag{11.60}
$$

If $Y_i(\mathbf{t}_i) = Y_i^*(\mathbf{t}_i)$ in (11.9), then, as pointed out by Laird and Ware (1982, Section 3.2), $\widehat{\mathbf{D}}_0$, $\widehat{\sigma}$, $\widehat{\gamma}_0(\widehat{\mathbf{D}}_0, \widehat{\sigma})$ and $\widehat{\gamma}_{0i}(\widehat{\mathbf{D}}_0, \widehat{\sigma})$ can be simultaneously obtained by maximizing the joint likelihood function based on the marginal distribution of $\{Y_i(\mathbf{t}_i) : i = 1, \ldots, n\}$. Here, since $Y_i(\mathbf{t}_i) \approx Y_i^*(\mathbf{t}_i)$ in (11.9), we can think of $\widehat{\mathbf{D}}_0$, $\widehat{\sigma}$, $\widehat{\gamma}_0(\widehat{\mathbf{D}}_0, \widehat{\sigma})$ and $\widehat{\gamma}_{0i}(\widehat{\mathbf{D}}_0, \widehat{\sigma})$ as the approximated maximize likelihood estimators of $\mathbf{D}_0$, σ, γ_0 and γ_{0i}, respectively.

Estimation of the approximate variances of $\widehat{\gamma}_0(\widehat{\mathbf{D}}_0, \widehat{\sigma})$, $\widehat{\gamma}_{0i}(\widehat{\mathbf{D}}_0, \widehat{\sigma})$ and $[\widehat{\gamma}_{0i}(\widehat{\mathbf{D}}_0, \widehat{\sigma}) - \gamma_{0i}]$ can be naturally achieved by substituting $\mathbf{D}_0$ and $\mathbf{V}_i$ in (11.55), (11.56) and (11.57) with their estimates $\widehat{\mathbf{D}}_0$ and $\widehat{\mathbf{V}}_i$. In this case, the estimators of $Var\left[\widehat{\gamma}_0(\widehat{\mathbf{D}}_0, \widehat{\sigma})\right]$, $Var\left[\widehat{\gamma}_{0i}(\widehat{\mathbf{D}}_0, \widehat{\sigma})\right]$ and $Var\left[\widehat{\gamma}_{0i}(\widehat{\mathbf{D}}_0, \widehat{\sigma}) - \gamma_{0i}\right]$ have the expressions

$$Var\left[\widehat{\gamma}_0(\widehat{\mathbf{D}}_0, \widehat{\sigma})\right] = \left[\sum_{i=1}^{n} B_0(\mathbf{t}_i)^T \widehat{\mathbf{V}}_i^{-1} B_0(\mathbf{t}_i)\right]^{-1}, \tag{11.61}$$

$$Var\left[\widehat{\gamma}_{0i}(\widehat{\mathbf{D}}_0, \widehat{\sigma})\right] = \widehat{\mathbf{D}}_0 B_0^*(\mathbf{t}_i)^T \left\{\widehat{\mathbf{V}}_i^{-1} - \widehat{\mathbf{V}}_i^{-1} B_0(\mathbf{t}_i)\right. \tag{11.62}$$
$$\times \left[\sum_{i=1}^{n} B_0(\mathbf{t}_i)^T \widehat{\mathbf{V}}_i^{-1} B_0(\mathbf{t}_i)\right]^{-1} B_0(\mathbf{t}_i)^T \widehat{\mathbf{V}}_i^{-1}\left.\right\} B_0^*(\mathbf{t}_i) \widehat{\mathbf{D}}_0,$$

and

$$Var\left[\widehat{\gamma}_{0i}(\widehat{\mathbf{D}}_0, \widehat{\sigma}) - \gamma_{0i}\right] = \widehat{\mathbf{D}}_0 - \widehat{\mathbf{D}}_0 B_0^*(\mathbf{t}_i)^T \widehat{\mathbf{V}}_i^{-1} B_0^*(\mathbf{t}_i) \widehat{\mathbf{D}}_0$$
$$+ \widehat{\mathbf{D}}_0 B_0^*(\mathbf{t}_i)^T \widehat{\mathbf{V}}_i^{-1} B_0(\mathbf{t}_i) \left[\sum_{i=1}^{n} B_0(\mathbf{t}_i)^T \widehat{\mathbf{V}}_i^{-1} B_0(\mathbf{t}_i)\right]^{-1}$$
$$\times B_0(\mathbf{t}_i)^T \widehat{\mathbf{V}}_i^{-1} B_0^*(\mathbf{t}_i) \widehat{\mathbf{D}}_0, \tag{11.63}$$

respectively. The appropriateness of these variance estimators depend strongly on the adequacy of $\widehat{\mathbf{D}}_0$ and $\widehat{\mathbf{V}}_i$.

Computation of the approximated maximum likelihood estimators $\widehat{\mathbf{D}}_0$, $\widehat{\sigma}$, $\widehat{\mathbf{V}}_i$, $\widehat{\gamma}_0(\widehat{\mathbf{D}}_0, \widehat{\sigma})$ and $\widehat{\gamma}_{0i}(\widehat{\mathbf{D}}_0, \widehat{\sigma})$ as shown in (11.58), (11.59) and (11.60) can be accomplished using the expectation-maximization (EM) algorithm as described in Laird and Ware (1982, Section 4). Because our B-spline approximated model relies on $Y_i(t) \approx Y_i^*(t)$ in (11.5), these approximated maximum likelihood estimators computed through the EM algorithm are specific for the pre-specified B-spline basis functions $B_0(\mathbf{t}_i)$ and $B_0^*(\mathbf{t}_i)$ with fixed numbers of basis terms K_0 and K_0^*.

11.3.3   Individual Trajectories

As discussed above, we use the estimators $\widehat{\gamma}_0$ and $\widehat{\gamma}_{0i}$ defined in (11.53) and (11.54), if the covariance components $\mathbf{D}_0$ and σ are known, and we use the estimators (11.59) and (11.60), if $\mathbf{D}_0$ and σ are unknown. The corresponding estimate of the ith individual's outcome trajectory based on the B-spline approximated model (11.5) is then the smooth curve

$$\widehat{Y}_i(t) = \sum_{s=1}^{K_0} \widehat{\gamma}_{0s} B_{0s}(t) + \sum_{s=1}^{K_0^*} \widehat{\gamma}_{0si} B_{0s}(t) = B_0(t)\widehat{\gamma}_0 + B_0^*(t)\widehat{\gamma}_{0i}, \tag{11.64}$$

which we refer to herein as the B-spline BLUP trajectory of $Y_i(t)$. We note that (11.64) combines information from the entire sample and from the individual subject in that it uses the population covariance structure to estimate the B-spline coefficients and shrinks the curve toward the population mean. The

BLUP trajectory (11.64) is well defined even when the observations from a particular subject are too sparse to support an ordinary least square fit.

When the local pattern of a subject-specific trajectory at a time point t is of interest, the derivative of $\widehat{Y}_i(t)$ with respect to t, $\widehat{Y}_i'(t) = d\widehat{Y}_i(t)/dt$ which represents the ith subject's local slope at t, can be easily computed from (11.64) and has the expression

$$\widehat{Y}_i'(t) = \sum_{s=1}^{K_0} \widehat{\gamma}_{0s} B_{0s}'(t) + \sum_{s=1}^{K_0^*} \widehat{\gamma}_{0si} B_{0s}'(t) = B_0'(t)\,\widehat{\gamma}_0 + B_0^{*\prime}(t)\,\widehat{\gamma}_{0i}, \qquad (11.65)$$

which, similar to (11.64), is referred to as the BLUP derivative trajectory of $Y_i(t)$, where $B_{0s}'(t) = dB_{0s}(t)/dt$,

$$B_0'(t) = \left(B_{01}'(t), \ldots, B_{0K_0}'(t)\right)^T \text{ and } B_0'(t) = \left(B_{01}'(t), \ldots, B_{0K_0^*}'(t)\right)^T.$$

A positive or negative derivative trajectory $\widehat{Y}_i'(t)$ would give indications of whether the ith subject's outcome trajectory is locally increasing or decreasing, respectively, at t. On the other hand, a zero derivative of $\widehat{Y}_i(t)$ at t would suggest that the ith subject's outcome trajectory $Y_i(t)$ has a local maximum, a local minimum or a saddle point at t.

11.3.4   Cross-Validation Smoothing Parameters

For practical applications of (11.64) and (11.65), the number and locations of the knots for the splines corresponding to the population mean function and the subject-specific random effects have to be specified. The number and locations of the knots are determined by the B-spline basis functions $\{B_{0s}(t) : s = 1, 2, \ldots\}$ and the numbers of terms K_0 and K_0^*. The equation (1.37) of Section 1.5.2 gives the exact expressions of B-spline basis expansions. In general, it is computationally intensive to use B-splines with data-driven choices of number and locations of the knots. As discussed in Sections 4.1.2 and 9.2.4, for computational convenience, we often use subjectively chosen knot locations and only select the number of knots through a data-driven procedure. In practice, we can consider several sets of knot locations and choose the one which gives the "best" results. A possible option is to place more knots at time points where there are more observations or where there is large curvature in the response curve. Another subjective choice of knots location is to simply use "equally spaced knots" $\{\xi_1, \ldots, \xi_K\}$ where $\xi_l - \xi_{l-1} = \xi_{l+1} - \xi_l > 0$ for all $l = 1, \ldots, K$. We assume here that the B-splines have equally spaced knots, so that the numbers of basis terms K_0 and K_0^* in (11.64) determine the numbers and locations of the knots.

1. Leave-One-Observation-Out Cross-Validation

We presented in Chapters 4 and 9 the leave-one-subject-out cross-validation (LSCV) for selecting the smoothing parameters in B-spline estimators of the population-mean coefficient curves, in which the LSCV scores were constructed by deleting all the repeated measurements from each subject one at a time. Under the current framework of mixed-effects models, Shi, Weiss and Taylor (1996, Section 3) suggest a natural cross-validation procedure which evaluates the differences between the observed values Y_{ij} with the BLUP trajectories $\widehat{Y}_i(t_{ij})$ of (11.64) for all $i = 1, \ldots, n$ and $j = 1, \ldots, n_i$. Since $\widehat{Y}_i(t_{ij})$ depends on the observations of the ith subject, this cross-validation procedure is computed by deleting the subject's observations at each of the time points.

Let $\widehat{Y}^{(-ij)}_{iK_0K_0^*}(t_{ij})$ be the B-spline BLUP trajectory of $Y_i(t)$ at time point $t = t_{ij}$ computed from (11.64) with the observation Y_{ij} deleted. The cross-validation score based on $\widehat{Y}^{(-ij)}_{iK_0K_0^*}(t_{ij})$ is defined by

$$CV(K_0, K_0^*) = \sum_{i=1}^{n} \sum_{j=1}^{n_i} w_i \left[Y_{ij} - \widehat{Y}^{(-ij)}_{iK_0K_0^*}(t_{ij}) \right]^2, \tag{11.66}$$

which we refer to as the leave-one-observation-out cross-validation (LOCV), where the weight w_i satisfies $\sum_{i=1}^{n} \sum_{j=1}^{n_i} w_i = 1$ and, in practice, can be simply chosen to be $w_i = 1/N$. The cross-validated smoothing parameters are denoted by $(K_{0,cv}, K_{0,cv}^*)$ which minimizes the LOCV score (11.66), assuming that the minimizer of $CV(K_0, K_0^*)$ uniquely exists. The advantage of using (11.66) is that $(K_{0,cv}, K_{0,cv}^*)$ intuitively leads to the best BLUP trajectories among all choices of (K_0, K_0^*).

We note that, since $\widehat{Y}^{(-ij)}_{iK_0K_0^*}(t_{ij})$ still depends on the observations of the ith subject at time points other than t_{ij}, the LOCV score (11.66) depends on the squared difference of the correlated Y_{ij} and $\widehat{Y}^{(-ij)}_{iK_0K_0^*}(t_{ij})$. It is still not well understood whether the potential correlations between Y_{ij} and $\widehat{Y}^{(-ij)}_{iK_0K_0^*}(t_{ij})$ could significantly influence the theoretical properties of the LOCV smoothing parameters $(K_{0,cv}, K_{0,cv}^*)$. A main advantage of using (11.66) is that it is computationally straightforward as $CV(K_0, K_0^*)$ is equivalent to the cross-validation score for a linear model.

2. Likelihood-Based Leave-One-Subject-Out Cross-Validation

Since our general expectation for a suitable cross-validation is that the deleted data and the remaining data in the cross-validation scores should be uncorrelated, we present here a likelihood-based leave-one-subject-out cross-validation (LSCV) procedure, which is suggested by Rice and Wu (2001) for computing the data-driven choices of K_0 and K_0^*. Under the B-spline approximated model (11.9) and (11.10), the joint log-likelihood function for $\{Y_i(\mathbf{t}_i) : i = 1, \ldots, n\}$ is given by

$$\ell\big(K_0, K_0^*, \gamma_0, \sigma, \mathbf{D}_0\big)$$

$$= -\frac{1}{2} \sum_{i=1}^{n} \left[n_i \log(2\pi) + \log |\mathbf{V}_i| \right]$$

$$-\frac{1}{2} \sum_{i=1}^{n} \left[Y_i(\mathbf{t}_i) - B_0(\mathbf{t}_i)\gamma_0 \right]^T \mathbf{V}_i^{-1} \left[Y_i(\mathbf{t}_i) - B_0(\mathbf{t}_i)\gamma_0 \right]. \qquad (11.67)$$

To cross-validate the joint likelihood $\ell(K_0, K_0^*, \gamma_0, \sigma, \mathbf{D}_0)$, we denote by $\widehat{\gamma}_0^{(-i)}$, $\widehat{\sigma}^{(-i)}$ and $\widehat{\mathbf{D}}_0^{(-i)}$ the estimators of γ_0, σ and $\mathbf{D}_0$, respectively, obtained by applying the procedures of Section 11.3.1 to the remaining data with the entire observations of the ith subject deleted. The likelihood-based LSCV smoothing parameters $(K_{0,LSCV}, K_{0,LSCV}^*)$ are the maximizers of the LSCV likelihood function

$$\ell_{LSCV}(K_0, K_0^*) = \sum_{i=1}^{n} \ell\left(K_0, K_0^*, \widehat{\gamma}_0^{(-i)}, \widehat{\sigma}^{(-i)}, \widehat{\mathbf{D}}_0^{(-i)}\right), \qquad (11.68)$$

where $\ell\left(K_0, K_0^*, \widehat{\gamma}_0^{(-i)}, \widehat{\sigma}^{(-i)}, \widehat{\mathbf{D}}_0^{(-i)}\right)$ is computed with the observations of the ith subject deleted and the summation up to $n-1$ at the right side of (11.67). In practice, we can speed up the computation by considering M-fold cross-validation of (11.68), which is achieved by deleting M subjects from the data within each cross-validation step.

3. Other Variable Selection Procedures

Since the model (11.5) depends on approximation through a linear space spanned by the B-spline basis functions, the selection of (K_0, K_0^*) can also be viewed as a model selection problem in linear models. By treating (11.5) as a linear model, other variable selection methods for linear models can also be applied to select the suitable (K_0, K_0^*). In particular, the Akaike Information Criterion (AIC) and the Bayesian Information Criterion (BIC) have been used in Rice and Wu (2001) to select (K_0, K_0^*). We do not repeat the technical details of these methods here because they are all well-known standard variable selection methods for the linear regression models. However, the theoretical properties of these variable selection methods under the current longitudinal data with B-spline approximated models (11.5) have not been systematically investigated in the literature. The numerical results of Rice and Wu (2001) gave similar results for the cross-validation approaches of (11.66) and (11.68) and the variable selection approaches based on the AIC and the BIC.

11.4   Functional Principal Components Analysis

The estimation methods of the previous section all depend on approximation through pre-specified basis functions. This approach may involve a large number of parameters, which can lead to two potential problems. First, for sparse longitudinal data when the numbers of repeated measurements n_i are small, the estimation procedures of Section 11.3 may not be computed because the number of parameters is too large. Second, even if the parameter estimated, the large number of parameters may lead to over-fitting of the data, which lead to large variations of the parameter estimates. In order to address this problem of over-fitting, a number of authors, i.e., James, Hastie and Sugar (2000), Rice and Wu (2001) and Yao, Müller and Wang (2005a, 2005b), proposed a functional principal components analysis (FPCA) method based on the well-known Karhunen-Loève expansion which can be used to efficiently extract the features of the time-trends from the data. We describe here the FPCA method and its applications for the nonparametric mixed-effects model (11.1).

11.4.1   The Reduced Rank Model

Given that the main source of over-fitting is the excess number of parameters in the B-spline approximated mixed-effects model (11.5), a useful modification is to summarize the random effect term of (11.1) by a few principal component functions. This produces a low-rank and low-frequency approximation to the covariance structure through eigenfunction decomposition as described in Rice and Silverman (1991) without requiring the data to be regularly spaced. Eigenfunctions of the estimated covariance functions can be easily computed and can provide insight into the modes of variability present among the individual curves.

We assume that, instead of the approximation (11.6), the covariance function $Cov[Y(s), Y(t)]$ of $Y_i(t)$ in (11.1) can be expressed as an orthogonal expansion in terms of eigenfunctions $\phi_l(t)$ and non-increasing eigenvalues λ_l, such that

$$Cov[Y(s), Y(t)] = \sum_{l=1}^{\infty} \lambda_l \phi_l(s) \phi_l(t) \tag{11.69}$$

for all $s \neq t \in \mathcal{T}$, where $\sum_{l=1}^{\infty} \lambda_l < \infty$, $\lambda_1 \geq \lambda_2 \geq \cdots$,

$$\int \phi_l^2(t) dt = 1, \qquad \int \phi_{l_1}(t) \phi_{l_2}(t) dt = 0 \quad \text{for any } l_1 \neq l_2,$$

and λ_l and $\phi_l(t)$ are unknown and need to be estimated from the data. It directly follows from (11.69) that, for any $1 \leq l \leq \infty$, $\lambda_l \phi_l(t)$ is the projection of $Cov[Y(s), Y(t)]$ to the eigenfunction $\phi_l(t)$ in the sense that

$$\int_{\mathcal{T}} Cov[Y(s), Y(t)] \times \phi_l(s) ds = \lambda_l \phi_l(t). \tag{11.70}$$

Associated with the eigenfunction decomposition (11.69) is that the subject-specific deviation curve $\beta_{0i}(t)$ of (11.1) can be expressed as $\beta_{0i}(t) = \sum_{l=1}^{\infty} \xi_{il} \phi_l(t)$, so that (11.1) becomes

$$Y_i(t) = \beta_0(t) + \sum_{l=1}^{\infty} \xi_{il} \phi_l(t) + \varepsilon_i(t), \tag{11.71}$$

where ξ_{il} are uncorrelated random variables with $E(\xi_{il}) = 0$ and variance $E(\xi_{il}^2) = \lambda_l$, and ξ_{il} and $\varepsilon_i(t)$ are independent.

The main advantage of the above eigenfunction decompositions is that in practice it is only necessary to use the first few eigenfunctions to capture the main features in (11.71). The remaining terms can then be ignored because they have ignorable contributions to the time-trends of $Y(t)$. If we use the first $K_0^{**} \geq 1$ terms of eigenfunction decompositions in (11.69) and (11.70), we then obtain the reduced rank model for (11.70)

$$
\begin{cases}
Y_i(t) & = \ \beta_0(t) + \sum_{l=1}^{K_0^{**}} \xi_{il} \, \phi_l(t) + \varepsilon_i(t), \\
Cov\big[Y_i(s), Y_i(t)\big] & = \ \sum_{l=1}^{K_0^{**}} \lambda_l \, \phi_l(s) \, \phi_l(t), \ s \neq t \in \mathscr{T},
\end{cases}
\qquad (11.72)
$$

where K_0^{**} may be either subjectively selected or chosen from the data. By using the first few eigenfunctions and eigenvalues, we can reduce the model complexity from (11.70) to (11.72), extract the main features of the individual curves, and gain insights into the modes of variability present among the individual curves. Interpretations of the eigenfunctions and eigenvalues depend on the outcomes of the study. In general, the first eigenfunction corresponds to an overall shift of the outcome level, the second to a time-trend, and the third to a change of the time-trend. Based on these interpretations, we can characterize the features of an individual curve through its corresponding eigenvalues.

11.4.2 Estimation of Eigenfunctions and Eigenvalues

There are two broad approaches for the estimation of the eigenfunctions and eigenvalues of (11.72) in the literature. The first approach relies on substituting $Cov[Y(s), Y(t)]$ of (11.70) with a nonparametric estimator and computing the estimated eigenfunctions $\{\phi_l(t) : l = 1, \dots, K_0^{**}\}$ on a fine grid of time points. The eigenfunctions at any time $t \in \mathscr{T}$ are then estimated by the linear interpolations of the estimated values at the grid points. The second approach relies on approximating the eigenfunctions in (11.72) with some spline expansions and estimating the corresponding coefficients through a maximum likelihood procedure. The estimated eigenfunctions are obtained by plugging the estimated coefficients into the spline expansions. We now describe the details of these two estimation approaches and discuss their advantages and disadvantages in real applications.

1. Projection Estimators

In this approach, we first obtain a nonparametric smoothing estimator $\widehat{Cov}[Y(s), Y(t)]$ of $Cov[Y(s), Y(t)]$ for any $s \neq t \in \mathscr{T}$. Using the B-spline approximation (11.6) and the approximated maximum likelihood estimator $\widehat{\mathbf{D}}_0$ used in (11.58), the B-spline estimator of $Cov[Y(s), Y(t)]$ is

$$
\widehat{Cov}[Y(s), Y(t)] = \sum_{s_1=1}^{K_0^*} \sum_{s_2=1}^{K_0^*} \big[\widehat{d}_{0s_1 s_2} B_{0s_1}(s) B_{0s_2}(t) \big],
\qquad (11.73)
$$

where $\widehat{d}_{0s_1s_2}$ is the (s_1, s_2)th element of $\widehat{\mathbf{D}}_0$. Using the projection (11.70), estimate the eigenfunctions $\phi_l(t)$ and eigenvalues λ_l by the solutions $\widehat{\phi}_l(t)$ and $\widehat{\lambda}_l$ of the projection equation

$$\int_{\mathscr{T}} \widehat{Cov}[Y(s), Y(t)] \times \widehat{\phi}_l(s) \, ds = \widehat{\lambda}_l \, \widehat{\phi}_l(t). \tag{11.74}$$

Computation of (11.74) is carried out by discretizing the left-side integral into a summation over a fine grid of time points $\mathscr{T}_{grid}$ in $\mathscr{T}$ and obtaining the numerical solutions $\widehat{\lambda}_l$ and $\widehat{\phi}_l(t)$ over $t \in \mathscr{T}_{grid}$. The values of $\widehat{\phi}_l(t)$ at $t \notin \mathscr{T}_{grid}$ are the values of linear interpolations computed using $\{\widehat{\phi}_l(t) : t \in \mathscr{T}_{grid}\}$. The B-splines projection estimators (11.74) were described in Rice and Wu (2001).

In reality, spline estimators of $Cov[Y(s), Y(t)]$ are not the only choice, and other types of smoothing estimators can also be used in (11.74). Yao, Müller and Wang (2005a, 2005b) studied the projection estimators (11.74) based on a local linear surface smoothing estimator of $Cov[Y(s), Y(t)]$ and derived their asymptotic properties including consistency, convergence rates and asymptotic distributions. Since the projection methods based on local linear smoothing estimators are different from the basis approximation methods described in this chapter, we refer to Yao, Müller and Wang (2005a, 2005b) for the details of the local smoothing projection methods.

2. Basis Approximated Eigenfunction Estimators

The second approach for the estimation of $\phi_l(t)$ and λ_l is to consider B-spline approximations for the unknown functions of the reduced rank model (11.72). This approach was suggested by James, Hastie and Sugar (2000) as a viable alternative to the projection method (11.74) because of two reasons. First, the projection method (11.74) depends on nonparametric smoothing estimators of $Cov[Y(s), Y(t)]$ without effectively utilizing the fact that only the first K_0^{**} terms in the orthogonal eigenfunction expansions are needed in the reduced rank model (11.72). Second, the final product obtained from the projection method (11.74) is a unsmoothed linear interpolation estimator of $\phi_l(t)$ based on a subjectively chosen fine grid of time points. The approach here is to approximate the eigenfunctions $\phi_l(t)$ by B-splines so that smoothed eigenfunction estimators can be computed directly from the reduced rank model (11.72).

Following the procedure described in James, Hastie and Sugar (2000), we consider a spline basis $b(t)$ and parameters

$$\left\{ \begin{array}{rcl} b(t) & = & \left(b_1(t), \ldots, b_{K_0}(t)\right)^T, \\ \gamma & = & \left(\gamma_1, \ldots, \gamma_{K_0}\right)^T, \quad \xi_i = \left(\xi_{i1}, \ldots, \xi_{iK_0^{**}}\right)^T, \\ \theta_l & = & \left(\theta_{l1}, \ldots, \theta_{lK_0^{**}}\right)^T \text{ for } l = 1, \ldots, K_0, \\ \theta & = & \left(\theta_1^T, \ldots, \theta_{K_0}^T\right)^T, \end{array} \right. \tag{11.75}$$

which satisfy the following constraints

$$\boldsymbol{\theta}^T \boldsymbol{\theta} = I, \int_{\mathcal{T}} b^T(t) b(t) \, dt = 1 \text{ and } \int\int_{\mathcal{T}\times\mathcal{T}} b^T(t) b(s) \, dt \, ds = 0, \qquad (11.76)$$

where I is the $K_0^{**} \times K_0^{**}$ identity matrix. We can then approximate $\beta_0(t)$, $\phi_l(t)$ and $Cov[Y(s), Y(t)]$ by

$$\begin{cases} \beta_0(t) & \approx & \gamma^T b(t), \quad \phi_l(t) \approx \theta_l^T b(t), \\ Cov[Y_i(s), Y_i(t)] & \approx & b(s) \boldsymbol{\theta} \mathbf{D} \boldsymbol{\theta}^T b^T(t) + \sigma^2 1_{[s=t]}, \end{cases} \qquad (11.77)$$

respectively, where $\mathbf{D} = diag\{\lambda_1, \dots, \lambda_{K_0^{**}}\}$ is the diagonal matrix for the variances of ξ_i. The B-spline approximated reduced rank model has the form

$$\begin{cases} Y_i(t) & \approx & Y_i^*(t), \\ Y_i^*(t) & = & \gamma^T b(t) + \xi_i^T \boldsymbol{\theta}^T b(t) + \varepsilon_i(t), \\ \xi_i & \sim & N(\mathbf{0}, \mathbf{D}), \\ \varepsilon_i(t) & \sim & N(0, \sigma^2). \end{cases} \qquad (11.78)$$

Using the representation similar to (11.9), we have

$$Y_i(\mathbf{t}_i) \sim N\left(b(\mathbf{t}_i)\gamma, \ b(\mathbf{t}_i)\boldsymbol{\theta} \mathbf{D} \boldsymbol{\theta}^T b^T(\mathbf{t}_i) + \sigma^2 I\right), \qquad (11.79)$$

where $b(\mathbf{t}_i) = (b_1(\mathbf{t}_i), \dots, b_{K_0}(\mathbf{t}_i))$ and $b_l(\mathbf{t}_i) = (b_l(t_{i1}), \dots, b_l(t_{in_i}))^T$ for $l = 1, \dots, K_0$. Let $f[Y_i(\mathbf{t}_i); \gamma, \boldsymbol{\theta}, \mathbf{D}, \sigma, K_0, K_0^{**}]$ be the density of $Y_i(\mathbf{t}_i)$. The joint log-likelihood function for the data is

$$L(\gamma, \boldsymbol{\theta}, \mathbf{D}, \sigma, K_0, K_0^{**}) = \sum_{i=1}^n \log f[Y_i(\mathbf{t}_i); \boldsymbol{\theta}, \mathbf{D}, \sigma, K_0, K_0^{**}]. \qquad (11.80)$$

The B-spline approximated maximum likelihood estimators $\{\widehat{\gamma}, \widehat{\boldsymbol{\theta}}, \widehat{\mathbf{D}}, \widehat{\sigma}\}$ are then maximizers of (11.80), i.e.,

$$L(\widehat{\gamma}, \widehat{\boldsymbol{\theta}}, \widehat{\mathbf{D}}, \widehat{\sigma}, K_0, K_0^{**}) = \max_{\gamma, \boldsymbol{\theta}, \mathbf{D}, \sigma} L(\gamma, \boldsymbol{\theta}, \mathbf{D}, \sigma, K_0, K_0^{**}), \qquad (11.81)$$

provided that (11.80) can be uniquely maximized. Consequently, the estimated eigenfunctions and eigenvalues are

$$\begin{cases} \widehat{\phi}_l(t) & = & \widehat{\theta}_l \, b(t), \quad l = 1, \dots, K_0^{**}, \\ \widehat{\lambda}_l & = & \text{the } l\text{th diagonal element of } \widehat{\mathbf{D}}. \end{cases} \qquad (11.82)$$

By focusing only on the first K_0^{**} main components of $Cov[Y(s), Y(t)]$, the approximated maximum likelihood method (11.82) has the advantage over the projection method (11.74) of avoiding estimating the entire surface $Cov[Y(s), Y(t)]$. However, it has the main drawback of being computationally intensive because the maximization of (11.80) over $\{\gamma, \boldsymbol{\theta}, \mathbf{D}, \sigma\}$ is generally a difficult nonconvex optimization problem. Thus, both the projection method (11.74) and the approximated maximum likelihood method (11.82) may be used in practice.

11.4.3   Model Selection of Reduced Ranks

In practice, we would like to select K_0^{**} so that the model (11.72), on one hand, captures the main features of variation and, on the other, is succinct with fewer parameters. Since (11.72) relies on linear expansions, the selection of K_0^{**} is a model selection problem in linear models. We present here two practical approaches for the selection of K_0^{**}. The first approach is based on evaluating the proportion of the total variation explained by the corresponding eigenfunction, which gives a heuristic interpretation of the model. The second approach is variable selection through AIC, which evaluates the contribution of each eigenfunction to the likelihood function.

1. Proportion of the Total Variation

Since, for a given K_0^{**} in (11.69) and (11.72), the eigenvalues $\lambda_1 \geq \lambda_2 \geq \cdots \geq \lambda_{K_0^{**}}$ determine the covariance functions of $\{Y_i(s), Y_i(t)\}$, the proportion of λ_l over $\sum_{l=1}^{K_0^{**}} \lambda_l$,

$$PTV_l = \frac{\lambda_l}{\sum_{l=1}^{K_0^{**}} \lambda_l}, \tag{11.83}$$

is a natural measure of the proportion of the total variation contributed by the eigenfunction $\phi_l(t)$. The main features of the individual curve of $Y_i(t)$ are captured by the first few terms with large PTV_l values, while the remaining terms can be ignored because they only have minor contributions to the total variation. In practice, we can start by fitting a model with a relative larger value of K_0^{**}, examine the values of $PTV_1 \geq PTV_2 \geq \cdots \geq PTV_{K_0^{**}}$, and select a submodel with a smaller value of K_0^{**}. The numerical applications of James, Hastie and Sugar (2000) and Rice and Wu (2001) demonstrate that this approach, although somewhat subjective, provides useful insights into the features of individual curves.

2. Model Selection by AIC

This approach is built on the idea that (11.78) is approximately a linear model, so that the likelihood function (11.81) is nondecreasing as K_0^{**} increases. The contribution of an additional eigenfunction can then be examined by the increase of this likelihood function when K_0^{**} is increased by one. This suggests that the well-known model selection procedures can be applied to select an appropriate K_0^{**}. When the Akaike Information Criterion (AIC), e.g., Akaike (1970), is used to (11.79), we would like to find K_0^{**} which maximizes

$$AIC(K_0^{**}) = -L(\widehat{\gamma}, \widehat{\theta}, \widehat{\mathbf{D}}, \widehat{\sigma}, K_0, K_0^{**}) + K_0^{**}, \tag{11.84}$$

where $L(\widehat{\gamma}, \widehat{\theta}, \widehat{\mathbf{D}}, \widehat{\sigma}, K_0, K_0^{**})$ is given in (11.81). Although the theoretical properties of (11.84) are still not well understood, the AIC approach has been shown by Rice and Wu (2001) and Yao, Müller and Wang (2005a) to lead to interpretable results in real applications.

11.5   Estimation and Prediction with Covariates

We now consider the estimation based on the B-spline approximated models of Section 11.2.3 with a univariate time-varying covariate. The case of multiple covariates in Section 11.2.4 is a direct extension of the method presented here. We omit the detailed description of the multiple covariate case, since it is straightforward and requires more complex notation. We note that it is still not well understood how the functional principal components analysis methods of Section 11.4 could be adequately extended to the models with time-varying covariates.

11.5.1   Models without Covariate Measurement Error

As a simple extension of the model (11.1), we first consider the estimation of the population-mean and subject-specific coefficient curves $\beta_0(t)$, $\beta_1(t)$ and $\beta_{1i}(t)$ of the model (11.11) or, equivalently, (11.14). Since in (11.14) the covariate $X_i(t_i)$ is observed without measurement error, it is suggested in Liang, Wu and Carroll (2003, Section 3.2) that $\beta_0(t)$, $\beta_1(t)$ and $\beta_{1i}(t)$ can be estimated by applying the procedures of Section 11.3.1 to the B-spline approximated model (11.20). Following the framework of (11.20), let

$$
\begin{cases}
\mathscr{X}_i(t) & = & \left(B_{01}(t), \ldots, B_{0K_0}(t), B_{11}(t)X_i(t), \ldots, B_{1K_1}(t)X_i(t)\right)^T, \\
\mathscr{Z}_i(t) & = & \left(B_{11}(t)X_i(t), \ldots, B_{1K_1^*}(t)X_i(t)\right)^T, \\
\gamma & = & (\gamma_0^T, \gamma_1^T)^T.
\end{cases}
\tag{11.85}
$$

We can rewrite the approximated model (11.20) as

$$
\begin{cases}
Y_i(t) & \approx & \gamma^T \mathscr{X}_i(t) + \gamma_{1i}^T \mathscr{Z}_i(t) + \varepsilon_i(t), \\
\gamma_{1i} & \sim & N(\mathbf{0}, \mathbf{D}_1), \\
\varepsilon_i(t) & \sim & N(0, \sigma^2).
\end{cases}
\tag{11.86}
$$

At the observed time points $\mathbf{t}_i$, we define

$$
\begin{cases}
\mathscr{X}_i(\mathbf{t}_i) & = & \left(\mathscr{X}_i(t_{i1}), \ldots, \mathscr{X}_i(t_{in_i})\right)^T, \\
\mathscr{Z}_i(\mathbf{t}_i) & = & \left(\mathscr{Z}_i(t_{i1}), \ldots, \mathscr{Z}_i(t_{in_i})\right)^T, \\
\varepsilon_i(\mathbf{t}_i) & = & \left(\varepsilon_i(t_{i1}), \ldots, \varepsilon_i(t_{in_i})\right),
\end{cases}
\tag{11.87}
$$

and write the model (11.86) as

$$
\begin{cases}
Y_i(\mathbf{t}_i) & \approx & \mathscr{X}_i(\mathbf{t}_i)\gamma + \mathscr{Z}_i(\mathbf{t}_i)\gamma_{1i} + \varepsilon_i(\mathbf{t}_i), \\
\gamma_{1i} & \sim & N(\mathbf{0}, \mathbf{D}_1), \\
\varepsilon_i(t_{ij}) & \sim & N(0, \sigma^2), \quad j = 1, \ldots, n_i.
\end{cases}
\tag{11.88}
$$

The covariance matrix of $Y_i(\mathbf{t}_i)$ under (11.86) can be approximated by

$$
Cov\left[Y_i(\mathbf{t}_i)\right] \approx \mathscr{V}_i = \mathscr{Z}_i(\mathbf{t}_i)\mathbf{D}_1 \mathscr{Z}_i(\mathbf{t}_i)^T + \sigma^2 I_{n_i}.
\tag{11.89}
$$

If $\mathbf{D}_1$ and σ are known $\mathscr{V}_i$ is nonsingular, the same approach as (11.53) and (11.54) leads to the population-mean estimator

$$\widehat{\gamma} = \left[\sum_{i=1}^{n} \mathscr{X}_i(\mathbf{t}_i)^T \mathscr{V}_i^{-1} \mathscr{X}_i(\mathbf{t}_i)\right]^{-1} \left[\sum_{i=1}^{n} \mathscr{X}_i(\mathbf{t}_i) \mathscr{V}_i^{-1} Y_i(\mathbf{t}_i)\right] \qquad (11.90)$$

and the BLUP random-effect estimator

$$\widehat{\gamma}_{1i} = \mathbf{D}_1 \mathscr{Z}_i(\mathbf{t}_i)^T \mathscr{V}_i^{-1} \left[Y_i(\mathbf{t}_i) - \mathscr{X}_i(\mathbf{t}_i)\widehat{\gamma}\right]. \qquad (11.91)$$

Under the general situation that the covariance structures $\mathbf{D}_1$ and σ are unknown, the parameters in (11.89) can be estimated by maximizing the joint likelihood function based on the marginal distribution of $\{Y_i(\mathbf{t}_i) : i = 1, \ldots, n\}$, as illustrated in Laird and Ware (1982, Section 3.2). Let $\widehat{\mathbf{D}}_1$ and $\widehat{\sigma}$ be the consistent estimators of $\mathbf{D}_1$ and σ, respectively, obtained by maximizing the joint likelihood function of (11.89). We can express the maximum likelihood estimator of $\mathscr{V}_i$ as

$$\widehat{\mathscr{V}}_i = \mathscr{Z}_i(\mathbf{t}_i) \widehat{\mathbf{D}}_1 \mathscr{Z}_i(\mathbf{t}_i)^T + \widehat{\sigma}^2 I_{n_i}. \qquad (11.92)$$

Substituting $\mathscr{V}_i$ of (11.90) and (11.91) with $\widehat{\mathscr{V}}_i$, we obtain the maximum likelihood estimator of the population-mean and random effect as

$$\widehat{\gamma}(\widehat{\mathbf{D}}_1, \widehat{\sigma}) = \left[\sum_{i=1}^{n} \mathscr{X}_i(\mathbf{t}_i)^T \widehat{\mathscr{V}}_i^{-1} \mathscr{X}_i(\mathbf{t}_i)\right]^{-1} \left[\sum_{i=1}^{n} \mathscr{X}_i(\mathbf{t}_i) \widehat{\mathscr{V}}_i^{-1} Y_i(\mathbf{t}_i)\right] \qquad (11.93)$$

and

$$\widehat{\gamma}_{1i}(\widehat{\mathbf{D}}_1, \widehat{\sigma}) = \widehat{\mathbf{D}}_1 \mathscr{Z}_i(\mathbf{t}_i)^T \widehat{\mathscr{V}}_i^{-1} \left[Y_i(\mathbf{t}_i) - \mathscr{X}_i(\mathbf{t}_i)\widehat{\gamma}(\widehat{\mathbf{D}}_1, \widehat{\sigma})\right], \qquad (11.94)$$

respectively.

11.5.2   Models with Covariate Measurement Error

For the case that the covariate $X_i(t)$ is measured with error, we consider the estimation of the population-mean and subject-specific coefficient curves $\beta_0(t)$, $\beta_1(t)$ and $\beta_{1i}(t)$ of the model (11.27) or, equivalently, (11.30), which includes the model (11.11) as a special case. Using the B-spline approximated model (11.34), a practical method for the estimation of $\beta_0(t)$, $\beta_1(t)$ and $\beta_{1i}(t)$, which is suggested by Liang, Wu and Carroll (2003, Section 3.2), is to use the two-step procedure described below.

Two-Step B-Spline Estimation Procedure:

(a) **Measurement Error Calibration.** *Applying the estimation procedures (11.59) and (11.60) to the B-spline approximated model $X_i(t) = X_i^*(t) \approx \xi^T B_x(t) + \xi_i^T B_x^*(t)$ in (11.34), we estimate the population-mean and subject-specific coefficients ξ and ξ_i by*

$$\widehat{\xi}(\widehat{\Sigma}_\xi, \widehat{\sigma}_u) = \left[\sum_{i=1}^{n} B_x(\mathbf{t}_i)^T \widehat{\mathbf{V}}_{X_i}^{-1} B_x(\mathbf{t}_i)\right]^{-1} \left[\sum_{i=1}^{n} B_x(\mathbf{t}_i)^T \widehat{\mathbf{V}}_{X_i}^{-1} Y_i(\mathbf{t}_i)\right], \qquad (11.95)$$

and

$$\widehat{\xi}_i(\widehat{\Sigma}_\xi, \widehat{\sigma}_u) = \widehat{\Sigma}_\xi \, B_x^*(\mathbf{t}_i)^T \, \widehat{\mathbf{V}}_{X_i}^{-1} \left[Y_i(\mathbf{t}_i) - B_x(\mathbf{t}_i) \, \widehat{\xi}(\widehat{\Sigma}_\xi, \widehat{\sigma}_u) \right], \qquad (11.96)$$

respectively, where, by (11.58),

$$\widehat{\mathbf{V}}_{X_i} = B_x^*(\mathbf{t}_i) \, \widehat{\Sigma}_\xi \, B_x^*(\mathbf{t}_i)^T + \widehat{\sigma}_u^2 \, I_{n_i} \qquad (11.97)$$

is the estimator of the covariance matrix $\mathbf{V}_{X_i} = Cov\left[X_i^*(\mathbf{t}_i)\right]$ *and* $\widehat{\Sigma}_\xi$ *and* $\widehat{\sigma}_u$ *are the consistent estimators of* Σ_ξ *and* σ_u *as described in Section 11.3.1. Using the estimators of (11.85) and (11.86), we compute the estimated individual's trajectory*

$$\widehat{X}_i(t) = B_x(t) \, \widehat{\xi} + B_x^*(t) \, \widehat{\xi}_i, \qquad (11.98)$$

of $X_i(t)$ *using the BLUP formula (11.64).*

(b) Regression Curve Estimation. *Substituting* $X_i^*(t)$ *of (11.34) with* $\widehat{X}_i(t)$*, we approximate (11.34) by the approximated model*

$$\begin{cases} Y_i(t) & \approx \gamma_0^T \, B_0(t) + \left[\gamma_1^T \, B_1(t) + \gamma_{1i}^T B_1^*(t) \right] \widehat{X}_i(t) + \varepsilon_i(t), \\ \gamma_{1i} & \sim N(\mathbf{0}, \mathbf{D}_1), \\ \varepsilon_i(t) & \sim N(0, \sigma^2). \end{cases} \qquad (11.99)$$

Similar to (11.85) and (11.86), we define

$$\begin{cases} \widehat{\mathscr{X}}_i(t) & = \left(B_{01}(t), \ldots, B_{0K_0}(t), B_{11}(t) \widehat{X}_i(t), \ldots, B_{1K_1}(t) \widehat{X}_i(t) \right)^T, \\ \widehat{\mathscr{Z}}_i(t) & = \left(B_{11}(t) \widehat{X}_i(t), \ldots, B_{1K_1^*}(t) \widehat{X}_i(t) \right)^T, \\ \gamma & = \left(\gamma_0^T, \gamma_1^T \right)^T, \end{cases}$$
$$\qquad (11.100)$$

and rewrite the approximated model (11.99) as

$$\begin{cases} Y_i(t) & \approx \gamma^T \, \widehat{\mathscr{X}}_i(t) + \gamma_{1i}^T \, \widehat{\mathscr{Z}}_i(t) + \varepsilon_i(t), \\ \gamma_{1i} & \sim N(\mathbf{0}, \mathbf{D}_1), \\ \varepsilon_i(t) & \sim N(0, \sigma^2). \end{cases} \qquad (11.101)$$

If $\mathbf{D}_1$ *and* σ *are known, we substitute* $\mathscr{X}_i(t)$ *and* $\mathscr{Z}_i(t)$ *in (11.90) and (11.91) with* $\widehat{\mathscr{X}}_i(t)$ *and* $\widehat{\mathscr{Z}}_i(t)$*, and obtain the population-mean and BLUP estimators*

$$\begin{cases} \widehat{\gamma} & = \left[\sum_{i=1}^n \widehat{\mathscr{X}}_i(\mathbf{t}_i)^T \, \mathscr{V}_i^{-1} \, \widehat{\mathscr{X}}_i(\mathbf{t}_i) \right]^{-1} \left[\sum_{i=1}^n \widehat{\mathscr{X}}_i(\mathbf{t}_i) \, \mathscr{V}_i^{-1} \, Y_i(\mathbf{t}_i) \right], \\ \widehat{\gamma}_{1i} & = \mathbf{D}_1 \, \widehat{\mathscr{Z}}_i(\mathbf{t}_i)^T \, \mathscr{V}_i^{-1} \left[Y_i(\mathbf{t}_i) - \widehat{\mathscr{X}}_i(\mathbf{t}_i) \, \widehat{\gamma} \right]. \end{cases} \qquad (11.102)$$

If $\mathbf{D}_1$ *and* σ *are unknown, we compute the consistent estimators* $\widehat{D}_1$ *and* $\widehat{\sigma}$ *of* $\mathbf{D}_1$ *and* σ*, respectively, by maximizing the joint likelihood function of*

(11.101), and obtain the population-mean and BLUP estimators

$$
\left\{
\begin{aligned}
\widehat{\mathscr{V}}_i &= \mathscr{L}_i(\mathbf{t}_i)\,\widehat{\mathbf{D}}_1\,\mathscr{L}_i(\mathbf{t}_i)^T + \widehat{\sigma}^2 I_{n_i}, \\
\widehat{\gamma}(\widehat{\mathbf{D}}_1, \widehat{\sigma}) &= \left[\sum_{i=1}^n \widehat{\mathscr{X}}_i(\mathbf{t}_i)^T\,\widehat{\mathscr{V}}_i^{-1}\,\widehat{\mathscr{X}}_i(\mathbf{t}_i)\right]^{-1} \\
&\quad \times \left[\sum_{i=1}^n \widehat{\mathscr{X}}_i(\mathbf{t}_i)\,\widehat{\mathscr{V}}_i^{-1}\,Y_i(\mathbf{t}_i)\right], \\
\widehat{\gamma}_{1i}(\widehat{\mathbf{D}}_1, \widehat{\sigma}) &= \widehat{\mathbf{D}}_1\,\widehat{\mathscr{X}}_i(\mathbf{t}_i)^T\,\widehat{\mathscr{V}}_i^{-1}\left[Y_i(\mathbf{t}_i) - \widehat{\mathscr{X}}_i(\mathbf{t}_i)\,\widehat{\gamma}(\widehat{\mathbf{D}}_1, \widehat{\sigma})\right],
\end{aligned}
\right.
\tag{11.103}
$$

by substituting $\mathscr{V}_i$ of (11.101) with $\widehat{\mathscr{V}}_i$. □

The main idea of the above procedure is to first substitute the $X_i(t)$ in (11.34) with its estimated subject-specific trajectory and then estimate the coefficient curves for $Y_i(t)$ using the maximum likelihood or restricted maximum likelihood approach. Extension of this procedure to the mixed-effects varying-coefficient model (11.42) with multiple time-varying covariates is straightforward. In this case, we compute the subject-specific trajectory for each of the covariate curve $X_i^{(l)}(t)$, $1 \le l \le k$, and proceed to step (b) by substituting $\{X_i^{(l)}(t) : l = 1, \ldots, k\}$ with their estimated trajectories.

11.6 R Implementation

11.6.1 The BMACS CD4 Data

The BMACS CD4 data has been described in Section 1.2 and analyzed in previous chapters. In Sections 3.5.2 and 4.3.2, we applied the unstructured local smoothing method and the B-spline based global smoothing method to estimate the population mean time curve of CD4 percentage after the HIV infection, but the subject-specific time-trends of CD4 percentage were not considered. In this section, we use this dataset to illustrate how to model the subject-specific CD4 percentage curves using B-spline functions with random coefficients. Consequently, as shown in Shi, Weiss and Taylor (1996) and Rice and Wu (2001), we estimate the variance and covariance functions using their approximations by tensor products of B-splines and demonstrate the usefulness of these variance-covariance estimates in real applications.

As shown in Figure 1.1 of Section 1.2.2, the CD4 percentage observations are noisy and sparse, and the dynamic features are not visually apparent except for a decreasing trend over time since HIV infection. Therefore, it is desirable to apply nonparametric methods, such as a spline approximated nonparametric mixed-effects model, since there are no natural *a priori* parametric models for estimating the mean, variance and covariance functions of these repeatedly measured curves. We consider in this example the simple model (11.1) without covariates other than time. The model with time-varying covariates is considered in the next example.

Under the assumption that both the population-mean CD4 percentage curve $\beta_0(t)$ and subject-specific deviation curve $\beta_{0i}(t)$ of the model (11.1) are smooth functions of time t, we use B-spline basis with fixed knots to approximate $\beta_0(t)$ and $\beta_{0i}(t)$ nonparametrically. The B-spline approximated nonparametric mixed-effects model is given by (11.5), where the covariance structure of the random curve $Y(t)$ modeled by (11.6) and the subject-specific trajectories estimated by (11.64). When K_0 and K_0^* are fixed, the model (11.5) is a classical linear mixed-effects model with the design matrix formed by B-splines. Using the methodology and algorithm developed for the linear mixed-effects models (e.g., Bates et al., 2015), we can then obtain the estimates of the spline coefficients, the variance-covariance parameters, and the best linear unbiased prediction (BLUP) estimates of the random effect and the outcome trajectory of individual subjects. For simplicity, we used B-splines with equally spaced knots in this example with the degree and number of knots chosen by model selection criteria, such as the Akaike Information Criterion (AIC). Other methods based on the cross-validation procedures may also be used.

The following R commands are used to compute the estimated fixed- and random-effects based on the quadratic spline with two equally spaced interior knots, which were selected by the AIC:

```
> library(lme4)
> data(BMACS)
> T.range<- range(BMACS$Time)
> Nk <- 4
> KN <- seq(from=T.range[1], to=T.range[2], length=Nk)[-c(1,Nk)]
> bs.time <- bs(BMACS$Time, knots=KN, degree=2, intercept=T)
> fmCD4   <- lmer(CD4 ~ 0+bs.time+(0+bs.time|ID), data=BMACS)
> summary(fmCD4)

Linear mixed model fit by REML ['lmerMod']
Formula: CD4 ~ 0 + bs.time + (0 + bs.time | ID)
   Data: BMACS

REML criterion at convergence: 12046.4

Scaled residuals:
    Min      1Q  Median      3Q     Max
-4.6213 -0.4912 -0.0250  0.4826  4.4676

Random effects:
 Groups   Name      Variance Std.Dev. Corr
 ID       bs.time1   94.04    9.697
          bs.time2   93.40    9.664   0.53
          bs.time3  127.84   11.307   0.46 0.59
          bs.time4  178.23   13.350   0.25 0.54 0.80
```

```
          bs.time5 198.19    14.078    0.20 0.59 0.81 0.99
 Residual               20.34    4.510
Number of obs: 1817, groups:  ID, 283

Fixed effects:
         Estimate Std. Error t value
bs.time1  36.4274     0.7116   51.19
bs.time2  32.3948     0.7340   44.14
bs.time3  25.3571     0.8417   30.13
bs.time4  22.4103     1.1668   19.21
bs.time5  21.4717     1.3303   16.14


Correlation of Fixed Effects:
          bs.tm1 bs.tm2 bs.tm3 bs.tm4
bs.time2 0.170
bs.time3 0.358  0.203
bs.time4 0.091  0.394  0.387
bs.time5 0.096  0.278  0.531  0.548
```

Using the model output for the estimated fixed- and random-effects param-eters, we then compute the curve for the mean CD4 percentage, and estimate the covariance, the corresponding eigenvalues, eigenfunctions and the individual trajectories on a given grid of time points (Sections 11.3.3 and 11.4.2). These estimates are obtained using the following R commands:

```
> N <- 50
> Tgrid <- seq(from=T.range[1], to=T.range[2], length=N)
> BS <- bs(Tgrid, knots=KN, degree=2,  intercept=T)
# obtain the mean curve estimate
> mean.hat <- BS %*% fixef(fmCD4)


# obtain the estimate of covariance for random effects and
# outcome
> GAMMA0 <- matrix(NA, 5, 5)
> VC <- as.data.frame(VarCorr(fmCD4))
> diag(GAMMA0) <- VC$vcov[1:5]
> GAMMA0[lower.tri(GAMMA0)] <- VC$vcov[6:15]
> GAMMA <- t(GAMMA0)
> GAMMA[lower.tri(GAMMA)] <- GAMMA0[lower.tri(GAMMA0)]


> COV.MAT <- matrix(NA, N, N)
> Index <- expand.grid(1:N, 1:N)
> f <- function(i){sum( GAMMA * (BS[Index[i,1],]
                         %o% BS[Index[i,2],]))}
```

```
> COV.MAT <- matrix(do.call('rbind', lapply(1:(N^2),f)),
                    nrow=N, byrow=F)

# obtain the eigenvalues and eigenvectors
> Eigenfun <-  eigen(COV.MAT)

> Nsub <- 283
> IDlevel<-rownames(coef(fmCD4)[[1]])
> BLUP.est <- matrix(NA, nrow= Nsub, ncol= N)
> Proj1 <- Proj2  <- numeric(Nsub)
> for (i in 1:Nsub)
  {
    Datai          <- BMACS[BMACS$ID==IDlevel[i],]
    BLUP.est[i,] <- BS %*% t(as.vector(coef(fmCD4)[[1]][i,]))
    Proj1[i] <- BLUP.est[i,] %*% Eigenfun$vectors[,1]
    Proj2[i] <- BLUP.est[i,] %*% Eigenfun$vectors[,2]
  }
```

Figure 11.1(A) shows the estimated population mean curve along with the observed individual trajectories of CD4 percentages over time. The estimated covariance function and the corresponding first and second eigenfunctions are shown in Figures 11.1(B)-(D), respectively. Interpretation for the eigenfunctions have been discussed in Section 11.4 and the literature, e.g., Rice and Silverman (1991) and Rice and Wu (2001). In this case, we note that the first eigenfunction corresponds to the overall level of CD4 percentage, while the second eigenfunction corresponds to the general trend and direction for the change of CD4 percentage over time. Since the first four eigenfunctions in our numerical results account for 85%, 11%, 2% and 1% of the total variability, we only consider the first two eigenfunctions, which explain most of the variability.

In practice, the eigenvalues can be useful to explore and identify the extreme cases explained by the corresponding eigenfunctions. Here, we calculate the individual projections into the directions of the eigenfunctions by the inner products of the BLUP estimates and the eigenfunctions. By identifying the extreme cases or outliers of the obtained eigenvalues, subjects with unusual CD4 percentage trajectories can be quickly singled out from a group of study subjects. Without examining the eigenvalues, it can be very challenging to visually identify the subjects with "unusual" outcome trajectories from the observation plots because of the irregular sampling and substantial noise among a large number of curves.

Figure 11.2 shows four subjects with extreme projections on the first two eigenfunctions. Although the population mean curve of CD4 percentage shows a decreasing trend over time with relatively low curvature, the BLUP estimates of the individual smoothed curves based on the nonparametric mixed-effects

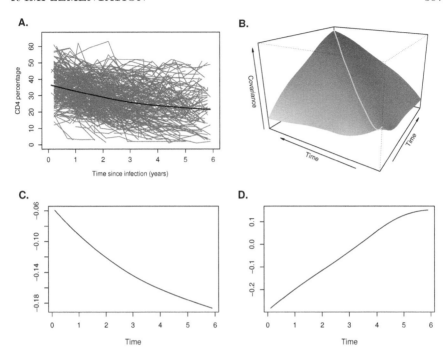

Figure 11.1 *(A) Estimated mean function and individual trajectories of 283 sequences of CD4 percentages. (B) Estimated covariance function for CD4 percentages. The first eigenfunction (C) and the second eigenfunction (D) of the estimated covariance function.*

model capture the subject-specific deviations from the mean CD4 percentage curve and fit the subjects' observations quite well. For example, the top row of Figure 11.2 shows two most extreme cases in the direction of the first eigenfunction. The subject shown in Figure 11.2(A) has the smallest projection, which is associated with the subject's very high levels of CD4 percentage over time, while the subject shown in Figure 11.2(B) has extremely large projection, which is associated with the corresponding low levels of CD4 percentage. Compared with the estimated population mean curve of CD4 percentage, the CD4 percentage trajectories of these two subjects have dramatic shifts from the population mean curve in opposite directions. The outcome trajectories in Figures 11.2(C)-(D) are from two subjects with extreme projection on the second eigenfunction. The subject in Figure 11.2(C) has extremely large projection (on the second eigenfunction), which, contrary to most other subjects, gives an increasing trend in the subject's outcome trajectory. On the other hand, the subject in Figure 11.2(D) has the smallest projection, which leads to a much faster declining CD4 percentage trend compared to other subjects in the sample. Because the levels of CD4 percentage are associated with the strength and functioning of the subject's immune system, the two sub-

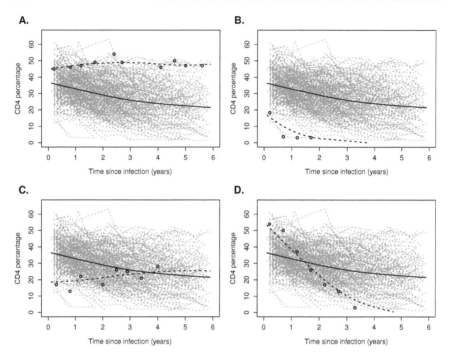

Figure 11.2 *The observations (dots) and smoothed curves (dashed lines) of four subjects with extreme projections on the first two eigenfunctions. Then mean curve (solid line) and all individual trajectories (gray dashed lines) are also displayed in each case.*

jects shown in Figures 11.2(B) and (D), who have either very low levels of CD4 percentage or a rapidly declining CD4 percentage time-trend, would be at high risk of worse clinical outcomes. Thus, these flexible nonparametric mixed-effects modeling and the associated curve exploration technique may provide helpful insight into disease risk classification and options for individual treatment decisions.

11.6.2   The NGHS BP Data

We illustrate here how to apply the nonparametric mixed-effects models of Section 11.2 to evaluate the covariate effects and estimate the subject-specific smoothed curves using spline approximation. The NGHS data has been described in Section 1.2. In Chapters 7 to 9, we applied the varying-coefficient model and the smoothing methods to estimate the baseline curve of systolic blood pressure (SBP) and the time-varying covariate effects. These models did not address the individual SBP trajectories. We present the special case of the model (11.51) with SBP as the outcome variable and two covariates, race and height percentiles, in which the actual height measurements are subject to measurement error.

For simplicity in the modeling and interpretation, we consider the special case of (11.51) with only random effect or subject-specific variations for the baseline curve. The model with subject-specific variation for the covariate effects can be easily applied by including additional random terms in the model. For the ith NGHS participant at t years of age, we denote by $Y_i(t)$, $X_i^{(1)}$, $X_i^{(2)}(t)$ and $\varepsilon_i(t)$ the subject's SBP level, race, height percentile and measurement error, respectively. Here, $X_i^{(1)} = 0$ if the girl is Caucasian, $X_i^{(1)} = 1$ if she is African American, and $X_i^{(2)}(t)$ is the age-adjusted height percentile as calculated in Section 7.4.1. Our varying-coefficient mixed-effects model, as a special case of (11.46), for $\{Y_i(t), X_i^{(1)}, X_i^{(2)}(t)\}$ is

$$\begin{cases} Y_i(t) &= \beta_0(t) + \beta_{0i}(t) + \beta_1(t)X_i^{(1)} + \beta_2(t)X_i^{(2)}(t) + \varepsilon_i(t), \\ W_i(t) &= X_i^{(2)}(t) + u_i, \end{cases} \quad (11.104)$$

where u_i is the mean zero measurement error and $W_i(t)$ is the observed age-adjusted height percentile. The B-spline approximations of (11.104) become

$$\begin{cases} X_i^{(2)}(t) &\approx \sum_{s=1}^{p} \xi_s B_s(t) + \sum_{s=1}^{p} \xi_{si} B_s(t), \\ W_i(t) &= X_i^{(2)} + u_i, \ u_i \sim N(0, \sigma_u^2), \\ Y_i(t) &= \sum_{s=1}^{K_0} \gamma_{0s} B_s(t) + \sum_{s=1}^{K_0^*} \gamma_{0si} B_s(t) \\ &\quad + \sum_{s=1}^{K_1} \gamma_{1s} B_s(t) X_i^{(1)} + \sum_{s=1}^{K_2} \gamma_{2s} B_s(t) X_i^{(2)}(t) + \varepsilon_i(t), \\ \varepsilon_i(t) &\sim N(0, \sigma^2), \ \gamma_{0i}^T = (\gamma_{01i}, \dots, \gamma_{0K_0^* i})^T \sim N(\mathbf{0}, \mathbf{D}). \end{cases} \quad (11.105)$$

Applying the procedures in Section 11.3, we use the following two-step procedure to correct the measurement error of height percentile in (11.105). First, we use the B-spline approximated model for $\{X_i^{(2)}(t), t\}$ and (11.64) to obtain the BLUP individual trajectories $\widehat{X}_i^{(2)}(t)$. We use the cubic spline with three equally spaced interior knots at 11.5, 14.5 and 16.5 years of age (the two boundary knots are at 9 and 19 years of age). Because the cubic spline fits the observed height percentiles very well, the BLUP height percentile trajectories closely match the individual observations and changes over time. Second, we substitute the $X_i^{(2)}(t)$ with the individual trajectory $\widehat{X}_i^{(2)}(t)$ in (11.105) to estimate the covariate effects $\{\beta_l(t), \ \beta_{li}(t) : l = 0, 1, 2; i = 1, \dots, n\}$ and the smoothed individual SBP trajectories.

The calculations for the coefficient curves and individual trajectories can be carried out using following R commands:

```
> KN1 <- seq(from=9, to=19, length=5)[-c(1,5)]
> Bs.age <- bs(NGHS$AGE, knots=KN1)
> fm.Ht <- lmer(HTPCT ~ 1+ Bs.age +(1+ Bs.age|ID), data=NGHS)
> NGHS$Htfitted <-  fitted(fm.Ht ) #BLUP estimate for height
```

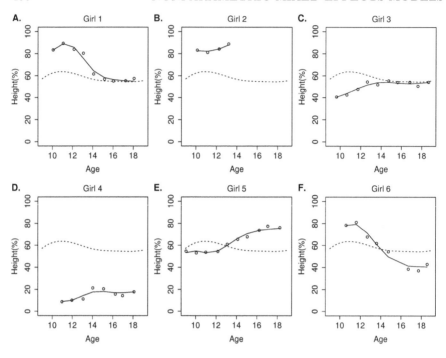

Figure 11.3 *The longitudinal height percentiles for three girls from the NGHS with estimated subject-specific curves and population-mean curves plotted in solid and dashed lines.*

```
> KN2 <- seq(from=9, to=19, length=4)[-c(1,4)]
> Bs.Age <- bs(NGHS$AGE, knots=KN2, intercept=T)
> Bs.Race <- bs(NGHS$AGE, knots=NULL,intercept=T)*NGHS$Black
> Bs.Ht <- bs(NGHS$AGE, knots=NULL,intercept=T)
                *(NGHS$Htfitted-50)
> fm.SBP <- lmer(SBP~ 0+ Bs.Age+ Bs.Race+ Bs.Ht+(0+ Bs.Age|ID),
                data=NGHS)
```

Figure 11.3 shows the observed height percentiles for six girls from the NGHS and the individual trajectories obtained from (11.5) and (11.64). Although the estimated mean curve $E\left[X_i^{(2)}(t)|t\right]$ stays close to 60% over the adolescent years, the trajectories $\widehat{X}_i^{(2)}(t)$ have considerable variations. These plots suggest that it is reasonable to use the estimated trajectories in the model for $Y_i(t)$ to correct the measurement errors of $X_i^{(2)}(t)$.

Figure 11.4 shows the mean baseline curve $\beta_0(t)$, and the two mean coefficient curves for race and height percentile $\beta_1(t)$ and $\beta_2(t)$, respectively, estimated by the cubic spline with two equally spaced interior knots at 12.3

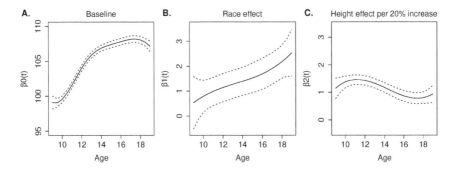

Figure 11.4 *The mean baseline curve $\beta_0(t)$, and two coefficient curves for race and height percentile $\beta_1(t)$ and $\beta_2(t)$ with 95% bootstrap pointwise confidence interval based on the model (11.105) with the cubic B-spline basis approximation.*

and 15.7 years of age chosen by the BIC. These curves suggest that the mean SBP increase with age, and the SBP measurements also depend on race and height percentile. The positive race coefficient curve suggests that the African American girls tend to have higher SBP levels than the Caucasian girls in this age range, and this racial difference in SBP increases with age. The height percentile coefficient curve indicates that the SBP levels increase with the height percentile for these adolescent girls, but this effect of height percentile on SBP tapers off at the older ages towards adulthood.

Furthermore, Figure 11.5 shows the observed longitudinal SBP measurements for three randomly selected subjects, with the estimated baseline mean curves and the subject-specific SBP trajectories based on the above B-spline approximated mixed-effects model (11.105). These scatter plots and the BLUP trajectory curves suggest that the model provide a fairly good fit to the SBP data.

11.7   Remarks and Literature Notes

The statistical models discussed in this chapter can be naturally viewed as "extended linear mixed-effects models." These models are attractive because, unlike the population based models of Chapters 3 to 10, they have built-in subject-specific time-trends, which capture the intra-subject correlation structures at different time points. Because the population-mean and subject-specific time-trends are nonparametric, these models are more flexible than the parametric mixed-effects models. But, on the other hand, we utilize the linear model structures by allowing the nonparametric time-trends to be approximated by expansions of some basis functions. This modeling approach allows for simple interpretations of the intra-subject correlations through an additive subject-specific deviation from the population time-trends.

The estimation methods of this chapter are natural extensions of the well-

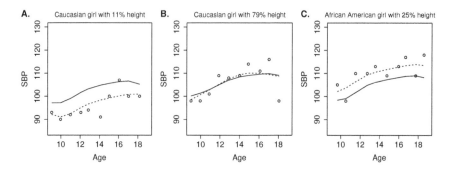

Figure 11.5 *The longitudinal SBP measurements for three NGHS girls with the estimated subject-specific curves and the mean population curves plotted in solid and dashed lines, respectively. The average height percentiles at the beginning of the study for the three girls were 11%, 79%, and 25%.*

known least squares and maximum likelihood methods in parametric mixed-effects models. By using basis expansions, we can approximate different shapes of the population average and subject-specific time-trends as well as the co-variance surfaces. Comparing with the models and methods of the previous chapters, the methods in this chapter have the advantage of providing individual outcome trajectory predictions, which can be used to cluster subjects with different outcome trajectories. We further elaborate the use of individual outcome trajectories in the applications of Chapter 15.

   The methods of this chapter can be found from a large number of publications in the "functional data analysis" literature. Our presentation of this chapter is mostly summarized from Shi, Weiss and Taylor (1996), James, Hastie and Sugar (2000), Rice and Wu (2001), Liang, Wu and Carroll (2003), Wu and Liang (2004), and Yao, Müller and Wang (2005a, 2005b). In order to limit our focus, the results of this chapter represent only a fraction of the general area of functional data analysis. Other smoothing methods, such as the local polynomial estimation for the nonparametric mixed-effects models developed by Wu and Zhang (2002), are not included because of our intention to limit the scope.

Part V

Nonparametric Models for Distributions

Chapter 12

Unstructured Models for Distributions

The statistical methods described in the previous chapters are all based on modeling the conditional means of the response variables with various longitudinal variance-covariance structures given time and a set of covariates. The conditional means and the longitudinal variance-covariance structures can be modeled either parametrically or nonparametrically. Although the conditional mean based models are popular in practice, they may be inadequate when the scientific objectives of the study require the evaluation of the conditional distribution functions. We present in this and the subsequent chapters a series of nonparametric models and estimation methods based on the conditional distribution functions. To start from a simple case, this chapter is focused on longitudinal data with time-invariant and categorical covariates, in which case the conditional distribution functions of the outcome variable can be directly estimated through a kernel smoothing method with an unstructured nonparametric model. The more complicated cases for the estimation of conditional distribution functions with time-dependent covariates require some modeling structures which are discussed in Chapters 13, 14 and 15. Throughout these chapters, we define a statistical index, the Rank-Tracking Probability, to measure the temporal tracking ability of a longitudinal variable, and discuss its estimation under various conditional distribution based models.

12.1 Objectives and General Setup

12.1.1 Objectives

Longitudinal analysis based on the conditional distribution functions is useful in practice because it can be used to evaluate two primary objectives.

(1) **Temporal Trends of Conditional Distributions:** When the outcome variable has non-Gaussian or skewed distributions, the temporal trends of the outcome variable and the covariate effects may be better described through the time-varying patterns of conditional distributions or conditional quantiles. The conditional-mean based regression models, on the other hand, do not lead to useful inferences about the distribution functions or quantiles, when the distributions of the error terms are non-Gaussian or

unknown. However, we should note that the estimation with conditional-distribution based models generally requires a larger sample size than the estimation with conditional-mean based models.

(2) **Developing Statistical Indices for Tracking:** An important objective in longitudinal study is to evaluate the *tracking ability* of a time-varying variable of interest. The concept of tracking is originated from the need of predicting future values of risk factors from serial measurements in epidemiological studies or longitudinal clinical trials, and has been studied by Clark, Woolson and Schrott (1976), Ware and Wu (1981), Foulkes and Davis (1981) and McMahan (1981) under two general definitions and a number of model formulations. As discussed in Foulkes and Davis (1981), the first definition of tracking is concerned with the ability to predict the future value of an outcome variable from its repeated measurements in the past, while the second definition relating to tracking concerns the maintenance over time of relative ranking with the response distribution among a group of peers. Despite the different technical definitions of tracking, the main objective in biomedical studies is to identify persistent disease risk factors which may eventually lead to disease events.   □

12.1.2   Applications

Applications of conditional distribution functions and tracking are widely available in biomedical studies. For example, Webber et al. (1991) described tracking of coronary heart disease risk factors in children in The Bogalusa Heart Study, and Wilsgaard et al. (2001) analyzed the tracking of cardiovascular risk factors in The Tromso Study. In these studies, the scientific objective is to determine whether an individual with unfavorable levels of cardiovascular risk factors, such as blood pressure, body mass index, and serum lipids, at younger ages is more likely to have unfavorable levels of the same risk factors at older ages.

We demonstrate the applications of the methods of this chapter using the NGHS study described in Section 1.2. This study is a typical example exhibiting the potential shortcomings of the conditional mean based regression method and the practical needs for statistical inferences from the conditional distribution based models. The NGHS study is appropriate for conditional distribution based regression because of the following features:

(1) **Relevant Scientific Objectives**: The scientific questions of the NGHS study require the evaluation of the conditional distribution functions. Among the publications based on the NGHS data, Daniels et al. (1998) studied the longitudinal mean effects of race, height and body mass index on the levels of systolic and diastolic blood pressures, Thompson et al. (2007) investigated the associations between childhood overweight and cardiovascular disease risk factors, and Obarzanek et al. (2010) evaluated

the prevalence and incidence of hypertension and their relationships with childhood obesity. These results all depend on modeling the conditional means of the outcome variables of interest, which are not sufficient to describe the time-dependent conditional distributions. An important objective for the NGHS is to determine whether the conditional distributions of the cardiovascular risk factors, such as blood pressure and body mass index change over time, and whether a girl's risk factor at an earlier age has any significant tracking ability to influence her risk factor at a later age. The importance of evaluating the risk factor's temporal patterns and tracking abilities in the NGHS study can be seen from the discussions in Thompson et al. (2007) and Obarzanek et al. (2010). More generally, Kavey et al. (2003) illustrates the importance of tracking the risk factors within subjects over various age ranges in pediatric cardiovascular studies.

(2) **Sufficiently Large Sample Size**: The sample size of the study is sufficiently large to warrant adequate estimation of the conditional distribution functions. As a typical large epidemiological study, the NGHS has 2379 subjects (1213 African American girls and 1166 Caucasian girls) and there are sufficient observations spread out across the age range. This large sample size provides sufficient data to model and estimate the conditional distributions and conditional quantiles of the cardiovascular risk factors, such as blood pressure, body mass index and serum lipids, during adolescence. Clinical classifications of blood pressure and body mass index for children or adolescents are defined through their conditional quantiles (NGHSRG, 1992, and NHBPEP, 2004).

(3) **Adequate Numbers of Repeated Measurements**: The study design has up to 10 annual follow-up visits for each participant. Although, as in almost all practical studies, some subjects have early withdraws or missing data on some of the planned visits, the numbers of repeated measurements still provide sufficient information to evaluate the population-wide and subject-specific temporal trends of the outcome variables of interest. These repeated measurements make it possible to estimate the *tracking abilities* of the risk factors.

(4) **Unknown Distribution Functions**: The conditional distributions of many of the cardiovascular risk factors in the NGHS study are unknown, non-Gaussian and skewed given age, race and other covariates. Since the conditional mean based regression models may not give appropriate statistical inferences on the conditional distribution functions if the distributions of the error terms are unknown, conditional distribution based regression models are more appropriate to answer the relevant scientific questions.

The above features are common in most epidemiological studies, which usually have large sample sizes and sufficiently long follow-up periods. When the studies have small to moderate sample sizes, nonparametric estimation of the conditional distribution functions is difficult and unreliable. In such cases, additional assumptions on the distribution functions, such as assumptions of parametric families, are required to accommodate the lack of information associated with the small to moderate sample sizes. Since the focus of this book is on nonparametric models and their corresponding estimation methods, these parametric models and estimation methods are omitted from our presentations, and we refer to the original papers, such as Ware and Wu (1981), Foulkes and Davis (1981) and McMahan (1981).

12.1.3 Estimation of Conditional Distributions

Nonparametric estimation methods for the conditional distribution functions have been studied in the literature under a number of settings. For the case with independent identically distributed data or a time series sample, Hall, Wolff and Yao (1999) presented a nonparametric smoothing method to estimate the conditional cumulative distribution functions (CDF) using either a local logistic estimator or an adjusted Nadaraya-Watson estimator, and Hall, Racine and Li (2004) proposed a cross-validation smoothing method to estimate the conditional probability densities. For longitudinal samples with time-dependent covariates, Wu, Tian and Yu (2010) and Wu and Tian (2013b) suggested a time-varying transformation model and a two-step smoothing method for estimating the temporal trends of the conditional distribution functions and the covariate effects.

When the covariates in the longitudinal sample are categorical and time-invariant, Wu and Tian (2013a) suggested a kernel smoothing method to estimate the conditional distribution functions without assuming any modeling structures. Since this chapter is only concerned with longitudinal samples with categorical and time-invariant covariates, the estimation procedures along with their applications and theoretical properties discussed in this chapter follow the ones described in Wu and Tian (2013a).

12.1.4 Rank-Tracking Probability

With respect to the two concepts of *tracking ability* described in Section 12.1.1, namely "predicting the future outcome" and "maintaining relative ranking," our focus in this chapter is on the definition and nonparametric estimation of the tracking index *rank-tracking probability* (RTP) suggested by Wu and Tian (2013a), which is essentially a nonparametric statistical index measuring the "maintaining relative ranking" ability. Definition and nonparametric estimation of the RTP for samples with time-dependent covariates require additional model structure assumptions, which are presented in Chapters 13, 14 and 15.

The RTP defined in this chapter is similar to the *tracking index* suggested by Foulkes and Davis (1981) as they both are indices measuring the "maintaining relative ranking" ability of the outcome variable. The only difference is that the *tracking index* defined and estimated by Foulkes and Davis (1981) is under the parametric assumption that the longitudinal outcomes and covariates follow a mixed-effects model. Under the parametric families of mixed-effects models, the "maintaining relative ranking" *tracking index* of Foulkes and Davis (1981) is consistent with the "predicting the future outcome" tracking definitions presented by Ware and Wu (1981) and McMahan (1981), which rely on the serial correlations of the repeated measurements. This is because, under the parametric assumption of the mixed-effects models, the serial correlation of the repeated measurements can be used to construct the *tracking index* of Foulkes and Davis (1981). In this sense, the serial correlation can be viewed as an indirect measure for the "maintaining relative ranking" ability.

Among other methods for *tracking index* in the literature, estimation of serial correlations in longitudinal analysis with nonparametric models can be found, for example, in Wu and Pourahmadi (2003) and Fan and Wu (2008). However, regression models based on serial correlations or conditional variance and covariance structures are generally insufficient to provide an intuitive quantitative measure of the "tracking ability" in many biomedical studies similar to the NGHS, particularly when the parametric families of the conditional distributions are unknown. The RTP, on the other hand, is a more direct and intuitive index for measuring the tracking ability under the setting of nonparametric models.

12.2   Data Structure and Conditional Distributions

12.2.1   Data Structure

The main results of this chapter are based on the following data structure, which is a special case of the longitudinal data with categorical and time-invariant covariates. This data structure is motivated by the practicality and clinical implications of the NGHS design (NGHSRG, 1992). We start from this simple special case because of its simplicity in mathematical expressions and straightforward interpretations in many biomedical studies.

Sample with Categorical and Time-Invariant Covariates:

(a) *The sample has n independent subjects, and the ith subject has n_i observations at time points $\{t_{ij} \in \mathcal{T}; j = 1, \ldots, n_i\}$, where $\mathcal{T}$ is the time interval of interest. The total number of observations is $N = \sum_{i=1}^{n} n_i$.*

(b) *At any time point $t \in \mathcal{T}$, $Y(t)$ is the real-valued outcome variable. The jth observation of $Y(t)$ for the ith subject, i.e., the ith subject's observed outcome at time t_{ij}, is given by Y_{ij}.*

(c) *The covariate X is time-invariant and categorical with $x \in \{1, \ldots, K\}$. The ith subject's observed covariate is X_i.*

(d) *The longitudinal sample for $\big(Y(t), X, t\big)$ is*

$$\mathscr{L} = \big\{ (Y_{ij}, X_i, t_{ij}) : 1 \le i \le n; 1 \le j \le n_i \big\},$$

which is a much simplified special case of commonly seen longitudinal data (e.g., Section 11.2.1).   □

In the application to the NGHS study of this chapter, we assume that $Y(t)$ is non-negative because all the important risk factors considered in the NGHS study are non-negative. But the methods and theory presented here allow for any $Y(t)$ on the real line.

The assumption of categorical and time-invariant covariate X is made for the simplicity of mathematical expressions and biological interpretations. When there are continuous or time-dependent covariates, nonparametric estimation of the conditional distribution functions and other related quantities requires multivariate smoothing methods, which could be computationally infeasible and difficult to interpret in practice. For the next two chapters, we present a useful dimension reduction approach based on the time-varying transformation models of Wu, Tian and Yu (2010) and Wu and Tian (2013a, 2013b), which can be used to estimate the conditional distribution functions with continuous and time-varying covariates. But this extension requires structural assumptions for the conditional distribution functions and their estimation procedures. The nonparametric estimation method of this chapter does not require these structural modeling assumptions.

The above data structure is consistent with the data formulation used in the NGHS publications, such as Daniels et al., (1998), Obarzanek et al. (2010) and Wu, Tian and Yu (2010). In these NGHS publications, the longitudinal sample has a set of $J > 1$ design time points $\mathbf{t} = \left(t_{(1)}, \dots, t_{(J)}\right)^T$, which contains the distinct values of all possible visit times. Each of the n independent subjects has actual visit times within a subset of $\mathbf{t}$. If the ith subject is observed at time point $t_{(j^*)}$, the corresponding outcome variable is $Y_i(t_{(j^*)})$, which may not be the same as the Y_{ij} in the data structure (2) above, since $t_{(j^*)}$ and t_{ij} are not necessarily the same. To see the connection between $Y_i(t_{(j^*)})$ and Y_{ij} for a given time point $t_{(j^*)} \in \mathbf{t}$, we define

$$\mathscr{S}_{j^*} = \{\text{ the set of subjects with outcome observed at } t_{(j^*)}\}. \qquad (12.1)$$

When $i \in \mathscr{S}_{j^*}$, the observed outcome for the subject at $t_{(j^*)}$ is $Y_i(t_{(j^*)})$. Thus, for each $t_{ij} \in \mathscr{T}$ given in our data structure (1), we can find an integer j^* such that $t_{ij} = t_{(j^*)}$, so that $i \in \mathscr{S}_{j^*}$ and $Y_i(t_{(j^*)}) = Y_{ij}$. Using the design time points $\mathbf{t}$ and (12.1), the numbers of observations at each time point in $\mathbf{t}$ are given by

$$n_{j^*} = \#\{i \in \mathscr{S}_{j^*}\} \quad \text{and} \quad n_{j_1^* j_2^*} = \#\{i \in \mathscr{S}_{j_1^*} \cap \mathscr{S}_{j_2^*}\}, \qquad (12.2)$$

which are the number of subjects in $\mathscr{S}_{j^*}$ and the number of subjects in both $\mathscr{S}_{j_1^*}$ and $\mathscr{S}_{j_2^*}$, respectively. Clearly, $n_{j_1^* j_2^*} \leq \min\{n_{j_1^*}, n_{j_2^*}\}$.

It is often the case in practice that there are a large number of distinct time points, so that the numbers of subjects observed at each of these time points are small. In such cases, adjacent time points may need to be pooled together to create a reasonable set of design time points $\mathbf{t}$. In Obarzanek et al. (2010), Wu, Tian and Yu (2010) and Wu and Tian (2013a, 2013b), the design time points of the NGHS study are specified by rounding up the age of the participants at the first decimal place, which is chosen because one decimal point is clinically sufficiently accurate.

12.2.2   Conditional Distribution Functions

Because the objective is to investigate the temporal trends of the distribution functions, we would like to estimate the conditional distributions of $Y(t)$ given $X = x$ at any time point t. These conditional distributions are treated as smooth functions of t within the time range $\mathscr{T}$. For a given $t \in \mathscr{T}$ and $X = x$, the conditional probability that $Y(t)$ belongs to a set $A(x,t)$, which may change with (x,t), is

$$P_A(x,t) = P\big[Y(t) \in A(x,t)\big|X = x, t\big]. \tag{12.3}$$

The estimation and statistical inferences of $P_A(x,t)$ depend on the shapes of $A(x,t)$ and whether $A(x,t)$ is known or has to be estimated from the data. In some cases, $A(x,t)$ is known from the scientific objectives or assumptions. In other cases, $A(x,t)$ is only assumed to satisfy some conditions, but its actual values depend on unknown quantities which have to be estimated from the data. For both cases of $A(x,t)$ known or unknown, our statistical objective is to estimate $P_A(x,t)$ using the sample $\mathscr{Z}$ of Section 12.2.1.

The choices of $A(x,t)$ depend on the scientific objectives of the analysis. In particular, if $A(x,t) = \big(0, y(x,t)\big]$ for a given function $y(x,t)$ of (x,t), $P_A(x,t)$ is the conditional cumulative distribution function (CDF) of $Y(t)$ given (x,t),

$$F_t\big[y(x,t)\big|x\big] = P\big[Y(t) \leq y(x,t)\big|x, t\big]. \tag{12.4}$$

Note that the above conditional CDF allows $y(x,t)$ to change with (x,t). A special case of (12.4) is to assume that $y(x,t) = y$, so that the statistical objective is to estimate the simple conditional CDF $F_t[y|x]$ as a smooth function of $t \in \mathscr{T}$. Although it is simple to estimate $F_t[y|x]$ for a fixed y, it is often useful in practice to allow $A(x,t)$ to change with both x and t. For example, the ranges of blood pressure status for children and adolescents are defined by risk categories determined by gender and age (e.g., NHBPEP, 2004). Thus, in pediatric studies, it is often meaningful to evaluate the conditional CDF defined by (12.2) with $y(x,t)$ being a pre-determined gender- and age-specific risk threshold curve.

12.2.3   Conditional Quantiles

Functionals of $P_A(x,t)$ may also be of interest and need to be estimated in practice. A particularly useful class of functionals of $F_t\big[y(x,t)\big|x\big]$ of (12.4), which has been used for defining pediatric hypertension status (e.g., NHBPEP, 2004), is the conditional quantiles. If, for any given on $X = x$ and t, there is a unique inverse of $F_t\big[y(x,t)\big|x\big] = \alpha$ for some $0 < \alpha < 1$, the $(100 \times \alpha)$th conditional quantile given (x,t) is the inverse of $F_t\big[y(x,t)\big|x\big] = \alpha$ denoted by

$$y_\alpha(x,t) = F_t^{-1}(\alpha|x). \tag{12.5}$$

When x and a suitable α are given, $y_\alpha(x,t)$ describes the $(100 \times \alpha)$th quantile as a function of t for the population characterized by $X = x$. For example, classifications of blood pressure and hypertension status for children and adolescents are defined in NHBPEP (2004) by the quantiles conditioning on the individual's age and gender. It is reasonable to assume in most biomedical studies that $y_\alpha(x, t)$ is a smooth function of t for any $\alpha \in (0, 1)$ and x.

Further functionals of the conditional quantiles $y_\alpha(x, t)$ maybe also of interest in a number of applications. A useful functional of the conditional quantiles $y_\alpha(x, t)$ is the *Inter-Quantile Range* (IQR) for (α_1, α_2) with $0 < \alpha_2 < \alpha_1 < 1$ defined by

$$\Delta_{\alpha_1, \alpha_2}(x, t) = y_{\alpha_1}(x, t) - y_{\alpha_2}(x, t), \qquad (12.6)$$

which is assumed to be a smooth function of $t \in \mathcal{T}$ for the population characterized by $X = x$. Examples of the IQR $\Delta_{\alpha_1, \alpha_2}(x, t)$ may be the blood pressure range between different hypertension categories defined for children and adolescents. More generally, it may be also useful to evaluate a *Generalized Inter-Quantile Range* (GIQR) defined to be the differences of conditional quantiles under different values of (x_1, t_1) and (x_2, t_2),

$$\Delta_{\alpha_1, \alpha_2}\big[(x_1, t_1), (x_2, t_2)\big] = y_{\alpha_1}(x_1, t_1) - y_{\alpha_2}(x_2, t_2), \qquad (12.7)$$

which, for any (α_1, x_1) and (α_2, x_2), are assumed to be smooth functions of (t_1, t_2). Similar to the IQR of (12.6), examples of the GIQR $\Delta_{\alpha_1, \alpha_2}\big[(x_1, t_1), (x_2, t_2)\big]$ may be the difference in blood pressure levels for Caucasian and African American girls with the same hypertension categories.

Functionals of $y_\alpha(x, t)$, other than (12.6) and (12.7), may also be defined based on the specific scientific questions. But, as general statistical quantities of interest, the IQR and GIQR defined in (12.6) and (12.7) are two of the most frequently used functionals of $y_\alpha(x, t)$ in real applications, so that our conditional quantile estimation methods of this chapter are limited to the quantities of (12.5), (12.6) and (12.7).

12.2.4   Rank-Tracking Probabilities

Given the importance of measuring the *tracking ability* among subjects with certain health status (e.g., Kavey et al., 2003; Thompson et al., 2007; Obarzanek et al., 2010), a direct index for measuring *tracking* can be defined through a conditional probability of the outcome variable $Y(t)$ at different time points.

1. Definition of Rank-Tracking Probabilities

Suppose that, for any $X = x$ and $t \in \mathcal{T}$, the health status of a subject at time t is determined by whether $Y(t) \in A(x, t)$ for a pre-determined set $A(x, t)$. Then, the tracking ability of $Y(t)$ at any two time points $s_1 < s_2$ can be measured by the conditional probability of $Y(s_2) \in A(x, s_2)$ given $Y(s_1) \in A(x, s_1)$ and

$X = x$, which is referred in Wu and Tian (2013a) as the RTP based on $A(\cdot, \cdot)$ at $s_1 < s_2$,

$$RTP_A(x, s_1, s_2) = P\big[Y(s_2) \in A(x, s_2)\big|Y(s_1) \in A(x, s_1), X = x\big]. \qquad (12.8)$$

Clearly, the interpretations of RTP depend on the choices and interpretations of $A(\cdot, \cdot)$, and the choice of $A(\cdot, \cdot)$ depends on the study questions and scientific objectives. In biomedical studies, the subject's health status at time t is often specified by the conditional quantiles of $Y(t)$, so that a common choice of $A(x, t)$ is the quantile-based interval

$$A_\alpha(x, t) = \big(y_\alpha(x, t), \infty\big), \qquad (12.9)$$

where $y_\alpha(x, t)$ is the $(100 \times \alpha)$th quantile of $Y(t)$ given $X = x$. As a special case of (12.8), we can define the quantile based RTP as

$$RTP_{\alpha_1, \alpha_2}(x, s_1, s_2) = P\big[Y(s_2) > y_{\alpha_2}(x, s_2)\big|Y(s_1) > y_{\alpha_1}(x, s_1), X = x\big], \quad (12.10)$$

which is the probability that $Y(t)$ is above the $(100 \times \alpha_2)$th quantile at time $t = s_2$ given that $X = x$ and $Y(t)$ is already above the $(100 \times \alpha_1)$th quantile at time $t = s_1$. For the special case that the covariate X is not included, the RTP based on a time-dependent outcome set $A(t)$ is

$$RTP_A(s_1, s_2) = P\big[Y(s_2) \in A(s_2)\big|Y(s_1) \in A(s_1)\big], \qquad (12.11)$$

and similarly its quantile-based analogue is

$$RTP_{\alpha_1, \alpha_2}(s_1, s_2) = P\big[Y(s_2) > y_{\alpha_2}(s_2)\big|Y(s_1) > y_{\alpha_1}(s_1)\big], \qquad (12.12)$$

where $y_\alpha(t)$ is the $(100 \times \alpha)$th quantile of $Y(t)$.

2. Interpretations of Rank-Tracking Probabilities

Although the value of $RTP_A(x, s_1, s_2)$ is within $[0, 1]$, the strength of *tracking* for $Y(t)$ is actually measured by the value of $RTP_A(x, s_1, s_2)$ relative to the probability $P\big[Y(s_2) \in A(x, s_2)\big|X = x\big]$. We can then distinguish *tracking ability* under the following situations:

(a) No Tracking Ability: If

$$RTP_A(x, s_1, s_2) = P\big[Y(s_2) \in A(x, s_2)\big|X = x\big], \qquad (12.13)$$

then knowing $Y(s_1) \in A(x, s_1)$ does not increase the chance of $Y(s_2) \in A(x, s_2)$, which implies that $Y(s_1) \in A(x, s_1)$ has no *tracking ability* for $Y(s_2) \in A(x, s_2)$.

(b) **Positive or Negative Tracking Ability:** On the other hand, $Y(s_1) \in A(x, s_1)$ is said to have positive or negative *tracking ability* for $Y(s_2) \in A(x, s_2)$, if

$$RTP_A(x, s_1, s_2) > P\big[Y(s_2) \in A(s_2, x)\big|X = x\big] \qquad (12.14)$$

or

$$RTP_A(x, s_1, s_2) < P\big[Y(s_2) \in A(s_2, x)\big|X = x\big], \qquad (12.15)$$

respectively. $\qquad \Box$

It easily follows that $Y(s_1)$ has no "tracking ability" for $Y(s_2)$ if $Y(s_1)$ and $Y(s_2)$ are conditionally independent given X.

12.2.5 Rank-Tracking Probability Ratios

Since the tracking index $RTP_A(x, s_1, s_2)$ is only meaningful when it is compared with the probability $P\big[Y(s_2) \in A(s_2, x)\big|X = x\big]$, a modified index for measuring the *relative tracking ability* for $Y(t)$ at time points $s_1 < s_2$ based on the RTP is the *Rank-Tracking Probability Ratio* (RTPR) defined by

$$RTPR_A(x, s_1, s_2) = \frac{RTP_A(x, s_1, s_2)}{P\big[Y(s_2) \in A(s_2, x)\big|X = x\big]}. \qquad (12.16)$$

Intuitively, $RTPR_A(x, s_1, s_2)$ describes the relative strength of the conditional probability $RTP_A(s_1, s_2, x)$ with the information of $Y(s_1) \in A(x, s_1)$ incorporated over the conditional probability $P\big[Y(s_2) \in A(s_2, x)\big|X = x\big]$ without incorporating the information of $Y(s_1) \in A(x, s_1)$.

When knowing $Y(s_1) \in A(x, s_1)$ or not at an earlier time point s_1 does not affect the probability of $Y(s_2) \in A(x, s_2)$ at a later time point $s_2 > s_1$, the equality of $RTP_A(s_1, s_2, x)$ and $P\big[Y(s_2) \in A(s_2, x)\big|X = x\big]$ suggests that $Y(s_1)$ has no *tracking ability* for $Y(s_2)$. The strength of positive or negative *tracking ability* can then be measured by comparing $RTPR_A(x, s_1, s_2)$ with 1. Specifically, we can define that

$$\begin{cases} RTPR_A(x, s_1, s_2) = 1, \\ \quad \text{if } Y(s_1) \text{ has no tracking ability for } Y(s_2); \\ RTPR_A(x, s_1, s_2) > 1, \\ \quad \text{if } Y(s_1) \text{ has positive tracking ability for } Y(s_2); \\ RTPR_A(x, s_1, s_2) < 1, \\ \quad \text{if } Y(s_1) \text{ has negative tracking ability for } Y(s_2). \end{cases} \qquad (12.17)$$

Special cases of the quantile-based RTPRs can also be defined by selecting $A(\cdot, \cdot)$ to be the interval $A_\alpha(\cdot, \cdot)$ of (12.9). Using the RTP of (12.10), the quantile-based RTPR of $Y(t)$ with (α_1, α_2) at time points (s_1, s_2) given $X = x$ is

$$RTPR_{\alpha_1, \alpha_2}(x, s_1, s_2) = \frac{RTP_{\alpha_1, \alpha_2}(x, s_1, s_2)}{P\big[Y(s_2) > y_{\alpha_2}(x, s_2)\big|X = x\big]}, \tag{12.18}$$

where $y_\alpha(x, t)$ is the $(100 \times \alpha)$th quantile of $Y(t)$ given $X = x$. The advantage of using the quantile-based RTPR is its simplicity in scientific interpretations. The application to the NGHS study presented in Section 12.4 is using the quantile-based RTPR defined above.

12.2.6   Continuous and Time-Varying Covariates

Since the data structure of this chapter (Section 12.2.1) only allows for time-invariant and categorical covariates, the conditional distributions and rank-tracking probability defined above, although having simple and straightforward interpretations, cannot be applied to the more general case with continuous and time-varying covariates. This is because the changing covariates at different time points have not been taken into account in the definitions of $P_A(x, t)$ of (12.3) and $RTP_A(x, s_1, s_2)$ of (12.8). We give here some brief comments on conceptual issues of this extension. Details of the actual approaches, including dimension reduction strategies through structural nonparametric models for distribution functions, are described in the next three chapters.

1. Conditional Distribution Functions with Time-Dependent Covariates

When there are continuous and time-varying covariates, we can denote the covariate vector by $X(t)$, and given the covariate values $\mathbf{X}(t) = \mathbf{x}$, the conditional probability of (12.3) is easily extended to

$$P_A(\mathbf{x}, t) = P\big[Y(t) \in A(\mathbf{x}, t)\big|\mathbf{X}(t) = \mathbf{x}, t\big]. \tag{12.19}$$

Unstructured nonparametric estimation of $P_A(\mathbf{x}, t)$ in (12.19) would require a multivariate smoothing method over both $\mathbf{x}$ and t, which could be difficult to compute, particularly when the dimensionality of $\mathbf{X}(t)$ is high. A dimension reduction alternative, which was proposed by Wu, Tian and Yu (2010), is to model $P_A(\mathbf{x}, t)$ using a class of time-varying transformation models, so that the dependence of $P_A(\mathbf{x}, t)$ on $\mathbf{x}$ at each given t is characterized through a structured nonparametric model which is determined by a set of coefficient curves. The models and estimation method of Wu, Tian and Yu (2010) are described in Chapter 13.

2. Tracking with Time-Dependent Covariates

Different definitions of the RTPs may be used depending on how the continuous and time-varying covariates are incorporated and the practical objectives of the tracking index. A definition of RTP, which may be close to the practical interpretations suggested by Kavey et al. (2003), is to consider the probability of $Y(s_2) \in A[\mathbf{X}(s_2), s_2]$ at time s_2 given $Y(s_1) \in A[\mathbf{X}(s_1), s_1]$ and $\mathbf{X}(s_1) \in B(s_1)$ for some set $B(s_1)$ of the covariates at time s_1. This consideration motivates the following definition of the "Rank-Tracking Probability" based on $\{A(\cdot, \cdot), B(\cdot)\}$ at time points $s_1 < s_2$,

$$
\begin{aligned}
RTP_A(B, s_1, s_2) & \\
= \quad & P\{Y(s_2) \in A[\mathbf{X}(s_2), s_2] \,|\, Y(s_1) \in A[\mathbf{X}(s_1), s_1], \mathbf{X}(s_1) \in B(s_1)\}.
\end{aligned}
\tag{12.20}
$$

In real applications, specific choices of $A(\cdot, \cdot)$ and $B(\cdot)$ have to be determined in order to have meaningful interpretations for $RTP_A(B, s_1, s_2)$. Other special versions of RTP and their corresponding RTPRs can be analogously defined. Similar to the estimation of $P_A(\mathbf{x}, t)$ in (12.19), the multivariate smoothing estimation of $RTP_A(B, s_1, s_2)$, without any structured modeling assumptions, is often infeasible in practice. Structured nonparametric methods for the estimation of (12.20) are presented in Chapter 14.

12.3   Estimation Methods

We present in this section two kernel smoothing methods for the estimation of conditional distribution functions and their useful functionals, such as conditional quantiles, RTPs and RTPRs. There are a few potential alternative smoothing estimation methods, such as the local polynomial estimators, the local logistic estimators, the adjusted Nadaraya-Watson estimators, or the global smoothing methods using basis approximations, which could be extended to the situations considered in this chapter. But, as for now, these alternative methods have not yet been systematically investigated in the literature.

12.3.1   Conditional Distribution Functions

By the definition of $P_A(x, t)$ in (12.3), we can rewrite $P_A(x, t)$ as

$$
P_A(x, t) = E\{1_{[Y(t) \in A(x,t), X=x]} \,|\, X = x, t\},
\tag{12.21}
$$

where, with a pre-specified set $A(x, t)$, $1_{[Y(t) \in A(x,t), X=x]} = 1$ if $Y(t) \in A(x,t)$ and $X = x$, and 0 otherwise. Since $x \in \{1, \ldots, K\}$ and K is a finite integer, which is often small in practice, we consider two different kernel smoothing approaches to estimate $P_A(x, t)$.

Let $K_x(\cdot)$ be a kernel function, which is often taken to be a probability density function, and let h_x be a positive bandwidth for a given $x \in \{1, \ldots, K\}$. The local sum of squares error of any estimator $p_A(x, t)$ of $P_A(x, t)$ based on $1_{[Y_{ij} \in A(x, t_{ij}), X_i = x]}$, $K_x(\cdot)$ and the subject uniform weight $(n n_i)^{-1}$ is

$$
L(x, t) = \sum_{i=1}^{n} \sum_{j=1}^{n_i} \left\{ \left(\frac{1}{n n_i} \right) \left[1_{[Y_{ij} \in A(x, t_{ij}), X_i = x]} - p_A(x, t) \right]^2 K_x\left(\frac{t - t_{ij}}{h_x} \right) \right\}.
\tag{12.22}
$$

Effectively $(n\,n_i)^{-1}$ in (12.22) weighs the observations from subjects with small numbers of repeated measurements more heavily than the observations from subjects with large numbers of repeated measurements. Minimizing (12.22) with respect to $p_A(x, t)$, the kernel estimator of $P_A(x, t)$ is given by

$$\widehat{P}_A(x, t) = \frac{\sum_{i=1}^{n} \left\{ n_i^{-1} \sum_{j=1}^{n_i} \left[1_{[Y_{ij} \in A(x, t_{ij}), X_i = x]} K_x((t - t_{ij})/h_x) \right] \right\}}{\sum_{i=1}^{n} \left\{ n_i^{-1} \sum_{j=1}^{n_i} [K_x((t - t_{ij})/h_x)] \right\}}. \qquad (12.23)$$

If, as an alternative to (12.22), the measurement uniform weight $N^{-1} = \left(\sum_{i=1}^{n} n_i \right)^{-1}$ is assigned to each of the observations, we can minimize the local sum of squares error

$$L^*(x, t) = \sum_{i=1}^{n} \sum_{j=1}^{n_i} \left\{ \left(\frac{1}{N} \right) \left[1_{[Y_{ij} \in A(x, t_{ij}), X_i = x]} - p_A(x, t) \right]^2 K_x \left(\frac{t - t_{ij}}{h_x} \right) \right\} \qquad (12.24)$$

with respect to $p_A(x, t)$. The minimizer of (12.24) leads to the kernel estimator

$$\widehat{P}_A^*(x, t) = \frac{\sum_{i=1}^{n} \sum_{j=1}^{n_i} \left[1_{[Y_{ij} \in A(x, t_{ij}), X_i = x]} K_x((t - t_{ij})/h_x) \right]}{\sum_{i=1}^{n} \sum_{j=1}^{n_i} [K_x((t - t_{ij})/h_x)]}. \qquad (12.25)$$

The above two estimators are special cases of the kernel estimators with the general weight w_i

$$L_w(x, t) = \sum_{i=1}^{n} \sum_{j=1}^{n_i} \left\{ w_i \left[1_{[Y_{ij} \in A(x, t_{ij}), X_i = x]} - p_A(x, t) \right]^2 K_x \left(\frac{t - t_{ij}}{h_x} \right) \right\}, \qquad (12.26)$$

which has the expression

$$\widehat{P}_{A,w}(x, t) = \frac{\sum_{i=1}^{n} \sum_{j=1}^{n_i} \left[w_i \, 1_{[Y_{ij} \in A(x, t_{ij}), X_i = x]} K_x((t - t_{ij})/h_x) \right]}{\sum_{i=1}^{n} \sum_{j=1}^{n_i} \left[w_i K_x((t - t_{ij})/h_x) \right]}. \qquad (12.27)$$

Similar to the discussion of weight choices in the previous chapters, e.g., Chapters 7, 8 and 9, the special cases of $w_i = (1/n\,n_i)$ and $w_i = 1/N$ are the most commonly used weights in practice. Thus, our presentation of the kernel estimators for $P_A(x, t)$ are also limited to these two weight choices.

We note that, in (12.23) and (12.25), different kernel functions $K_x(\cdot)$ and bandwidths h_x are allowed for different values x of the covariate X. This is only possible when the number of possible values for X is not too large. The possibility of using different kernel functions and bandwidths allows for different smoothness adjustments in $\widehat{P}_{A,w}(x, t)$ for different x. Because of the local smoothing nature of (12.23) and (12.25), the potential correlations of the repeated measurements are not used. This phenomenon is consistent with other local smoothing approaches in longitudinal analysis, such as the methods of Chapters 6 through 9. The potential correlations of the measurements, however, affect the values of RTPs and their smoothing estimators. In particular, nonparametric estimators for the RTPs cannot be obtained from cross-sectional i.i.d. data.

When all the subjects have similar numbers of repeated measurements, i.e., the n_i's are similar for all $i = 1, \ldots, n$, $\widehat{P}_A(x, t)$ and $\widehat{P}_A^*(x, t)$ should be approximately the same, although $\widehat{P}_A^*(x, t)$ is more influenced by subjects with more repeated measurements. The estimator $\widehat{P}_{A,w}(x, t)$ shares some similarities with the componentwise kernel estimators of Chapter 6 for the conditional mean based varying-coefficient models. The weight choices between $w_i = (1/nn_i)$ and $w_i = (1/N)$ have been shown in Section 6.8 to lead to kernel estimators with different asymptotic properties. The asymptotic properties between $\widehat{P}_A(x, t)$ and $\widehat{P}_A^*(x, t)$, which are presented in Section 12.6, are also potentially different, while neither one is uniformly superior to the other.

Although it is known in the literature that the Nadaraya-Watson kernel approach may have excessive biases at the boundary points compared with the local polynomial smoothing method, it has the main advantages that it is computationally simple and produces estimates of the conditional distributions with values between 0 and 1. The local polynomial smoothing estimator using $1_{[Y_{ij} \in A(x,t), X_i = x]}$ does not always lead to an estimator of $P_A(x, t)$ with values between 0 and 1. Under the cross-sectional i.i.d. data or time series data, Hall, Wolff and Yao (1999) proposed to estimate the conditional distribution functions by the local logistic method (LLM) or the adjusted Nadaraya-Watson Estimator (ANWE), and showed that the estimators obtained from LLM and ANWE have good asymptotic properties with values between 0 and 1. However, LLM and ANWE are computationally intensive, and their extension to the current longitudinal data structure has not been investigated in the literature, although this extension warrants substantial future research.

12.3.2 Conditional Cumulative Distribution Functions

To estimate the conditional CDF $F_t[y(x, t)|x]$, we need to specify whether the threshold $y(x, t)$ is known in advance. When $y(x, t)$ is pre-specified and known, the estimation of $F_t[y(x, t)|x]$ is a direct application of the estimators in (12.23), (12.25) and (12.27). In some situations, however, $y(x, t)$ is not given and has to be estimated first before applying the methods in (12.23), (12.25) and (12.27).

1. Conditional CDF Estimation with Known Threshold

Applying (12.23) and (12.25) to the estimation of $F_t[y(x, t)|x]$ for a given threshold value $y(x, t)$, the corresponding kernel estimators based on the weights $(1/nn_i)$ and $(1/N)$ are

$$\widehat{F}_t[y(x, t)|x] = \frac{\sum_{i=1}^n \left\{ n_i^{-1} \sum_{j=1}^{n_i} \left[1_{[Y_{ij} \leq y(x, t_{ij}), X_i = x]} K_x((t - t_{ij})/h_x) \right] \right\}}{\sum_{i=1}^n \left\{ n_i^{-1} \sum_{j=1}^{n_i} \left[K_x((t - t_{ij})/h_x) \right] \right\}} \qquad (12.28)$$

and

$$\widehat{F}_t^* \big[y(x, t)|x\big] = \frac{\sum_{i=1}^{n} \sum_{j=1}^{n_i} \left[1_{[Y_{ij} \leq y(x, t_{ij})], X_i = x]} K_x((t - t_{ij})/h_x)\right]}{\sum_{i=1}^{n} \sum_{j=1}^{n_i} \left[K_x((t - t_{ij})/h_x)\right]}, \qquad (12.29)$$

respectively. The kernel conditional CDF estimator for a general weight function w_i based on (12.27) is

$$\widehat{F}_{t,w} \big[y(x, t)|x\big] = \frac{\sum_{i=1}^{n} \left\{ w_i \sum_{j=1}^{n_i} \left[1_{[Y_{ij} \leq y(x, t_{ij})], X_i = x]} K_x((t - t_{ij})/h_x)\right] \right\}}{\sum_{i=1}^{n} w_i \left\{ \sum_{j=1}^{n_i} \left[K_x((t - t_{ij})/h_x)\right] \right\}}. \qquad (12.30)$$

These estimators in (12.28), (12.29) and (12.30) have the attractive property of having non-decreasing values between 0 and 1 as $y(x, t)$ increases.

If $K_x(\cdot)$ is chosen to be the uniform density function on $(-1/2, 1/2)$, then $\widehat{F}_t \big[y(x, t)|x\big]$, $\widehat{F}_t^* \big[y(x, t)|x\big]$ and $\widehat{F}_{t,w} \big[y(x, t)|x\big]$ are the local empirical conditional CDFs constructed based on the observed indicators for $t_{ij} \in (t - h_x, t + h_x)$,

$$\mathscr{I}(t, h_x) = \left\{ 1_{[Y_{ij} \leq y(x, t_{ij})], X_i = x]} : |t_{ij} - t| < h_x \right\}. \qquad (12.31)$$

For more general kernel functions, the estimators in (12.28), (12.29) and (12.30) are some weighted local empirical distribution functions based on some observed indicator data similar to $\mathscr{I}(t, h_x)$.

2. Conditional CDF Estimation with Estimated Threshold

When the threshold $y(x, t)$ is unknown, we have to obtain a consistent estimator of $y(x, t)$ before applying the kernel estimators (12.23), (12.25) and (12.27). This can be done by splitting the sample into two sub-samples, so that $y(x, t)$ can be estimated from the first sub-sample and $F_t \big[y(x, t)|x\big]$ can be estimated from the second sub-sample with $y(x, t)$ replaced by its consistent estimator. The main reason of using sample splitting is to ensure that the estimator for $y(x, t)$ and the estimator for $F_t \big[y(x, t)|x\big]$ are constructed from two independent datasets. For the ease of notation, we assume here that $\widehat{y}(x, t)$ is a consistent estimator of $y(x, t)$ computed from a sample (or a sub-sample) that is independent from our dataset $\mathscr{Z}$ specified in Section 12.2.1. Then, the corresponding kernel estimators based on the weights $(1/nn_i)$, $(1/N)$ and w_i are

$$\widehat{F}_t \big[\widehat{y}(x, t)|x\big] = \frac{\sum_{i=1}^{n} \left\{ n_i^{-1} \sum_{j=1}^{n_i} \left[1_{[Y_{ij} \leq \widehat{y}(x, t_{ij})], X_i = x]} K_x((t - t_{ij})/h_x)\right] \right\}}{\sum_{i=1}^{n} \left\{ n_i^{-1} \sum_{j=1}^{n_i} \left[K_x((t - t_{ij})/h_x)\right] \right\}}, \qquad (12.32)$$

$$\widehat{F}_t^* \big[\widehat{y}(x, t)|x\big] = \frac{\sum_{i=1}^{n} \sum_{j=1}^{n_i} \left[1_{[Y_{ij} \leq \widehat{y}(x, t_{ij})], X_i = x]} K_x((t - t_{ij})/h_x)\right]}{\sum_{i=1}^{n} \sum_{j=1}^{n_i} \left[K_x((t - t_{ij})/h_x)\right]}, \qquad (12.33)$$

and

$$\widehat{F}_{t,w} \big[\widehat{y}(x, t)|x\big] = \frac{\sum_{i=1}^{n} \left\{ w_i \sum_{j=1}^{n_i} \left[1_{[Y_{ij} \leq \widehat{y}(x, t_{ij})], X_i = x]} K_x((t - t_{ij})/h_x)\right] \right\}}{\sum_{i=1}^{n} w_i \left\{ \sum_{j=1}^{n_i} \left[K_x((t - t_{ij})/h_x)\right] \right\}}, \qquad (12.34)$$

respectively. Since $\hat{y}(x, t)$ is a consistent estimator of $y(x, t)$, these estimators have values between 0 and 1 and are asymptotically non-decreasing as $y(x, t)$ increases.

12.3.3 Conditional Quantiles and Functionals

A direct application of the above conditional CDF estimators is that they can be used to construct the conditional quantile estimators. Suppose that, for any $0 < \alpha < 1$, $t \in \mathscr{T}$ and $X = x$, $y_\alpha(x, t)$ is the unique $(100 \times \alpha)$th conditional quantile of $F_t[y(x, t)|x]$ defined in (12.5). A simple estimator of $y_\alpha(x, t)$ based on $\widehat{F}_t[y(x, t)|x]$ can be obtained by

$$\hat{y}_\alpha(x, t) = \frac{1}{2}\left\{\inf_y\left[y : \widehat{F}_t(y|x) \geq \alpha\right] + \sup_y\left[y : \widehat{F}_t(y|x) < \alpha\right]\right\}. \tag{12.35}$$

Similarly, replacing $\widehat{F}_t(y|x)$ of (12.35) with $\widehat{F}_t^*(y|x)$ of (12.29) or $\widehat{F}_{t,w}(y|x)$ of (12.30), we obtain a simple estimator $\hat{y}_\alpha^*(x, t)$ based on $\widehat{F}_t^*(y|x)$, such that,

$$\hat{y}_\alpha^*(x, t) = \frac{1}{2}\left\{\inf_y\left[y : \widehat{F}_t^*(y|x) \geq \alpha\right] + \sup_y\left[y : \widehat{F}_t^*(y|x) < \alpha\right]\right\} \tag{12.36}$$

or $\hat{y}_\alpha(x, t)$ based on $\widehat{F}_{t,w}(y|x)$, such that,

$$\hat{y}_{\alpha,w}(x, t) = \frac{1}{2}\left\{\inf_y\left[y : \widehat{F}_{t,w}(y|x) \geq \alpha\right] + \sup_y\left[y : \widehat{F}_{t,w}(y|x) < \alpha\right]\right\}. \tag{12.37}$$

Since $y_\alpha(x, t)$ is nondecreasing as α increases, $\hat{y}_\alpha(x, t)$, $\hat{y}_\alpha^*(x, t)$ and $\hat{y}_{\alpha,w}(x, t)$ are all non-decreasing functions of α.

Functionals of the conditional quantiles can be estimated by substituting $y_\alpha(x, t)$ with $\hat{y}_\alpha(x, t)$, $\hat{y}_\alpha^*(x, t)$ or $\hat{y}_{\alpha,w}(x, t)$. In particular, $\Delta_{\alpha_1, \alpha_2}\left[(x_1, t_1), (x_2, t_2)\right]$ of (12.7) can be estimated by

$$\widehat{\Delta}_{\alpha_1, \alpha_2}\left[(x_1, t_1), (x_2, t_2)\right] = \hat{y}_{\alpha_1}(x_1, t_1) - \hat{y}_{\alpha_2}(x_2, t_2), \tag{12.38}$$

$$\widehat{\Delta}_{\alpha_1, \alpha_2}^*\left[(x_1, t_1), (x_2, t_2)\right] = \hat{y}_{\alpha_1}^*(x_1, t_1) - \hat{y}_{\alpha_2}^*(x_2, t_2) \tag{12.39}$$

or

$$\widehat{\Delta}_{\alpha_1, \alpha_2, w}\left[(x_1, t_1), (x_2, t_2)\right] = \hat{y}_{\alpha_1, w}(x_1, t_1) - \hat{y}_{\alpha_2, w}(x_2, t_2). \tag{12.40}$$

For example, we can estimate the conditional inter-quartile range given (x, t) by

$$\widehat{\Delta}_{0.75, 0.25}(x, t) = \hat{y}_{0.75}(x, t) - \hat{y}_{0.25}(x, t). \tag{12.41}$$

Substituting $\hat{y}_\alpha(x, t)$ with $\hat{y}_\alpha^*(x, t)$ or $\hat{y}_{\alpha,w}(x, t)$ for $\alpha = 0.75$ and 0.25, we can estimate $\Delta_{0.75, 0.25}(x, t)$ by $\widehat{\Delta}_{0.75, 0.25}^*(x, t)$ or $\widehat{\Delta}_{0.75, 0.25, w}(x, t)$.

12.3.4 Rank-Tracking Probabilities

For the estimation of the rank-tracking probability $RTP_A(x, s_1, s_2)$ with $s_1 < s_2$ and $x \in \{1, \ldots, K\}$, we define the indicator function

$$I_{A(x,\cdot)}[Y(s_2), Y(s_1), X] = 1_{[Y(s_2) \in A(x, s_2), Y(s_1) \in A(x, s_1), X=x]}, \tag{12.42}$$

and use the conditional probabilities

$$
\begin{aligned}
P_A(x, s_1, s_2) &= P\big[Y(s_2) \in A(x, s_2), Y(s_1) \in A(x, s_1) \big| X = x\big] \\
&= E\big[I_{A(x,\cdot)}[Y(s_2), Y(s_1), X] \big| X = x\big]
\end{aligned} \tag{12.43}
$$

and

$$P\big[Y(s_1) \in A(x, s_1) \big| X = x\big] = E\big[1_{[Y(s_1) \in A(x, s_1), X=x]} \big| X = x\big]. \tag{12.44}$$

Thus, it follows from (12.8), (12.34), (12.35) and (12.36) that the RTP-based on $A(\cdot, \cdot)$ at $s_1 < s_2$ can be rewritten as

$$RTP_A(x, s_1, s_2) = \frac{E\big[I_{A(x,\cdot)}[Y(s_2), Y(s_1), X] \big| X = x\big]}{E\big[1_{[Y(s_1) \in A(x, s_1), X=x]} \big| X = x\big]}. \tag{12.45}$$

The estimation of $RTP_A(x, s_1, s_2)$ can be naturally carried out by estimating the numerator $P_A(x, s_1, s_2)$ and denominator $P\big[Y(s_1) \in A(x, s_1) \big| X = x\big]$ in (12.45) separately, and then calculating their ratios.

1. Estimation of Joint Probabilities

Similar to (12.23), we can estimate $P_A(x, s_1, s_2)$ by minimizing the weighted local square error with the $1/(n n_i)$ weight

$$
\begin{aligned}
L(x, s_1, s_2) = \sum_{i=1}^{n} \sum_{j_1 \neq j_2=1}^{n_i} \Bigg\{ &\left(\frac{1}{n n_i}\right) \Big[I_{A(x,\cdot)}[Y_{ij_2}, Y_{ij_1}, X_i] - p_A(x, s_1, s_2)\Big]^2 \\
&\times K_x\left(\frac{s_1 - t_{ij_1}}{h_{x,1}}, \frac{s_2 - t_{ij_2}}{h_{x,2}}\right) \Bigg\}
\end{aligned} \tag{12.46}
$$

with respect to $p_A(x, s_1, s_2)$, so that the resulting kernel estimator of $P_A(x, s_1, s_2)$, which minimizes (12.46) is

$$\widehat{P}_A(x, s_1, s_2) \tag{12.47}$$

$$= \frac{\sum_{i=1}^{n} \sum_{j_1 \neq j_2=1}^{n_i} \left\{(n n_i)^{-1} I_{A(x,\cdot)}[Y_{ij_2}, Y_{ij_1}, X_i] K_x\left(\frac{s_1 - t_{ij_1}}{h_{x,1}}, \frac{s_2 - t_{ij_2}}{h_{x,2}}\right)\right\}}{\sum_{i=1}^{n} \sum_{j_1 \neq j_2=1}^{n_i} \left\{(n n_i)^{-1} K_x\left(\frac{s_1 - t_{ij_1}}{h_{x,1}}, \frac{s_2 - t_{ij_2}}{h_{x,2}}\right)\right\}},$$

where, for each $x \in \{1, \ldots, K\}$, $K_x(\cdot, \cdot)$ is a non-negative kernel function on the plane $R \times R$, and $h_{x,1}$ and $h_{x,2}$ are the corresponding bandwidths.

When the $1/N$ weight is used, we estimate $P_A(x, s_1, s_2)$ by minimizing the weighted local square error

$$L^*(x, s_1, s_2) = \sum_{i=1}^{n} \sum_{j_1 \neq j_2 = 1}^{n_i} \left\{ \left(\frac{1}{N}\right) \left[I_{A(x, \cdot)}[Y_{ij_2}, Y_{ij_1}, X_i] - p_A(x, s_1, s_2) \right]^2 \right. $$
$$\left. \times K_x \left(\frac{s_1 - t_{ij_1}}{h_{x,1}}, \frac{s_2 - t_{ij_2}}{h_{x,2}} \right) \right\} \tag{12.48}$$

with respect to $p_A(x, s_1, s_2)$, and obtain the kernel estimator

$$\widehat{P}_A^*(x, s_1, s_2) \tag{12.49}$$
$$= \frac{\sum_{i=1}^{n} \sum_{j_1 \neq j_2 = 1}^{n_i} \left\{ I_{A(x, \cdot)}[Y_{ij_2}, Y_{ij_1}, X_i] K_x \left(\frac{s_1 - t_{ij_1}}{h_{x,1}}, \frac{s_2 - t_{ij_2}}{h_{x,2}} \right) \right\}}{\sum_{i=1}^{n} \sum_{j_1 \neq j_2 = 1}^{n_i} \left\{ K_x \left(\frac{s_1 - t_{ij_1}}{h_{x,1}}, \frac{s_2 - t_{ij_2}}{h_{x,2}} \right) \right\}},$$

where $K_x(\cdot, \cdot)$, $h_{x,1}$ and $h_{x,2}$ are the same as in (12.46) and (12.47). More generally, for the weight function w_i, the weighted local square error is

$$L_w(x, s_1, s_2) = \sum_{i=1}^{n} \sum_{j_1 \neq j_2 = 1}^{n_i} \left\{ w_i \left[I_{A(x, \cdot)}[Y_{ij_2}, Y_{ij_1}, X_i] - p_A(x, s_1, s_2) \right]^2 \right. $$
$$\left. \times K_x \left(\frac{s_1 - t_{ij_1}}{h_{x,1}}, \frac{s_2 - t_{ij_2}}{h_{x,2}} \right) \right\}, \tag{12.50}$$

and, by minimizing (12.50) with respect to $p_A(x, s_1, s_2)$, we obtain the kernel estimator

$$\widehat{P}_{A,w}(x, s_1, s_2) \tag{12.51}$$
$$= \frac{\sum_{i=1}^{n} \sum_{j_1 \neq j_2 = 1}^{n_i} \left\{ w_i I_{A(x, \cdot)}[Y_{ij_2}, Y_{ij_1}, X_i] K_x \left(\frac{s_1 - t_{ij_1}}{h_{x,1}}, \frac{s_2 - t_{ij_2}}{h_{x,2}} \right) \right\}}{\sum_{i=1}^{n} \sum_{j_1 \neq j_2 = 1}^{n_i} \left\{ w_i K_x \left(\frac{s_1 - t_{ij_1}}{h_{x,1}}, \frac{s_2 - t_{ij_2}}{h_{x,2}} \right) \right\}},$$

where $K_x(\cdot, \cdot)$, $h_{x,1}$ and $h_{x,2}$ are the same as in (12.46) and (12.47).

2. Estimation of RTP through Ratios of Probability Estimators

Substituting both expected values of the numerator and denominator in (12.45) with their corresponding estimators, the kernel estimators of $RTP_A(x, s_1, s_2)$ based on the weight functions $1/(nn_i)$, $1/N$ and w_i are, respectively,

$$\widehat{RTP}_A(x, s_1, s_2) = \frac{\widehat{P}_A(x, s_1, s_2)}{\widehat{P}_A(x, s_1)}, \tag{12.52}$$

where $\widehat{P}_A(x, s_1)$ is defined in (12.23),

$$\widehat{RTP}_A^*(x, s_1, s_2) = \frac{\widehat{P}_A^*(x, s_1, s_2)}{\widehat{P}_A^*(x, s_1)}, \tag{12.53}$$

where $\widehat{P}_A^*(x, s_1)$ is defined in (12.25), and

$$\widehat{RTP}_{A,w}(x, s_1, s_2) = \frac{\widehat{P}_{A,w}(x, s_1, s_2)}{\widehat{P}_{A,w}(x, s_1)}, \tag{12.54}$$

where $\widehat{P}_{A,w}(x, s_1)$ is defined in (12.27). The bandwidths h_x, $h_{x,1}$ and $h_{x,2}$ in (12.52), (12.53) and (12.54) may not be the same in general. However, for computational simplicity, the same bandwidth h_x for $h_{x,1}$ and $h_{x,2}$ may be used in practice.

As known in the literature, under-smoothing or over-smoothing of kernel estimators are mainly caused by inappropriate bandwidth choices, but are rarely influenced by the shapes of the kernel functions (e.g., Härdle, 1990, Sec. 4.5). Thus, a number of kernel functions may be used for the computation of $\widehat{P}_A(x, t)$, $\widehat{P}_A^*(x, t)$, $\widehat{P}_A(x, s_1, s_2)$ and $\widehat{P}_A^*(x, s_1, s_2)$. For the numerical results of this chapter, we use the Epanechnikov kernel

$$K_E(u) = (3/4)\left(1 - u^2\right) 1_{[|u|<1]} \tag{12.55}$$

for the univariate smoothing estimators $\widehat{P}_A(x, t)$ and $\widehat{P}_A^*(x, t)$, and the multiplicative Epanechnikov kernel

$$K_E(u_1, u_2) = (3/4)^2 \left(1 - u_1^2\right)\left(1 - u_2^2\right) 1_{[|u_1|<1, |u_2|<1]} \tag{12.56}$$

for $\widehat{P}_A(x, s_1, s_2)$ and $\widehat{P}_A^*(x, s_1, s_2)$.

12.3.5   Cross-Validation Bandwidth Choices

We present here two cross-validation approaches for the selection of data-driven bandwidths. These data-driven bandwidths are for the conditional distribution functions. Other quantities, such as the conditional quantiles and the rank-tracking probabilities, are estimated from the plug-in estimators of the conditional distribution functions with the corresponding data-driven bandwidths. For local smoothing estimators of the conditional means, sensible bandwidth choices may often be determined by comparing the plots of the observations and the fitted curves. For the estimation of conditional distribution functions and their functionals, the observations are local indicators, such as (12.42), with values 0 or 1, so that it is difficult to visualize the appropriateness of a bandwidth through a plot. A data-driven bandwidth would be helpful to adjust the fitness of the smoothing estimators in practice.

1. Leave-One-Subject-Out Cross-Validation

The first approach for the selection of data-driven bandwidths is to use the leave-one-subject-out cross-validation (LSCV), which has been used for smoothing with conditional-mean based regression models in Chapters 6 and 7. This bandwidth selection method is based on deleting all the observations of a subject one at a time, and has been investigated by Wu, Tian and Yu (2010), Wu and Tian (2013a, 2013b) for conditional-distribution based regression models.

LSCV Bandwidth for Estimator with a Single Time:

Applying this cross-validation procedure to the kernel estimator $\widehat{P}_{A,w}(x, t)$ based on the w_i weight (12.27), we compute the cross-validated bandwidth $h_{x,LSCV}$ using the following two steps:

(a) *Compute the leave-one-subject-out estimator* $\widehat{P}_{A,w,-i}(x, t_{ij}; h_x)$ *at all the time points* $\{t_{ij} : i = 1, \ldots, n; j = 1, \ldots, n_i\}$ *using the remaining data from the sample with the ith subject's observations deleted.*

(b) *Compute the following LSCV score based on* $\widehat{P}_{A,w,-i}(x, t_{ij}; h_x)$,

$$LSCV_{P_{A,w}}(h_x) = \sum_{i=1}^{n} \sum_{j=1}^{n_i} \left\{ w_i \left[1_{[Y_{ij} \in A(x, t_{ij}), X_i = x]} - \widehat{P}_{A,w,-i}(x, t_{ij}; h_x) \right]^2 \right\}. \quad (12.57)$$

If the right side of (12.57) can be uniquely minimized with respect to h_x, the minimizer of $LSCV_{P_{A,w}}(h_x)$ is the LSCV bandwidth $h_{x,LSCV}$. □

LSCV Bandwidth for Estimator with Bivariate Time:

For the bivariate smoothing estimator $\widehat{P}_{A,w}(x, s_1, s_2)$ of (12.51), a pair of bandwidths $(h_{x,1}, h_{x,2})$ need to be selected for the two time points (s_1, s_2). The LSCV bandwidth selection procedure can be computed as follows:

(a) *Compute the leave-one-subject-out estimator* $\widehat{P}_{A,w,-i}(x, t_{ij_1}, t_{ij_2}; h_{x,1}, h_{x,2})$ *at all the time points* $t_{ij_1} < t_{ij_2}$ *using the remaining data from the sample with the ith subject's observations deleted.*

(b) *Compute the following LSCV score based on* $\widehat{P}_{A,w,-i}(x, t_{ij_1}, t_{ij_2}; h_{x,1}, h_{x,2})$

$$LSCV_{P_A,w}(h_{x,1}, h_{x,2}) = \sum_{i=1}^{n} \sum_{j_1 \neq j_2 = 1}^{n_i} \left\{ w_i \left[I_{A(x,\cdot)}[Y_{ij_2}, Y_{ij_1}, X_i] \right. \right. \quad (12.58)$$

$$\left. \left. - \widehat{P}_{A,w,-i}(x, t_{ij_1}, t_{ij_2}; h_{x,1}, h_{x,2}) \right]^2 1_{[n_i \geq 2]} \right\}.$$

If the right side of (12.58) can be uniquely minimized with respect to the bandwidth pair $(h_{x,1}, h_{x,2})$, the minimizer of $LSCV_{P_{A,w}}(h_{x,1}, h_{x,2})$ is the LSCV bandwidth pair $(h_{x,1,LSCV}, h_{x,2,LSCV})$. □

For the computation of $h_{x,1,LSCV}$ and $h_{x,2,LSCV}$, the subjects with observations at only one time point (i.e., the subjects with $n_i = 1$) are omitted in the minimization of (12.58). For simplicity, we may take $h_{x,1} = h_{x,2}$ in (12.58), so that a single LSCV bandwidth $h_{x,1,LSCV} = h_{x,2,LSCV}$ is used for both time points s_1 and s_2. The LSCV bandwidths obtained from (12.57) and (12.58) can be used in (12.54) for the estimation of $RTP_A(x, s_1, s_2)$.

Although theoretical properties of the LSCV bandwidths under the current context require further systematic investigation, the appropriateness of (12.57) and (12.58) in practice may be seen heuristically from some of their expansions. We illustrate this heuristic justification using the univariate case of (12.57) with the $w_i = 1/(nn_i)$ weight. The bivariate case of (12.58) can be justified similarly. A direct expansion of (12.57) for $w_i = 1/(nn_i)$ shows that

$$
\begin{aligned}
LSCV_{P_A}(h_x) &= \sum_{i=1}^{n}\sum_{j=1}^{n_i}\left\{\left(\frac{1}{nn_i}\right)\left[1_{[Y_{ij}\in A(x,t_{ij}),X_i=x]} - P_A\left(x, t_{ij}\right)\right]^2\right\} \\
&+ \sum_{i=1}^{n}\sum_{j=1}^{n_i}\left\{\left(\frac{1}{nn_i}\right)\left[P_A\left(x, t_{ij}\right) - \widehat{P}_{A,-i}\left(x, t_{ij}; h_x\right)\right]^2\right\} \\
&+ 2\sum_{i=1}^{n}\sum_{j=1}^{n_i}\left\{\left(\frac{1}{nn_i}\right)\left[1_{[Y_{ij}\in A(x,t_{ij}),X_i=x]} - P_A\left(x, t_{ij}\right)\right]\right. \\
&\qquad\qquad\left. \times\left[P_A\left(x, t_{ij}\right) - \widehat{P}_{A,-i}\left(x, t_{ij}; h_x\right)\right]\right\}. \quad (12.59)
\end{aligned}
$$

The first term at the right side of (12.59) does not depend on the bandwidth. Since the ith subject is not included in the estimator $\widehat{P}_{A,-i}(x, t_{ij}; h_x)$, the expected value of the third term at the right side of (12.59) is zero. Since the expectation of the second term at the right side of (12.59) is approximately the expectation of the average squared error

$$
ASE\left[\widehat{P}_A(x, \cdot)\right] = \sum_{i=1}^{n}\sum_{j=1}^{n_i}\left\{\left(\frac{1}{nn_i}\right)\left[P_A\left(x, t_{ij}\right) - \widehat{P}_{A,-i}\left(x, t_{ij}; h_x\right)\right]^2\right\}, \quad (12.60)
$$

we observe that, by minimizing $LSCV_{1,P_A}(x)$, the LSCV bandwidth approximately minimizes the expectation of $ASE\left[\widehat{P}_A(x, \cdot)\right]$.

2. Leave-One-Time-Point-Out Cross-Validation

A potential drawback of the LSCV approach above is that minimizing the cross-validation scores (12.57) and (12.58) can be computationally intensive. To alleviate the computational burden, an alternative cross-validation approach, which mimics the cross-validation with cross-sectional data and is suggested by Wu, Tian and Yu (2010), is by deleting the distinct design time points $\mathbf{t} = \left(t_{(1)}, \ldots, t_{(J)}\right)^T$ one at a time. We refer to this approach as the *leave-one-time-point-out cross-validation* (LTCV).

Notation for Distinct Time Points:
We first recall the notation for distinct time points. Since $\{t_{(1)}, \ldots, t_{(J)}\}$ contains all the distinct possible time points after proper round-up or binning (e.g., Wu, Tian and Yu, 2010, Remark 2 and Sec. 4), the time points of the

ith subject, for each $1 \leq i \leq n$, $\{t_{i1}, \ldots, t_{in_i}\}$ are only a subset of $\mathbf{t}$, and, on the other hand, each $t_{(j^*)}$ contains the observations from n_{j^*} different subjects. Then, for each time point $t_{(j^*)} \in \mathbf{t}$, the set of subjects with observations at $t_{(j^*)}$ is $\mathscr{S}_{j^*}$. For each $i \in \mathscr{S}_{j^*}$, the observed outcome at time point $t_{(j^*)}$ is $Y_i(t_{(j^*)})$. $\square$

The LTCV bandwidths for the kernel estimators of the conditional probabilities with univariate time and bivariate time $P_A(x, t)$ and $P_A(x, s_1, s_2)$, respectively, based on the general w_i weight function are computed using the following steps.

LTCV Bandwidth for Estimator with a Single Time:

(a) *Compute the leave-one-time-point-out estimator* $\widehat{P}_{A,w}^{-(j^*)}(x, t_{(j^*)}; h_x)$ *using (12.27) with the remaining data after deleting all the observations at time point $t_{(j^*)}$, i.e., with $\{Y_i(t_{(j^*)}); i \in \mathscr{S}_{j^*}\}$ deleted.*

(b) *Compute the following LTCV score based on* $\widehat{P}_{A,w}^{-(j^*)}(x, t_{(j^*)}; h_x)$,

$$
LTCV_{P_{A,w}}(h_x) \quad = \quad \sum_{j^*=1}^{J} \sum_{i \in \mathscr{S}_{j^*}} \Big\{ 1_{[Y_i(t_{(j^*)}) \in A(x, t_{(j^*)}), X_i = x]}
$$

$$
- \widehat{P}_{A,w}^{-(j^*)}(x, t_{(j^*)}; h_x) \Big\}^2. \qquad (12.61)
$$

If the right side of (12.61) can be uniquely minimized with respect to h_x, the minimizer of $LTCV_{P_{A,w}}(h_x)$ is the LTCV bandwidth $h_{x, LTCV}$. $\square$

LTCV Bandwidth for Estimator with Bivariate Time:

(a) *Compute the leave-one-time-point-out estimator* $\widehat{P}_{A,w}^{-(j_1^*, j_2^*)}\big[x, t_{(j_1^*)}, t_{(j_2^*)}; h_{x,1}, h_{x,2}\big]$ *using (12.51) with the remaining data after deleting all the observations at time points $\{t_{(j_1^*)}, t_{(j_2^*)}\}$, i.e., with $\{(Y_i(t_{(j_1^*)}), Y_i(t_{(j_2^*)})); i \in \mathscr{S}_{j_1^*} \cap \mathscr{S}_{j_2^*}\}$ deleted.*

(b) *Compute the following LTCV score based on* $\widehat{P}_{A,w}^{-(j_1^*, j_2^*)}\big[x, t_{(j_1^*)}, t_{(j_2^*)}; h_{x,1}, h_{x,2}\big]$,

$$
LTCV_{P_{A,w}}(h_{x,1}, h_{x,2}) \quad = \quad \sum_{j_1^* \neq j_2^* = 1}^{J} \sum_{i \in \mathscr{S}_{j_1^*} \cap \mathscr{S}_{j_2^*}} \Big\{ 1_{A(x, \cdot)}\big[Y_i(t_{(j_2^*)}), Y_i(t_{(j_1^*)}), X_i = x\big]
$$

$$
- \widehat{P}_A^{-(j_1^*, j_2^*)}\big[x, t_{(j_1^*)}, t_{(j_2^*)}; h_{x,1}, h_{x,2}\big] \Big\}^2. \qquad (12.62)
$$

If the right side of (12.62) can be uniquely minimized with respect to the bandwidth pair $(h_{x,1}, h_{x,2})$, the minimizer of $LTCV_{P_{A,w}}(h_{x,1}, h_{x,2})$ is the LTCV bandwidth pair $(h_{x,1,LTCV}, h_{x,2,LTCV})$. $\square$

Similar to the LSCV score in (12.58), the subjects with observations at only one time point are omitted in (12.62), and the computation for min-imizing $LTCV_{P_{A,w}}(h_{x,1}, h_{x,2})$ may be simplified by setting $h_{x,1} = h_{x,2}$, so that $h_{x,1,LTCV} = h_{x,2,LTCV}$. Unlike the LSCV bandwidths, which rely on deleting independent subjects one at a time, the LTCV bandwidths of (12.61) and (12.62) rely on deleting the design time points in $\mathbf{t}$ one at a time, and ignores the potential intra-subject correlations across the time points. Therefore, the heuristic justification for the LSCV bandwidths may not be directly applied to the LTCV bandwidths. A potential advantage of using the LTCV bandwidths instead of the LSCV bandwidths is that minimizing (12.61) and (12.62) is often computationally simpler than minimizing (12.57) and (12.58). The sim-ulation results of Wu, Tian and Yu (2010) show that the LTCV approach may often lead to adequate bandwidth choices under various longitudinal designs similar to the NGHS data. The computational simplicity makes LTCV an attractive approach in practice compared with the LSCV approach.

12.3.6 Bootstrap Pointwise Confidence Intervals

Explicit expressions of the asymptotic distributions for the estimators in Sec-tions 12.3.1 to 12.3.5 have not been described in the literature. Wu and Tian (2013a) suggest to use the same resampling subject bootstrap method as in Section 8.3.1 to construct the pointwise confidence intervals for these estima-tors. This resampling subject bootstrap is based on the practical suitability of the bootstrap confidence intervals shown in the simulation study of Wu and Tian (2013a). Following the spirit of the resampling bootstrap procedure of Section 8.3.1, we construct the bootstrap pointwise confidence intervals for $P_A(x, t)$ and $P_A(x, s_1, s_2)$ using the following steps.

Bootstrap Pointwise Confidence Intervals with a Single Time:

(a) **Bootstrap Estimators:** *Draw B resampling-subject bootstrap samples*

$$\mathscr{L}^b = \left\{ (Y_{ij}^b, X_i^b, t_{ij}^b) : 1 \leq i \leq n; 1 \leq j \leq n_i \right\}, \ b = 1, \ldots, B, \qquad (12.63)$$

and compute the bootstrap estimators $\left\{ \widehat{P}_{A,w}^b(x, t) : b = 1, \ldots, B \right\}$ us-ing (12.27).

(b) **Approximated Bootstrap Confidence Intervals:** *Calculate the lower and upper $[100 \times (1 - \alpha/2)]$th percentiles of the B bootstrap estimators $\left\{ \widehat{P}_{A,w}^b(x, t) : b = 1, \ldots, B \right\}$, and denote the corresponding percentiles by $L_{P_{A,w}, \alpha/2}(t)$ and $U_{P_{A,w}, \alpha/2}(t)$. The $[100 \times (1 - \alpha)]$th bootstrap percentile point-wise confidence interval for $P_A(x, t)$ is*

$$\left(L_{P_{A,w}, \alpha/2}(t), U_{P_{A,w}, \alpha/2}(t) \right). \qquad (12.64)$$

The $[100 \times (1 - \alpha/2)]$th normal approximated pointwise confidence interval for $P_A(x, t)$ is

$$\widehat{P}_{A,w}(x, t) \pm z_{1-\alpha/2} \, \widehat{se}\left[\widehat{P}_{A,w}^b(x, t) \right], \qquad (12.65)$$

where $\widehat{se}\big[\widehat{P}^b_{A,w}(x,t)\big]$ is the sample standard error of the bootstrap estimators $\big\{\widehat{P}^b_{A,w}(x,t) : b = 1, \ldots, B\big\}$. □

The steps for constructing the bootstrap confidence intervals for $P_A(x, s_1, s_2)$ are essentially the same as above with the minor modifications that the estimators of (12.47) based on the original sample and the bootstrap samples have to be used.

Bootstrap Pointwise Confidence Intervals with Bivariate Time:

(a) **Bootstrap Estimators:** *Generate the bootstrap samples as in step (a) above and obtain the bootstrap estimators* $\big\{\widehat{P}^b_{A,w}(x, s_1, s_2) : b = 1, \ldots, B\big\}$.

(b) **Approximated Bootstrap Confidence Intervals:** *The* $[100 \times (1 - \alpha)]\%$ *bootstrap percentile pointwise confidence interval for* $P_A(x, s_1, s_2)$ *is*

$$\big(L_{P_{A,w},\,\alpha/2}(s_1, s_2),\ U_{P_{A,w},\,\alpha/2}(s_1, s_2)\big), \tag{12.66}$$

where $L_{P_{A,w},\,\alpha/2}(s_1, s_2)$ *and* $U_{P_{A,w},\,\alpha/2}(s_1, s_2)$ *are the lower and upper* $[100 \times (1 - \alpha/2)]th$ *percentiles of the bootstrap estimators. The* $[100 \times (1 - \alpha/2)]\%$ *normal approximated pointwise confidence interval for* $P_A(x, s_1, s_2)$ *is*

$$\widehat{P}_{A,w}(x, s_1, s_2) \pm z_{1-\alpha/2}\,\widehat{se}\big[\widehat{P}^b_{A,w}(x, s_1, s_2)\big], \tag{12.67}$$

where $\widehat{se}\big[\widehat{P}^b_{A,w}(x, s_1, s_2)\big]$ *is the sample standard error of the bootstrap estimators.* □

Similar to the discussion of Section 7.3.1, the normal approximated intervals of (12.65) and (12.67) are not rigorously proper asymptotically $[100 \times (1 - \alpha)]th$ confidence intervals because they do not adjust for the asymptotic biases of $\widehat{P}_{A,w}(x, t)$ and $\widehat{P}_{A,w}(x, s_1, s_2)$. Theoretically, we may consider a plug-in approach to adjust the potential bias using an estimator of the asymptotic bias. But, since it is usually difficult to obtain an appropriate estimate of the asymptotic bias, such plug-in type approximate confidence intervals may not have practical advantages over (12.65) and (12.67), which completely ignore the asymptotic biases. Bootstrap pointwise confidence intervals for the functionals of $P_A(x,t)$ and $P_A(x, s_1, s_2)$ can be constructed using similar procedures as above. For example, the bootstrap confidence intervals for the conditional quantiles $y_\alpha(x, t)$ can be constructed using $\widehat{y}_{\alpha,w}(x, t)$, and the bootstrap confidence intervals for the rank-tracking probability $RTP_A(x, s_1, s_2)$ and the rank-tracking probability ratio $RTPR_A(x, s_1, s_2)$ can be constructed using their corresponding estimators (12.54) and (12.55).

12.4 R Implementation

12.4.1 The NGHS BMI Data

The NGHS body mass index (BMI) data has been described in Sections 1.2 and 5.2.2. The study had high retention rate during 10 years of follow-up. The median visits for the individual girls is 9, with range 1 to 10 and inter-quartile range of 8 to 10. Among all the important risk factors that have been studied by the NGHS investigators, childhood overweight and obesity measured by BMI have been linked by a number of NGHS publications, e.g., Daniels et al. (1998), Kimm et al. (2000, 2001, 2002), Thompson et al. (2007) and Obarzanek et al. (2010), to cardiovascular disease risk factors including hypertension and unhealthy lipid levels. Applying the procedures of Section 12.3 to the NGHS BMI data, the kernel estimates are obtained for the conditional probabilities of overweight and obesity, as well as the RTPs for BMI given race and various ages for this population of adolescent girls. These estimates, particularly the estimated RTPs, illustrate the importance of tracking BMI at an early age for the prevention of later incidence of overweight and obesity for this population.

Because the entry age starts at 9, the observed age in our analysis is limited to $\mathscr{T} = [9, 19)$ and rounded up to the first decimal point. This age round-up has the required clinical accuracy for age (Obarzanek et al., 2010), which leads to $J = 100$ distinct time-design points $\{t_{(1)} = 9.0, t_{(2)} = 9.1, \ldots, t_{(100)} = 18.9\}$. For a given $1 \leq j^* \leq J$, $i \in \mathscr{S}_{j^*}$ denotes that the ith girl has a BMI observation $Y_i(t_{(j^*)})$ at age $t_{(j^*)}$, $X_i = 0$ if she is Caucasian, and $X_i = 1$ if she is African American. Clearly, $Y_i(t_{(j^*)}) = Y_{ij}$, if $t_{(j^*)} = t_{ij}$ is the age of the ith girl at the jth BMI measurement, $1 \leq j \leq n_i$. The random variables for BMI at age $t \in \mathscr{T}$ and race are $Y(t)$ and X, respectively. Since a main objective of this study is to evaluate the patterns and temporal trends of overweight and obesity for this cohort of girls (e.g., Kimm et al., 2002), we define

$$A(x, t) = \{(Y(t), X) : Y(t) > y(t), X = x\},$$

where $y(t)$ is a given BMI quantile determined by the CDC BMI growth chart for girls at age t (Kuczmarski, et al., 2002), so that

$$\left\{ \begin{array}{rcl} P_A(x, t) & = & P[Y(t) > y(t) | X = x, t], \\ P_A(x, s_1, s_2) & = & P[Y(s_1) > y(s_1), Y(s_2) > y(s_2) | X = x, s_1, s_2], \quad (12.68) \\ RTP_A(x, s_1, s_2) & = & P[Y(s_2) > y(s_2) | Y(s_1) > y(s_1), X = x]. \end{array} \right.$$

In particular, we choose $y(t)$ in (12.68) to be the 85th percentile of the CDC BMI growth chart, which was used by Obarzanek et al. (2010) to define overweight and obese girls at age t.

For the kernel estimation of $P_A(x, t)$ and $P_A(x, s_1, s_2)$, we use the corresponding univariate and bivariate Epanechnikov kernels with both $(n n_i)^{-1}$ and N^{-1} weights and the LSCV and LTCV bandwidths, where the same bandwidth $h_{x,1} = h_{x,2}$ is used in the estimation of $P_A(s_1, s_2, x)$. Since both $(n n_i)^{-1}$ and N^{-1}

weights give similar estimators (as most girls in NGHS have n_i between 8 and 10) and the bandwidths obtained from LSCV and LTCV are comparable, we only present the estimation results from the N^{-1} weight with the LTCV bandwidths. Computed separately for Caucasian and African American girls, the univariate and bivariate LTCV bandwidths for the estimates of $P_A(x, t)$ and $P_A(x, s_1, s_2)$ are $h_{0,LTCV} = 3.5$, $h_{1,DTCV} = 2.6$, $h_{0,1,LTCV} = h_{0,2,LTCV} = 0.8$ and $h_{1,1,LTCV} = h_{1,2,LTCV} = 0.9$. The estimators for

$$
\begin{cases}
RTP_A(x, s_1, s_2) & = P\big[Y(s_2) > y(s_2)\big|Y(s_1) > y(s_1), X = x\big], \\
RTPR_A(x, s_1, s_2) & = RTP_A(x, s_1, s_2)/P\big[Y(s_2) > y(s_2)\big|X = x\big]
\end{cases}
\tag{12.69}
$$

based on (12.68) are computed by substituting their components with the corresponding kernel estimators. The 95% pointwise confidence intervals are computed from the resampling-subject bootstrap empirical quantile confidence intervals with $B = 500$ bootstrap samples.

We compute the kernel estimates of the $P_A(x, t)$ and $P_A(x, s_1, s_2)$ with $s_2 = s_1 + 2$ and $A(\cdot)$ including those girls with BMI percentile greater than the CDC age-adjusted $85th$ percentile for the African American girls. Then, we use these kernel estimates to compute their RTP and RTPR of overweight and obesity. Similarly, we can compute the conditional probabilities and tracking indices for the Caucasian girls. The R code for the computation is given below:

```
> library(npmlda)
> NGHS.B <- NGHS[NGHS$RACE==2,]
> IDD <- unique(NGHS.B$ID)
> nID <- length(IDD)
> nID # no. of subjects

[1] 1213

> nrow(NGHS.B) # no. of visits

[1] 10028

> Grid  <- 0.2
> S1cat <- seq(9.0, 18.8, by=Grid)
> S12cat <- seq(9.0, 16.8, by=Grid)
## Compute the PA(x,s1 )##
> KK1 <- length(S1cat)
> Yvec<- (NGHS.B$BMIPCT>=85)*1
> Xvec<- NGHS.B$agebin
> Prob.S1 <- numeric(KK1)
> for(k in 1:KK1)
  {
   Prob.S1[k]<-NW.Kernel(Xvec, Yvec, S1cat[k], Bndwdth=2.6)
  }
```

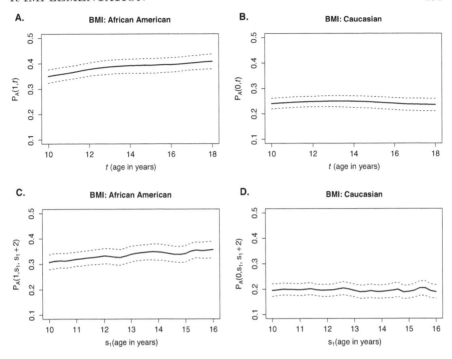

Figure 12.1 *The estimated conditional probability curves and their bootstrap 95% empirical quantile pointwise CIs.*

```
## Compute the PA(x,s1,s2) ##
> KK2 <- length(S12cat)
> Prob.S1S2 <- numeric(KK2)
> IDD <- NGHS.B$ID

> for(k in 1:KK2)
  {
    Prob.S1S2[k] <- Kernel2D(IDD, Xvec, Yvec, X01=S12cat[k],
                 X02=S12cat[k]+2, Bndwdth1=0.9, Bndwdth2=0.9)
  }
## Compute RTP and RTPR ##
> IND1 <- (S1cat >=9 & S1cat <=16.8)
> IND2 <- (S1cat >=11 & S1cat <=18.8)
> RTP   <- Prob.S1S2/(Prob.S1[IND1])
> RTPR <- RTP/(Prob.S1[IND2])
```

Figure 12.1 shows the estimated probabilities,

$$
\begin{cases}
P_A(x, t) &= P\big[Y(t) > y(t) \big| X = x, t\big] \quad \text{for } t \in [10, 18], \\
P_A(x, s_1, s_1 + 2) &= P\big[Y(s_1) > y(s_1), Y(s_1 + 2) > y(s_1 + 2) \\
&\quad \big| X = x, s_1, s_1 + 2\big] \quad \text{for } s_1 \in [10, 16]
\end{cases}
\tag{12.70}
$$

and their corresponding 95% pointwise confidence intervals, where $y(t)$ is the 85th percentile of the CDC BMI growth chart for girls at age t. In (12.70), $P_A(x,t)$ with $x = 0$ or 1 represents the probability of a Caucasian or an African American girl being overweight or obese at age t, and $P_A(x, s_1, s_1 + 2)$ with $x = 1$ or 0 represents the joint probability of an African American or a Caucasian girl being overweight or obese at both age s_1 and age $s_1 + 2$. The estimated $P_A(x,t)$ for both racial groups are higher than 15%, suggesting that this cohort of girls have higher overweight or obese probabilities than the general population of girls described in the CDC BMI growth chart. The African American girls also tend to have higher overweight or obese probability than the Caucasian girls. The probability of overweight and obesity for African American girls appears to be gradually increasing over age, but the probability for Caucasian girls stays constant across the age range.

Figure 12.2 shows the estimated $RTP_A(x, s_1, s_1 + 2)$ and $RTPR_A(x, s_1, s_1 + 2)$ for $s_1 \in [10, 16]$ years and their corresponding 95% pointwise confidence intervals. Here $RTP_A(x, s_1, s_1 + 2)$ with $x = 1$ or 0 represents the probability of an African American or Caucasian girl being overweight or obese at age $s_1 + 2$ given that she is already overweight or obese at age s_1. The top panels of Figure 12.2 show that, given that a girl is already overweight or obese at age s_1, her overweight or obese probability at age $s_1 + 2$ is approximately 90% or 80% if she is African American or Caucasian, respectively. The bottom panels of Figure 12.2 show that $RTPR_A(x, s_1, s_1 + 2)$ is approximately between 2.1 and 2.5 for African American girls, and between 3.2 and 3.6 for Caucasian girls. These RTPR estimates suggest that BMI has high degree of "positive tracking ability" for both racial groups.

12.5 Asymptotic Properties

We establish the asymptotic biases, variances and mean squared errors of the kernel estimators $\widehat{P}_A(x, t)$ and $\widehat{P}_A^*(x, t)$ defined in (12.23) and (12.25), respectively. The results of this section suggest that, since these estimators rely on the two different weights $w_i = 1/(n n_i)$ and $w_i = 1/N$, their asymptotic properties are also different. We pay special attention to these two weights because, in the absence of the actual intra-subject correlation structures, they are the most commonly used weight choices in real applications. The asymptotic properties of $\widehat{P}_{A,w}(x, t)$ of (12.27) with a general weight w_i can be derived using the same steps in the proofs of the theorems of this section, but are omitted due to the lack of interest for using weights other than $1/(n n_i)$ and $1/N$.

Asymptotic properties for the other estimators of Section 12.3, such as the

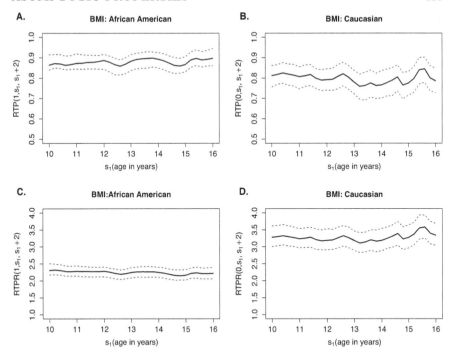

Figure 12.2 *The estimated RTP and RTP curves and their bootstrap 95% empirical quantile pointwise CIs for BMI over the 85th population BMI percentile at the corresponding age.*

estimators of the conditional distribution functions with bivariate time and the rank-tracking probabilities, can be derived using similar approaches. But, the details of these extensions, which require tedious computations, have not been explicitly developed in the literature. Although our theoretical development of this section is limited to the estimators $\widehat{P}_A(x, t)$ and $\widehat{P}_A^*(x, t)$ under a single time, the technique of this section is applicable to kernel estimators of other conditional distribution functions and their functionals.

12.5.1   Asymptotic Assumptions

The following assumptions are assumed throughout this chapter:

(a) *For all $t \in \mathscr{T}$ and x, there are a density function $f(x, t) > 0$ and an integer $d_x \geq 0$, such that $f(x, t)$ is continuously differentiable with respect to t and $P_A(x, t)$ is $(d_x + 2)$ time continuously differentiable with respect to t.*

(b) *For each x, $K_x(\cdot)$ is a compactly supported $(d_x + 2)$th order kernel function* *in the sense that, for all* $1 \le k < d_x + 2$, $K_x(\cdot)$ *satisfies*

$$\begin{cases} \int K_x(u)\,du = 1, & \int u^k K_x(u)\,du = 0, \\ \mu_{(d_x+2)} = \int u^{d_x+2} K_x(u)\,du < \infty, & R(K_x) = \int K_x^2(u)\,du. \end{cases} \tag{12.71}$$

Without loss of generality, the support of $K_x(\cdot)$ *is assumed to be* $[-1,1]$.

(c) *For each* x, *the bandwidth* h_x *satisfies* $h_x \to 0$ *and* $n h_x \to \infty$ *as* $n \to \infty$.

(d) *For all* $i = 1, \dots, n$ *and any* $j_1 \ne j_2$, $|t_{ij_1} - t_{ij_2}| > h_x$.   □

The above assumptions are consistent with the standard assumptions used for kernel smoothing estimators in conditional mean based regression models, e.g., Sections 6.6.2 and 7.5.1. In particular, Assumption (d), which is used in the derivations of Section 12.5.3, is appropriate in practice, because in most longitudinal studies the subject's visit times are spread apart by design, so that meaningful scientific information may be obtained at different visits.

12.5.2 Asymptotic Mean Squared Errors

1. Asymptotic Expressions with the Subject Uniform Weight

We present first the approximate mean squared errors of $\widehat{P}_A(x,t)$ in (12.23), which has the subject uniform weight $w_i = 1/(n n_i)$ weight. To consider the numerator and the denominator at the right side of (12.23) separately, we define the numerator and denominator to be

$$\widehat{p}_A(x,t) = \sum_{i=1}^{n} \sum_{j=1}^{n_i} \left\{ \left(\frac{1}{n n_i h_x} \right) \left[1_{[Y_{ij} \in A(x,t_{ij}), X_i=x]} K_x \left(\frac{t - t_{ij}}{h_x} \right) \right] \right\} \tag{12.72}$$

and

$$\widehat{f}(x,t) = \sum_{i=1}^{n} \sum_{j=1}^{n_i} \left[\left(\frac{1}{n n_i h_x} \right) K_x \left(\frac{t - t_{ij}}{h_x} \right) \right], \tag{12.73}$$

respectively, such that

$$\widehat{P}_A(x,t) = \frac{\widehat{p}_A(x,t)}{\widehat{f}(x,t)}. \tag{12.74}$$

Using the same approximation argument as (3.37) and (3.38), we note that

$$\widehat{f}(x,t) \to f(x,t) \quad \text{in probability as } n \to \infty \text{ and } h_x \to 0, \tag{12.75}$$

for any t in the support of $f(x,t)$, and, by (12.72) through (12.75),

$$[1 - \delta(x,t)] [\widehat{P}_A(x,t) - P_A(x,t)] = [f(x,t)]^{-1} [\widehat{p}_A(x,t) - P_A(x,t)\widehat{f}(x,t)], \tag{12.76}$$

where $\delta(x,t) = 1 - [\widehat{f}(x,t)/f(x,t)]$. Consequently, by $f(x,t) > 0$, (12.75) and (12.76), we have that $\delta(x,t) = o_p(1)$ and

$$[1 + o_p(1)] [\widehat{P}_A(x,t) - P_A(x,t)] = [f(x,t)]^{-1} \widehat{R}_A(x,t), \tag{12.77}$$

where $\widehat{R}_A(x,t) = \widehat{p}_A(x,t) - P_A(x,t)\widehat{f}(x,t)$.

Using the right side of (12.77), we define the approximate mean squared error (MSE) and the approximate mean integrated squared error (MISE) of $\widehat{P}_A(x,t)$ to be

$$
MSE\left[\widehat{P}_A(x,t), P_A(x,t)\right]
$$

$$
= E\left\{\left[f^{-1}(x,t)\widehat{R}_A(x,t)\right]^2\right\}
$$

$$
= \left\{E\left[f^{-1}(x,t)\widehat{R}_A(x,t)\right]\right\}^2 + V\left\{\left[f(x,t)\right]^{-1}\widehat{R}_A(x,t)\right\} \qquad (12.78)
$$

and

$$
MISE\left[\widehat{P}_A(x), P_A(x)\right] = \int MSE\left[\widehat{P}_A(x,t), P_A(x,t)\right]\pi(t)\,dt, \qquad (12.79)
$$

where $E(\cdot)$ and $V(\cdot)$ are the conditional expectation and variance given x, and $\pi(t)$ is a pre-specified non-negative weight function on $\mathscr{T}$. Following the decomposition of (12.78), we refer to

$$
\begin{cases}
Bias\left[\widehat{P}_A(x,t)\right] &= E\left[f^{-1}(x,t)\widehat{R}_A(x,t)\right], \\
Var\left[\widehat{P}_A(x,t)\right] &= V\left[f^{-1}(x,t)\widehat{R}_A(x,t)\right]
\end{cases} \qquad (12.80)
$$

as the bias and variance terms of $\widehat{P}_A(x,t)$. The following theorem summarizes the asymptotic expression for the MSE of $\widehat{P}_A(x,t)$.

Theorem 12.1. *When the number of subjects n is large, t is an interior point of the support of $f(x,\cdot)$ and Assumptions (a)-(d) of Section 12.5.1 are satisfied, the following asymptotic expressions for $\widehat{P}_A(x,t)$ hold:*

(a) *The asymptotic bias of (12.80) is*

$$
Bias\left[\widehat{P}_A(x,t)\right] = h_x^{(d_x+2)}\mu_{(d_x+2)}\left[\frac{P_A^{(d_x+2)}(x,t)}{(d_x+2)!} + \frac{P_A^{(d_x+1)}(x,t)f'(x,t)}{(d_x+1)!f(x,t)}\right]
$$

$$
+o\left(h_x^{(d_x+2)}\right). \qquad (12.81)
$$

(b) *The asymptotic variance of (12.80) is*

$$
Var\left[\widehat{P}_A(x,t)\right] = \left[nh_xf(x,t)\right]^{-1}\left[\sum_{i=1}^{n}(nn_i)^{-1}\right]\left[1 - P_A(x,t)\right]P_A(x,t)R(K_x)
$$

$$
+o\left(h_x^{-1}n^{-2}\sum_{i=1}^{n}n_i^{-1}\right). \qquad (12.82)
$$

(c) *The asymptotic mean squared error of (12.78) is*

$$
MSE\left[\widehat{P}_A(x,t), P_A(x,t)\right] = \left\{Bias\left[\widehat{P}_A(x,t)\right]\right\}^2 + Var\left[\widehat{P}_A(x,t)\right]
$$

$$
+o\left(h_x^{2(d_x+2)} + h_x^{-1}n^{-2}\sum_{i=1}^{n}n_i^{-1}\right), \qquad (12.83)
$$

where the asymptotic expressions of $Bias[\widehat{P}_A(x,t)]$ *and* $Var[\widehat{P}_A(x,t)]$ *are given by (12.81) and (12.82), respectively.* ∎

Proof of Theorem 12.1 is given in Section 12.5.3.

In particular, when $P_A(x,t)$ is twice continuously differentiable, i.e., $d_x = 0$, the asymptotic expressions of the bias, variance and MSE in (12.81), (12.82) and (12.83) are simplified to

$$Bias[\widehat{P}_A(x,t)] = h_x^2 \mu_{(2)} \left[\frac{P_A''(x,t)}{2} + \frac{P_A'(x,t)f'(x,t)}{f(x,t)} \right] + o(h_x^2), \tag{12.84}$$

$$Var[\widehat{P}_A(x,t)] = [nh_x f(x,t)]^{-1} \left[\sum_{i=1}^n (nn_i)^{-1} \right] [1 - P_A(x,t)] P_A(x,t) R(K_x)$$
$$+ o\left(h_x^{-1} n^{-2} \sum_{i=1}^n n_i^{-1} \right) \tag{12.85}$$

and

$$MSE[\widehat{P}_A(x,t), P_A(x,t)] = \left\{ Bias[\widehat{P}_A(x,t)] \right\}^2 + Var[\widehat{P}_A(x,t)]$$
$$+ o\left(h_x^4 + h_x^{-1} n^{-2} \sum_{i=1}^n n_i^{-1} \right), \tag{12.86}$$

respectively.

2. Asymptotic Expressions with the Measurement Uniform Weight

The asymptotic expressions of the bias, variance and MSE for $\widehat{P}_A^*(x,t)$ of (12.25) with the measurement uniform weight $w_i = 1/N$ weight can be derived using the same approach as above. In this case, we define

$$\widehat{p}_A^*(x,t) = \sum_{i=1}^n \sum_{j=1}^{n_i} \left\{ \left(\frac{1}{Nh_x} \right) \left[1_{[Y_{ij} \in A(x,t_{ij}), X_i = x]} K_x \left(\frac{t - t_{ij}}{h_x} \right) \right] \right\} \tag{12.87}$$

and

$$\widehat{f}^*(x,t) = \sum_{i=1}^n \sum_{j=1}^{n_i} \left\{ \left(\frac{1}{Nh_x} \right) K_x \left(\frac{t - t_{ij}}{h_x} \right) \right\}, \tag{12.88}$$

so that,

$$\widehat{P}_A^*(x,t) = \frac{\widehat{p}_A^*(x,t)}{\widehat{f}^*(x,t)}. \tag{12.89}$$

Following the same arguments in (12.75) to (12.76), we have the approximation

$$[1 + o_p(1)][\widehat{P}_A^*(x,t) - P_A(x,t)] = f^{-1}(x,t)\widehat{R}_A^*(x,t), \tag{12.90}$$

where $\widehat{R}_A^*(x,t) = \widehat{p}_A^*(x,t) - P_A(x,t)\widehat{f}^*(x,t)$. The approximate mean squared errors are

$$MSE\left[\widehat{P}_A^*(x,t), P_A(x,t)\right]$$
$$= \left\{E\left[f^{-1}(x,t)\widehat{R}_A^*(x,t)\right]\right\}^2 + V\left[f^{-1}(x,t)\widehat{R}_A^*(x,t)\right] \qquad (12.91)$$

and

$$MISE\left[\widehat{P}_A^*(x), P_A(x)\right] = \int MSE\left[\widehat{P}_A^*(x,t), P_A(x,t)\right]\pi(t)\,dt. \qquad (12.92)$$

The bias and variance for $\widehat{P}_A^*(x,t)$ are

$$\begin{cases} Bias\left[\widehat{P}_A^*(x,t)\right] &= E\left[f^{-1}(x,t)\widehat{R}_A^*(x,t)\right], \\ Var\left[\widehat{P}_A^*(x,t)\right] &= V\left[f^{-1}(x,t)\widehat{R}_A^*(x,t)\right]. \end{cases} \qquad (12.93)$$

The following theorem shows the asymptotic MSE of $\widehat{P}_A^*(x,t)$.

Theorem 12.2. *When the number of subjects n is large, t is an interior point of the support of $f(x, \cdot)$ and Assumptions (a)-(d) of Section 12.5.1 are satisfied, the following asymptotic expressions for $\widehat{P}_A^*(x,t)$ hold:*

(a) *The asymptotic bias of (12.93) is*

$$Bias\left[\widehat{P}_A^*(x,t)\right] = h_x^{(d_x+2)}\mu_{(d_x+2)}\left[\frac{P_A^{(d_x+2)}(x,t)}{(d_x+2)!} + \frac{P_A^{(d_x+1)}(x,t)\,f'(x,t)}{(d_x+1)!\,f(x,t)}\right]$$
$$+o\left(h_x^{(d_x+2)}\right). \qquad (12.94)$$

(b) *The asymptotic variance of (12.93) is*

$$Var\left[\widehat{P}_A^*(x,t)\right] = \left[Nh_xf(x,t)\right]^{-1}\left[1-P_A(x,t)\right]P_A(x,t)R(K_x)$$
$$+o\left[(Nh_x)^{-1}\right]. \qquad (12.95)$$

(c) *The asymptotic mean squared error of (12.93) is*

$$MSE\left[\widehat{P}_A^*(x,t), P_A(x,t)\right] = \left\{Bias\left[\widehat{P}_A^*(x,t)\right]\right\}^2 + Var\left[\widehat{P}_A^*(x,t)\right]$$
$$+o\left[(Nh_x)^{-1} + h_x^{2(d_x+2)}\right], \qquad (12.96)$$

where the asymptotic expressions of $Bias\left[\widehat{P}_A^*(x,t)\right]$ and $Var\left[\widehat{P}_A^*(x,t)\right]$ are given by (12.94) and (12.95), respectively. ∎

Proof of Theorem 12.2 is given in Section 12.5.3.

For the case that $P_A(x,t)$ is twice continuously differentiable, i.e., $d_x = 0$, the asymptotic expressions of the bias, variance and MSE are

$$Bias\left[\widehat{P}_A^*(x,t)\right] = h_x^2\mu_{(2)}\left[\frac{P_A''(x,t)}{2} + \frac{P_A'(x,t)f'(x,t)}{f(x,t)}\right] + o\left(h_x^4\right), \qquad (12.97)$$

$$Var\left[\widehat{P}_A^*(x,t)\right] = \left[Nh_x f(x,t)\right]^{-1}\left[1 - P_A(x,t)\right]P_A(x,t)R(K_x) + o\left[(Nh_x)^{-1}\right] \quad (12.98)$$

and

$$
\begin{aligned}
MSE\left[\widehat{P}_A^*(x,t), P_A(x,t)\right] &= \left\{Bias\left[\widehat{P}_A^*(x,t)\right]\right\}^2 + Var\left[\widehat{P}_A^*(x,t)\right] \\
&\quad + o\left[(Nh_x)^{-1} + h_x^4\right].
\end{aligned}
\quad (12.99)
$$

The main implication of Theorems 12.1 and 12.2 is that both kernel estimators $\widehat{P}_A(x,t)$ and $\widehat{P}_A^*(x,t)$ have the same asymptotic biases, but their asymptotic variances are different because of the different weight choices. These asymptotic variances depend on the number of subjects n as well as the numbers of repeated measurements n_i. If all the subjects have similar numbers of repeated measurements, e.g., $n_i \approx m$ for all $i = 1, \ldots, n$, then $\widehat{P}_A(x,t)$ and $\widehat{P}_A^*(x,t)$ are asymptotically equivalent in the sense that they have approximately the same asymptotic variances because $n^{-2}\sum_{i=1}^{n} n_i^{-1} \approx N$. On the other hand, if the number of repeated measurements n_i vary significantly across subjects, the asymptotic variances of $\widehat{P}_A(x,t)$ and $\widehat{P}_A^*(x,t)$ could be very different.

12.5.3 Theoretical Derivations

We now give the proofs of Theorems 12.1 and 12.2. The derivations of both proofs follow the same technical approach, although the convergence rates are differences. These derivations also bear some resemblance to the proofs for the time-varying coefficient models given in Section 7.5.

Proof of Theorem 12.1:
By (12.72), (12.73) and the change of variables, we have that, for any x,

$$
\begin{aligned}
&E\left[f(x,t)^{-1}\widehat{R}_A(x,t)\right] \\
&= \frac{1}{nh_x f(x,t)}\sum_{i=1}^{n}\sum_{j=1}^{n_i}\int \frac{1}{n_i}\left\{E\left[1_{[Y_{ij}\in A(x,t_{ij}),X_i=x]}\big|t_{ij}=s\right] - P_A(x,t)\right\} \\
&\qquad\qquad\qquad\qquad\qquad \times K_x\left(\frac{t-s}{h_x}\right)f(x,s)\,ds \quad (12.100) \\
&= f(x,t)^{-1}\int \left[P_A(x,t-h_x u) - P_A(x,t)\right]f(x,t-h_x u)K_x(u)\,du.
\end{aligned}
$$

Then, it follows from the Assumptions (a)-(d) of Section 12.5.1 and the Taylor's expansions of $P_A(x, t - h_x u)$ at $P_A(x,t)$ and $f(x, t - h_x u)$ at $f(x,t)$ that

$$
E\left[f(x,t)^{-1}\widehat{R}_A(x,t)\right] \quad (12.101)
$$

$$
= h_x^{d_x+2}\mu_{(d_x+2)}\left[\frac{P_A^{(d_x+2)}(x,t)}{(d_x+2)!} + \frac{P_A^{(d_x+1)}(x,t)f'(x,t)}{(d_x+1)!f(x,t)}\right](1+o(1)),
$$

which shows (12.81).

To derive the expression of $V\left[f(x,t)^{-1}\widehat{R}_A(x,t)\right\}$ in (12.82), we define

$$Z_{ij}(x,t) = \frac{1}{n_i}\left[1_{[Y_{ij}\in A(x,t_{ij}),X_i=x]} - P_A(x,t)\right],\qquad(12.102)$$

so that, direct expansions based on (12.72), (12.73) and (12.77) show that

$$\left[f(x,t)^{-1}\widehat{R}_A(x,t)\right]^2 = A_1(x,t) + A_2(x,t) + A_3(x,t),\qquad(12.103)$$

where it follows from (12.102) that

$$A_1(x,t) = \left[\frac{1}{f(x,t)\,n\,h_x}\right]^2 \sum_{i=1}^{n}\sum_{j=1}^{n_i}\left[Z_{ij}^2(x,t)K_x^2\left(\frac{t-t_{ij}}{h_x}\right)\right],$$

$$A_2(x,t) = \left[\frac{1}{f(x,t)\,n\,h_x}\right]^2 \sum_{i=1}^{n}\sum_{j_1\neq j_2}\left[Z_{ij_1}(x,t)Z_{ij_2}(x,t)K_x\left(\frac{t-t_{ij_1}}{h_x}\right)K_x\left(\frac{t-t_{ij_2}}{h_x}\right)\right]$$

and

$$A_3(x,t) = \left[\frac{1}{f(x,t)\,n\,h_x}\right]^2 \sum_{i_1\neq i_2}\sum_{j_1,j_2}\left[Z_{i_1j_1}(x,t)Z_{i_2j_2}(x,t)K_x\left(\frac{t-t_{i_1j_1}}{h_x}\right)K_x\left(\frac{t-t_{i_2j_2}}{h_x}\right)\right].$$

Direct calculation for the expected value of $A_1(x,t)$ shows that

$$\begin{aligned}
E\left[A_1(x,t)\right] &= \left[\frac{1}{f(x,t)\,n\,h_x}\right]^2 \sum_{i=1}^{n}\sum_{j=1}^{n_i}\int E\left[Z_{ij}^2(x,t)|t_{ij}=s\right]K_x^2\left(\frac{t-s}{h_x}\right)f(s)\,ds,\\
&= \left[\frac{1}{f(x,t)\,n\,h_x}\right]^2 \sum_{i=1}^{n}\sum_{j=1}^{n_i}\left\{\left(\frac{1}{n_i^2}\right)\int\left[P_A(x,s) - 2P_A(x,s)P_A(x,t)\right.\right.\\
&\qquad\qquad\qquad\qquad \left.\left.+P_A(x,t)\right]K_x^2\left(\frac{t-s}{h_x}\right)f(s)\,ds\right\}\qquad(12.104)\\
&= \frac{1}{f(x,t)\,n\,h_x}\left[\sum_{i=1}^{n}\left(\frac{1}{n\,n_i}\right)\right][1-P_A(x,t)]\,P_A(x,t)R(K_x)[1+o(1)].
\end{aligned}$$

For the expectation of $A_2(x,t)$, we have a bounded function $c(x,t)$ so that

$$\begin{aligned}
&E\left[A_2(x,t)\right]\\
&= \left[\frac{1}{f(x,t)\,n\,h_x}\right]^2 \sum_{i=1}^{n}\sum_{j_1\neq j_2}\left\{\int\int E\left[Z_{ij_1}(x,t)Z_{ij_2}(x,t)|t_{ij_1}=s_1, t_{ij_2}=s_2\right]\right.\\
&\qquad\qquad\qquad\quad\left.\times K_x\left(\frac{t-s_1}{h_x}\right)K_x\left(\frac{t-s_2}{h_x}\right)f(s_1)\,f(s_2)\,ds_1\,ds_2\right\}\\
&= 0,\qquad(12.105)
\end{aligned}$$

since, by Assumption (d), $|s_1 - s_2| > \delta$, the $[-1,1]$ support of $K_x(\cdot)$ implies that, for all $t \in \mathcal{T}$, $K_x[(t - s_1)/h_x] \times K_x[(t - s_2)/h_x] = 0$ when h_x is sufficiently small.

For the expectation of $A_3(x,t)$, it can be shown by the independence of subjects i_1 and i_2 that

$$E\big[A_3(x, t)\big] = E^2\big[f(x, t)^{-1} \widehat{R}_A(x, t)\big]. \tag{12.106}$$

Combining the results of (12.103) through (12.106), we have that

$$V\big[f(x, t)^{-1} \widehat{R}_A(x, t)\big] \tag{12.107}$$
$$= \frac{1}{f(x, t)n h_x} \left[\sum_{i=1}^{n} \left(\frac{1}{n n_i}\right)\right] [1 - P_A(x, t)] P_A(x, t) R(K_x)[1 + o(1)].$$

The conclusion of (12.82) follows from (12.103) and (12.107). The conclusion of (12.83) is a direct consequence of (12.78), (12.81) and (12.82). The conclusions of (12.84), (12.85) and (12.86) follow from (12.81), (12.82) and (12.83) by taking $d_x = 0$. This completes the proof of the theorem. ∎

Proof of Theorem 12.2:

To show the asymptotic bias (12.94), direct calculation using $N = \sum_{i=1}^{n} m_i$, (12.87) and (12.88) shows that, for any x,

$$E\big[f(x, t)^{-1} \widehat{R}_A^*(x, t)\big]$$
$$= \left[\frac{1}{N h_x f(x, t)}\right] \sum_{i=1}^{n} \sum_{j=1}^{n_i} \int \left\{ E\big[1_{[Y_{ij} \in A(x, t_{ij}), X_i = x]} \big| t_{ij} = s\big] - P_A(x, t) \right\}$$
$$\times K_x\left(\frac{t - s}{h_x}\right) f(x, s) \, ds \tag{12.108}$$
$$= f(x, t)^{-1} \int [P_A(x, t - h_x u) - P_A(x, t)] f(x, t - h_x u) K_x(u) \, du,$$

which is the same as $E\big[f(x, t)^{-1} \widehat{R}_A(x, t)\big]$ in (12.101). Consequently, (12.94) follows from (12.81).

For the variance term $V\big[f(x, t)^{-1} \widehat{R}_A^*(x, t)\big]$, we define

$$Z_{ij}^*(x, t) = 1_{[Y_{ij} \in A(x, t_{ij}), X_i = x]} - P_A(x, t), \tag{12.109}$$

so that, by (12.109) and the same expansion as (12.103), we have

$$\big[f(x, t)^{-1} \widehat{R}_A^*(x, t)\big]^2 = A_1^*(x, t) + A_2^*(x, t) + A_3^*(x, t), \tag{12.110}$$

where

$$A_1^*(x, t) = \left[\frac{1}{f(x, t) N h_x}\right]^2 \sum_{i=1}^{n} \sum_{j=1}^{n_i} \left\{ [Z_{ij}^*(x, t)]^2 K_x^2\left(\frac{t - t_{ij}}{h_x}\right) \right\},$$

$$A_2^*(x,t) = \left[\frac{1}{f(x,t)Nh_x}\right]^2 \sum_{i=1}^n \sum_{j_1 \neq j_2} \left[Z_{ij_1}^*(x,t)Z_{ij_2}^*(x,t)K_x\left(\frac{t-t_{ij_1}}{h_x}\right)K_x\left(\frac{t-t_{ij_2}}{h_x}\right)\right]$$

and

$$A_3^*(x,t) = \left[\frac{1}{f(x,t)Nh_x}\right]^2 \sum_{i_1 \neq i_2} \sum_{j_1,j_2} \left[Z_{i_1 j_1}^*(x,t)Z_{i_2 j_2}^*(x,t)K_x\left(\frac{t-t_{i_1 j_1}}{h_x}\right)K_x\left(\frac{t-t_{i_2 j_2}}{h_x}\right)\right].$$

Direct calculation similar to (12.104), (12.105) and (12.106) shows that

$$E\left[A_1^*(x,t)\right] = \frac{1}{f(x,t)Nh_x}\left[1 - P_A(x,t)\right]P_A(x,t)R(K_x)[1+o(1)], \qquad (12.111)$$

and, by Assumptions (b), (c) and (d) of Section 12.5.1, i.e., the compact support of $K_x(\cdot)$ and $h_x \to 0$ as $n \to \infty$, we have

$$E\left[A_2^*(x,t)\right]$$

$$= \left[\frac{1}{f(x,t)Nh_x}\right]^2 \sum_{i=1}^n \sum_{j_1 \neq j_2} \left\{\int E\left[Z_{ij_1}(x,t)Z_{ij_2}(x,t)\big|t_{ij_1} = s_1, t_{ij_2} = s_2\right]\right.$$

$$\left. \times K_x\left(\frac{t-s_1}{h_x}\right)K_x\left(\frac{t-s_2}{h_x}\right)f(s_1)f(s_2)ds\right\}$$

$$= 0, \qquad (12.112)$$

and

$$E\left[A_3(x,t)\right] = \left\{E\left[f(x,t)^{-1}\widehat{R}_A(x,t)\right]\right\}^2. \qquad (12.113)$$

The conclusion of (12.95) follows from (12.110) through (12.113). The conclusion of (12.96) directly follows from (12.94) and (12.95). The conclusions of (12.97) to (12.98) are special cases of (12.94) to (12.86), respectively, with $d_x = 0$. The proof is completed. ∎

12.6   Remarks and Literature Notes

As discussed in the previous chapters, the conditional-mean based regression models have been well established in the literature, for example, Hart and Wehrly (1993), Shi, Weiss and Taylor (1996), Hoover et al. (1998), Fan and Zhang (2000), Lin and Carroll (2001), Rice and Wu (2001), James, Hastie and Sugar (2000), Diggle et al. (2002), Molenberghs and Verbeke (2005), Sentürk and Müller (2006), Zhou, Huang and Carroll (2008), and Fitzmaurice et al. (2009). The theory and methods presented in this chapter, which is adapted from Wu and Tian (2013a), provide a comprehensive statistical tool for evaluating the conditional distribution functions and their functionals in a longitudinal study. Such methods are usually appropriate for long-term follow-up studies, which have a large number of subjects and sufficient numbers of repeated measurements over time. Our application to the NGHS BMI data demonstrates that the "Rank-Tracking Probability" is a useful nonparametric index for the tracking ability of a health outcome over time. Other useful statistical indices for the tracking ability in the literatures, such as Ware and Wu (1981), Foulkes and Davis (1981) and McMahan (1981), depend on the parametric mixed-effects models. Although our methodology and theoretical results are limited to the kernel smoothing estimators under two different weight schemes and a set of asymptotic assumptions, they provide some useful insight into the accuracy of the statistical results under different repeated measurement scenarios.

There are a number of theoretical and methodological aspects that warrant further investigation. First, further theoretical and simulation studies are needed to investigate the properties of other smoothing methods, such as global smoothing methods through splines and other basis approximations, and their corresponding asymptotic inference procedures. Second, many longitudinal studies have multivariate outcome variables, so that appropriate statistical models and estimation methods for multivariate conditional distribution functions deserve to be systematically investigated. The extension of flexible conditional distribution based statistical models incorporating both time-dependent and time-invariant covariates are discussed in Chapters 13 and 14.

Chapter 13

Time-Varying Transformation Models - I

The conditional distribution functions and their functionals discussed in Chapter 12 are only limited to the situations with time-invariant and categorical covariates. For this reason, nonparametric estimators of the conditional distribution functions can be constructed using the kernel smoothing methods without imposing any modeling structures between the response variable and the covariates. This unstructured estimation approach may not work well when there are time-varying and continuous covariates, because unstructured estimation in such situations requires high-dimensional multivariate kernel smoothing that may only work when the sample size is unusually large. In this chapter, we introduce a class of structured nonparametric models, namely the time-varying transformation models, for the conditional distribution functions, which can incorporate the time-varying and continuous covariates. This class of models, which was first suggested by Wu, Tian and Yu (2010), provides a simple and flexible framework for connecting the well-known regression models in survival analysis with the time-varying random variables.

13.1 Overview and Motivation

As discussed in Section 12.1.1, the regression methods for longitudinal data presented in Chapters 6 through 11 primarily focus on evaluating the time-varying mean response curves and their relationships with the covariates. These conditional mean-based regression models could be inadequate when the scientific objective is to evaluate the conditional distribution functions and the outcome variables have skewed or non-Gaussian distributions. A well-known approach for skewed random variables in practice is to apply a transformation, such as the Box-Cox transformation, to these variables so that the transformed variables have approximately Gaussian conditional distributions, e.g., Lipsitz, Ibrahim and Molenberghs (2000). A limitation of this approach is that the form of the transformation has to be time-invariant and known in advance. For longitudinal studies, the outcome variable $Y(t)$ may have different skewed distributions over time t, so that a fixed and known transformation for $Y(t)$ may not lead to an approximately Gaussian distribution for all t within the range of interest.

The approach of this chapter directly models the conditional distributions of the outcome variable $Y(t)$ at a time point t giving the covariates. This approach is motivated by the attempt for modeling the covariate effects on the distributions of blood pressure and other cardiovascular risk factors in the National Growth and Health Study (NGHS). Since a main objective of the NGHS is to evaluate the patterns of obesity and cardiovascular risk factors during adolescence, prior publications of this study, such as Daniels et al. (1998) and Thompson et al. (2007), studied the longitudinal effects of childhood obesity on the levels of several cardiovascular risk factors, such as blood pressure, lipoprotein cholesterol (LDL) and triglyceride (TG), using the conditional means or the probabilities of unhealthy risk levels specified by certain threshold values. These results, although informative to some extent, do not give a clear indication of whether similar associations between childhood obesity and cardiovascular risks still hold if other definitions of unhealthy risk levels are used. It is desirable to have a flexible regression method to describe the covariate effects on the entire distribution of a risk variable.

In Chapters 6 through 8, we have seen the varying-coefficient models as a useful dimensional reduction strategy, because these models retain a simple parametric structure at each time point and preserve the nonparametric flexibility when the time point varies. Extending the idea of varying coefficients to the conditional distribution functions, a natural approach is to consider a class of time-varying transformation models. The transformation models with cross-sectional i.i.d. data have been studied extensively in survival analysis, for example, Cheng, Wei and Ying (1995), Lu and Ying (2004), Lu and Tsiatis (2006) and Zeng and Lin (2006). The extension to longitudinal data can be established by allowing the coefficients of the transformation models to be nonparametric functions of time. Our focus in this chapter is on the estimation of the coefficient curves through a two-step procedure. This two-step estimation procedure is computationally simple and can automatically adjust the smoothing parameters for different coefficient curves.

13.2 Data Structure and Model Formulation

13.2.1 Data Structure

For simplicity, we use a set of slightly different notation to describe the same longitudinal samples from the previous chapters, such as Sections 6.1 and 12.2. Similar notation has been used in Section 7.4 for describing the two-step estimation procedure of Fan and Zhang (2000) for conditional means.

Population Random Variables and Distribution Functions:
The real-valued outcome variable of interest at time t is $Y(t)$, and the covariate vector at time $t \in \mathscr{T}$ is $\mathbf{X}(t) = \left(X_1(t), \ldots, X_P(t)\right)^T$, where, for any $1 \leq p \leq P$, $X_p(t)$ is allowed to be time-varying and may be either continuous or discrete. The conditional cumulative distribution function (CDF) of $Y(t)$ at

time t given $\mathbf{X}(t) = \mathbf{x}$ is

$$F_t(y|\mathbf{x}) = P\big[Y(t) \le y | \mathbf{X}(t) = \mathbf{x}, t\big]. \tag{13.1}$$

The statistical objectives are to model the relationship between $F_t(y|\mathbf{x})$ and $\mathbf{x}$, and to estimate the effects of $X_p(t)$, $1 \le p \le P$, on $F_t(y|\mathbf{x})$ for any $t \in \mathcal{T}$.

Longitudinal Data with Time-Varying Covariates:

(a) *The sample has n independent subjects, and each subject is observed at a randomly selected subset of $J > 1$ distinct design time points $\mathbf{t} = (t_1, \ldots, t_J)^T$.*

(b) *Since not all the subjects are observed at every t_j, we denote by $\mathscr{S}_j$ the set of subjects whose observations are available at time t_j.*

(c) *Let $Y_i(t_j)$ and $\mathbf{X}_i(t_j) = \big(X_{i1}(t_j), \ldots, X_{iP}(t_j)\big)^T$ be the real-valued outcome and the $P \times 1$ covariate vector, respectively, at t_j for the ith subject when $i \in \mathscr{S}_j$. The longitudinal sample of $\{Y(t), \mathbf{X}(t), t \in \mathcal{T}\}$ is*

$$\left\{ \begin{array}{rcl} \mathscr{L} & = & \{Y_i(t_j), \mathbf{X}_i(t_j), t_j : i \in \mathscr{S}_j; j = 1, \ldots, J\}, \\ \mathscr{D} & = & \{\mathbf{X}_i(t_j), t_j : i \in \mathscr{S}_j; j = 1, \ldots, J\} \end{array} \right. \tag{13.2}$$

with $\mathscr{D}$ being the set of observed covariates.

(d) *Denote the number of subjects in $\mathscr{S}_j$ and the number of subjects in both $\mathscr{S}_{j_1}$ and $\mathscr{S}_{j_2}$ when $j_1 \ne j_2$ by*

$$n_j = \#\{i \in \mathscr{S}_j\} \quad and \quad n_{j_1 j_2} = \#\{i \in \mathscr{S}_{j_1} \bigcap \mathscr{S}_{j_2}\}, \tag{13.3}$$

respectively. It clearly follows that $n_{j_1 j_2} \le \min\{n_{j_1}, n_{j_2}\}$. □

13.2.2 The Time-Varying Transformation Models

As a flexible dimension reduction alternative to unstructured $F_t(y|\mathbf{x})$, we consider a class of linear transformation models, which adopt the following structured time-varying covariate effects on $F_t(y|\mathbf{x})$ at each $t \in \mathcal{T}$,

$$\left\{ \begin{array}{rcl} g\big[S_t(y|\mathbf{X})\big] & = & h(y, t) + \mathbf{X}^T(t)\,\beta(t), \\ \beta(t) & = & (\beta_1(t), \ldots, \beta_P(t))^T, \\ \beta_p(t) & = & \text{smooth function on } \mathcal{T}, \; p = 1, \ldots, p, \end{array} \right. \tag{13.4}$$

where $g(\cdot)$ is a known decreasing link function, $S_t(y|\mathbf{X}) = 1 - F_t(y|\mathbf{X})$ is the probability of $Y(t) > y$, $h(\cdot, \cdot)$ is an unspecified and strictly increasing function in y, and $\beta_p(t)$ describes the effect of the covariate $X_p(t)$ on the "survival function" $S_t(y|\mathbf{X})$. When t is fixed, (13.4) is a semiparametric transformation model with $\beta_p(t)$ representing the change of $g\big[S_t(y|\mathbf{X})\big]$ associated with a unit increase of $X_p(t)$. Under this framework of longitudinal data, we are interested in the time-trends of the covariate effects. Hence, the estimation and inferences then focus on the coefficient curves $\{\beta_p(t) : p = 1, \ldots, P\}$ over $t \in \mathscr{T}$.

Useful special cases of (13.4) can be obtained by specifying the forms of $g(\cdot)$. In particular, the *time-varying proportional hazards model* is given by (13.4) with

$$g\big[S_t(y|\mathbf{X})\big] = \log\big\{ -\log\big[S_t(y|\mathbf{X})\big]\big\}, \tag{13.5}$$

and the *time-varying proportional odds model* is given by (13.4) with

$$g\big[S_t(y|\mathbf{X})\big] = -\log\big[S_t(y|\mathbf{X})/F_t(y|\mathbf{X})\big]. \tag{13.6}$$

There is no theoretical guideline on which link function $g(\cdot)$ should be used for a given longitudinal sample. In practice, $g(\cdot)$ is often selected by examining the fitness of the data and evaluating the scientific or clinical interpretations of the results. Our application of (13.4) to the NGHS blood pressure data (Section 13.4) is based on the time-varying proportional odds model (13.6).

In survival analysis, most estimation methods for the linear transformation models are developed for failure times with random censoring. Examples of the linear transformation models involving censored failure time data include Cheng, Wei and Ying (1995, 1997), Lu and Ying (2004) and Zeng and Lin (2006). Although censoring remains a theoretical possibility in longitudinal settings, none of the repeatedly measured variables in the NGHS data involve censoring. Thus, nonparametric estimation of $\beta(t)$ in (13.4) with $Y_i(t_j)$ subject to random censoring is only theoretically interesting. For practical considerations, our estimation methods of this chapter are developed for (13.4) without censoring.

13.3 Two-Step Estimation Method

We describe here a two-step estimation method similar to the one described in Section 8.2 for the time-varying coefficient models. In this approach, we first obtain the raw estimates of the coefficient curves $\beta(t) = \big(\beta_1(t), \ldots, \beta_P(t)\big)^T$ of (13.4) at the design time points $\mathbf{t} = (t_1, \ldots, t_J)^T$, and then apply a nonparametric smoothing method to estimate $\beta(t)$ at any $t \in \mathscr{T}$ by smoothing over the raw estimates. Similar to the time-varying coefficient models of Chapter 8, the smoothing step serves two purposes: First, it reduces the variability of the curve estimates at $t \in \mathscr{T}$ by borrowing the information obtained from the raw estimates at the design time points which are adjacent to t. Second, it produces the curve estimates of $\beta(t)$ for any time point $t \in \mathscr{T}$. A main reason for this two-step approach is its computational simplicity, since, for both the raw estimator step and the smoothing step, we can use the existing methods which are known to have good statistical properties.

13.3.1 Raw Estimates of Coefficients

The coefficients $\beta(t_j)$ of (13.4) can be estimated by adapting the estimating equations of Cheng, Wei and Ying (1995) to the observations at t_j. Using equation (1.4) of Cheng, Wei and Ying (1995), we can define

$$\varepsilon_{ij} = g\{S_{t_j}[Y_i(t_j)|\mathbf{X}_i(t_j)]\}, \tag{13.7}$$

and, by

$$P[\varepsilon_{ij} \leq u|\mathbf{X}_i(t_j), t_j] = P\{S_{t_j}[Y_i(t_j)|\mathbf{X}_i(t_j)] \geq g^{-1}(u)|\mathbf{X}_i(t_j), t_j\}, \tag{13.8}$$

we can verify, as in equation (1.4) of Cheng, Wei and Ying (1995), that (13.4) is equivalent to

$$h[Y_i(t_j), t_j] = -\mathbf{X}_i^T(t_j)\boldsymbol{\beta}(t_j) + \varepsilon_{ij}, \tag{13.9}$$

where ε_{ij} are defined in (13.7) and have distribution function $G(\cdot) = 1 - g^{-1}(\cdot)$.

For a given time design point t_j, we define

$$\begin{cases} Z_{i_1,i_2}(t_j) &= 1_{[Y_{i_1}(t_j) \geq Y_{i_2}(t_j)]}, \\ \mathbf{X}_{i_1,i_2}(t_j) &= \mathbf{X}_{i_1}(t_j) - \mathbf{X}_{i_2}(t_j), \\ \xi(s) &= \int_{-\infty}^{\infty}[1 - G(t+s)]\,dG(t), \end{cases} \tag{13.10}$$

and, by taking the conditional expectation of $Z_{i_1,i_2}(t_j)$ given $\mathbf{X}_{i_1}$ and $\mathbf{X}_{i_2}$, it follows from (13.7) and (13.10) that

$$\begin{aligned} E[Z_{i_1,i_2}(t_j)|\mathbf{X}_{i_1}, \mathbf{X}_{i_2}, t_j] &= P\{h[Y_{i_1}(t_j), t_j] \geq h[Y_{i_2}(t_j), t_j]|\mathbf{X}_{i_1}, \mathbf{X}_{i_2}, t_j\} \\ &= P[\varepsilon_{i_1,j} - \varepsilon_{i_2,j} \geq \mathbf{X}_{i_1,i_2}^T(t_j)\boldsymbol{\beta}(t_j)] \\ &= \xi[\mathbf{X}_{i_1,i_2}^T(t_j)\boldsymbol{\beta}(t_j)]. \end{aligned} \tag{13.11}$$

Then it follows from (13.10) and (13.11) that, conditioning on $\{\mathbf{X}_{i_1}, \mathbf{X}_{i_2}\}$, the expected value of

$$\Delta_{i_1,i_2}[\mathbf{X}_{i_1,i_2}(t_j)] = Z_{i_1,i_2}(t_j) - \xi[\mathbf{X}_{i_1,i_2}^T(t_j)\boldsymbol{\beta}(t_j)] \tag{13.12}$$

is zero, and, at the time design point t_j, $\boldsymbol{\beta}(t_j)$ can be estimated using a moment estimation approach by setting a weighted sample mean of (13.12) to zero. Following this moment estimation approach, we can apply the estimation method of Cheng, Wei and Ying (1995) for observations at the time design point t_j, which leads to the raw estimator for $\boldsymbol{\beta}(t_j)$ as a solution $\widetilde{\boldsymbol{\beta}}(t_j)$ to the estimating equation

$$\begin{cases} \sum_{i_1 \neq i_2 \in \mathscr{S}_j} U_{i_1 i_2}[\widetilde{\boldsymbol{\beta}}(t_j)] &= 0, \\ U_{i_1 i_2}[\widetilde{\boldsymbol{\beta}}(t_j)] &= w[\mathbf{X}_{i_1,i_2}^T(t_j)\widetilde{\boldsymbol{\beta}}(t_j)]\,\mathbf{X}_{i_1,i_2}(t_j)\Delta_{i_1,i_2}[\mathbf{X}_{i_1,i_2}(t_j)], \end{cases} \tag{13.13}$$

where $w(\cdot)$ is a pre-specified weight function.

The choice of $w(\cdot)$ in (13.13) is subjective and has some potential to affect the statistical properties of the raw estimator $\widetilde{\boldsymbol{\beta}}(t_j)$. However, the influence of

$w(\cdot)$ on the final smoothing estimators is fairly limited, because many simple choices of $w(\cdot)$ can lead to appropriate raw estimators $\widetilde{\beta}(t_j)$, at least in the sense of asymptotic convergence rates, and the smoothing step further reduces the variability of the final estimators of $\beta(t)$. Since for each fixed t_j, the estimating equation (13.13) is identical to the estimating equation (2.2) of Cheng, Wei and Ying (1995), their conclusions of uniqueness and asymptotic uniqueness of the solutions also hold for (13.13). Thus, by the conclusions in Section 2 and Appendix 1 of Cheng, Wei and Ying (1995), we can show that, if $w(\cdot)$ is positive, the estimating equation (13.13) has *asymptotically a unique solution*, and, furthermore, when $w(\cdot) = 1$ and the matrix $\sum_{i_1 \neq i_2 \in \mathscr{S}_j} \mathbf{X}_{i_1,i_2}(t_j) \mathbf{X}_{i_1,i_2}^T(t_j)$ is positive definite, the estimating equation (13.13) has a unique solution. Cheng, Wei and Ying (1995) also illustrated through examples that the estimation procedure of (13.13) with $w(\cdot) = 1$ worked well for the proportional odds model and the model with standard normal error.

The assumption of having J distinct design time points $\mathbf{t} = (t_1, \ldots, t_J)^T$ is a simplification for the theoretical discussions. In practice, the design time points are usually not prespecified. In such situations, the dataset may have a large number of distinct time points, and the raw estimate $\widetilde{\beta}(t_j)$ may not exist when there are very few subjects observed at t_j. If there are only a few such time points, we may simply leave the raw estimates missing at these points, and compute the smoothing estimates using the raw estimates obtained at other time points. A more practical approach is to group the observed time points into small time bins, so that the raw estimates can be computed at each bin. When a binning method is used, we need to have small bin sizes, so that the raw estimates are undersmoothed relative to the smoothing parameters used in the smoothing step.

13.3.2   Bias, Variance and Covariance of Raw Estimates

Explicit expressions for the finite sample mean and variance of $\widetilde{\beta}(t_j)$ in (13.13) are generally not available. So, we illustrate the asymptotic properties of $\widetilde{\beta}(t_j)$ by establishing a large sample approximation for $[\widetilde{\beta}(t_j) - \beta(t_j)]$, defining the bias, variance and covariance matrix for $\widetilde{\beta}(t_j)$, and deriving the asymptotic expressions for the bias, variance and covariance matrix. Since, for a fixed t_j, the raw estimator $\widetilde{\beta}(t_j)$ is the special case of the estimator in Cheng, Wei and Ying (1995) without censoring, the asymptotic properties of $\widetilde{\beta}(t_j)$ are essentially the same as theirs, with the exception that, when different time points are involved, the raw estimators are correlated because of the intra-subject correlations of the data.

1. Large Sample Approximation

Because the raw estimator $\widetilde{\beta}(t_j)$ does not have a simple explicit expression, the asymptotic properties of the error term $[\widetilde{\beta}(t_j) - \beta(t_j)]$ have to be investigated by an asymptotically equivalent term which has tractable asymptotic expressions. For this reason, the next lemma gives an asymptotic approximation for the raw estimators, which can be used to study the asymptotic expressions of the mean, covariance and variance of $\widetilde{\beta}(t_j)$.

Lemma 13.1. *For any t_j in the time design points* $\mathbf{t} = (t_1, \ldots, t_J)^T$, *the raw estimator* $\widetilde{\beta}(t_j) = (\widetilde{\beta}_1(t_j), \ldots, \widetilde{\beta}_P(t_j))^T$ *of* $\beta(t_j) = (\beta_1(t_j), \ldots, \beta_P(t_j))^T$ *satisfy the following asymptotic approximation when n and n_j are sufficiently large,*

$$
\begin{cases}
A(t_j) & = (A_1(t_j), \ldots, A_P(t_j))^T \\
& = [I + o_p(I)] [\widetilde{\beta}(t_j) - \beta(t_j)] \\
& = [n_j(n_j - 1)]^{-1} \Lambda^{-1}(t_j) \sum_{i_1 \neq i_2 \in \mathscr{S}_j} U_{i_1, i_2}[\beta(t_j)], \\
\Lambda(t_j) & = -E\left\{ \partial U_{i_1 i_2}[\beta(t_j)] / \partial \beta(t_j) \big| \mathscr{D} \right\},
\end{cases}
\tag{13.14}
$$

where I is the identity matrix. In particular, we have

$$
\Lambda(t_j) = E\left\{ w\left[\mathbf{X}_{i_1, i_2}^T(t_j)\beta(t_j)\right] \xi'\left[\mathbf{X}_{i_1, i_2}^T(t_j)\beta(t_j)\right] \mathbf{X}_{i_1, i_2}(t_j) \mathbf{X}_{i_1, i_2}^T(t_j) \right\}, \tag{13.15}
$$

when $w(\cdot)$ is a constant. ∎

Proof of Lemma 13.1 is given in Section 13.5.4.

Using the approximation (13.14), we essentially simplify the intractable error term $[\widetilde{\beta}(t_j) - \beta(t_j)]$ by its approximation $A(t_j)$ which is specified by the random term $\sum_{i_1 \neq i_2 \in \mathscr{S}_j} U_{i_1, i_2}[\beta(t_j)]$ and the non-random term $\Lambda(t_j)$.

2. Bias, Variance and Covariance Matrix

Given the observed covariates $\mathscr{D}$, we define the following bias, variance and covariance matrix of $\widetilde{\beta}(t_j)$ based on its approximation $A(t_j)$ of (13.14).

Bias: For $p = 1, \ldots, P$ and fixed t_j, the bias of $\widetilde{\beta}_p(t_j)$ is defined by

$$
Bias[\widetilde{\beta}_p(t_j)] = E[A_p(t_j)|\mathscr{D}]. \tag{13.16}
$$

A raw estimator $\widetilde{\beta}(t_j)$ is asymptotically unbiased or simply unbiased at t_j if $Bias[\widetilde{\beta}_p(t_j)] = 0$.

Variance: For $p = 1, \ldots, P$ and fixed t_j, the variance of $\widetilde{\beta}_p(t_j)$ is defined by

$$
Var[\widetilde{\beta}_p(t_j)|\mathscr{D}] = Var[A_p(t_j)|\mathscr{D}]. \tag{13.17}
$$

Variance-Covariance Matrix: For any $\{j_1, j_2\}$, the variance-covariance matrix of $\widetilde{\beta}(t_{j_1})$ and $\widetilde{\beta}(t_{j_2})$ is defined by

$$Cov\left[\widetilde{\beta}(t_{j_1}), \widetilde{\beta}(t_{j_2})\middle|\mathscr{D}\right] = Cov\left[A(t_{j_1}), A(t_{j_2})\middle|\mathscr{D}\right]. \tag{13.18}$$

For the special case of $j_1 = j_2 = j$, the diagonal elements of (13.18) are $\left\{Var\left[A_1(t_j)\middle|\mathscr{D}\right], \ldots, Var\left[A_P(t_j)\middle|\mathscr{D}\right]\right\}$. $\square$

3. Asymptotic Properties

Asymptotic properties of the raw estimator $\widetilde{\beta}(t_j)$ can be summarized into the following three aspects.

(1) Asymptotic Unbiasedness

The raw estimator $\widetilde{\beta}(t_j)$ is asymptotically unbiased for $\beta(t_j)$. This can be seen from (13.11), (13.12) and the definition of $U_{i_1 i_2}\left[\beta(t_j)\right]$ in (13.13), in which we have

$$E\left\{U_{i_1 i_2}\left[\beta(t_j)\right]\middle|\mathscr{D}\right\} = 0. \tag{13.19}$$

It then easily follows from (13.14), (13.16) and (13.19) that the raw estimator $\widetilde{\beta}(t_j)$ of (13.13) is asymptotically unbiased at t_j because

$$E\left[A(t_j)\middle|\mathscr{D}\right] = \left[n_j\left(n_j - 1\right)\right]^{-1}\Lambda^{-1}(t_j) \sum_{i_1 \neq i_2 \in \mathscr{S}_j} E\left\{U_{i_1,i_2}\left[\beta(t_j)\right]\middle|\mathscr{D}\right\} = 0. \tag{13.20}$$

(2) Asymptotic Covariance Matrix

Using the definitions (13.14) and (13.18), the next lemma summarizes the asymptotic expression of the variance-covariance matrix of $\widetilde{\beta}(t_{j_1})$ and $\widetilde{\beta}(t_{j_2})$.

Lemma 13.2. *The variance-covariance matrix of $\widetilde{\beta}(t_{j_1})$ and $\widetilde{\beta}(t_{j_2})$ for any $\{j_1, j_2\}$ has the following asymptotic expression, when n_{j_1}, n_{j_2} and $n_{j_1 j_2}$ are sufficiently large,*

$$Cov\left[A(t_{j_1}), A(t_{j_2})\middle|\mathscr{D}\right]\left[I + o_p(I)\right] = r_{j_1 j_2}\Sigma(t_{j_1}, t_{j_2}), \tag{13.21}$$

where, for all $\{i_1, i_2, i_3\}$ such that $i_1 \neq i_2 \neq i_3$,

$$\Sigma(t_{j_1}, t_{j_2}) = \Lambda^{-1}(t_{j_1})\rho(t_{j_1}, t_{j_2})\left[\Lambda^{-1}(t_{j_2})\right]^T, \tag{13.22}$$

$$\rho(t_{j_1}, t_{j_2}) = E\left\{U_{i_1 i_2}\left[\beta(t_{j_1})\right]U_{i_2 i_3}\left[\beta(t_{j_2})\right]\middle|\mathscr{D}\right\} \tag{13.23}$$

and

$$r_{j_1 j_2} = \left[\frac{4 n_{j_1 j_2}}{n_{j_1} n_{j_2}\left(n_{j_1} - 1\right)\left(n_{j_2} - 1\right)}\right]\left\{\left(n_{j_1} - n_{j_1 j_2}\right)\left(n_{j_2} - n_{j_1 j_2}\right)\right.$$
$$\left. + \frac{1}{2}\left(n_{j_1 j_2} - 1\right)\left[\left(n_{j_1} + n_{j_2} - 2 n_{j_1 j_2}\right) + \frac{1}{3}\left(n_{j_1 j_2} - 2\right)\right]\right\}. \tag{13.24}$$

When $j_1 \neq j_2$, (13.21) gives the covariance matrix at two different time design points $t_{j_1} \neq t_{j_2}$. ∎

Proof of Lemma 13.2 is given in Section 13.5.4.

We refer to $r_{j_1 j_2} \Sigma(t_{j_1}, t_{j_2})$ as the asymptotic variance-covariance matrix of $\widetilde{\beta}(t_{j_1})$ and $\widetilde{\beta}(t_{j_2})$ because, by (13.21), $Cov[A(t_{j_1}), A(t_{j_2})|\mathscr{D}]$ is asymptotically equivalent to $r_{j_1 j_2} \Sigma(t_{j_1}, t_{j_2})$. Since $n_{j_1 j_2} \leq \min\{n_{j_1}, n_{j_2}\}$, it is clear from (13.24) that the rate for $r_{j_1 j_2} \Sigma(t_{j_1}, t_{j_2})$ converging to zero depends on how fast n_{j_1}, n_{j_2} and $n_{j_1 j_2}$ converging to infinity.

(3) Asymptotic Variance

When $j_1 = j_2 = j$ and $n_{jj} = n_j$ is large, the same calculations as in the proof of Lemma 13.2 suggests that (13.24) becomes

$$r_{jj} = \frac{2(n_j - 2)}{3n_j(n_j - 1)} \approx 2/(3n_j). \tag{13.25}$$

Let $\sigma_p^2(t_j)$ be the pth diagonal element of $\Sigma(t_j, t_j)$. The asymptotic variance of $\widetilde{\beta}_p(t_j)$ defined in (13.17) is

$$Var[A_p(t_j)|\mathscr{D}] = \frac{2}{3n_j}\sigma_p^2(t_j)[1 + o_p(1)]. \tag{13.26}$$

These asymptotic variance and covariance expressions are useful for computing the asymptotic biases and variances for the smoothing estimators.

13.3.3 Smoothing Estimators

For the second step of the estimation procedure, we need to obtain a smoothing estimator for $\beta(t) = (\beta_1(t), \ldots, \beta_P(t))^T$ of (13.4) at any time point $t \in \mathscr{T}$ by applying a smoothing method to the raw estimates $\{\widetilde{\beta}(t_j) : j = 1, \ldots, J\}$ obtained in (13.13). The smoothing step can reduce the variation by sharing information from the adjacent time points.

1. General Expressions of Linear Smoothing Estimators

By treating the raw estimates $\{\widetilde{\beta}_p(t_j) : j = 1, \ldots, J\}$ as the pseudo-observations for $\beta_p(t)$, $p = 1, \ldots, P$, a "local" smoothing estimator of $\beta_p(t)$ can be expressed as a linear estimator of the form

$$\widehat{\beta}_p(t) = \sum_{j=1}^{J} w(t_j, t) \widetilde{\beta}_p(t_j), \tag{13.27}$$

where $w(t_j, t)$ is a smoothing weight function which gives more weight if the time point t_j is close to t. If $w(t_j, t)$ is formed by a kernel function $K(\cdot)$, which is often chosen as a probability density function, and a bandwidth h, such that

$$w(t_j, t) = K\left(\frac{t_j - t}{h}\right) \Big/ \sum_{j=1}^{J} K\left(\frac{t_j - t}{h}\right),$$

then $\widehat{\beta}_p(t)$ is given by

$$\widehat{\beta}_p(t) = \sum_{j=1}^{J} \left[\widetilde{\beta}_p(t_j) K\left(\frac{t_j - t}{h}\right)\right] \Big/ \sum_{j=1}^{J} K\left(\frac{t_j - t}{h}\right), \tag{13.28}$$

which is a kernel smoothing estimator of $\beta_p(t)$.

More generally, the qth derivatives of $\beta_p(t)$ with respect to t, which is denoted by $\beta_p^{(q)}(t)$, can also be estimated by a linear estimator based on $\{\widetilde{\beta}_p(t_j) : j = 1, \ldots, J\}$. The special case of $q = 0$ corresponds to $\beta_p(t)$ itself, that is $\beta_p^{(0)}(t) = \beta_p(t)$. Suppose that there is an integer $Q \geq 0$ such that, for a $1 \leq p \leq P$, $\beta_p(t)$ is $Q + 1$ times differentiable with respect to t. The qth derivative of $\beta_p(t)$, $\beta_p^{(q)}(t)$, for any integer $0 \leq q \leq Q + 1$ can be estimated by

$$\widehat{\beta}_p^{(q)}(t) = \sum_{j=1}^{J} w_{q, Q+1}(t_j, t) \widetilde{\beta}_p(t_j), \tag{13.29}$$

where the weight function $w_{q, Q+1}(t_j, t)$ is determined by the smoothing method as well as the values of q and Q. When $q = 0$, $\widehat{\beta}_p(t) = \widehat{\beta}_p^{(0)}(t)$ is an estimator of $\beta_p(t)$. Specific choices of $w_{q, Q+1}(t_j, t)$ can be constructed by splines, local polynomials or other smoothing methods.

2. Local Polynomial Estimators

We focus in this chapter on the local polynomial estimators. Using the same smoothing method of Section 7.2.3, we define, for any time design point $t_j \in \mathbf{t}$, $t \in \mathscr{T}$, the bandwidth and kernel pair $\{h, K(\cdot) : h > 0\}$ and $K_h(t) = K(t/h)/h$,

$$\begin{cases} C_j & = \ (1, t_j - t, \ldots, (t_j - t)^Q)^T, \ j \ = \ 1, \ldots, J, \\ \mathbf{C} & = \ (C_1, \ldots, C_J)^T, \\ W_j & = \ K_h(t_j - t) \ \text{and} \ \mathbf{W} \ = \ \text{diag}(W_1, \ldots, W_J). \end{cases} \tag{13.30}$$

Following (13.29), the Qth order local polynomial estimator of the qth derivative $\beta_p^{(q)}(t)$, $0 \leq q < Q + 1$, of $\beta_p(t)$ with respect to t uses the weight function

$$w_{q, Q+1}(t_j, t; h) = q! e_{q+1, Q+1}^T (\mathbf{C}^T \mathbf{W} \mathbf{C})^{-1} C_j W_j, \ j = 1, \ldots, J, \tag{13.31}$$

and has the expression

$$\widehat{\beta}_p^{(q)}(t) = \sum_{j=1}^{J} \left\{ [q! e_{q+1, Q+1}^T (\mathbf{C}^T \mathbf{W} \mathbf{C})^{-1} C_j W_j] \widetilde{\beta}_p(t_j) \right\}, \tag{13.32}$$

where $e_{q+1,Q+1} = (0,\ldots,0,1,0,\ldots,0)^T$ is the $(Q+1)$ column vector with 1 at its $(q+1)$th place and 0 elsewhere. By taking $q=0$, the Qth order local polynomial estimator of $\beta_p(t)$ has the expression

$$\widehat{\beta}_p^{(0)}(t) = \sum_{j=1}^{J} \left\{ [q! e_{1,Q+1}^T (\mathbf{C}^T \mathbf{W} \mathbf{C})^{-1} C_j W_j] \widetilde{\beta}_p(t_j) \right\}, \tag{13.33}$$

where $e_{1,Q+1} = (1,0,\ldots,0)^T$. The special case of local linear estimator is obtained by setting $q = 0$ and $Q = 1$ in (13.33). In this case, the local linear estimator $\widehat{\beta}_p^L(t)$ of $\beta_p(t)$ has the expression

$$\widehat{\beta}_p^L(t) = \sum_{j=1}^{J} \left\{ [e_{1,2}^T (\mathbf{C}^T \mathbf{W} \mathbf{C})^{-1} C_j W_j] \widetilde{\beta}_p(t_j) \right\}, \tag{13.34}$$

where $e_{1,2} = (1,0)^T$. By taking $e_{2,2} = (0,1)^T$ in (13.32), the derivative $\beta_p'(t)$ of $\beta_p(t)$ with respect to t can be estimated by the local linear estimator

$$\widehat{\beta}_p^{L,(1)}(t) \sum_{j=1}^{J} \left\{ [e_{2,2}^T (\mathbf{C}^T \mathbf{W} \mathbf{C})^{-1} C_j W_j] \widetilde{\beta}_p(t_j) \right\}. \tag{13.35}$$

In Chapter 7, the coefficient curves in the conditional mean based time-varying coefficient models can also be estimated by a "one-step smoothing method", such as the least squares based smoothing methods of Section 7.2. Similar one-step smoothing methods are still not available for the current conditional distribution based time-varying transformation models, because we are currently lacking a rank based smoothing method that avoids the initial raw estimation. The two-step approach, on the other hand, utilizes the existing estimation methods both in the raw estimation step and the smoothing step. This approach is computationally simple and has the additional advantage of automatically adjusting different smoothing needs for different coefficient curves. However, it is important to note that the two-step approach often requires large sample sizes. When binning is used, it also requires the bin sizes to be small.

An important application of the time-varying transformation model (13.4) is to estimate and predict the conditional distributions and quantiles of the response variable $Y(t)$ given the covariates $\mathbf{X}(t)$. Under this circumstance, we would need to construct a nonparametric estimator of $h(y,t)$ that is monotone increasing or "order-preserving" in y for all $t \in \mathscr{T}$. In principle, $h(y,t)$ can be estimated by first obtaining a set of raw estimators at the distinct time design points $\{t_j; j = 1,\ldots,J\}$ using the approach described in Cheng, Wei and Ying (1997) and then smoothing the raw estimates over the time range $\mathscr{T}$. The methods for this estimation problem are discussed in Chapter 14.

13.3.4 Bandwidth Choices

Similar to the two-step smoothing estimation of Chapter 7, the choice of bandwidth h of (13.33) is crucial for obtaining an appropriate smoothing estimate.

We describe here two cross-validation methods for selecting the bandwidths based on the data.

1. Leave-One-Subject-Out Cross-Validation

For the purpose of preserving the correlation structure of the data, the first data-driven bandwidth choice is the leave-subject-out cross-validation (LSCV) described in Section 7.2.5, which depends on deleting the entire observations of a subject each time. Extending this approach to the local polynomial estimator (13.33), we first compute the local polynomial estimator $\widehat{\beta}_{p,-(i_1,i_2)}^{(0)}(t;h_p)$ for any pairs of subjects $\{(i_1,i_2) : i_1 \neq i_2; i_1 = 1, \ldots, n; i_2 = 1, \ldots, n\}$ from (13.33) with bandwidth h_p using the remaining data with all the observations from the subject pair (i_1,i_2) deleted. Let $\mathbf{h} = (h_1, \ldots, h_P)^T$ be the vector of bandwidths, where, for $p = 1, \ldots, P$, h_p is the bandwidth in $\widehat{\beta}_{p,-(i_1,i_2)}^{(0)}(t;h_p)$, and let

$$\widehat{\beta}_{-(i_1,i_2)}(t;\mathbf{h}) = \left(\widehat{\beta}_{1,-(i_1,i_2)}^{(0)}(t;h_1), \ldots, \widehat{\beta}_{P,-(i_1,i_2)}^{(0)}(t;h_P)\right)^T \tag{13.36}$$

be the local polynomial estimator for $\beta(t) = (\beta_1(t), \ldots, \beta_P(t))^T$. The LSCV cross-validation score is

$$LSCV(\mathbf{h}) = \sum_{j=1}^{J} \sum_{i_1 \neq i_2 \in \mathscr{S}_j} \left\{ Z_{i_1,i_2}(t_j) - \xi\left[\mathbf{X}_{i_1,i_2}^T(t_j)\widehat{\beta}_{-(i_1,i_2)}(t_j;\mathbf{h})\right] \right\}^2, \tag{13.37}$$

where $Z_{i_1,i_2}(t_j)$ and $\xi(\cdot)$ are defined in (13.10). If (13.37) can be uniquely minimized with respect to $\mathbf{h}$, the LSCV bandwidth

$$\mathbf{h}_{LSCV} = \left(h_{1,LSCV}, \ldots, h_{P,LSCV}\right)^T \tag{13.38}$$

is then the unique minimizer of (13.37).

When minimizing (13.37) is computationally intensive, we can replace (13.37) by a "M-fold LSCV" to speed up the computation, which is calculated by deleting a block of subjects each time. In this approach, we randomly divide the subjects into M blocks $\{b(m) : m = 1, \ldots, M\}$ and compute the M-fold LSCV score

$$MLSCV(\mathbf{h}) = \sum_{j=1}^{J} \sum_{m=1}^{M} \sum_{i_1 \neq i_2 \in \mathscr{S}_j; (i_1,i_2) \in b(m)}$$

$$\left\{ Z_{i_1,i_2}(t_j) - \xi\left[\mathbf{X}_{i_1,i_2}^T(t_j)\widehat{\beta}_{-b(m)}(t_j;\mathbf{h})\right] \right\}^2, \tag{13.39}$$

where $\widehat{\beta}_{-b(m)}(t_j;\mathbf{h})$ is the local polynomial estimator of $\beta(t) = (\beta_1(t), \ldots, \beta_P(t))^T$ computed from (13.33) with the bandwidth vector $\mathbf{h} = (h_1, \ldots, h_P)^T$ and the remaining data with all the observations in the entire block $b(m)$ deleted. The M-fold LSCV bandwidth is then the minimizer

$$\mathbf{h}_{MLSCV} = \left(h_{1,MLSCV}, \ldots, h_{P,MLSCV}\right)^T \tag{13.40}$$

of (13.39), assuming that the unique minimizer of (13.39) exists.

The minimization of (13.37) and (13.39) is computed over the multivariate $\mathbf{h} = (h_1, \ldots, h_P)^T$. This suggests that, when the dimensionality P is large, the LSCV bandwidths (13.38) and (13.40) can be difficult to compute even though the minimizers of (13.37) and (13.39) exist and are unique.

2. Leave-One-Time-Point-Out Cross-Validation

To speed up the computation, it is preferable to consider a component-wise cross-validation score which can be minimized by a univariate bandwidth. Intuitively, since the smoothing estimator $\widehat{\beta}_p^{(0)}(t)$ of (13.33) is obtained by treating the raw estimates $\{\widetilde{\beta}_p(t_j) : j = 1, \ldots, J\}$ as the pseudo-observations, we can consider a bandwidth that leads to "proper" smoothness over these pseudo-observations. The leave-one-time-point-out cross-validation (LTCV) is performed separately for each component curve $\beta_p(t)$ with $p = 1, \ldots, P$ by deleting the pseudo-observations $\{\widetilde{\beta}_p(t_j) : j = 1, \ldots, J\}$ one at a time or a block at a time. To do this, we first compute the local polynomial estimator $\widehat{\beta}_{p,-j}^{(0)}(t; h_p)$ from (13.33) with bandwidth h_p using the remaining pseudo-observations with $\widetilde{\beta}_p(t_j)$ deleted, and then compute the cross-validation score

$$LTCV(h_p) = \sum_{j=1}^{J} \left[\widetilde{\beta}_p(t_j) - \widehat{\beta}_{p,-j}^{(0)}(t_j; h_p) \right]^2. \tag{13.41}$$

If (13.41) can be uniquely minimized with respect to h_p, the LTCV bandwidths are given by

$$\begin{cases} \mathbf{h}_{LTCV} &= \left(h_{1,LTCV}, \ldots, h_{P,LTCV} \right)^T, \\ h_{p,LTCV} &= \arg\min_{h_p} LTCV(h_p) \quad \text{for} \quad p = 1, \ldots, P. \end{cases} \tag{13.42}$$

To speed up the computation, the above LTCV can be computed by grouping the time points $\{t_1, \ldots, t_J\}$ into M blocks $\{b(m) : m = 1, \ldots, M\}$ and minimizing the M-fold LTCV score

$$MLTCV(h_p) = \sum_{m=1}^{M} \left[\widetilde{\beta}_p(t_j) - \widehat{\beta}_{p,-b(m)}^{(0)}(t_j; h_p) \right]^2 1_{[t_j \in b(m)]} \tag{13.43}$$

with respect to h_p, where $\widehat{\beta}_{p,-b(m)}^{(0)}(t_j; h_p)$ is the local polynomial estimator of $\beta_p(t_j)$ computed from (13.33) with bandwidth h_p using the remaining pseudo-observations with $\{\widetilde{\beta}_p(t_j) : t_j \in b(m)\}$ deleted and $1_{[\cdot]}$ is the indicator function. Assuming that (13.43) can be uniquely minimized, the M-fold LTCV bandwidth vector is

$$\begin{cases} \mathbf{h}_{MLTCV} &= \left(h_{1,MLTCV}, \ldots, h_{P,MLTCV} \right)^T, \\ h_{p,MLTCV} &= \arg\min_{h_p} MLTCV(h_p) \quad \text{for} \quad p = 1, \ldots, P. \end{cases} \tag{13.44}$$

Since the LTCV score (13.41) is minimized for each p separately, it has the potential to save computing time compared with the LSCV score (13.37). An advantage of using the LSCV is that it has the potential to preserve the intra-subject correlation structure of the data (e.g., Section 7.2). Although the LTCV score (13.41) ignores the potential intra-subject correlations, it has been shown to have similar performance compared to the LSCV methods by simulation in Wu, Tian and Yu (2010).

13.3.5 Bootstrap Confidence Intervals

We have demonstrated previously in Section 7.3 that a "$\pm Z_{1-\alpha/2}$ standard error band," which ignores the bias, can often lead to an approximate $[100 \times (1-\alpha)]\%$ confidence interval for a smoothing estimator. Since biases of the smoothing estimators are often difficult to estimate, we consider here the approximate pointwise confidence intervals of the local polynomial estimator $\widehat{\beta}_p^{(0)}(t)$ for $\beta_p(t)$ at any $t \in \mathscr{T}$ based on the resampling subject bootstrap without bias correction. Because of the unknown correlation structures of the data, it is difficult to implement in practice an approximate confidence interval based on the asymptotic distributions of $\widehat{\beta}_p^{(0)}(t)$.

Approximate Bootstrap Pointwise Confidence Intervals:

(a) **Computing Bootstrap Estimators.** *Generate B bootstrap samples using the resampling-subject bootstrap procedure of Section 3.4.1, denote the resulting bootstrap samples by*

$$\left\{ \begin{array}{rcl} \mathscr{X}^b & = & \{Y_i^b(t_j^b), \mathbf{X}_i^b(t_j^b), t_j^b : i \in \mathscr{S}_j, j = 1, \ldots, J\}, \\ \mathscr{X}^{Boot} & = & \{\mathscr{X}^b : b = 1, \ldots, B\}, \end{array} \right. \tag{13.45}$$

and obtain B bootstrap estimators $\{\widehat{\beta}_{p,b}^{(0)}(t) : b = 1, \ldots, B\}$.

(b) **Approximate Bootstrap Confidence Intervals.** *Denote by $l_{\alpha/2}^p(t)$ and $u_{\alpha/2}^p(t)$ the lower and upper $[100 \times (\alpha/2)]th$ percentiles of the B bootstrap estimators. The $[100 \times (1-\alpha)]\%$ bootstrap percentile pointwise confidence intervals for $\beta_p^{(0)}(t)$ is*

$$\left(l_{\alpha/2}^p(t), u_{\alpha/2}^p(t) \right). \tag{13.46}$$

Let $\widehat{se}_p^{Boot}(t; \widehat{\beta}_p^{(0)})$ be the sample standard deviation of the B bootstrap estimators, i.e.,

$$\widehat{se}_p^{Boot}\left(t; \widehat{\beta}_p^{(0)}\right) = \left\{ \frac{1}{B-1} \sum_{b=1}^B \left[\widehat{\beta}_{p,b}^{(0)}(t) - \frac{1}{B} \sum_{b'=1}^B \widehat{\beta}_{p,b'}^{(0)}(t) \right]^2 \right\}^{1/2}.$$

The normal approximate pointwise confidence interval for $\beta_p(t)$ based on the bootstrap samples $\mathscr{X}^{Boot}$ is

$$\widehat{\beta}_p^{(0)}(t) \pm Z_{1-\alpha/2} \widehat{se}_p^{Boot}(t), \tag{13.47}$$

where z_a is the $[100 \times a]$th quantile of the standard normal distribution. □

   The above pointwise confidence intervals are potentially useful for the construction of simultaneous confidence bands for all or a subset of the coefficient curves $\beta(t) = \left(\beta_1(t), \ldots, \beta_P(t)\right)^T$ at a collection of the time points. The generalization to simultaneous confidence bands in principle can be carried out using the similar methods discussed in Section 7.3.2. However, the problem of simultaneous statistical inferences for the current time-varying transformation models has not been investigated in the literature.

13.4   R Implementation

13.4.1   The NGHS Data

The NGHS has been described in Section 1.2. Recall that the NGHS is a multicenter population-based cohort study aimed at evaluating the longitudinal changes and racial difference in cardiovascular risk factors between Caucasian and African American girls during childhood and adolescence. We illustrate here how to apply the time-varying transformation models to the NGHS data to evaluate the age-specific covariate effects on the distribution functions of four important outcome variables, namely, systolic and diastolic blood pressures (SBP and DBP), low density lipoprotein (LDL) cholesterol, and triglyceride (TG).

   In a series of preliminary goodness-of-fit tests for normality, such as the Shapiro-Wilk tests and the Kolmogorov-Smirnov tests, and visual inspections of the quantile-quantile plots, we observed that SBP, DBP, LDL and TG for this population of girls conditioning on various covariate values of age, race, height and body mass index (BMI) were clearly not normally distributed. In particular, many of these conditional distributions were skewed, and were rejected for normality by the goodness-of-fit tests at 5% significance level. These preliminary findings suggest that the conditional mean-based results, such as obtained in Daniels et al. (1998), may not give an adequate description of the covariate effects on the overall conditional distributions of the blood pressure and lipid levels, and the conclusions in Thompson (2007), based on specific threshold choices for the outcome variables, may not hold if other threshold values were used. However, the time-varying transformation models have certain major advantages over the existing analyses. With a flexible and parsimonious structure, the time-varying transformation models summarize the age-specific effects of these covariates on the overall distributions of the outcome variables. Moreover, statistical inferences for the age-dependent coefficient curves have meaningful biological interpretations and do not reply on the normality assumption for the outcome variables or the threshold values for defining unhealthy risk levels.

   Since the children's blood pressures are increasing with age and height percentiles (NHBPEP, 2004; James et al., 2014), in our exploratory analysis, we examine the odds-ratio plots of SBP and DBP under various age, race and height strata. These odds-ratio plots suggest that the proportional odds models give reasonable approximations to the relationships between the conditional distribution functions of these outcomes and the covariates. Using the notation of Section 13.2.1, SBP is the outcome $Y(t)$, and race and height percentile are the covariates, $X^{(1)}$ and $X^{(2)}(t)$, respectively. We fit the following time-varying proportional odds model to evaluate the age-specific covariate effects of race and height on the distribution functions of SBP,

$$logit\left[P\left(Y(t) > y_1 | t, X^{(1)}, X^{(2)}(t)\right)\right] = h(t) + \beta_1(t)X^{(1)} + \beta_2(t)X^{(2)}(t).$$

Here the visit time t (age) ranges from 9 to 19 years for the girls and the time-varying covariate, height, is measured by the age-adjusted percentile computed from the CDC population growth chart.

Because the entry age starts at 9 years, we use the age range $\mathcal{T} = [9, 19)$, and round off the age to one decimal place in years, which leads to $J = 100$ equally spaced age bins $[9.0, 9.1), \ldots, [18.9, 19.0)$ with time design points $\{t_1 = 9.0, t_2 = 9.1, \ldots, t_{100} = 18.9\}$. If the ith girl has observations within $[t_j, t_{j+1})$, $Y_i(t_j)$ and $X_i^{(2)}(t_j)$ are the ith girl's observed blood pressure and height percentile at age t_j, $X_i^{(1)} = 1$ for an African American girl, and $X_i^{(1)} = 0$ for a Caucasian girl. A positive (negative) value for $\beta_1(t)$ suggests that African American girls tend to have higher (lower) SBP levels than Caucasian girls at age t, and $\beta_2(t)$ represents the changes of the log-odds of $Y(t) > y$ associated with a unit increase in height percentile $X^{(2)}(t)$ at age t.

We carry out the two-step procedure of Section 13.3 for estimating the coefficient curves of SBP. First, the raw estimates $\tilde{\beta}_1(t_j)$ and $\tilde{\beta}_2(t_j)$ are computed at the time design points with weight $w(\cdot) = 1$ for SBP. Then, in the second step, the smoothing estimates, $\hat{\beta}_1(t)$ and $\hat{\beta}_2(t)$, are computed using the local linear method with the Epanechnikov kernel and the cross-validated bandwidths. The covariate effects of race and height on the outcome DBP can be estimated similarly. We use the following R code:

```
> library(npmlda)
> NGHS.sbp <- NGHS[!is.na(NGHS$SBP) & !is.na(NGHS$HTPCT),]
> Agebins <- seq(90, 189, by=1) #100 bins
# Raw estimates 100*2 matrix
> attach(NGHS.sbp)
> OR.SBP <- TVtrans.fit(Agebins, Y=SBP, X1=(RACE==2)*1,
                        X2=HTPCT)
# local linear smoothing estimates
> SBP.beta1.lm <- LocalLm(Agebins, Agebins, OR.SBP[,1], bw=16)
> SBP.beta2.lm <- LocalLm(Agebins, Agebins, OR.SBP[,2], bw=24)
```

Figure 13.1 shows the two-step local linear estimators of $\beta_1(t)$ and $\beta_2(t)$

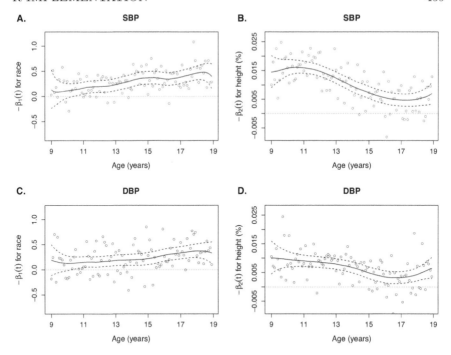

Figure 13.1 *The systolic blood pressure (A)-(B) and diastolic blood pressure (C)-(D). Each row shows the raw estimates (circles), and the two-step local linear estimates (solid lines) computed with the LTCV bandwidths, and their bootstrap standard error bands (dashed lines) for the covariate effects of race and height.*

computed from the LTCV cross-validated bandwidths and the "±1.96 standard error bands" from 1000 bootstrap repetitions. To ease the computational burden, the cross-validated bandwidths obtained from the original sample are used for the bootstrap standard error bands. Similar results are obtained from the bandwidths based on the LSCV method or the bootstrap percentiles. For both SBP and DBP, the effects of race and height clearly change over time. The racial difference is close to zero and slightly increases with age, suggesting that African American girls tend to have greater odds of high blood pressure than Caucasian girls during the later adolescent years. The positive but decreasing estimate of $\beta_2(t)$ suggests that height contributes positively to the odds of high SBP or DBP, but the height effect diminishes with age.

As reported in Thompson (2007), childhood overweight is significantly associated with unhealthful lipid levels. Hence we apply the time-varying proportional odds models for LDL and TG levels to evaluate the covariate effects of race and BMI. Figure 13.2 shows the coefficient curves based on the two-step local linear estimators for LDL and TG, respectively. These results suggest that there is no difference in the two racial groups in LDL, but there is signif-

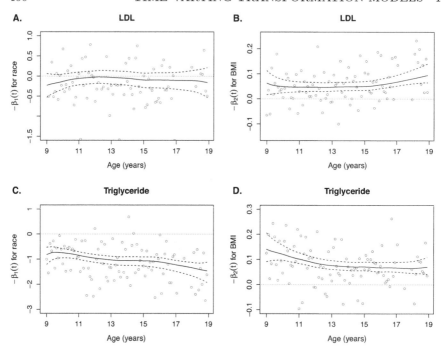

Figure 13.2 *The low density lipoprotein (LDL) cholesterol (A)-(B) and triglyceride (C)-(D). Each row shows the raw estimates (circles), and the two-step local linear estimates (solid lines) computed with the LTCV bandwidths, and their ±1.96 bootstrap standard error bands (dashed lines) for the covariate effects of race and BMI.*

icant race effect on the triglyceride levels, and this race effect varies with age. As expected, there is also a strong BMI effect on the lipids. For both LDL and TG, higher BMI is associated with greater odds of having high lipid levels during the adolescent years for the NGHS girls.

13.5   Asymptotic Properties

We derive in this section the large sample properties of the local polynomial estimator $\widehat{\beta}_p^{(q)}(t)$ of (13.32) at any time point $t \in \mathcal{T}$. The special case of $q = 0$ corresponds to the asymptotic properties for the local polynomial estimator of $\beta_p(t)$. The derivation is followed by two main steps. First, we approximate the estimator as a weighted linear combination of the dominating terms in (13.14). Second, we compute the asymptotic errors using the asymptotic properties of the raw estimator $\widetilde{\beta}(t_j)$ developed in Section 13.3.2.

13.5.1   Conditional Mean Squared Errors

We consider the following conditional bias, variance and mean squared errors (MSE) of $\widehat{\beta}_p^{(q)}(t)$ given the observed covariates $\mathscr{D}$ in (13.2):

$$
\begin{cases}
B\big[\widehat{\beta}_p^{(q)}(t)\big|\mathscr{D}\big] & = \ E\big[\widehat{\beta}_p^{(q)}(t) - \beta_p^{(q)}(t)\big|\mathscr{D}\big], \\[2mm]
Var\big[\widehat{\beta}_p^{(q)}(t)\big|\mathscr{D}\big] & = \ E\big\{\big[\widehat{\beta}_p^{(q)}(t) - E\big(\widehat{\beta}_p^{(q)}(t)\big|\mathscr{D}\big)\big]^2\big|\mathscr{D}\big\}, \\[2mm]
MSE\big[\widehat{\beta}_p^{(q)}(t)\big|\mathscr{D}\big] & = \ E\big\{\big[\widehat{\beta}_p^{(q)}(t) - \beta_p^{(q)}(t)\big]^2\big|\mathscr{D}\big\} \\[2mm]
& = \ B^2\big[\widehat{\beta}_p^{(q)}(t)\big|\mathscr{D}\big] + Var\big[\widehat{\beta}_p^{(q)}(t)\big|\mathscr{D}\big].
\end{cases}
\tag{13.48}
$$

Although the quantities in (13.48) are only defined for the pth component curve $\widehat{\beta}_p^{(q)}(t)$, they can be straightforwardly generalized to the vector

$$
\widehat{\beta}^{(q)}(t) = \Big(\widehat{\beta}_1^{(q)}(t), \dots, \widehat{\beta}_P^{(q)}(t)\Big)^T.
\tag{13.49}
$$

If we assign equal weight to each of the component in $\widehat{\beta}^{(q)}(t)$, a simple extension of the MSE to (13.49) is

$$
\begin{aligned}
MSE\big[\widehat{\beta}^{(q)}(t)\big|\mathscr{D}\big] & = \ \textstyle\sum_{p=1}^{P} E\big\{\big[\widehat{\beta}_p^{(q)}(t) - \beta_p^{(q)}(t)\big]^2\big|\mathscr{D}\big\} \\[2mm]
& = \ \textstyle\sum_{p=1}^{P} B^2\big[\widehat{\beta}_p^{(q)}(t)\big|\mathscr{D}\big] + \sum_{p=1}^{P} Var\big[\widehat{\beta}_p^{(q)}(t)\big|\mathscr{D}\big].
\end{aligned}
\tag{13.50}
$$

Clearly, (13.50) can be generalized by giving different weights to different components. Since the MSE in (13.50) and its generalizations depend on the asymptotic expressions of the terms in (13.48). The asymptotic developments of this section are focused on (13.48).

13.5.2   Asymptotic Assumptions

We make the following assumptions throughout this section. These assumptions are similar to the ones given in Section 7.5.1 for the conditional mean based time-varying coefficient models.

(a) *When $n \to \infty$, $n_j \approx c_j n$ for some constant $0 < c_j < 1$, and the bandwidth h satisfies $h \to 0$, $n^{1/2} h^{Q-q+1} \to \infty$, $nJh^{2q+1} \to \infty$ and $nh/(Jn_{j_1 j_2}) \to \infty$.*

(b) *The design time points $\{t_j \in \mathscr{T} : j = 1, \dots, J\}$ are independent and identically distributed with the density function $\pi(t)$, and $\mathscr{T}$ is the support of $\pi(\cdot)$.*

(c) *The coefficient curves $\{\beta_p(t) : 1 \le p \le P\}$ of (13.4) are $Q+1$ times continuously differentiable with respect to t.*

(d) *The matrix $\Lambda(t_j)$ of (13.14) is nonsingular and $\Sigma_{p_1, p_2}(t_1, t_2)$, the (p_1, p_2)th element, are $Q+1$ times continuously differentiable with respect to (t_1, t_2).*

(e) *The kernel $K(\cdot)$ is a symmetric probability density on a bounded set.*   □

Assumption (a) serves the purpose of producing a consistent local polyno-mial estimator $\widehat{\beta}_p^{(q)}(t)$ when n and n_j tend to infinity. In particular, $h \to 0$, $n^{1/2} h^{Q-q+1} \to \infty$ and $nJh^{2q+1} \to \infty$ guarantee that both $B[\widehat{\beta}_p^{(q)}(t)|\mathscr{D}]$ and $Var[\widehat{\beta}_p^{(q)}(t)|\mathscr{D}]$ tend to zero in probability, while $nh/(Jn_{j_1 j_2}) \to \infty$ ensures that the asymptotic expression of $Var[\widehat{\beta}_p^{(q)}(t)|\mathscr{D}]$ is not affected by the covariance structure of the data. We use these assumptions for the asymptotic proper-ties of $\widehat{\beta}_p^{(q)}(t)$ because they represent a large number of situations in practice. Other asymptotic properties of $\widehat{\beta}_p^{(q)}(t)$ may be derived if Assumption (a) is replaced by another set of conditions.

13.5.3 Asymptotic Risk Expressions

For any Q and q, we define the following functionals of the kernel function $K(\cdot)$:

$$
\begin{cases}
S &= \left(s_{j_1 j_2}\right)_{j_1, j_2 = 0, \ldots, Q}, \\
s_{j_1 j_2} &= \int K(u) u^{j_1 + j_2} du, \\
K_{q, Q+1}(t) &= e_{q+1, Q+1}^T S^{-1} \left(1, t, \cdots, t^Q\right)^T K(t), \\
B_{Q+1}(K) &= \int K(u) u^{Q+1} du, \\
V(K) &= \int K^2(u) du.
\end{cases}
\tag{13.51}
$$

We denote by $a_n \approx b_n$ that both a_n and b_n converge to the same limit in probability as $n \to \infty$. The next theorem shows the asymptotic approximations for $B[\widehat{\beta}_p^{(q)}(t)|\mathscr{D}]$, $V[\widehat{\beta}_p^{(q)}(t)|\mathscr{D}]$, and $MSE[\widehat{\beta}_p^{(q)}(t)|\mathscr{D}]$.

Theorem 13.1. *Under Assumptions (a) to (e), the asymptotic expressions of the conditional bias, variance and MSE as $n \to \infty$ are*

$$
B[\widehat{\beta}_p^{(q)}(t)|\mathscr{D}] = \frac{h^{Q-q+1}(q!)}{(Q+1)!} \beta_p^{(Q+1)}(t) B_{Q+1}\left(K_{q, Q+1}\right) [1 + o(1)],
\tag{13.52}
$$

$$
Var[\widehat{\beta}_p^{(q)}(t)|\mathscr{D}] = \frac{2(q!)^2}{3cnJh^{2q+1}\pi(t)} \sigma_p^2(t) V\left(K_{q, Q+1}\right) [1 + o(1)],
\tag{13.53}
$$

and

$$
\begin{aligned}
MSE[\widehat{\beta}_p^{(q)}(t)|\mathscr{D}\} &= \frac{h^{2(Q-q+1)}(q!)^2}{[(Q+1)!]^2} \left[\beta_p^{(Q+1)}(t) B_{Q+1}\left(K_{q, Q+1}\right)\right]^2 \\
&\quad + \frac{2(q!)^2}{3cnJh^{2q+1}\pi(t)} \sigma_p^2(t) V\left(K_{q, Q+1}\right) \\
&\quad + o\left[h^{2(Q-q+1)} + \left(nJh^{2q+1}\right)^{-1}\right]
\end{aligned}
\tag{13.54}
$$

for some constant $0 < c < 1$ given in Assumption (a), where $\sigma_p^2(t)$ is the pth diagonal element of $\Sigma(t, t)$ defined in (13.22). ∎

Proof of Theorem 13.1 is given in Section 13.5.4.

The theoretically optimal bandwidths for $\widehat{\beta}_p^{(q)}(t)$ can be derived by minimizing the mean squared errors $MSE\left[\widehat{\beta}_p^{(q)}(t)\big|\mathscr{D}\right]$ with respect to h. The next theorem gives the expression of the asymptotically optimal bandwidth for $\widehat{\beta}_p^{(0)}(t)$.

Theorem 13.2. *If the assumptions of Theorem 13.1 are satisfied, when $q = 0$ and $Q = 1$, the asymptotic expressions of the bias, variance and MSE for $\widehat{\beta}_p^{(0)}(t)$ are*

$$B\left[\widehat{\beta}_p^{(0)}(t)\big|\mathscr{D}\right] = (h^2/2)\,\beta_p''(t)\,B_2(K_{0,2})\,[1+o(1)], \tag{13.55}$$

$$V\left[\widehat{\beta}_p^{(0)}(t)\big|\mathscr{D}\right] = \frac{2}{3\,cnJh\pi(t)}\,\sigma_p^2(t)\,V(K_{0,2})\,[1+o(1)], \tag{13.56}$$

$$MSE\left[\widehat{\beta}_p^{(0)}(t)\big|\mathscr{D}\right\} = (h^4/4)\left[\beta_p''(t)\,B_2(K_{0,2})\right]^2$$
$$+\frac{2}{3\,cnJh\pi(t)}\,\sigma_p^2(t)\,V(K_{0,2}) \tag{13.57}$$
$$+o\left[h^4 + (nJh)^{-1}\right],$$

where $K_{0,2}(\cdot)$, $B_2(K_{0,2})$ and $V(K_{0,2})$ are defined in (13.51). The theoretically optimal pointwise bandwidth is

$$h_{opt}\left[\beta_p(t)\right] = \left\{\frac{2\,\sigma_p^2(t)\,V(K_{0,2})}{3\,cnJ\,\pi(t)\left[\beta_p''(t)\,B_2(K_{0,2})\right]^2}\right\}^{1/5}, \tag{13.58}$$

which minimizes the right side of (13.57). ∎

Proof of Theorem 13.2 is given in Section 13.5.4.

Since the expression of $h_{opt}\left[\beta_p(t)\right]$ in (13.58) depends on $t \in \mathscr{T}$ and the unknown functions $\sigma_p^2(t)$, $\beta_p''(t)$ and $\pi(t)$, it only gives a theoretical sense of the ideal bandwidth and cannot be directly used in practice. If we ignore the constant term of (13.58) and focus on its convergence rate, an implication of (13.58) is that an appropriate bandwidth h_p should converge to zero in the $(nJ)^{1/5}$ rate, i.e., $(nJ)^{1/5}h_p \to C_p(t)$ as $n \to \infty$ and $J \to \infty$ for some $C_p(t) > 0$. Another implication of (13.58) is that, if the smoothness assumptions of two different component curves $\beta_{p_1}(t)$ and $\beta_{p_2}(t)$ for $p_1 \neq p_2$ are different, the appropriate bandwidths h_{p_1} and h_{p_2} should also be different. Since the two-step smoothing procedure of this chapter depends on component-wise smoothing of the raw estimates for each $p = 1, \ldots, P$, Theorem 13.2 suggests that the appropriate bandwidth can be obtained for each $\beta_p(t)$.

13.5.4   Theoretical Derivations

This section gives the proofs of Lemma 13.1, Lemma 13.2, Theorem 13.1 and Theorem 13.2.

Proof of Lemma 13.1:

Using equation (13.13) and the Taylor's expansion, we have the following approximation when n and n_j are sufficiently large,

$$\sum_{i_1 \neq i_2 \in \mathscr{S}_j} U_{i_1 i_2}[\beta(t_j)] = \sum_{i_1 \neq i_2 \in \mathscr{S}_j} \left\{ \frac{\partial U_{i_1 i_2}[\beta(t_j)]}{\partial \beta(t_j)} \right\}$$
$$\times \left[\tilde{\beta}(t_j) - \beta(t_j) \right] \left[\mathbf{I} + o_p(\mathbf{I}) \right]. \tag{13.59}$$

By the law of large numbers for U-statistics, e.g., Serfling (1980, Sec. 5.2), we have that, as $n \to \infty$,

$$\left[\frac{1}{n_j(n_j - 1)} \right] \sum_{i_1 \neq i_2 \in \mathscr{S}_j} \left\{ \frac{\partial U_{i_1 i_2}[\beta(t_j)]}{\partial \beta(t_j)} \right\} \to \Lambda(t_j) \quad \text{in probability.} \tag{13.60}$$

Since $\Lambda(t_j)$ is nonsingular and $A(t_j)$ is defined as

$$A(t_j) = \left[\frac{1}{n_j(n_j - 1)} \right] \Lambda^{-1}(t_j) \sum_{i_1 \neq i_2 \in \mathscr{S}_j} U_{i_1 i_2}[\beta(t_j)],$$

the approximation of (13.14) follows from multiplying $[n_j(n_j - 1)]^{-1} \Lambda^{-1}(t_j)$ to both sides of (13.59).   ∎

Proof of Lemma 13.2:

By the definition of $U_{i_1 i_2}[\beta(t_j)]$ and (13.14), we have $E[A(t_j)|\mathscr{D}] = 0$ and

$$Cov\left[A(t_{j_1}), A(t_{j_2})|\mathscr{D}\right]$$
$$= \left[n_{j_1}(n_{j_1} - 1) n_{j_2}(n_{j_2} - 1) \right]^{-1} \Lambda^{-1}(t_{j_1})$$
$$\times \sum_{i_1 \neq i_2 \in \mathscr{S}_{j_1}} \sum_{i_3 \neq i_4 \in \mathscr{S}_{j_2}} E\left\{ U_{i_1 i_2}[\beta(t_{j_1})] U_{i_3 i_4}[\beta(t_{j_2})] \Big| \mathscr{D} \right\}$$
$$\times \left[\Lambda^{-1}(t_{j_2}) \right]^T. \tag{13.61}$$

To evaluate the summation term at the right side of (13.61), we consider two sets of distinct integers $\{i_1, i_2\}$ and $\{i_3, i_4\}$, and let c be the number of common integers in these two sets. The values of c are 0, 1 and 2, since $\{i_1, i_2\}$ and $\{i_3, i_4\}$ may have 0, 1 or 2 integers in common.

We now evaluate the right side of (13.61) for the three separate cases of $c = 0$, 1 and 2. When $c = 0$, we have

$$E\{U_{i_1 i_2}[\beta(t_{j_1})] U_{i_3 i_4}[\beta(t_{j_2})]|\mathscr{D}\} = 0 \quad \text{for } i_1 \neq i_2 \neq i_3 \neq i_4. \tag{13.62}$$

Thus, when computing the right side of (13.61), the summation terms with $i_1 \neq i_2 \neq i_3 \neq i_4$, i.e., $c = 0$, can be ignored.

When $c = 1$, it can be directly verified by the symmetry of $U_{i_1 i_2}[\beta(t_j)]$ that

$$\rho(t_{j_1}, t_{j_2}) = E\left\{ U_{i_1 i_2}[\beta(t_{j_1})] U_{i_3 i_4}[\beta(t_{j_2})] \Big| \mathscr{D} \right\} \tag{13.63}$$

for any $\{i_1, i_2\}$ and $\{i_3, i_4\}$ that have only one integer in common. To compute the number of ways of selecting one integer in common, there are four situations which need to be considered:

(a) $i_1 = i_3 \in \mathscr{S}_{j_1} \cap \mathscr{S}_{j_2}$, $i_2 \in \mathscr{S}_{j_1} \backslash \mathscr{S}_{j_2}$, $i_4 \in \mathscr{S}_{j_2} \backslash \mathscr{S}_{j_1}$;

(b) $i_1 = i_3 \in \mathscr{S}_{j_1} \cap \mathscr{S}_{j_2}$, $i_2 \in \mathscr{S}_{j_1} \cap \mathscr{S}_{j_2}$, $i_4 \in \mathscr{S}_{j_2} \backslash \mathscr{S}_{j_1}$;

(c) $i_1 = i_3 \in \mathscr{S}_{j_1} \cap \mathscr{S}_{j_2}$, $i_2 \in \mathscr{S}_{j_1} \backslash \mathscr{S}_{j_2}$, $i_4 \in \mathscr{S}_{j_2} \cap \mathscr{S}_{j_1}$;

(d) $i_1 = i_3 \in \mathscr{S}_{j_1} \cap \mathscr{S}_{j_2}$, $i_2 \in \mathscr{S}_{j_1} \cap \mathscr{S}_{j_2}$, $i_4 \in \mathscr{S}_{j_2} \cap \mathscr{S}_{j_1}$, $i_3 \neq i_4$.
$$\tag{13.64}$$

Here $i \in \mathscr{S}_{j_1} \backslash \mathscr{S}_{j_2}$ means $i \in \mathscr{S}_{j_1}$ and $i \notin \mathscr{S}_{j_2}$. The number of ways of selecting one common integer corresponding to the situations in (13.64) is

(a) $n_{j_1 j_2}(n_{j_1} - n_{j_1 j_2})(n_{j_2} - n_{j_1 j_2})$;

(b) $n_{j_1 j_2}(n_{j_1 j_2} - 1)(n_{j_2} - n_{j_1 j_2})/2$;

(c) $n_{j_1 j_2}(n_{j_1 j_2} - 1)(n_{j_1} - n_{j_1 j_2})/2$;

(d) $n_{j_1 j_2}(n_{j_1 j_2} - 1)(n_{j_1 j_2} - 2)/6$.
$$\tag{13.65}$$

When $c = 2$, we can assume by symmetry that $i_1 = i_3$ and $i_2 = i_4$, and define

$$\rho^*(t_{j_1}, t_{j_2}) = E\left\{ U_{i_1 i_2}[\beta(t_{j_1})] U_{i_1 i_2}[\beta(t_{j_2})] \Big| \mathscr{D} \right\}. \tag{13.66}$$

The number of ways of selecting two common integers from $\{i_1, i_2\} \in \mathscr{S}_{j_1}$ and $\{i_3, i_4\} \in \mathscr{S}_{j_2}$ is $n_{j_1 j_2}(n_{j_1 j_2} - 1)/2$.

Adding up all the terms for $c = 1$ and $c = 2$ using (13.63) through (13.66), Assumption (a) shows that the summation term at the right side of (13.61) is

$$\sum_{i_1 \neq i_2 \in \mathscr{S}_{j_1}} \sum_{i_3 \neq i_4 \in \mathscr{S}_{j_2}} E\left\{ U_{i_1 i_2}[\beta(t_{j_1})] U_{i_3 i_4}[\beta(t_{j_2})] \Big| \mathscr{D} \right\}$$

$$= \sum_{c=1} 4\rho(t_{j_1}, t_{j_2}) + \sum_{c=2} \rho^*(t_{j_1}, t_{j_2})$$

$$= 4n_{j_1 j_2}\left[(n_{j_1} - n_{j_1 j_2})(n_{j_2} - n_{j_1 j_2}) + \frac{1}{2}(n_{j_1 j_2} - 1)(n_{j_1} + n_{j_2} - 2n_{j_1 j_2}) \right.$$

$$\left. + \frac{1}{6}(n_{j_1 j_2} - 1)(n_{j_1 j_2} - 2) \right] \rho(t_{j_1}, t_{j_2}) + \frac{n_{j_1 j_2}(n_{j_1 j_2} - 1)}{2} \rho^*(t_{j_1}, t_{j_2})$$

$$= 4n_{j_1 j_2}\left[(n_{j_1} - n_{j_1 j_2})(n_{j_2} - n_{j_1 j_2}) + \frac{1}{2}(n_{j_1 j_2} - 1)(n_{j_1} + n_{j_2} - 2n_{j_1 j_2}) \right.$$

$$\left. + \frac{1}{6}(n_{j_1 j_2} - 1)(n_{j_1 j_2} - 2) \right] \rho(t_{j_1}, t_{j_2})[1 + o_p(1)], \tag{13.67}$$

since $(n_{j_1 j_2} - 1) = o\big[(n_{j_1 j_2} - 1)(n_{j_1 j_2 - 2})\big]$. Substituting (13.67) into (13.61), we get the desired result of (13.21). This completes the proof of Lemma 13.2. ∎

Proof of Theorem 13.1:
Assuming that Assumption (a) is satisfied, the following four equations have been shown in Lemma 1 and Lemma 2 of Fan and Zhang (2000, Appendix A) to hold for the local polynomial weight function $w_{q,Q+1}(t_j, t; h)$ defined in (13.31):

$$w_{q,Q+1}(t_j, t; h) = \frac{q!}{J h^{q+1} \pi(t)} K_{q,Q+1}\left(\frac{t_j - t}{h}\right)[1 + o_p(1)], \quad j = 1, \ldots, J; \quad (13.68)$$

$$\sum_{j=1}^{J} w_{q,Q+1}(t_j, t; h)(t_j - t)^k = q! \, 1_{[k=q]}, \quad k = 0, \ldots, P; \quad (13.69)$$

$$\sum_{j=1}^{J} w_{q,Q+1}(t_j, t; h)(t_j - t)^{Q+1} = q! \, h^{Q-q+1} B_{Q+1}(K_{q,Q+1})[1 + o_p(1)]; \quad (13.70)$$

$$\sum_{j=1}^{J} w_{q,Q+1}^2(t_j, t; h) = \frac{(q!)^2}{J h^{2q+1} \pi(t)} V(K_{q,Q+1})[1 + o_p(1)]; \quad (13.71)$$

where $K_{q,Q+1}(t) = e_{q+1,Q+1}^T S^{-1}(1, t, \ldots, t^Q)^T K(t)$, which is defined in (13.51), is the equivalent kernel of the local polynomial fit with $S = (s_{kl})_{k,l=0,1,\ldots,Q}$ and $s_{kl} = \int K(u) u^{k+l} \, du$. We do not repeat the proofs of (13.68) through (13.71) here, and simply use these equations as facts for our derivations. Details for the derivations of (13.68) through (13.71) can be found in Fan and Zhang (2000) and the references cited therein.

To derive the bias term (13.52) of $\widehat{\beta}_p^{(q)}(t)$, we note the decomposition

$$
\begin{aligned}
B\big[\widehat{\beta}_p^{(q)}(t)\big|\mathscr{D}\big] &= E\big[\widehat{\beta}_p^{(q)}(t)\big|\mathscr{D}\big] - \beta_p^{(q)}(t) \\
&= \sum_{j=1}^{J}\left\{w_{q,Q+1}(t_j, t; h) E\big[\widetilde{\beta}_p(t_j)\big|\mathscr{D}\big]\right\} - \beta_p^{(q)}(t). \quad (13.72)
\end{aligned}
$$

Now, let

$$W_1 = \sum_{j=1}^{J}\left\{w_{q,Q+1}(t_j, t; h) E\big[\widetilde{\beta}_p(t_j) - \beta_p(t_j)\big|\mathscr{D}\big]\right\} \quad (13.73)$$

and

$$W_2 = \sum_{j=1}^{J}\left[w_{q,Q+1}(t_j, t; h) \beta_p(t_j)\right] - \beta_p^{(q)}(t). \quad (13.74)$$

It follows from (13.72) that $B\big[\widehat{\beta}_p^{(q)}(t)\big|\mathscr{D}\big] = W_1 + W_2$, and it suffices to evaluate W_1 and W_2 separately.

Using the fact that $n_j \approx cn$ for some $0 < c < 1$ in Assumption (a) and $\tilde{\beta}_p(t_j) - \beta_p(t_j) = O_p(n_j^{-1/2})$, it then follows from (13.73) that

$$W_1 = \sum_{j=1}^{J} \left[w_{q,Q+1}(t_j, t) \times O_p(n_j^{-1/2}) \right] = O_p(n^{-1/2}), \qquad (13.75)$$

where the last equality holds because, by (13.68), $\sum_{j=1}^{J} w_{q,Q+1}(t_j, t)$ is bounded in probability. To compute W_2, we have, by the Taylor's expansion of $\beta_p(t_j)$ at t in (13.74),

$$
\begin{aligned}
W_2 &= \sum_{j=1}^{J} w_{q,Q+1}(t_j, t) \left\{ \sum_{k=0}^{Q+1} \beta_p^{(k)}(t) \frac{(t_j - t)^k}{k!} + o_p\left[(t_j - t)^{Q+1} \right] \right\} - \beta_p^{(q)}(t) \\
&= \sum_{k=0}^{Q+1} \beta_p^{(k)}(t) \left[\sum_{j=1}^{J} w_{q,Q+1}(t_j, t) \frac{(t_j - t)^k}{k!} \right] - \beta_p^{(q)}(t) \\
&\quad + \sum_{j=1}^{J} \left\{ w_{q,Q+1}(t_j, t) \times o_p\left[(t_j - t)^{Q+1} \right] \right\} \\
&= \beta_p^{(q)}(t) - \beta_p^{(q)}(t) + \frac{(q!)\, h^{Q-q+1}}{(Q+1)!} \beta_p^{(Q+1)}(t) B_{Q+1}\left(K_{q,Q+1} \right) \left[1 + o_p(1) \right] \\
&\quad + \sum_{j=1}^{J} \left\{ w_{q,Q+1}(t_j, t) \times o_p\left[(t_j - t)^{Q+1} \right] \right\} \\
&= \frac{(q!)\, h^{Q-q+1}}{(Q+1)!} \beta_p^{(Q+1)}(t) B_{Q+1}\left(K_{q,Q+1} \right) \left[1 + o_p(1) \right], \qquad (13.76)
\end{aligned}
$$

where the last two equality signs follow from (13.68), (13.69) and (13.70). By the assumption that $\lim_{n \to \infty} n^{1/2} h^{Q-q+1} \to \infty$ in Assumption (a), (13.75) and (13.76) then imply that $W_1 = o_p(W_2)$, so that, consequently, by (13.72),

$$B\left[\hat{\beta}_p^{(q)}(t) \big| \mathscr{D} \right] = W_2 + o_p(W_2). \qquad (13.77)$$

The expression of the asymptotic bias (13.52) is a direct consequence of (13.76) and (13.77).

For the derivation of the asymptotic variance (13.53), we note that

$$
\begin{aligned}
Var\left[\hat{\beta}_p^{(q)}(t) \big| \mathscr{D} \right] &= E\left\{ \sum_{j=1}^{J} w_{q,Q+1}(t_j, t) \tilde{\beta}_p(t_j) - E\left[\hat{\beta}_p^{(q)}(t) \big| \mathscr{D} \right] \right\}^2 \\
&= E\left\{ \sum_{j=1}^{J} w_{q,Q+1}(t_j, t) \left[\tilde{\beta}_p(t_j) - E\left[\tilde{\beta}_p(t_j) \big| \mathscr{D} \right] \right] \right\}^2 \\
&= W_3 + W_4, \qquad (13.78)
\end{aligned}
$$

where, by (13.17) and (13.18),

$$
\begin{cases}
W_3 &= \sum_{j=1}^{J} w_{q,Q+1}^2(t_j, t)\, Var\left[A_p(t_j) \big| \mathscr{D} \right], \\
W_4 &= \sum_{j_1 \neq j_2} w_{q,Q+1}(t_{j_1}, t)\, w_{q,Q+1}(t_{j_2}, t)\, Cov\left[A_p(t_{j_1}) A_p(t_{j_2}) \big| \mathscr{D} \right].
\end{cases}
\qquad (13.79)
$$

It suffices to evaluate W_3 and W_4 separately.

For W_3, by Assumption (a), (13.71) and $V[A_p(t_j)|\mathscr{D}] \approx [2/(3n_j)]\sigma_p^2(t)$ in (13.26),

$$W_3 = \frac{2(q!)^2}{3cnJh^{2q+1}\pi(t)} \sigma_p^2(t) V(K_{q,Q+1})[1 + o_p(1)]. \qquad (13.80)$$

For W_4, we consider the convergence rate $r_{j_1 j_2}$ of $Cov[A_p(t_{j_1})A_p(t_{j_2})|\mathscr{D}]$ given in (13.21) and (13.24) for two distinct time points $t_{j_1} \neq t_{j_2}$. Substituting (13.21) into W_4 in (13.79), it follows from (13.68) that W_4 has the convergence rate

$$W_4 = O_p\left(\frac{r_{j_1 j_2}}{h^{2q+2}}\right). \qquad (13.81)$$

Since $n_{j_1 j_2} \leq \min\{n_{j_1}, n_{j_2}\}$ for any $j_1 \neq j_2$, we can directly verify from (13.24) that

$$r_{j_1 j_2} = O\left(\frac{n_{j_1 j_2}}{n_{j_1} n_{j_2}}\right), \qquad (13.82)$$

where $n_{j_1 j_2}$ may or may not tend to infinity as $n \to \infty$. It then follows from (13.81) and (13.82) that

$$W_4 = O_p\left(\frac{n_{j_1 j_2}}{n_{j_1} n_{j_2} h^{2q+2}}\right). \qquad (13.83)$$

Comparing (13.80) with (13.83), the assumptions of $\lim_{n\to\infty}[nh/(Jn_{j_1 j_2})] = \infty$ and $n_j \approx c_j n$ in Assumption (a) implies that

$$W_4 = o_p(W_3). \qquad (13.84)$$

The conclusion of (13.53) directly follows from (13.78), (13.80) and (13.84). Finally, the conclusion of (13.54) is obtained by (13.48) with $B^2[\widehat{\beta}_p^{(q)}(t)|\mathscr{D}]$ and $Var[\widehat{\beta}_p^{(q)}(t)|\mathscr{D}]$ replaced by the dominating terms at the right sides of (13.52) and (13.53), respectively. ∎.

Proof of Theorem 13.2:

When $Q = 1$, Assumption (c) implies that $\beta_p(t)$ is twice differentiable with respect to t, and its second derivative $\beta''(t)$ is a continuous function of $t \in \mathscr{T}$. The conclusions of (13.55), (13.56) and (13.57) follow from (13.52), (13.53) and (13.54), respectively, by setting $q = 0$ and $Q = 1$. Minimizing the right side of (13.57) with respect to h, the theoretically optimal pointwise bandwidth satisfies the normal equation

$$h_{opt}^3[\beta_p(t)]\left[\beta_p''(t)B_2(K_{0,2})\right]^2 = \frac{2}{3cnJ\pi(t)} h_{opt}^{-2}[\beta_p(t)]\sigma_p^2(t)V(K_{0,2}), \qquad (13.85)$$

which then leads to the conclusion of (13.58). ∎

13.6 Remarks and Literature Notes

This chapter summarizes a time-varying nonparametric approach for modeling the conditional distribution functions and their temporal trends. As a special case of the conditional distribution based regression models, the time-varying transformation models have practical advantages over the well-established conditional mean-based regression models in longitudinal analysis when the scientific objective is better achieved by analyzing the conditional distribution functions. The application to the NGHS data illustrates the advantages of modeling the conditional distribution functions in a typical biomedical study. Similar to the conditional mean-based varying-coefficient models of Chapters 6 to 8, the time-varying transformation models effectively reduce the estimation complexity by utilizing a regression structure which retains a high degree of model flexibility.

The main scope of this chapter focuses on the two-step estimation method described in Wu, Tian and Yu (2010). This estimation method, which is based on smoothing over the raw estimates obtained from the estimating equations of Cheng, Wei and Ying (1995), is limited to the covariate effects of the target distribution functions. The advantage of this two-step estimation method is its conceptual simplicity and flexibility of providing different bandwidths for different component curves. However, in view that the method of Cheng, Wei and Ying (1995) is not the only available method for the transformation models, alternative estimation methods other than the ones described in Wu, Tian and Yu (2010) are possible but have not been systematically studied. Existing possibilities include using alternative raw estimators of the transformation models or alternative smoothing methods in the smoothing step. The question of whether the two-step estimation method can be replaced by a "one-step" estimation approach is not well understood and deserves further investigation. This question is perhaps more interesting from the theoretical and methodological point of view, because, judging from the NGHS application of Section 13.4 and simulation results, the two-step estimation method of Section 13.3 performs very well in practice. Further research based on the established premises is warranted in several fronts. These include developing appropriate goodness-of-fit tests to formally examine the adequacy of the time-varying models under appropriate statistical hypotheses, extending the time-varying models to incorporate the joint distributions of a multivariate response variable, and comparing the time-varying transformation models to potential quantile regression models for longitudinal data.

Chapter 14

Time-Varying Transformation Models - II

In addition to the estimation of covariate effects, the time-varying transformation models of Chapter 13 can be further used to predict the ranking of a subject's health status at one or multiple time points. We describe in this chapter the methods for estimating the conditional distribution functions and some of their useful functionals based on the models and estimation methods of Chapter 13. These estimators lead to useful statistical indices to quantitatively track the changes of health status in biomedical studies.

14.1 Overview and Motivation

In Chapter 12, we discussed two primary objectives for the longitudinal analysis of conditional distributions, namely estimation of temporal trends of the conditional distributions and the rank-tracking probabilities. Since Chapter 12 considers the simple data structure which contains only categorical and time-invariant covariates, its estimation methods can be carried out with a simple unstructured kernel smoothing method. When the data contain time-dependent covariates, the unstructured model and kernel smoothing methods are difficult to use because of the multivariate smoothing. In this chapter, we present a two-step method to estimate the conditional distributions and the rank-tracking probabilities using time-varying covariates and the models of Chapter 13.

1. Motivation

The NGHS data, which has been repeatedly analyzed in the previous chapters, is again an excellent motivating example for the methods of this chapter. Because of the methodological limitations, two important questions of the NGHS have not been answered in Chapters 12 and 13:

(a) *What are the conditional distributions of blood pressure, either systolic blood pressure (SBP) or diastolic blood pressure (DBP), given a set of time-dependent covariates?*

(b) *Are the study subjects with a set of time-varying covariates and high blood pressure levels at a younger age more likely to have high blood pressure levels at an older age?*

The first question is raised because the estimated curves of Chapter 13 only describe the covariate effects on the outcome variable's conditional distributions but not the actual predicted values of the distributions. The second question is discussed in Kavey et al. (2003), Thompson et al. (2007) and Obarzanek et al. (2010), where these authors illustrate the need to measure the "tracking ability" of a cardiovascular risk factor within subjects over different ages. Appropriate answers for both questions depend on the models and estimation methods for the conditional distribution functions. Given the simple and flexible modeling structure in Chapter 13, it is natural to expect that the above two questions could be investigated using the time-varying transformation models with time-dependent covariates. This motivates the estimation methods of this chapter.

2. Conditional Distribution Estimators

We present in this chapter a two-step smoothing method, which was developed by Wu and Tian (2013b), to estimate the conditional distribution functions and longitudinal tracking abilities. Similar to the tracking concept in Chapter 12, we consider two useful functionals of the conditional distributions, the rank-tracking probability (RTP) and the rank-tracking probability ratio (RTPR), based on the time-varying transformation models, and illustrate the interpretations of the RTP and the RTPR through the NGHS data. We estimate the conditional distribution functions using the similar smoothing estimators as Chapter 13, in which we first compute the raw estimates at a set of distinct time points and then smooth the raw estimates through a smoothing method. In the theoretical development, we derive the asymptotic biases, variances and mean squared errors for the two-step local polynomial estimators of the conditional distribution functions, and show that the smoothing step may reduce the variances of the estimators. For the choices of data-driven bandwidths, we present two cross-validation methods based on "leave subject out one at a time" or "leave time point out one at a time." Since high dimensional smoothing is used to estimate the RTPs and the RTPRs, statistical properties of the RTP and RTPR estimators are investigated by a simulation study. We construct the pointwise confidence intervals for the smoothing estimators using the resampling-subject bootstrap. The application to the NGHS blood pressure (BP) data shows that the time-varying transformation models and the two-step estimators can lead to useful scientific insights.

14.2   Data Structure and Distribution Functionals

14.2.1   Data Structure

The data considered in this chapter has exactly the same structure as Chapter 13. To briefly reiterate the notation and data structure of Section 13.2, we consider the stochastic processes indexed by the time point $t \in \mathcal{T}$, where $\mathcal{T}$ is a bounded subset of $[0, \infty)$. At any given $t \in \mathcal{T}$, $Y(t) \in R$ is the real-valued outcome variable, and $\mathbf{X}(t) = \left(X_1(t), \ldots, X_P(t) \right)^T$, $X_p(t) \in R$ for $1 \leq p \leq P$, is the R^P-valued covariate vector. The longitudinal sample of $\{Y(t), \mathbf{X}(t) : t \in \mathcal{T}\}$ contains n independent subjects, and each subject is observed at a randomly selected subset of $J > 1$ distinct time points $\mathbf{t} = (t_1, \ldots, t_J)^T \in \mathcal{T}$.

In a real study, the number of distinct time points J may be large, and not all the subjects are observed at every t_j. When the observed time points are not exactly contained in $\mathbf{t}$, it is common to round-off the observed time points into $\mathbf{t}$ provided that the round-off has meaningful scientific interpretations. At each t_j, $\mathcal{S}_j$ is the set of subjects with observations,

$$\mathcal{L} = \left\{ Y_i(t_j), \mathbf{X}_i(t_j), t_j : i \in \mathcal{S}_j, j = 1, \ldots, J \right\}$$

is the sample of $\{Y(t), \mathbf{X}(t) : t \in \mathcal{T}\}$, and

$$\mathcal{D} = \left\{ \mathbf{X}_i(t_j), t_j : i \in \mathcal{S}_j, j = 1, \ldots, J \right\}$$

is the set of observed covariates, where $Y_i(t_j)$ and $\mathbf{X}_i(t_j) = \left(X_{i1}(t_j), \ldots, X_{iP}(t_j) \right)^T$ are the outcome and covariates for the ith subject when $i \in \mathcal{S}_j$. We denote by $n_j = \#\{i \in \mathcal{S}_j\}$ the number of subjects in $\mathcal{S}_j$ and $n_{j_1 j_2} = \#\{i \in \mathcal{S}_{j_1} \cap \mathcal{S}_{j_2}\}$ the number of subjects in both $\mathcal{S}_{j_1}$ and $\mathcal{S}_{j_2}$, where $n_{j_1 j_2} \leq \min\{n_{j_1}, n_{j_2}\}$. Implications and justifications for this data structure have already been discussed in Section 13.2.

14.2.2   Conditional Distribution Functions

The statistical objective is to estimate the conditional distribution functions and the tracking ability of $\{Y(t) : t \in \mathcal{T}\}$ given $\{\mathbf{X}(t), t\}$ based on the sample $\mathcal{L}$. Given any covariates $\mathbf{X}(t) = \mathbf{x}(t)$ at time $t \in \mathcal{T}$ and any time-varying set of outcomes $A[\mathbf{x}(t), t] \subseteq R$ which depends on both t and $\mathbf{x}(t)$, the conditional probability of $Y(t)$ belonging to $A[\mathbf{x}(t), t]$, i.e., $Y(t) \in A[\mathbf{x}(t), t]$, is

$$P_A[\mathbf{x}(t), t] = P\left\{ Y(t) \in A[\mathbf{x}(t), t] \,\big|\, \mathbf{X}(t) = \mathbf{x}(t), t \right\}. \tag{14.1}$$

The choice of $A[\mathbf{x}(t), t]$ can be either known or estimated from the data, and is specified by the study objectives. For the interesting special case of

$$A[\mathbf{x}(t), t] = \left(-\infty, y[\mathbf{x}(t), t] \right] \tag{14.2}$$

for some pre-specified function $y[\mathbf{x}(t), t]$, $P_A[\mathbf{x}(t), t]$ reduces to the conditional cumulative distribution function (CDF), which, by (14.1), is

$$F_t\{y[\mathbf{x}(t), t] | \mathbf{x}(t)\} = P\{Y(t) \le y[\mathbf{x}(t), t] | \mathbf{X}(t) = \mathbf{x}(t), t\}. \tag{14.3}$$

Other useful special cases of (14.1) may include different intervals, such as

$$A[\mathbf{x}(t), t] = (y_1[\mathbf{x}(t), t], y_2[\mathbf{x}(t), t]] \tag{14.4}$$

for some $y_1[\mathbf{x}(t), t] < y_2[\mathbf{x}(t), t]$, which lead to

$$P_{(y_1[\mathbf{x}(t), t], y_2[\mathbf{x}(t), t]]}[\mathbf{x}(t), t]$$
$$= P\{y_1[\mathbf{x}(t), t] < Y(t) \le y_2[\mathbf{x}(t), t] | \mathbf{X}(t) = \mathbf{x}(t), t\}. \tag{14.5}$$

The probability defined in (14.5) may be used to define different categories of health status or disease risks.

Although we assume in (14.1) the general situation that $A[\mathbf{x}(t), t]$ depends on the values of the covariates $\mathbf{x}(t)$ at time t, disease progression or health status in real applications may not depend on covariates. For situations where the set of outcomes may change with time t but does not depend on the covariates, we have

$$A[\mathbf{x}(t), t] = A(t). \tag{14.6}$$

The particular cases for the CDFs and their functionals based on (14.3) and (14.4) are given by

$$F_t[y(t) | \mathbf{x}(t)] = P\left[Y(t) \le y(t) | \mathbf{X}(t) = \mathbf{x}(t), t\right] \tag{14.7}$$

and

$$P_{(y_1(t), y_2(t))}[\mathbf{x}(t), t] = P\left[y_1(t) < Y(t) \le y_2(t) | \mathbf{X}(t) = \mathbf{x}(t), t\right]. \tag{14.8}$$

The conditional CDF in $F_t[y(t) | \mathbf{x}(t)]$ reduces to the conditional CDF of (12.4) when the covariate is time-invariant and categorical, i.e., $\mathbf{x}(t) = x$.

14.2.3 Conditional Quantiles

The conditional quantiles of $Y(t)$ at time t given $\mathbf{X}(t) = \mathbf{x}(t)$ can be naturally defined using (14.2) and (14.3). If, for any time $t \in \mathcal{T}$ and given $\mathbf{X}(t) = \mathbf{x}(t)$, there is a unique quantity $y_\alpha[\mathbf{x}(t), t]$ such that

$$F_t\{y_\alpha[\mathbf{x}(t), t] | \mathbf{x}(t)\} = \alpha \quad \text{for some } 0 < \alpha < 1, \tag{14.9}$$

then $y_\alpha[\mathbf{x}(t), t]$ is the $(100 \times \alpha)$th conditional quantile given $\{\mathbf{x}(t), t\}$, which is defined to be the inverse of $F_t\{y[\mathbf{x}(t), t] | \mathbf{x}(t)\} = \alpha$ and is denoted by

$$y_\alpha[\mathbf{x}(t), t] = F_t^{-1}[\alpha | \mathbf{x}(t)]. \tag{14.10}$$

Useful functionals of the conditional quantiles $y_\alpha\left[\mathbf{x}(t),t\right]$ include the conditional Inter-Quantile Range (IQR) defined by

$$\Delta_{\alpha_1,\alpha_2}\left[\mathbf{x}(t),t\right] = y_{\alpha_1}\left[\mathbf{x}(t),t\right] - y_{\alpha_2}\left[\mathbf{x}(t),t\right] \quad \text{for any } 0 \le \alpha_2 < \alpha_1 < 1 \quad (14.11)$$

or, when different covariate and time values $\{\mathbf{x}_1(t_1),t_1\}$ and $\{\mathbf{x}_2(t_2),t_2\}$ are included,

$$\Delta_{\alpha_1,\alpha_2}\left\{\left[\mathbf{x}_1(t_1),t_1\right],\left[\mathbf{x}_2(t_2),t_2\right]\right\} = y_{\alpha_1}\left[\mathbf{x}_1(t_1),t_1\right] - y_{\alpha_2}\left[\mathbf{x}_2(t_2),t_2\right]. \quad (14.12)$$

The generalized version (14.12) allows for quantile differences under different sub-groups specified by the covariates and the different time points. For example, when $\alpha_1 = \alpha_2 = \alpha$ and $t_1 = t_2 = t$, (14.12) reduces to

$$\Delta_\alpha\left\{\left[\mathbf{x}_1(t),t\right],\left[\mathbf{x}_2(t),t\right]\right\} = y_\alpha\left[\mathbf{x}_1(t),t\right] - y_\alpha\left[\mathbf{x}_2(t),t\right], \quad (14.13)$$

which represents the difference of the $(100 \times \alpha)$th quantiles at time $t \in \mathscr{T}$ between the subjects with $\mathbf{x}_1(t)$ and the subjects with $\mathbf{x}_2(t)$. If two different time points $t_1 \ne t_2$ are used in (14.13), then

$$\Delta_\alpha\left\{\left[\mathbf{x}_1(t_1),t_1\right],\left[\mathbf{x}_2(t_2),t_2\right]\right\} = y_\alpha\left[\mathbf{x}_1(t_1),t_1\right] - y_\alpha\left[\mathbf{x}_2(t_2),t_2\right] \quad (14.14)$$

represents the difference of the $(100 \times \alpha)$th quantiles between the subjects at time $t_1 \in \mathscr{T}$ with $\mathbf{x}_1(t_1)$ and the subjects at time $t_2 \in \mathscr{T}$ with $\mathbf{x}_2(t_2)$.

14.2.4   Rank-Tracking Probabilities

As shown in Section 12.2.4, the tracking ability can be quantified by evaluating whether a subject's outcome at an earlier time point affects the distribution of its outcome at a later time point. For the current situation with time-varying covariates, a general index for measuring tracking ability should take the covariates into account. Let $A\left[\mathbf{X}(t),t\right] \subseteq R$ be a pre-specified set of possible outcome values, which describes the health status of the subject at time t and may depend on the covariates $\mathbf{X}(t)$. Let $B(t) \subseteq R^P$ be a set of the possible values for the covariates $\mathbf{X}(t)$. The tracking ability of $Y(s)$ at time points $s_1 < s_2$ can be measured by the Rank-Tracking Probability (RTP)

$$RTP_{s_1,s_2}(A,B) \;=\; P\Big\{Y(s_2) \in A\left[\mathbf{X}(s_2),s_2\right]\Big|$$

$$Y(s_1) \in A\left[\mathbf{X}(s_1),s_1\right], \mathbf{X}(s_1) \in B(s_1)\Big\}. \quad (14.15)$$

In the RTP defined above, the outcome set $A\left[\mathbf{X}(t),t\right]$ is interpreted as a covariate-specific health status indicator given that the covariates $\mathbf{X}(t)$ are pre-determined. Once the scientific interpretations of $A\left[\mathbf{X}(t),t\right]$ are well defined, $RTP_{s_1,s_2}(A,B)$ measures the tracking ability of $Y(t)$ at $t = s_1$ and s_2 through the conditional probability of $Y(s_2) \in A\left[\mathbf{X}(s_2),s_2\right]$ at time point s_2 given that $Y(s_1) \in A\left[\mathbf{X}(s_1),s_1\right]$ and $\mathbf{X}(s_1) \in B(s_1)$ at time point s_1. Thus, a large value of $RTP_{s_1,s_2}(A, B)$ would suggest strong tracking ability of $Y(t)$ for the given health status indicator $A(\cdot, \cdot)$ at time points $s_1 < s_2$ among the subjects with $\mathbf{X}(s_1) \in B(s_1)$. On the other hand, a small value of $RTP_{s_1,s_2}(A, B)$ suggests weak tracking ability of $Y(t)$ for $A(\cdot, \cdot)$ at $s_1 < s_2$. If $A[\mathbf{X}(t), t]$ is the interval given in (14.2) for some $y[\mathbf{X}(t), t]$, the RTP of (14.15) has the expression

$$RTP_{s_1,s_2}(A, B) = P\Big\{Y(s_2) \leq y[\mathbf{X}(s_2), s_2]\Big|$$
$$Y(s_1) \leq y[\mathbf{X}(s_1), s_1], \mathbf{X}(s_1) \in B(s_1)\Big\}. \quad (14.16)$$

When the disease progression and health status are determined without depending on the covariates, a covariate independent set $A(t) \subseteq R$ defined in (14.6) should be used. The RTP of (14.15) at time points $s_1 < s_2$ can be simplified to

$$RTP_{s_1,s_2}(A, B) = P\Big[Y(s_2) \in A(s_2)\Big|Y(s_1) \in A(s_1), \mathbf{X}(s_1) \in B(s_1)\Big]. \quad (14.17)$$

If $A(t)$ is given by (14.6) for some $y(t)$, then

$$RTP_{s_1,s_2}(A, B) = P\Big[Y(s_2) \leq y(s_2)\Big|Y(s_1) \leq y(s_1), \mathbf{X}(s_1) \in B(s_1)\Big]. \quad (14.18)$$

Examples of covariate independent set $A(t)$ include hypertension and elevated blood pressure levels defined for adult populations (James et al., 2014), where elevated levels of systolic blood pressure or diastolic blood pressure are defined for all adults without depending on other covariates.

14.2.5 Rank-Tracking Probability Ratios

The relative tracking strength for $Y(s)$ with $A[\mathbf{X}(t), t]$ at $s_1 < s_2$ given $\mathbf{X}(s_1) \in B(s_1)$ can be measured by the Rank-Tracking Probability Ratio (RTPR)

$$RTPR_{s_1,s_2}(A, B) = \frac{RTP_{s_1,s_2}(A, B)}{P\{Y(s_2) \in A[\mathbf{X}(s_2), s_2]|\mathbf{X}(s_1) \in B(s_1)\}}, \quad (14.19)$$

which essentially compares $RTP_{s_1,s_2}(A, B)$ with the conditional probability of $Y(s_2) \in A[\mathbf{X}(s_2), s_2]$ at time s_2 given $\mathbf{X}(s_1) \in B(s_1)$ at time s_1. As in Section 12.2.5, the $RTPR_{s_1,s_2}(A, B)$ can be used to distinguish the tracking ability of $Y(t)$ under the following situations:

(a) **No Tracking Ability.** If knowing $Y(s_1) \in A[\mathbf{X}(s_1), s_1]$ does not change the probability of $Y(s_2) \in A[\mathbf{X}(s_2), s_2]$ among the subjects with $\mathbf{X}(s_1) \in B(s_1)$, then

$$RTPR_{s_1,s_2}(A, B) = 1, \quad (14.20)$$

so that $Y(s)$ has no tracking ability based on $A[\mathbf{X}(t), t]$ at the two time points $s_1 < s_2$. Clearly, $Y(s_1)$ has no tracking ability for $Y(s_2)$ if $Y(s_1)$ and $Y(s_2)$ are conditionally independent given $\mathbf{X}(s_1) \in B(s_1)$.

(b) Positive or Negative Tracking Ability If

$$RTPR_{s_1,s_2}(A, B) > 1 \quad \text{or} \quad RTPR_{s_1,s_2}(A, B) < 1, \tag{14.21}$$

then $Y(t)$ has positive or negative, respectively, tracking ability based on $A[\mathbf{X}(t), t]$ at the two time points $s_1 < s_2$.

For the special case of (14.2), $RTPR_{s_1,s_2}(A, B)$ has the expression

$$RTPR_{s_1,s_2}(A, B) = \frac{RTP_{s_1,s_2}(A, B)}{P\{Y(s_2) \leq y[\mathbf{X}(s_2), s_2] \mid \mathbf{X}(s_1) \in B(s_1)\}}, \tag{14.22}$$

where $RTP_{s_1,s_2}(A, B)$ is given by (14.16).   □

Another useful special case is the RTPs and RTPRs defined by (14.15) and (14.19) based on quantiles, i.e.,

$$A[\mathbf{X}(t), t] = A(t) = (y_\alpha(t), \infty), \tag{14.23}$$

where $y_\alpha(t)$ for any $0 < \alpha < 1$ is the $(100 \times \alpha)$th quantile of $Y(t)$. Using (14.23), the RTP and RTPR at time points $s_1 < s_2$ are

$$\begin{cases} RTP_{s_1,s_2}[(y_{\alpha_1}, y_{\alpha_2}), B] &= P[Y(s_2) > y_{\alpha_2}(s_2) \mid \\ & \quad Y(s_1) > y_{\alpha_1}(s_1), \mathbf{X}(s_1) \in B(s_1)], \\ RTPR_{s_1,s_2}[(y_{\alpha_1}, y_{\alpha_2}), B] &= \dfrac{RTP_{s_1,s_2}[(y_{\alpha_1},y_{\alpha_2}),B]}{P[Y(s_2)>y_{\alpha_2}(s_2) \mid \mathbf{X}(s_1) \in B(s_1)]}. \end{cases} \tag{14.24}$$

In practice, $RTP_{s_1,s_2}[(y_{\alpha_1}, y_{\alpha_2}), B]$ and $RTPR_{s_1,s_2}[(y_{\alpha_1}, y_{\alpha_2}), B]$ may be used to evaluate the tracking ability of elevated levels of an outcome variable from s_1 to s_2. Implications of (14.24) for the NGHS blood pressure data are illustrated in Section 14.4.

14.2.6   The Time-Varying Transformation Models

Nonparametric estimation of $P_A(\mathbf{x}, t)$, $RTP_{s_1,s_2}(A, B)$ and $RTPR_{s_1,s_2}(A, B)$ defined in (14.1), (14.15) and (14.19), respectively, can be computationally infeasible and difficult to interpret, when the covariates $\mathbf{X}(t)$ are continuous and time-varying. As a practical compromise between model flexibility and computational feasibility, the time-varying transformation models given in (13.4) can be used to construct a class of structured nonparametric estimators for these quantities.

Recall that the model (13.4) assumes the following flexible structure for the time-varying outcome and covariates $\{Y(t), \mathbf{X}(t), t\}$,

$$g\{S_t[y \mid \mathbf{X}(t)]\} = h(y, t) + \mathbf{X}^T(t)\beta(t), \tag{14.25}$$

where $g(\cdot)$ is a known decreasing link function, $S_t[y|\mathbf{X}(t)] = 1 - F_t[y|\mathbf{X}(t)]$, $h(\cdot, \cdot)$ is an unknown baseline function strictly increasing in y, $\beta(t) = (\beta_1(t), \ldots, \beta_P(t))^T$, and $\beta_p(t)$ are smooth functions of $t \in \mathscr{T}$. Useful special cases of (14.25) include the proportional hazard model and the proportional odds model defined in (13.5) and (13.6). If we denote the inverse of $g(\cdot)$ by

$$\phi(u) = g^{-1}(u) \quad \text{for any } 0 \le u \le 1, \tag{14.26}$$

it is shown in equation (13.9) that, for any subject with observations at time t_j, (14.25) is equivalent to

$$h[Y_i(t_j), t_j] = -\mathbf{X}_i^T(t_j)\beta(t_j) + \varepsilon_{ij}, \tag{14.27}$$

where $\varepsilon_{ij} = g\{S_{t_j}[Y_i(t_j)|\mathbf{X}_i(t_j)]\}$ are random errors with distribution

$$G(\cdot) = 1 - \phi(\cdot).$$

Under (14.25), the conditional CDF of $Y(t)$ given $\mathbf{X}(t)$ at time t is

$$F_t[y|\mathbf{X}(t)] = 1 - \phi[h(y, t) + \mathbf{X}^T(t)\beta(t)]. \tag{14.28}$$

The conditional probability of $Y(t) \in A[\mathbf{X}(t), t]$ can be computed from the expression of $F_t[y|\mathbf{X}(t)]$ in (14.28) and $A[\mathbf{X}(t), t]$. For example, if $A(t) = (y(t), \infty)$ and, for time points $s_1 < s_2$, $y(s_1) = y_1$, $y(s_2) = y_2$, and $B(s_1) = \mathbf{x}(s_1)$ for some fixed $\mathbf{x}(s_1)$, then it follows that, under (14.28),

$$
\begin{cases}
RTP_{s_1,s_2}[(y_1, y_2), \mathbf{x}] &= \dfrac{S_{s_1,s_2}[y_1, y_2|\mathbf{x}(s_1)]}{\phi[h(y_1, s_1) + \mathbf{x}^T(s_1)\beta(s_1)]}, \\[2ex]
RTPR_{s_1,s_2}[(y_1, y_2), \mathbf{x}] &= \dfrac{S_{s_1,s_2}[y_1, y_2|\mathbf{x}(s_1)]}{\phi[h(y_1, s_1) + \mathbf{x}^T(s_1)\beta(s_1)]\, S_{s_2}[y_2|\mathbf{x}(s_1)]},
\end{cases}
\tag{14.29}
$$

where $S_{s_1,s_2}[y_1, y_2|\mathbf{x}(s_1)]$ and $S_{s_2}[y_2|\mathbf{x}(s_1)]$ are the conditional probabilities defined by

$$
\begin{cases}
S_{s_1,s_2}[y_1, y_2|\mathbf{x}(s_1)] &= P[Y(s_2) > y_2, Y(s_1) > y_1|\mathbf{X}(s_1) = \mathbf{x}(s_1)], \\
S_{s_2}[y_2|\mathbf{x}(s_1)] &= P[Y(s_2) > y_2|\mathbf{X}(s_1) = \mathbf{x}(s_1)].
\end{cases}
\tag{14.30}
$$

It is important to note that (14.25) only assumes the structure between $\mathbf{X}(t)$ and $Y(t)$ at a given time $t \in \mathscr{T}$. Since the structure between $Y(s_2)$ and $\{Y(s_1), \mathbf{X}(s_1)\}$ for any $s_1 < s_2$ is not specified, (14.25) may not be used to estimate the conditional probabilities in (14.30). There are two reasons for leaving the correlation structure at two time points unspecified. First, the dependence structures between the observations at different time points are usually completely unknown, so that unstructured smoothing estimators of (14.28) are preferred in an exploratory analysis. Second, the unstructured smoothing estimators are always needed because any appropriate models for the correlation structure across time points need to be compared with the unstructured smoothing estimators.

14.3 Two-Step Estimation and Prediction Methods

Using the similar estimation approach of Chapter 13, we present a two-step method for the nonparametric estimation and prediction of the conditional distribution functions and their corresponding RTPs and RTPRs based on the time-varying transformation models (14.25). In this method, we first obtain the raw estimates of $\beta(t_j)$, $h(\cdot, t_j)$ and $F_{t_j}(\cdot|\cdot)$ at each distinct time point t_j and then compute their smoothing estimates at any $t \in \mathcal{T}$ based on the available raw estimates. This two-step method is used because it is generally difficult to compute any potentially likelihood-based estimators for (14.25) without assumptions on the correlation structures for $\mathcal{Z}$.

14.3.1 Raw Estimators of Distribution Functions

We construct the raw estimators of the conditional CDF $F_t[y|\mathbf{X}(t)]$ of (14.25) at a time design point t_j, $1 \le j \le J$, using the following three steps:

(a) estimating the coefficients $\beta(t_j)$;

(b) estimating the baseline function $h(y, t_j)$;

(c) estimating the conditional CDF $F_{t_j}[y|\mathbf{X}(t_j)]$ itself.

For step (a), a class of raw estimators of $\beta(t_j)$ have already been established in Section 13.3.1 by applying the estimating equations of Cheng, Wei and Ying (1995) to the subjects with observations at the time point t_j, i.e., subjects in $\mathcal{S}_j$. Let $\widetilde{\beta}(t_j)$ be the raw estimator of $\beta(t_j)$ obtained by solving the estimating equation (13.13) using the observations from the subjects in $\mathcal{S}_j$. The statistical properties of $\widetilde{\beta}(t_j)$ have been derived in Section 13.3.2.

For the estimation of $h(y, t_j)$ in step (b), we substitute $\beta(t_j)$ with $\widetilde{\beta}(t_j)$ and observe from (14.25) that

$$S_{t_j}[y|\mathbf{X}_i(t_j)] = 1 - F_{t_j}[y|\mathbf{X}_i(t_j)] = \phi[h(y, t_j) + \mathbf{X}_i^T(t_j)\beta(t_j)], \qquad (14.31)$$

a raw estimator $\widetilde{h}(y, t_j)$ can be obtained by solving the estimating equation

$$V[\widetilde{h}(y, t_j)] = \frac{1}{n_j} \sum_{i \in \mathcal{S}_j} \left\{ 1_{[Y_i(t_j)>y]} - \phi[\widetilde{h}(y, t_j) + \mathbf{X}_i^T(t_j)\widetilde{\beta}(t_j)] \right\} = 0. \qquad (14.32)$$

The above estimating equation (14.32) is an extension of the estimating equation of Cheng, Wei and Ying (1997), where the data are from a cross-sectional i.i.d. sample.

For the estimation in step (c), we substitute $h(\cdot, t_j)$ and $\beta(t_j)$ in (14.28) with $\widetilde{h}(\cdot, t_j)$ and $\widetilde{\beta}(t_j)$, and obtain the raw estimator $\widetilde{F}_{t_j}(y|\mathbf{x})$ of $F_{t_j}(y|\mathbf{x})$ at time point t_j given $\mathbf{X}(t) = \mathbf{x}(t)$ by

$$\widetilde{F}_{t_j}(y|\mathbf{x}) = 1 - \widetilde{S}_{t_j}(y|\mathbf{x}) = 1 - \phi[\widetilde{h}(y, t_j) + \mathbf{x}^T(t_j)\widetilde{\beta}(t_j)]. \qquad (14.33)$$

For a new subject with covariates $\mathbf{x}(t)$ at time $t = t_j$, $\widetilde{F}_{t_j}(y|\mathbf{x})$ is used to predict the subject's chance of $Y(t) \leq y$ at time $t = t_j$.

There are two issues which we need to consider for the estimators in (14.32) and (14.33):

(a) **Order-Preserving Property.** Since both the baseline function $h(y, t)$ and the conditional CDF $F_t(y|\mathbf{x})$ are nondecreasing functions in y for any $t \in \mathscr{T}$, i.e., $h(y_1, t) \leq h(y_2, t)$ and $F_t(y_1|\mathbf{x}) \leq F_t(y_2|\mathbf{x})$ for any $y_1 \leq y_2$ and $t \in \mathscr{T}$, we prefer that the estimators of these two functions have the order-preserving property in the sense that these estimators are also nondecreasing functions in y for any $t \in \mathscr{T}$. Indeed, it has been shown by Cheng, Wei and Ying (1997) that both $\widetilde{h}(y, t_j)$ and $\widetilde{F}_{t_j}(y|\mathbf{x})$ are nondecreasing in y for any given t_j, hence, $\widetilde{h}(y, t_j)$ and $\widetilde{F}_{t_j}(y|\mathbf{x})$ are order-preserving.

(b) **Bin Sizes.** The raw estimators of $h(\cdot, t_j)$ and $F_{t_j}(\cdot|\cdot)$ require the number of observations n_j at t_j to be sufficiently large, so that (14.32) and (14.33) can be solved numerically. When the sample size n_j is not sufficiently large, we can round off or group some of the adjacent time points into small bins, and compute the raw estimates within each bin (Fan and Zhang, 2000; Wu, Tian and Yu, 2010). In biomedical studies, the unit of time is often rounded off with an acceptable precision. The effects of rounding off or binning on the asymptotic properties of the smoothed estimators have not been studied in the literature and are beyond the scope of this chapter.   □

14.3.2 Smoothing Estimators for Conditional CDFs

Using the above raw estimators as pseudo-observations, our smoothing step here is to construct local polynomial estimators for the baseline function $h(y, t)$ and the conditional CDF $F_t(y|\mathbf{x})$ at any time point $t \in \mathscr{T}$. A basic assumption for this smoothing step is that $h(y, t)$ and $F_t(y|\mathbf{x})$ are smooth functions of $t \in \mathscr{T}$. Statistical properties of the smoothing estimators of $h(y, t)$ and $F_t(y|\mathbf{x})$ depend on the smoothness assumptions for $h(y, t)$ and $F_t(y|\mathbf{x})$. This smoothing step is used to

(a) provide estimators of $h(y, t)$ and $F_t(y|\mathbf{x})$ for any t in $\mathscr{T}$ including these time points not included in the design time points $\mathbf{t} = (t_1, \ldots, t_J)^T$;

(b) reduce the variances of the estimators by borrowing the information from the unsmoothed raw estimators at the design time points adjacent to t. □

1. General Expressions of Linear Smoothing Estimators

We consider here the same commonly used smoothness families as the ones in Chapter 13. Suppose that $h(y, t)$ and $F_t(y|\mathbf{x})$ are $(Q+1)$ times continuously differentiable with respect to $t \in \mathscr{T}$ for some $Q \geq 0$. Given the unsmoothed raw estimators $\widetilde{h}(y, t_j)$ and $\widetilde{F}_{t_j}(y|\mathbf{x})$ for all $t_j \in \mathbf{t}$ obtained from (14.32) and (14.33), we can construct the smoothing estimators of $h(y, t)$ and $F_t(y|\mathbf{x})$ for any $t \in \mathcal{T}$ through linear estimators of the form

$$\begin{cases} \widehat{h}(y, t) &= \sum_{j=1}^{J} w(t_j, t)\, \widetilde{h}(y, t_j), \\ \widehat{F}_t(y|\mathbf{x}) &= \sum_{j=1}^{J} w(t_j, t)\, \widetilde{F}_{t_j}(y|\mathbf{x}), \end{cases} \tag{14.34}$$

where $w(t_j, t)$ is a known weight function. If $w(t_j, t)$ is a weight function determined by a kernel function $K(\cdot)$ and bandwidth $b > 0$ of the form

$$w(t_j, t) = K\left(\frac{t_j - t}{b}\right) \Big/ \sum_{j=1}^{J} K\left(\frac{t_j - t}{b}\right),$$

the two-step kernel estimators of $h(y, t)$ and $F_t(y|\mathbf{x})$ obtained from (14.34) are given by

$$\begin{cases} \widehat{h}_K(y, t) &= \sum_{j=1}^{J} \left[\widetilde{h}(y, t_j) K\left(\frac{t_j - t}{b}\right)\right] \Big/ \sum_{j=1}^{J} K\left(\frac{t_j - t}{b}\right), \\ \widehat{F}_{t,K}(y|\mathbf{x}) &= \sum_{j=1}^{J} \left[\widetilde{F}_{t_j}(y|\mathbf{x}) K\left(\frac{t_j - t}{b}\right)\right] \Big/ \sum_{j=1}^{J} K\left(\frac{t_j - t}{b}\right). \end{cases} \tag{14.35}$$

To estimate the derivatives of $h(y, t)$ and $F_t(y|\mathbf{x})$, let $h^{(q)}(y, t)$ and $F_t^{(q)}(y|\mathbf{x})$, $0 \le q \le Q$, be the qth derivatives of $h(y, t)$ and $F_t(y|\mathbf{x})$ with respect to t. The two-step smoothing estimators of the qth derivatives $h^{(q)}(y, t)$ and $F_t^{(q)}(y|\mathbf{x})$ for any $t \in \mathcal{T}$ can be constructed through linear estimators of the form

$$\begin{cases} \widehat{h}^{(q)}(y, t) &= \sum_{j=1}^{J} w_{q, Q+1}(t_j, t)\, \widetilde{h}(y, t_j), \\ \widehat{F}_t^{(q)}(y|\mathbf{x}) &= \sum_{j=1}^{J} w_{q, Q+1}(t_j, t)\, \widetilde{F}_{t_j}(y|\mathbf{x}), \end{cases} \tag{14.36}$$

where $w_{q, Q+1}(\cdot, \cdot)$ is a known weight function. It follows from (14.38) and (14.39) that the estimators of $h(y,t)$ and $F_t(y|\mathbf{x})$ in (14.34) are given by

$$\begin{cases} \widehat{h}(y, t) &= \widehat{h}^{(0)}(y,t) = \sum_{j=1}^{J} w_{0, Q+1}(t_j, t)\, \widetilde{h}(y, t_j), \\ \widehat{F}_t(y|\mathbf{x}) &= \widehat{F}_t^{(0)}(y|\mathbf{x}) = \sum_{j=1}^{J} w_{0, Q+1}(t_j, t)\, \widetilde{F}_{t_j}(y|\mathbf{x}), \end{cases} \tag{14.37}$$

which are obtained by (14.36) with $q = 0$.

2. Local Polynomial Estimators

Different choices of $w_{q, Q+1}(t_j, t)$ lead to different smoothing estimators of (14.36). Let $K(\cdot)$ be a non-negative kernel function and $b > 0$ be a bandwidth. It follows from (14.36) that the Qth order local polynomial estimators $\widehat{h}^{(q)}(y, t)$ and $\widehat{F}_t^{(q)}(y|\mathbf{x})$ for $h^{(q)}(y, t)$ and $F_t^{(q)}(y|\mathbf{x})$ are given by

$$\begin{cases} \widehat{h}^{(q)}(y, t) &= \sum_{j=1}^{J} w_{q, Q+1}(t_j, t; b)\, \widetilde{h}(y, t_j), \\ \widehat{F}_t^{(q)}(y|\mathbf{x}) &= \sum_{j=1}^{J} w_{q, Q+1}(t_j, t; b)\, \widetilde{F}_{t_j}(y|\mathbf{x}), \\ w_{q, Q+1}(t_j, t; b) &= q!\, e_{q+1, Q+1}^T \left(\mathbf{C}^T \mathbf{W} \mathbf{C}\right)^{-1} C_j W_j, \quad j = 1, \dots, J, \end{cases} \tag{14.38}$$

where the weight function $w_{q,Q+1}(t_j, t; b)$ is the same as (13.31), $e_{q+1,Q+1}$ is the $(Q+1)$th column vector with 1 at its $(q+1)$th row and 0 everywhere else, C_j, $\mathbf{C}$, W_j and $\mathbf{W}$ are defined in (13.30).

For the special case of $q = 0$ and $Q = 1$, the weight function in (14.38) is $w_{0,2}(t_j, t; b)$ and the two-step local linear estimators of $h(y, t)$ and $F_t(y|\mathbf{x})$ are given by

$$
\begin{cases}
\widehat{h}^L(y, t) &= \sum_{j=1}^{J} \left[e_{1,2}^T \left(\mathbf{C}^T \mathbf{W} \mathbf{C} \right)^{-1} C_j W_j \right] \widetilde{h}(y, t_j), \\
\widehat{F}_t^L(y|\mathbf{x}) &= \sum_{j=1}^{J} \left[e_{1,2}^T \left(\mathbf{C}^T \mathbf{W} \mathbf{C} \right)^{-1} C_j W_j \right] \widetilde{F}_{t_j}(y|\mathbf{x}).
\end{cases}
\tag{14.39}
$$

Because of their mathematical simplicity and numerical stability, the above two-step local linear estimators of $h(y, t)$ and $F_t(y|\mathbf{x})$ are commonly used in real applications.

The following two comments are about the order-preserving properties of the kernel and local polynomial estimators:

(a) **Order-Preserving of Kernel Estimators.** Since, by the discussion of Section 14.3.1, $\widetilde{h}(y, t_j)$ and $\widetilde{F}_{t_j}(y|\mathbf{x})$ are order-preserving in y for any t_j, it easily follows from (14.35) that, because $K(u) \geq 0$ for all u and $K(u) > 0$ for some u,

$$
\widehat{h}_K(y_1, t) \leq \widehat{h}_K(y_2, t) \quad \text{and} \quad \widehat{F}_{t,K}(y_1|\mathbf{x}) \leq \widehat{F}_{t,K}(y_2|\mathbf{x})
\tag{14.40}
$$

for all $y_1 \leq y_2$ and all $t \in \mathcal{T}$, so that the kernel estimators $\widehat{h}_K(y, t)$ and $\widehat{F}_{t,K}(y|\mathbf{x})$ are order-preserving in y for any t.

(b) **Asymptotic Order-Preserving of Local Polynomials.** Because the weight functions of the local polynomial estimators are possible to have negative values, the local polynomial estimators $\widehat{h}(y, t)$ and $\widehat{F}_t(y|\mathbf{x})$ may not have the order-preserving property for any finite sample size n. However, as we will see later in the asymptotic properties of Section 14.5.2, we can conclude that, for any $y_1 \leq y_2$ and all $t \in \mathcal{T}$,

$$
\widehat{h}^{(0)}(y_1, t) \leq \widehat{h}^{(0)}(y_2, t) \quad \text{and} \quad \widehat{F}_t^{(0)}(y_1|\mathbf{x}) \leq \widehat{F}_t^{(0)}(y_2|\mathbf{x}) \quad \text{in probability,}
\tag{14.41}
$$

when n and n_j, $j = 1, \ldots, J$, are sufficiently large, which we refer to herein as the asymptotically order-preserving property in y.   □

Other than the kernel estimators (14.35), further research is needed to develop two-step local smoothing estimators with the order-preserving property for finite sample sizes. Hall and Müller (2003) suggests a useful order-preserving method for the estimation of conditional distributions and quantiles. But, their method cannot be directly extended to the current models with longitudinal data.

14.3.3 Smoothing Estimators for Quantiles

Useful functionals of $F_t(y|\mathbf{x})$ under the time-varying transformation models (14.25), such as the conditional quantile

$$y_\alpha(t, \mathbf{x}) = F_t^{-1}(\alpha|\mathbf{x}) = \left\{ y : 1 - \phi\left[h(y, t) + \mathbf{X}^T(t)\beta(t)\right] = \alpha \right\} \qquad (14.42)$$

obtained (14.28), may be estimated from $\widehat{F}_t(y|\mathbf{x})$. In particular, a simple estimator of $y_\alpha(t, \mathbf{x})$ based on $\widehat{F}_t(y|\mathbf{x})$ can be given by

$$\widehat{y}_\alpha(t, \mathbf{x}) = \frac{1}{2}\left\{ \inf_y \left[y : \widehat{F}_t(y|\mathbf{x}) \geq \alpha\right] + \sup_y \left[y : \widehat{F}_t(y|\mathbf{x}) \leq \alpha\right]\right\}, \qquad (14.43)$$

where $\widehat{F}_t(y|\mathbf{x})$ can be constructed using the smoothing estimators of Section 14.3.2, such as (14.35), (14.37) and (14.39).

Since, for any fixed $\{t, \mathbf{x}\}$, the conditional quantile $y_\alpha(t, \mathbf{x})$ of (14.42) is nondecreasing when α decreases, it is generally preferred to use an order-preserving estimator $\widehat{F}_t(y|\mathbf{x})$ in (14.43). The two-step kernel estimator $\widehat{F}_{t,K}(y|\mathbf{x})$ is a good choice for (14.43) because of its order-preserving property (14.40). The two-step local linear estimator $\widehat{F}_t^L(y|\mathbf{x})$, on the other hand, is asymptotically order-preserving in the sense of (14.41), although not necessarily order-preserving for a given finite sample size, which suggests that $\widehat{F}_t^L(y|\mathbf{x})$ is a reasonable choice for (14.43) when n and n_j are sufficiently large.

Nonparametric quantile regression methods using statistical models other than (14.25) have been studied in the literature. For example, Wei et al. (2006) studied a quantile regression method based on an unstructured nonparametric model, and Wang, Zhu and Zhou (2009) studied a quantile regression in partially linear varying coefficient models. However, asymptotic properties of nonparametric quantile estimators based on (14.25) have not been derived, which, of course, is a worthwhile topic for future research.

14.3.4 Estimation of Rank-Tracking Probabilities

To estimate the rank-tracking probability $RTP_{s_1,s_2}\left[(y_1, y_2), \mathbf{x}\right]$ defined in (14.29), we note that, by (14.25) and (14.29),

$$RTP_{s_1,s_2}\left[(y_1, y_2), \mathbf{x}\right] = \frac{S_{s_1,s_2}\left[y_1, y_2|\mathbf{x}(s_1)\right]}{S_{s_1}\left[y_1|\mathbf{x}(s_1)\right]}, \qquad (14.44)$$

with

$$\left\{ \begin{array}{rcl} S_{s_1,s_2}\left[y_1, y_2|\mathbf{x}(s_1)\right] & = & P\left[Y(s_2) > y_2, Y(s_1) > y_1 \big| \mathbf{X}(s_1) = \mathbf{x}(s_1)\right], \\ S_{s_1}\left[y_1|\mathbf{x}(s_1)\right] & = & P\left[Y(s_1) > y_1 \big| \mathbf{X}(s_1) = \mathbf{x}(s_1)\right]. \end{array} \right. \qquad (14.45)$$

If there exist consistent estimators

$$\widehat{S}_{s_1,s_2}\left[y_1, y_2|\mathbf{x}(s_1)\right] \quad \text{and} \quad \widehat{S}_{s_1}\left[y_1|\mathbf{x}(s_1)\right] \qquad (14.46)$$

of (14.45), then a straightforward plugging type estimator of $RTP_{s_1,s_2}\big[(y_1,y_2),\mathbf{x}\big]$ is to substitute $S_{s_1,s_2}\big[y_1,y_2\big|\mathbf{x}(s_1)\big]$ and $S_{s_1}\big[y_1\big|\mathbf{x}(s_1)\big]$ of (14.44) with the corresponding estimators in (14.46), so that, the RTP can be estimated by

$$\widehat{RTP}_{s_1,s_2}\big[(y_1,y_2),\mathbf{x}\big] = \frac{\widehat{S}_{s_1,s_2}\big[y_1,y_2\big|\mathbf{x}(s_1)\big]}{\widehat{S}_{s_1}\big[y_1\big|\mathbf{x}(s_1)\big]}. \tag{14.47}$$

The estimators given below can be used to estimate $S_{s_1}\big[y_1\big|\mathbf{x}(s_1)\big]$ and $S_{s_1,s_2}\big[y_1,y_2\big|x(s_1)\big]$:

(a) **Smoothing Estimators of** $S_{s_1}\big[y_1\big|x(s_1)\big]$: Using the time-varying transformation model (14.28), the conditional CDF $F_{s_1}(y_1|\mathbf{x})$ can be estimated by any of the special cases of the two-step linear smoothing estimators $\widehat{F}_{s_1}(y_1|\mathbf{x})$ given in (14.34). Thus, $S_{s_1}\big[y_1\big|\mathbf{x}(s_1)\big]$ can be simply estimated by

$$\widehat{S}_{s_1}\big[y_1\big|\mathbf{x}(s_1)\big] = 1 - \widehat{F}_{s_1}(y_1|\mathbf{x}), \tag{14.48}$$

where special choices of $\widehat{F}_{s_1}(y_1|\mathbf{x})$ include the kernel estimator $\widehat{F}_{s_1,K}(y_1|\mathbf{x})$ of (14.35) and the local linear estimator $\widehat{F}_{s_1}^{L}(y_1|\mathbf{x})$ of (14.39).

(b) **Kernel Estimators of** $S_{s_1,s_2}\big[y_1,y_2\big|x(s_1)\big]$: Given that we do not impose any correlation assumptions on $\{Y(t),\mathbf{X}(t)\}$ between s_1 and s_2, unstructured smoothing estimators of $S_{s_1,s_2}\big[y_1,y_2\big|\mathbf{x}(s_1)\big]$ should be used in (14.47). Since the covariate vector $\mathbf{X}(t)$ may contain time-varying components, the estimation of $S_{s_1,s_2}\big[y_1,y_2\big|\mathbf{x}(s_1)\big]$ depends on the dimensionality P of $\mathbf{X}(t)$ and whether $X_p(t)$, $p=1,\cdots,P$, are discrete or continuous. In real applications, when the components of $\mathbf{X}(t)$ are continuous, useful unstructured smoothing estimators $\widehat{S}_{s_1,s_2}\big[y_1,y_2\big|\mathbf{x}(s_1)\big]$ may be obtained only when the dimensionality P is small, such as $P=1$ or 2. If $P=1$, i.e., $\mathbf{X}(t)=X(t)$, and $X(t)$ is a continuous random variable at each $t\in\mathscr{T}$, we can estimate $S_{s_1,s_2}\big[y_1,y_2\big|x(s_1)\big]$ using the following kernel smoother. Let

$$a_{1,j_1}=\frac{t_{j_1}-s_1}{b_1}, \quad a_{2,j_2}=\frac{t_{j_2}-s_2}{b_2} \quad \text{and} \quad a_{3,i,j_1}=\frac{X_i(t_{j_1})-x(s_1)}{b_3}, \tag{14.49}$$

where $\{b_1,b_2,b_3\}$ are positive bandwidths, and let $\pi(\cdot,\cdot,\cdot)$ be a weight function on R^3 with bandwidths $\{b_1,b_2,b_3\}$. A simple kernel estimator of $S_{s_1,s_2}\big[y_1,y_2\big|x(s_1)\big]$ is

$$\widehat{S}_{s_1,s_2}\big[y_1,y_2\big|x(s_1)\big] = \sum_{j_1,j_2,i} \pi\big(a_{1,j_1},a_{2,j_2},a_{3,i,j_1}\big)\,1_{\big[Y_i(t_{j_2})>y_2,\,Y_i(t_{j_1})>y_1\big]}, \tag{14.50}$$

where the summation is over the set

$$\{j_1,j_2,i\}\in\{j_1\neq j_2,\,1\leq j_1,\,j_2\leq J,\,i\in\mathscr{S}_{j_1}\cap\mathscr{S}_{j_2}\}.$$

Useful choices of $\pi(\cdot,\cdot,\cdot)$ include kernel weights for multivariate kernel estimators or the weight function for multivariate local linear estimators. $\square$

14.3.5   Estimation of Rank-Tracking Probability Ratios

By the definition of RTPR given in (14.24) and (14.29), we note that

$$RTPR_{s_1,s_2}\left[(y_1,y_2),\mathbf{x}\right] = \frac{S_{s_1,s_2}\left[y_1,y_2|\mathbf{x}(s_1)\right]}{S_{s_1}\left(y_1|\mathbf{x}\right) S_{s_2}\left[y_2|\mathbf{x}(s_1)\right]}. \tag{14.51}$$

Following the plug-in approach for the RTP, we estimate $RTPR_{s_1,s_2}\left[(y_1,y_2),\mathbf{x}\right]$ by

$$\widehat{RTPR}_{s_1,s_2}\left[(y_1,y_2),\mathbf{x}\right] = \frac{\widehat{S}_{s_1,s_2}\left[y_1,y_2|\mathbf{x}(s_1)\right]}{\widehat{S}_{s_1}\left(y_1|\mathbf{x}\right) \widehat{S}_{s_2}\left[y_2|\mathbf{x}(s_1)\right]}, \tag{14.52}$$

where $\widehat{S}_{s_1}(y_1|\mathbf{x})$ and $\widehat{S}_{s_1,s_2}\left[y_1,y_2|\mathbf{x}(s_1)\right]$ are given by (14.48) and (14.50), respectively, and $\widehat{S}_{s_2}\left[y_2|\mathbf{x}(s_1)\right]$, which depends on the covariate vector at time s_1 and the outcome variable at time s_2, can be estimated by a kernel smoothing method similar to $\widehat{S}_{s_1,s_2}\left[y_1,y_2|\mathbf{x}(s_1)\right]$.

A simple special case of $\widehat{S}_{s_2}\left[y_2|\mathbf{x}(s_1)\right]$ is described in the following. We consider the special case that $P=1$ and $\mathbf{X}(t)=X(t)$ is a continuous random variable on the real line at each $t \in \mathscr{T}$. Let $\pi(\cdot,\cdot,\cdot)$ be the same weight on R^3 function as defined in (14.50) which depends on $\{a_{1,j_1}, a_{2,j_2}, a_{3,i,j_1}\}$ of (14.49) and the bandwidths $\{b_1, b_2, b_3\}$. The kernel estimator of $S_{s_2}\left[y_2|x(s_1)\right]$ has the expression

$$\widehat{S}_{s_2}\left[y_2|x(s_1)\right] = \sum_{j_1,j_2,i} \pi(a_{1,j_1}, a_{2,j_2}, a_{3,i,j_1})\, 1_{\left[Y_i(t_{j_2})>y_2\right]} \tag{14.53}$$

with the summation over the set

$$\{j_1, j_2, i\} \in \{j_1 \neq j_2, 1 \le j_1, j_2 \le J, i \in \mathscr{S}_{j_1} \cap \mathscr{S}_{j_2}\}.$$

Similar to the estimation of $S_{s_1,s_2}\left[y_1,y_2|\mathbf{x}(s_1)\right]$, useful choices of $\pi(\cdot,\cdot,\cdot)$ may include kernel weights for multivariate kernel estimators or the weight function for multivariate local linear estimators. Note that, because we do not impose any correlation assumptions on $\{Y(t),\mathbf{X}(t)\}$ in (14.53) between any two different time points s_1 and s_2, (14.53) is a completely unstructured smoothing estimators of $S_{s_2}\left[y_2|\mathbf{x}(s_1)\right]$. Although (14.53) has the attractive feature of flexibility, a potential drawback of using an unstructured smoothing estimator is that it may only be applied to situations of low-dimensional covariate vectors, such as $P=1$ or 2.

14.3.6   Bandwidth Choices

Data-driven bandwidth choices for the smoothing estimators of conditional distribution functions can be obtained using the similar cross-validation approaches in Chapter 12. Our data-driven bandwidth choices in this section are focused on the smoothing estimators of $h(y,t)$ and $F_t\left[y|\mathbf{X}(t)\right]$ of the model (14.25). These include the leave-one-subject-out cross-validation (LSCV) and the leave-one-time-point-out cross-validation (LTCV). Since the RTPs and RTPRs are estimated in Sections 14.3.4 and 14.3.5 by plugging in the smoothing estimators of their corresponding components, some subjectively chosen bandwidths may have to be used in the kernel estimators of $S_{s_1,s_2}[y_1, y_2|\mathbf{x}(s_1)]$ and $S_{s_2}[y_2|\mathbf{x}(s_1)]$.

1. Subjective Bandwidth Choices

The smoothing estimators of $S_{s_1,s_2}[y_1, y_2|\mathbf{x}(s_1)]$ and $S_{s_2}[y_2|\mathbf{x}(s_1)]$ depend on the dimensionality of $\mathbf{X}(t)$ and the bandwidths for $\mathbf{X}_i(t_j)$. In the estimators $\widehat{S}_{s_1,s_2}[y_1, y_2|x(s_1)]$ and $\widehat{S}_{s_2}[y_2|x(s_1)]$ given in (14.50) and (14.53), respectively, $\mathbf{X}(t) = X(t)$ is a one-dimensional random variable for each given $t \in \mathcal{T}$, so that, in addition to the bandwidths b_1 and b_2 used for the time points t_{j_1} and t_{j_2}, an additional bandwidth $b_3 > 0$ is used in (14.49) to restrict the estimators using the values of $X_i(t_{j_1})$ within a neighborhood of $x(s_1)$. When the dimensionality of $\mathbf{X}(t)$ is greater than one, a bandwidth vector should be used for $\mathbf{X}_i(t_j)$. Data-driven bandwidth choices for $\mathbf{X}_i(t_j)$, such as b_3 for the one-dimensional case, have not been investigated in the literature. Thus, in practice, subjectively chosen bandwidths are needed for the continuous components of $\mathbf{X}_i(t_j)$. The appropriateness of these subjectively chosen bandwidths can be visually inspected by evaluating the smoothness of the estimators $S_{s_1,s_2}[y_1, y_2|\mathbf{x}(s_1)]$ and $S_{s_2}[y_2|\mathbf{x}(s_1)]$.

2. Leave-One-Subject-Out Cross-Validation

We demonstrate the bandwidth selection procedures using the two-step local linear estimators $\widehat{h}^L(y,t)$ and $\widehat{F}_t^L(y|\mathbf{x})$ of (14.39). However, the cross-validation procedures here apply to the general two-step smoothing estimators.

LSCV Bandwidths for $\widehat{h}^L(y,t)$ and $\widehat{F}_t^L(y|\mathbf{x})$:

(a) Let $\widehat{h}_{b,-i}^L(y,t)$ and $\widehat{F}_{t,b,-i}^L(y|\mathbf{x})$ be the smoothing estimators of $h(y,t)$ and $F_t(y|\mathbf{x})$ computed from (14.37) using the remaining data with all the observations from the ith subject deleted.

(b) When the goal is to estimate $h(y,t)$ and $F_t[y|\mathbf{X}(t)]$ at a single value $y \in R$, the LSCV bandwidth $b_{h,y,LSCV}$ for $\widehat{h}^L(y,t)$ is the minimizer of

$$LSCV_{h,y}(b) = \sum_{j=1}^{J} \sum_{i \in \mathscr{S}_j} \frac{1}{Jn_j} \left\{ 1_{[Y_{ij}>y]} - \phi\left[\widehat{h}_{b,-i}^L(y,t_j) + \mathbf{X}_i^T(t_j)\widetilde{\beta}(t_j)\right] \right\}^2, \quad (14.54)$$

where $\widetilde{\beta}(t_j)$ is the raw estimator of $\beta(t_j)$ used in (14.32), and the LSCV bandwidth $b_{F,y,LSCV}$ for $\widehat{F}_t^L(y|\mathbf{x})$ is the minimizer of

$$LSCV_{F,y}(b) = \sum_{j=1}^{J} \sum_{i \in \mathscr{S}_j} \frac{1}{Jn_j} \left\{ 1_{[Y_{ij} \leq y]} - \widehat{F}_{t_j,b,-i}^L[y|\mathbf{X}_i^T(t_j)] \right\}^2. \quad (14.55)$$

(c) *When the goal is to estimate $h(y,t)$ and $F_t[y|\mathbf{x}(t)]$ over the entire range of $y \in R$ based on a known weight function $\pi(y)$, the LSCV bandwidth $b_{h,LSCV}$ for $\widehat{h}^L(\cdot, t)$ is the minimizer of*

$$LSCV_h(b) = \sum_{j=1}^{J} \sum_{i \in \mathscr{S}_j} \frac{1}{Jn_j} \int \left\{ 1_{[Y_{ij} > y]} - \right.$$

$$\left. \phi \left[\widehat{h}^L_{b,-i}(y,t_j) + \mathbf{X}_i^T(t_j) \widetilde{\beta}(t_j) \right] \right\}^2 \pi(y)\, dy, \quad (14.56)$$

where $\widetilde{\beta}(t_j)$ is the raw estimator of $\beta(t_j)$ used in (14.32), and the LSCV bandwidth $b_{F,LSCV}$ for $\widehat{F}^L_t(\cdot|\mathbf{x})$ is the minimizer of

$$LSCV_F(b) = \sum_{j=1}^{J} \sum_{i \in \mathscr{S}_j} \frac{1}{Jn_j} \int \left\{ 1_{[Y_{ij} \le y]} - \widehat{F}^L_{t_j,b,-i}\left[y|\mathbf{X}_i^T(t_j)\right] \right\}^2 \pi(y)\, dy. \quad (14.57)$$

The choice of $\pi(y) = 1$ for all $y \in R$ corresponds to the situation of assigning a uniform weight to all the values of $y \in R$ for $h(y,t)$ and $F_t(y|\mathbf{x})$. □

The reason for using different data-driven bandwidth choices in steps (b) and (c) above is the different estimation objectives. In step (b), the objective is to find bandwidths which are best for the estimation of $h(y,t)$ and $F_t(y|\mathbf{x})$ at the pre-specified value of y, so that the LSCV scores (14.54) and (14.55) are only focused on the squared errors of the smoothing estimators at y, and the cross-validated bandwidths are the ones that minimize the estimators' squared errors at y only. In step (c), on the other hand, the objective is to find bandwidths which are suitable for the estimation of $h(y,t)$ and $F_t(y|\mathbf{x})$ for the entire range of $y \in R$. Consequently, the cross-validated bandwidth obtained in step (c) minimizes the integrated squared errors of the smoothing estimators.

3. Leave-One-Time-Point-Out Cross-Validation

The second cross-validation procedure, as described in Section 12.3.5, depends on deleting the raw estimates at the time points $\{t_1, \ldots, t_J\}$ one at a time. This leave-one-time-point-out cross-validation (LTCV) is computationally simpler than the LSCV described above.

LTCV Bandwidths for $\widehat{h}^L(y,t)$ and $\widehat{F}^L_t(y|\mathbf{x})$:

(a) *Let $\widehat{h}^L_{b,-t_j}(y,t)$ and $\widehat{F}^L_{t,b,-t_j}(y|\mathbf{x})$ be the local polynomial estimators of $h(y,t)$ and $F_t(y|\mathbf{x})$ computed with the corresponding raw estimators at the time point t_j deleted.*

(b) *When the goal is to estimate $h(y,t)$ and $F_t[y|\mathbf{X}(t)]$ at a single value $r \in R$, the LTCV bandwidth $b_{h,y,LTCV}$ for $\widehat{H}^L(y,t)$ is the minimizer of*

$$LTCV_{h,y}(b) = (1/J) \sum_{j=1}^{J} \left[\widetilde{h}(y,t_j) - \widehat{h}^L(y,t_j) \right]^2, \quad (14.58)$$

and the LTCV bandwidth $b_{F,y,LTCV}$ for $\widehat{F}_t^L(y|\mathbf{x})$ is the minimizer of

$$LTCV_{F,y}(b) = (1/J) \sum_{j=1}^{J} \left\{ \widetilde{F}_{t_j}\left[\mathbf{X}_i(t_j)\right] - \widehat{F}_{t_j,b,-t_j}^L \left[y\big|\mathbf{X}_i(t_j)\right] \right\}^2, \qquad (14.59)$$

where $\widetilde{h}(y,t_j)$ and $\widetilde{F}_{t_j}\left[y\big|\mathbf{X}_i(t_j)\right]$ are the raw estimators of (13.32) and (13.33).

(c) When the goal is to estimate $h(y,t)$ and $F_t\left[y\big|\mathbf{x}(t)\right]$ over the entire range of $y \in R$ based on a known weight function $\pi(y)$, the LTCV bandwidth $b_{h,LTCV}$ for $\widehat{h}^L(\cdot,t)$ is the minimizer of

$$LTCV_{(h)}(b) = (1/J) \sum_{j=1}^{J} \int \left[\widetilde{h}(y,t_j) - \widehat{h}_{t_j,b,-t_j}^L(y,t_j)\right]^2 \pi(y)\,dy, \qquad (14.60)$$

and the minimizer of

$$LTCV_{(F)}(b) = (1/J) \sum_{j=1}^{J} \int \left\{ \widetilde{F}_{t_j}\left[y\big|\mathbf{X}_i(t_j)\right] - \widehat{F}_{t_j,b,-t_j}\left[y\big|\mathbf{X}_i(t_j)\right] \right\}^2 \pi(y)\,dy.$$
$$(14.61)$$

is the LTCV bandwidth $b_{F,LTCV}$.   □

The bandwidths $b_{h,LTCV}$ and $b_{F,LTCV}$ ignore the possible correlations of the raw estimates, and are computationally simpler than $b_{h,LSCV}$ and $b_{F,LSCV}$. Asymptotic properties of these two cross-validation methods under the current models have not been investigated in the literature. Practical appropriateness of $b_{h,LTCV}$ and $b_{F,LTCV}$ have been investigated through a simulation study of Wu and Tian (2013b).

14.4   R Implementation

14.4.1   Conditional CDF for the NGHS SBP Data

The NGHS data has been described in Section 1.2 and analyzed in several previous chapters. In particular, we have discussed in Chapter 13 the estimation of the time-varying effects of covariates, such as race and height, on several cardiovascular risk factors, without estimating the conditional distributions of these outcomes. We illustrate here how to estimate the conditional distributions, the rank-tracking probability (RTP) and the rank-tracking probability ratio (RTPR) based on the time-varying transformation models (14.25).

Using the same logistic model as in Section 13.4, the conditional distribution of SBP has the expression

$$-\log\left\{ \frac{1 - F_t\left[y\big|X^{(1)}, X^{(2)}(t)\right]}{F_t\left[y\big|X^{(1)}, X^{(2)}(t)\right]} \right\} = h(y,t) + \beta_1(t)X^{(1)} + \beta_2(t)X^{(2)}(t), \qquad (14.62)$$

where $Y(t)$ is the SBP at age t, $F_t\left[y\big|\mathbf{X}(t)\right]$ is the conditional CDF for $Y(t)$

given $\mathbf{X}(t) = \left(X^{(1)}, X^{(2)}(t)\right)^T$ with $X^{(1)}$ and $X^{(2)}(t)$ denoting the race and height percentile, respectively. Here $X_i^{(1)} = 1$ if the ith girl is African American, and $X_i^{(1)} = 0$ if she is Caucasian. Using the two-step estimation procedure in Section 14.3, we estimate $S_t \left[y|\mathbf{x}(t)\right] = 1 - F_t \left[y|\mathbf{X}(t)\right]$, $RTP_{s_1, s_2} \left[(y_1, y_2), \mathbf{x}\right]$ and $RTPR_{s_1, s_2} \left[(y_1, y_2), \mathbf{x}\right]$ based on (14.62) over a set of $\{y, y_1, y_2, t, s_1, s_2\}$ values for $x^{(1)} = 0$ or 1, and a range of $x^{(2)}(t)$ values.

For the estimation of $S_t \left[y|\mathbf{x}(t)\right]$ in (??), we first compute the raw estimates $\widetilde{h}(y, t_j)$ and $\widetilde{S}_{t_j} \left[y|\mathbf{x}(t_j)\right]$ at the 100 equally spaced time design points $\{t_1 = 9.0, t_2 = 9.1, \ldots, t_{100} = 18.9\}$ using (13.13), (14.32) and (14.33) with $w(\cdot) = 1$, and then calculate the local linear estimators $\widehat{S}_t \left[y|\mathbf{x}(t)\right]$ using (14.39) with $q = 0$, $Q = 1$, the Epanechnikov kernel, and the LSCV and LTCV bandwidths. We compute the 95% pointwise confidence intervals for the estimators using the percentile bootstrap approach, described in Section 13.3.5.

We use the following R code to compute the two-step smoothing estimator of the conditional probability of "SBP > 100 mmHg" for an African American girl with median (50%) height:

```
# NGHS.sbp and Agebins are given in Sec 13.4
# Obtain raw estimate of probability given Race=1, HTPCT=50%,
> attach(NGHS.sbp)
> Agebins <- seq(90, 189, by= 1)
> Prob.Y100 <- Cond.Prob(Agebins, Y=SBP, X1= (RACE==2)*1,
                    X2=HTPCT, Y0=100, X10=1, X20=50)
# Local linear smoothing estimate
> Prob.Y100.lm <- LocalLm(Agebins, Agebins, Prob.Y100, bw=16)
```

The conditional probabilities for other SBP values given different covariates can be computed similarly.

Figure 14.1 shows the $\widehat{S}_t \left[100|\mathbf{x}(t)\right]$ computed by the LTCV bandwidths for the two races, $x^{(1)} = 0$ or 1, at five height percentile values, $x^{(2)}(t) = 5$, 25, 50, 75 and 95. These curves suggest that the conditional probabilities of "SBP > 100 mmHg" increase with age, but the slopes of these curves taper off after 13 to 15 years of age. The differences of $\widehat{S}_t \left[100|\mathbf{x}(t)\right]$ among different height percentiles diminish when age t increases, thus the effect of height on SBP gradually decreases with age. Comparing the predicted probabilities at the same height percentiles between two races, African American girls tend to have higher probabilities of "SBP > 100 mmHg" than Caucasian girls and these differences become more evident when the girls are getting older, so that race is a significant factor affecting the SBP during middle to late adolescence. Similar conclusions are obtained from the results of the LSCV bandwidths and SBP values other than 100 mmHg or results based on other smoothing methods such as kernel estimators.

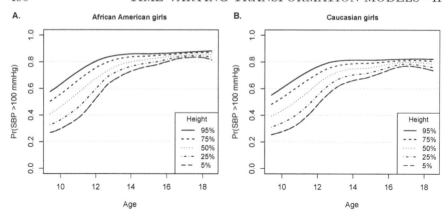

Figure 14.1 *The predicted curves* $\widehat{S}_t\left[100\big|x^{(1)}, x^{(2)}(t)\right] = 1 - \widehat{F}_t\left[100\big|x^{(1)}, x^{(2)}(t)\right]$ *of "SBP > 100 mmHg", computed using the LTCV bandwidths, over age* $9 \le t \le 19$ *years for (A) African American* ($x^{(1)} = 1$) *girls and (B) Caucasian* ($x^{(1)} = 0$) *girls at height percentile values,* $x^{(2)}(t) = 5, 25, 50, 75$ *and 95.*

14.4.2 RTP and RTPR for the NGHS SBP Data

To compute the estimators of $RTP_{s_1,s_2}\left[(y_1, y_2), \mathbf{x}\right]$ and $RTPR_{s_1,s_2}\left[(y_1, y_2), \mathbf{x}\right]$ in (14.47) and (14.52), we first compute the estimator $\widehat{S}_{s_1}\left[y_2\big|\mathbf{x}(s_1)\right]$ as above, and then compute $\widehat{S}_{s_1,s_2}\left[y_1, y_2\big|\mathbf{x}(s_1)\right]$ and $\widehat{S}_{s_2}\left[y_2\big|\mathbf{x}(s_1)\right]$ separately for the two races, $x^{(1)} = 0$ and 1, using (14.50) and (14.53) with the product Epanechnikov kernel $\pi(\cdot, \cdot, \cdot)$.

Let $y_t(q, a)$ be the qth SBP percentile for girls with ath height percentile at age t given in NHBPEP (2004). We calculate the estimates for $RTP_{s_1,s_2}\left[(y_1, y_2), \mathbf{x}\right]$, $RTPR_{s_1,s_2}\left[(y_1, y_2), \mathbf{x}\right]$ and their bootstrap 95% pointwise percentile confidence intervals over age $9 \le s_1 \le 14$ for $s_2 = s_1 + 3$, $y_1 = y_{s_1}(75, 50)$, $y_2 = y_{s_2}(75, 50)$, and Caucasian and African American girls with the 50th and 75th percentiles $(x^{(2)}(s_1) = 50, 75)$ in height. Some of the R code for these estimators is given below:

```
# For African American girls, with S1=10, height(S1)=75%
> NGHS.B <- subset(NGHS.sbp, RACE==2)
> S1 <- 10; S2 <- 10 + 3;
> attach(NGHS.B)
# BP75IND is the indicator of Y(t)> y(75,50) quantile at t
# agebin ranges from 90-189 to indicate age-bins at 9.0, 9.1 ..
> Pr.S12 <- Kernel3D(ID, Y=BP75IND, Time=agebin, X=HTPCT,
              T1=S1*10, T2=S2*10, X0=75, Bndwdth1=15,
              Bndwdth2=15, Bndwdth3=25)
```

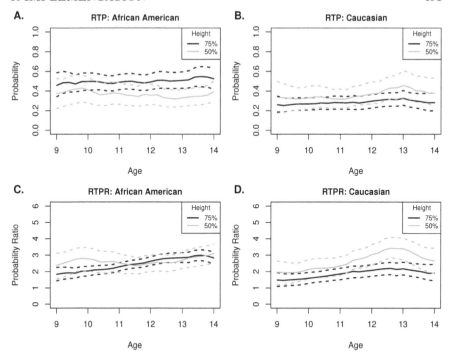

Figure 14.2 *The solid lines are the estimators* $\widehat{RTP}_{s_1,s_1+3}\left[(y_1, y_2), (x^{(1)}, x^{(2)}(s_1))\right]$ *(A)-(B) and* $\widehat{RTPR}_{s_1,s_1+3}\left[(y_1, y_2), (x^{(1)}, x^{(2)}(s_1))\right]$ *(C)-(D) at* $y_1 = y_{s_1}(75, 50)$ *and* $y_2 = y_{s_1+3}(75, 50)$ *for African American* $(x^{(1)} = 1,$ *A and C) girls and Caucasian* $(x^{(1)} = 0,$ *B and D) girls with the median* $(x^{(2)}(s_1) = 50,$ *gray lines) and the 75th percentile* $(x^{(2)}(s_1) = 75,$ *dark lines) height. The dashed lines are the corresponding bootstrap 95% pointwise percentile confidence intervals.*

```
> Pr.S2 <- Kernel3D.S2(ID, Y=BP75IND, Time=agebin, X=HTPCT,
         T1=S1*10, T2=S2*10, X0=75, Bndwdth1=15,
         Bndwdth2=15, Bndwdth3=25)
```

Figure 14.2 shows these estimated 3-year rank-tracking quantities. These RTPs and RTPRs describe the likelihood for those girls to have SBP values greater than the age- and height-adjusted 75th SBP percentile, $y_{s_1+3}(75, 50)$ at age $s_1 + 3$ given that, at age s_1, their SBP values are greater than $y_{s_1}(75, 50)$ and their height is at the median and the 75th percentile at s_1, respectively. The estimated RTPRs in Figures 14.2(C)-(D) are all greater than one, indicating that SBP has positive tracking abilities. This suggests that the girls with high SBP values at age s_1 are more likely to have high SBP values three years later, compared to a randomly selected girl in this population with elevated SBP value without the knowledge of her SBP status at a younger age. Since the estimated rank-tracking quantities in Figure 14.2 vary with different races and height percentiles, the tracking abilities of SBP also appear to be affected by these variables.

The smoothing estimators of Figures 14.1 and 14.2 provide useful insight into the SBP distributions for adolescent girls, which have not been previously investigated. The results of Figure 14.2 suggest that the multivariate smoothing method in Section 14.3 gives appropriate RTP and RTPR estimators for a typical longitudinal study when the dimensionality of the covariate vector $\mathbf{X}(t)$ is low. These RTPs and RTPRs may be used as an exploratory analysis tool for evaluating the correlation structures of the longitudinal variables. However, when the dimensionality of $\mathbf{X}(t)$ is high, the smoothing approach may not be feasible for the estimation of RTPs and RTPRs, so that appropriate longitudinal models across different time points and their goodness-of-fit tests need to be developed.

14.5   Asymptotic Properties

We present the asymptotic properties of both the raw and smoothing estimators of $h(y, t)$ and $F_t(y|\mathbf{x})$. For the raw estimators, we derive the asymptotic expressions of their biases and variances at each of the design time points $\{t_1, \ldots, t_J\}$ and their covariances between two different design time points t_{j_1} and t_{j_2}. These asymptotic expressions are the key components for deriving the asymptotic mean squared errors of the smoothing estimators.

14.5.1   Asymptotic Assumptions

Although the asymptotic derivations can be carried out for various types of smoothing estimators based on the general expression of linear estimators in (14.34), our derivations are specifically given for the local polynomial estimators (14.38). Since the asymptotic derivations require some complex notation, we first define, for any $t_j \in \{t_1, \ldots, t_J\}$,

$$\begin{cases} V[h(y, t_j)] & = (1/n_j) \sum_{i \in \mathscr{S}_j} v_i(y, t_j), \\ v_i(y, t_j) & = 1_{[Y_{ij} \geq y]} - \phi[h(y, t_j) + \mathbf{X}_i^T(t_j)\beta(t_j)], \end{cases} \qquad (14.63)$$

where $\phi(u)$ is defined in (14.26). The derivatives of $V[h(y, t_j)]$ with respect to $h(y, t_j)$ and the elements of $\beta(t_j)$ are defined by

$$V_1[h(y, t_j)] = \frac{\partial V[h(y, t_j)]}{\partial h(y, t_j)} \quad \text{and} \quad V_2[h(y, t_j)] = \frac{\partial V[h(y, t_j)]}{\partial \beta(t_j)}. \qquad (14.64)$$

Since, by (14.32), the raw estimator $\widetilde{h}(y, t_j)$ satisfies

$$V[\widetilde{h}(y, t_j)] = 0 \quad \text{for any } t_j \in \{t_1, \ldots, t_J\},$$

it follows from (14.63), (14.64) and the Taylor's expansion that, when n is sufficiently large,

$$
\begin{aligned}
V\left[h(y, t_j)\right] &= -V_1\left[h(y, t_j)\right]\left[\tilde{h}(y, t_j) - h(y, t_j)\right] - V_2\left[h(y, t_j)\right]\left[\tilde{\beta}(t_j) - \beta(t_j)\right] \\
&+ o_p\left[\left|\tilde{h}(y, t_j) - h(y, t_j)\right|\right] + o_p\left\{\left|\sum_{p=1}^{P}\left[\tilde{\beta}_p(t_j) - \beta_p(t_j)\right]\right|\right\}.
\end{aligned} \tag{14.65}
$$

In order to take the variation of $\left[\tilde{\beta}(t_j) - \beta(t_j)\right]$ into account, we consider $U_{i_1 i_2}\left[\beta(t_j)\right]$ of (13.13) and define

$$
\Lambda(t_j) = -E\left\{\frac{\partial U_{i_1 i_2}[\beta(t_j)]}{\partial \beta(t_j)}\bigg| \mathscr{D}\right\}. \tag{14.66}
$$

We make the following asymptotic assumptions throughout this chapter for the local polynomial estimators (14.38).

Asymptotic Assumptions:

(a) *For any q and Q, the bandwidth b and number of design time points J satisfy $b \to 0$, $n^{1/2} b^{Q-q+1} \to \infty$, $Jb \to \infty$ and $nJb^{2q+1} \to \infty$ as $n \to \infty$.*

(b) *The design time points $\{t_j \in \mathscr{T} : j = 1, \ldots, J\}$ are i.i.d. with density function $\pi(t)$ and support $\mathscr{T}$. For all j, j_1 and j_2, there are known constants $0 < c_j \le 1$ and $0 < c_{j_1 j_2} \le 1$, such that $\lim_{n \to \infty} (n_j/n) = c_j$ and $\lim_{n \to \infty} (n_{j_1 j_2}/n) = c_{j_1 j_2}$.*

(c) *For any $1 \le p \le P$ and y, $\beta_p(t)$ and $h(y, t)$ of (14.25) are $Q+1$ times continuously differentiable with respect to t. The inverse of the link function, $\phi(s) = g^{-1}(s)$, is $Q+1$ times continuously differentiable with respect to s.*

(d) *The matrices $E\{V_1\left[h(y, t_j)\right]|\mathscr{D}\}$ and $\Lambda(t_j)$ are nonsingular.*

(e) *The kernel $K(\cdot)$ is a bounded symmetric probability density function.* □

14.5.2 Raw Baseline and Distribution Function Estimators

Since the asymptotic properties of the raw estimator $\tilde{\beta}(t_j)$ have already been established in Section 12.5, our focus is on the asymptotic properties of the raw baseline estimator $\tilde{h}(t, t_j)$. The asymptotic properties of the raw CDF estimator $\tilde{F}_{t_j}(y|\mathbf{x})$ are derived from the asymptotic properties of $\tilde{\beta}(t_j)$ and $\tilde{h}(y, t_j)$ using the model (14.31).

1. Baseline Estimators

We now show that $\tilde{h}(y, t_j)$ of (14.32) is asymptotically unbiased for the baseline function $h(y, t_j)$, and establish its asymptotic variance-covariance matrix.

We define

$$
\begin{cases}
W(y, t_j) &= \left[n_j(n_j-1)\right]^{-1} \sum_{i_1 \neq i_2 \in \mathscr{S}_j} \left[v_{i_1}^*(y, t_j) + u_{i_1 i_2}^*(y, t_j)\right], \\
B_1(y, t_j) &= \left\{E\left[V_1(h(y, t_j))\big|\mathscr{D}\right]\right\}^{-1}, \\
B_2(y, t_j) &= B_1(y, t_j) E\left\{V_2\left[h(y, t_j)\right]\big|\mathscr{D}\right\}, \\
v_i^*(y, t_j) &= -B_1(y, t_j) v_i(y, t_j), \\
u_{i_1 i_2}^*(y, t_j) &= B_1(y, t_j)\Lambda^{-1}(t_j) U_{i_1 i_2}\left[\beta(t_j)\right].
\end{cases}
\tag{14.67}
$$

In addition, with c_j and $c_{j_1 j_2}$ of Assumption (b), we define $a_{j_1 j_2}$ and $\rho^{(k)}(y, t_{j_1}, t_{j_2})$ for $k = 0, 1, 2$ by

$$
\begin{cases}
a_{j_1 j_2} &= c_{j_1 j_2}\left[3(c_{j_2}-c_{j_1 j_2})(2c_{j_1}-c_{j_1 j_2})\right. \\
&\quad \left. +c_{j_1 j_2}(3c_{j_1}-2c_{j_1 j_2})\right]/(6c_{j_1}^2 c_{j_2}^2), \\
\rho^{(0)}(y, t_{j_1}, t_{j_2}) &= E\left[v_{i_1}^*(y, t_{j_1}) v_{i_1}^*(y, t_{j_2})\big|\mathscr{D}\right], \\
\rho^{(1)}(y, t_{j_1}, t_{j_2}) &= E\left[v_{i_1}^*(y, t_{j_1}) u_{i_1 i_3}^*(y, t_{j_2})\big|\mathscr{D}\right], \\
\rho^{(2)}(y, t_{j_1}, t_{j_2}) &= E\left[u_{i_1 i_2}^*(y, t_{j_1}) u_{i_1 i_3}^*(y, t_{j_2})\big|\mathscr{D}\right]
\end{cases}
\tag{14.68}
$$

for any $\{j_1, j_2\}$ and

$$
\left\{(i_1, i_2, i_3) : i_1 \neq i_2 \neq i_3;\ i_1 \in \mathscr{S}_{j_1} \cap \mathscr{S}_{j_2};\ i_2 \in \mathscr{S}_{j_1};\ i_3 \in \mathscr{S}_{j_2}\right\}.
\tag{14.69}
$$

The following lemma shows the asymptotic approximation, asymptotic unbiasedness and asymptotic variance-covariance matrix for $\widetilde{h}(y, t_j)$.

Lemma 14.1. *If Assumptions (b) and (d) are satisfied, the following conclusions hold when n is sufficiently large:*

(a) *The baseline error $\widetilde{h}(y, t_j) - h(y, t_j)$ has the approximation*

$$
\begin{aligned}
&\left[\widetilde{h}(y, t_j) - h(y, t_j)\right]\left[1 + o_p(1)\right] \\
&= W(y, t_j) + o_p\left\{\left|\sum_{p=1}^{P}\left[\widetilde{\beta}_p(t_j) - \beta_p(t_j)\right]\right|\right\}.
\end{aligned}
\tag{14.70}
$$

(b) *The baseline estimator $\widetilde{h}(y, t_j)$ is asymptotically unbiased for $h(y, t_j)$ in the sense that*

$$
E\left[W(y, t_j)\big|\mathscr{D}\right] = 0 \quad \text{for any } t_j \in \mathscr{T}.
\tag{14.71}
$$

(c) *For any $(t_{j_1}, t_{j_2}) \in \mathscr{T} \times \mathscr{T}$, the asymptotic covariance of $\widetilde{h}(y, t_j)$ is*

$$
\begin{aligned}
&\text{Cov}\left[W(y, t_{j_1}), W(y, t_{j_2})\big|\mathscr{D}\right] \\
&= n^{-1} a_{j_1 j_2}\left[\rho^{(0)}(y, t_{j_1}, t_{j_2}) + 2\rho^{(1)}(y, t_{j_1}, t_{j_2})\right. \\
&\quad \left. +2\rho^{(1)}(y, t_{j_2}, t_{j_1}) + 4\rho^{(2)}(y, t_{j_1}, t_{j_2})\right].
\end{aligned}
\tag{14.72}
$$

(d) *For any $t_j \in \mathcal{T}$, the asymptotic variance of $\widetilde{h}(y, t_j)$ is given by*

$$Var\left[W(y, t_j)\big|\mathcal{D}\right] \quad = \quad Cov\left[W(y, t_j), W(y, t_j)\big|\mathcal{D}\right] \tag{14.73}$$
$$= \quad \left[\rho^{(0)}(y, t_j) + 4\rho^{(1)}(y, t_j) + 4\rho^{(2)}(y, t_j)\right]/(6c_j n).$$

which is obtained by taking $j_1 = j_2 = j$ in (14.72), $\rho^{(k)}(y, t_j) = \rho^{(k)}(y, t_j, t_j)$ for $k = 0, \ldots, 3$, $c_{j_1} = c_{j_2} = c_{j_1 j_2} = c_j$, $a_{jj} = 1/(6c_j)$. ∎

Proof of Lemma 14.1 is given in Section 14.5.4.

We clarify a number of implications of the above lemma. The asymptotic approximation of (14.70) implies that, when n and n_j are sufficiently large, $\left[\widetilde{h}(y, t_j) - h(y, t_j)\right]$ is approximated by $W(y, t_j)$ plus a smaller order term of $\left|\Sigma_{p=1}^{P}\left[\widetilde{\beta}_p(t_j) - \beta_p(t_j)\right]\right|$. Because of (14.70), this implies that on average $\widetilde{h}(y, t_j)$ is close to $h(y, t_j)$, hence is asymptotically unbiased, although the conditional expectation of $\widetilde{h}(y, t_j)$ given $\mathcal{D}$ may not necessarily equal $h(y, t_j)$ for finite sample size situations. The asymptotic covariance (14.72) and variance (14.73) depend on the limits of c_j, $c_{j_1 j_2}$ and $a_{j_1 j_2}$ as $n \to \infty$. Consequently, various special cases can be derived depending on the choices of c_j, $c_{j_1 j_2}$ and other quantities in (14.72) and (14.73). For example, if we consider the special case of an "ideal" longitudinal design with $c_j = c_{j_1 j_2} = 1$, i.e., all the subjects have observations at all the time design points, the asymptotic covariance of $\widetilde{h}(y, t_{j_1})$ and $\widetilde{h}(y, t_{j_2})$ is given by (14.72) with $a_{j_1 j_2} = 1/6$.

Comparing (14.72) with (14.73), if $c_{j_1 j_2}$ is ignorable relative to c_{j_1} and c_{j_2}, the asymptotic covariance $Cov\left[W(y, t_{j_1}), W(y, t_{j_2})\big|\mathcal{D}\right]$ can be ignored relative to $Var\left[W(y, t_j)\big|\mathcal{D}\right]$. This comparison suggests that the correlation structures of the data only affect the asymptotic properties of $\widetilde{h}(y, t_j)$ under the dense longitudinal data in the sense that $c_{j_1 j_2}$ cannot be ignored relative to c_{j_1} and c_{j_2}. Because of various practical concerns, such as cost and logistical issues, dense longitudinal data are rarely used in real clinical studies. Thus, it is often reasonable to ignore the contributions of the asymptotic covariance $Cov\left[W(y, t_{j_1}), W(y, t_{j_2})\big|\mathcal{D}\right]$ when we evaluate the asymptotic mean squared errors of a local polynomial smoothing estimator in a real application.

2. Distribution Function Estimators

We assume that $\phi(s)$ is a continuously differentiable function with derivative $\phi'(s)$. Based on (14.33), the asymptotic approximations of $\widetilde{S}_{t_j}(y|\mathbf{x}) = 1 - \widetilde{F}_{t_j}(y|\mathbf{x})$ can be expressed as

$$\widetilde{S}_{t_j}(y|\mathbf{x}) - S_{t_j}(y|\mathbf{x})$$
$$= \quad \phi'\left[h(y, t_j) + \mathbf{x}^T(t_j)\beta(t_j)\right]\left\{\widetilde{h}(y, t_j) - h(y, t_j) + \mathbf{x}^T(t_j)\left[\widetilde{\beta}(t_j) - \beta(t_j)\right]\right\}$$
$$+ o_p\left[\left|\widetilde{h}(y, t_j) - h(y, t_j)\right|\right] + o_p\left[\sum_{p=1}^{P}\left|\widetilde{\beta}_p(t_j) - \beta_p(t_j)\right|\right]. \tag{14.74}$$

Following the functions defined in (14.65) and (14.66), we define that, for any $\{j_1, j_2\}$ and (i_1, i_2, i_3) satisfying (14.69),

$$
\begin{cases}
u_{i_1 i_2}^{**}(y, \mathbf{x}, t_j) &= \left[B_1(y, t_j) + \mathbf{x}^T(t_j)\right] \Lambda^{-1}(t_j) U_{i_1 i_2}\left[\beta(t_j)\right], \\
\rho^{(1*)}(y, \mathbf{x}, t_{j_1}, t_{j_2}) &= E\left[v_{i_1}^*(y, t_{j_1}) u_{i_1 i_3}^*(y, \mathbf{x}, t_{j_2}) \big| \mathscr{D}\right], \\
\rho^{(2*)}(y, \mathbf{x}, t_{j_1}, t_{j_2}) &= E\left[u_{i_1 i_2}^*(y, \mathbf{x}, t_{j_1}) u_{i_1 i_3}^*(y, \mathbf{x}, t_{j_2}) \big| \mathscr{D}\right], \\
\mathscr{A}_{i_1 i_2}(y, \mathbf{x}, t_j) &= \phi'\left[h(y, t_j) + \mathbf{x}^T(t_j)\beta(t_j)\right] \\
&\quad \times \left[v_{i_1}^*(y, t_j) + u_{i_1 i_2}^{**}(y, \mathbf{x}, t_j)\right].
\end{cases}
\tag{14.75}
$$

The next lemma summarizes the asymptotic expressions of the bias, covariance and variance for $\widetilde{S}_{t_j}(y|\mathbf{x})$.

Lemma 14.2. *If Assumptions (b) and (d) are satisfied, then, as $n \to \infty$,*

$$
\begin{aligned}
\left[\widetilde{S}_{t_j}(y|\mathbf{x}) - S_{t_j}(y|\mathbf{x})\right]\left[1 + o_p(1)\right] &= \mathscr{A}(y, \mathbf{x}, t_j) \tag{14.76} \\
&= \left[n_j(n_j - 1)\right]^{-1} \sum_{i_1 \neq i_2 \in \mathscr{S}_j} \mathscr{A}_{i_1 i_2}(y, \mathbf{x}, t_j),
\end{aligned}
$$

and the following conclusions hold:

(a) $\widetilde{S}_{t_j}(y|\mathbf{x})$ *is asymptotically unbiased for $S_{t_j}(y|\mathbf{x})$ in the sense that*

$$
E\left[\mathscr{A}(y, \mathbf{x}, t_j) \big| \mathscr{D}\right] = 0 \quad \text{for any } t_j \in \mathscr{T}.
\tag{14.77}
$$

(b) *For any $(t_{j_1}, t_{j_2}) \in \mathscr{T} \times \mathscr{T}$, the asymptotic covariance of $\widetilde{S}_{t_j}(y|\mathbf{x})$ is*

$$
\begin{aligned}
&Cov\left[\mathscr{A}(y, \mathbf{x}, t_{j_1}), \mathscr{A}(y, \mathbf{x}, t_{j_2}) \big| \mathscr{D}\right] \\
&= n^{-1} a_{j_1 j_2} \phi'\left[h(y, t_{j_1}) + \mathbf{x}^T(t_{j_1})\beta(t_{j_1})\right] \phi'\left[h(y, t_{j_2}) + \mathbf{x}^T(t_{j_2})\beta(t_{j_2})\right] \\
&\quad \times \left[\rho^{(0)}(y, t_{j_1}, t_{j_2}) + 2\rho^{(1*)}(y, \mathbf{x}, t_{j_1}, t_{j_2}) + 2\rho^{(1*)}(y, \mathbf{x}, t_{j_2}, t_{j_1})\right. \\
&\quad \left. + 4\rho^{(2*)}(y, \mathbf{x}, t_{j_1}, t_{j_2})\right],
\end{aligned}
\tag{14.78}
$$

where $a_{j_1 j_2}$ and $\rho^{(0)}(\cdot)$ are defined in (14.68), and $\rho^{(1*)}(\cdot)$ and $\rho^{(2*)}(\cdot)$ defined in (14.75).

(c) *For any $t_j \in \mathscr{T}$, the asymptotic variance of $\widetilde{S}_{t_j}(y|\mathbf{x})$ is given by (14.78) with $j_1 = j_2 = j$, that is,*

$$
\begin{aligned}
Var\left[\mathscr{A}(y, \mathbf{x}, t_j) \big| \mathscr{D}\right] &= \frac{1}{6c_j n}\left\{\phi'\left[h(y, t_j) + \mathbf{x}^T(t_j)\beta(t_j)\right]\right\}^2 \tag{14.79} \\
&\quad \times \left[\rho^{(0)}(y, t_j) + 4\rho^{(1*)}(y, \mathbf{x}, t_j) + 4\rho^{(2*)}(y, \mathbf{x}, t_j)\right],
\end{aligned}
$$

where $\rho^{(0)}(y, t_j) = \rho^{(0)}(y, t_j, t_j)$ and $\rho^{(k*)}(y, \mathbf{x}, t_j) = \rho^{(k*)}(y, \mathbf{x}, t_j, t_j)$ for $k = 1$ and 2. ∎

Proof of Lemma 14.2 is given in Section 14.5.4.

Since $E\left[\mathscr{A}(y,\mathbf{x},t_j)|\mathscr{D}\right]=0$, Lemma 14.2 suggests that $\widetilde{S}_{t_j}(y|\mathbf{x})$ is on average close to $S_{t_j}(y|\mathbf{x})$. The asymptotic covariance (14.78) depends on c_j, $c_{j_1 j_2}$ and $a_{j_1 j_2}$. If $c_{j_1 j_2}$ can be ignored relative to c_{j_1} and c_{j_2}, then it follows from (14.68), (14.78) and (14.79) that $Cov\left[\mathscr{A}(y,\mathbf{x},t_{j_1}),\mathscr{A}(y,\mathbf{x},t_{j_2})|\mathscr{D}\right]$ can be ignored relative to $Var\left[\mathscr{A}(y,\mathbf{x},t_j)|\mathscr{D}\right]$.

14.5.3   Local Polynomial Smoothing Estimators

We establish the asymptotic biases, variances and mean-squared errors of the local polynomial estimators $\widehat{h}^{(q)}(y,t)$ and $\widehat{F}_t^{(q)}(y|\mathbf{x})$ of (14.38). Let $\widehat{\psi}_t(y,\mathbf{x})$ be a local polynomial estimator of $\psi_t(y,\mathbf{x})$, which may be either $h^{(q)}(y,t)$ or $F_t^{(q)}(y|\mathbf{x})$. We define the conditional bias, variance and mean squared error (MSE) for $\widehat{\psi}_t(y,\mathbf{x})$ given $\mathscr{D}$ by

$$
\left\{
\begin{array}{rcl}
Bias\left[\widehat{\psi}_t(y,\mathbf{x})|\mathscr{D}\right] & = & E\left[\widehat{\psi}_t(y,\mathbf{x})-\psi_t(y,\mathbf{x})|\mathscr{D}\right], \\
Var\left[\widehat{\psi}_t(y,\mathbf{x})|\mathscr{D}\right] & = & E\left\{\left[\widehat{\psi}_t(y,\mathbf{x})-E\left[\widehat{\psi}_t(y,\mathbf{x})|\mathscr{D}\right]\right]^2|\mathscr{D}\right\}, \\
MSE\left[\widehat{\psi}_t(y,\mathbf{x})|\mathscr{D}\right] & = & Bias\left[\widehat{\psi}_t(y,\mathbf{x})|\mathscr{D}\right]^2+Var\left[\widehat{\psi}_t(y,\mathbf{x})|\mathscr{D}\right].
\end{array}
\right. \tag{14.80}
$$

These biases, variances and MSEs are all pointwise in the sense that they are specific for the pair (y,t). When the adequacy of $\widehat{\psi}_t(y,\mathbf{x})$ is considered over a range $\mathscr{R}$ of (y,t), we can generalize $MSE\left[\widehat{\psi}_t(y,\mathbf{x})|\mathscr{D}\right]$ to the following integrated MSEs

$$
IMSE_w\left[\widehat{\psi}_{\cdot}(\cdot,\mathbf{x})|\mathscr{D}\right]=\int_{\mathscr{R}} MSE\left[\widehat{\psi}_t(y,\mathbf{x})|\mathscr{D}\right]w(y,t)\,dy\,dt,
$$

where $w(y,t)$ is some pre-specified non-negative weight function on $\mathscr{R}$.

We present here the cases for $\widehat{\psi}_t(y,\mathbf{x})=\widehat{h}^{(q)}(y,t)$ and $\widehat{\psi}_t(y,\mathbf{x})=\widehat{F}_t^{(q)}(y,t)$. The asymptotic expressions for $IMSE_w\left[\widehat{\psi}_{\cdot}(\cdot,\mathbf{x})|\mathscr{D}\right]$ can be directly expressed using $MSE\left[\widehat{\psi}_t(y,\mathbf{x})|\mathscr{D}\right]$ and $w(y,t)$, hence, are omitted.

1. Smoothing Baseline Estimators

For a given kernel function $K(\cdot)$, we define

$$
\left\{
\begin{array}{rcl}
S & = & \left(s_{j_1 j_2}\right)_{j_1,j_2=0,\ldots,Q}, \text{ matrix with elements } s_{j_1 j_2}, \\
s_{j_1 j_2} & = & \int K(u)u^{j_1+j_2}\,du, \\
K_{q,Q+1}(u) & = & e_{q+1,Q+1}^T S^{-1}\left(1,u,\cdots,u^Q\right)^T K(u), \\
B_{Q+1}(K) & = & \int K(u)u^{Q+1}\,du, \\
V(K) & = & \int K^2(u)\,du.
\end{array}
\right. \tag{14.81}
$$

The asymptotic expressions for $\widehat{h}^{(q)}(y,t)$ are given in the following theorem.

Theorem 14.1. *If Assumptions (a) to (e) are satisfied with $c_j = c$ for all $j = 1,\ldots,J$ and some $0 < c \le 1$, the following conclusions hold when n is sufficiently large:*

(a) *The asymptotic bias for $\widehat{h}^{(q)}(y, t)$ satisfies*

$$Bias\left[\widehat{h}^{(q)}(y, t)\big|\mathscr{D}\right] = b^{(Q-q+1)}\mathscr{B}_h(y, t)\left[1 + o_p(1)\right], \qquad (14.82)$$

where $b > 0$ is the bandwidth and

$$\mathscr{B}_h(y, t) = [q!/(Q+1)!]\, h^{(Q+1)}(y, t)\, B_{Q+1}\left(K_{q,Q+1}\right). \qquad (14.83)$$

(b) *The asymptotic variance for $\widehat{h}^{(q)}(y, t)$ satisfies*

$$Var\left[\widehat{h}^{(q)}(y, t)\big|\mathscr{D}\right] = (nJ)^{-1}b^{-2q-1}\mathscr{V}_h(y, t)\left[1 + o_p(1)\right], \qquad (14.84)$$

where, with $\pi(t)$ being the density of design time points defined in Assumption (b),

$$\begin{cases} \mathscr{V}_h(y, t) &= \{(q!)^2/[6c\,\pi(t)]\}V\left(K_{q,Q+1}\right)\sigma_h^2(y, t), \\ \sigma_h^2(y, t) &= \rho^{(0)}(y, t) + 4\rho^{(1)}(y, t) + 2\rho^{(2)}(y, t). \end{cases} \qquad (14.85)$$

(c) *It follows from (a) and (b) above that, for sufficiently large n,*

$$\begin{aligned} MSE\left[\widehat{h}^{(q)}(y, t)\big|\mathscr{D}\right] &= \left[b^{2(Q-q+1)}\mathscr{B}_h^2(y, t) + (nJ)^{-1}b^{-2q-1}\mathscr{V}_h(y, t)\right] \\ &\quad \times \left[1 + o_p(1)\right]. \end{aligned} \qquad (14.86)$$

where the bias and variance components $\mathscr{B}_h^2(y, t)$ and $\mathscr{V}_h(y, t)$ have the expressions in (14.83) and (14.85), respectively. ∎

Proof of Theorem 14.1 is given in Section 14.5.4.

It is clear from Theorem 14.1(a) that a small bandwidth b leads to a small absolute value of the asymptotic bias of $\widehat{h}^{(q)}(y, t)$, while, on the other hand, it is seen from (14.84) that a small bandwidth b leads to a large asymptotic variance of $\widehat{h}^{(q)}(y, t)$. Thus, the mean squared error of $\widehat{h}^{(q)}(y, t)$ converges to zero if and only if both $b^{2(Q-q+1)}$ and $(nJ)^{-1}b^{-2q-1}$ converge to zero. For the special case of estimating $h(y, t)$ using the local linear estimator $\widehat{h}^L(y, t)$ defined in (14.39), Theorem 14.1 leads to the following corollary by setting $Q = 1$ and $q = 0$ in (14.82) through (14.86).

Corollary 14.1. *If the assumptions of Theorem 14.1 are satisfied and Assumption (c) holds for $Q = 1$, the following conclusions hold for the local linear estimator $\widehat{h}^L(y, t)$ of $h(y, t)$ when n is sufficiently large:*

(a) *The asymptotic bias for $\widehat{h}^L(y, t)$ satisfies*

$$Bias\left[\widehat{h}^L(y, t)\big|\mathscr{D}\right] = b^2\,\mathscr{B}_h^*(y, t)\left[1 + o_p(1)\right], \qquad (14.87)$$

where $b > 0$ is the bandwidth,

$$\mathscr{B}_h^*(y, t) = (1/2) h''(y, t) B_2(K_{0,2}) \tag{14.88}$$

and $h''(y, t)$ is the second derivative of $h(y, t)$ with respect to t.

(b) *The asymptotic variance for $\widehat{h}^L(y, t)$ satisfies*

$$Var[\widehat{h}^L(y, t)|\mathscr{D}] = (nJb)^{-1} \mathscr{V}_h^*(y, t) [1 + o_p(1)], \tag{14.89}$$

where, with $\pi(t)$ being the density of design time points defined in Assumption (b),

$$\begin{cases} \mathscr{V}_h(y, t) &= \{1/[6c\,\pi(t)]\} V(K_{0,2}) \sigma_h^2(y, t), \\ \sigma_h^2(y, t) &= \rho^{(0)}(y, t) + 4\rho^{(1)}(y, t) + 2\rho^{(2)}(y, t). \end{cases} \tag{14.90}$$

(c) *It follows from (a) and (b) above that, for sufficiently large n,*

$$\begin{aligned} MSE[\widehat{h}^L(y, t)|\mathscr{D}] &= \left\{ b^4 [\mathscr{B}_h^*(y, t)]^2 + (nJb)^{-1} \mathscr{V}_h^*(y, t) \right\} \\ &\times [1 + o_p(1)], \end{aligned} \tag{14.91}$$

where the bias and variance components $[\mathscr{B}_h^*(y, t)]^2$ and $\mathscr{V}_h^*(y, t)$ have the expressions in (14.88) and (14.90), respectively. ∎

We can derive a number of theoretical conclusions based on the asymptotic expressions of the mean squared errors in (14.86) and (14.91). First, by setting the derivative of the dominating part of (14.86) with respect to t to zero, the asymptotically optimal bandwidth $b_{opt, h^{(q)}}$, which minimizes the dominating part of the right side of (14.86), is given by

$$b_{opt, h^{(q)}} = \left[\frac{(2q + 1) \mathscr{V}_h(y, t)}{2nJ(Q - q + 1) \mathscr{B}_h^2(y, t)} \right]^{1/(2Q+3)}. \tag{14.92}$$

Substituting b of (14.86) with the above $b_{opt, h^{(q)}}$, we can show that, under the assumptions of Theorem 14.1, the asymptotically optimal mean squared error of $\widehat{h}^{(q)}(y, t)$ is

$$\begin{aligned} &MSE_{opt}[\widehat{h}^{(q)}(y, t)|\mathscr{D}] \\ &= (nJ)^{-\frac{2(Q-q+1)}{2Q+3}} [\mathscr{B}_h(y, t)]^{\frac{2(2q+1)}{2Q+3}} [\mathscr{V}_h(y, t)]^{\frac{2(Q-q+1)}{2Q+3}} \\ &\times \left\{ \left[\frac{2q+1}{2(Q-q+1)} \right]^{\frac{2(Q-q+1)}{2Q+3}} + \left[\frac{2q+1}{2(Q-q+1)} \right]^{-\frac{q+1}{2Q+3}} \right\} \\ &\times [1 + o_p(1)]. \end{aligned} \tag{14.93}$$

The right side of (14.93) suggests that, when the ideal bandwidth $b_{opt, h^{(q)}}$ is used, the optimal convergence rate for the mean squared error of $\widehat{h}^{(q)}(y,t)$ under the assumptions of Theorem 14.1 is $(nJ)^{[-2(Q-q+1)]/(2Q+3)}$. However, because the ideal bandwidth of (14.92) depends on the unknown quantities $\mathscr{B}_h(y,t)$ and $\mathscr{V}_h(y,t)$, the expression at the right side of (14.93) only gives some theoretical insight into the appropriateness of the estimator $\widehat{h}^{(q)}(y,t)$, which may not be actually used in practice.

The asymptotically optimal bandwidth and mean squared error of the local linear estimator $\widehat{h}^L(y,t)$ under the assumptions of Corollary 14.1 are given by (14.92) and (14.93), respectively, with $Q=1$ and $q=0$. As a result, the optimal convergence rate for the mean squared error of $\widehat{h}^L(y,t)$ is $(nJ)^{-4/5}$.

2. Smoothing Distribution Function Estimators

Building on the asymptotic representations of $\widehat{h}^{(q)}(y,t)$, the next theorem summarizes the asymptotic biases, variances and mean squared errors of the conditional CDF estimator $\widehat{F}_t^{(q)}(y|\mathbf{x})$.

Theorem 14.2. *If Assumptions (a) through (e) are satisfied with $c_j = c$ for all $j = 1, \ldots, J$ and some $0 < c \le 1$, the following conclusions hold when n is sufficiently large:*

(a) *The asymptotic bias of $\widehat{F}_t^{(q)}(y|\mathbf{x})$ satisfies*

$$Bias\left[\widehat{F}_t^{(q)}(y|\mathbf{x})\big|\mathscr{D}\right] = b^{(Q-q+1)}\,\mathscr{B}_F(y,\mathbf{x},t)\left[1+o_p(1)\right], \qquad (14.94)$$

where $b > 0$ is the bandwidth and

$$\mathscr{B}_F(y,t) = \left[q!/(Q+1)!\right]F_t^{(Q+1)}(y|\mathbf{x})\,B_{Q+1}\big(K_{q,Q+1}\big). \qquad (14.95)$$

(b) *The asymptotic variance of $\widehat{F}_t^{(q)}(y|\mathbf{x})$ satisfies*

$$Var\left[\widehat{F}_t^{(q)}(y|\mathbf{x})\big|\mathscr{D}\right] = (nJ)^{-1}\,b^{-2q-1}\mathscr{V}_F(y,\mathbf{x},t)\left[1+o_p(1)\right], \qquad (14.96)$$

where, with $\pi(t)$ being the density of time design points defined in Assumption (b),

$$\begin{cases} \mathscr{V}_F(y,\mathbf{x},t) &= \left\{(q!)^2/[6c\,\pi(t)]\right\}V\big(K_{q,Q+1}\big)\,\sigma_F^2(y,\mathbf{x},t), \\ \sigma_F^2(y,\mathbf{x},t) &= \left\{\phi'[h(y,t)+\mathbf{x}^T\beta(t)]\right\}^2 \\ & \quad \times\left[\rho^{(0)}(y,t)+4\rho^{(1*)}(y,\mathbf{x},t)+4\rho^{(2*)}(y,\mathbf{x},t)\right]. \end{cases} \qquad (14.97)$$

(c) *It follows from (a) and (b) above that, for sufficiently large n,*

$$MSE\left[\widehat{F}_t^{(q)}(y,\mathbf{x})\big|\mathscr{D}\right] = \left\{b^{2(Q-q+1)}\left[\mathscr{B}_F(y,\mathbf{x},t)\right]^2 \qquad (14.98)$$

$$+(nJ)^{-1}b^{-2q-1}\,\mathscr{V}_F(y,\mathbf{x},t)\right\}\left[1+o_p(1)\right],$$

where the bias and variance components $\mathscr{B}_F(y,\mathbf{x},t)$ and $\mathscr{V}_F(y,\mathbf{x},t)$ have the expressions in (14.95) and (14.97), respectively. ∎

Proof of Theorem 14.2 is given in Section 14.5.4.

Comparing Theorem 14.2(c) with Theorem 14.1(c), we can see that the mean squared errors of $\widehat{h}^{(q)}(y,t)$ and $\widehat{F}_t^{(q)}(y,\mathbf{x})$ converge to zero in the same rate as $b^{2(Q-q+1)}$ and $(nJ)^{-1}b^{-2q-1}$ converging to zero. For the special case of estimating $F_t(y|\mathbf{x})$ with the local linear estimator $\widehat{F}_t^L(y|\mathbf{x})$ of (14.39), Theorem 14.2 reduces to the following corollary with $Q=1$ and $q=0$.

Corollary 14.2. *If the assumptions of Theorem 13.2 are satisfied and Assumption (c) holds for $Q=1$, the following conclusions hold when n is sufficiently large:*

(a) *The asymptotic bias of $\widehat{F}_t^L(y|\mathbf{x})$ satisfies*

$$Bias\left[\widehat{F}_t^L(y|\mathbf{x})\big|\mathscr{D}\right] = b^2\,\mathscr{B}_F^*(y,\mathbf{x},t)\left[1+o_p(1)\right], \tag{14.99}$$

where $b>0$ is the bandwidth,

$$\mathscr{B}_F^*(y,t) = (1/2)\,F_t''(y|\mathbf{x})\,B_2\big(K_{0,2}\big). \tag{14.100}$$

and $F_t''(y|\mathbf{x})$ is the second derivative with respect to t.

(b) *The asymptotic variance of $\widehat{F}_t^L(y|\mathbf{x})$ satisfies*

$$Var\left[\widehat{F}_t^L(y|\mathbf{x})\big|\mathscr{D}\right] = (nJb)^{-1}\,\mathscr{V}_F^*(y,\mathbf{x},t)\left[1+o_p(1)\right], \tag{14.101}$$

where, with $\pi(t)$ being the density of design time points defined in Assumption (b),

$$\left\{\begin{array}{rcl} \mathscr{V}_F^*(y,\mathbf{x},t) &=& \left\{1/\left[6c\,\pi(t)\right]\right\}V\left(K_{0,2}\right)\sigma_F^2(y,\mathbf{x},t), \\[4pt] \sigma_F^2(y,\mathbf{x},t) &=& \left[\rho^{(0)}(y,t)+4\rho^{(1*)}(y,\mathbf{x},t)+4\rho^{(2*)}(y,\mathbf{x},t)\right] \\[4pt] && \times\left\{\phi'\left[h(y,t)+\mathbf{x}^T\beta(t)\right]\right\}^2. \end{array}\right. \tag{14.102}$$

(c) *It follows from (a) and (b) above that, for sufficiently large n,*

$$MSE\left[\widehat{F}_t^L(y,\mathbf{x})\big|\mathscr{D}\right] \tag{14.103}$$
$$= \left\{b^4\left[\mathscr{B}_F^*(y,\mathbf{x},t)\right]^2+(nJb)^{-1}\,\mathscr{V}_F^*(y,\mathbf{x},t)\right\}\left[1+o_p(1)\right],$$

where the bias and variance components $\mathscr{B}_F^*(y,\mathbf{x},t)$ and $\mathscr{V}_F^*(y,\mathbf{x},t)$ have the expressions in (14.100) and (14.102), respectively. ∎

Similar to the comments following Theorem 14.1 and Corollary 14.1, we can derive the same theoretical conclusions based on the asymptotic expressions of the mean squared errors in (14.98) and (14.103). Setting the derivative of the dominating part of (14.98) with respect to t to zero, the asymptotically optimal bandwidth $b_{opt,F^{(q)}}$ is

$$b_{opt,F^{(q)}} = \left[\frac{(2q+1)\,\mathscr{V}_F(y,t)}{2nJ(Q-q+1)\,\mathscr{B}_F^2(y,t)}\right]^{1/(2Q+3)}. \tag{14.104}$$

Substituting b of (14.98) with the above $b_{opt,F^{(q)}}$, the asymptotically optimal mean squared error of $\widehat{F}^{(q)}(y,t)$ under the assumptions of Theorem 14.2 is

$$MSE_{opt}\left[\widehat{F}^{(q)}(y,t)\Big|\mathscr{D}\right]$$

$$= (nJ)^{-\frac{2(Q-q+1)}{2Q+3}}\left[\mathscr{B}_F(y,t)\right]^{\frac{2(2q+1)}{2Q+3}}\left[\mathscr{V}_F(y,t)\right]^{\frac{2(Q-q+1)}{2Q+3}}$$

$$\times\left\{\left[\frac{2q+1}{2(Q-q+1)}\right]^{\frac{2(Q-q+1)}{2Q+3}}+\left[\frac{2q+1}{2(Q-q+1)}\right]^{-\frac{q+1}{2Q+3}}\right\}$$

$$\times\left[1+o_p(1)\right], \tag{14.105}$$

which has the optimal convergence rate $(nJ)^{[-2(Q-q+1)]/(2Q+3)}$ under the assumptions of Theorem 14.2. Since (14.104) depends on the unknown quantities $\mathscr{B}_F(y,t)$ and $\mathscr{V}_F(y,t)$, $b_{opt,F^{(q)}}$, (14.105) only gives some theoretical insights into the appropriateness of $\widehat{F}^{(q)}(y,t)$. Under the assumptions of Corollary 14.2, the asymptotically optimal bandwidth and mean squared errors of the local linear estimator $\widehat{F}^L(y,t)$ are given by (14.104) and (14.105), respectively, with $Q=1$ and $q=0$, and the optimal convergence rate for the mean squared error of $\widehat{F}^L(y,t)$ is $(nJ)^{-4/5}$.

3. Justifications of Smoothing Step

Justifications of the smoothing step can be seen from the asymptotic variances and MSEs between the raw estimators $\widetilde{h}(y,t_j)$ and $\widetilde{F}_{t_j}(y|\mathbf{x})$ and the smoothing estimators $\widehat{h}(y,t)$ and $\widehat{F}_t(y|\mathbf{x})$. We illustrate this justification here using a simple special case. Other scenarios can be similarly evaluated.

Suppose that $t=t_j$ for some $1\le j\le J$. Lemmas 14.1 and 14.2 imply that $\widetilde{h}(y,t_j)$ and $\widetilde{F}_{t_j}(y|\mathbf{x})$ are asymptotically unbiased and the asymptotic variances of $\widetilde{h}(y,t_j)$ and $\widetilde{F}_{t_j}(y|\mathbf{x})$ are $O(n^{-1})$. Suppose that $q=0$, $Q=1$, and $\widehat{h}(y,t_j)$ and $\widehat{F}_{t_j}(y|\mathbf{x})$ are the local linear estimators. Theorems 14.1 and 14.2 imply that the asymptotic variances of $\widehat{h}(y,t_j)$ and $\widehat{F}_{t_j}(y|\mathbf{x})$ are $O[(nJb)^{-1}]$, which, by Assumption (a), is of smaller order than $O(n^{-1})$. The asymptotic unbiasedness of $\widetilde{h}(y,t_j)$ and $\widetilde{F}_{t_j}(y|\mathbf{x})$ implies that the asymptotic MSEs of $\widetilde{h}(y,t_j)$ and $\widetilde{F}_{t_j}(y|\mathbf{x})$ are also $O(n^{-1})$. When the asymptotic bias terms are included, the asymptotic MSEs of $\widehat{h}(y,t_j)$ and $\widehat{F}_{t_j}(y|\mathbf{x})$ are $O[b^4+(nJb)^{-1}]$, whose minimum is attained at $O[(nJ)^{-4/5}]$ by taking $b=O[(nJ)^{-1/5}]$.

Comparing $O[(nJ)^{-4/5}]$ with $O(n^{-1})$, we observe that the asymptotic MSEs of $\widehat{h}(y,t_j)$ and $\widehat{F}_{t_j}(y|\mathbf{x})$ converge to zero in a faster rate than that of $\widetilde{h}(y,t_j)$ and $\widetilde{F}_{t_j}(y|\mathbf{x})$ if $\lim_{n\to\infty}Jn^{-1/4}=\infty$. This suggests that the smoothing step reduces the MSEs if J is large relative to $n^{1/4}$. However, the smoothing step may not lead to better estimators if J is small relative to $n^{1/4}$.

14.5.4   Theoretical Derivations

We provide in this section the proofs of Lemmas 14.1 and 14.2, and Theorems 14.1 and 14.2.

1. Proofs for Raw Estimators

Proof of Lemma 14.1:

Following (14.32), (14.64), (14.65) and the law of large numbers, we first observe the approximation

$$E\{V_1\left[h(y, t_j)\right]|\mathscr{D}\}\left[\widetilde{h}(y, t_j) - h(y, t_j)\right]\left[1 + o_p(1)\right]$$
$$= -V\left[h(y, t_j)\right] - E\left[V_2(y, t_j)|\mathscr{D}\right]\left[\widetilde{\beta}(t_j) - \beta(t_j)\right]. \qquad (14.106)$$

If $\Lambda(t_j)$ is nonsingular, we also know from Lemma 12.1 and (13.14) that

$$A(t_j) = \left[I + o_p(I)\right]\left[\widetilde{\beta}(t_j) - \beta(t_j)\right]$$
$$= \left[n_j(n_j - 1)\right]^{-1}\Lambda^{-1}(t_j)\sum_{i_1 \neq i_2 \in \mathscr{S}_j} U_{i_1 i_2}\left[\beta(t_j)\right]. \qquad (14.107)$$

Since $E\{V_1\left[h(y, t_j)\right]|\mathscr{D}\}$ is assumed to be nonsingular, we have, by (14.32), (14.106), (14.107) and the definitions of $v_i^*(y, t_j)$, $u_{i_1 i_2}^*(y, t_j)$ and $W(y, t_j)$ in (14.67), that

$$\left[\widetilde{h}(y, t_j) - h(y, t_j)\right]\left[1 + o_p(1)\right] = W(y, t_j)$$
$$+ o_p\left\{\left\|\sum_{p=1}^{P}\left[\widetilde{\beta}_p(t_j) - \beta_p(t_j)\right]\right\|\right\},$$

which gives the conclusion (14.70) of the lemma.
    Next, using the definitions given in (14.63) and (14.67), we observe that

$$\begin{cases} E\left[v_i(y, t_j)|\mathscr{D}\right] = E\{U_{i_1 i_2}\left[\beta(t_j)\right]|\mathscr{D}\} = 0, \\ E\left[v_{i_1}^*(y, t_j)|\mathscr{D}\right] = E\{u_{i_1 i_2}^*\left[\beta(t_j)\right]|\mathscr{D}\} = 0. \end{cases} \qquad (14.108)$$

Substituting (14.108) into the definition of $W(y, t_j)$ given in (14.67), we observe that

$$E\left[W(y, t_j)|\mathscr{D}\right]$$
$$= \frac{1}{n_j(n_j - 1)}\sum_{i_1 \neq i_2 \in \mathscr{S}_j}\left\{E\left[v_{i_1}^*(y, t_j)|\mathscr{D}\right] + E\left[u_{i_1 i_2}^*(\beta(t_j))|\mathscr{D}\right]\right\}$$
$$= 0.$$

which gives the assertion (14.71) of the lemma.
    The next task is to show Lemma 14.1(c). To do this, we first observe

from (14.67) and (14.108) that the covariance of $W(y, t_{j_1})$ and $W(y, t_{j_2})$ has the following expression

$$
\begin{aligned}
&Cov\left[W(y, t_{j_1}), W(y, t_{j_2})\big|\mathscr{D}\right] \\
&= \left[\frac{1}{n_{j_1}(n_{j_1} - 1)}\right]\left[\frac{1}{n_{j_2}(n_{j_2} - 1)}\right] \\
&\quad \times \sum_{i_1 \neq i_2 \in \mathscr{S}_{j_1}} \sum_{i_3 \neq i_4 \in \mathscr{S}_{j_2}} E\left[w^*_{i_1 i_2}(y, t_{j_1}) w^*_{i_3 i_4}(y, t_{j_2})\big|\mathscr{D}\right],
\end{aligned}
\tag{14.109}
$$

where

$$
w^*_{i_1 i_2}(y, t_j) = v^*_{i_1}(y, t_j) + u^*_{i_1 i_2}(y, t_j).
\tag{14.110}
$$

Writing out the expansion of (14.109) from (14.110), we need to evaluate the summation of

$$
\begin{aligned}
&E\left[w^*_{i_1 i_2}(y, t_{j_1}) w^*_{i_3 i_4}(y, t_{j_2})\big|\mathscr{D}\right] \\
&= E\left[v^*_{i_1}(y, t_{j_1}) v^*_{i_3}(y, t_{j_2})\big|\mathscr{D}\right] + E\left[v^*_{i_1}(y, t_{j_1}) u^*_{i_3 i_4}(y, t_{j_2})\big|\mathscr{D}\right] \\
&\quad + E\left[v^*_{i_3}(y, t_{j_2}) u^*_{i_1 i_2}(y, t_{j_1})\big|\mathscr{D}\right] + E\left[u^*_{i_1 i_2}(y, t_{j_1}) u^*_{i_3 i_4}(y, t_{j_2})\big|\mathscr{D}\right]
\end{aligned}
$$

over $\{(i_1, i_2) : i_1 \neq i_2\}$ and $\{(i_3, i_4) : i_3 \neq i_4\}$.

Since, for any $\{i_1, i_2, i_3, i_4\}$, we can derive from (14.67) that

$$
\begin{cases}
E\left[v^*_{i_1}(y, t_{j_1}) v^*_{i_3}(y, t_{j_2})\big|\mathscr{D}\right] = 0, & \text{if } i_1 \neq i_3, \\
E\left[v^*_{i_1}(y, t_{j_1}) u^*_{i_3 i_4}(y, t_{j_2})\big|\mathscr{D}\right] = 0, & \text{if } i_1 \neq i_3 \text{ and } i_1 \neq i_4, \\
E\left[u^*_{i_1 i_2}(y, t_{j_1}) u^*_{i_3 i_4}(y, t_{j_2})\big|\mathscr{D}\right] = 0, & \text{if } i_1 \neq i_2 \neq i_3 \neq i_4,
\end{cases}
\tag{14.111}
$$

and $u^*_{i_1 i_2}(y, t_{j_1})$ is symmetric in (i_1, i_2), it suffices to evaluate the non-zero terms defined in (14.109). To do this, we consider the different situations where the two pairs of integers $\{(i_1, i_2) : i_1 \neq i_2\}$ and $\{(i_3, i_4) : i_3 \neq i_4\}$ contain different numbers of integers in common.

Let $C = 0, 1, 2$ be the numbers of integers in common in $\{(i_1, i_2) : i_1 \neq i_2\}$ and $\{(i_3, i_4) : i_3 \neq i_4\}$. If $C = 0$, then $\{(i_1, i_2) : i_1 \neq i_2\}$ and $\{(i_3, i_4) : i_3 \neq i_4\}$ do not contain any common integers, i.e., $i_1 \neq i_2 \neq i_3 \neq i_4$, and it easily follows from (14.110) that

$$
\begin{aligned}
E\left[w^*_{i_1 i_2}(y, t_{j_1}) w^*_{i_3 i_4}(y, t_{j_2})\big|\mathscr{D}\right] &= E\left[w^*_{i_1 i_2}(y, t_{j_1})\big|\mathscr{D}\right] \times E\left[w^*_{i_3 i_4}(y, t_{j_2})\big|\mathscr{D}\right] \\
&= 0.
\end{aligned}
\tag{14.112}
$$

If $C = 1$, $\mathscr{C}_1 = \{(i_1, i_2, i_3, i_4) : i_1 \neq i_2; i_3 \neq i_4; C = 1\}$ is the union of the disjoint subsets

$$
\begin{cases}
\mathscr{C}_{1a} = \{(i_1, i_2, i_3, i_4) : i_1 = i_3; i_1 \neq i_2 \neq i_4\}, \\
\mathscr{C}_{1b} = \{(i_1, i_2, i_3, i_4) : i_1 = i_4; i_1 \neq i_2 \neq i_3\}, \\
\mathscr{C}_{1c} = \{(i_1, i_2, i_3, i_4) : i_2 = i_3; i_1 \neq i_2 \neq i_4\}, \\
\mathscr{C}_{1d} = \{(i_1, i_2, i_3, i_4) : i_2 = i_4; i_1 \neq i_2 \neq i_3\}.
\end{cases}
\tag{14.113}
$$

Then, using the definitions given in (14.67) and (14.68), we can directly compute from (14.110) and (14.113) that

$$E\left[w^*_{i_1 i_2}(y, t_{j_1})\, w^*_{i_3 i_4}(y, t_{j_2})\,\big|\,\mathscr{D}\right] \tag{14.114}$$

$$= \begin{cases}
\begin{aligned}
&\rho^{(0)}(y, t_{j_1}, t_{j_2}) + \rho^{(1)}(y, t_{j_1}, t_{j_2}) \\
&\quad + \rho^{(1)}(y, t_{j_2}, t_{j_1}) + \rho^{(2)}(y, t_{j_1}, t_{j_2}),
\end{aligned} & \text{if } (i_1, i_2, i_3, i_4) \in \mathscr{C}_{1a}, \\[2mm]
\rho^{(1)}(y, t_{j_1}, t_{j_2}) + \rho^{(2)}(y, t_{j_1}, t_{j_2}), & \text{if } (i_1, i_2, i_3, i_4) \in \mathscr{C}_{1b}, \\[2mm]
\rho^{(1)}(y, t_{j_2}, t_{j_1}) + \rho^{(2)}(y, t_{j_1}, t_{j_2}), & \text{if } (i_1, i_2, i_3, i_4) \in \mathscr{C}_{1c}, \\[2mm]
\rho^{(2)}(y, t_{j_1}, t_{j_2}), & \text{if } (i_1, i_2, i_3, i_4) \in \mathscr{C}_{1d}.
\end{cases}$$

Next we need to compute the numbers of terms in $\mathscr{C}_{1a}$ through $\mathscr{C}_{1d}$. We first recognize that $\mathscr{C}_{1a}$ is the union of four disjoint subsets, i.e.,

$$\mathscr{C}_{1a} = \mathscr{C}_{1a1} \cup \mathscr{C}_{1a2} \cup \mathscr{C}_{1a3} \cup \mathscr{C}_{1a4}, \tag{14.115}$$

where

$$\begin{aligned}
\mathscr{C}_{1a1} &= \big\{(i_1, i_2, i_3, i_4) \in \mathscr{C}_{1a} : i_1 = i_3 \in \mathscr{S}_{j_1} \cap \mathscr{S}_{j_2}, \\
&\qquad\qquad i_2 \in \mathscr{S}_{j_1} \backslash \mathscr{S}_{j_2},\, i_4 \in \mathscr{S}_{j_2} \backslash \mathscr{S}_{j_1}\big\}, \\
\mathscr{C}_{1a2} &= \big\{(i_1, i_2, i_3, i_4) \in \mathscr{C}_{1a} : i_1 = i_3 \in \mathscr{S}_{j_1} \cap \mathscr{S}_{j_2}, \\
&\qquad\qquad i_2 \in \mathscr{S}_{j_1} \cap \mathscr{S}_{j_2},\, i_4 \in \mathscr{S}_{j_2} \backslash \mathscr{S}_{j_1}\big\}, \\
\mathscr{C}_{1a3} &= \big\{(i_1, i_2, i_3, i_4) \in \mathscr{C}_{1a} : i_1 = i_3 \in \mathscr{S}_{j_1} \cap \mathscr{S}_{j_2}, \\
&\qquad\qquad i_2 \in \mathscr{S}_{j_1} \backslash \mathscr{S}_{j_2},\, i_4 \in \mathscr{S}_{j_2} \cap \mathscr{S}_{j_1}\big\}, \\
\mathscr{C}_{1a4} &= \big\{(i_1, i_2, i_3, i_4) \in \mathscr{C}_{1a} : i_1 = i_3 \in \mathscr{S}_{j_1} \cap \mathscr{S}_{j_2}, \\
&\qquad\qquad i_2 \in \mathscr{S}_{j_1} \cap \mathscr{S}_{j_2},\, i_4 \in \mathscr{S}_{j_1} \cap \mathscr{S}_{j_2}\big\},
\end{aligned}$$

where $\mathscr{S}_{j_1} \backslash \mathscr{S}_{j_2} = \mathscr{S}_{j_1} \cap \mathscr{S}^c_{j_2}$ denotes the set of subjects in $\mathscr{S}_{j_1}$ but not in $\mathscr{S}_{j_2}$. The numbers of terms in $\mathscr{C}_{1ak}$, $k = 1, 2, 3, 4$, are

$$\begin{cases}
m_{1a1} &= n_{j_1 j_2}(n_{j_1} - n_{j_1 j_2})(n_{j_2} - n_{j_1 j_2}), \\
m_{1a2} &= n_{j_1 j_2}(n_{j_1 j_2} - 1)(n_{j_2} - n_{j_1 j_2})/2, \\
m_{1a3} &= n_{j_1 j_2}(n_{j_1 j_2} - 1)(n_{j_1} - n_{j_1 j_2})/2, \\
m_{1a4} &= n_{j_1 j_2}(n_{j_1 j_2} - 1)(n_{j_1 j_2} - 2)/6.
\end{cases} \tag{14.116}$$

Collecting all the terms from above, we obtain the number of terms in $\mathscr{C}_{1a}$, such that

$$\begin{aligned}
m_{1a} &= m_{1a1} + m_{1a2} + m_{1a3} + m_{1a4} \tag{14.117} \\
&= n_{j_1 j_2}\left[\left(n_{j_2} - n_{j_1 j_2}\right)\left(n_{j_1} - \frac{n_{j_1 j_2}}{2} - \frac{1}{2}\right) + (n_{j_1 j_2} - 1)\left(\frac{n_{j_1}}{2} - \frac{n_{j_1 j_2}}{3} - \frac{1}{3}\right)\right].
\end{aligned}$$

Applying the same approach to $\mathscr{C}_{1b}$, $\mathscr{C}_{1c}$ and $\mathscr{C}_{1d}$, we can divide $\mathscr{C}_{1b}$, $\mathscr{C}_{1c}$

and $\mathscr{C}_{1d}$ into similar disjoint subsets as for $\mathscr{C}_{1a}$ in (14.115), and compute the corresponding numbers of terms m_{1b}, m_{1c} and m_{1d} using the calculations in (14.116) and (14.117). We omit these calculations here because they follow the same approach as for m_{1a}. These calculations show that the numbers of terms m_{1b}, m_{1c} and m_{1d} for $\mathscr{C}_{1b}$, $\mathscr{C}_{1c}$ and $\mathscr{C}_{1d}$, respectively, are also given by (14.117). The summation of (14.114) for terms in $\mathscr{C}_1$ is then given by

$$\sum_{(i_1,i_2,i_3,i_4)\in\mathscr{C}_1} E\left[w^*_{i_1 i_2}(y, t_{j_1})\, w^*_{i_3 i_4}(y, t_{j_2})\middle|\mathscr{D}\right] \tag{14.118}$$

$$= m_{1a}\left[\rho^{(0)}(y, t_{j_1}, t_{j_2}) + 2\rho^{(1)}(y, t_{j_1}, t_{j_2}) + 2\rho^{(1)}(y, t_{j_2}, t_{j_1}) + 4\rho^{(2)}(y, t_{j_1}, t_{j_2})\right],$$

where m_{1a} is given by (14.117).

For $C = 2$, we recognize that $\mathscr{C}_2 = \{(i_1, i_2, i_3, i_4) : i_1 \neq i_2, i_3 \neq i_4, C = 2\}$ is the union of two disjoint subsets, i.e.,

$$\mathscr{C}_2 = \mathscr{C}_{2a} \cup \mathscr{C}_{2b}, \tag{14.119}$$

where

$$\begin{aligned}
\mathscr{C}_{2a} &= \big\{(i_1, i_2, i_3, i_4) : i_1 \neq i_2, i_3 \neq i_4, \\
&\qquad i_1 = i_3 \in \mathscr{S}_{j_1} \cap \mathscr{S}_{j_2}, i_2 = i_4 \in \mathscr{S}_{j_1} \cap \mathscr{S}_{j_2}\big\}, \\
\mathscr{C}_{2b} &= \big\{(i_1, i_2, i_3, i_4) : i_1 \neq i_2, i_3 \neq i_4, \\
&\qquad i_1 = i_4 \in \mathscr{S}_{j_1} \cap \mathscr{S}_{j_2}, i_2 = i_3 \in \mathscr{S}_{j_1} \cap \mathscr{S}_{j_2}\big\}.
\end{aligned}$$

Direct calculation shows that the numbers of terms in $\mathscr{C}_{2a}$ and $\mathscr{C}_{2b}$ are

$$m_{2a} = m_{2b} = n_{j_1 j_2}\left(n_{j_1 j_2} - 1\right)/2. \tag{14.120}$$

Furthermore, using (14.68), (14.110) and the definitions of $\mathscr{C}_{2a}$ and $\mathscr{C}_{2b}$, we can compute that

$$E\left[w^*_{i_1 i_2}(y, t_{j_1})\, w^*_{i_3 i_4}(y, t_{j_2})\middle|\mathscr{D}\right] \tag{14.121}$$

$$= \begin{cases}
\begin{aligned}
&\rho^{(0)}(y, t_{j_1}, t_{j_2}) + \rho^{(1)}(y, t_{j_1}, t_{j_2}) \\
&\quad + \rho^{(1)}(y, t_{j_2}, t_{j_1}) + \rho^{(3)}(y, t_{j_1}, t_{j_2}),
\end{aligned} & \text{if } (i_1, i_2, i_3, i_4) \in \mathscr{C}_{2a}, \\[2ex]
\begin{aligned}
&\rho^{(1)}(y, t_{j_1}, t_{j_2}) + \rho^{(1)}(y, t_{j_2}, t_{j_1}) \\
&\quad + \rho^{(3)}(y, t_{j_1}, t_{j_2}),
\end{aligned} & \text{if } (i_1, i_2, i_3, i_4) \in \mathscr{C}_{2b},
\end{cases}$$

where

$$\rho^{(3)}(y, t_{j_1}, t_{j_2}) = E\left[u^*_{i_1 i_2}(y, t_{j_1})\, u^*_{i_1 i_2}(y, t_{j_2})\middle|\mathscr{D}\right] \text{ with } i_2 \in \mathscr{S}_{j_1} \cap \mathscr{S}_{j_2}. \tag{14.122}$$

Combining (14.120), (14.121) and (14.122), the summation of (14.121) terms in $\mathscr{C}_2$ is

$$\sum_{(i_1,i_2,i_3,i_4)\in\mathscr{C}_2} E\left[w^*_{i_1 i_2}(y, t_{j_1})\, w^*_{i_3 i_4}(y, t_{j_2})\middle|\mathscr{D}\right]$$

$$= \frac{n_{j_1 j_2}\left(n_{j_1 j_2} - 1\right)}{2}\Big\{\rho^{(0)}(y, t_{j_1}, t_{j_2}) + 2\rho^{(1)}(y, t_{j_1}, t_{j_2})$$

$$+ 2\rho^{(1)}(y, t_{j_2}, t_{j_1}) + 2\rho^{(3)}(y, t_{j_1}, t_{j_2})\Big\}. \tag{14.123}$$

The summation of (14.118) and (14.123) then gives

$$\sum_{i_1 \neq i_2 \in \mathcal{S}_{j_1}} \sum_{i_3 \neq i_4 \in \mathcal{S}_{j_2}} E\left[w_{i_1 i_2}^*(y, t_{j_1}) w_{i_3 i_4}^*(y, t_{j_2}) \middle| \mathcal{D} \right]$$

$$= \left[m_{1a} + \frac{n_{j_1 j_2}(n_{j_1 j_2} - 1)}{2} \right] \left[\rho^{(0)}(y, t_{j_1}, t_{j_2}) \right. \tag{14.124}$$

$$\left. + 2\rho^{(1)}(y, t_{j_1}, t_{j_2}) + 2\rho^{(1)}(y, t_{j_2}, t_{j_1}) \right]$$

$$+ 4 m_{1a} \rho^{(2)}(y, t_{j_1}, t_{j_2}) + n_{j_1 j_2}(n_{j_1 j_2} - 1) \rho^{(3)}(y, t_{j_1}, t_{j_2}),$$

which, by (14.117) and ignoring the smaller order term $n_{j_1 j_2}(n_{j_1 j_2} - 1)$, gives (14.69). The conclusion in (14.73) directly follows from (14.69) through (14.72) by setting $j_1 = j_2 = j$. ∎

Proof of Lemma 14.2:

The proof is analogous to proof of Lemma 14.1, so we only outline the main steps. First, by (13.14) in Lemma 13.1 and (14.70) in Lemma 14.1, the dominating term for $\widetilde{h}(y, t_j) - h(y, t_j)$ is $W(y, t_j)$, and the dominating term for $\widetilde{\beta}(t_j) - \beta(t_j)$ is $A(t_j)$. Substituting $\widetilde{h}(y, t_j) - h(y, t_j)$ and $\widetilde{\beta}(t_j) - \beta(t_j)$ with $W(y, t_j)$ and $A(t_j)$, respectively, we can directly verify from (14.74) and (14.75) that

$$\mathscr{A}(y, \mathbf{x}, t_j) = \phi'\left[h(y, t_j) + \mathbf{x}^T(t_j) \beta(t_j) \right] \left[W(y, t_j) + \mathbf{x}^T(t_j) A(t_j) \right], \tag{14.125}$$

which implies (14.76). By $E\left[W(y, t_j) \middle| \mathcal{D} \right] = E\left[A(t_j) \middle| \mathcal{D} \right] = 0$, it then follows from (14.125) that

$$E\left[\mathscr{A}(y, \mathbf{x}, t_j) \middle| \mathcal{D} \right] = 0.$$

To compute the covariance, we use the same approach as (14.109) and consider

$$Cov\left[\mathscr{A}(y, \mathbf{x}, t_{j_1}), \mathscr{A}(y, \mathbf{x}, t_{j_2}) \middle| \mathcal{D} \right]$$

$$= \left[\frac{1}{n_{j_1}(n_{j_1} - 1)} \right] \left[\frac{1}{n_{j_2}(n_{j_2} - 1)} \right] \tag{14.126}$$

$$\times \sum_{i_1 \neq i_2 \in \mathcal{S}_{j_1}} \sum_{i_3 \neq i_4 \in \mathcal{S}_{j_2}} E\left[\mathscr{A}_{i_1 i_2}(y, \mathbf{x}, t_{j_1}) \mathscr{A}_{i_3 i_4}(y, \mathbf{x}, t_{j_2}) \middle| \mathcal{D} \right],$$

where, with $i_1 \neq i_2$ and $i_3 \neq i_4$,

$$E\left[\mathscr{A}_{i_1 i_2}(y, \mathbf{x}, t_{j_1}) \mathscr{A}_{i_3 i_4}(y, \mathbf{x}, t_{j_2}) \middle| \mathcal{D} \right]$$

$$= \phi'\left[h(y, t_{j_1}) + \mathbf{x}^T(t_{j_1}) \beta(t_{j_1}) \right] \phi'\left[h(y, t_{j_2}) + \mathbf{x}^T(t_{j_2}) \beta(t_{j_2}) \right]$$

$$\times \left\{ E\left[v_{i_1}^*(y, t_{j_1}) v_{i_3}^*(y, t_{j_2}) \middle| \mathcal{D} \right] + E\left[v_{i_1}^*(y, t_{j_1}) u_{i_3 i_4}^*(y, t_{j_2}) \middle| \mathcal{D} \right] \right.$$

$$\left. + E\left[v_{i_3}^*(y, t_{j_2}) u_{i_1 i_2}^*(y, t_{j_1}) \middle| \mathcal{D} \right] + E\left[u_{i_1 i_2}^*(y, t_{j_1}) u_{i_3 i_4}^*(y, t_{j_2}) \middle| \mathcal{D} \right] \right\}.$$

Using the same derivations as in (14.111) through (14.123) with $w^*_{i_1 i_2}(\cdot)$ and $w^*_{i_3 i_4}(\cdot)$ replaced by $\mathscr{A}_{i_1 i_2}(\cdot)$ and $\mathscr{A}_{i_3 i_4}(\cdot)$, respectively, we get

$$\sum_{i_1 \neq i_2 \in \mathscr{S}_{j_1}} \sum_{i_3 \neq i_4 \in \mathscr{S}_{j_2}} E\left[\mathscr{A}_{i_1 i_2}(y, \mathbf{x}, t_{j_1}) \mathscr{A}_{i_3 i_4}(y, \mathbf{x}, t_{j_2}) \middle| \mathscr{D}\right]$$

$$= \phi'\left[h(y, t_{j_1}) + \mathbf{x}^T(t_{j_1}) \beta(t_{j_1})\right] \phi'\left[h(y, t_{j_2}) + \mathbf{x}^T(t_{j_2}) \beta(t_{j_2})\right]$$

$$\times \left\{\left[m_{1a} + \frac{n_{j_1 j_2}(n_{j_1 j_2} - 1)}{2}\right]\left[\rho^{(0)}(y, t_{j_1}, t_{j_2})\right.\right.$$

$$\left.+ 2\rho^{(1*)}(y, \mathbf{x}, t_{j_1}, t_{j_2}) + 2\rho^{(1*)}(y, \mathbf{x}, t_{j_2}, t_{j_1})\right]$$

$$\left.+ 4 m_{1a} \rho^{(2*)}(y, \mathbf{x}, t_{j_1}, t_{j_2}) + n_{j_1 j_2}(n_{j_1 j_2} - 1) \rho^{(3*)}(y, \mathbf{x}, t_{j_1}, t_{j_2})\right\},$$

where, for $i_2 \in \mathscr{S}_{j_1} \cap \mathscr{S}_{j_2}$,

$$\rho^{(3*)}(y, \mathbf{x}, t_{j_1}, t_{j_2}) = E\left[u^*_{i_1 i_2}(y, \mathbf{x}, t_{j_1}) u^*_{i_1 i_2}(y, \mathbf{x}, t_{j_2}) \middle| \mathscr{D}\right].$$

By (14.117), (14.126) and ignoring the smaller order term $n_{j_1 j_2}(n_{j_1 j_2} - 1)$, the above summation gives (14.78). The conclusion of (14.79) is a direct consequence of (14.78) with $j_1 = j_2 = j$.   ∎

2. Proofs for Smoothing Estimators

Proof of Theorem 14.1:

We first note that, under Assumption (a), the weight function $w_{q,Q+1}(t_j, t; b)$ in (14.38) satisfies equations (13.68) through (13.71). Since the bias of $\widehat{h}^{(q)}(y, t)$ has the decomposition

$$\begin{cases} Bias\left[\widehat{h}^{(q)}(y, t) \middle| \mathscr{D}\right] = E\left[\widehat{h}^{(q)}(y, t) \middle| \mathscr{D}\right] - h^{(q)}(y, t) = \mathscr{W}_1 + \mathscr{W}_2, \\ \mathscr{W}_1 = \sum_{j=1}^{J} w_{q,Q+1}(t_j, t; b)\left\{E\left[\widetilde{h}(y, t_j) \middle| \mathscr{D}\right] - h(y, t_j)\right\} \\ \mathscr{W}_2 = \sum_{j=1}^{J} w_{q,Q+1}(t_j, t; b) h(y, t_j) - h^{(q)}(y, t), \end{cases} \quad (14.127)$$

it suffices to evaluate the asymptotic expressions of $\mathscr{W}_1$ and $\mathscr{W}_2$. By Assumption (a), equation (13.69) and Lemma 13.1, we observe that

$$\mathscr{W}_1 = \sum_{j=1}^{J} \frac{q!}{J b^{q+1} \pi(t)} K_{q,Q+1}\left(\frac{t_j - t}{b}\right)[1 + o_p(1)] O_p(n^{-1/2})$$

$$= O_p(n^{-1/2}), \quad (14.128)$$

where the second equality sign holds because

$$\sum_{j=1}^{J} \left| q! \left[J b^{q+1} \pi(t)\right]^{-1} K_{q,Q+1}\left[(t_j - t)/b\right] \right|$$

is bounded. By Assumption (c), equations (13.69) and (13.70), and the Taylor expansions for $h(y, t_j)$, we have

$$
\mathscr{W}_2 = \sum_{j=1}^{J} w_{q,Q+1}(t_j, t; b) \left\{ \sum_{k=0}^{Q+1} \left[h^{(k)}(y, t) \frac{(t_j - t)^k}{k!} \right] + o_p \left[(t_j - t)^{Q+1} \right] \right\}
$$
$$
- h^{(q)}(y, t)
$$
$$
= \frac{(q!) b^{Q-q+1}}{(Q+1)!} h^{(Q+1)}(y, t) B_{Q+1}(K_{q,Q+1}) \left[1 + o_p(1) \right]. \tag{14.129}
$$

It follows from Assumption (a) and (14.128) that (14.129) is the dominating term, so that the conclusion of (14.82) holds.

To compute the variance of $\widehat{h}^{(q)}(y, t)$, we observe that, by the definition of $\widehat{h}^{(q)}(y, t)$ in (14.38),

$$
Var\left[\widehat{h}^{(q)}(y, t) \big| \mathscr{D} \right] = E \left\{ \left[\sum_{j=1}^{J} w_{q,Q+1}(t_j, t; b) \left[\widetilde{h}(y, t_j) - h^*(y, t_j) \right] \right]^2 \big| \mathscr{D} \right\}, \tag{14.130}
$$

where $h^*(y, t_j) = E\left[\widetilde{h}(y, t_j) | \mathscr{D} \right]$.

By Assumption (a), equations (13.68) and (13.71), Lemma 13.1, $c_j = c$ and $\lim_{t_j \to t} \sigma^2(y, t_j) = \sigma^2(y, t)$, the right side of (14.130) is the sum of $\mathscr{W}_3$ and $\mathscr{W}_4$, where

$$
\mathscr{W}_3 = \sum_{j=1}^{J} w_{q,Q+1}^2(t_j, t; b) E\left\{ \left[\widetilde{h}(y, t_j) - h^*(y, t_j) \right]^2 \big| \mathscr{D} \right\}
$$
$$
= \frac{(q!)^2}{6 c n J b^{2q+1} \pi(t)} V(K_{q,Q+1}) \sigma^2(y, t) \left[1 + o_p(1) \right] \tag{14.131}
$$

$$
\mathscr{W}_4 = \sum_{j_1 \neq j_2} \left\{ w_{q,Q+1}(t_{j_1}, t; b) w_{q,Q+1}(t_{j_2}, t; b) \right.
$$
$$
\times E\left[\left[\widetilde{h}(y, t_{j_1}) - h^*(y, t_{j_1}) \right] \left[\widetilde{h}(y, t_{j_2}) - h^*(y, t_{j_2}) \right] \big| \mathscr{D} \right] \right\}
$$
$$
= \sum_{j_1 \neq j_2} \left\{ w_{q,Q+1}(t_{j_1}, t; b) w_{q,Q+1}(t_{j_2}, t; b) Cov\left[W(y, t_{j_1}), W(y, t_{j_2}) | \mathscr{D} \right] \right\}
$$
$$
\times \left[1 + o_p(1) \right]
$$
$$
= o_p\left[\left(n J b^{2q+1} \right)^{-1} \right]. \tag{14.132}
$$

It follows from (14.131) and (14.132) that $Var\left[\widehat{h}^{(q)}(y, t) | \mathscr{D} \right] = \mathscr{W}_3 \left[1 + o_p(1) \right]$, so that (14.84) holds. The conclusion of (14.86) is a simple consequence of (14.82) and (14.84). ∎

Proof of Theorem 14.2:

The derivations follow the same arguments in the proof of Theorem 14.1 by substituting $\widetilde{h}(y,t)$, $h(y,t)$ and $W(y,t_j)$ with $\widetilde{F}_t(y|\mathbf{x})$, $F_t(y|\mathbf{x})$ and $\mathscr{A}(y,\mathbf{x},t)$, respectively. Thus, the expressions of (14.94) and (14.96) are the same as (14.82) and (14.84) with $h^{(Q+1)}(y,t)$ and $Var\big[W(y,t)\big|\mathscr{D}\big]$ replaced by $F_t^{(Q+1)}(y|\mathbf{x})$ and $Var\big[\mathscr{A}(y,\mathbf{x},t)\big|\mathscr{D}\big]$. We omit the details to avoid repetition. ∎

14.6 Remarks and Literature Notes

The models and estimation methods presented in this chapter are mostly adapted from the results of Wu and Tian (2013b). These models and estimation methods provide a useful tool for evaluating the conditional distribution functions and rank-tracking abilities measured by the Rank-Tracking Probability (RTP) and the Rank-Tracking Probability Ratio (RTPR). Since the time-varying transformation models considered in this chapter do not incorporate any correlation or dependence structures between distribution functions at different time points, the estimation methods for the RTPs and the RTPRs depend on smoothing over bivariate time scales. These estimation methods are flexible but generally require a large sample size and reasonable numbers of repeated measurements. The application to the NGHS Blood Pressure data demonstrates that the RTPs and the RTPRs are useful quantitative tools for tracking the temporal trends of health outcomes in long-term longitudinal studies. Although our asymptotic results are limited to the conditional distribution functions, they provide insight into the accuracy of the smoothing estimators under typical longitudinal settings.

There are a number of theoretical and methodological issues that warrant further investigation. First, systematic investigations are warranted to derive the asymptotic properties of the smoothing RTP and RTPR estimators, and suitable models should be developed to describe the correlation structures across different time points. Second, theoretical and simulation studies are needed to investigate the properties of smoothing methods other than the two-step local polynomial estimators in Section 14.3. Third, we need to develop some goodness-of-fit tests and variable selection methods for evaluating the time-varying transformation models of Section 14.2.6. Finally, in studies with moderate sample sizes, reliable raw estimates may be difficult to obtain, so that, as a potentially useful alternative, it may be necessary to develop some one-step estimation methods without relying on the initial raw estimators of Section 14.3.1.

Chapter 15

Tracking with Mixed-Effects Models

We present in this chapter an alternative method for estimating the conditional distributions, the Rank-Tracking Probabilities (RTP) and the Rank-Tracking Probability Ratios (RTPR). This method is motivated by the recognition that the estimation method of Chapter 13 has two limitations in practice. First, when constructing the raw estimators at any two time points $s_1 < s_2$, we need a sufficiently large number of subjects with observations at time points around (s_1, s_2). Second, the smoothing step requires bivariate smoothing, which again requires a large sample. These limitations are caused by the fact that the time-varying transformation models (14.25) do not take into account the dependence structure between any two time points $s_1 < s_2$. To alleviate these drawbacks, the alternative approach here relies on estimating the conditional distributions using two simple steps: (a) predicting the subjects' trajectory curves from the nonparametric mixed-effects models of Chapter 11; (b) constructing the conditional distribution estimators based on the predicted outcome trajectories. The trajectory prediction step in (a) is crucial for the accuracy of the estimators in (b). Unlike the modeling approaches of Chapters 13 and 14, the trajectory prediction gives a natural link between the conditional-mean based mixed-effects models and the conditional-distribution based estimation methods. The estimation method of this chapter has the advantage of evaluating the conditional means, conditional distributions and tracking abilities of an outcome variable under a unified regression framework.

15.1 Data Structure and Models

15.1.1 Data Structure

The data structure is the same as in Chapter 11. In summary, we assume that our longitudinal sample contains n independent subjects and the ith subject has n_i number of visits at time points $\{t_{ij} \in \mathscr{T} : j = 1, \ldots, n_i\}$, where $\mathscr{T}$ is the time interval of the study. At any time $t \in \mathscr{T}$, $Y(t)$ is the real-valued outcome variable and $\mathbf{X}(t) = \left(X^{(1)}(t), \ldots, X^{(k)}(t)\right)^T$ is the R^k-valued covariate vector, which may include both time-invariant and time-dependent covariates. The ith subject's outcome and covariate curves over $t \in \mathscr{T}$ are $\left\{\left(Y_i(t), t, \mathbf{X}_i^T(t)\right)^T\right.$:

$t \in \mathscr{T}$}. The observed longitudinal sample is

$$\left\{ (Y_{ij}, t_{ij}, \mathbf{X}_{ij}^T) : i = 1, \ldots, n; \, j = 1, \ldots, n_i \right\},$$

where $Y_{ij} = Y_i(t_{ij})$ and $\mathbf{X}_{ij} = \mathbf{X}_i(t_{ij}) = \left(X_i^{(1)}(t_{ij}), \ldots, X_i^{(k)}(t_{ij}) \right)^T$.

15.1.2   The Nonparametric Mixed-Effects Models

Using the modeling framework of Chapter 11, we consider here that the stochastic processes $Y(t)$ and $\mathbf{X}(t)$ satisfy the mixed-effects varying-coefficient models (11.11) or (11.37), so that a series of global smoothing methods through basis approximations can be used to estimate the coefficient curves and predict the outcome trajectories. For the simple case of evaluating $\{(Y(t), t)^T : t \in \mathscr{T}\}$ without covariates, $Y_i(t)$ at time $t \in \mathscr{T}$ satisfies nonparametric mixed-effects model (11.1), which, with a slight change of notation for the clarity of presentation, we rewrite as follows

$$Y_i(t) = \mu(t) + \zeta_i(t) + \varepsilon_i(t), \tag{15.1}$$

where $\mu(t)$ is the population mean curve of $Y(t)$, $\zeta_i(t)$ is the ith subject's random departure from $\mu(t)$ which satisfies $E[\zeta_i(t)] = 0$, and $\varepsilon_i(t)$ is the mean zero measurement error satisfying

$$E[\varepsilon_i(t)] = 0, \; Var[\varepsilon_i(t)] = \sigma^2 \; \text{ and } \; Cov[\varepsilon_i(s), \varepsilon_i(t)] = 0 \; \text{ if } s \neq t.$$

When the covariate $\mathbf{X}(t)$ is incorporated, we also consider the following varying-coefficient mixed-effects model for $\{(Y_i(t), t, \mathbf{X}_i(t))^T : t \in \mathscr{T}\}$, which is a special case of the model (11.37),

$$\left\{ \begin{array}{rcl} Y_i(t) & = & \beta_{0i}(t) + \mathbf{X}_i^T(t)\,\beta_{1i}(t) + \varepsilon_i(t), \\ \beta_{0i}(t) & = & \beta_0(t) + \gamma_{0i}(t), \\ \beta_{1i}(t) & = & \beta_1(t) + \gamma_{1i}(t), \\ \beta_1(t) & = & \left(\beta_{11}(t), \ldots, \beta_{1k}(t) \right)^T, \\ \gamma_{1i}(t) & = & \left(\gamma_{1i1}(t), \ldots, \gamma_{1ik}(t) \right)^T, \\ \varepsilon_i(t) & \sim & N(0, \sigma^2) \; \text{ and } \; Cov[\varepsilon_i(s), \varepsilon_i(t)] = 0 \; \text{ if } s \neq t, \end{array} \right. \tag{15.2}$$

where $\{\beta_0(t), \beta_1(t)\}$ are smooth functions of t, $\{\gamma_{0i}(t), \gamma_{1i}(t)\}$ are the mean zero stochastic processes that represent the ith subject's random deviations from the population, $\{\varepsilon_i(t), \gamma_{0i}(t), \gamma_{1i}(t)\}$ are mutually independent for any given i and $\{\varepsilon_{i_1}(t), \gamma_{0i_1}(t), \gamma_{1i_1}(t)\}$ and $\{\varepsilon_{i_2}(t), \gamma_{0i_2}(t), \gamma_{1i_2}(t)\}$ are independent for all $i_1 \neq i_2$.

15.1.3   Conditional Distributions and Tracking Indices

The objective here is to estimate the conditional distributions, the RTPs and the RTPRs defined in Section 14.2 based on $\{Y(t), t, \mathbf{X}(t)\}$ satisfying the models (15.1) or (15.2). This means that, for a given set $A[\mathbf{X}(t), t]$ of the outcome values and a given set $B(s_1)$ of the covariate vectors, we are interested in estimating the conditional distribution function,

$$P_{s_1,s_2}(A, B) = P\{Y(s_2) \in A[\mathbf{X}(s_2), s_2] | \mathbf{X}(s_1) \in B(s_1)\}, \tag{15.3}$$

the rank-tracking probability,

$$\begin{aligned} RTP_{s_1,s_2}(A, B) &= P\{Y(s_2) \in A[\mathbf{X}(s_2), s_2] | \\ &\quad Y(s_1) \in A[\mathbf{X}(s_1), s_1], \mathbf{X}(s_1) \in B(s_1)\}, \end{aligned} \tag{15.4}$$

and the rank-tracking probability ratio

$$RTPR_{s_1,s_2}(A, B) = \frac{RTP_{s_1,s_2}(A, B)}{P_{s_1,s_2}(A, B)} \tag{15.5}$$

at any two time points $s_1 \leq s_2$. Interpretations of (15.3), (15.4) and (15.5) as measures of the tracking abilities of the outcome variable $Y(t)$ have already been discussed in Section 14.2. By assuming the varying-coefficient mixed-effects model (15.2) to the variables $\{Y(t), t, \mathbf{X}(t)\}$, we can estimate the effects of $\mathbf{X}(t)$ on the conditional means of $Y(t)$ as well as the tracking indices (15.3), (15.4) and (15.5) under one modeling framework. In contrast, the time-varying transformation models of Chapters 13 and 14, which are based on modeling the conditional distributions, are very different from the mixed-effects models (15.1) and (15.2), which are mainly used to evaluate the conditional means of the outcome variable.

Special cases of (15.4) and (15.5) may also be considered in some applications. When the covariates are not considered in (15.4) and (15.5), i.e., $B(s_1)$ is chosen to be the entire space of the covariates at time s_1, the statistical objective is to estimate the RTP,

$$RTP_{s_1,s_2}(A) = P\{Y(s_2) \in A[\mathbf{X}(s_2), s_2] | Y(s_1) \in A[\mathbf{X}(s_1), s_1]\}, \tag{15.6}$$

and the RTPR,

$$RTPR_{s_1,s_2}(A) = \frac{RTP_{s_1,s_2}(A)}{P_{s_2}(A)}, \tag{15.7}$$

where

$$P_{s_2}(A) = P\{Y(s_2) \in A[\mathbf{X}(s_2), s_2]\}. \tag{15.8}$$

Other special cases of (15.4) and (15.5) may be considered by choosing specific forms of $A(\cdot, \cdot)$, such as $A[\mathbf{X}(t), t] = A(t)$. Although the choices of $A(\cdot, \cdot)$ depend on the specific scientific objectives, it is common in biomedical studies to define the health status at time t based on the conditional quantiles of $Y(t)$ given $\mathbf{X}(t)$, so that a useful choice of $A[\mathbf{X}(t), t]$ is

$$A_\alpha[\mathbf{X}(t), t] = \{Y(t) : Y(t) > q_\alpha[t, \mathbf{X}(t)]\}, \tag{15.9}$$

where $q_\alpha[t, \mathbf{X}(t)]$ is the $(100 \times \alpha)$th quantile of $Y(t)$ given $\mathbf{X}(t)$.

15.2 Prediction and Estimation Methods

We present a class of smoothing methods based on B-spline approximations for the estimation and inferences of the statistical indices defined in (15.3), (15.4) and (15.5). Although other basis functions, such as Fourier bases and Wavelet bases, may also be considered, we focus on B-splines because of their good numerical properties and simplicity in practical implementation.

15.2.1 B-spline Prediction of Trajectories

We first present a B-spline approximation method to estimate the coefficient curves and predict the outcome trajectories based on the varying-coefficient mixed-effects model (15.2), which includes (15.1) as a special case.

1. B-spline Approximations, Estimation and Prediction

For simplicity, we illustrate the case of a single covariate $\mathbf{X}(t) = X(t)$ with $P = 1$. The case of multivariate covariates can be extended analogously. When B-splines are used for (15.2), we have the following approximations

$$\begin{cases} \beta_0(t) &\approx b_0^T(t)\,\xi_0, & \beta_1(t) &\approx b_1^T(t)\,\xi_1, \\ \gamma_{0i}(t) &\approx b_0^T(t)\,\eta_i, & \gamma_{1i}(t) &\approx b_1^T(t)\,\phi_i \end{cases} \tag{15.10}$$

based on the B-spline basis functions

$$b_0(t) = \big(b_{01}(t), \ldots, b_{0m}(t)\big)^T \quad \text{and} \quad b_1(t) = \big(b_{11}(t), \ldots, b_{1q}(t)\big)^T \tag{15.11}$$

for some integers $m, q > 0$, where

$$\xi_0 = \big(\xi_{01}, \ldots, \xi_{0m}\big)^T \quad \text{and} \quad \xi_1 = \big(\xi_{11}, \ldots, \xi_{1q}\big)^T \tag{15.12}$$

are the vectors of coefficients for the fixed-effects components,

$$\eta_i = \big(\eta_{i1}, \ldots, \eta_{im}\big)^T \quad \text{and} \quad \phi_i = \big(\phi_{i1}, \ldots, \phi_{ip}\big)^T \tag{15.13}$$

are the vectors of coefficients for the subject-specific normal random components with mean zero and covariance matrices

$$\Gamma = Cov(\eta_i) \quad \text{and} \quad \Phi = Cov(\phi_i). \tag{15.14}$$

If we denote the observed outcome, covariates and spline bases functions at time points $\{t_{i1}, \ldots, t_{in_i}\}$ by

$$\begin{cases} Y_i &= \big(Y_{i1}, \ldots, Y_{in_i}\big)^T, \ X_i &= \big(X_{i1}, \ldots, X_{in_i}\big)^T, \\ B_{0i} &= \big(b_0(t_1), \ldots, b_0(t_{n_i})\big)^T, \ B_{1i} &= \big(b_1(t_1), \ldots, b_1(t_{n_i})\big)^T, \\ \varepsilon_i &= \big(\varepsilon_{i1}, \ldots, \varepsilon_{in_i}\big)^T, \ \varepsilon_{ij} &= \varepsilon_i(t_{ij}), \\ E\big(\varepsilon_{ij}\big) &= 0, \ Cov(\varepsilon_i) &= \Sigma = \sigma^2 I_{n_i \times n_i} \end{cases} \tag{15.15}$$

where B_{0i} and B_{1i} are $n_i \times m$ and $n_i \times q$ matrices, respectively, and $I_{n_i \times n_i}$ is the $n_i \times n_i$ identity matrix, the B-spline approximation for (15.2) is

$$Y_i \approx B_{0i} \left(\xi_0 + \eta_i \right) + B_{1i} \left(\xi_1 + \phi_i \right) X_i + \varepsilon_i. \qquad (15.16)$$

When the covariate is not included, the B-spline approximation for (15.1) is

$$Y_i \approx B_{0i} \left(\xi_0 + \eta_i \right) + \varepsilon_i, \qquad (15.17)$$

which is a special case of (15.16). Since ε_i has a mean zero Gaussian distribution, we have shown in Chapter 11 that the parameters $\{\xi_0, \xi_1, \Sigma, \Gamma, \Phi\}$ of (15.2), (15.12), (15.14) and (15.15) can be estimated by the maximum likelihood estimators (MLE) or the restricted MLEs. We denote by $\{\widehat{\xi}_0, \widehat{\xi}_1, \widehat{\Sigma}, \widehat{\Gamma}, \widehat{\Phi}\}$ either the MLE or restricted MLE of $\{\xi_0, \xi_1, \Sigma, \Gamma, \Phi\}$ for (15.16), where $\widehat{\Sigma} = \widehat{\sigma}^2 I_{n_i \times n_i}$.

The best linear unbiased predictors (BLUPs) of the random effects $\widehat{\eta}_i$ and $\widehat{\phi}_i$ can be computed by the EM algorithm as in Chapter 11. By plugging in the coefficient estimates $\{\widehat{\xi}_0, \widehat{\xi}_1, \widehat{\eta}_i, \widehat{\phi}_i\}$ into (15.16), the B-spline predicted outcome trajectory curve, or BLUP trajectory, for the ith subject with covariate $X_i(t)$ at any time point t is

$$\widehat{Y}_i(t) = b_0^T(t) \left(\widehat{\xi}_0 + \widehat{\eta}_i \right) + b_1^T(t) \left(\widehat{\xi}_1 + \widehat{\phi}_i \right) X_i(t). \qquad (15.18)$$

We note that (15.18) is only useful when $X_i(t)$ is observed and measured without error. For example, $X_i(t) = X_i$ is a time-invariant categorical variable, such as race and gender. For situations where $X_i(t)$ is measured with error and needs to be predicted from the data, the predicted subject-specific outcome in (15.18) has to be modified by substituting the unknown $X_i(t)$ with its predicted value. For the special case (15.17), the predicted subject-specific outcome trajectory curve is

$$\widehat{Y}_i(t) = b_0^T(t) \left(\widehat{\xi}_0 + \widehat{\eta}_i \right). \qquad (15.19)$$

In this case, only ξ_0 and η_i need to be estimated.

2. Best Linear Unbiased Prediction for Covariate and Outcome

When $X_i(t)$ is not directly observed or measured with error, we have to predict $X_i(t)$ from the observations $\{X_{ij} : j = 1, \ldots, n_i\}$. To do this, we assume that the time-varying covariate $X_i(t)$ is measured with errors and satisfies the following nonparametric mixed-effects model, which has the same structure as (15.1),

$$X_i(t) = \mu_x(t) + \zeta_i(t) + u_i(t), \qquad (15.20)$$

where $\mu_x(t)$ is the population mean curve of $X(t)$, $\zeta_i(t)$ is the ith subject's random departure from $\mu_x(t)$ which satisfies $E\left[\zeta_i(t)\right] = 0$, and $u_i(t)$ is the mean zero measurement error which satisfies

$$E\left[u_i(t)\right] = 0, \quad Var\left[u_i(t)\right] = \sigma_x^2 \quad \text{and} \quad Cov\left[u_i(s), u_i(t)\right] = 0 \quad \text{if } s \neq t.$$

Applying the similar B-spline basis approximations as (15.17) to (15.20), we can obtain the B-spline estimator $\widehat{\mu}_x(t)$ of the mean curve $\mu_x(t)$ and the BLUP predicted subject-specific curve $\widehat{\zeta}_i(t)$ of $\zeta_i(t)$. The BLUP predictor of the subject-specific trajectory curve, or BLUP trajectory, $\widehat{X}_i(t)$ for $X_i(t)$ is

$$\widehat{X}_i(t) = b_*^T(t)\left(\widehat{\mu}_x + \widehat{\zeta}_i\right), \tag{15.21}$$

where $b_*(t)$ is the spline basis function used in the approximation of (15.20). Instead of using (15.18), the predicted subject-specific trajectory curves $\widehat{X}_i(t)$ should be used in the prediction of $\widehat{Y}_i(t)$ when $X_i(t)$ is not directly observed and/or measured with error. Substituting $X_i(t)$ of (15.18) with $\widehat{X}_i(t)$, the predicted trajectory curve of $Y_i(t)$ is

$$\widehat{Y}_i(t) = b_0^T(t)\left(\widehat{\xi}_0 + \widehat{\eta}_i\right) + b_1^T(t)\left(\widehat{\xi}_1 + \widehat{\phi}_i\right)\widehat{X}_i(t). \tag{15.22}$$

For the general case of multivariate covariates $\mathbf{X}(t)$, we can construct the predicted value $\widehat{X}_i(t)$ of the ith subject using a multivariate linear mixed-effects model, i.e., a multivariate generalization of (15.20). For simplicity, we omit the details of the multivariate $\mathbf{X}(t)$ case here, as this generalization is straightforward and requires more complex notation.

3. Predicted Observations

The predicted outcome trajectories obtained above are not suitable for the estimation of the conditional distribution functions, the RTPs and the RTPRs. This is because $\widehat{Y}_i(t)$ given in (15.18), (15.19) and (15.22) are the subject-specific mean curves for the ith subject at time t, which is not necessarily its potential observations at time t. The distributions of the subject-specific mean curves have smaller variations than the distributions of the subject-specific random processes $\{Y_i(t) : i = 1, \ldots, n; t \in \mathscr{T}\}$. Thus, using $\widehat{Y}_i(t)$ instead of the observations of $Y_i(t)$ could lead to biased estimators of the conditional distribution functions.

In order to obtain unbiased distribution estimators, we need to add the subject-specific random errors or measurement errors back to the predicted subject-specific trajectories. Following (15.2), (15.18), (15.19) and (15.20), a pseudo-estimator of the ith subject's random error $\varepsilon_i(t)$ is

$$\widehat{\varepsilon}_i(t) = Y_i(t) - \widehat{Y}_i(t). \tag{15.23}$$

Note that, because $Y_i(t)$ is not observed if t is not one of the time points in $\{t_{ij} : j = 1, \ldots, n_i\}$, $\widehat{\varepsilon}_i(t)$ of (15.23) may not be computed directly, hence a pseudo-estimator. The observed residuals of (15.18), (15.19) and (15.22) are

$$\widehat{\varepsilon}_{ij} = \widehat{\varepsilon}_i(t_{ij}) = Y_{ij} - \widehat{Y}_i(t_{ij}) \quad \text{for } i = 1, \ldots, n, \ j = 1, \ldots, n_i. \tag{15.24}$$

Because (15.2) assumes that $\varepsilon_i(t)$ does not depend on t and has the mean zero normal distribution with variance σ^2, the observed residuals $\widehat{\varepsilon}_{ij}$ of (15.24) are assumed to satisfy the assumption that

$$\left\{\widehat{\varepsilon}_{ij} : i = 1, \ldots, n; j = 1, \ldots, n_i\right\} \quad \text{are independent and} \quad \widehat{\varepsilon}_{ij} \sim N\left(0, \sigma^2\right). \quad (15.25)$$

Under (15.25), the subject-specific random error $\widetilde{\varepsilon}_i(t)$ of $\widehat{Y}_i(t)$ can be generated by one of the two following approaches:

Estimated Random Error for $\widehat{Y}_i(t)$:

(a) $\widetilde{\varepsilon}_i(t)$ *is a randomly selected value from the residuals* $\left\{\widehat{\varepsilon}_{ij} : i = 1, \ldots, n; j = 1, \ldots, n_i\right\}$.

(b) $\widetilde{\varepsilon}_i(t)$ *is a randomly selected value from the distribution* $N\left(0, \widehat{\sigma}^2\right)$, *where* $\widehat{\sigma}^2$ *is the MLE or restricted MLE of* σ^2. □

Using an estimated random error $\widetilde{\varepsilon}_i(t)$ obtained from either one of the two approaches above, we can compute an estimated observation $\widetilde{Y}_i(t)$ of $Y_i(t)$ based on any one of the models (15.18), (15.19) and (15.22). Applying the same approach as (a) or (b) above to the covariate $X_i(t)$ under the model (15.20), we can compute an estimated random error $\widetilde{u}_i(t)$ and then an estimated observation $\widetilde{X}_i(t)$. If multivariate covariates $\mathbf{X}_i(t)$ are involved, the multivariate generalization of (15.20) and (15.21) are used to compute the multivariate random error $\widetilde{u}_i(t)$ and the predicted observation $\widetilde{\mathbf{X}}_i(t)$. The resulting predicted observations of $\left\{Y_i(t), \mathbf{X}_i(t)\right\}$ are then denoted by

$$\begin{cases} \widetilde{Y}_i(t) & = & \widehat{Y}_i(t) + \widetilde{\varepsilon}_i(t), \\ \widetilde{\mathbf{X}}_i(t) & = & \widehat{\mathbf{X}}_i(t) + \widetilde{u}_i(t). \end{cases} \quad (15.26)$$

We note that $\widetilde{\mathbf{X}}_i(t)$ is used in the next section for the estimation of the conditional distributions, the RTPs and the RTPRs, only if it is subject to measurement errors. In case that the measurement errors of $\mathbf{X}_i(t)$ can be ignored, we would prefer to use $\widehat{\mathbf{X}}_i(t)$ instead of $\widetilde{\mathbf{X}}_i(t)$.

15.2.2 Estimation with Predicted Outcome Trajectories

We construct the estimators of the conditional distribution functions, the RTPs and the RTPRs under three scenarios: the nonparametric mixed-effects model (15.1) without covariates; the mixed-effects varying-coefficients model (15.2); the combined unstructured mixed-effect models of (15.1) and (15.20). Because different predicted trajectory curves are used under each of these scenarios, the estimation methods are different.

1. Estimation without Covariates

We first consider the estimation of the RTPs and the RTPRs based on the

observations $\{Y_{ij} = Y_i(t_{ij}) : j = 1, \ldots, n_i; i = 1, \ldots, n\}$ which satisfy the model
(15.19). By (15.6), the RTP for $Y(t) \in A(t)$ at time points $s_1 < s_2$ is

$$RTP_{s_1,s_2}(A) = \frac{E\{1_{[Y(s_2) \in A(s_2), Y(s_1) \in A(s_1)]}\}}{E\{1_{[Y(s_1) \in A(s_1)]}\}}, \tag{15.27}$$

where $1_{[\cdot]}$ is the indicator function and $A(t)$ is a pre-specified and known risk
set at time t. If we use the set of quantiles

$$A(t) = A_\alpha(t) = \{Y(t) : Y(t) > q_\alpha(t), 0 < \alpha < 1\} \tag{15.28}$$

which is a special case of (15.9) with $q_\alpha(t)$ being the known $(100 \times \alpha)$th quantile of $Y(t)$, the RTP defined in (15.27) then measures the tracking probability
of $Y(t)$ greater than its $(100 \times \alpha)$th quantile, such that

$$RTP_{s_1,s_2}(A_\alpha) = \frac{E\{1_{[Y(s_2) > q_\alpha(s_2), Y(s_1) > q_\alpha(s_1)]}\}}{E\{1_{[Y(s_1) > q_\alpha(s_1)]}\}}. \tag{15.29}$$

The RTP defined in (15.6) can be estimated using the predicted observa-
tions of $Y_i(t)$ of (15.26). By (15.27), an estimator $\widehat{RTP}_{s_1,s_2}(A)$ of $RTP_{s_1,s_2}(A)$
can be obtained by substituting the expectations

$$E\{1_{[Y(s_2) \in A(s_2), Y(s_1) \in A(s_1)]}\} \quad \text{and} \quad E\{1_{[Y(s_1) \in A(s_1)]}\}$$

of (15.27) by their corresponding empirical estimators

$$\begin{cases} \widehat{E}\{1_{[Y(s_2) \in A(s_2), Y(s_1) \in A(s_1)]}\} &= (1/n) \sum_{i=1}^{n} 1_{[\widetilde{Y}_i(s_2) \in A(s_2), \widetilde{Y}_i(s_1) \in A(s_1)]}, \\ \widehat{E}\{1_{[Y(s_1) \in A(s_1)]}\} &= (1/n) \sum_{i=1}^{n} 1_{[\widetilde{Y}_i(s_1) \in A(s_1)]}, \end{cases} \tag{15.30}$$

where $\widetilde{Y}(t)$ is the estimated observation of $Y_i(t)$ in (15.26). The estimator of
$RTP_{s_1,s_2}(A)$ is

$$\widehat{RTP}_{s_1,s_2}(A) = \frac{\sum_{i=1}^{n} 1_{[\widetilde{Y}_i(s_2) \in A(s_2), \widetilde{Y}_i(s_1) \in A(s_1)]}}{\sum_{i=1}^{n} 1_{[\widetilde{Y}_i(s_1) \in A(s_1)]}}. \tag{15.31}$$

In particular, by (15.28) and (15.29), the estimator of $RTP_{s_1,s_2}(A_\alpha)$ is

$$\widehat{RTP}_{s_1,s_2}(A_\alpha) = \frac{\sum_{i=1}^{n} 1_{[\widetilde{Y}_i(s_2) > q_\alpha(s_2), \widetilde{Y}_i(s_1) > q_\alpha(s_1)]}}{\sum_{i=1}^{n} 1_{[\widetilde{Y}_i(s_1) > q_\alpha(s_1)]}}. \tag{15.32}$$

To estimate the RTPR of (15.7), we first recognize that, by (15.7), (15.8)
and (15.27),

$$RTPR_{s_1,s_2}(A) = \frac{E\{1_{[Y(s_2) \in A(s_2), Y(s_1) \in A(s_1)]}\}}{E\{1_{[Y(s_1) \in A(s_1)]}\} E\{1_{[Y(s_2) \in A(s_2)]}\}}, \tag{15.33}$$

and, when $A_\alpha(t)$ of (15.28) is used,

$$RTPR_{s_1,s_2}(A_\alpha) = \frac{E\left\{1_{[Y(s_2)>q_\alpha(s_2),Y(s_1)>q_\alpha(s_1)]}\right\}}{E\left\{1_{[Y(s_1)>q_\alpha(s_1)]}\right\}E\left\{1_{[Y(s_2)>q_\alpha(s_2)]}\right\}}. \tag{15.34}$$

Similar to (15.30), we can estimate $E\left\{1_{[Y(s_2)\in A(s_2)]}\right\}$ by

$$E\left\{1_{[Y(s_2)\in A(s_2)]}\right\} = (1/n)\sum_{i=1}^{n}1_{[\tilde{Y}_i(s_2)\in A(s_2)]}. \tag{15.35}$$

Using (15.31), (15.34) and (15.35), the estimator of $RTPR_{s_1,s_2}(A)$ is

$$\widehat{RTPR}_{s_1,s_2}(A) = \frac{n\sum_{i=1}^{n}1_{[\tilde{Y}_i(s_2)\in A(s_2),\tilde{Y}_i(s_1)\in A(s_1)]}}{\left\{\sum_{i=1}^{n}1_{[\tilde{Y}_i(s_1)\in A(s_1)]}\right\}\left\{\sum_{i=1}^{n}1_{[\tilde{Y}_i(s_2)\in A(s_2)]}\right\}}, \tag{15.36}$$

and the estimator of $RTPR_{s_1,s_2}(A_\alpha)$ is

$$\widehat{RTPR}_{s_1,s_2}(A) = \frac{n\sum_{i=1}^{n}1_{[\tilde{Y}_i(s_2)>q_\alpha(s_2),\tilde{Y}_i(s_1)>q_\alpha(s_2)]}}{\left\{\sum_{i=1}^{n}1_{[\tilde{Y}_i(s_1)>q_\alpha(s_2)]}\right\}\left\{\sum_{i=1}^{n}1_{[\tilde{Y}_i(s_2)>q_\alpha(s_2)]}\right\}}. \tag{15.37}$$

When $A(t)$ is unknown, we can estimate it from the same sample that is used to estimate the RTPs and the RTPRs. For example, the $(100 \times \alpha)$th quantile $q_\alpha(t)$ used in $A_\alpha(t)$ may not be known for a given study population and has to be estimated from the predicted trajectories. We describe in Section 15.3.2 a split sample approach for dealing with the situation that $A(t)$ is unknown and has to be estimated from the sample.

2. Estimation with the Varying-Coefficient Mixed-Effects Models

We now consider the estimation of conditional distribution function (15.3), the RTP (15.4) and the RTPR (15.5) under the varying-coefficient mixed-effects model (15.2). We assume the general case that the covariates $X_i(t)$ are measured with error, so that the predicted observations of $\{Y_i(t), X_i(t)\}$ in (15.26) are used for the estimation.

For any given sets $\{A[X(t),t], B(t) : t \in \mathcal{T}\}$ and any $s_1, s_2 \in \mathcal{T}$, let

$$\left\{\begin{array}{rcl} E_1(s_1) & = & E\left\{1_{[X(s_1)\in B(s_1)]}\right\}, \\ E_2(s_1,s_2) & = & E\left\{1_{[Y(t_{s_2})\in A[X(s_2),s_2],X(s_1)\in B(s_1)]}\right\}, \\ E_3(s_1,s_2) & = & E\left\{1_{[Y(t_{s_2})\in A[X(s_2),s_2],Y(t_{s_1})\in A[X(s_1),s_1],X(s_1)\in B(s_1)]}\right\}, \end{array}\right. \tag{15.38}$$

and denote the quantities in (15.3), (15.4) and (15.5) by

$$P_{s_1,s_2}(A,B) = \frac{E_2(s_1,s_2)}{E_1(s_1)}, \tag{15.39}$$

$$RTP_{s_1,s_2}(A, B) = \frac{E_3(s_1, s_2)}{E_2(s_1, s_1)}, \tag{15.40}$$

and

$$RTPR_{s_1,s_2}(A, B) = \frac{E_3(s_1, s_2) \times E_1(s_1)}{E_2(s_1, s_1) \times E_2(s_1, s_2)}. \tag{15.41}$$

Using the predicted observations $\{\widetilde{Y}_i(t), \widetilde{\mathbf{X}}_i(t) : t \in \mathcal{T}\}$ computed from (15.20), (15.22) and (15.26), we can estimate the expectations in (15.38) by

$$\begin{cases} \widehat{E}_1(s_1) &= (1/n) \sum_{i=1}^{n} 1_{[\widetilde{\mathbf{X}}_i(s_1) \in B(s_1)]}, \\ \widehat{E}_2(s_1, s_2) &= (1/n) \sum_{i=1}^{n} 1_{[\widetilde{Y}_i(t_{s_2}) \in A[\widetilde{\mathbf{X}}_i(s_2), s_2], \widetilde{\mathbf{X}}_i(s_1) \in B(s_1)]}, \\ \widehat{E}_3(s_1, s_2) &= (1/n) \times \\ & \quad \sum_{i=1}^{n} 1_{[\widetilde{Y}_i(t_{s_2}) \in A[\widetilde{\mathbf{X}}_i(s_2), s_2], \widetilde{Y}_i(t_{s_1}) \in A[\widetilde{\mathbf{X}}_i(s_1), s_1], \widetilde{\mathbf{X}}_i(s_1) \in B(s_1)]} \end{cases} \tag{15.42}$$

and, consequently, estimate the conditional probability (15.39), the RTP (15.40) and the RTPR (15.41) by

$$\widehat{P}_{s_1,s_2}(A, B) = \frac{\widehat{E}_2(s_1, s_2)}{\widehat{E}_1(s_1)}, \tag{15.43}$$

$$\widehat{RTP}_{s_1,s_2}(A, B) = \frac{\widehat{E}_3(s_1, s_2)}{\widehat{E}_2(s_1, s_1)}, \tag{15.44}$$

and

$$\widehat{RTPR}_{s_1,s_2}(A, B) = \frac{\widehat{E}_3(s_1, s_2) \times \widehat{E}_1(s_1)}{\widehat{E}_2(s_1, s_1) \times \widehat{E}_2(s_1, s_2)}, \tag{15.45}$$

respectively.

3. Estimation Based on Unstructured Mixed-Effects Models

The estimators presented above, i.e., (15.42) through (15.45), depend strongly on the varying-coefficient mixed-effects model (15.2), and utilize the predicted observations $\{\widetilde{Y}_i(t), \widetilde{\mathbf{X}}_i(t) : i = 1, \ldots, n\}$ computed based on the predicted subject-specific mean curves (15.22). When the outcome and covariates $\{Y_i(t), \mathbf{X}_i(t) : t \in \mathcal{T}, i = 1, \ldots, n\}$ do not satisfy the model (15.2), the predicted observations computed based on (15.22) could be inadequate for the estimation of (15.39), (15.40) and (15.41). In case that (15.2) is not necessarily satisfied, an alternative approach is to use predicted subject-specific observations of $Y_i(t)$ computed from the observations $\{Y_{ij} : i = 1, \ldots, n; j = 1, \ldots, n_i\}$ alone. We refer to this approach as the estimation based on unstructured mixed-effects models, which is computed using the following two steps.

Estimation with Unstructured Mixed-Effects Models:

(a) *Compute the predicted observations $\{\widetilde{Y}_i(t), \widetilde{\mathbf{X}}_i(t) : t \in \mathscr{T}, i = 1, \ldots, n\}$ based on (15.19), (15.20) and (15.26).*

(b) *Compute the estimators of $P_{s_1, s_2}(A, B)$, $RTP_{s_1, s_2}(A, B)$ and $RTPR_{s_1, s_2}(A, B)$ of (15.39), (15.40) and (15.41), respectively, using (15.43), (15.44) and (15.45) with $\widetilde{Y}_i(t)$ and $\widetilde{\mathbf{X}}_i(t)$ computed from (a).*   □

Notice that, since the predicted subject-specific observations $\widetilde{Y}_i(t)$ in (a) above are computed without using any regression models for $Y(t)$ and $\mathbf{X}(t)$, the estimators given in (b) depend on the unstructured mixed-effects model of $Y_i(t)$, as opposed to the structured mixed-effects model for $Y_i(t)$ and $\mathbf{X}_i(t)$ in (15.2). It is then reasonable to expect that the unstructured estimation method in (a) and (b) above can be more generally applied than the structured method of the previous section. But, on the other hand, when (15.2) is a reasonable model for $\{Y_i(t), \mathbf{X}_i(t) : t \in \mathscr{T}; i = 1, \ldots, n\}$, the unstructured approach may lead to estimators with larger variances than the estimators obtained from the structured approach based on (15.2).

15.2.3 *Estimation Based on Split Samples*

An important assumption for the estimation methods of Section 15.3.2 is that the outcome and covariate status sets $A[\mathbf{X}(t), t]$ and $B(t)$ used in (15.3), (15.4) and (15.5) are known. For example, in (15.9) the conditional quantiles $q_\alpha[t, \mathbf{X}(t)]$ are assumed to be a known function of $\{t, \mathbf{X}(t)\}$. In practice, $B(t)$ is usually known, but $A[\mathbf{X}(t), t]$ is possibly unknown, which can be estimated from the sample. In such situations, a practical approach is to randomly split the subjects into two sub-samples, so that one sub-sample is used to estimate $A[\mathbf{X}(t), t]$, while the other is used to estimate the conditional distributions. We proceed with this approach through the following three steps:

Estimation with Sample Splitting:

(a) *Randomly split the sample $\{Y_{ij}, \mathbf{X}_{ij} : j = 1, \ldots, n_i; i = 1, \ldots, n\}$ into sub-samples I_1 and I_2 with sample sizes n_{I_1} and n_{I_2}, such that $n_{I_1} + n_{I_2} = n$.*

(b) *The first sub-sample I_1 is used to estimate $A[\mathbf{X}(t), t]$. Let $\widehat{A}[\mathbf{X}(t), t]$ be the estimated set of $A[\mathbf{X}(t), t]$ obtained based on the sample I_1.*

(c) *Compute the estimators of $P_{s_1, s_2}(A, B)$, $RTP_{s_1, s_2}(A, B)$ and $RTPR_{s_1, s_2}(A, B)$ in (15.3), (15.4) and (15.5) by applying the estimation methods of Section 15.3.2 to the sub-sample I_2 with $\widehat{A}[\widetilde{\mathbf{X}}_i(t), t]$ in place of $A[\widetilde{\mathbf{X}}_i(t), t]$.*   □

The estimated set $\widehat{A}[\mathbf{X}(t), t]$ depends on the specific choice of $A[\mathbf{X}(t), t]$ and has to be constructed on a case-by-case basis. For the conditional quantile based set $A_\alpha[\mathbf{X}(t), t]$ of (15.9), we can use the estimated set given by

$$\widehat{A}_\alpha[\mathbf{X}(t), t] = \left\{ Y(t) : Y(t) > \widehat{q}_\alpha[t, \mathbf{X}(t)] \right\}, \qquad (15.46)$$

where $\widehat{q}_\alpha[t, \mathbf{X}(t)]$ is the estimated $(100 \times \alpha)$th conditional quantile of $Y(t)$ given $\mathbf{X}(t)$ based on the sub-sample I_1.

15.2.4 Bootstrap Pointwise Confidence Intervals

Similar to the inference procedures of previous chapters, such as Chapters 13 and 14, we use the following resampling-subject bootstrap method to construct the pointwise confidence intervals for the conditional distribution functions, RTPs and RTPRs.

Approximate Bootstrap Pointwise Confidence Intervals:

(a) Bootstrap Estimators: *Generate B bootstrap samples from the original dataset* $\{Y_{ij}, \mathbf{X}_{ij} : j = 1, \ldots, n_i; \ i = 1, \ldots, n\}$ *and denote the resulting bootstrap samples by*

$$\begin{cases} \mathscr{L}^{(b)} &= \left\{ Y_{ij}^{(b)}, \mathbf{X}_{ij}^{(b)}, t_{ij}^{(b)} : i = 1, \ldots, n; j = 1, \ldots, n_{ij} \right\} \\ \mathscr{L}^{boot} &= \left\{ \mathscr{L}^{(b)} : b = 1, \ldots, B \right\}. \end{cases} \qquad (15.47)$$

Compute the estimators $\widehat{P}_{s_1, s_2}^{(b)}(A, B)$, $\widehat{RTP}_{s_1, s_2}^{(b)}(A, B)$ *and* $\widehat{RTPR}_{s_1, s_2}^{(b)}(A, B)$ *using the bth bootstrap sample* $\mathscr{L}^{(b)}$ *and methods of Section 15.3.2, such as (15.43), (15.44) and (15.45), and obtain*

$$\begin{cases} \left\{ \widehat{P}_{s_1, s_2}^{(b)}(A, B) : b = 1, \ldots, B \right\}, \\ \left\{ \widehat{RTP}_{s_1, s_2}^{(b)}(A, B) : b = 1, \ldots, B \right\}, \\ \left\{ \widehat{RTPR}_{s_1, s_2}^{(b)}(A, B) : b = 1, \ldots, B \right\}. \end{cases} \qquad (15.48)$$

(b) Bootstrap Percentile Confidence Intervals: *Let* $l_{\alpha/2}^p(s_1, s_2)$ *and* $u_{\alpha/2}^p(s_1, s_2)$ *be the lower and upper* $[100 \times (\alpha/2)]$th *percentiles of the corresponding B bootstrap estimators given in (15.48). The* $[100 \times (\alpha/2)]\%$ *bootstrap percentile pointwise confidence interval for the corresponding estimator in (15.48) is*

$$\left(l_{\alpha/2}^p(s_1, s_2), u_{\alpha/2}^p(s_1, s_2) \right). \qquad (15.49)$$

(c) Bootstrap Normal Approximated Confidence Interval: *Let* $\widehat{\xi}_{s_1, s_2}^{(b)}(A, B)$ *be any of the estimators given in (15.48) computed using the bootstrap sample* $\mathscr{L}^{(b)}$, *and* $\widehat{se}_{s_1, s_2}^{boot}(A, B)$ *be the corresponding sample standard deviations of these B bootstrap estimators, i.e.,*

$$\widehat{se}_{s_1, s_2}^{boot}(A, B) = \left\{ \frac{1}{B-1} \sum_{b=1}^{B} \left[\widehat{\xi}_{s_1, s_2}^{(b)}(A, B) - \frac{1}{B} \sum_{b=1}^{B} \widehat{\xi}_{s_1, s_2}^{(b)}(A, B) \right]^2 \right\}^{1/2}. \qquad (15.50)$$

The $[100 \times (\alpha/2)]\%$ normal approximated pointwise confidence interval for the corresponding estimator in (15.48) is

$$\widehat{\xi}_{s_1,s_2}^{(b)}(A, B) \pm z_{1-\alpha/2}\widehat{se}_{s_1,s_2}^{boot}(A, B), \tag{15.51}$$

where $z_{1-\alpha/2}$ is the $[100 \times (\alpha/2)]th$ quantile of the standard normal distribution. □

We note that, because the potential biases of the estimators of (15.48) have been ignored, (15.49) and (15.51) are in fact only some error bands. These error bands will have approximately adequate coverage probabilities for the quantities being estimated if the bias of the estimator is small. In practice, we can reduce the bias of the estimator by increasing the number of basis functions used in the B-spline approximations.

15.3   R Implementation with the NGHS Data

The NGHS data has been described in Section 1.2. We have discussed in Chapters 13 and 14 the estimation of the time-varying effects of covariates, the conditional distributions and the rank-tracking probabilities of several cardiovascular risk factors using the time-varying transformation models. Here we illustrate how to model and estimate these quantities based on the nonparametric mixed-effects models to track body mass index (BMI) and systolic blood pressure (SBP) with the NGHS data.

15.3.1   Rank-Tracking for BMI

Based on Obarzanek et al. (2010), a lower bound of the age- and sex-adjusted 85th percentile, $q_{0.85}(t)$, from the Centers for Disease Control and Prevention (CDC) BMI growth chart is used to define overweight and obese status for girls at a given age t. We fit separately the nonparametric mixed-effects models (15.1) to the NGHS BMI data by the two racial groups, each model with cubic B-spline approximation and four equally spaced knots. We then compute the subject-specific BMI trajectory curves over 9 to 19 years of age, and estimate $RTP_A(x, s_1, s_1 + \delta)$ and $RTPR_A(x, s_1, s_1 + \delta)$ of BMI for Caucasian $(x = 0)$ and African American $(x = 1)$ girls over the age range $9 \leq s_1 \leq 16$ and $\delta = 3$, where $A_{0.85}(t)$ is the set of subjects whose BMI values at age t are greater than $q_{0.85}(t)$. Some example R code for fitting the model and estimating the tracking indices is given below:

```
> library(npmlda)
> NGHS.B <- NGHS[NGHS$RACE==2,]
> nID <- length(unique(NGHS.B$ID)) #1213
# fit the linear mixed model with cubic B-splines
```

```
> KN1 <- seq(from=9, to=18.9, length=4)[-c(1,4)]
> Bs.age <- bs(NGHS.B$AGE, knots=KN1)
> fm1 <- lmer(BMI ~ 1+ Bs.age +(1+ Bs.age|ID), data=NGHS.B)
# generate the BLUP predictions on a grid
> T2<- seq(from=9, to=18.8, by=0.2)
> S.BS <- bs(T2, knots=KN1 )
> IDlevel<- rownames(coef(fm1)[[1]])
> NGHSped<- data.frame(ID= NA, AGE=rep(T2, nID), BMI=NA)
> nT2<- 50
> for (i in 1:nID)
  {
   KK <- i-1
   Datai <- NGHS.B[NGHS.B$ID==IDlevel[i],]
   mean.pred <-  cbind(1,S.BS) %*% t(as.vector(coef(fm1)[[1]][i,]))
   NGHSped[(KK*nT2+1):(KK*nT2+nT2),]$ID  <- IDlevel[i]
   NGHSped[(KK*nT2+1):(KK*nT2+nT2),]$BMI <- mean.pred
  }
>  NGHSped$BMIp <- NGHSped$BMI + rnorm(nrow(NGHSped) ,
                    mean = 0, sd = sigma(fm1)  )
## Compute the PA(x,s1)##
> NN <- nrow(NGHSped)
> S1cat <- seq(9.0, 18.8, by=0.2)
> S12cat <- seq(9.0, 15.8, by=0.2)
> KK1 <- length(S1cat)
> KK2 <- length(S12cat)
> Prob.S1 <- numeric(KK1 )
> for (i in 1:KK1)
  {
    SeqKK1 <- seq(from=i, to =NN-nT2+i, by=nT2)
    Datai <- NGHSped[SeqKK1,]
    Prob.S1.f1[i]  <- mean(Datai$BMIp>=BMIq85[i])
   # BMIq85 is the CDC percentile curve
  }
## Compute the PA(x,s1,s2)##
> Prob.S1S2 <- numeric( KK2 )
> for (i in 1:KK2)
  {
   SeqKK1 <- seq(from=i, to =NN-nT2+i, by=nT2)
   Datai <- cbind(NGHSped[SeqKK1,]$BMIp,
                  NGHSped[SeqKK1+15,]$BMIp)
   Prob.S1S2[i]<- mean((Datai[,1]>=BMIq85[i])&
                       (Datai[,2]>=BMIq85[i+15]))
 }
# Then follow the same code in Chapter 12 to calculate RTP, RTPR
```

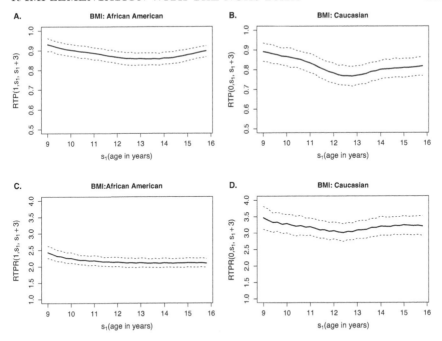

Figure 15.1 *The estimated BMI $RTP_A(x, s_1, s_1 + 3)$, $RTPR_A(x, s_1, s_1 + 3)$ and their 95% bootstrap percentile confidence intervals for Caucasian $(x = 0)$ and African American $(x = 1)$ girls.*

Figure 15.1 shows the estimated 3-year RTP and RTPR curves for both African American and Caucasian girls with their bootstrap 95% pointwise percentile confidence intervals based on $B = 500$ bootstrap replications. Figures 15.1 (A)-(B) show that the conditional probabilities of being overweight or obese are 84%-94% and 74%-90% for those girls who were already overweight or obese 3 years earlier, respectively. This suggests that the African American girls are more likely to remain in the undesirable overweight status compared to the Caucasian girls. Since the NGHS participants are slightly more overweight than the general population, we examine the relative strength of the BMI tracking ability through the RTPR curves displayed in Figures 15.1 (C)-(D). For both racial groups, RTPR curves are significantly greater than 1, which indicates that knowing a girl's overweight status at an earlier age increases the chance of being overweight at a later age about two to three times compared to the probability of her being overweight without knowing her previous weight status. Our spline-based estimation results are similar to the results based on the kernel estimation in Section 12.4, which suggests that BMI has high rank-tracking ability for adolescent girls.

15.3.2   Rank-Tracking for SBP

Based on NHBPEP (2004), the age, sex, and height specific conditional percentiles for the blood pressure are used to define pre-hypertension and hypertension in children. With the NGHS SBP data, we can estimate the RTPs and RTPRs of SBP based on the predicted SBP trajectory curves obtained either from the model (15.2) or the separate univariate mixed-effects models (15.1) and (15.20).

Using the framework of (15.2), the mixed-effects varying-coefficient model for the data is

$$Y_i(t_{ij}) = \beta_{0i}(t_{ij}) + X_{1i}\beta_{1i}(t_{ij}) + X_{2i}(t_{ij})\beta_{2i}(t_{ij}) + \varepsilon_i(t_{ij}), \tag{15.52}$$

where $Y_i(t)$, X_{1i} and $X_{2i}(t)$ are ith girl's SBP, race and height percentile at t years of age, with $X_{1i} = 0$ if the girl is Caucasian and $X_{1i} = 1$ if she is African American. We have described the R implementation of the similar model fitting and prediction in Section 11.6 for this data.

Since the conditional 90th percentile $q_{0.9}[t, X_1, X_2(t)]$ is not known, we randomly split the subjects into two sub-samples with approximately equal sample sizes, and compute the estimates of $q_{0.9}[t, X_1, X_2(t)]$ using the first sub-sample. We then use the cubic B-spline predicted SBP trajectories values from the second sub-sample to estimate $RTP_A(x_1, s_1, s_1 + 2)$ and $RTPR_A(x_1, s_1, s_1 + 2)$, the 2-year rank-tracking probabilities of SBP at ages s_1 and $s_1 + 2$ for both racial groups.

Figure 15.2 shows the estimated tracking indices and their 95% pointwise bootstrap percentile confidence intervals obtained with $B = 500$ bootstrap samples. The RTPs in Figures 15.2 (A)-(B) are 30%-40% for both African-American and Caucasian girls with $s_1 \in [9, 17]$ years. These values are significantly larger than 10%, which is roughly the probability of SBP greater than 90th percentiles at $t = s_1 + 2$. To evaluate the relative strength of the "rank-tracking ability" of SBP for these girls, we examine the estimates of $RTPR_A(x_1, s_1, s_1 + 2)$ in Figures 15.2 (C)-(D) for those girls. The estimated RTPRs are approximately 2.9 to 3.1 for African American girls and 2.6 to 3.6 for Caucasian girls. That is, compared to the probability of a girl having elevated SBP without knowing the SBP status at an earlier age, knowing that a girl already had elevated SBP two years earlier increases the likelihood of her having elevated SBP at a later age more than two times. The estimates of these tracking indices suggest that SBP has high degree of risk tracking for both African American and Caucasian girls within this age range. We have also estimated RTPs and RTPRs based on the separate univariate mixed-effects models (15.1) and (15.20), and the estimate curves are very similar to the results presented in Figure 15.2.

15.4   Remarks and Literature Notes

This chapter summarizes a class of global smoothing methods for estimating the conditional distribution functions, the RTPs and the RTPRs, which

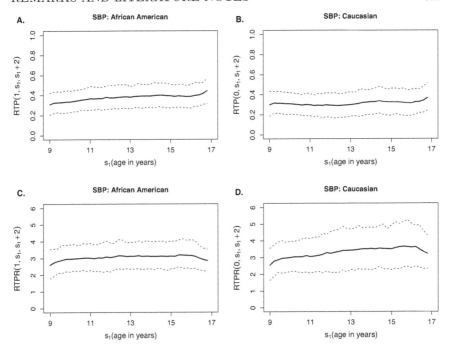

Figure 15.2 *The estimated SBP $RTP_A(x_1, s_1, s_1 + 2)$, $RTPR_A(x_1, s_1, s_1 + 2)$ and their 95% bootstrap percentile confidence intervals for Caucasian $(x = 0)$ and African American $(x = 1)$ girls.*

are quantitative indices of the tracking abilities for a time-dependent outcome variable. These estimators incorporate the subject's covariates through a class of flexible varying-coefficient mixed-effects models. The inherent model flexibility and straightforward interpretations enable these estimators to serve as a convenient statistical tool to identify disease risk factors that track over time. The estimation methods of this chapter are based on the B-spline approximations. Depending on the scientific questions and the nature of the data, a number of other estimation methods, such as approximations through other basis functions, may be considered in given situations.

The modeling and estimation methods of this chapter are developed by Shi, Weiss and Taylor (1996), Rice and Wu (2001), Liang, Wu and Carroll (2003) and Wu and Tian (2013b). Specifically, Shi, Weiss and Taylor (1996) and Rice and Wu (2001) established the basis approximation estimation and prediction methods for nonparametric mixed-effects models with time-invariant covariates, Liang, Wu and Carroll (2003) suggested the mixed-effects varying-coefficient models with measurement errors to incorporate time-dependent covariates, and Wu and Tian (2013b) developed the conditional distribution, the RTP and the RTPR estimators based on the predicted subject-specific outcome and covariate values.

Bibliography

Akaike, H. Statistical predictor identification, *Annals of the Institute of Statistical Mathematics*, 22:203–217, 1970.

Altman, N. S. Kernel smoothing of data with correlated errors. *Journal of the Americal Statistical Association*, 85:749–759, 1990.

Bang, H. and Robins, J. Doubly robust estimation in missing data and causal inference models. *Biometrics*, 61:962–972, 2005.

Bates, D. M. and Pinheiro, J. C. *Mixed Effects Models in S*. Springer, New York, 1999.

Bates, D., Mächler, M., Bolker, B. M. and Walker S. C. Fitting linear mixed-effects models using lme4. *Journal of Statistical Software*, 67(1), *doi:10.18637/jss.v067.i01*, 2015.

Bickel, P. J. One-step Huber estimates in linear models. *Journal of the American Statistical Association*, 34:584–653, 1975.

Brauer, F. and Nohel, J. A. *Ordinary Differential Equations: A First Course*, (2nd Ed.), W. A. Benjamin, Menlo Park, CA, USA, 1973.

Bretz, F., Hothorn, T. and Westfall, P. *Multiple Comparisons Using R*. Chapman & Hall/CRC Press, Boca Raton, FL, 2011.

Cheng, S. C., Wei, L. J. and Ying, Z. Analysis of transformation models with censored data. *Biometrika*, 82:835–845, 1995.

Cheng, S. C., Wei, L. J. and Ying, Z. Predicting survival probabilities with semiparametric transformation models. *Journal of the American Statistical Association*, 92:227–235, 1997.

Chiang, C.-T., Rice, J. A. and Wu, C. O. Smoothing spline estimation for varying coefficient models with repeatedly measured dependent variable. *Journal of the American Statistical Association*, 96(454):605–619, 2001.

Clark, W., Woolson, R. and Schrott, H. Tracking of blood pressure serum lipids and obesity in children: The Muscatine Study. Abstract. *Circulation*, 23, Supplement II, 53–54, 1976.

Cook, T. D. and DeMets, D. L. *Introduction to Statistical Methods for Clinical Trials*. Chapman & Hall/CRC Press, Boca Raton, FL, USA, 2008.

Dalgaard, P. *Introductory Statistics with R, 2nd edition*. Springer, New York, 2008.

Daniels, S. R., McMahon, R. P., Obarzanek, E., Waclawiw, M. A., Similo,

S. L., Biro, F. M., Schreiber, G. B., Kimm, S. Y. S., Morrison, J. A. and Barton, B. A. Longitudinal correlates of change in blood pressure in adolescent girls. *Hypertension*, 31:97–103, 1998.

Davidian, M. and Giltinan, D.M. *Nonlinear Models for Repeated Measurement Data*. Chapman & Hall, New York, 1995.

Davidian, M. and Giltinan, D.M. Non-linear models for repeated measurement data: An overview and update. *Journal of Agricultural, Biological, and Environmental Statistics*, 8:387–419, 2003.

de Boor, C. *A Practical Guide to Splines*. Springer, New York, 1978.

Diggle, P. J. An approach to the analysis of repeated measurements. *Biometrics*, 44:959–971, 1988.

Diggle, P. J., Heagerty, P., Liang, K.-Y. and Zeger, S. L. *Analysis of Longitudinal Data, Second Edition*. Oxford University Press, Oxford, England, 2002.

Diggle, P. J. and Verbyla, A. Nonparametric estimation of covariance structure in longitudinal data. *Biometrics*, 54:401–415, 1998.

Efron, B. and Tibshirani, R. J. *An Introduction to the Bootstrap*. Chapman & Hall, Boca Raton, FL, 1993.

ENRICHD Investigators. Enhancing recovery in coronary heart disease patients (ENRICHD): study intervention rationale and design. *Psychosomatic Medicine*, 63:747–755, 2001.

ENRICHD Investigators. Enhancing recovery in coronary heart disease patients (ENRICHD): the effects of treating depression and low perceived social support on clinical events after myocardial infarction. *Journal of the American Medical Association*, 289:3106–3116, 2003.

Eubank, R. L. *Nonparametric Regression and Spline Smoothing, Second Edition*. Marcel Dekker, Inc., New York, 1999.

Eubank, R. L. and Speckman, P. L. Confidence bands in nonparametric regression. *Journal of the American Statistical Association*, 88:1287–1301, 1993.

Fahrmeir, L. and Lang, S. Bayesian inference for generalised additive mixed models based on Markov random field priors. *Applied Statistics*, 50:201–220, 2000.

Fan, J. Design-adaptive nonparametric regression. *Journal of the American Statistical Association*, 87:998–1004, 1992.

Fan, J. and Gijbels, I. *Local Polynomial Modeling and Its Applications*. Chapman & Hall, London, United Kingdom, 1996.

Fan, J. and Marron, J. S. Fast implementations of nonparametric curves estimators. *Journal of Computation and Graphical Statistics*, 3:35–56, 1994.

Fan, J., Wu, Y. Semiparametric estimation of covariance matrices for longitudinal data. *Journal of the American Statistical Association*, 103:1520–

1533, 2008.

Fan, J. and Zhang, J.-T. Functional linear models for longitudinal data. *Journal of the Royal Statistical Society, B*, 62:303–322, 2000.

Fitzmaurice, G., Davidian, M., Verbeke G. and Molenberghs G. *Longitudinal Data Analysis*. Chapman & Hall/CRC, Boca Raton, Florida, 2009.

Follmann, D. and Wu, M. C. An approximate generalized linear model with random effects for informative missing data. *Biometrics*, 51:151–168, 1995.

Foulkes, M. A. and Davis, C. E. An index of tracking for longitudinal data. *Biometrics*, 37(3):439–446, 1981.

Friedman, L. M., Furberg, C. D., DeMets, D., Reboussin, D. M. and Granger, C. B. *Fundamentals of Clinical Trials, Fifth Edition*. Springer, Switzerland, 2015.

Green, P. J., and Silverman, B. W. *Nonparametric Regression and Generalized Linear Models, A roughness penalty approach*. Chapman & Hall/CRC, Boca Raton, FL, 1994.

Guo, W. Functional mixed effects models. *Biometrics*, 58:121–128. 2002.

Hall, P. and Müller, H.G. Order-preserving nonparametric regression, with applications to conditional distributions and quantile function estimation. *Journal of the American Statistical Association*, 98:598–608, 2003.

Hall, P., Racine, J. and Li, Q. Cross-validation and the estimation of conditional probability densities. *Journal of the American Statistical Association*, 99:1015–1026, 2004.

Hall, P. and Titterington, D. M. On confidence bands in nonparametric density estimation and regression. *Journal of Multivariate Analysis*, 27:228–254, 1988.

Hall, P. and Wehrly, T.E. A geometrical method for removing edge effects from kernel-type nonparametric regression estimators. *Journal of the American Statistical Association*, 86:665–672, 1991.

Hall, P., Wolff, R. C. L. and Yao, Q. Methods for estimating a conditional distribution function. *Journal of the American Statistical Association*, 94:154–163, 1999.

Hansen, M. H. and Koopergerg, C. Spline adaptation in extended linear models. *Statistical Science*, 17:2–51, 2002.

Härdle, W. *Applied Nonparametric Regerssion*. Cambridge University Press, Cambridge, United Kingdom, 1990.

Härdle, W. and Marron, J. S. Bootstrap simultaneous error bars for nonparametric regression. *The Annals of Statistics*, 19:778–796, 1991.

Hart, J. D. Kernel regression estimation with time series errors. *Journal of the Royal Statistical Society, Ser. B*, 53:173–187, 1991.

Hart, J. D. *Nonparametric Smoothing and Lack-of-Fit Tests*. Springer, New

York, 1997.

Hart, J. D. and Wehrly, T. E. Kernel regression estimation using repeated measurements data. *Journal of the American Statistical Association*, 81:1080–1088, 1986.

Hart, J. D. and Wehrly, T. E. Consistency of cross-validation when the data are curves. *Stochastic Processes and Their Applications*, 45:351–361, 1993.

Harville, D. A. Bayesian inference for variance components using only error contrasts. *Biometrika*, 61:383–385, 1974.

Hastie, T. J. and Tibshirani, R. J. Varying-coefficient models. *Journal of the Royal Statistical Society, Ser. B*, 55:757–796, 1993.

Hoover, D. R., Rice, J. A., Wu, C. O. and Yang, L.-P. Nonparametric smoothing estimates of time-varying coefficient models with longitudinal data. *Biometrika*, 85:809–822, 1998.

Huang, J. Z. Projection estimation in multiple regression with application to functional ANOVA models. *The Annals of Statistics*, 26:242–272, 1998.

Huang, J. Z. Concave extended linear modeling: A theoretical synthesis. *Statistica Sinica*, 11:173–197, 2001.

Huang, J. Z. Local asymptotics for polynomial spline regression. *The Annals of Statistics*, 31:1600-1635, 2003.

Huang, J. Z. and Stone, C. J. Extended linear modeling with splines. In: *Nonlinear Estimation and Classification (Denison, D. D., Hansen, M. H., Holmes, C. C., Mallick, B. and Yu, B., eds)*. Lecture Notes in Statistics, vol 171. Springer, New York, NY, 2003.

Huang, J., Wu, C. O. and Zhou, L. Varying coefficient models and basis function approximations for the analysis of repeated measurements. *Biometrika*, 89:111–128, 2002.

Huang, J. Z., Wu, C. O. and Zhou, L. Polynomial spline estimation and inference for varying coefficient models with longitudinal data. *Statistica Sinica*, 14:763–788, 2004.

James, G. M., Hastie, T. J. and Sugar, C. A. Principal component models for sparse functional data. *Biometrika*, 87:587–602, 2000.

James, P. A., Oparil, S., Carter, B. L., et al. Evidence-based guideline for the management of high blood pressure in adults: Report from the panel members appointed to the Eighth Joint National Committee (JNC 8). *Journal of the American Medical Association*, 311(5):507–520, 2014.

James, G., Witten, D., Hastie, T., and Tibshirani, R. *An Introduction to Statistical Learning*. Springer, New York, 2013.

Jiang, J. *Linear and Generalized Linear Mixed Models and Their Applications*. Springer, New York, 2007.

Jones, M. C. On higher order kernels. *Journal of Nonparametric Statistics*,

5(2):215–221, 1995.

Jones, R. H. and Ackerson, L. M. Serial correlation in unequally spaced longitudinal data. *Biometrika*, 77:721–731, 1990.

Jones, R. H. and Boadi-Boteng, F. Unequally spaced longitudinal data with serial correlation. *Biometrics*, 47:161–175, 1991.

Kaslow, R. A., Ostrow, D. G., Detels, R., Phair, J. P., Polk, B. F. and Rinaldo, C. R. The Multicenter AIDS Cohort Study: rationale, organization and selected characteristics of the participants. *American Journal of Epidemiology*, 126:310–318, 1987.

Kavey, R. E. W., Daniels, S. R., Lauer, R. M., Atkins, D. L., Hayman, L. L. and Taubert, K. American Heart Association guidelines for primary prevention of atherosclerotic cardiovascular disease beginning in childhood. *Circulation*, 107:1562–1566, 2003.

Kimm, S. Y., Barton, B. A., Obarzanek, E., McMahon, R. P., Sabry, Z. I., Waclawiw, M. A., et al. Racial divergence in adiposity during adolescence: the NHLBI Growth and Health Study. *Pediatrics*, 107:E34, 2001.

Kimm, S. Y., Barton, B. A., Obarzanek, E., McMahon, R. P., Kronsberg, S. S., Waclawiw, M. A., et al. Obesity development during adolescence in a biracial cohort: the NHLBI Growth and Health Study. *Pediatrics*, 110:E54, 2002.

Kimm, S. Y., Glynn, N. W., Kriska, A. M., Fitzgerald, S. L., Aaron, D. J., Similo, S. L., et al. Longitudinal changes in physical activity in a biracial cohort during adolescence. *Medicine & Science in Sports & Exercise*, 32:1445–1454, 2000.

Knafl, G., Sacks, J. and Ylvisaker, D. Confidence bands for regression functions. *Journal of the American Statistical Association*, 80:683–691, 1985.

Kuczmarski, R. J., Ogden, Guo, S. S., Grummer-Strawn L. M., Flegal K. M., Mei, Z., Wei, R., Curtin, L. R., Roche, A. F. and Johnson, C. L. 2000 CDC growth charts for the United States: methods and development. *Vital and Health Statistics, Series 11*, 246:1–190, 2002.

Laird, N. M. and Ware, J. H. Random-effects models for longitudinal data. *Biometrics*, 38:963–974, 1982.

Li, Z., Wang Y., Wu, P., Xu, W. and Zhu, L. Tests for variance components in varying coefficient mixed models. *Statistica Sinica*, 22:123–148, 2012.

Li, Z., Xu, W. and Zhu, L. Influence diagnostics and outlier tests for varying coefficient mixed models. *Journal of Multivariate Analysis*, 100:2002–2017, 2009.

Liang, H., Wu, H. and Carroll R. J. The relationship between virologic and immunologic responses in AIDS clinical research using mixed-effects varying-coefficient models with measurement error. *Biostatistics*, 4:297–312, 2003.

Liang, K.-Y. and Zeger, S. L. Longitudinal data analysis using generalized linear models. *Biometrika*, 73:13–22, 1986.

Lin, X. and Carroll, R. J. Nonparametric function estimation for clustered data when the predictor is measured without/with error. *Journal of the American Statistical Association*, 95:520–534, 2000.

Lin, X. and Carroll, R. J. Semiparametric regression for clustered data using generalized estimating equations. *Journal of the American Statistical Association*, 96:1045–1056, 2001.

Lin, X. and Carroll, R. J. Semiparametric estimation in general repeated measures problems. *Journal of the Royal Statistical Society, Series B*, 68:69–88, 2006.

Lin, X., Wang, N., Welsh, A. and Carroll, R. J. Equivalent kernels of smoothing splines in nonparametric regression for clustered data. *Biometrika*, 91:177–193, 2004.

Lin, D. and Ying, Z. Semiparametric and nonparametric regression analysis of longitudinal data. *Journal of the American Statistical Association*, 96:103–126, 2001.

Lin, X. and Zhang, D. Inference in generalized additive mixed models by using smoothing splines. *Journal of the Royal Statistical Society, Ser. B*, 61(2):381–400, 1999.

Lipsitz, S. R., Ibrahim, J. and Molenberghs, G. Using a Box-Cox transformation in the analysis of longitudinal data with incomplete response. *Journal of the Royal Statistical Society, Ser. C (Applied Statistics)*, 49(2):287–296, 2000.

Lu, W. and Tsiatis, A. A. Semiparametric transformation models for the case-cohort study. *Biometrika*, 93:207–214, 2006.

Lu, W. and Ying, Z. On semiparametric transformation cure models. *Biometrika*, 91:331–343, 2004.

Luo Z., Zhu, L. and Zhu, H. Single-index varying coefficient model for functional responses. *Biometrics*, DOI: 10.1111/biom.12526, 2016.

McMahan, C. A. An index of tracking. *Biometrics*, 37(3):447–455, 1981.

Melenhorst, J.J., Tian, X., Xu, D., Sandler, N.G., Scheinberg, P., Biancotto, A., Scheinberg, P., McCoy, J.P. Jr, Hensel, N.F., McIver, Z., Douek, D.C. and Barrett, A.J. Cytopenia and leukocyte recovery shape cytokine fluctuations after myeloablative allogeneic hematopoietic stem cell transplantation. *Haematologica*, 97(6):867–73, 2012.

Messer, K. A comparison of a spline estimate to its equivalent kernel estimate. *The Annals of Statistics*, 19:817–829, 1991.

Messer, K. and Goldstein, L. A new class of kernels for nonparametric curve estimation. *The Annals of Statistics*, 21:179–195, 1993.

Molenberghs, G. and Verbeke, G. *Models for Discrete Longitudinal Data.*

Springer, New York, 2005.

Moyeed, R. A. and Diggle, P. J. Rates of convergence in semiparametric modeling of longitudinal data. *Australian Journal of Statistics*, 36:75–93, 1994.

Müller, H.-G. *Nonparametric Regression Analysis of Longitudinal Data*. Lecture Notes in Statistics, 46. Springer, New York, 1988.

Müller, H.-G. On the boundary kernel method for non-parametric curve estimation near endpoints. *Scandinavian Journal of Statistics*, 20:313–328, 1993.

National High Blood Pressure Education Program Working Group on High Blood Pressure in Children and Adolescents, (NHBPEP Working Group). The fourth report on the diagnosis, evaluation, and treatment of high blood pressure in children and adolescents. *Pediatrics*, 114:555–576, 2004.

National Heart, Lung, and Blood Institute Growth and Health Research Group (NGHSRG). Obesity and cardiovascular disease risk factors in black and white girls: the NHLBI Growth and Health Study. *American Journal of Public Health*, 82:1613–1620, 1992.

Nychka, D. Splines as local smoothers. *The Annals of Statistics*, 23:1175–1197, 1995.

Obarzanek, E., Wu, C. O., Cutler, J. A., Kavey, R. W., Pearson, G. D. and Daniels, S. R. Prevalence and incidence of hypertension in adolescent girls. *The Journal of Pediatrics*, 157(3):461–467, 2010.

Pantula, S. G. and Pollock, K. H. Nested analysis of variance with autocorrelated errors. *Biometrics*, 41:909–920, 1985.

Patterson, H. D. and Thompson, R. Recovery of inter-block information when block sizes are unequal. *Biometrika*, 58:545–554, 1971.

Pepe, M. S. and Anderson, G. A cautionary note on inference of marginal regression models with longitudinal data and general correlated response data. *Communications in Statistics: Simulation and Computation*, 23:939–951, 1994.

Pinheiro, J. C. and Bates, D. M. *Mixed-Effects Models in S and S-Plus*. Springer, New York, 2000.

Pinheiro J., Bates, D., DebRoy, S., Sarkar, D., R Core Team. *NLME: Linear and Nonlinear Mixed Effects Models*. R package version 3.1-131.1, URL http://CRAN.R-project.org/package=nlme, 2018.

R Core Team. R: A language and environment for statistical computing. *R Foundation for Statistical Computing*, Vienna, Austria, 2017 (https://www.R-project.org/).

Ramsay, J. O. and Silverman, B. W. *Functional Data Analysis, Second Edition*. Springer, New York, 2005.

Rice, J. A. Boundary modification for kernel regression. *Communication in*

Statistics - Theory and Methods, 13:893–900, 1984.

Rice, J. A. and Silverman, B. W. Estimating the mean and covariance structure nonparametrically when the data are curves. *Journal of the Royal Statistical Society, Ser. B*, 53:233–243, 1991.

Rice, J. A. and Wu, C. O. Nonparametric mixed effects models for unequally sampled noisy curves. *Biometrics*, 57:253–259, 2001.

Robinson, G. K. That BLUP is a good thing: The estimation of random effects. *Statistical Science*, 6:15–51, 1991.

Rosenblatt, M. Conditional probability density and regression estimators. In *Multivariate Analysis 2*, Ed. P. R. Krishnaiah, pp.25–31. North-Holland, Amsterdam, 1969.

Rosenbaum, P. R. *Observational Studies, Second Edition*. Springer, New York, 2002.

Ruppert, D. and Wand, M. P. Multivariate weighted least squares regression. *The Annals of Statistics*, 22:1346–1370, 1994.

Schumaker, J. *Spline Functions: Basic Theory*. John Wiley & Sons, New York, 1981.

Schwarz, G. Estimating the dimension of a model. *The Annals of Statistics*, 6:461–464, 1978.

Sentürk, D. and Müller, H. G. Inference for covariante adjusted regression via varying coefficient models. *The Annals of Statistics*, 34:654–679, 2006.

Serfling, R. J. *Approximation Theorems of Mathematical Statistics*. John Wiley & Sons, New York, 1980.

Shi, M., Weiss, R. E. and Taylor, J. M. G. An analysis of paediatric CD4 counts for acquired immune deficiency syndrome using flexible random curves. *Journal of the Royal Statistical Society. Series C (Applied Statistics)*, 45:151–163, 1996.

Shibata, R. An optimal selection of regression variables. *Biometrika*, 68:45–54, 1981.

Silverman, B. W. *Density Estimation for Statistics and Data Analysis*. Chapman & Hall/CRC, Boca Raton, FL, 1986.

Stone, C. J. Optimal global rates of convergence for nonparametric regression. *The Annals of Statistics*, 10:1040–1053, 1982.

Stone, C. J. The use of polynomial splines and their tensor products in multivariate function estimation. *The Annals of Statistics*, 22:118–171, 1994.

Stone, C. J., Hansen, M., Kooperberg, C. and Truong, Y. Polynomial splines and their tensor products in extended inear modeling (with discussion). *The Annals of Statistics*, 25:1371–1470, 1997.

Taylor, C. B., Youngblood, M. E., Catellier, D., Veith, R. C., Carney, R. M., Burg, M. M., Kaufmann, P., Shuster, J., Mellman, T., Blumenthal, J. A., Krishnan, R. and Jaffe, A. S. Effects of antidepressant medication on morbidity and mortality in depressed patients after myocardial infarction. *Archives of General Psychiatry*, 62:792–298, 2005.

Thompson, D. R., Obarzanek, E., Franko, D. L., Barton, B. A., Morrison, J., Biro, F. M., Daniels, S. R. and Striegel-Moore, R. H. Childhood overweight and cardiovascular disease risk factors: The National Heart, Lung, and Blood Institute Growth and Health Study. *Journal of Pediatrics*, 150:18–25, 2007.

Venables, W.N., and Ripley, B.D. *Modern Applied Statistics with S, 4th edition*. Springer, New York, 2002.

Verbeke, G. and Molenberghs, G. *Linear Mixed Models for Longitudinal Data*. Springer, New York, 2000.

Vonesh, E. F. and Chinchilli, V. M. *Linear and Nonlinear Models for the Analysis of Repeated Measurements*. Marcel Dekker, New York, 1997.

Wahba, G. Smoothing noisy data with spline functions. *Numerische Mathematik*, 24:383–393, 1975.

Wahba, G. *Spline Models for Observational Data*. Society for Industrial and Applied Mathematics, Philadelphia, PA, USA, 1990.

Wang, H. J., Zhu, Z. and Zhou, J. Quantile regression in partially linear varying coefficient models. *The Annals of Statistics*, 37:3841–3866, 2009.

Wang, N. Marginal nonparametric kernel regression accounting for within-subject correlation. *Biometrika*, 90:43–52, 2003.

Wang, N., Carroll, R. J. and Lin, X. Efficient semiparametric marginal estimation for longitudinal/clustered data. *Journal of the American Statistical Association*, 100:147–157, 2005.

Ware, J. H. Linear models for the analysis of longitudinal studies. *The American Statistician*, 39:95–101, 1985.

Ware, J. H. and Wu M. C. Tracking: prediction of future values from serial measurements. *Biometrics*, 37(3):427–437, 1981.

Webber L. S., Srinivasan S. R., Wattigney, W. A. and Berenson G. S. Tracking of serum lipids and lipoproteins from childhood to adulthood. *American Journal of Epidemiology*, 133(9):884–899, 1991.

Wei, Y., Pere, A., Koenker, R. and He, X. Quantile regression methods for reference growth charts. *Statistics in Medicine*, 25:1369–1382, 2006.

Wickham, H., RStudio. Tidy data. *Journal of Statistical Software*, 59(10), 2014. http://www.jstatsoft.org/

Wickham, H. *R Packages: Organize, Test, Document, and Share Your Code, 1st Edition*. O'Reilly Media, California, 2015.

Wickham H. and Grolemund G. *R for Data Science: Import, Tidy, Transform, Visualize, and Model Data*. O'Reilly Media, California, 2017.

Wilsgaard, T., Jacobsen, B. K., Schirmer, H., Thune, I., Løchen, M-L., Njølstad, I., and Arnesen E. Tracking of cardiovascular risk factors. *American*

Journal of Epidemiology, 154(5):418–426, 2001.

Wu, B. and Pourahmadi, M. Nonparametric estimation of large covariance matrices of longitudinal data. *Biometrika*, 90:831–844, 2003.

Wu, C. O. and Chiang, C.-T. Kernel smoothing on varying coefficient models with longitudinal dependent variable. *Statistica Sinica*, 10:433–456, 2000.

Wu, C. O., Chiang, C.-T. and Hoover, D. R. Asymptotic confidence regions for kernel smoothing of a varying-coefficient model with longitudinal data. *Journal of the American Statistical Association*, 93:1388–1402, 1998.

Wu, C. O. and Tian, X. Nonparametric estimation of conditional distributions functions and rank-tracking probabilities with longitudinal data. *Journal of Statistical Theory and Practice*, 7:259–284, 2013a.

Wu, C. O. and Tian, X. Nonparametric estimation of conditional distributions and rank-tracking probabilities with time-varying transformation models in longitudinal studies. *Journal of the American Statistical Association*, 108:971–982, 2013b.

Wu, C. O., Tian, X. and Bang, H. A varying-coefficient model for the evaluation of time-varying concomitant intervention effects in longitudinal studies. *Statistics in Medicine*, 27:3042–3056, 2008.

Wu, C. O., Tian, X. and Jiang, W. A shared parameter model for the estimation of longitudinal concomitant intervention effects. *Biostatistics*, 12(4):737–749, 2011.

Wu, C. O., Tian, X. and Yu, J. Nonparametric estimation for time-varying transformation models with longitudinal data. *Journal of Nonparametric Statistics*, 22:133–147, 2010.

Wu, C. O., Yu, K. F. and Chiang, C.-T. A two-step smoothing method for varying-coefficient models with repeated measurements. *Annals of the Institute of Statistical Mathematics*, 52:519–543, 2000.

Wu, C. O., Yu, K. F. and Yuan, V. W. S. Large sample properties and confidence bands for component-wise varying-coefficient regression with longitudinal dependent variable. *Communication in Statistics, Theory and Methods*, 29:1017–1037, 2000.

Wu, H. and Liang, H. Backfitting random varying-coefficient models with time-dependent smoothing covariates. *Scandinavian Journal of Statistics*, 31:3–19, 2004.

Wu, H. and Zhang, J.-T. Local polynomial mixed-effects models for longitudinal data. *Journal of the American Statistical Association*, 97:883–897, 2002.

Xu, W. and Zhu, L. Goodness-of-fit testing for varying-coefficient models. *Metrika*, 68:129–146, 2008.

Xu, W. and Zhu, L. Testing the adequacy of varying coefficient models with missing responses at random. *Metrika*, 76:53–69, 2013.

Xue, L. and Zhu, L. Empirical likelihood for a varying coefficient model with longitudinal data. *Journal of the American Statistical Association*, 102:642–654, 2007.

Yao, F., Müller, H. G. and Wang, J. L. Functional data analysis for sparse longitudinal data. *Journal of the American Statistical Association*, 100:577–590, 2005a.

Yao, F., Müller, H. G. and Wang, J. L. Functional linear regression analysis for longitudinal data. *The Annals of Statistics*, 33:2873–2903, 2005b.

Zeger, S. L. and Diggle, P. J. Semiparametric models for longitudinal data with application to CD4 cell numbers in HIV seroconverters. *Biometrics*, 50:689–699, 1994.

Zeger, S. L., Liang, K.-Y. and Albert, P. S. Models for longitudinal data: a generalized estimating equation approach. *Biometrics*, 44:1049–1060, 1988.

Zeng, D. and Lin, D. Y. Efficient estimation of semiparametric transformation models for counting processes. *Biometrika*, 93:627–640, 2006.

Zhang, D., Lin, X., Raz, J. and Sowers, M. Semiparametric stochastic mixed models for longitudinal data. *Journal of the American Statistical Association*, 93:710–719, 1998.

Zhou, L., Huang, J. Z. and Carroll, R. J. Joint modeling of paired sparse functional data using principal components. *Biometrika*, 95:601–619, 2008.

Index